Random Signals and Systems

PRENTICE HALL SIGNAL PROCESSING SERIES

Alan V. Oppenheim, Editor

ANDREWS AND HUNT *Digital Image Restoration*
BRIGHAM *The Fast Fourier Transform*
BRIGHAM *The Fast Fourier Transform and Its Applications*
BURDIC *Underwater Acoustic System Analysis*
CASTLEMAN *Digital Image Processing*
COWAN AND GRANT *Adaptive Filters*
CROCHIERE AND RABINER *Multirate Digital Signal Processing*
DUDGEON AND MERSEREAU *Multidimensional Digital Signal Processing*
HAMMING *Digital Filters, 3/E*
HAYKIN, ED. *Array Signal Processing*
JAYANT AND NOLL *Digital Coding of Waveforms*
KAY *Modern Spectral Estimation*
KINO *Acoustic Waves: Devices, Imaging, and Analog Signal Processing*
LEA, ED. *Trends in Speech Recognition*
LIM *Two-Dimensional Signal and Image Processing*
LIM, ED. *Speech Enhancement*
LIM AND OPPENHEIM, EDS. *Advanced Topics in Signal Processing*
MARPLE *Digital Spectral Analysis with Applications*
McCLELLAN AND RADER *Number Theory in Digital Signal Processing*
MENDEL *Lessons in Digital Estimation Theory*
OPPENHEIM, ED. *Applications of Digital Signal Processing*
OPPENHEIM, WILLSKY, WITH YOUNG *Signals and Systems*
OPPENHEIM AND SCHAFER *Digital Signal Processing*
OPPENHEIM AND SCHAFER *Discrete-Time Signal Processing*
PICINBONO *Random Signals and Systems*
QUACKENBUSH ET AL. *Objective Measures of Speech Quality*
RABINER AND GOLD *Theory and Applications of Digital Signal Processing*
RABINER AND SCHAFER *Digital Processing of Speech Signals*
ROBINSON AND TREITEL *Geophysical Signal Analysis*
STEARNS AND DAVID *Signal Processing Algorithms*
TRIBOLET *Seismic Applications of Homomorphic Signal Processing*
WIDROW AND STEARNS *Adaptive Signal Processing*

Random Signals and Systems

BERNARD PICINBONO

Université de Paris-Sud
École Supérieure d'Électricité

PRENTICE HALL, Englewood Cliffs, NJ 07632

Library of Congress Cataloging-in-Publication Data

Picinbono, Bernard
Random signals and systems / Bernard Picinbono
p. cm.
Includes bibliographical references and index.
1. Signal processing--Digital techniques. 2. Signal processing--Statistical methods. 3. Stochastic processes. I. Title.
TK5102.5.P54 1993
621.382'23--dc20
92-44386
CIP

Acquisitions editor: **MARCIA HORTON**
Editorial/production supervision: **JENNIFER WENZEL**
Copy editor: **LYNN McARTHUR**
Cover design: **BRUCE KENSELAAR**
Prepress buyer: **LINDA BEHRENS**
Manufacturing buyer: **DAVID DICKEY**
Editorial assistant: **DOLORES MARS**
Supplements editor: **ALICE DWORKIN**

Printed in the United States of America

10 9 8 7 6 5 4 3 2 1

ISBN 0-13-752270-3

Prentice-Hall International (UK) Limited, London
Prentice-Hall of Australia Pty. Limited, Sydney
Prentice-Hall Canada Inc., Toronto
Prentice-Hall Hispanoamericana, S.A., Mexico
Prentice-Hall of India Private Limited, New Delhi
Prentice-Hall of Japan, Inc., Tokyo
Simon & Schuster Asia Pte. Ltd., Singapore
Editora Prentice-Hall do Brasil, Ltda., Rio de Janeiro

To Béatrice

Contents

Preface

Le hasard sait toujours trouver ceux qui savent s'en servir.

Romain Rolland

This book is the result of long experience in teaching many different groups of students. The reaction of this audience has directly influenced its writing and helped me to clarify the main principles of its organization.

First, a few words about the choice of title. Looking at the content, one could conclude that the book is mainly devoted to probability and stochastic processes. These terms refer to mathematically oriented material, and students not majoring in mathematics may have doubts concerning its interest in applied physics or the engineering sciences. By using the title *Random Signals and Systems*, my intention is to indicate clearly that this material covers the broad field of the statistical description of signals and methods of statistical signal processing. In fact, it is virtually impossible today to have a good grasp of information transmission or communication without a sound knowledge of transmission signals. These signals can be either deterministic or random, each aspect of which demands a specific presentation. To emphasize this, the terms "random functions" and "stochastic processes" are very rarely used. The question is not one of terminology, but more of fundamental approach.

On the other hand, if you leaf through this book you will find a large number of equations. The reason for this is that many concepts in signal representation cannot be understood without some basis of mathematical knowledge. This is true both for deterministic and random signals. Take, for example, the idea of frequency, an everyday concept for users of an FM radio receiver. To understand this, and its use in signal transmission and modulation, we must be able to use the Fourier representation of random signals analyzed in Chapter 6. Similarly, the concept of correlation is often used, as, for example, when we say that there is no correlation between the weather in Paris and Sydney. Yet, to understand this fully and to use it correctly requires a precise definition, which is impossible without resorting to mathematical expressions.

At this stage the question arises as to what level of knowledge and mathematical methods is required to approach this book. The situation is similar to that encountered in politics: whatever your position, you will always find others to the right and left of you. In the description of random signals, you will always find those who say that it is only possible with very abstract and sophisticated means, and others who feel that for engineers only intuitive concepts are necessary.

It is clear that the probability theory is a very broad field of pure mathematical science, and can be presented using the measure theory, functional analysis, and abstract integration, but these concepts have no place in the basic background of those studying engineering or physics. Does this mean that they cannot study and use probabilistic concepts? The answer is clearly no, as can be seen by a glance at those working in statistical signal processing or statistical physics. The purpose of this book is precisely to discuss a large number of important questions in signal theory without introducing the abstract concepts which require long explanations. This means that the level of mathematics needed to work through this text is the standard level of students of physics and electrical engineering. This would include the basic concepts of linear algebra, and vectors and matrices, which are commonly used in signal processing, a good knowledge of Fourier and Laplace transforms such as that introduced in courses on deterministic signals, and the fundamental principles of mathematical analysis.

Special mention must be made of Hilbert spaces. These are the basis of geometric representations of random variables, and the geometric approach is fundamental to the presentation of topics such as estimation and prediction. As Hilbert spaces are sometimes not included in mathematics courses, and in order to unify the notation throughout the book, a detailed review of Hilbert spaces is presented in Chapter 3. It contains all the material necessary for the understanding of the geometric approach widely used in the text.

Let us now discuss some points of content. As the presentation is oriented towards signal theory, the discussion of probability within the framework of combinatorics has been cut to a minimum. It seems that playing with colored balls in urns is not an everyday activity of those working in signal processing. It is, however, essential to have a good training in the description of multidimensional random variables to understand the problems in signal estimation and prediction. It is for this reason that a whole chapter is devoted to the second order properties of random vectors.

For teaching use, this book can be divided into three parts. The first, made up of Chapters 1 to 4, can be taught at undergraduate level. This part deals with the basic elements of probabilities and random variables needed to understand the concept of random signals. Obviously, points can be selected where desired, especially in Chapter 4, where various examples are discussed. Particular attention is given to complex normal random variables, showing clearly that passing from real to complex random variables is not an obvious task. Furthermore, as mentioned above, geometric presentation is of great interest when discussing random variables, and the orthogonality properties of the regression and the geometric representation of sample mean and variance are given as examples.

The second part corresponds to Chapters 5 to 9, dealing with random signals. This can be taught at both undergraduate and graduate level, depending on the organisation of studies. The basic concept to be retained in these chapters is that of the correlation function and power spectrum. Even for students not majoring in signal processing, a minimum knowledge of random signals or noise should include that of power and its description in the time and frequency domains. However, these chapters present some new fields of research, such as higher-order statistics. Poisson processes and their consequences are also described in some detail.

The third part, definitely at graduate level, corresponds to Chapters 10 to 14, and deals with the basic ideas of estimation, prediction, Wiener and Kalman filtering and matched filters. In these chapters geometric presentation is widely used, simplifying and unifying the discussion, and opening new concepts on constrained estimation and prediction.

Problems are set out at the end of each chapter, some requiring the direct application of concepts just introduced, some at a more advanced level. These problems are presented according to the French tradition: a problem is like a story requiring various steps, the first elementary and the last often complex. It is, however, possible to leave the story at any step and to return to finalise it later.

I would now like to thank all those who have contributed to the different stages of this book. I owe a very special debt to Professor André Blanc-Lapierre, who was my first teacher in this field and later became a collaborator and co-author of many books and papers. I would also like to thank many colleagues at the Laboratoire des Signaux et Systèmes and at the École Supérieure d'Électricité and especially C. Bendjaballah, P. Boelcke, P. Bondon, P. Combettes, M. Jaïdane, H. Krim, P. Regalia, and M. Savio. Parts of the text were written as a direct consequence of discussions with these colleagues. Thanks are also due to J.H. McClellan for reviewing the manuscript.

Finally, I would particularly like to thank two people who have contributed to the realization of the manuscript. The text was edited and typed in Australia by Mrs. Eve Salinas, whose contribution was essential to the quality of the final text. The equations and final production, particularly onerous tasks, were realized by Mrs. Marie-Thérèse Guerry. It was a privilege for me to benefit from their cooperation.

Chapter 1

Introduction to Random Signals and Probability

1.1 INTRODUCTION

The purpose of this chapter is to explain, without using concepts which are too abstract, why randomness must be introduced into problems of signals and systems. Even though the arguments used in this chapter are rather intuitive, leaving the detailed analysis for the following chapters, we recommend that the reader take good note of them. In fact, a good understanding of the practical situations in which randomness appears is necessary to understand the interest and limitations of more abstract theories.

Let us start with very simple examples, and consider the two signals represented in Figures 1.1 and 1.2. The signal $x(t)$ in Figure 1.1 is delivered by a frequency generator and can be considered approximately as a deterministic signal with a single frequency f, written as

$$x(t) = a \cos(2\pi f t + \phi) \qquad (1.1)$$

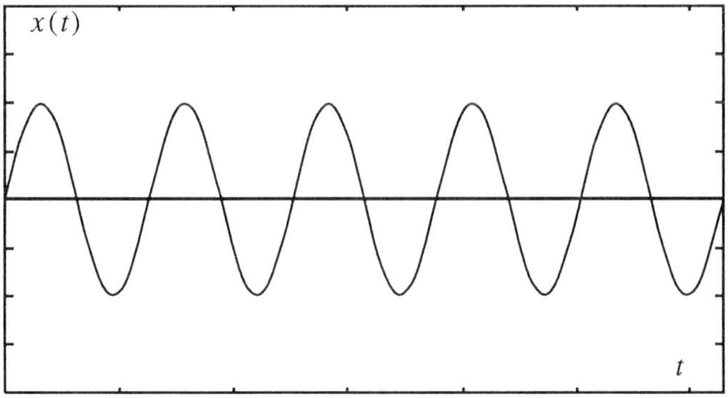

Figure 1.1 Single frequency signal.

Its amplitude and frequency are a and f respectively, while ϕ represents the phase at the time $t = 0$. This signal appears to be extremely regular, and we have the impression from its observation during an arbitrary time interval that it is possible to deduce its value at any time instant. This is quite true in (1.1), which defines a periodic signal equal to its repetition, which means that it is completely known, when it is known only on a time interval of one period.

On the other hand, the signal $x(t)$ in Figure 1.2 delivered by a "noise generator" has a completely different structure, and we will try to describe intuitively its main features. The first idea we have of this signal is its irregular behavior, which is exactly the opposite of what was seen in Figure 1.1. In particular, we have the impression that it is impossible to determine the values of the signal outside the limits of the figure. In other words, this signal appears "unpredictable," and, as seen later, *prediction* is one of the important topics in the study of random signals. Other terms used to describe the general aspect of this signal are "irregular," "non deterministic," "chaotic," and, of course, "random."

In fact it is necessary to analyze this situation in more detail. The randomness of a signal can be an illusion due to our lack of knowledge. This is especially true of the signal in Figure 1.2, which has been obtained by a computer program that is completely deterministic. In other words, for those knowing the program, the signal is completely known even though it has a disordered behavior.

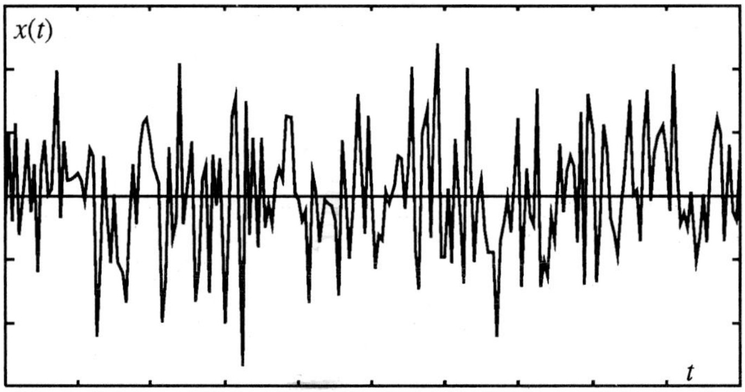

Figure 1.2 Signal $x(t)$ delivered by a noise generator.

This has a relationship with the origin of *randomness*, a very interesting subject in the philosophy of science but not discussed in detail here. To explain this point, consider another example of a random signal appearing in *meteorology*. Suppose that the temperature is taken at a given location every day at the same hour for a very long period of time (say one hundred years). The values of this temperature can be considered as a discrete-time signal, which for obvious reasons is often called a "time series." The observation of this signal exhibits two facts: the first is a regular and periodic mean behavior due to the effect of the seasons, with clearly a period of one year equal to the period of the earth's rotation. The second is that, after the suppression of this regular part, a discrete-time signal which seems unpredictable, or random, remains. But the value of the temperature is the result of physical phenomena described by deterministic laws. With a detailed knowledge of these phenomena, it would be possible to predict exactly the temperature at the same time next year, for example. However, this is impossible because of our lack of knowledge, and the randomness appears to be due to this imperfection. On the other hand, on the scale of microphysics, there is a *fundamental randomness* which is a consequence of the principles of quantum mechanics, although this will not be discussed here. Finally, let us note that in recent years there has been increasing interest in phenomena called *deterministic chaos*. This is a field of research that is expanding rapidly at the present time. Roughly speaking, it corresponds to signals with a random behavior but which are completely deterministic. Even though the origin is quite different, there is some analogy with the signal in Figure 1.2 given by a deterministic computer program. The expression of "pseudorandom signal" is sometimes used.

The term "random signal" is often associated with the term "noise" or "random noise". In fact, if we amplify the signal in Figure 1.2 and pass it through a loudspeaker, we will have the impression of a *noise*. Conversely, the same method applied to the signal in Figure 1.1 will give the impression of a pure tone. In general it is quite different from the sound of any musical instrument because the quality of a sound is due to harmonic frequencies, which are not present in (1.1).

The term "noise" can have a pejorative sense, because in everyday life noise is usually disturbing. This aspect also appears somewhat in signal problems, and a large part of work in signal processing is devoted to fighting against noise effects. For example, random noise can introduce significant perturbations in communication systems, and there are very sophisticated methods in the category of "noise cancellation." Some of these will be studied in this book.

But noise can also be interesting in itself; for example, the noise can be the interesting signal, and not the disturbance. This is especially the case in passive sonar systems where the noise generated by a ship is the means of localizing it. Similarly, a good knowledge of the random aspects of traffic problems allows us to improve traffic capacity. Other examples could also be given.

1.2 SOME EXAMPLES OF CONTINUOUS-TIME RANDOM SIGNALS

A scalar valued continuous-time random signal is usually written as

$$y = s(t;\, \omega) \tag{1.2}$$

where ω is related to randomness as specified below. Note that we do not differentiate between the expressions "random signal," "random process," and "stochastic process," which are often used in this context. The signal y is a function of time by the continuous parameter t and also of an event specified by the parameter ω. For a specified value of ω, say ω_0, $y = s(t;\, \omega_0)$ is a deterministic signal. This signal is sometimes called a realization or a sample function or a *trajectory* of the random signal. To explain this point, let us consider some examples in more detail.

Suppose we have an ensemble of N identical resistors R_i, $1 \leq i \leq N$. Following the Ohm law, the voltage observed at each resistor is null. But if we measure this voltage after amplification, we can observe a signal similar to that in Figure 1.2. The mean value is null but it presents fluctuations which are due to the thermal motion of the electrons in the metal. In other words, all the resistors are macroscopically identical, and this appears in the mean which is null according to the Ohm law, but all the voltages at the same time instant t are different, which shows a microscopic difference. Let us call $s(t;\ i)$ the voltage at this time instant t of the resistor R_i. In this expression i represents the event of specifically choosing this resistor, and corresponds to the parameter ω in $s(t;\ \omega)$. The main point to note is that in this example ω is an element of a finite set, or at least countable, if there is an infinity of resistors. There are more complex situations where ω can belong to a continuous set.

Consider now another physical device, often used in optics or in nuclear physics, called a *photomultiplier*. This system transfers an optical field into an electric current by means of the photoelectric effect. Roughly speaking, the photons of the optical field are absorbed and transformed into photoelectrons. With a very specific electronic device it is possible to identify with a good approximation the time instants at which photoelectrons are emitted, and the output signal appears as a random sequence of identical pulses. An example of such a sequence is represented in Figure 1.3. This sequence can be considered as a particular trajectory of a random signal. As the amplitude of the pulses are the same, the only feature of the signal is the set of time instants at which photoelectrons are emitted. This kind of phenomenon is called a *point stochastic process*, because it is a random distribution of points in the time axis. It is of course possible to introduce two-dimensional point processes, as for example in image processing.

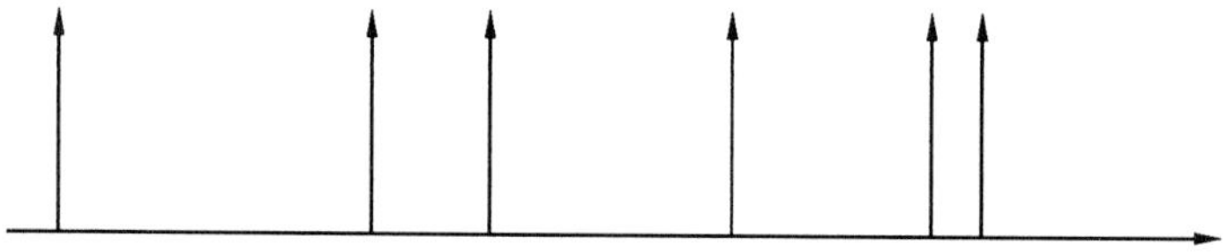

Figure 1.3 Sequence of pulses at the output of a photomultiplier.

Point stochastic processes also appear on a macroscopic scale in traffic problems. Consider the case of a telephone switchboard. The time instants at which telephone calls are received constitute a point process and in a first approach we can consider this process as random. Of course, some general features can be predicted in advance. For example, without a specific and unpredictable event, the mean number of telephone calls processed at midnight is smaller than at noon. The *duration* of each call is also unpredictable, and can thus be considered as random. Consequently the number $N(t)$ of calls processed at the time instant t is a random signal taking only integer non negative values.

When a call is registered, $N(t)$ increases by one, and decreases by one at the end of each call. Assuming that two calls cannot begin or end at the same instant, the function $N(t)$ has random variations of ± 1 at time instants at the beginning and end of calls. A trajectory of this function, which is of course deduced from a point process, is given in Figure 1.4.

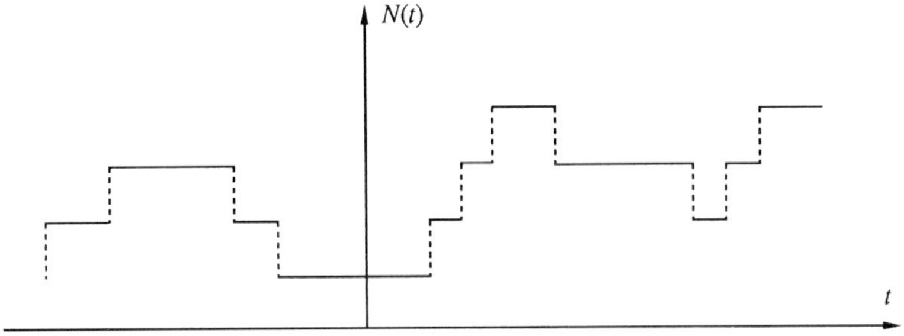

Figure 1.4 Number of calls processed by a switchboard.

It is important to note that point processes are not discrete-time signals. For example, the point process describing the emission of photoelectrons can be described by (1.2) where, as for the resistors, ω represents the experiment. For a given value of ω, i.e., for a given experiment represented by ω_0 the function $s(t; \omega_0)$ is non null only at the time instants t_k where an electron is emitted. But these instants appear in continuous time.

When the parameter ω takes its values in a discrete set, the study of continuous-time random signals is equivalent to that of a set of a countable number of deterministic signals, which presents no great difficulty. On the other hand, the problem becomes much more complex when parameter ω is continuous.

A rigorous approach to such problems would require a good knowledge of *measure theory*, which is not necessarily an acquisition of researchers in signal processing. Fortunately, almost all problems in this field can be correctly treated without this knowledge, although some results are not rigorously proved. A similar situation is found in the case of deterministic signals when distributions are introduced without all the mathematical arguments.

Finally, let us introduce the concept of the family of finite distribution, which allows a transition to the discrete time case, and will be explained more precisely later. Consider signal (1.2) at a given time instant, say t_1. The value of y is now only a function of the parameter ω, and is called a one-dimensional random variable (RV). The study of these RVs is the topic of most books on elementary probability. In order to present consistent material here, we shall recall the most important properties of RVs in the next chapters but those readers interested in the foundations of probability theory should consult more specialized works.

An RV X is usually described by its distribution function $F(x)$ defined by

$$F(x) = P[X \le x] \tag{1.3}$$

where P is the probability. The same can be said of a multidimensional RV, or of a set of n one-dimensional RVs.

To introduce the concept of a family of finite distributions, consider now, instead of one instant t_1, a set of n arbitrary time instants $t_1, t_2, \ldots, t_n$. The values of the signal $s(t_1), s(t_2), \ldots, s(t_n)$ are now an n-dimensional RV described by its n-dimensional distribution function. The collection of all these functions for arbitrary t_i and arbitrary n constitutes the family of finite distributions, whose knowledge is sufficient for the description of all random signals considered hereafter. By this method we reduce the study of continuous-time random signals to that of multidimensional RVs.

1.3 DISCRETE-TIME RANDOM SIGNALS

We have already presented the concepts of time series by discussing the case of temperature measurement. Certainly, there are many other examples of time series and especially in Economics. For example, the values or prices of certain objects can be measured every day and common sense indicates that they contain some unpredictable components or some randomness. A possible approach is to represent these values or prices as a random time series or a discrete-time signal. But discrete-time signals can also be obtained by sampling continuous-time signals. In this case the randomness of the result can be a consequence of the randomness either of the sampling procedure or of the continuous-time signal. Sampling methods are extensively discussed for deterministic signals, but the randomness introduces, of course, new problems. For a general presentation, a discrete-time random signal can be written as

$$y = s[n; \omega] \tag{1.4}$$

where n is an integer, and it is in reality a sequence of random variables. All problems related to continuous time completely disappear, and the study of the sequence of RVs is a part of the study of RVs themselves. In other words, the use of discrete-time random signals does not introduce *a priori* new concepts, which greatly simplifies the situation.

However, they do introduce new problems which are not studied within the framework of RVs. These problems are due to the time character of the parameter n in (1.4) above which justifies the expression of time series. Let us present two examples. As time is a dual aspect of *frequency*, it is important to expand a signal as a sum of complex exponential signals. This is the principle of the harmonic representation of time series and if the signal is random, the expansion introduces random coefficients. This is a very important topic, which will be discussed in detail later. The interest of harmonic representation of signals is widely described in the context of deterministic signals [Oppenheim], [Picinbono]. The main reason is that exponential signals are eigensignals of convolution, or of linear time-invariant systems. This property remains the same in a random context and linear filtering as well as harmonic representation of disrete-time signals is a very important topic.

The second example is related to prediction. In this problem we observe a discrete-time random signal at the time instants $t_1, t_2, ..., t_n$, and we want to estimate its future, or its value at time instants t_{n+1}, t_{n+2}, etc. In estimation problems the distinction between past and future is essential and these concepts are meaningless for a sequence of RVs without any relationship with time. All these questions will be carefully analyzed in the following chapters.

1.4 FUNDAMENTAL PRINCIPLES OF PROBABILITY SPACES

The purpose of this section is to present the fundamental ideas introducing the concept of probability spaces. These ideas are given largely intuitively and as simply as possible, without the proofs which require the use of measure theory.

Consider an experiment the result of which is not known in advance, as, for example, throwing a die. The basic concept in this experiment is that if it is repeated under apparently the same conditions the result has no reason to be the same, a fact which can be verified by observation. We resume this property by saying that the result of this experiment is *random.*

Let us analyze in more detail the experiment of *throwing a die*, which is a very simple situation allowing the introduction of almost all the concepts used in probability theory.

The realization of an experiment is called a *trial* and the result of this trial is an *experimental outcome* or an *elementary event.* Thus in our particular experiment there are six elementary events, and the set Ω of elementary events is

$$\Omega = \{1, 2, 3, 4, 5, 6\} \tag{1.5}$$

But there are many other possible *events*, for example, that of obtaining an even number which is realized if the trial gives either 2 or 4 or 6. So the subset of Ω

$$A \triangleq \{2, 4, 6\} \tag{1.6}$$

is a possible event, and by generalization any set of elementary events is an event. We thus arrive at the conclusion that any set of elements of Ω can be considered as an event. The introduction of set terminology leads to the application of the fundamental concepts of set theory.

To each event, say A, we can associate the *complementary* event, denoted A^*. This is the event in which A fails to occur. For example, the complementary event to obtaining a one at the throw of a die is to obtain a result other than one, i.e., 2, 3, 4, 5 or 6.

Consider two events A and B. The event that *at least either* A or B occurs is the *sum* or the *union* of A and B, noted either $A + B$ or $A \cup B$. For example, the event of an even number is the union of the events 2, 4 or 6.

The same can be said for the *product* or *intersection* of events. The event $A.B$ or $A \cap B$ means that both A and B occur in the experiment. For example, if A is the event of an even number and B that of a number greater than 3, $A.B$ is realized if 4 or 6 occur.

Finally, an impossible event is an event which cannot occur, as for example obtaining a 7. Similarly, a *sure* event will always occur, as for example obtaining a number smaller than 7.

In conclusion, we see that starting from the elementary events of the throw of a die, we can construct a family, or field, of events which is closed with respect to the three operations introduced above: complement, sum and product. This is the starting point for the introduction of a probability space.

Roughly speaking, the procedure is as follows. We start from an arbitrary set Ω with elements ω called elementary events. Any set of elements of Ω is called an *event*. By all the operations introduced above we construct a family $\mathscr{F}$ of events. This family is called an σ-field or a Borel field. To each element of $\mathscr{F}$, say A, we associate a probability measure $P(A)$ satisfying

$$0 \le P(A) \le 1 \quad ; \quad P(\Omega) = 1 \quad ; \quad P(\varnothing) = 0 \tag{1.7}$$

and the property of additivity meaning that if $A \cap B = A.B = \varnothing$,

$$P(A \cup B) = P(A + B) = P(A) + P(B) \tag{1.8}$$

The triplet $(\Omega, \mathscr{F}, P)$ defines a *probability space*. Consequently the set Ω is a sure event and the empty set is an impossible event. Furthermore if $P(A) = 0$, the event A is said to be an event with zero probability.

The example of throwing a die was a good way of introducing the basic concepts of probability space, but it is insufficient to explain the difficulties which appear when the set Ω is continuous. Let us take a specific example.

By using a radar system it is possible to measure the distance between the antenna and a target. Suppose that the electronic device is realized in such a way that the target is localized by a point on a screen. If there is only one target, there is one point on the screen which can be represented by a circle, and, assuming that the target is randomly located, we can consider that the experimental outcome is a mathematical point M inside a circle. This point can be characterized by its polar coordinates r, θ. The possible events are sets of points inside the circle. So, let us consider an arbitrary domain D inside the circle. We can associate the event $A(D)$ which is the set of points M belonging to D. In order to simplify the discussion, suppose that point M is *uniformly distributed*. This means that if the area of the domain D is $S(D)$, the probability of the event A is

$$P(A) = S/\pi\rho^2 \tag{1.9}$$

where ρ is the radius of the circle limiting the screen. The consequence of this assumption is that any experimental outcome, or any elementary event, has a zero probability, because the area associated with a point is null. Thus, any elementary event is an event of zero probability, which is different to the case of throwing a die, where Ω is a finite set. The calculation of the probability of an arbitrary event requires more advanced mathematical methods.

We shall not focus our attention in this book on the properties of probability spaces, which are analyzed in books on measure theory. Probability spaces will only be an introductory means to the study of random variables and processes, which are the main subject of our attention.

However, from the previous considerations we can deduce various results which are often used. Let A be an arbitrary event with probability $P(A)$. The complementary event A^* introduced previously satisfies

$$A \cap A^* = \varnothing \quad ; \quad A \cup A^* = \Omega \tag{1.10}$$

From (1.8) we see that

$$P(A^*) = 1 - P(A) \tag{1.11}$$

Furthermore (1.8) can be extended to the case of three events A, B and C. In fact, the sum or union of sets is associative, and

$$A + B + C = A + D \tag{1.12}$$

$$D = B + C \tag{1.13}$$

By applying (1.8) twice we immediately deduce that if A, B, and C are mutually exclusive events, or if $A.B = B.C = C.A = \varnothing$, then

$$P(A + B + C) = P(A) + P(B) + P(C) \tag{1.14}$$

This relation can of course be extended to the sum of an arbitrary number of events.

Let us now calculate the probability in (1.8) in the case where A and B are not mutually exclusive. This situation is represented in Figure 1.5.

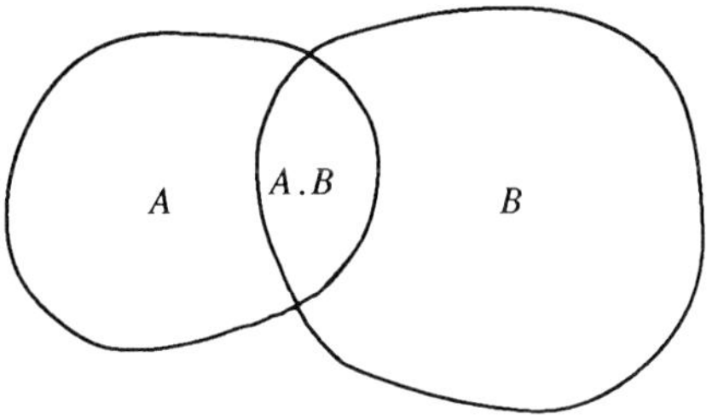

Figure 1.5 Non exclusive sets A and B.

In order to apply (1.8) we must decompose $A + B$ in a sum of mutually exclusive sets. This results from

$$A + B = A + A^*.B \tag{1.15}$$

which gives

$$P(A+B) = P(A) + P(A^*.B) \tag{1.16}$$

Furthermore, as $A + A^* = \Omega$, we have

$$B = A.B + A^*.B \tag{1.17}$$

and as the events $A.B$ and $A^*.B$ are mutually exclusive, we obtain

$$P(B) = P(A.B) + P(A^*.B) \tag{1.18}$$

Combining (1.16) and (1.18), we obtain

$$P(A + B) = P(A) + P(B) - P(A.B) \tag{1.19}$$

and the interpretation of this result in Figure 1.5 is quite evident.

1.5 CONDITIONAL PROBABILITIES, INDEPENDENCE

Let us consider again the random experiment of throwing a die. There are six possible elementary events defined by (1.5). Suppose that A is the event defined by (1.6) and that B is the event of obtaining a number higher than 3. The subset of Ω defining B is then

$$B \triangleq \{4, 5, 6\} \tag{1.20}$$

As a result, the event $A.B$ or $A \cap B$ is defined by the subset

$$A.B = \{4, 6\} \tag{1.21}$$

Suppose now that the die is without any default, which means that the six elementary events have the same probability equal to 1/6. Consequently, the probabilities of A, B and $A.B$ are equal to

$$P(A) = 1/2\,;\; P(B) = 1/2\,;\; P(A.B) = 1/3 \tag{1.22}$$

Now suppose that we know that A is realized. In this case the random experiment can only provide the three numbers 2, 4 or 6 with the same probabilities, which are then equal to 1/3. As a result, we can introduce the conditional probability of obtaining B, knowing that A is realized, which is the probability of obtaining 4 or 6. The corresponding probability, noted $P(B \mid A)$, is then 2/3 and we have

$$P(A.B) = P(A)\,.\,P(B \mid A) \tag{1.23}$$

This result, introduced intuitively, is taken as the *definition* of the conditional probability.

More precisely, let us consider an arbitrary probability space, $(\Omega, \mathcal{F}, P)$ and two events A and B defined on this space. By definition, the *conditional probability* of B with respect to A, or given A, is

$$P(B \mid A) \triangleq \frac{P(A.B)}{P(A)} \tag{1.24}$$

It is the probability of the event B relative to the hypothesis that A has occurred. From this equation we deduce, of course, (1.23) which was introduced intuitively. It is clear that $P(A \mid B)$ can be defined by only permuting A and B, which gives

$$P(A.B) = P(A)\, P(B \mid A) = P(B)P(A \mid B) \tag{1.25}$$

From this equation we deduce the so-called *Bayes rule*, written as

$$P(B \mid A) = \frac{P(B)P(A \mid B)}{P(A)} \tag{1.26}$$

Using the conditional probabilities we can now introducc thc concept of independence.

The concept of independence. Two events A and B are independent if

$$P(A \mid B) = P(A) \tag{1.27}$$

This definition is symmetric because (1.26) and (1.27) give, of course,

$$P(B \mid A) = P(B) \tag{1.28}$$

This can also be written in the form

$$P(A.B) = P(A)P(B) \tag{1.29}$$

The physical meaning of independence between A and B is that the realization of B does not give any information on the probability of obtaining A. The simplest example appears with the die experiment again. Suppose that we now play with two dice. If there is no mechanical relation between them, it appears obvious that the probability of obtaining an arbitrary pair (a, b) is the product of the two probabilities equal to 1/6.

Some elementary properties of conditional probabilities. The conditional probabilities satisfy the same rule as the probabilities introduced in the previous section. For example, let us consider the situation used in (1.8), i.e., the case of two mutually exclusive events A and B. From the relation $A.B = \varnothing$, we deduce that for any event M we have

$$(A.M)(B.M) = \varnothing \tag{1.30}$$

Furthermore, the union of sets is associative, which means that

$$(A + B).M = A.M + B.M \tag{1.31}$$

This is pictured in Figure 1.6. As a result we have

$$P[(A + B).M] = P(A.M + B.M) \tag{1.32}$$

Now, because of (1.30) we can apply (1.8) which gives

$$P[(A + B).M] = P(A.M) + P(B.M) \tag{1.33}$$

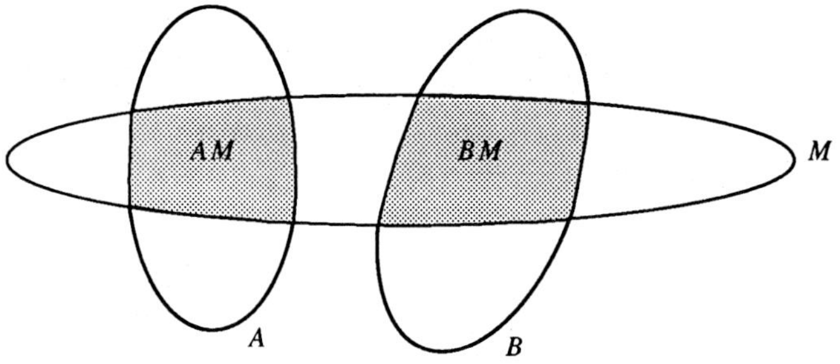

Figure 1.6 Union and product of sets.

Using the rule (1.25) in (1.33) we obtain

$$P(M)\,P(A+B \mid M) = P(M)\,[P(A \mid M) + P(B \mid M)]$$

which gives

$$P(A+B \mid M) = P(A \mid M) + P(B \mid M) \tag{1.34}$$

This is the extension of (1.8) to conditional probabilities.

However, there is another additive rule which is of great interest. Suppose that A and B are a *partition* of Ω. This means that

$$A.B = \varnothing \text{ and } A + B = \Omega \tag{1.35}$$

Let M be an arbitrary event. We then have

$$P(M) = P(M \mid A)\,P(A) + P(M \mid B)\,P(B) \tag{1.36}$$

In order to prove this relation, we note that the second relation (1.35) implies that

$$M = M.(A+B) = M.A + M.B$$

Furthermore, the first relation (1.35) implies that $M.A$ and $M.B$ are mutually exclusive. Applying (1.8) and (1.25), we arrive at (1.36).

This expression can be generalized to a *partition* of Ω by an arbitrary number of sets A_i. Such a partition is characterized by

$$A_i.A_j = \varnothing,\ i \neq j \text{ and } \sum_{i=1}^{n} A_i = \Omega \tag{1.37}$$

and (1.36) becomes

$$P(M) = \sum_{i=1}^{n} P(M|A_i)\,P(A_i) \tag{1.38}$$

From this we can deduce the *Bayes theorem* or the relation

$$P(A_i|M) = \frac{P(M|A_i)\,P(A_i)}{\sum_{i=1}^{n} P(M|A_i)\,P(A_i)} \tag{1.39}$$

This is a direct consequence of the Bayes rule (1.26) where the denominator $P(M)$ is replaced by (1.38).

1.6 SOME ELEMENTARY PROBLEMS OF PROBABILITY

In signal processing problems we are mainly interested in the description of properties of random signals. These descriptions are usually obtained by using some expected values such as, for example, the correlation function. The calculation which directly concerns probabilities is less important than in other fields of science or engineering. We shall therefore discuss these questions only briefly in this introductory chapter.

Let us start with an example where only probability and non expected values are used. Signals are often used to detect an event. For example, from a radar signal we must decide whether or not there is a target. This type of problem is analyzed in *detection theory*. In order to detect a target we can use one radar system, but many modern detection systems make use of multiple sensors, which introduces the concept of decentralized detection. Suppose, for example, that we use n detection systems $S_1, S_2, \ldots S_n$, working independently, and that, for given conditions, they are characterized by an alarm probability p_i. The global alarm can be decided under various conditions, and we shall give the most classical here. We can decide the global alarm if all the detection systems are in the alarm position (a very restrictive condition), or if there are at least p alarms ($1 \leq p < n$). It is clear that the probability of global alarm depends a great deal on the chosen strategy, sometimes called the *fusion rule*. It is this type of problem which will be analyzed in this section. As there is no general method available, we shall discuss some examples, while others are given as problems at the close of the chapter.

1.6.1 Calculation of the Probability of Events

We shall restrict this section to the case where the set of elementary events is *finite*. In order to calculate the probability of an event A the most elementary method consists in decomposing A as a sum of elementary events and applying (1.8). This is, however, not necessarily the simplest method, as we shall see below.

Example 1.1 Consider the random experiment with two dice. The set Ω of elementary events is the set of pairs of integers $[a, b]$ defined by

$$\Omega = \{[a, b] | 1 \leq a \leq 6,\ 1 \leq b \leq 6\} \tag{1.40}$$

There are of course $6^2 = 36$ distinct elementary events. As they are assumed to have the same probability, this is $1/36$. Let A be the event when the two dice give the same result, which can be written $[a, a]$, where $1 \le a \le 6$. It can then be decomposed as the sum of 6 distinct elementary events. Applying (1.14) we find that $P(A) = 1/6$.

Let B be the event when the two dice give a different result. This event is the sum of all the events $[a, b]$ where $a \ne b$. In order to decompose B as a sum of elementary events, we can operate as follows. Choosing a first, we must choose b such that $b \ne a$. There are then five different choices, and as there are six possibilities of choosing a, we find that

$$P(B) = (6.5)/36 = 5/6 \tag{1.41}$$

But there is a more direct way of calculating this probability. If we note that $B = A^*$, and use (1.11), we obtain (1.41).

Example 1.2 Let us now extend the previous example to the case of three dice. The set Ω is now the set of triplets of integers defined as in (1.40) by

$$\Omega = \{[a, b, c] | 1 \le a \le 6,\ 1 \le b \le 6,\ 1 \le c \le 6\} \tag{1.42}$$

There are $6^3 = 216$ elementary events with the same probability 1/216. Let us now consider various events which can be deduced from a random experiment with three dice.

Suppose that A is the event of obtaining the same results for three dice. This can be written as $[a, a, a]$, and six different situations are possible. Consequently $P(A) = 6/6^3 = 1/36$.

Let B be the event when the three outcomes are different. This can be written as $[a, b, c]$, where $a \ne b \ne c \ne a$. Using the same procedure as for two dice, we find that

$$P(B) = 6.\ 5.\ 4/6^3 = 5/9 \tag{1.43}$$

Let C be the event when at least two outcomes are equal. It is possible to calculate the number of situations that give this result, but it is simpler to note that the complementary event C^* of C is B. Applying (1.11), we obtain

$$P(C) = 1 - 5/9 = 4/9 \tag{1.44}$$

It is a little more difficult to calculate the probability of event D, which is when two dice out of three give the same result. The corresponding events take the form $[a, a, b]$, $[a, b, a]$ or $[b, a, a]$, where $a \ne b$. The number of distinct events in the form $[a, a, b]$ is of course $6.5 = 30$. Consequently, the probability of event D is

$$P(D) = 90/216 = 5/12 \tag{1.45}$$

We verify that $P(A) + P(B) + P(D) = 1$, and it is obvious that any elementary event $[a, b, c]$ belongs to the sum $A + B + D$. It is clear that the general procedure for the type of problems considered above consists in calculating the number of distinct elementary events the sum of which results in a given event. The general method for doing this belongs to a field of mathematical science known as *combinatorics*, about which there are numerous specialized texts for interested readers.

Example 1.3 The alarm problem. Consider the alarm problem introduced above in connection with distributed detection. Suppose that we have n alarm systems S_i with the same alarm probability α. We want to calculate the global alarm probability for various fusion rules.

The most secure fusion rule consists in deciding the global alarm when each detection system is in the alarm position. This fusion rule F_0, sometimes called the "and" fusion rule, gives a global alarm probability

$$P_0(A) = \alpha^n \tag{1.46}$$

which decreases whenever n increases. Note that this is the unanimity rule when a group of people must decide by vote, and it is well known that for large groups it is extremely difficult to reach a decision. However, when a decision is taken its meaning is very important, as it reflects the unanimity of the group.

A less restrictive fusion rule F_1 is to decide the global alarm when at least one sensor is in alarm position. In the case of two sensors this introduces the "or" rule. In order to calculate $P_1(A)$ we note that A^* is realized when all the sensors are in the non-alarm position, an event with probability $(1 - \alpha)^n$. Using (1.11), we deduce

$$P_1(A) = 1 - (1 - \alpha)^n \tag{1.47}$$

It is easy to show that for $n > 1$ we have $P_1(A) > P_0(A)$. Furthermore, for large values of n, $P_1(A)$ tends to one. This means that while F_0 is too restrictive, almost never giving the global alarm, F_1 is too supple, almost always giving the alarm for large values of n. It is then necessary to introduce less extreme fusion rules. A possible way is to decide the alarm when *at least* k systems, among the n in operation, are in the alarm position. To calculate the probability $P_k(A)$ of alarm with this decision rule, let us introduce the events E_p defined by the fact that p sensors are in the alarm position while $n - p$ are not in this position. Using the classical results of combinatorics, it is well known that starting from a set of n elements (here the set of n sensors), the number of distinct subsets containing p distinct elements is

$$\binom{n}{p} = \frac{n!}{p!(n - p)!} \tag{1.48}$$

As a result the probability $P(E_p)$ of E_p is

$$P(E_p) = \binom{n}{p} \alpha^p (1 - \alpha)^{n-p} \tag{1.49}$$

For the chosen fusion rule, the event of alarm A is obviously

$$A = E_k + E_{k+1} + \dots + E_n \tag{1.50}$$

and, as all these events are mutually exclusive, we obtain

$$P_k(A) = \sum_{p=k}^{n} \binom{n}{p} \alpha^p (1 - \alpha)^{n-p} \tag{1.51}$$

For $k = 1$, we again find (1.47) because of the well-known expression

$$(a+b)^n = \sum_{p=0}^{n} \binom{n}{p} a^p b^{n-p} \tag{1.52}$$

Using $a = \alpha$ and $b = 1 - \alpha$, we obtain $a + b = 1$, and this gives

$$1 = \binom{n}{0}(1-\alpha)^n + P_1(A) \tag{1.53}$$

Using (1.48), we immediately deduce $P_1(A)$ given by (1.47).

1.6.2 Repeated Trials

The concept of randomness was introduced in the previous section by referring to an experiment whose result is not known in advance, or whose result is not the same even though it is repeated under the same conditions. This is especially obvious in the experiment of throwing a die. Thus the idea of repeated trials is essential in probability. This can be introduced in two ways.

To explain the problem let us take the same die throwing example. If we repeat the experiment n times and register n results we can say that this experiment is equivalent to the other consisting in one trial with n dice, although it is simpler to throw the die 30 times rather than to throw 30 dice simultaneously.

From a *mathematical* point of view this can be explained by using the concept of the Cartesian product. Consider two sets Ω_1 and Ω_2 with elements ω_1 and ω_2. The Cartesian product $\Omega_1 \times \Omega_2$ is the set whose elements are pairs such as $[\omega_1, \omega_2]$. It can be written formally as

$$\Omega_1 \times \Omega_2 = \{[\omega_1, \omega_2] | \omega_1 \in \Omega_1 \text{ and } \omega_2 \in \Omega_2\} \tag{1.54}$$

Note that this is exactly the same structure as in (1.40) in the case of two dice, and can be extended to an arbitrary number of sets by an obvious generalization. Thus (1.42) corresponds to the Cartesian product of three sets. Let us explain here how to introduce a probability space corresponding to a repeated trial of n experiments. Starting from a probability space $(\Omega, \mathcal{F}, P)$ we want to construct the probability space $(\Omega_n, \mathcal{F}_n, P_n)$ corresponding to the n times repeated experiment. The set Ω_n is simply the Cartesian product of Ω n times by itself, or the nth order Cartesian power. The family $\mathcal{F}_n$ of events can be deduced immediately. Any event A of $\mathcal{F}_n$ is then the Cartesian product of $A_1 \times A_2 \times \ldots \times A_n$ where the A_is are events, or sets, of $\mathcal{F}$. If we assume that the repeated trials are independent, the probability measure of A is simply

$$P(A) = P(A_1) \,.\, P(A_2) \ldots P(A_n) \tag{1.55}$$

From a *physical* point of view the use of repeated trials introduces the concept of *relative frequency*. We shall first explain this with the elementary case of the die experiment. Let A be the event of obtaining the result equal to one. Repeating the experiment n

times, we introduce $n(A)$, which is the number of experiments giving the event A. The relative frequency is the ratio

$$\pi_n(A) = \frac{n(A)}{n} \tag{1.56}$$

Anyone with a die can verify that when n increases, $\pi_n(A)$ tends to 1/6, which means that the relative frequency converges to the probability. The definition (1.56) can be applied to any event of an arbitrary probability space, and if the repeated trials are independent the relative frequency converges to the probability $P(A)$. The proof of this result will appear later, when the concepts of convergence are introduced. The relation between probability and relative frequencies was an important question in the early stages of probability theory, and is certainly the link between physical experiments and mathematical concepts.

PROBLEMS

1.1 Show the following properties of events or of sets of elements belonging to an arbitrary set Ω:

$$A + A = A.A = A$$
$$A + B = B + A$$
$$A.B = B.A$$
$$(A + B) + C = A + (B + C)$$
$$(A.B).C = A.(B.C)$$
$$A.(B + C) = A.B + A.C$$
$$A + A^* = \Omega$$
$$A.A^* = \varnothing$$
$$A + \Omega = \Omega$$
$$A.\Omega = A$$
$$A + \varnothing = A$$
$$A.\varnothing = \varnothing$$

1.2 In a random experiment with four dice several events can appear. Calculate the probabilities of the following events:

- A : all the results are different
- B_2: two dice give the same result, the other two give different results which are also different from that given by the dice giving the same result
- B'_2: two dice give the same result, the other two give the same result which is not the same as that of the other two dice
- B_3 : three dice give the same result and the other a different result
- C : the four dice give the same result.

Calculate $P(A + B_2 + B_2' + B_3 + C)$ and explain the result.

1.3 Taking the same experiment as in the previous problem, calculate the probabilities of the following events:

D_2 : at least two dice give the same result

D_3 : at least three dice give the same result.

1.4 Suppose that we throw a die four times successively. Let us call A the event that the result of the first throw is a, $1 \le a \le 6$. Consider the following events concerning the last three throws:

B_0 : the result a does not appear

B_1 : the result a appears once

B_2 : the result a appears twice

B_3 : the result a appears three times.

Calculate the conditional probabilities $P(B_j | A)$, $0 \le j \le 3$. Calculate the sum of these probabilities and interpret the result.

1.5 Suppose that we throw one die n times. Calculate the probability that the same result appears at least twice for $1 \le n \le 10$. Interpret the result for $n > 6$.

1.6 In the same experiment as in the previous problem, calculate the probability that the same result appears at least three times for $1 \le n \le 15$. Interpret the result for $n > 12$.

1.7 A group of n people are eating together. Introducing some appropriate assumptions, what is the probability of hearing "Happy Birthday"?

1.8 In the same situation as in the previous problem, calculate the probability of hearing "Happy Birthday" for at least two people.

1.9 Show the chain rule

$$P(A.B.C) = P(A)\, P(B | A)\, P(C | A.B)$$

Generalize this chain rule to n events and calculate $P(A_1.A_2. \ldots A_n)$.

1.10 Suppose that we play with three dice and that the faces of these dice are distinguished by numbers 1, 2, 3, 4, 5, and 6. Calculate the probability that the sum of the results appearing in the three dice is equal to n. What are the possible values of n?

1.11 Using the same notations as in the previous problem, let A_n be the event that the first die gives the result n and B_m the event that the sum of the three results is m. Calculate $P(A_n)$, $P(B_m)$, and the conditional probability $P(A_n | B_n)$. Explain why the two events A_n and B_m are not independent.

1.12 Show that $P(A + B) \le P(A) + P(B)$ and extend the result in the form

$$P\left(\sum_{i=1}^{n} A_i\right) \le \sum_{i=1}^{n} P(A_i)$$

which is called the union bound.

1.13 In order to interpret the results of the experiment with a point process such as that presented in Figure 1.3, we introduce the probability $p[n]$ to observe n points in an arbitrary time interval T. In many experiments this probability can be written

$$p[n] = \exp(-m)\,\frac{m^n}{n!}$$

where $m = \lambda T$. Show that $\sum p[n] = 1$. Calculate the probability of recording an even number of points.

1.14 Consider the same situation as in the previous problem and suppose that T_1 and T_2 are two non overlapping time intervals of duration T_1 and T_2. In many cases one can show that any event concerning the interval T_1 is independent of any event concerning the interval T_2. Calculate the probability to observe p points in the interval $T_1 + T_2$.

1.15 Let us call A_n the event of recording n points in T_1 and B_p the event of recording p points in $T_1 + T_2$. Calculate the conditional probability $P(A_n | B_p)$, $n < p$ and show that it does not depend on λ.

1.16 Let E_i $(1 \le i \le n)$ be a partition of Ω and p_i the probability of the event E_i. Calculate $\sum p_i$. The entropy associated with the probabilities p_i is defined by

$$H = -\sum_{i=1}^{n} p_i \ln p_i$$

(a) Calculate this entropy when $n = 2$ and find the probabilities p_i for which H is maximum.

(b) Calculate the entropy when $p_i = 1/n$.

1.17 **Failure probability.** A device is in operation at time $t = 0$ and can fail at a posterior time instant T. This time instant is random and can have any non-negative value. In many applications it is necessary to calculate the probability $F(t)$ of the event $T \le t$, which is then the probability that the failure appears before a given instant t.

(a) Let E_t be the event that the failure appears after t. Calculate $P(E_t)$ in terms of the function $F(t)$.

(b) Let $I_{t,dt}$ be the event that the failure appears between t and $t + dt$, $dt > 0$. Express this event in terms of the events E_t and E_{t+dt}.

(c) Deduce $P(I_{t,dt})$.

(d) The *conditional failure rate* $\beta(t)$ is a function such that

$$\beta(t)\,dt = P(I_{t,dt} | E_t)$$

which is the conditional probability that the failure appears in the interval $[t, t + dt]$ given that the failure appears after t. Show that

$$\beta(t) = \frac{F'(t)}{1 - F(t)}$$

where $F'(t)$ is the derivative of $F(t)$.

(e) Deduce the value $F(0)$ from the fact that the device is in operation at $t = 0$.

(f) Calculate $F(t)$ in terms of $\beta(t)$.

(g) In many physical processes $\beta(t) = k$. This means that the device does not age. With this assumption deduce the function $F(t)$.

1.18 Suppose that the probability of recording k points of a point process in an interval T is $p_k = ca^k$, $k \geq 0$. Determine the value of the constant c. Calculate the probability of finding less than K points in this interval.

Chapter 2

A Review of Scalar Random Variables

2.1 THE CONCEPT OF A RANDOM VARIABLE

In most physics experiments or signal processing problems it is necessary to assign a numerical value to the result of a random experiment. For example, the signal $x(t)$ in Figure 1.2 can represent the voltage of the thermal noise measured with a resistor, as explained on p. 4. So $x(t)$ is a numerical value, measured for example in volts. Similarly, the value of the temperature at a given location is also a number, measured in degrees. We also associate a number to the experiment of throwing dice. However, this number was not used in the previous chapter, and the results would be the same if the faces of the dice were marked with letters, colors, or any other distinctive marks.

In a random experiment the experimental outcomes, or the elementary events introduced above are the elements ω of a set Ω. To introduce the concept of a random variable (RV) we must assign a numerical value to each element ω. In other words, we must introduce a function $X(\omega)$ taking real values, but this function is not completely arbitrary.

Consider an arbitrary probability space $(\Omega, \mathscr{F}, P)$ and a real function $X(\omega)$ taking its values in the real line $R = (-\infty, +\infty)$. To be considered as an RV, the function $X(\omega)$ must be *measurable*. Roughly speaking, this can be explained as follows. Consider the set $A(x)$ of points ω of Ω such that $X(\omega) \le x$, where x is arbitrary, or

$$A(x) = \{\omega \mid X(\omega) \le x\} \tag{2.1}$$

If this set is well defined and has a probability measure $P(A)$, the function $X(\omega)$ is said to be measurable and is then an RV. Furthermore, the probability $P(A)$ is called the distribution function (DF) of X defined by

$$F(x) = P[A(x)] \tag{2.2}$$

It is clear that if Ω contains a finite number of elements, any function $X(\omega)$ is measurable, and the only possible values of the RV are $X_i = X(\omega_i)$. In this case the RV is said to be discrete.

As the DF will play a crucial role in what follows, it is important to study its properties in more detail.

2.2 DISTRIBUTION FUNCTIONS OF RANDOM VARIABLES

2.2.1 Definitions

The DF defined by (2.2) can also be written as

$$F(x) = P[X(\omega) \leq x] \tag{2.3}$$

Note that this definition implies that $X(\omega)$ is real. It is also possible to introduce complex RVs, the simplest way being to write $X(\omega) = A(\omega) + j\,B(\omega)$ where $A(\omega)$ and $B(\omega)$ are a pair of real RVs defined on the same probability space. In this chapter the RVs are real unless otherwise specified. For all the following, an RV is completely characterized by its distribution function, although this is not perfectly exact from a strictly mathematical point of view. Knowing this function $F(x)$, we can completely ignore the probability space $(\Omega, \mathcal{F}, P)$ and simply write X instead of $X(\omega)$. Note that in order to avoid any confusion between the notation of the RV and its possible values, we write the RV in capital letters and its value in lower-case letters. It would in fact be necessary to write the DF as $F_X(x)$ to indicate that it is the DF of the RV X taking the possible values x. Where no confusion is possible we will use only the simpler form $F(x)$.

Let us now give an example of two distinct RVs $X(\omega)$ and $Y(\omega)$ with the same DF. Let us call I and D the sets of points ω defined by

$$I = \{\omega \mid X(\omega) = Y(\omega)\}$$

$$D = \{\omega \mid X(\omega) \neq Y(\omega)\}$$

This means that I is the set of points ω for which $X(\omega)$ and $Y(\omega)$ are equal. Furthermore, I and D are a partition of Ω, or

$$I \cap D = \varnothing \ ; \ I \cup D = \Omega \tag{2.4}$$

Suppose now that $P(I) = 1$. In this case it is said that $X(\omega)$ and $Y(\omega)$ are equal with probability one. This does not mean that the two RVs $X(\omega)$ and $Y(\omega)$ are equal. In order to verify this point, consider the following example. Suppose that Ω is the set of real numbers satisfying $0 < \omega \leq 1$. Let $A(\omega_1, \omega_2)$ be the set of points ω satisfying $\omega_1 < \omega \leq \omega_2$ and suppose that the probability measure of $A(\omega_1, \omega_2)$ is simply $\omega_2 - \omega_1$. It is then said that the points ω are *uniformly distributed* in (0, 1]. As a consequence, the probability that $\omega = \omega_1$, where ω_1 is given, is equal to zero. Suppose now that $X(\omega) = 0$, whatever ω, and that $Y(\omega) = 0$ if $\omega \neq \omega_1$ and $Y(\omega_1) = 1$. It follows that the probability that $X(\omega) \neq Y(\omega)$ is null, or $P(D) = 0$, and consequently $P(I) = 1$. So $X(\omega)$ and $Y(\omega)$ are different but equal with probability one.

Let us now show that if, as in the previous example, $P(I) = 1$, the DFs of $X(\omega)$ and $Y(\omega)$ are equal. For this purpose it suffices to apply (1.35), noting that I and D satisfy the same relations as A and B appearing in this equation. Let us call M the event that $Y(\omega) \leq y$. We can then apply (1.36) and write

$$P(M) = P(M \mid I)\, P(I) + P(M \mid D) P(D)$$

However, as $P(I) = 1$, we deduce from (2.4) that $P(D) = 0$. Consequently we have

$$P(M) = P(M \mid I) = P[Y(\omega) \leq y \mid X(\omega) = Y(\omega)]$$

But if $X(\omega) = Y(\omega)$, the last term is equal to $P[X(\omega) \leq y]$, which finally gives

$$F_Y(y) = F_X(y)$$

It is clear that this property is quite general and that two different RVs which are equal with probability one have the same DF. This is why the DF does not completely define an RV. As this is of no importance in what follows, we do not distinguish two RVs that are equal with probability one.

2.2.2 Properties of the Distribution Function

Consider two real numbers a and b, $a < b$, and the two events $E_1 = [X \leq a]$ and $E_2 = [a < X \leq b]$. It is clear that $E_1 . E_2 = \varnothing$ and that

$$E_1 + E_2 = [X \leq b] \tag{2.5}$$

Applying (1.8), we deduce that

$$F(a) + P(a < X \leq b) = F(b)$$

or

$$P(a < X \leq b) = F(b) - F(a) \tag{2.6}$$

As a result of this relation we obtain $F(b) - F(a) \geq 0$ as being the probability of an event. Then, if $a < b$, $F(a) \leq F(b)$, which characterizes a *non-decreasing function*. Assuming that X takes its value in $(-\infty, +\infty)$, we deduce

$$F(-\infty) = 0 \quad ; \quad F(+\infty) = 1 \tag{2.7}$$

If $F(b) = F(a)$, then the event $a < X \leq b$ has a zero probability: it is called an almost impossible event. If $F(a) = 0$ and $F(b) = 1$, the same event has a probability of one, and is called an almost sure event. In this case it is clear that X takes its values in the set $a < x \leq b$ with probability one.

An example of distribution function is presented in Figure 2.1. This function is discontinuous at the point x_0 and the variation of F is p_0. From (2.6) this means that $P(X = x_0) = p_0$.

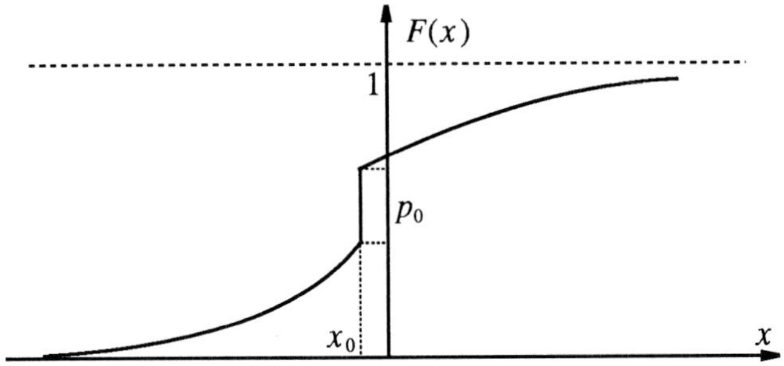

Figure 2.1 Example of distribution function.

2.2.3 Discrete Random Variables

Consider an RV $X(\omega)$ taking only the values x_i with the probabilities p_i respectively. It is obvious that these probabilities must satisfy

$$\sum_i p_i = 1 \tag{2.8}$$

Let us now deduce the form of its DF. For this purpose, suppose that the possible values x_i of X are ordered in such a way that $x_i < x_{i+1}$. Consider a real number x satisfying $x_k < x < x_{k+1}$. It is clear that the event $X \le x$ is the sum of the exclusive events $X = x_1$, $X = x_2$, ..., $X = x_k$. Consequently the DF $F(x)$ is given by

$$F(x) = \sum_{i=1}^{k} p_i \tag{2.9}$$

As a result $F(x)$ is a constant function when x belongs to the interval (x_k, x_{k+1}). Suppose now that $x = x_k$. As the event $X \le x_k$ is still the sum of the same exclusive events, we have

$$(X \le x_k) = (X = x_1) + (X = x_2) + \ldots + (X = x_k)$$

and $F(x_k)$ is still given by (2.9). The result of this discussion is that $F(x)$ is a step-wise function which only varies by jumps of amplitudes p_i at the points x_i. So $F(x)$ is discontinuous at the points x_i, which means that the limit of $F(x_i + \varepsilon)$ when $\varepsilon \to 0$ depends on the sign of ε. More precisely, we have

$$\lim_{|\varepsilon| \to 0} F(x_i + |\varepsilon|) = F(x_i)$$

which is expressed by saying that $F(x)$ is continuous to the right at $x = x_i$. Of course, we have

$$\lim_{|\varepsilon| \to 0} F(x_i - |\varepsilon|) = F(x_i) - p_i$$

which means that $F(x)$ is not continuous to the left at $x = x_i$. The general form of this kind of DF is represented in Figure 2.2.

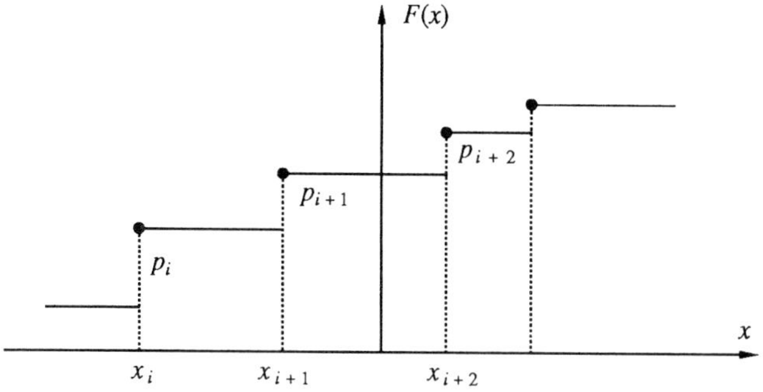

Figure 2.2 Form of the DF of a discrete RV.

2.2.4 Continuous Random Variables

Suppose now that $F(x)$ is continuous and has a derivative whatever the value of x. Let us call this derivative $p(x)$, defined by

$$p(x) \triangleq \frac{dF(x)}{dx} = \frac{dF}{dx} \tag{2.10}$$

In this case it is clear that (2.6) can be written as

$$F(b) - F(a) = \int_a^b p(x)\, dx \tag{2.11}$$

where this integral is the classical Riemann integral. Furthermore, as $F(x)$ is a non-decreasing function, its derivative is non-negative. Finally (2.7) takes the form

$$\int_{-\infty}^{+\infty} p(x)\, dx = 1 \tag{2.12}$$

The function $p(x)$ which is non-negative and has an integral satisfying (2.12) is called the *probability density function* (PDF) of the RV $X(\omega)$.

An RV with a PDF defined whatever x is called a *continuous* RV. It must be noted that if $X(\omega)$ is a continuous RV, the event $X(\omega) = x$ has a zero probability. In order to verify this point it suffices to start from (2.11) and to note that

$$P(x < X \leq x + \Delta x) = F(x + \Delta x) - F(x)$$

It is clear that this expression tends to zero when $\Delta x \to 0$ because $F(x)$ is continuous.

The case of discrete RVs can be discussed in the framework of continuous RVs by using the Dirac distribution, or the *unit impulse function*. More precisely, the RV discussed in Section 2.2.3 can be described by the PDF

$$p(x) = \sum_i p_i \, \delta(x - x_i) \tag{2.13}$$

This expression is a consequence of the fact that $\delta(x)$ can be considered as the derivative of the unit step function $u(x)$ (See [Oppenheim], p. 22 or [Picinbono], p. 43).

2.2.5 General Case

The DF represented in Figure 2.1 is neither continuous whatever x, nor a stepwise function. It is possible to show that any DF $F(x)$ can be decomposed as a sum of three terms and written as

$$F(x) = F_c(x) + F_d(x) + F_s(x) \tag{2.14}$$

The first term of this sum has a derivative $f(x)$ whatever x and corresponds to the case presented in 2.2.4. The second term is a stepwise function similar to the DF represented in Figure 2.2. The last term is a singular function which is continuous but without derivative. It is called the *singular* component of the DF. It has no practical application and in all the following we shall assume that $F_s(x) = 0$. Furthermore, the two equations (2.8) and (2.12) now become

$$\int f(x)\, dx + \sum p_i = 1 \tag{2.15}$$

If $F_d(x) = 0$, we return to the case of a continuous RV. If the two components $F_c(x)$ and $F_d(x)$ are present in $F(x)$, this function is only continuous to the right, as is $F_d(x)$. It is the situation shown in Figure 2.1.

2.3 EXPECTATION AND MOMENTS

2.3.1 Expectation

We have previously noted that an RV is completely described by its DF defined by (2.3). Furthermore, the knowledge of this DF is equivalent to that of the PDF $p(x)$ if the RV is continuous or of the probabilities p_i and the values x_i if the RV is discrete. In many applications a lower level of information is quite sufficient. For example, taking again a situation introduced in the first chapter, the thermal noise appearing in a resistor can be described with sufficient accuracy by its mean value and its mean power. We have noted that the mean value is null, which is the Ohm law. Similarly, it is often sufficient to know the mean temperature in January in a particular city, without knowing the DF of this temperature. It is then necessary to give a more precise definition of the intuitive concept of mean value. This is the purpose of this section. In order to avoid any confusion we shall proceed by successive steps, depending on the discrete or continuous nature of the RV.

First, consider a discrete RV X taking the values x_i with the probabilities p_i. The mean value or expectation, or expected value of X, written as $E(X)$, is defined by

$$E(X) = \sum x_i \, p_i \tag{2.16}$$

The interpretation of this expression is quite simple. Each possible value x_i is weighted by its probability and we take the sum of the results. There is also a mechanical analogy of $E(X)$. Suppose that p_i is a mass located at the point x_i. This mass is sometimes called probability mass. Then $E(X)$ corresponds to the center of gravity, or center of mass. The physical interpretation is that $E(X)$ is also the expected value of an RV Y taking only the value $E(X)$ with the probability one.

Suppose now that X is a continuous RV defined by the PDF $p(x)$. In this case the expected value is defined by

$$E(X) = \int x \, p(x) \, dx \tag{2.16'}$$

where the integral is calculated from $-\infty$ to $+\infty$. The same definition is used in mechanics for the definition of the center of mass of a continuous distribution with a density.

We can pass from (2.16) to (2.16') by saying that

$$E(X) = \lim_{\Delta x_i \to 0} \sum x_i \, p(x_i) \, \Delta x_i$$

where $p(x_i)\Delta x_i$ is the probability associated with the interval $(x_i, x_i + \Delta x_i)$. It is clear that the integral in (2.16') is the classical Riemann integral and the range of integration is from $-\infty$ to $+\infty$, when it is not otherwise specified.

In case the RV X is neither continuous nor discrete but contains a discrete and a continuous part, we can use an expression like (2.15) combining the two previous equations and giving

$$E(X) = \int x f(x) \, dx + \sum x_i \, p_i \tag{2.16''}$$

But there is another way to write one equation only, directly using the DF $F(x)$. This can be expressed as

$$E(X) = \int x \, dF(x) \tag{2.17}$$

which replaces (2.16") by introducing the so-called *Stieltjes integral*. If $F(x)$ has a derivative, $dF(x) = p(x)dx$, and we return to (2.16'). If $F(x)$ is a stepwise function, $dF(x)$, which is the differential increment of $F(x)$, is null for $x \neq x_i$ and $dF(x_i) = p_i$, which again gives (2.16). In the general case, (2.17) is equivalent to (2.16").

In the following we will systematically use the Stieltjes integral, and those unfamiliar with this concept may translate the equations either into sums, as (2.16), into Riemann integrals, as (2.16'), or into combinations, as (2.16").

Describing an RV by its expectation $E(X)$ only is of course quite insufficient in many applications. So it is necessary to introduce other expected values. For this purpose, consider an arbitrary function $y = h(x)$. If x is replaced by the RV X, y becomes an RV noted $h(X)$. For example, if X is a discrete RV, Y is also discrete and takes the values $y_i = h(x_i)$ with the probabilities p_i. The expectation of this RV is defined by

$$E\{h(X)\} = \int h(x)\, dF(x) \tag{2.18}$$

which is, as in (2.17), a Stieltjes integral. For discrete RVs this gives

$$E\{h(X)\} = \sum h(x_i)\, p_i \tag{2.18'}$$

and for continuous RVs we have

$$E\{h(X)\} = \int h(x)\, p(x)\, dx \tag{2.18''}$$

which is a Riemann integral. Three points should be noticed immediately. First, the function $h(x)$ is quite general and can, in particular, be complex. Secondly, (2.18) can be infinite, which means that $h(X)$ has no mean value. This can even happen for $E(X)$ of (2.16'), when $p(x)$ is such that (2.12) holds but decreases too slowly at infinity in such a way that $xp(x)$ has no finite integral. Finally, if $h(x) = 1$, the mean value is also equal to one because the integral (2.18) becomes $F(+\infty) - F(-\infty) = 1$, as a result of (2.7). More generally, if $h(x) = c$, we have $E[h(X)] = c$.

2.3.2 Moments

Among all the possible functions $h(x)$ the monomial functions such as x^k are the simplest, and probably the most important, introducing the concept of *moment*. By definition the kth-order moment of X is

$$m_k \triangleq E(X^k) = \int x^k\, dF(x) \tag{2.19}$$

It is clear that this integral can be infinite and it is easy to construct examples of DFs $F(x)$ such that m_k is finite if $k \leq N$ and infinite otherwise.

For $k = 1$, we obtain the mean value $E(X)$ again, defined by (2.17). For many distributions we have $m_1 = 0$. In this case it is said that X is a *zeromean* RV. Note that if $m_1 \neq 0$, the RV $Y = X - m_1$ is zeromean-valued. In fact, replacing $h(x)$ by $x - m_1$ in (2.18), we immediately obtain

$$E(X - m_1) = 0$$

This leads to the concept of central moments. The *central moment of order* k is defined by

$$\mu_k = E[(X - m_1)^k] \tag{2.20}$$

It is of course the moment of order k of the RV Y introduced above. By construction $\mu_1 = 0$. It is clear that $(X - m_1)^k$ can be expanded as a polynomial of degree k in X. Inserting this expression in (2.18) we obtain a sum of terms using the moments of m_i up to the order k. In other words, the central moments can be expressed in terms of the moments.

This can especially be applied to $k = 2$, which gives

$$\mu_2 = E(X^2 - 2m_1X + m_1^2) = m_2 - (m_1)^2$$

This moment is particularly important for applications and is called the *variance*, generally noted as σ^2. Thus we have

$$\sigma^2 = E[(X - m_1)^2] = m_2 - (m_1)^2 \tag{2.21}$$

The square root of the variance, σ, is called the *standard deviation* of the RV X.

The variance plays a very important role in all that follows. Using the mechanical analogy presented above, the variance represents the moment of inertia of the mass distribution with respect to the center of mass. In statistical problems the standard deviation gives an idea of the fluctuations of the RV around its mean value. In particular if $\sigma = 0$, the randomness disappears. To verify this, we note that (2.21) can be written as (2.18) in the form

$$\sigma^2 = \int (x - m_1)^2 \, dF(x) \tag{2.22}$$

Suppose now that $\sigma = 0$. The integral must then be equal to zero. But $(x - m_1)^2 \geq 0$ and $dF(x) \geq 0$, because $F(x)$ is a non-decreasing function. So we must have $dF(x) = 0$ when $(x - m_1) \neq 0$ and thus $dF(x) = 1$ for $x = m_1$. This means that $F(x) = 0$ for $x < m_1$ and $F(x) = 1$ for $x \geq m_1$, which characterizes an RV equal to m_1. So $X = m_1$, and is no longer random. This property of the variance can also be shown by means of the Tchebyshev inequality.

The Tchebyshev Inequality. We have

$$P(|X - m_1| \geq \varepsilon) \leq \frac{\sigma^2}{\varepsilon^2} \tag{2.23}$$

Proof. Let us call D the set $\{x \mid |x - m_1| \geq \varepsilon\}$. We deduce from (2.22) that

$$\sigma^2 \geq \int_D (x - m_1)^2 \, dF(x)$$

But as in D we have $(x - m_1)^2 \geq \varepsilon^2$, it follows that

$$\sigma^2 \geq \varepsilon^2 \int_D dF(x) = \varepsilon^2 P(X \in D) = \varepsilon^2 P(|X - m_1| \geq \varepsilon)$$

which leads to (2.23).

Let us discuss the consequences of this inequality. If $\sigma = 0$, we again find the result indicated above that X is no longer random. In fact, (2.23) shows that the only possible value of X is m_1. On the other hand, if we take $\varepsilon = k\sigma$, we obtain

$$P(|X - m_1| \geq k\sigma) \leq 1/k^2$$

which gives a physical interpretation of the standard deviation σ. This inequality gives a limit of the probability that X deviates from its mean value m_1 in terms of the variance.

In more specific problems it is of interest to use moments of a higher order than 2. The *skewness* of an RV is defined by

$$sk = \mu_3 / \sigma^3 \tag{2.24}$$

which is a generalization of (2.21) to the order 3. The *kurtosis* is defined by

$$\chi = (\mu_4 - 3\sigma^4)/\sigma^4 \tag{2.25}$$

We will see later that it is null for the normal distribution.

2.4 EXAMPLES OF RANDOM VARIABLES

In this section we will present the most common types of RVs used in applications. In each case we will calculate the mean value and the variance, leaving for other sections the calculation of other expected values. In the case of continuous RVs it is worth noting that if the PDF is an even, or symmetric function, which is characterized by the relation $p(x) = p(-x)$, we have

$$m_{2k+1} = 0 \tag{2.26}$$

which is a direct consequence of (2.19) where $dF(x)$ is replaced by $p(x)dx$.

Example 2.1 Uniform distribution on an interval (a, b)

This is a PDF defined by

$$p(x) = \frac{1}{b - a} r(x;\, a,\, b) \tag{2.27}$$

where $r(x; a, b)$ is the rectangular function equal to 1 for $a < x < b$ and zero otherwise. The PDF and the DF of a uniform distribution are represented in Figure 2.3.

If $b \to a$, the mean value m_1 tends to a and the variance to 0. The function $F(x)$ tends to have a step of amplitude 1 at $x = a$. The PDF becomes singular, and in reality tends to the Dirac distribution. It is very simple to calculate the mean and variance corresponding to (2.27) which are

$$m_1 = (1/2)\,(a + b) \tag{2.28}$$

$$\sigma^2 = (1/12)\,(a - b)^2 \tag{2.29}$$

Its DF, integral of $p(x)$, is of course,

$$F(x) = \frac{x - a}{b - a} \quad \text{for } a \le x \le b$$

and

$$F(x) = 0 \text{ if } x \le a \quad ; \quad F(x) = 1 \text{ if } x \ge b$$

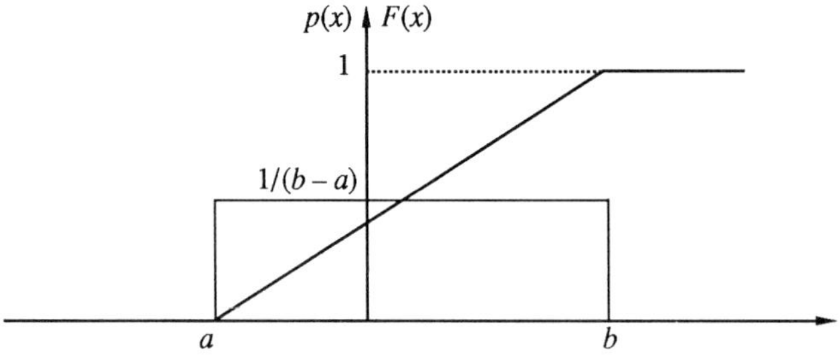

Figure 2.3 Uniform distribution.

Example 2.2 One-sided exponential distribution

This is the PDF defined by

$$p(x) = u(x)\, a \exp(-ax)\,, \quad a > 0 \tag{2.30}$$

where $u(x)$ is the unit step function equal to 0 for $x < 0$ and to 1 otherwise. Its mean and variance are

$$m_1 = 1/a \tag{2.31}$$

$$\sigma^2 = 1/a^2 \tag{2.32}$$

This function is of course defined by one parameter only.

Example 2.3 Two-sided exponential distribution

This is defined by

$$p(x) = (a/2) \exp(-a|x|), \quad a > 0 \tag{2.33}$$

This is an even function, which means that all the odd moments are null. The variance is

$$\sigma^2 = m_2 = 2/a^2 \tag{2.34}$$

It is also defined by one parameter only.

Example 2.4 Normal distribution

An RV X is said to be normal, or Gaussian, if its PDF is

$$\boxed{p(x) = (2\pi\sigma^2)^{-1/2} \exp\left\{-\frac{(x - m)^2}{2\sigma^2}\right\}} \tag{2.35}$$

This is sometimes written $N(m, \sigma^2)$. As this is an even function of $(x - m)^2$, its mean value is of course m. The form of the PDF of the distribution $N(0,1)$ is presented in Figure 2.4.

The calculation of its variance is a little more complex and is discussed in Problem 2.1. We will reconsider it later within the framework of characteristic functions. Because of the notation, it is clear that this variance is σ^2. The normal distribution is defined by the two parameters m and σ^2. It plays an *essential role* in what follows, and we will discover many of its properties below.

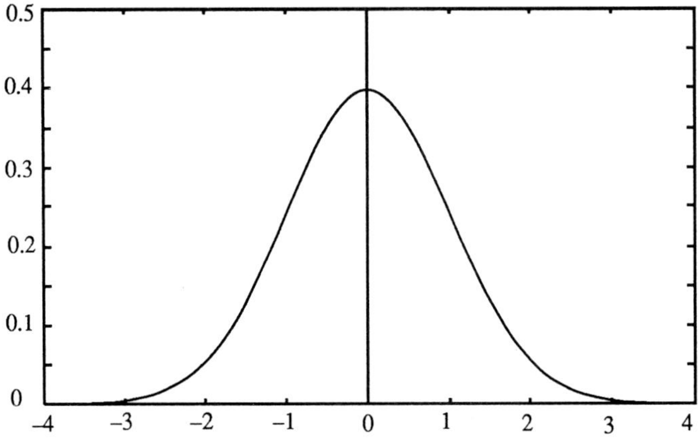

Figure 2.4 Probability density function of the distribution $N(0, 1)$.

Example 2.5 The heads or tails RV

By tossing a coin we can realize two events. Suppose that heads and tails correspond to 0 and 1 respectively. We then have an RV taking only two values, and we let p equal $P(X = 1)$. Of course $P(X = 0) = 1 - p = q$. We obviously have $X^k = X$ and so

$$E(X^k) = p \tag{2.36}$$

As a result the variance is

$$\sigma^2 = p - p^2 = pq \tag{2.37}$$

Example 2.6 Binomial distribution

Consider an RV X taking the values 0, 1, ..., n with the probabilities

$$p_k = \binom{n}{k} p^k (1 - p)^{n-k} = \binom{n}{k} p^k q^{n-k} \tag{2.38}$$

This is a binomial distribution, and starting from the expression of the binomial

$$(p + q)^n = \sum_{k=0}^{n} \binom{n}{k} p^k q^{n-k} \tag{2.39}$$

we see that (2.8) is verified because $p + q = 1$. Using the properties of the binomial coefficients, we encourage the reader to verify that

$$E(X) = n\,p \tag{2.40}$$

$$\sigma^2 = n\,p\,q \tag{2.41}$$

We will prove these relations later by another method.

Example 2.7 Poisson distribution

A Poisson-distributed RV X takes the integer values 0, 1, 2, ... with the probabilities

$$p_k = P(X = k) = \exp(-m)\,\frac{m^k}{k!} \tag{2.42}$$

As

$$\exp(m) = \sum_{k=0}^{\infty} \frac{m^k}{k!} \tag{2.43}$$

we see that (2.8) is verified. The expected value m_1 is given by

$$m_1 = \sum_{k=0}^{\infty} k\,\exp(-m)\,\frac{m^k}{k!} \tag{2.44}$$

which can be written as

$$m_1 = m\,\exp(-m) \sum_{k=1}^{\infty} \frac{m^{k-1}}{(k-1)!} = m \tag{2.45}$$

The parameter m defining the Poisson distribution (2.42) is then the expected value of X. We can calculate m_2 by the same procedure, which gives

$$m_2 = m + m^2 \tag{2.46}$$

As a result we obtain

$$\sigma^2 = m_2 - (m_1)^2 = m \tag{2.47}$$

which shows that the mean and the variance of a Poisson distribution are equal.

For some applications it is interesting to calculate the probabilities P_+ and P_- of the events A_+ and A_- to obtain an even number or an odd number respectively. It is obvious that these two events cannot appear simultaneously and satisfy (1.10). Applying (1.8), we obtain

$$P(A_+ + A_-) = P_+ + P_- = 1 \tag{2.48}$$

Furthermore, by again applying (1.8), we have

$$P_+ = \exp(-m) \sum_{k=0}^{\infty} \frac{m^{2k}}{(2k)!} \tag{2.49}$$

and the same kind of expression for P_-, which gives

$$P_+ - P_- = \exp(-m) \sum_{k=0}^{\infty} \frac{(-m)^k}{k!} = \exp(-2m) \tag{2.50}$$

Using (2.48) and (2.50) we find

$$P_+ = (1/2)\ \{1 + \exp(-2m)\} = \exp(-m)\ \text{ch}m \tag{2.51}$$

$$P_- = (1/2)\ \{1 - \exp(-2m)\} = \exp(-m)\ \text{sh}m \tag{2.52}$$

2.5 CHARACTERISTIC FUNCTIONS

The characteristic function of an RV is the expectation of $h(X) = \exp(juX)$. It is then a function $\phi(u)$ defined by (2.18) or

$$\phi(u) \triangleq E\{\exp(juX)\} = \int \exp(jux)\, dF(x) \tag{2.53}$$

In the continuous case it can be written

$$\phi(u) = \int \exp(jux)\, p(x)\, dx \tag{2.54}$$

and in the discrete case

$$\phi(u) = \sum_i \exp(jux_i)\, p_i \tag{2.55}$$

The characteristic function is of course a *complex function* of the real variable u. As $|\exp(jux)| \leq 1$, we deduce that

$$|\phi(u)| \leq \phi(0) = 1 \tag{2.56}$$

and then the integral or the series defining $\phi(u)$ are always convergent.

The characteristic function has a very important property: it is a *non-negative definite function.* This is specified by the inequality

$$\sum_{k=1}^{N} \sum_{l=1}^{N} \lambda_k \lambda_l^* \phi(u_k - u_l) \geq 0 \tag{2.57}$$

for every set of complex numbers λ_i and real numbers u_i, $1 \leq i \leq N$, and arbitrary N. To show this property, let us consider the RV Z defined by

$$Z \triangleq \sum_{k=1}^{N} \lambda_k \exp(ju_k X) \tag{2.58}$$

We have of course

$$|Z|^2 = ZZ^* = \sum_{k=1}^{N} \sum_{l=1}^{N} \lambda_k \lambda_l^* \exp\{j(u_k - u_l)X\} \tag{2.59}$$

As $|Z|^2 \geq 0$, we also have $E(|Z|^2) \geq 0$, which is (2.57).

If we are given an arbitrary function $f(u)$, it is in general very difficult to verify whether it is non-negative definite by using (2.57). Fortunately, there are other means of doing this, which will be discussed later.

Let us now present the most interesting properties of the characteristic function.

Characteristic functions and probability distributions. Suppose that X is a continuous RV. In this case $\phi(u)$ is given by (2.54) and appears as the Fourier transform (FT) of $p(x)$. Using the classical inversion formula of FT, we obtain

$$p(x) = \frac{1}{2\pi} \int \phi(u) \exp(-jux)\, du \tag{2.60}$$

This explains the term "characteristic function": there is the same information on the RV X in its PDF as in its characteristic function, although it is presented in a different way.

This allows us to understand the origin of (2.57). In fact, the function $p(x)$ is not arbitrary, but must satisfy $p(x) \geq 0$. This was the reason for the property $E(|Z|^2) \geq 0$ used to arrive at (2.57). Conversely, it is possible to show that if a complex function satisfies (2.57), its FT is non-negative. The proof is outside the scope of this discussion.

For those unfamiliar with these concepts it is important to recall that by Fourier transformation we pass from a *non-negative* function to a *non-negative definite* function. But we must not confuse these two expressions, which are unfortunately too similar. A non-negative definite function is in general complex, and thus has no reason to be non-negative itself.

Finally, as a result of (2.60) we find that the simplest way to verify if a given function is non-negative definite is to calculate its FT and verify that it is real and has no negative values.

Suppose now that X is discrete, or that $\phi(u)$ is given by (2.55). As $\delta(x)$ is the FT of the constant unit signal, we obtain

$$\delta(x) = \frac{1}{2\pi} \int \exp(-jux)\, du \tag{2.61}$$

Inserting this expression in (2.60), where $\phi(u)$ is given by (2.55), we obtain

$$p(x) = \sum_i \delta(x - x_i)\, p_i \tag{2.62}$$

which is (2.13). In conclusion, we obtain by Fourier transformation either $p(x)$ or p_i and x_i, which characterizes the probability distribution of the RV X.

Characteristic function and moments. A major interest of the characteristic function is that it gives very simply all the moments of the RV X. For this purpose, let us recall that $\exp(jux)$ is defined by

$$\exp(jux) = \sum_k \frac{j^k u^k}{k!} x^k \tag{2.63}$$

Replacing this expression in (2.53) and using (2.19), we obtain

$$\phi(u) = \sum_k \frac{j^k}{k!} m_k u^k \tag{2.64}$$

This is the McLaurin expansion of $\phi(u)$, written as

$$\phi(u) = \sum_k \frac{\phi^{(k)}[0]}{k!} u^k \tag{2.65}$$

where $\phi^{(k)}[0]$ is the value for $u = 0$ of the kth derivative of $\phi(u)$. By comparing (2.64) and (2.65) we obtain

$$m_k = j^{-k} \phi^{(k)}[0] \tag{2.66}$$

These calculations are effected provided that all the series are convergent, which requires at least that the moments be finite. As some probability distributions have no moments, some care is necessary in the use of (2.66).

Example 2.8 Some characteristic functions

We will simply take all the previous examples, calculate the corresponding characteristic functions and deduce the moments.

The characteristic function of the uniform distribution presented in Example 2.1 is

$$\phi(u) = \frac{1}{b-a} \frac{\exp(jub) - \exp(jua)}{ju} \tag{2.67}$$

In order to calculate the moments it is simpler to write the expansion of $\phi(u)$ directly than to calculate its derivatives for (2.66). Using (2.63), we obtain

$$\exp(jub) - \exp(jua) = \sum_{k=1}^{\infty} \frac{(ju)^k}{k!} (b^k - a^k) \tag{2.68}$$

The series begins from $k = 1$ because the term corresponding to $k = 0$ is null. Noting that

$$b^k - a^k = (b-a) P_k \tag{2.69}$$

with

$$P_k = a^{k-1} + a^{k-2} b + \ldots + ab^{k-2} + b^{k-1} \tag{2.70}$$

and $P_1 = 1$, we obtain

$$\phi(u) = \sum_{k=0}^{\infty} \frac{(ju)^k}{k!} \frac{P_{k+1}}{k+1} \tag{2.71}$$

Comparing this with (2.64) yields

$$m_k = \frac{P_{k+1}}{k+1} \tag{2.72}$$

With these values we again find (2.28) and (2.29).

Consider now the one-sided exponential distribution. Its characteristic function is

$$\phi(u) = \frac{a}{a - ju} \tag{2.73}$$

and we verify that this is effectively a complex function. By a geometric series expansion we obtain

$$\phi(u) = \sum_{k=0}^{\infty} \frac{j^k u^k}{a^k} \tag{2.74}$$

and, comparing this with (2.64), we obtain

$$m_k = \frac{k!}{a^k} \tag{2.75}$$

which leads to (2.31) and (2.32).

For the two-sided exponential distribution, we obtain

$$\phi(u) = \frac{a^2}{a^2 + u^2} \tag{2.76}$$

which has a Lorentzian shape. By a geometric series expansion we obtain

$$\phi(u) = \frac{1}{1 + (u^2/a^2)} = \sum_{k=0}^{\infty} (-1)^k \frac{u^{2k}}{a^{2k}} \tag{2.77}$$

Again using (2.64), we obtain

$$m_{2k+1} = 0 \quad ; \quad m_{2k} = \frac{(2k)!}{a^{2k}} \tag{2.78}$$

We verify that the odd moments are null, as they must be for any symmetric distribution. This is a consequence of the fact that $\phi(u)$ is only a function of u^2. Furthermore, the even moments of the two exponential distributions are the same, which is also normal because of the symmetry. Finally, we note that $\phi(u)$ defined by (2.76) is a positive function. It is thus simultaneously positive and positive definite. The same is, of course, true for its Fourier transform (2.33).

Consider now the very important case of *normal distribution*. The corresponding characteristic function is the Fourier transform of the probability density given by (2.35). We already know that the result is still a Gaussian function. But because of its importance, we shall now present a complete calculation. Taking $y = (x - m)/\sigma$, and after simple algebra, we find

$$\phi(u) = \exp(jum).(2\pi)^{-1/2} \int \exp\{-(1/2)(y^2 - 2ju\sigma y)\}\, dy \tag{2.79}$$

Writing

$$y^2 - 2ju\sigma y = (y - ju\sigma)^2 + u^2\sigma^2 \tag{2.80}$$

we obtain

$$\phi(u) = \exp\{jum - (1/2)\sigma^2 u^2\}\ I \tag{2.81}$$

where

$$I = \frac{1}{\sqrt{2\pi}} \int \exp\{-(1/2)(y - ju\sigma)^2\}\ dy \tag{2.82}$$

Taking $w = y - ju\sigma$, we obtain the integral of a distribution $N(0,1)$, and $I = 1$. In fact the integral in w is an integral on a straight line of the complex plane parallel to the real axis. The proof thus needs a complement analyzed in Problem 2.2. Finally, we obtain

$$\phi(u) = \exp\{jmu - (1/2)\sigma^2 u^2\} \tag{2.83}$$

If $m = 0$, this is

$$\phi(u) = \exp\{-(1/2)\sigma^2 u^2\} \tag{2.84}$$

which is a very important expression. Its expansion in power of u^2 gives

$$\phi(u) = \sum_{k=0}^{\infty} (-1/2)^k \frac{\sigma^{2k}}{k!} u^{2k} \tag{2.85}$$

and, comparing this with (2.64), we obtain $m_{2k+1} = 0$ and

$$m_{2k} = \frac{(2k)!\sigma^{2k}}{2^k\ k!} = (2k - 1)!!\ \sigma^{2k} \tag{2.86}$$

where $(2k - 1)!! = 1.3.5.\ \ldots\ (2k - 1)$. As in the previous example, we note that the odd moments are null because of the symmetry of the PDF, and that $m_4 = 3\sigma^4$ which implies that the kurtosis defined by (2.25) is null. Furthermore (2.84) gives a function which is simultaneously definite positive and positive.

Let us now consider the case of the heads or tails RV. Its characteristic function is

$$\phi(u) = 1 + p\{\exp(ju) - 1\} \tag{2.87}$$

which immediately gives (2.36).

For a binomial distribution we have

$$\phi(u) = \sum_{k=0}^{n} \binom{n}{k} p^k q^{n-k} \exp(jku) \tag{2.88}$$

and as $\exp(jku) = \{\exp(ju)\}^k$, we obtain

$$\phi(u) = \{p\ \exp(ju) + q\}^n = [1 + p\{\exp(ju) - 1\}]^n \tag{2.89}$$

which is deduced from (2.87) by taking the nth power. The calculation of the corresponding moments is proposed as a problem.

Finally, let us consider the case of Poisson distribution. Starting from (2.42), we obtain

$$\phi(u) = \exp(-m) \sum_{k=0}^{\infty} \frac{m^k}{k!} \exp(jku) \tag{2.90}$$

which immediately gives

$$\phi(u) = e^{-m} \exp\{me^{ju}\} = \exp\{m(e^{ju} - 1)\} \tag{2.91}$$

Note that the term $\exp(ju) - 1$ also appears in (2.87), (2.89) and (2.91). The reason for this is discussed in another chapter.

There is no simple expression for the general moments, but for a given value of the order it is always possible to make a limited expansion of (2.91) to arrive at these moments. For example, m_1 and m_2 given by (2.45) and (2.46) are obtained by an expansion limited to the term in u^2.

Before ending this section, let us make some comments on the so-called *moment problem*. Up to now we have used the characteristic function in order to calculate some moments, either by using (2.66) directly or, more often, by taking an expansion of $\phi(u)$ as in (2.64). Consider now the inverse problem. Suppose that we are given a sequence of numbers m_k, and consider the function $\phi(u)$ constructed by the series (2.64). Of course, if this series is not convergent there is nothing more to say. We shall thus suppose that it is convergent. There is no reason in general that the resulting function should be a non-negative definite function, and the moment problem is precisely related to the conditions ensuring this property. In other words, there is no reason why a sequence of numbers should be a sequence of moments of any distribution function. It is impossible to develop this problem here, but we can say that we already know some necessary conditions. For example, we deduce from (2.21) that the first two moments must satisfy $m_2 \geq m_1^2$. Other inequalities are discussed as problems.

Finally, let us note that the characteristic function will play an important role in many other questions discussed below, such as the sum of independent RVs, the convergence of series of RVs, and so on. For these questions a knowledge of vector RVs, presented in the following chapters, is required.

2.6 THE SECOND CHARACTERISTIC FUNCTION

The second characteristic function, or log characteristic function, denoted $\psi(u)$, is defined by

$$\psi(u) = \ln\{\phi(u)\} \tag{2.92}$$

Note that, as $\phi(u)$ is a complex number, $\psi(u)$ is a multiform function and it is then necessary to choose an arbitrary determination.

This function exhibits some interesting properties, and has a particularly simple form for some distributions. Thus we deduce from (2.83) that the second characteristic function of a distribution $N(m, \sigma^2)$ is simply

$$\psi(u) = jmu - (1/2)\sigma^2 u^2 \tag{2.93}$$

Similarly, we obtain for a Poisson distribution the result deduced from (2.91)

$$\psi(u) = m\{\exp(ju) - 1\} \tag{2.94}$$

The McLaurin expansion of $\psi(u)$ introduces coefficients c_k defined as m_k in (2.64) and called the *cumulants* of the distribution. Thus the cumulants of a normal distribution are all null except for the first two. There is of course the same information in the cumulants as in the moments, but for some problems cumulants can be more interesting, as discussed later. Relations between first moments and cumulants are analyzed in Problem 2.3.

2.7 FUNCTIONS OF RANDOM VARIABLES

In many signal processing problems we must consider functions of an RV. For example a noise current, which is a signal $x(t)$ as presented in Figure 1.2, at a time instant t is an RV X. But this RV can be observed after a square law rectifier giving $Y = X^2$. Thus the square of an RV is an example of the function of this RV.

More precisely, consider an RV X and a function $y = h(x)$, as used in (2.18). It introduces an RV Y defined by

$$Y = h(X) \tag{2.95}$$

which is sometimes called a deterministic function of an RV.

There is no particular problem in calculating the expected values of Y. For example, the moments of Y are obtained from (2.18), where $h(x)$ is replaced by $[h(x)]^k$.

On the other hand, the calculation of the DF $F_Y(y)$ of Y from $F_X(x)$, DF of X, is a little more complex. This is due to the fact that the function $h(x)$ is not necessarily invertible. To explain this point, let us return to the square, or $Y = X^2$. By definition $F_Y(y) = P(Y \leq y)$. If $y < 0$, this probability is null. If $y > 0$, the event $Y \leq y$ is realized if $-\sqrt{y} \leq X \leq \sqrt{y}$. Using (1.8) and (2.6), we deduce

$$F_Y(y) = F_X(\sqrt{y}) - F_X(-\sqrt{y}) + P(X = -\sqrt{y}) \tag{2.96}$$

By derivation with respect to y we obtain the probability density.

As the difficulty comes from the possible absence of the inverse function of $h(x)$, let us first consider the simpler case, where $h(x)$ is a strictly monotonic transformation. Suppose for example that $h(x)$ is an increasing function, or that its derivative is always positive. Then the event $Y \leq y$, or $h(X) \leq y$, is equivalent to the event $X \leq h^{-1}(y)$. As a result we have

$$F_Y(y) = F_X\{h^{-1}(y)\} \tag{2.97}$$

where F_X and F_Y are the DFs of X and Y respectively. Now suppose that $h(x)$ is decreasing. In this case the event $Y \leq y$ appears if $X \geq h^{-1}(y)$. Using (1.8) we note that $P(X \geq x) = P(X > x) + P(X = x)$ and as $P(X > x) = 1 - P(X \leq x)$ or $1 - F_X(x)$, we deduce that

$$F_Y(y) = 1 - F_X\{h^{-1}(y)\} + P\{X = h^{-1}(y)\} \tag{2.98}$$

The last term of course disappears in the case of a continuous RV.

In this latter case it is sometimes more interesting to use the PDF instead of the DF. This density is obtained by the derivation of (2.97) or (2.98), which gives

$$p_Y(y) = \frac{p_X\{h^{-1}(y)\}}{|h'\{h^{-1}(y)\}|} \tag{2.99}$$

where h' is the derivative of h which is negative in the case resulting in (2.98).

When the transformation $y = h(x)$ is non-monotone, there is no general expression for finding $F_Y(y)$, but the general method is exactly the same. Let us call $D(y)$ the domain of the real line corresponding to the points x such that $h(x) \le y$. The event $Y \le y$ is the same as the event $X \in D(y)$, and we deduce

$$F_Y(y) = \int_{D(y)} dF_X(x) \tag{2.100}$$

Some examples are discussed as problems.

The results of this section can be interestingly applied in the generation of *random numbers* with a specific distribution. There are many computer programs for generating random numbers which are uniformly distributed and correspond to independent trials. This last property is rather difficult to show in practice, while distribution is easy to measure. Suppose, for simplicity, that these random numbers x_i are uniformly distributed over the interval [0, 1]. This means that their PDF is given by (2.27) where $a = 0$ and $b = 1$, represented in Figure 2.3. By using a transformation such as (2.95) it is possible to transform the sequence of random numbers X_i into another sequence Y_i with another PDF. The basic idea of this transformation is as follows. Let us suppose that in the transformation (2.95) the function $h(x)$ is simply $F_X(x)$, and assume that $F_X(x)$ is everywhere differentiable, which means that the RV X is continuous. As the derivative of $F_X(x)$ is the PDF, we deduce that the numerator and the denominator of (2.99) are equal, which means that $p_Y(y)$ is constant. But, as $0 \le F(x) \le 1$, because it is a probability, we deduce that the RV Y defined by (2.95) takes its values in the interval [0, 1]. So it is uniformly distributed in this interval with the PDF (2.27) where $a = 0$ and $b = 1$. In conclusion, any continuous PDF can be transformed into an RV uniformly distributed in [0, 1] by using the transformation $Y = F_X(X)$. Using the inverse transformation it is possible to transform a uniform distribution in [0, 1] into any other arbitrary distribution. This is the method applied to generate random numbers with arbitrary distributions from sequences of random numbers uniformly distributed.

PROBLEMS

2.1 Consider the PDF (2.35) where $m = 0$ and $\sigma^2 = 1$. Show that $p(x)$ has an integral equal to one and that the variance associated with $p(x)$ is also equal to one.

2.2 In order to calculate the integral I appearing in (2.82), let us call $f(z)$ the function $(2\pi)^{-1/2}\exp\{-(1/2)z^2\}$ and consider the straight lines, or sets of points $z = x + jy$ of the complex plane, defined by $D(a) = \{z \mid z = x, -a \le x \le a\}$, $D_-(a) = \{z \mid z = x - ju\sigma, -a \le x \le a\}$, $\Delta_+(a) = \{z \mid z = a - juy, 0 \le y \le \sigma\}$ and $\Delta_-(a) = \{z \mid z = -a - juy, 0 \le y \le \sigma\}$.

(a) Show that $D(a) + D_-(a) + \Delta_+(a) + \Delta_-(a)$ introduces a closed contour C.
(b) Show that the integral of $f(z)$ on C is null.
(c) Show that the two integrals of $f(z)$ on $\Delta_+(a)$ and $\Delta_-(a)$ tend to zero when a tends to infinity.
(d) Deduce that $I = 1$.

2.3 Relations between the first moments and cumulants. The moments m_k are defined from the expansion of $\phi(u)$ by (2.64). The cumulants c_k are defined by the same equation where $\phi(u)$ is replaced by $\psi(u)$ defined by (2.92) and m_k is replaced by c_k.
(a) Writing $\phi(u)$ as $\phi(u) = 1 + T$, expand $\psi(u)$ as a series in T.
(b) Using (2.64) limited to the term u^4, write an expression of $\psi(u)$ limited to the same term.
(c) Deduce the cumulants c_i, $i \leq 4$, in terms of the moments m_k, $k \leq 4$.
(d) Conversely express these moments in terms of these cumulants.
(e) Write the same expressions when $m_1 = 0$ or $c_1 = 0$.

2.4 Generating function. Consider an RV X taking only non-negative integer values with the probabilities $p_k = P(X = k)$. The generating function of X is defined by $g(s) = E(s^X)$, where s is a complex number.
(a) Calculate $g(1)$ and deduce that $g(s)$ is convergent inside the unit circle.
(b) By a McLaurin expansion of $g(s)$, show that p_k can be deduced from $g(s)$.
(c) The factorial moments f_k of X are defined by $f_k = E\{X(X-1)(X-2)...(X-k+1)\}$, $k \geq 1$, where k is integer. Calculate f_k in terms of $g(s)$.
(d) Express $g(s)$ in terms of the characteristic function $\phi(u)$.
(e) Calculate $g(s)$ and the factorial moments of a binomial and a Poisson RV.

2.5 Likelihood ratio. Let us consider a continuous RV X with the probability density $p_0(x)$ under the hypothesis H_0 and $p_1(x)$ under H_1. The likelihood ratio appearing in the decision between H_0 and H_1 is defined by $L(x) = p_1(x)/p_0(x)$. In what follows we call Y the RV $Y = L(X)$.
(a) Let us call m_k and μ_l the moments of Y when X has the density $p_0(x)$ and $p_1(x)$ respectively. Establish the relation between m_k and μ_l. Calculate μ_0.
(b) Let us call $\phi_0(u)$ and $\phi_1(u)$ the characteristic functions of the RV Y when X has the density $p_0(x)$ and $p_1(x)$ respectively. Express $\phi_1(u)$ in terms of $\phi_0(u)$.
(c) Deduce the relation between the probability densities $q_0(y)$ and $q_1(y)$ of Y corresponding to $\phi_0(u)$ and $\phi_1(u)$.

2.6 Moments and cumulants of a binomial distribution.
(a) Using the results of Problem 2.3, calculate the first four cumulants of the heads and tails RV. Verify the results by expanding the second characteristic function $\psi_1(u)$.
(b) Calculate the second characteristic function $\psi_2(u)$ of the binomial distribution defined by (2.38). Express $\psi_2(u)$ in terms of $\psi_1(u)$, the two functions being calculated for the same value p.
(c) Deduce the first four cumulants of the binomial distribution.
(d) Using the results of Problem 2.3 again, calculate the corresponding first four moments.

2.7 Calculate the moments m_i, $1 \le i \le 4$, of a Poisson RV with parameter m either from (2.42) or from (2.91). Give the expression of the moment m_{p+1} in terms of $m_1, m_2, \ldots, m_p$.

2.8 **Mixture of Poisson distributions.** Consider a discrete valued RV X taking the integer values k ($k \ge 0$) with the probabilities

$$p_k = P(X = k) = \int_0^{+\infty} \exp(-m) \frac{m^k}{k!} w(m)\, dm$$

where $w(m)$ is a weighting function or a mixture function. The RV X is sometimes called a Poisson compound RV.

(a) Find the condition on $w(m)$ ensuring that (2.8) is satisfied.
(b) Show that if $w(m)$ is a probability density, then $p_k > 0$.
(c) Express the characteristic and generating functions of X in terms of $w(m)$.
(d) Calculate the first two moments of X in terms of $w(m)$.
(e) Suppose that $w(m)$ is the probability density of an RV M with finite variance. Express the variance σ^2 of X in terms of the mean value and the variance σ_M^2 of M. Deduce that $\sigma^2 > E(X)$, which means that the variance is greater than the one corresponding to a Poisson distribution.
(f) Suppose that M has a one-sided exponential distribution or that $w(m)$ is given by (2.30). Calculate in this case p_k, the characteristic and generating functions and the factorial moments of X.
(g) Suppose that $w(m) = (\lambda + 1)a\exp(-ax) - \lambda b\exp(-bx)$, where $\lambda > 0$ and $0 < a < b$. Find the values of λ such that: (i) $w(m) > 0$, $\forall\, m$; (ii) $p_k > 0, \forall\, k$; (iii) $\sigma^2 < E(X)$.
(h) In the set of values of λ such that $p_k > 0$, determine the subsets corresponding to the three situations: $w(m) > 0$; $w(m) \not> 0$ and $\sigma^2 > E(X)$; $w(m) \not> 0$ and $\sigma^2 < E(X)$. Comment on these results.

2.9 **The error function.** This function erf x is defined by

$$\operatorname{erf} x = (2\pi)^{-1/2} \int_x^{+\infty} \exp(-t^2/2)\, dt$$

It is represented in Figure 2.5.

(a) Calculate erf$(-\infty)$, erf 0, and erf$(+\infty)$. Establish the relation between erf(x) and erf$(-x)$.
(b) Calculate the distribution function $F(x)$ of the RV $N(m, \sigma^2)$ in terms of erf$x$.
(c) By expanding $\exp(-t^2/2)$ in McLaurin series and expressing erf$x$ with an integral from 0 to x, find the series expansion of erf$x$.
(d) Show that the terms of this series can be calculated recursively.

2.10 Starting with a continuous RV X we construct the RV $Y = X^2$.
(a) Calculate the PDF of Y in terms of the PDF of X. What is the result when the PDF of X is symmetric?
(b) Apply this result to the case where X has the distribution $N(0, \sigma^2)$.
(c) Make the same calculation when the PDF of X is given by (2.30).

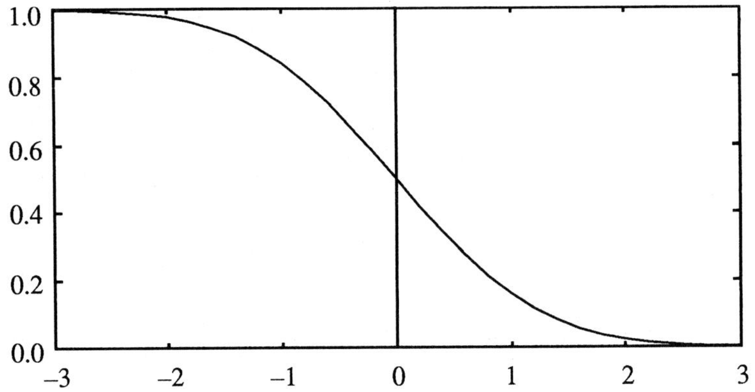

Figure 2.5 The error function erf x.

2.11 Consider the signal $X = \cos\phi$ where ϕ is a random phase uniformly distributed over the interval 0, 2π. Calculate the PDF $p_X(x)$ of X.

2.12 The PDF of a Cauchy distribution is given by $p(x) = a\,[\pi\,(a^2 + x^2)]^{-1}$. Calculate its characteristic function. What are the moments of this distribution?

2.13 By considering the RVs $(X^2 - m_2)^2$ and $(X^2 - X)^2$ find two inequalities between the moments m_2, m_3, and m_4.

Chapter 3

Second Order Random Vectors

3.1 INTRODUCTION

Up until now we have considered only one real scalar RV. For many purposes, however, and especially in the case of random signals, we are obliged to study many RVs simultaneously. As an example, let us take the situation described on p. 2. Instead of measuring the temperature at a given location every day at the same time, we can take the same measurement in two different locations. The result of the experiment is now composed of two numbers which can be viewed as a two-dimensional RV, studied in the next section. Common sense tells us that if the locations are very close, the temperatures can exhibit similar behavior, while in the reverse case there is almost no relation between the results of the random experiment. In the first case we shall say that the events are *highly correlated* while in the second case they are almost *independent*.

Take now a signal such as is presented in Figure 1.2. Instead of using one noise generator we can use two or many more generators. We then collect an ensemble of signals which can be considered as the components of a random vector signal. At a given time instant this vector signal is a random vector.

In the following we make no distinction between the term "random vector" and the term "multidimensional random variable." The latter expression is probably more commonly used when studying two-dimensional vectors, as discussed in the next section.

As indicated just above, an important point in the study of random vectors is the evaluation of the correlation between the components. This term has an intuitive meaning and is often used in everyday life when we try to establish a correlation between certain events. Of course, within the framework of signal theory, we need a precise mathematical definition of the concept of *correlation*. This is one of the main objectives of this chapter because second order random vectors and correlation properties are intimately connected.

To facilitate matters for those not so familiar with the concepts of *linear algebra*, we begin with a section devoted to two-dimensional random vectors. For these vectors the use of components is quite sufficient and vector notations are not necessary. However, these latter are so powerful in the case of general vectors that we do not hesitate to use them systematically. We also encourage the reader to use this notation, now commonly used by all those working in signal processing or in the communication field.

3.2 TWO-DIMENSIONAL RANDOM VARIABLES

Consider a pair of two real RVs $X(\omega)$ and $Y(\omega)$ defined on the probability space $(\Omega, \mathcal{F}, P)$. They can be considered as the components of a random vector $\mathbf{V}(\omega)$ of the space $\mathbb{R}^2$. Such a vector is presented in Figure 3.1. In this figure an arbitrary domain D also appears, and to any such domain we can associate the event $A(D)$ which is the event when the random vector $\mathbf{V}(\omega)$ lies inside this domain. It is also the event when the point $M(\omega)$ with coordinates $X(\omega)$ and $Y(\omega)$ belongs to D.

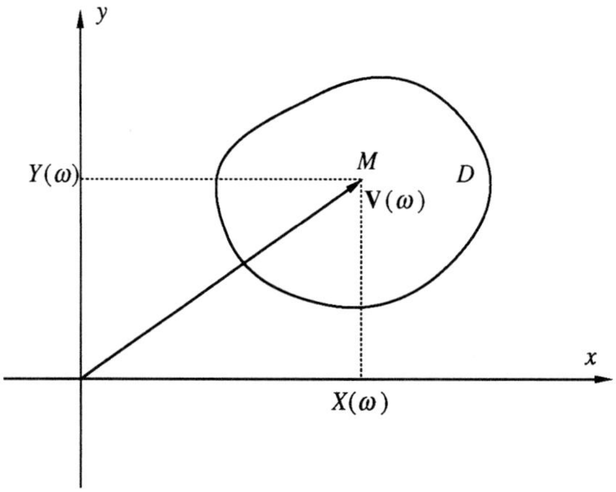

Figure 3.1 Two-dimensional random vector.

To any such event it is possible to assign a probability $P(D)$. This is the probability measure associated with the set of points ω defined by

$$A(D) = \{\omega | M(\omega) \in D\}$$

As in the scalar case, we can consider the set $A(x, y)$ of points ω such that

$$A(x, y) = \{\omega | X(\omega) \leq x \text{ and } Y(\omega) \leq y\} \tag{3.1}$$

The probability measure of this set is the *distribution function* $F(x, y)$ of the two-dimensional RV $X(\omega)$, $Y(\omega)$. For the following we shall ignore the probability space and work only with DFs. Let us first indicate their more important properties.

3.2.1 Distribution Functions

From the previous discussion the DF of a two-dimensional RV is defined by

$$F(x, y) = P\{(X \leq x) \,.\, (Y \leq y)\} \tag{3.2}$$

This DF is sometimes called a *bivariate* DF. It satisfies the same kind of general properties as those studied previously in 2.2.2. Among these, note that $F(x, y)$ is a non-decreasing function of x and y. Note also that if x and y become infinite, $F(x, y)$ is equal to one. Physically $F(x, y)$ is the probability associated with the domain D of the plane appearing in Figure 3.2.

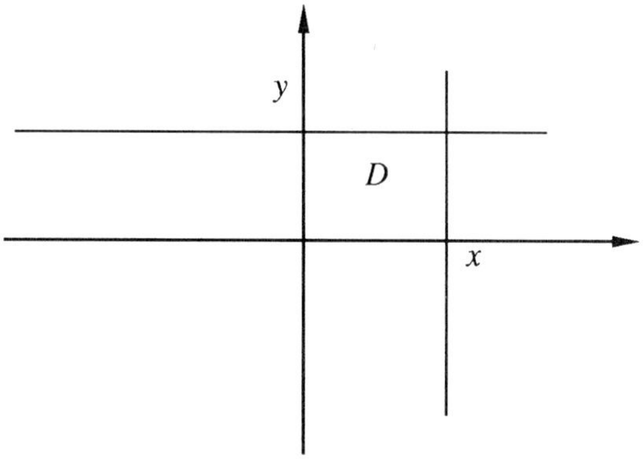

Figure 3.2 Domain D corresponding to $F(x, y)$.

More generally, the knowledge of $F(x, y)$ allows us to calculate the probability associated with any arbitrary domain D of the plane by the expression

$$P(D) = \iint_D d^2F(x, y) \tag{3.3}$$

which is the generalization of (2.100) for the two-dimensional case. If D is the whole plane we get $F(+\infty, +\infty)$ which is then equal to one.

The notation in (3.3) is a mathematical means of avoiding any differentiation between continuous and discrete RVs. This method was also used in the previous chapter when defining moments by (2.18). But, as before, it is also possible to treat the two cases separately.

The two-dimensional RV (X, Y) is continuous if $F(x, y)$ has a derivative or if we can write

$$p(x, y) = \frac{\partial^2 F(x, y)}{\partial x \partial y} \tag{3.4}$$

which is called the probability density function of (X, Y). In this case (3.3) becomes

$$P(D) = \iint_D p(x, y)\, dx\, dy \tag{3.5}$$

On the other hand, if $F(x, y)$ is a step-wise function with steps at points x_i, y_j of the plane, the two-dimensional RV (X, Y) is discrete and takes only the values x_i and y_j with the probabilities

$$p(i, j) = p_{i,j} = P\{(X = x_i) \,.\, (Y = y_j)\} \tag{3.6}$$

As in (3.5), the probability associated with the domain D is now

$$P(D) = \sum_D p(i, j) \tag{3.7}$$

where the sum is extended to all those points (x_i, y_j) belonging to D. Combinations of continuous and discrete distributions are, of course, easy to imagine.

Marginal distributions. Each component of the random vector $\mathbf{V}(\omega)$ is a scalar RV and can be considered separately. For example, if $X(\omega)$ and $Y(\omega)$ are the temperatures, measured at the same time in two different cities, it is in some cases possible to ignore $Y(\omega)$ and to study the temperatures in the first city only. This introduces the concept of marginal distribution.

More precisely, the knowledge of $F(x, y)$ defined by (3.2) is sufficient to introduce the DF of $X(\omega)$ alone, which is called the *marginal distribution function.* As this function is defined by

$$F_X(x) = P[X \leq x]$$

we see that it is deduced from (3.2) by replacing y by $+\infty$, or

$$F_X(x) = F(x, +\infty) \tag{3.8}$$

It corresponds to the domain D of the plane defined only by $X \leq x$. The same can be said concerning y. As a result we can see that the marginal distribution is simply obtained by replacing one of the two variables x or y by $+\infty$. In our example this means that we now consider X as a scalar RV, regardless of the value of Y, which is completely dropped. Note that the marginal distribution is sometimes called *a priori* distribution.

Let us apply this to the continuous and discrete cases successively. Using (3.4) we obtain

$$d^2F(x, y) = p(x, y)\, dx\, dy \tag{3.9}$$

which was applied in (3.5). Furthermore, (3.8) can be written

$$F_X(x) = \int_{-\infty}^{x} \int_{-\infty}^{+\infty} p(x', y')\, dx'\, dy' \tag{3.10}$$

and by derivation with respect to x we obtain the marginal PDF of X which is

$$p_X(x) = \int p(x, y)\, dy \tag{3.11}$$

Similarly by integration over x we obtain the marginal PDF $p_Y(y)$ of Y. The marginal PDF is then obtained simply by integration over the other variable.

The same is valid for the discrete case and, for example, the marginal probability p_i is given by

$$p_i = \sum_j p_{i,j} \tag{3.12}$$

3.2.2 Expectation

The relation (2.18) can be translated into the two-dimensional case and becomes

$$E\{h(X, Y)\} = \iint h(x, y)\, d^2F(x, y) \tag{3.13}$$

insofar as it exists. It is important to note that this integral is a two-dimensional *Stieltjes integral*, which is a very powerful tool to calculate the expectations when it is not known whether the random vector is a continuous or a discrete RV. As in the scalar case, those unfamiliar with this integral can use separate expressions for the two cases, which are

$$E\{h(X, Y)\} = \iint h(x, y)\, p(x, y)\, dx\, dy$$

or

$$E\{h(X, Y)\} = \sum_{i,j} h(x_i, x_j)\, p(i, j)$$

Let us now extract some interesting results from these relations. If $h(x, y)$ is only a function of x, say $h(x)$, we obtain

$$E\{h(X)\} = \iint h(x)\, d^2F(x, y) \tag{3.14}$$

But integration over the variable y gives the marginal distribution $F_X(x)$ and as a result (3.14) is equivalent to (2.18). This becomes particularly obvious in the continuous case, and using (3.9) and (3.11) we obtain

$$E\{h(X)\} = \iint h(x)\, p(x, y)\, dx\, dy = \int h(x)\, p_X(x)\, dx \tag{3.15}$$

This shows that in order to calculate the expectation of a function of X only, we can completely ignore the second RV Y, or the random vector, and consider X as a scalar RV defined by its marginal distribution. The same reasoning can obviously be applied to a function of Y only.

Suppose now that $h(x, y) = x + y$. Applying (3.14) twice, we obtain

$$E(X + Y) = E(X) + E(Y) \tag{3.16}$$

which means that the expectation of the sum of two RVs is the sum of their expectation, regardless of whether they are or are not independent.

Now consider the case of the product, or $h(x, y) = xy$. Applying (3.13) we see that in general

$$E(XY) \neq E(X)\,E(Y) \tag{3.17}$$

The consideration of the expectation of a product leads to a more careful investigation of the case of random vectors with *independent components*.

The concept of independent events A and B is characterized by (1.29). When applied to a pair of scalar RVs this allows us to define the concept of independent RVs. Two RVs X and Y are independent if, whatever x and y, we have

$$P[(X \le x) \,.\, (Y \le y)] = P(X \le x) \,.\, P(Y \le y)$$

In terms of distribution functions, this gives

$$F(x, y) = F_X(x)\,F_Y(y) \tag{3.18}$$

In the case of continuous RVs, it results from (3.4) that the *independence* is characterized by

$$p(x, y) = p_X(x)\,p_Y(y) \tag{3.18'}$$

For discrete RVs this becomes

$$p(i, j) = p_X(i)\,p_Y(j) \tag{3.18''}$$

and more generally (3.18) implies that $d^2F(x, y) = dF_X(x)\,dF_Y(y)$. Inserting this in (3.13) we deduce that if X and Y are *independent* we have

$$E[h(X)\,g(Y)] = E[h(X)]\,E[g(Y)]$$

whatever the functions $h(.)$ or $g(.)$, provided that the expected values do exist. This gives, more particularly,

$$E(XY) = E(X)\,E(Y) \tag{3.19}$$

and (3.17) does not hold for independent RVs.

It is important to note that the converse is not true and (3.19) can appear even for dependent RVs. To distinguish the two situations, we say that (3.19) characterizes *uncorrelated* RVs. As a consequence two independent RVs are uncorrelated, while two uncorrelated RVs can be dependent. Let us discuss this point using some examples.

Example 3.1 Uncorrelated and dependent RVs. Consider a PDF in the form

$$p(x, y) = f(x^2 + y^2) \tag{3.20}$$

where $f(.) \ge 0$ and such that the integral (3.5) over the whole plane is one. Applying (3.15) we obtain $E(X) = E(Y) = 0$, because $p(x, y) = p(-x, y) = p(x, -y) = p(-x, -y)$. Applying (3.13), we now obtain

$$E(XY) = \iint xy\ f(x^2 + y^2)\ dx\ dy = 0 \qquad (3.21)$$

In fact, if we begin by the integral over x, we obtain zero, always because of the parity of $f(.)$. We then have $E(XY) = E(X)E(Y) = 0$. But as $f(x^2 + y^2)$ is not a product as is (3.18'), the variables X and Y are not independent. However, if $f(.)$ takes an exponential form, we obtain a product which gives the independence. We shall see later that this characterizes two-dimensional Gaussian RVs, for which uncorrelation and independence are equivalent.

Now consider the two RVs X and Y defined by

$$X = \cos\phi \quad ; \quad Y = \sin\phi \qquad (3.22)$$

where ϕ is an RV uniformly distributed between 0 and 2π. This leads to

$$E(X) = E(Y) = E(XY) = 0 \qquad (3.23)$$

which means that X and Y are uncorrelated. But as $Y^2 = 1 - X^2$, these two RVs are not independent because if X is known, we can deduce the value of Y^2.

3.3 SECOND ORDER PROPERTIES

Consider a two-dimensional RV (X, Y) and let us call m_x and m_y the mean values of X and Y, or

$$m_x = E(X) \quad ; \quad m_y = E(Y) \qquad (3.24)$$

These mean values are calculated by (3.14) where $h(x)$ is simply replaced by x for m_x and by y for m_y. For continuous RVs it is simpler to use (3.15). If $m_x = m_y = 0$, the RV(X, Y) is said to be *zeromean*. If that is not the case, it is clear that the RV $(X - m_x, Y - m_y)$ is zeromean-valued.

The variances σ_x^2 and σ_y^2 of X and Y respectively are defined by (2.21) applied to X and Y successively and by replacing m_1 by m_x or m_y.

The *covariance* γ_{xy}, sometimes written as $\mathrm{Cov}(x, y)$, is defined by

$$\gamma_{xy} = E\{(X - m_x)\ (Y - m_y)\} \qquad (3.25)$$

By developing the product, we obtain

$$\gamma_{xy} = E(XY) - m_x\, m_y \qquad (3.26)$$

which is the extension of (2.21) for covariance. As a result of (3.19) and of the definition which follows, two uncorrelated RVs have a null covariance. This can introduce some confusion, as discussed in more detail further down, because the concepts of covariance and correlation are different. It is obvious that

$$\sigma_x^2 = \gamma_{xx} \qquad (3.27)$$

which is another form of (2.21). For zeromean RVs, the covariance is simply $E(XY)$.

The Schwarz inequality for random variables. Consider two RVs A and B and an arbitrary number λ. As $(\lambda A + B)^2 \geq 0$, we also have

$$E\{(\lambda A + B)^2\} = \lambda^2 E(A^2) + 2\lambda E(AB) + E(B^2) \geq 0 \tag{3.28}$$

As this polynomial in λ is always non-negative, it must have no real roots, or only a double root. Thus we have

$$\{E(AB)\}^2 \leq E(A^2) E(B^2) \tag{3.29}$$

which is the Schwarz inequality. If the *equality* holds, there is a double root λ_0, for which $E\{(\lambda_0 A + B)^2\} = 0$. If this is realized we conclude, using the same argument as after (2.22), that

$$\lambda_0 A + B = 0 \tag{3.30}$$

almost surely, or with probability one. This means that the two RVs A and B are almost surely proportional.

Let us apply this result to various situations. Suppose first that $A = X$ and $B = 1$. We deduce

$$\{E(X)\}^2 \leq E(X^2) \tag{3.31}$$

Comparing this with (2.22), this means that $\sigma_x^2 \geq 0$, a fact deduced directly from (2.21). If the equality holds, we have $X = m_x$, as indicated after (2.22). Furthermore, this shows that, if $E(X^2) < +\infty$, the mean value m_x is also finite. In this case we say that X is a *second order random variable.*

Suppose now that $A = X - m_x$ and $B = Y - m_y$. Using (3.29), (3.25) and (2.21), we obtain

$$|\gamma_{xy}| \leq \sigma_x \sigma_y \tag{3.32}$$

Introducing the *correlation coefficient* c defined by

$$c \triangleq \frac{\gamma_{xy}}{\sigma_x \sigma_y} \tag{3.33}$$

we deduce from (3.32) that $|c| \leq 1$.

If $c = 0$, X and Y are said to be *uncorrelated.* On the other hand, if $|c| = 1$, X and Y are almost surely proportional and are sometimes called fully correlated. Thus the correlation coefficient gives information on the statistical relation between X and Y. Note, as seen before, that $c = 0$ does not mean that X and Y are independent. Note also that if we replace X and Y by $\alpha(X - a)$ and $\beta(Y - b)$ respectively, where $\alpha > 0$ and $\beta > 0$, the correlation coefficient c is invariant. Finally, note that c can have negative values. That is, for example, the case if $Y = -X$, which gives $c = -1$.

The correlation coefficient appears in the calculation of the *variance of the sum* of two RVs. Suppose, for simplicity, that they are zeromean. We then have

$$\sigma_{x+y}^2 \triangleq E\{(X + Y)^2\} \tag{3.34}$$

By expanding this and using (3.33), we obtain

$$\sigma_{x+y}^2 = \sigma_x^2 + \sigma_y^2 + 2\, c\, \sigma_x\, \sigma_y \tag{3.35}$$

As $-1 \le c \le +1$, we have

$$(\sigma_x - \sigma_y)^2 \le \sigma_{x+y}^2 \le (\sigma_x + \sigma_y)^2 \tag{3.36}$$

and for $c = 0$ we have

$$\sigma_{x+y}^2 = \sigma_x^2 + \sigma_y^2 \tag{3.37}$$

Thus we have an additivity rule for the variance, similar to (3.16), but valid only for *uncorrelated RVs*. It is sometimes said that it is an incoherent sum, while the upper bound of (3.36) corresponds to a coherent sum, giving the maximum value of the variance of the sum.

Extensions to complex random variables. For many purposes discussed later, it is necessary to use *complex RVs*. This is especially the case when considering random currents or voltages for which the complex notation is particularly well adapted, as known in circuit theory. Note that complex RVs have already been used in (2.53) for the definition of characteristic functions.

As a complex number is nothing less than a pair of two real numbers, an RV $Z(\omega)$ is simply

$$Z(\omega) = X(\omega) + j\, Y(\omega) \tag{3.38}$$

where $X(\omega)$ and $Y(\omega)$, real and imaginary parts, constitute a two-dimensional RV. It is then completely described by a DF $F(x, y)$ similar to (3.2). However, note that this function is not necessarily a function of $z = x + jy$, and, in particular, there is no sense in talking about a probability such as $P(Z \le z)$, because the concept of inequality is not valid for complex numbers.

In other words, when discussing complex RVs the simplest way is to return to the real case by using (3.38). Nevertheless, this is not always absolutely necessary, especially when using only moments. For example, the mean value and the variance of Z are defined by

$$m_z = E(Z) \tag{3.39}$$

$$\sigma_z^2 = E\{|Z - m_z|^2\} \tag{3.40}$$

It is easy to verify that

$$\sigma_z^2 = E(|Z|^2) - |m_z|^2 \tag{3.41}$$

These moments can be expressed in terms of X and Y by

$$m_z = m_x + j m_y \tag{3.42}$$

$$\sigma_z^2 = E(X^2) + E(Y^2) - m_x^2 - m_y^2 \tag{3.43}$$

and using (3.26) and (3.27) this gives

$$\sigma_z^2 = \sigma_x^2 + \sigma_y^2 \tag{3.44}$$

Note that this expression is valid whatever the correlation between X and Y, and it is important to avoid any confusion with (3.37).

Let us now introduce the concept of covariance between two complex RVs. Suppose that we have a pair of complex RVs (U, V), defined as Z in (3.38). This corresponds to an ensemble of four real RVs defined by a distribution function as in (3.2). The mean values and variances of U and V are defined as above. The covariance, extension of (3.25) for the complex case, is defined by

$$\gamma_{uv} \triangleq E\{(U - m_u)(V - m_v)^*\} \tag{3.45}$$

where the star means the complex conjugate. If $U = V$, we find σ_u^2 defined again by (3.40). It is obvious that, as m_u and m_v, γ_{uv} is a complex number. It is also possible to return to the real and imaginary parts defined by

$$U = X + jY \quad ; \quad V = S + jT \tag{3.46}$$

where X, Y, S and T are real. After a simple calculation we obtain

$$\gamma_{uv} = \gamma_{xs} + \gamma_{yt} + j(\gamma_{ys} - \gamma_{xt}) \tag{3.47}$$

If $U = V$, we again find $\gamma_{uu} = \gamma_{xx} + \gamma_{yy}$ given by (3.44). It is important to note that γ_{uv} is, in general, a complex number. We also note that the knowledge of the complex covariance γ_{uv} is not sufficient to determine the covariances between the real and imaginary parts of U and V because we have only two equations to determine the four real covariances appearing in (3.47). This problem will be discussed in more detail later.

From (3.45) we immediately deduce that

$$\gamma_{uv} = \gamma_{vu}^* \tag{3.48}$$

Furthermore, the Schwarz inequality can be extended in the complex case, and instead of (3.29) we now have

$$|E(AB^*)|^2 \leq E(|A|^2)\ E(|B|^2) \tag{3.49}$$

From this we deduce that

$$|\gamma_{uv}|^2 \leq \gamma_{uu}\ \gamma_{vv} \tag{3.50}$$

and the whole discussion on the correlation coefficient can be transposed without difficulty to the complex case.

Finally, we can introduce a covariance matrix defined by

$$\Gamma = \begin{bmatrix} \gamma_{uu} & \gamma_{uv} \\ \gamma_{vu} & \gamma_{vv} \end{bmatrix} \tag{3.51}$$

We already know that it is Hermitian, because of (3.48), and that its determinant is non-negative, because of (3.50). This will be extensively studied in a following section.

3.4 RANDOM VECTORS

As stated in the introduction, this term is equivalent to that of multidimensional RVs, by an extension of the term of two-dimensional RVs used above.

Consider an ensemble of n real RVs $X_1(\omega)$, $X_2(\omega)$, ..., $X_n(\omega)$. These RVs can be considered as the components of a vector of $\mathbb{R}^n$ written $\mathbf{X}(\omega)$, which is then a random vector or a vector RV. The case $n = 2$ studied above corresponds to a random vector in a plane. Before continuing with the statistical description of random vectors, let us give some examples of physical situations in which they can be used. There are plenty of these, but we provide only a few for a better understanding of the practical importance of random vectors.

In many modern signal processors we use multiple sensors, each of which gives a signal as in Figure 1.2. This is especially the case of very large antennas, and the outputs of the sensors are recombined in an optimal way to achieve some performance. If there are n sensors, their signals at a given time instant constitute an n-dimensional random vector.

On the other hand, suppose that we have only one signal $s(t)$ and that we want to calculate its discrete Fourier transform. We take n samples $s(t_i)$ which can also be considered as the components of an n-dimensional vector. If the signal is random, this vector also becomes random. Its discrete Fourier transform, calculated for example using the fast Fourier transform techniques, is also a random vector. But in this case the components of this vector are in general complex, which introduces a complex random vector. The properties of such vectors will be analyzed in Section 3.5.

The passage from $n = 2$ to an arbitrary value does not change the situation significantly. For this reason we shall not present the general results with as much detail as for $n = 2$, leaving it to the reader to complete the proofs which are merely outlined.

3.4.1 Distribution Functions

As in the scalar case we admit hereafter that a random vector is fully described by its DF, an extension of (3.2) and written as

$$F(\mathbf{x}) = F(x_1, x_2, \ldots, x_n) = P\{(X_1 \le x_1) \, . \, (X_2 \le x_2) \ldots (X_n \le x_n)\} \quad (3.52)$$

From its definition it is clear that this DF is a monotonic function of all the variables x_i. Furthermore, if all the x_is tend to $-\infty$ or to $+\infty$, $F(\mathbf{x})$ becomes equal to zero or to one respectively. If now only k variables x_i tend to $+\infty$, $F(\mathbf{x})$ becomes the DF of a random vector of dimension $n - k$. This is the extension of the concept of marginal distribution.

From a strictly mathematical point of view, $F(\mathbf{x})$ can be decomposed into three terms by

$$F(\mathbf{x}) = F_c(\mathbf{x}) + F_d(\mathbf{x}) + F_s(\mathbf{x}) \tag{3.53}$$

where $F_c(\mathbf{x})$, $F_d(\mathbf{x})$ and $F_s(\mathbf{x})$ are the continuous, discrete and singular parts of the DF. In the following we assume that $F_s(\mathbf{x}) = 0$, and we then have only the two parts already introduced in 3.2.1. As in (3.4) the continuous part introduces a PDF defined by

$$p(\mathbf{x}) = \frac{\partial^n F}{\partial x_1\, \partial x_2 \ldots \partial x_n} \tag{3.54}$$

On the other hand, the discrete part is defined by probabilities p_i such that

$$p_i = P(\mathbf{X} = \mathbf{x}_i) \tag{3.55}$$

Note that the notation is different from that used in (3.6) for $n = 2$, to avoid the writing of n-indices. These indices appear in the components of the vectors $\mathbf{x}_i$. The normalization condition is, of course,

$$\int_{\mathbb{R}^n} d^n F(\mathbf{x}) = \int_{\mathbb{R}^n} p(\mathbf{x})\, d\mathbf{x} + \sum_i p_i = 1 \tag{3.56}$$

Henceforth every integral without indication will mean an integral on the whole space $\mathbb{R}^n$. This expression is quite general and allows us to avoid any distinction between continuous and discrete cases. For those not familiar with integrals using the term $d^n F(\mathbf{x})$, the simplest way is to use each term of the r.h.s. of (3.56) separately. Furthermore, the notation of the Riemann integral in (3.56) is a multiple integral, usually written as

$$\int p(\mathbf{x})\, d\mathbf{x} = \underbrace{\int \cdots \int}_{n} p(x_1, x_2, \ldots, x_n)\, dx_1\, dx_2 \ldots dx_n \tag{3.56'}$$

3.4.2 Expectation

Consider a vector function $\mathbf{h}(\mathbf{x})$ taking its values in $\mathbb{R}^m$, where m is arbitrary. The mean value of $\mathbf{h}(\mathbf{X})$ is, of course,

$$E\{\mathbf{h}(\mathbf{X})\} = \int \mathbf{h}(\mathbf{x})\, d^n F(\mathbf{x}) \tag{3.57}$$

The meaning of this expression must be clearly understood. If $F(\mathbf{x})$ is reduced to its first term of (3.53), we can use the PDF and the expectation becomes

$$E[\mathbf{h}(\mathbf{X})] = \int \mathbf{h}(\mathbf{x})\, p(\mathbf{x})\, d\mathbf{x}$$

where the integral has the same meaning as in (3.56'). It is important to note that $E[\mathbf{h}(\mathbf{X})]$ is a vector and that its m components are given by

$$E[h_i(\mathbf{X})] = \int h_i(\mathbf{x})\, p(\mathbf{x})\, d\mathbf{x}$$

The same can be said concerning $\mathbb{M}(\mathbf{x})$, a matrix valued function of $\mathbf{x}$.

Thus the mean value of $\mathbf{X}$ is a vector $\mathbf{m}$ defined by

$$E(\mathbf{X}) = \int \mathbf{x} \; d^n F(\mathbf{x}) = \mathbf{m} \tag{3.58}$$

and if $\mathbf{m} = \mathbf{0}$, the random vector is said to be zeromean. It is clear that the components of $E(\mathbf{X})$ are the mean values of those of $\mathbf{X}$, and can be calculated using the marginal distributions.

For the second order properties two objects are of particular interest: the inner and outer products $\mathbf{x}^T\mathbf{x}$ and $\mathbf{x}\mathbf{x}^T$.

The first one is a scalar defined by

$$\mathbf{x}^T\mathbf{x} = \sum_{i=1}^{n} x_i^2 \tag{3.59}$$

and is sometimes referred to as the energy of the vector. It is also the square of the norm, or the length, of the vector. A unit vector is characterized by the property $\mathbf{x}^T\mathbf{x} = 1$.

The second is an $n \times n$ matrix noted $\mathbf{x}\mathbf{x}^T$. This is sometimes called a *dyadic* matrix, and is the product of the column vector $\mathbf{x}$ by the row vector $\mathbf{x}^T$. It is clear that the matrix elements of this matrix are $x_i x_j$. It is important to note that this matrix is symmetric because $x_i x_j = x_j x_i$, and that it is a rank one matrix. To verify this let us apply this matrix to an arbitrary vector $\mathbf{u}$. Using the fact that a product of matrices is associative, we obtain

$$\mathbf{x}\mathbf{x}^T . \; \mathbf{u} = \mathbf{x} \; (\mathbf{x}^T\mathbf{u}) = (\mathbf{x}^T\mathbf{u}) \; \mathbf{x} \tag{3.60}$$

where $\mathbf{x}^T\mathbf{u}$ is the scalar product between $\mathbf{x}$ and $\mathbf{u}$. The matrix $\mathbf{x}\mathbf{x}^T$ applied to $\mathbf{u}$ then gives a vector proportional to $\mathbf{x}$, which is the definition of a matrix of rank one. Another way to show that $\mathbf{x}\mathbf{x}^T$ is of rank one is to note that all the columns are proportional, a characteristic of this matrix. In fact, column i of $\mathbf{x}\mathbf{x}^T$ is simply $x_i\mathbf{x}$, and all the columns are proportional to $\mathbf{x}$.

The *covariance matrix* $\mathbb{\Gamma}$ is defined by

$$\mathbb{\Gamma} \triangleq E(\mathbf{X} \; \mathbf{X}^T) - \mathbf{m}\mathbf{m}^T \tag{3.61}$$

This matrix is very important in what follows and is sometimes noted $\mathbb{R}$ or $\mathbb{C}$. The matrix elements of $\mathbb{\Gamma}$ noted $\Gamma_{i,\,j}$ or Γ_{ij} are given by

$$\Gamma_{i,\,j} = E(X_i X_j) - m_i \, m_j \tag{3.62}$$

which is exactly (3.26) where X_i and X_j replace X and Y respectively. Because of their importance, the main properties of this matrix will be studied in Section 3.5. Note that the diagonal elements of $\mathbb{\Gamma}$ are nothing less that the variances of the components of $\mathbf{X}$, which justifies the term, sometimes used, of *variance-covariance* matrix. Finally, it is clear that $\mathbb{\Gamma}$ can also be defined as

$$\mathbb{\Gamma} = E\{(\mathbf{X} - \mathbf{m})(\mathbf{X} - \mathbf{m})^T\} \tag{3.63}$$

We shall complete this section by noting a small point of terminology. The covariance matrix $\mathbb{\Gamma}$ defined by (3.61) appears as a difference of two terms. The first is sometimes called the *correlation matrix*, while the second is only due to the mean value. When it is null, which means that $\mathbf{X}$ is zeromean, there is no difference between covariance and correlation matrices.

Let us discuss this terminology in more detail. Suppose that the mean value $\mathbf{m}$ is non null. The covariance matrix is defined by (3.61), and the matrix $\mathbb{C}$ defined by

$$\mathbb{C} \triangleq E(\mathbf{X}\mathbf{X}^T) \tag{3.63'}$$

is then called the *correlation matrix*. This denomination is not at all satisfactory. To explain this, consider the element $i\,j$ of $\mathbb{C}$ which is $C_{ij} = E(X_i X_j)$. Using (3.26), we can say that (3.62) gives the *covariance* between X_i and X_j. If $\mathbb{C}$ is called the correlation matrix, it is logical to say that C_{ij} is the correlation between X_i and X_j. Suppose now that $E(X_i X_j) = 0$. The consequence of the above denomination is that the correlation is null, or that X_i and X_j are uncorrelated, *but this is not true*. In fact, uncorrelated random variables are defined after (3.33) and characterized by a null correlation coefficient. This means that their *covariance* is null. But as $m_i\, m_j \neq 0$, the property that $E(X_i X_j) = 0$ does not mean that (3.62) is null. We therefore arrive at the conclusion that X_i and X_j have a null correlation but are correlated, which explains why the terminology is not satisfactory. It is sometimes said that if $\Gamma_{ij} = 0$ the RVs X_i and X_j are *uncorrelated* and that if $C_{ij} = 0$ they are *orthogonal*. But this terminology does not prevent the name "correlation matrix" for $\mathbb{C}$ defined by (3.63') to be unsatisfactory, even though it is widely used. Of course all this is invalid if $\mathbf{m} = \mathbf{0}$, because then the matrices $\mathbb{C}$ and $\mathbb{\Gamma}$ are the same.

To avoid the contradiction described above, the matrix $\mathbb{C}$ defined by (3.63') will be called the *second order matrix* below. When $n = 1$ this gives the second moment defined by (2.19) for $k = 2$.

3.4.3 Vectors Sets

Instead of studying one vector $\mathbf{X}$ with its distribution function, it is sometimes necessary to study several similar vectors simultaneously. For simplification we will restrict our attention to the case of two vectors $\mathbf{X}$ and $\mathbf{Y}$, the extension to a larger number being straightforward. The distribution function of a pair of vectors $\mathbf{X}$ and $\mathbf{Y}$ with dimension m and n respectively is quite similar to (3.52) and can then be written as

$$F(\mathbf{x}, \mathbf{y}) = P\left\{\prod_{i=1}^{m} (X_i \leq x_i) \prod_{j=1}^{n} (Y_j \leq y_j)\right\} \tag{3.64}$$

If $m = n = 1$, we again find the case of a two-dimensional RV, studied above, and in fact most of the results obtained in this case can be easily extended for arbitrary values of m and n. We will now focus our attention on the second order properties studied in section 3.3.

The covariance γ_{xy} defined by (3.25) becomes a *crosscovariance matrix* $\Gamma_{x, y}$ defined by

$$\Gamma_{x, y} \triangleq E\{(\mathbf{X} - \mathbf{m}_x)(\mathbf{Y} - \mathbf{m}_y)^T\} = E(\mathbf{X}\mathbf{Y}^T) - \mathbf{m}_x \mathbf{m}_y^T \tag{3.65}$$

To explain this expression, let us recall that the matrix $\mathbf{u}\,\mathbf{v}^T$ associated with two vectors of dimensions m and n respectively is a matrix with matrix elements $u_i\, v_j$, $1 \leq i \leq m$, $1 \leq j \leq n$, where u_i and v_j are the components of the vectors $\mathbf{u}$ and $\mathbf{v}$ respectively. The outer product appearing in (3.60) is a particular case corresponding to $\mathbf{u} = \mathbf{v}$.

The term crosscovariance matrix is used to distinguish this case from that corresponding to (3.63), which by comparison introduces a matrix, sometimes written as Γ_x, and called the *autocovariance matrix*. From its definition it is clear that $\Gamma_{x, y}$ is a rectangular matrix of dimension $m \times n$. As for Γ, the matrix elements of $\Gamma_{x, y}$ are the covariances between the components of $\mathbf{X}$ and $\mathbf{Y}$. However, the striking difference is that no element of $\Gamma_{x, y}$ is a variance, as are the diagonal elements of Γ_x.

As for the scalar case, it is said that the two vectors $\mathbf{X}$ and $\mathbf{Y}$ are uncorrelated if the matrix $\Gamma_{x, y}$ is null. Furthermore, the term $E(\mathbf{X}\mathbf{Y}^T)$ in (3.65) is sometimes called the crosscorrelation matrix. For zeromean vectors there is no difference between crosscovariance and crosscorrelation matrices or between autocorrelation and autocovariance matrices. Furthermore, where no confusion is possible, the prefix "auto" is not used in either case.

3.5 SECOND ORDER RANDOM VECTORS, COVARIANCE MATRICES

A random vector is said to be second order if all its components are second order RVs. In this section we assume that the mean value is null. From the definition we obtain that all the diagonal elements of the covariance matrix Γ defined by (3.61) are finite. Using the Schwarz inequality (3.29) we find that all the matrix elements of Γ are also finite. In other words, a second order random vector has a finite covariance matrix.

In this section devoted to second order properties, it is convenient to work with *complex* random vectors, the real case appearing as a particular case.

The concept of a scalar complex RV was introduced by (3.38). A complex random vector $\mathbf{X}(\omega)$ is simply a vector the components of which are complex RVs. This vector is an element of the space $\mathbb{C}^n$.

Before developing the main results of this section we will briefly recall the most important concepts of *linear algebra* used below. We present here only the results which have a direct application to the study of second order random vectors, and especially to the properties of covariance matrices.

3.5.1 Reminder of Linear Algebra

Consider the space $\mathbb{C}^n$ of complex vectors of dimension n. To each vector $\mathbf{u}$ with components u_i we can associate its transpose $\mathbf{u}^T$, which is a row vector of dimension n, or a $1 \times n$ matrix. The complex conjugate of $\mathbf{u}$ is, of course, the vector the components of which are u_i^*. Most important is the Hermitian conjugate $\mathbf{u}^H$ which is the row vector with components u_i^* satisfying $\mathbf{u}^H = (\mathbf{u}^T)^*$.

A rectangular matrix $\mathbb{M}$ with m rows and n columns is defined by its matrix elements M_{ij} where i and j refer to rows and columns respectively. Thus we have $1 \le i \le m$ and $1 \le j \le n$. If $m = n$ we obtain a square matrix. It follows that column vectors of $\mathbb{C}^n$ are $n \times 1$ matrices while their transpose or their Hermitian conjugates are $1 \times n$ matrices.

The product of a matrix by a vector $\mathbf{u}$ is a vector $\mathbf{v}$, written as $\mathbf{v} = \mathbb{M}\mathbf{u}$. If $\mathbb{M}$ is an $m \times n$ matrix, $\mathbf{u}$ and $\mathbf{v}$ are vectors of dimension n and m respectively. The components of $\mathbf{v}$ are defined by

$$v_i = \sum_{j=1}^{n} M_{ij} u_j \tag{3.66}$$

Similarly the product of two $m \times p$ and $p \times n$ matrices $\mathbb{A}$ and $\mathbb{B}$ is an $m \times n$ matrix $\mathbb{M}$ with matrix elements defined by

$$M_{ij} = \sum_{k=1}^{p} A_{ik} B_{kj} \tag{3.67}$$

In the case of square matrices, it is often interesting to introduce bilinear forms defined by

$$\mathbf{u}^H \mathbb{M} \mathbf{v} \triangleq \sum_{i,j} M_{ij} u_i^* v_j \tag{3.68}$$

It is obvious that $\mathbf{u}^H \mathbb{M} \mathbf{v}$ is the scalar product between the vectors $\mathbf{u}$ and $\mathbb{M}\mathbf{v}$. If $\mathbf{u} = \mathbf{v}$, we obtain the *quadratic form* defined by $\mathbb{M}$ and $\mathbf{u}$ given by

$$\mathbf{u}^H \mathbb{M} \mathbf{u} = \sum_{i,j} M_{ij} u_i^* u_j \tag{3.69}$$

It is also possible to define the quantity $\mathbf{u}^T \mathbb{M} \mathbf{v}$ by using (3.68) where u_i^* is simply replaced by u_i.

As in the case of vectors, it is possible to introduce the transpose $\mathbb{M}^T$ and the Hermitian conjugate $\mathbb{M}^H$ of a matrix $\mathbb{M}$. The matrix elements of $\mathbb{M}^T$ and $\mathbb{M}^H$ are deduced from those of $\mathbb{M}$ by permuting rows and columns which gives

$$(\mathbb{M}^T)_{ji} = M_{ij} \quad ; \quad (\mathbb{M}^H)_{ji} = M_{ij}^* \tag{3.70}$$

If $\mathbb{M}$ is an $m \times n$ matrix, $\mathbb{M}^T$ and $\mathbb{M}^H$ are $n \times m$ matrices. Applying (3.68) and (3.70) we easily obtain

$$\mathbf{u}^H \mathbb{M} \mathbf{v} = (\mathbf{v}^H \mathbb{M}^H \mathbf{u})^* \quad ; \quad \mathbf{u}^T \mathbb{M} \mathbf{v} = \mathbf{v}^T \mathbb{M}^T \mathbf{u} \tag{3.71}$$

A square matrix $\mathbb{M}$ is *Hermitian* if $\mathbb{M} = \mathbb{M}^H$ and *symmetric* if $\mathbb{M} = \mathbb{M}^T$. The two concepts are equivalent for real matrices. The inverse $\mathbb{M}^{-1}$ of a square matrix $\mathbb{M}$ is defined, when $\mathbb{M}$ is invertible, by the relation $\mathbb{M}\mathbb{M}^{-1} = \mathbb{M}^{-1}\mathbb{M} = \mathbb{I}$, where $\mathbb{I}$ is the identity matrix. Applying (3.67) and the previous definitions we easily obtain

$$(\mathbb{A}\mathbb{B})^H = \mathbb{B}^H \mathbb{A}^H \quad ; \quad (\mathbb{A}\mathbb{B})^{-1} = \mathbb{B}^{-1} \mathbb{A}^{-1} \tag{3.72}$$

The last equation is, of course, valid only if $\mathbb{A}$ and $\mathbb{B}$ are invertible.

A large part of the following is devoted to *non-negative definite* or *positive definite* matrices that will now be defined.

A matrix $\mathbb{M}$ is said to be *positive definite*, which is written PD, or $\mathbb{M} > \mathbb{0}$, if for any non-zero vector $\mathbf{u}$ the quadratic form $\mathbf{u}^H\mathbb{M}\mathbf{u}$ defined by (3.69) is positive. Replacing positive by non-negative, we obtain the definition of a non-negative definite matrix, written as NND or $\mathbb{M} \geq \mathbb{0}$. It is interesting to note the analogy with the condition (2.57) introduced for characteristic functions.

The case of *real matrices* need to be considered carefully. If the property $\mathbf{u}^H\mathbb{M}\mathbf{u} \geq 0$ is valid only for real vectors, and therefore can be written as $\mathbf{u}^T\mathbb{M}\mathbf{u} \geq 0$, then $\mathbb{M}$ is said to be NND *for real vectors*. As we will see later, a real matrix $\mathbb{M}$ can be NND for real vectors, without having this property for complex vectors.

Let us now present the most significant properties of NND matrices. In the following, the term NND means non-negative definite for complex vectors, when it is not specified otherwise.

(a) Any NND matrix is Hermitian. Consider an NND matrix $\mathbb{M}$, characterized by the property $\mathbf{u}^H\mathbb{M}\mathbf{u} \geq 0$ for any complex vector $\mathbf{u}$. As $\mathbb{M}$ is generally complex, it can be written as

$$\mathbb{M} = \mathbb{A} + j\,\mathbb{B} \tag{3.73}$$

where $\mathbb{A}$ and $\mathbb{B}$ are real $n \times n$ matrices, real and imaginary parts of $\mathbb{M}$ respectively. First, let us calculate the quadratic form (3.69) for a *real* vector $\mathbf{a}$. This gives

$$\mathbf{a}^H\mathbb{M}\mathbf{a} = \mathbf{a}^T\mathbb{M}\mathbf{a} = \mathbf{a}^T\mathbb{A}\mathbf{a} + j\,\mathbf{a}^T\mathbb{B}\mathbf{a} \geq 0 \tag{3.74}$$

The last inequality shows that $\mathbf{a}^T\mathbb{B}\mathbf{a} = 0$ for any vector $\mathbf{a}$. Applying this property to the vector $\mathbf{a} + \mathbf{b}$, where $\mathbf{a}$ and $\mathbf{b}$ are arbitrary real vectors, we obtain

$$\mathbf{a}^T\mathbb{B}\mathbf{a} + \mathbf{b}^T\mathbb{B}\mathbf{b} + \mathbf{a}^T\mathbb{B}\mathbf{b} + \mathbf{b}^T\mathbb{B}\mathbf{a} = 0 \tag{3.75}$$

As the two first terms are null, we obtain with (3.71)

$$\mathbf{a}^T\mathbb{B}\mathbf{b} = -\,\mathbf{b}^T\mathbb{B}\mathbf{a} = -\,\mathbf{a}^T\mathbb{B}^T\mathbf{b} \tag{3.76}$$

for arbitrary $\mathbf{a}$ and $\mathbf{b}$. This shows that $\mathbf{a}^T(\mathbb{B} + \mathbb{B}^T)\mathbf{b} = 0$, or $\mathbb{B} = -\mathbb{B}^T$. In fact, if a real matrix $\mathbb{N}$ is such that for any real vectors $\mathbf{a}$ and $\mathbf{b}$ we have $\mathbf{a}^T\mathbb{N}\mathbf{b} = 0$, it follows that $\mathbb{N}\mathbf{b}$ is orthogonal to any vector of $\mathbb{R}^n$, and then $\mathbb{N}\mathbf{b} = \mathbf{0}$. But as this is valid for any vector $\mathbf{b}$, we deduce that $\mathbb{N} = \mathbb{0}$.

Let us now apply the property $\mathbf{u}^H \mathbb{M} \mathbf{u} \geq 0$ for a complex vector $\mathbf{u} = \mathbf{a} + j\mathbf{b}$, where $\mathbf{a}$ and $\mathbf{b}$ are real. Combining this with (3.73) and using the fact that $\mathbf{u}^H = \mathbf{a}^T - j\mathbf{b}^T$, we can calculate the imaginary part of $\mathbf{u}^H \mathbb{M} \mathbf{u}$, which is

$$\mathrm{Im}(\mathbf{u}^H \mathbb{M} \mathbf{u}) = \mathbf{a}^T \mathbb{B} \mathbf{a} + \mathbf{b}^T \mathbb{B} \mathbf{b} + \mathbf{a}^T (\mathbb{A} - \mathbb{A}^T) \mathbf{b} \tag{3.77}$$

As $\mathbb{M}$ is NND, this quantity must be null. However, as the two first terms are null, we obtain, by the same argument as before, $\mathbb{A} = \mathbb{A}^T$. Then, if $\mathbb{M}$ is NND, we have

$$\mathbb{A} = \mathbb{A}^T \quad ; \quad \mathbb{B} = -\mathbb{B}^T \tag{3.78}$$

However, as $\mathbb{M} = \mathbb{A} + j\,\mathbb{B}$, we deduce from (3.70) that

$$\mathbb{M}^H = \mathbb{A}^T - j\,\mathbb{B}^T \tag{3.79}$$

and (3.78) shows that $\mathbb{M} = \mathbb{M}^H$.

(b) Properties of the real and imaginary parts of a NND matrix. They are specified by (3.78). Furthermore, we deduce from the previous calculation that if $\mathbf{u}$ is a *real* vector, we have

$$\mathbf{u}^T \mathbb{A} \mathbf{u} \geq 0 \quad ; \quad \mathbf{u}^T \mathbb{B} \mathbf{u} = 0 \tag{3.80}$$

Combining (3.78) and (3.80) we see that the real part $\mathbb{A}$ is symmetric and NND for real vectors.

(c) Properties of eigenelements of NND matrices. Let us call λ_i and $\mathbf{u}_i$ the eigenelements of an NND matrix $\mathbb{M}$, satisfying $\mathbb{M}\mathbf{u}_i = \lambda_i \mathbf{u}_i$. From this equation we obtain

$$\mathbf{u}_i^H \mathbb{M} \mathbf{u}_i = \lambda_i \mathbf{u}_i^H \mathbf{u}_i \tag{3.81}$$

and as $\mathbf{u}_i^H \mathbf{u}_i$ is positive, we deduce from the property of definition $\mathbf{u}^H \mathbb{M} \mathbf{u} \geq 0$ that λ_i is non-negative. So the eigenvalues of an NND matrix are *non-negative.*

We shall now show that the eigenvectors of an NND matrix corresponding to distinct eigenvalues are *orthogonal.* For this purpose, note that as $\mathbb{M}$ is Hermitian, we deduce from (3.71) that $\mathbf{u}_i^H \mathbb{M} \mathbf{u}_j = (\mathbf{u}_j^H \mathbb{M} \mathbf{u}_i)^*$. Using the relation $\mathbb{M}\mathbf{u}_i = \lambda_i \mathbf{u}_i$ written for i and j we deduce that

$$\lambda_j \mathbf{u}_i^H \mathbf{u}_j = \lambda_i (\mathbf{u}_j^H \mathbf{u}_i)^* = \lambda_i \mathbf{u}_i^H \mathbf{u}_j \tag{3.82}$$

because λ_i is real. As $\lambda_i \neq \lambda_j$, we deduce that $\mathbf{u}_i$ and $\mathbf{u}_j$ are orthogonal. Let us now consider the case of *multiple* eigenvalues. An eigenvalue λ_i is said to be multiple, or degenerate, if there are at least two linearly independent vectors $\mathbf{v}_{i1}$ and $\mathbf{v}_{i2}$ which are eigenvectors of $\mathbb{M}$ with this eigenvalue λ_i. To simplify the discussion let us assume that there is no other vector $\mathbf{v}$, eigenvector of $\mathbb{M}$ with the eigenvalue λ_i, and such that $\mathbf{v}$, $\mathbf{v}_{i1}$, $\mathbf{v}_{i2}$ are linearly independent. In this case it is said that λ_i is a multiple eigenvalue of order 2. It is clear that any vector $\mathbf{v}$ which is a linear combination of $\mathbf{v}_{i1}$ and $\mathbf{v}_{i2}$ is eigenvector of $\mathbb{M}$ with the same eigenvalue λ_i. So the subspace E_i of $\mathbb{C}^n$ generated by $\mathbf{v}_{i1}$ and $\mathbf{v}_{i2}$ can be called the *eigensubspace* corresponding to the eigenvalue λ_i. This subspace is of course of dimension 2. Using the result shown above, we see that any vector $\mathbf{v}$ of E_i is orthogonal to any eigenvector $\mathbf{u}_j$ corresponding to an eigenvalue λ_j distinct from λ_i. This is expressed by saying that the subspace E_i is orthogonal to $\mathbf{u}_j$. Starting from $\mathbf{v}_{i1}$ and $\mathbf{v}_{i2}$ it is easy to construct two vectors $\mathbf{u}_{i1}$ and $\mathbf{u}_{i2}$ belonging to E_i and which are orthogonal. To do this we state $\mathbf{u}_{i1} = \mathbf{v}_{i1}$ and $\mathbf{u}_{i2} = \mathbf{v}_{i1} + \alpha \mathbf{v}_{i2}$ and we determine α in such a way that $\mathbf{u}_{i1}$ and $\mathbf{u}_{i2}$ are orthogonal. We can extend this procedure if λ_i is a multiple eigenvalue of arbitrary order, and repeat the same procedure for all multiple eigenvalues.

In conclusion, we see that starting from an NND matrix $\mathbb{M}$ we can find n orthogonal vectors $\mathbf{u}_i$ which are eigenvectors of $\mathbb{M}$ with the eigenvalues λ_i, whether or not those eigenvalues are distinct. Finally, by multiplying each vector $\mathbf{u}_i$ by a scalar α_i we can obtain a normalized vector. So the set of vectors $\mathbf{u}_i$ becomes a set of orthonormal vectors of $\mathbb{C}^n$. This set of orthonormal vectors constitutes a basis of $\mathbb{C}^n$ called an *eigenbasis*.

Let us explain this procedure with a few examples. Consider first the identity matrix $\mathbb{I}$. It is of course PD because the quadratic for (3.69) is positive. It has only one eigenvalue equal to one. It is clear that any orthonormal basis of $\mathbb{C}^n$ is an eigenbasis.

Consider now the dyadic matrix $\mathbb{M} = \mathbf{v}\mathbf{v}^H$, where $\mathbf{v}$ is a unit vector. This matrix is obviously NND and satisfies $\mathbb{M}^2 = \mathbb{M}$. It is then a projection matrix on the subspace generated by $\mathbf{v}$. It is clear that $\mathbb{M}\mathbf{v} = \mathbf{v}$ and that any vector $\mathbf{u}$ orthogonal to $\mathbf{v}$ satisfies $\mathbb{M}\mathbf{u} = 0\mathbf{u}$. So there are only two eigenvalues, eigenvalue 1 being distinct and eigenvalue 0 being multiple and of order $n-1$. The eigensubspace E_0 corresponding to the eigenvalue 0 is the subspace orthogonal to $\mathbf{v}$. An eigenbasis can be constructed by using a set of arbitrary orthonormal vectors $\mathbf{u}_{i0}$ belonging to E_0 and by adding $\mathbf{v}$ we obtain the basis.

(d) Spectral decomposition of an NND matrix. Consider the matrix $\mathbb{P}$ defined by

$$\mathbb{P} \triangleq \sum_{i=1}^{n} \lambda_i \mathbf{u}_i \mathbf{u}_i^H \tag{3.83}$$

where the λ_is and the $\mathbf{u}_i$s are the eigenvalues and an eigenbasis of $\mathbb{M}$ respectively. This gives

$$\mathbb{P}\mathbf{u}_i = \lambda_i \mathbf{u}_i \tag{3.84}$$

which means that λ_i and $\mathbf{u}_i$ are the eigenelements of $\mathbb{P}$. As a result the matrices $\mathbb{M}$ and $\mathbb{P}$ have the same eigenelements and are then equal, which gives the spectral decomposition, of $\mathbb{M}$ written as

$$\mathbb{M} = \sum_{i=1}^{n} \lambda_i \mathbf{u}_i \mathbf{u}_i^H \tag{3.85}$$

Furthermore as any vector $\mathbf{v}$ can be written in the form

$$\mathbf{v} = \sum_{j=1}^{n} \alpha_j \mathbf{u}_j \tag{3.86}$$

where the α_js are the components in the orthonormal eigenbasis $\{\mathbf{u}_i\}$, we obtain

$$\mathbf{v}^H \mathbb{M} \mathbf{v} = \sum_{i=1}^{n} \lambda_i \, |\alpha_i|^2 \geq 0 \qquad (3.86')$$

and we verify that $\mathbb{M}$ is an NND matrix. Another verification can be deduced from (3.85) because the spectral decomposition is a sum of n NND matrices, and this sum is also NND.

(e) Rank of an NND matrix. As the determinant of a square $n \times n$ matrix is the product of its eigenvalues, this determinant is non-negative for an NND matrix. If all the eigenvalues are positive, the rank is n. If the rank is $n - p$, this means that there are p null eigenvalues.

(f) The inverse of an PD matrix is also PD. For the proof of this property we start from (3.85) and show that the matrix $\mathbb{Q}$, which has the same structure but where the λ_is are replaced by λ_i^{-1}, is the inverse of $\mathbb{M}$.

(g) Characterization of an NND matrix. As it is sometimes difficult to characterize an NND matrix starting from (3.69), several other procedures are possible. The first consists of writing a given matrix in the form (3.83) and verifying that the $\mathbf{u}_i$s are orthogonal and the λ_i non-negative.

Another way is to use a result shown in Problem 3.10 which states that a Hermitian matrix is PD if all the principal minors are positive. There are also particular criteria when $\mathbb{M}$ has a specific structure, as seen later with Toeplitz matrices.

(h) Case of real NND matrices for real vectors. Such matrices are always defined by $\mathbf{u}^H \mathbb{M} \mathbf{u} \geq 0$ where $\mathbf{u}^H$ and u_i^* in (3.69) are replaced by $\mathbf{u}^T$ and u_i respectively, because $\mathbf{u}$ is a real vector.

Without any other specification, all the previous properties disappear. In fact, it is perfectly possible to construct examples of NND matrices for real vectors that are not symmetric and that do not have non-negative eigenvalues, as seen in Problems 3.1, 3.2, 3.3 and 3.4.

Let us now assume that $\mathbb{M}$ is not only NND for *real vectors*, but also *symmetric*. In this case it appears that $\mathbb{M}$ becomes NND for complex vectors and all the previous properties again become valid . To verify this, consider a complex vector $\mathbf{u} = \mathbf{a} + j\mathbf{b}$. As $\mathbf{u}^H = \mathbf{a}^T - j\mathbf{b}^T$, we can write

$$\mathbf{u}^H \mathbb{M} \mathbf{u} = \mathbf{a}^T \mathbb{M} \mathbf{a} + \mathbf{b}^T \mathbb{M} \mathbf{b} + j(\mathbf{a}^T \mathbb{M} \mathbf{b} - \mathbf{b}^T \mathbb{M} \mathbf{a}) \qquad (3.87)$$

which shows the real and imaginary parts of $\mathbf{u}^H \mathbb{M} \mathbf{u}$. But as $\mathbb{M} = \mathbb{M}^T$, we deduce from (3.71) that the imaginary part is null, and as $\mathbb{M}$ is NND for real vectors, such as $\mathbf{a}$ and $\mathbf{b}$, we obtain $\mathbf{u}^H \mathbb{M} \mathbf{u} \geq 0$.

The previous results can even be simplified when we note that the eigenvectors are real. In fact, as the eigenvalues are non-negative and therefore real, the eigenequation can be written

$$(\mathbb{M} - \lambda_i \mathbb{I})\, \mathbf{u}_i = \mathbf{0} \tag{3.88}$$

where $\mathbb{I}$ is the identity matrix. This means that $\mathbf{u}_i$ is in the kernel of the real matrix $\mathbb{M} - \lambda_i \mathbb{I}$, which is a set of real vectors.

Consequently, the spectral decomposition of $\mathbb{M}$ can now be written

$$\mathbb{M} = \sum_{i=1}^{n} \lambda_i \, \mathbf{u}_i \, \mathbf{u}_i^T \tag{3.89}$$

where the $\mathbf{u}_i$s are real orthonormal vectors satisfying $\mathbf{u}_i^T \mathbf{u}_j = \delta_{i,j}$.

3.5.2 Properties of Covariance Matrices

The covariance matrix of a complex random vector is defined from an extension of (3.61) or (3.63) by

$$\Gamma \triangleq E(\mathbf{X}\,\mathbf{X}^H) - \mathbf{m}\mathbf{m}^H = E[(\mathbf{X} - \mathbf{m})(\mathbf{X} - \mathbf{m})^H] \tag{3.90}$$

which is equal to $E(\mathbf{X}\,\mathbf{X}^H)$ if we assume, as stated before, that $\mathbf{m} = \mathbf{0}$. The basic property of Γ is that *it is an NND matrix.* This fact was partially indicated after (3.51).

In order to show that Γ is NND, sometimes written as $\Gamma \geq 0$, let us consider the quadratic form $\mathbf{u}^H \Gamma \mathbf{u}$. Starting from (3.90), we obtain for any vector $\mathbf{u}$

$$\mathbf{u}^H \Gamma \mathbf{u} = \mathbf{u}^H E[(\mathbf{X} - \mathbf{m})(\mathbf{X} - \mathbf{m})^H]\mathbf{u} = E(|\mathbf{u}^H(\mathbf{X} - \mathbf{m})|^2) \geq 0 \tag{3.91}$$

which shows the property. The same procedure can be used to show that the second order matrix defined by (3.63'), where $\mathbf{X}^T$ is replaced by $\mathbf{X}^H$, is also NND.

Conversely, any NND matrix $\mathbb{M}$ can be considered as the covariance matrix of at least one random vector $\mathbf{X}$. To prove this, let us start from (3.90). Consider a set of zeromean uncorrelated RVs X_i such that

$$E(|X_i|^2) = \lambda_i \quad ; \quad E(X_i X_j^*) = 0, \ i \neq j \tag{3.92}$$

and the random vector $\mathbf{V}$ defined by

$$\mathbf{V} \triangleq \sum_{i=1}^{n} X_i \, \mathbf{u}_i \tag{3.93}$$

where the $\mathbf{u}_i$s are the eigenvectors of $\mathbb{M}$. The covariance matrix Γ of $\mathbf{V}$ is

$$\Gamma = E(\mathbf{V}\mathbf{V}^H) = \sum_{i=1}^{n} \sum_{j=1}^{n} E(X_i X_j^*)\, \mathbf{u}_i \, \mathbf{u}_j^H \tag{3.94}$$

Using (3.92) and (3.85), we immediately obtain $\Gamma = \mathbb{M}$. The only constraint on the variables X_i is specified by (3.92) and there is still a large degree of freedom in choosing the set of variables. This can be used for some specific purposes.

The same arguments can be used when NND is replaced by PD. As we have seen in (e) of 3.5.1, an NND matrix is PD if it is full rank or if all of its eigenvalues are positive. So, a full rank covariance matrix is PD. Conversely, a PD matrix can always be considered as a full rank covariance matrix. For this purpose we again use (3.92) and $\mathbb{\Gamma}$ given by (3.94) is a covariance matrix without null eigenvalue or full rank.

Let us now investigate some particular consequences of these results.

(a) The sum of two covariance matrices is a covariance matrix. If $\mathbb{\Gamma}_1$ and $\mathbb{\Gamma}_2$ are two covariance matrices, they are NND. The sum $\mathbb{\Gamma} = \mathbb{\Gamma}_1 + \mathbb{\Gamma}_2$ is then still NND, and is thus a covariance matrix.

(b) The "direct product" of two covariance matrices is a covariance matrix. The direct product $\mathbb{\Gamma}$ is a matrix the elements of which are

$$\Gamma_{m,\,n} = \Gamma_{1;m,\,n} \cdot \Gamma_{2;m,\,n} \tag{3.95}$$

where $\Gamma_{i;m,\,n}$ are the matrix elements of $\mathbb{\Gamma}_i$. To show this we can use the same procedure as in (3.92) and (3.94) and then find two *independent* vectors $\mathbf{X}$ and $\mathbf{Y}$ with the correlation matrices $\mathbb{\Gamma}_1$ and $\mathbb{\Gamma}_2$ respectively. Now consider the vector $\mathbf{Z}$ with components Z_i defined by $X_i Y_i$. Using (3.19) generalized to complex RVs we get

$$E(Z_m Z_n^*) = E(X_m X_n^*)E(Y_m Y_n^*) \tag{3.96}$$

and it follows that $\mathbb{\Gamma}$ defined by (3.95) is the covariance matrix of $\mathbf{Z}$.

(c) If $\mathbb{A}$ is an arbitrary rectangular matrix, $\mathbb{A}\mathbb{\Gamma}\mathbb{A}^H$ is a covariance matrix. This is a consequence of the fact that $\mathbb{A}\mathbb{\Gamma}\mathbb{A}^H$ is an NND matrix, as directly seen from the definition. Furthermore, it is clear that if $\mathbb{\Gamma}$ is the covariance matrix of a random vector $\mathbf{X}$, $\mathbb{A}\mathbb{\Gamma}\mathbb{A}^H$ is the covariance matrix of $\mathbf{Y} = \mathbb{A}\mathbf{X}$. This is a result of the fact that $\mathbf{Y}\mathbf{Y}^H$ can be written as $\mathbb{A}\mathbf{X}\mathbf{X}^H\mathbb{A}^H$ and by taking the expectation we obtain $\mathbb{A}\mathbb{\Gamma}\mathbb{A}^H$.

(d) Rank of covariance matrices. A covariance matrix is full rank if all the eigenvalues λ_i are positive. On the other hand, if there are p null eigenvalues, the rank is $n-p$.

(e) Covariance basis. The eigenbasis of a covariance matrix $\mathbb{\Gamma}$ is also called the covariance basis. To understand the importance of this basis, let us consider a random vector $\mathbf{X}$ and its covariance matrix $\mathbb{\Gamma}$. This vector can be expanded in the covariance basis, as in (3.93). This gives

$$\mathbf{X} = \sum_{i=1}^{n} X_i \mathbf{u}_i \; ; \; X_i = \mathbf{u}_i^H \mathbf{X} \tag{3.97}$$

It is clear that in this "covariance basis" the components X_i are uncorrelated, because

$$E(X_i X_j^*) = \mathbf{u}_i^H \mathbb{\Gamma} \mathbf{u}_j = \lambda_i \; \delta_{i,\,j} \tag{3.98}$$

It is sometimes said that the expansion (3.97) is *doubly orthogonal*. This means that the vectors $\mathbf{u}_i$ are orthogonal, as shown above. But the random components are also said to be orthogonal because, as seen later, the covariance (3.98) can be interpreted as a scalar product. The expansion (3.97), where the $\mathbf{u}_i$s are eigenvectors of $\mathbb{F}$, is sometimes called the *Karhunen-Loève expansion* of the random vector $\mathbf{X}$, and the random components X_i are sometimes called the *principal components* of the random vector $\mathbf{X}$. However, if the vectors $\mathbf{u}_i$, always orthonormal, are no longer eigenvectors of $\mathbb{F}$, (3.97) is still valid, but the last term of (3.98) is not. In this case the expansion (3.97) is no longer doubly orthogonal.

Now, if the rank of $\mathbb{F}$ is smaller than n, this means that some eigenvalues are null or, from (3.98), that some components X_i have a null variance, and are then almost surely equal to zero, as seen on p. 29. In consequence, the random vector takes its value only in the subspace of $\mathbb{C}^n$ orthogonal to the subspace spanned by the eigenvectors corresponding to the null eigenvalues. When this situation appears, it is generally recommended to suppress this subspace in order to obtain a full rank covariance matrix.

3.5.3 Factorization of Covariance Matrices

Consider a random vector $\mathbf{X}$ with components X_i and suppose that these components are uncorrelated and with the same variance σ^2. For several reasons, explained later, such a vector is called a *white vector*. Its covariance matrix is, of course, equal to $\sigma^2\,\mathbb{I}$, where $\mathbb{I}$ is the unity matrix of $\mathbb{C}^n$. For simplification we also assume that $\sigma^2 = 1$, and in this case $\mathbf{X}$ is called a *white unit vector*. Suppose that this vector is transformed into another one by linear transformation defined by a square matrix giving $\mathbf{Y} = \mathbb{A}\mathbf{X}$. The covariance matrix $\mathbb{F}$ of $\mathbf{Y}$ is

$$\mathbb{F} = \mathbb{A}\mathbb{A}^H \tag{3.99}$$

and we have a *factorization* of $\mathbb{F}$, using the matrix $\mathbb{A}$, sometimes called a square root of $\mathbb{F}$. Conversely, starting from a given covariance matrix $\mathbb{F}$, it is of interest to write its factorization as in (3.99).

The first point to note is that this factorization has no reason to be unique. In fact, consider any matrix $\mathbb{B}$ such that $\mathbb{B}\,\mathbb{B}^H = \mathbb{I}$ and call $\mathbb{C}$ the matrix $\mathbb{A}\,\mathbb{B}$. It is clear that $\mathbb{F} = \mathbb{C}\,\mathbb{C}^H$, which is another factorization of $\mathbb{F}$. Furthermore, the matrix $\exp(j\,\varphi)\,\mathbb{A}$ gives the same factorization.

It is possible to reduce the ambiguity by imposing that $\mathbb{A}$ be a triangular matrix, which is related to the *causality* problem, extensively studied later in another context.

Suppose then that $\mathbb{A}$ is a lower triangular matrix with non-negative diagonal elements, and let us write $\mathbf{Y} = \mathbb{A}\mathbf{X}$ as

$$Y_i = \sum_{j=1}^{n} a_{ij}\, X_j \tag{3.100}$$

where $\mathbf{X}$ is a white unit vector. The assumption that $\mathbb{A}$ is lower triangular means that $a_{ij} = 0$ if $j > i$, which reduces the number of terms in (3.100) from n to i. As we have $Y_1 = a_{11}X_1$ and as $E(|Y_1|^2) = \Gamma_{11} = |a_{11}|^2$, we deduce from the assumption $a_{11} \geq 0$ that $a_{11} = (\Gamma_{11})^{1/2}$. Let us now show that $\mathbb{A}$ can be calculated recursively. In order to do this, suppose that the a_{ij}s are known for $1 \leq j \leq i$ and $1 \leq i \leq n-1$, and let us calculate a_{nm}. As the X_js are uncorrelated we have

$$\Gamma_{nk} = E(Y_n Y_k^*) = a_{n1} a_{k1}^* + a_{n2} a_{k2}^* + \dots + a_{nk} a_{kk}^* \tag{3.101}$$

Applying it to $k = 1$, we obtain a_{n1} because a_{11} is known. Similarly, for $k = 2$ we obtain a_{n2}, because a_{n1}, a_{21}^* , a_{22}^* are known, etc. The last step is

$$\Gamma_{nn} = |a_{n1}|^2 + |a_{n2}|^2 + \dots + |a_{n,n-1}|^2 + |a_{nn}|^2 \tag{3.102}$$

As only a_{nn} is unknown, we obtain its value without ambiguity by the condition $a_{nn} \geq 0$.

If all the diagonal elements of $\mathbb{A}$ are non-null, the matrix $\mathbb{A}$ has an inverse $\mathbb{A}^{-1}$ which is also lower triangular, and which gives

$$\mathbf{X} = \mathbb{A}^{-1}\,\mathbf{Y} \tag{3.103}$$

This operation is called a whitening of $\mathbf{Y}$ because it transforms an arbitrary random vector into a white unit vector. Let us note that $\mathbb{A}^{-1}$ appears in the factoring of $\mathbb{\Gamma}^{-1}$ because (3.99) and (3.72) also gives

$$\mathbb{\Gamma}^{-1} = (\mathbb{A}^H)^{-1}\,\mathbb{A}^{-1} \tag{3.104}$$

Another interesting factorization is the Cholesky factorization written as

$$\mathbb{\Gamma} = \mathbb{A}\,\mathbb{D}\,\mathbb{A}^H \tag{3.105}$$

where $\mathbb{A}$ is a lower triangular matrix with diagonal elements equal to one and $\mathbb{D}$ a diagonal matrix with non-negative elements. This factorization and some of its consequences are discussed in Problems 3.8 and 3.9.

Finally, we note that for real vectors $\mathbf{X}$, (3.90) can also be written by replacing H by T. As a result $\mathbb{\Gamma}$ is real, but also symmetric, according to its definition. Thus all the results in this section remain valid when H is replaced by T and the complex conjugates in all the previous equations are suppressed.

Example 3.2. Covariance matrix of signal plus noise. In many signal processing problems we encounter a situation where an observation vector $\mathbf{X}$ appears as the sum of a signal and a noise component. This situation can be described as

$$\mathbf{X} = \mathbf{N} + A\,\mathbf{s} \tag{3.106}$$

where $\mathbf{N}$ is the noise vector, $\mathbf{s}$ a deterministic signal and A its amplitude. Suppose that the random variables A and $\mathbf{N}$ are zeromean and uncorrelated. In this case $\mathbf{X}$ is also zeromean and its covariance matrix can be written

$$\mathbb{\Gamma}_X = \mathbb{\Gamma}_N + \sigma_A^2\,\mathbf{s}\mathbf{s}^H \tag{3.107}$$

where $\mathbb{\Gamma}_N$ is the covariance matrix of the noise and σ_A^2 the variance of the amplitude. An interesting situation appears when the noise is white, which is characterized by a covariance matrix equal to $\sigma_N^2\ \mathbb{I}$. In this case $\mathbb{\Gamma}_X$ can be written in the form $\sigma_N^2\ \mathbb{C}$ with

$$\mathbb{C} = \mathbb{I} + \rho\, \mathbf{s}\mathbf{s}^H \tag{3.108}$$

ρ being a signal-to-noise ratio coefficient. In order to simplify, suppose that $\mathbf{s}$ is a unit vector. It is clear that $\mathbf{s}$ is a normalized eigenvector of $\mathbb{C}$ with the eigenvalue $1 + \rho$, because $\mathbf{s}^H\mathbf{s} = 1$. Furthermore, any unit vector orthogonal to $\mathbf{s}$ is a normalized eigenvector of $\mathbb{C}$ with the eigenvalue 1. In other words the vector $\mathbf{s}$ is the eigenvector of $\mathbb{C}$ with the largest eigenvalue.

Example 3.3. Covariance matrices and the moment problem. We saw on p. 39 that the moments m_i of a DF must satisfy some inequalities, some of which were considered in Problem 2.13. By using the covariance matrix we will find more general results. To do this, we associate to an arbitrary RV X the random vector $\mathbf{X}$ with components X, X^2. Its mean value is the vector $[m_1, m_2]^T$ where m_1 and m_2 are the first two moments of X, in such a way that the dyadic matrices appearing in (3.61) are

$$\mathbf{X}\mathbf{X}^T = \begin{bmatrix} X^2 & X^3 \\ X^3 & X^4 \end{bmatrix} \qquad \mathbf{m}\mathbf{m}^T = \begin{bmatrix} m_1^2 & m_1 m_2 \\ m_1 m_2 & m_2^2 \end{bmatrix}$$

Taking the expectation, we obtain the covariance matrix expressed in terms of the moments of X by

$$\mathbb{\Gamma} = \begin{bmatrix} m_2 - m_1^2 & m_3 - m_1 m_2 \\ m_3 - m_1 m_2 & m_4 - m_2^2 \end{bmatrix} \tag{3.109}$$

We note that the diagonal elements of this matrix are the variances of X and X^2 respectively. These variances are of course non-negative. However, the matrix $\mathbb{\Gamma}$ must also be an NND matrix, which is ensured if its determinant is non-negative. By calculating this determinant we obtain

$$m_2\, m_4 + 2m_1\, m_2\, m_3 - m_2^3 - m_1^2\, m_4 - m_3^2 \ \geq 0 \tag{3.110}$$

which is especially interesting for zeromean RVs, for which we have

$$m_2\, m_4 - m_2^3 - m_3^2 \ \geq 0 \tag{3.111}$$

The situation which appears when the equality holds, is analyzed in Problem 3.19.

Another way to obtain the same result is to consider the second-order matrix (or correlation matrix) of the random vector $\mathbf{X}$ with components 1, X and X^2. This matrix takes the form

$$\mathbb{C} = E(\mathbf{X}\mathbf{X}^T) = \begin{bmatrix} 1 & m_1 & m_2 \\ m_1 & m_2 & m_3 \\ m_2 & m_3 & m_4 \end{bmatrix} \tag{3.109'}$$

and must be an NND matrix. This is ensured if the principal minors are non-negative. The first one gives $m_2 - m_1^2 \geq 0$, which is the variance of X, and the second is the determinant of $\mathbb{C}$. The calculation of this determinant gives exactly (3.110).

Example 3.4. Factoring some covariance matrices. Let **U** be a white unit vector and **X** the vector $\mathbb{M}\mathbf{U}$ where

$$\mathbb{M} = \begin{bmatrix} 1 & 1 & 0 \\ 1 & 1 & 1 \\ 0 & 1 & 1 \end{bmatrix}$$

It is clear that this matrix is symmetric. The covariance matrix of **X** is

$$\Gamma = \mathbb{M}\mathbb{M}^T = \mathbb{M}^2 = \begin{bmatrix} 2 & 2 & 1 \\ 2 & 3 & 2 \\ 1 & 2 & 2 \end{bmatrix}$$

which is also symmetric. Note that $\mathbb{M}$ is a Toeplitz matrix, which means that the elements M_{ij} are only functions of $|i - j|$. However, Γ is no longer a Toeplitz matrix, because all the diagonal elements are not equal. This shows that the square of a Toeplitz matrix is not necessarily Toeplitz.

The application of the factoring method introduced by (3.101) and (3.102) shows that Γ takes the form of (3.99) with the matrix

$$\mathbb{A} = 2^{-1/2} \begin{bmatrix} 2 & 0 & 0 \\ 2 & \sqrt{2} & 0 \\ 1 & \sqrt{2} & 1 \end{bmatrix}$$

which is lower triangular. It is easy to verify that $\Gamma = \mathbb{M}\mathbb{M}^T = \mathbb{A}\mathbb{A}^T$ and that matrices $\mathbb{A}$ and $\mathbb{M}$ are quite different. Let us repeat the same procedure with the matrix

$$\mathbb{M} = \begin{bmatrix} 0 & 1 & 0 \\ 0 & 0 & 1 \\ 1 & 0 & 0 \end{bmatrix}$$

which has a Toeplitz structure but is not symmetric. The corresponding covariance matrix is the unity matrix, or

$$\Gamma = \mathbb{M}\mathbb{M}^T = \mathbb{I}$$

This can be easily verified by a direct calculation, but there is also a physical interpretation. In fact, the application of $\mathbb{M}$ to vector **U** gives vector **X** with the components

$$X_1 = U_2 \quad ; \quad X_2 = U_3 \quad ; \quad X_3 = U_1$$

which means that $\mathbb{M}$ is a permutation matrix. As the components U_i are uncorrelated, the same is true for the components X_i of **X**, which is then also a white unit vector. Consequently, its covariance matrix is the unity matrix and the factorization (3.99) can be realized with the matrix $\mathbb{A}$, also equal to the unity matrix.

3.6 SEQUENCES OF SECOND ORDER RANDOM VARIABLES

The component X_k of a random vector can be considered as a finite sequence of RVs. However, in many signal processing problems it is necessary to use infinite sequences of RVs, which immediately introduces the concept of convergence. In this chapter, which is devoted to second order random vectors, we will restrict our attention to the concept of *convergence in the quadratic mean sense.* This concept uses only the second order properties of RVs, while different concepts calling for other knowledge are presented in the next chapter.

Definition. A sequence of random variables X_k defined on the same probability space is convergent in quadratic mean to a random variable X if

$$\lim_{k \to \infty} E[|X_k - X|^2] = 0 \tag{3.112}$$

This property is sometimes written as

$$\lim_{k \to \infty} \text{q.m. } X_k = X \tag{3.113}$$

and is also called *mean square convergence.*

Before presenting a more careful analysis of this kind of convergence, it is important to understand its physical meaning. In particular it must be noted that neither (3.112) nor (3.113) imply that the sequence of non random numbers $X_k(\omega_0)$ has the limit $X(\omega_0)$. Using the vocabulary of Section 1.2, this means that an arbitrary realization of the sequence X_k has no reason to be convergent to X.

To illustrate this point we will compare this mode of convergence with that used for deterministic signals when introducing the Fourier transformation (see [Picinbono], p.19). Using the probability measure P, (3.112) can be written as

$$\lim_{k \to \infty} \int_{\Omega} |X_k(\omega) - X(\omega)|^2 \, dP(\omega) = 0 \tag{3.114}$$

For signals $x(t)$ defined on $[0, 1]$ we can also obtain the relation

$$\lim_{k \to \infty} \int_0^1 |x_k(t) - x(t)|^2 \, dt = 0 \tag{3.115}$$

which does not mean that at any arbitrary time t the signal $x_k(t)$ converges to $x(t)$ if $k \to \infty$.

Let us briefly indicate the most important consequences of mean square convergence. We do not of course claim to present here an exhaustive discussion, as this can be found in books specializing in probability theory.

In classical analysis one of the most important problems is to specify the convergence without any knowledge of the limit. For this purpose there is a *Cauchy criterion* that states that X_k is mean square convergent if and only if

$$E[|X_k - X_l|^2] \to 0 \tag{3.116}$$

when k and l tend independently to infinity.

From this relation it is very easy to deduce the most interesting criterion of convergence.

Criterion. The sequence X_k is mean square convergent if and only if

$$E[X_k \; X_l^*] \to c \tag{3.117}$$

when k and l tend independently to infinity.

Proof. Suppose first that (3.117) is fulfilled. If we calculate the term of (3.116) we obtain

$$E[|X_k - X_l|^2] = E[X_k \; X_k^*] + E[X_l \; X_l^*] - E[X_k \; X_l^*] - E[X_l \, X_k^*] \tag{3.118}$$

which converges to zero from (3.117). Thus applying (3.116) we see that X_k is convergent in quadratic mean. The limit X is unknown, but it is clear that the expectation value $E\left[|X|^2\right]$ is equal to c. Conversely, if (3.112) is fulfilled, we deduce from the Schwarz inequality (3.49) applied to $(X_k - X)$ and $(X_l - X)$, or to $(X_k - X)$ and X, that $E[X_k \; X_l^*]$ converges to $E(|X|^2)$ which is (3.117).

One of the most important examples of the application of these results to signal processing problems appears when discussing the linear filtering of stochastic signals. It is well known that the input-output relationship of a causal linear filter is a convolution, or

$$y[k] = \sum_{l=0}^{+\infty} h[l] \; x[k-l] \tag{3.119}$$

The question is to know whether or not this series is mean square convergent when the input is a stochastic signal. To avoid the excess of indices, consider the problem for the series

$$S = \sum_{l=0}^{+\infty} h_l \, X_l \tag{3.120}$$

where X_l is a sequence of RVs and h_l a sequence of deterministic numbers. In order to simplify the discussion, which will be reintroduced later, let us assume that the RVs X_l are zeromean, uncorrelated and have the same variance σ^2. The partial sums S_k are given when l goes from zero to k. In order to use the criterion (3.117), we calculate $E(S_k \; S_l^*)$ which, using the property of the RVs X_l, gives

$$E(S_k S_l^*) = \sum_{i=0}^{k \wedge l} \sigma^2 |h_i|^2 \tag{3.121}$$

where $k \wedge l$ means the minimum of k and l. As a consequence we find that S is mean square convergent if the series $|h_l|^2$ is convergent, and the variance of the limit S is of course

$$E(|S|^2) = \sigma^2 \sum_{l=0}^{+\infty} |h_l|^2 \tag{3.122}$$

3.7 HILBERT SPACES OF SECOND ORDER RANDOM VARIABLES

In order to study second order RVs or random vectors, we shall introduce geometric concepts that can also be applied to many problems considered later. For those not familiar with the concepts of Hilbert spaces we *briefly* recall their principal properties.

3.7.1 Review of Hilbert Spaces

The purpose of this section is to present the principal definitions and properties concerning Hilbert spaces as simply and logically as possible. The concept of Hilbert space allows the introduction of many geometric reasonings in the framework of random signals and is therefore of great use. The ideas presented hereafter are very elementary and do not need much mathematical background. However, they are sufficient to understand all the problems which are discussed in this book.

(a) Linear space. Let S be a set of elements called "points" or "vectors" and noted by a lower-case letter such as x. Let us suppose that the sum of two vectors x and y is a vector of S noted $x + y$. This sum has all the classical properties of the addition, and in particular, it is commutative and associative and there is a null vector such that $x + 0 = x$. Let us also suppose that the product of a vector x by a complex number λ is a vector of S noted λx and that this product satisfies

$$\lambda(x + y) = \lambda x + \lambda y$$

The set S is a linear space, or a vector space, if, when x and y belong to S, $\lambda_1 x + \lambda_2 y$ belongs to S, whatever the complex numbers λ_1 and λ_2. If this property is only valid for real numbers λ, it is said that S is a real linear space.

Example 3.5. The above used space $\mathbb{R}^n$ is, of course, a real vector space and $\mathbb{C}^n$ is a complex vector space. On the other hand, let us consider the subset S of vectors of $\mathbb{C}^n$, such that $\mathbf{x}^H\mathbf{x} < 1$. It is the set of points of $\mathbb{C}^n$ located inside a sphere of radius 1. It is clear that this set S is not a linear space, because to any vector $\mathbf{x}$ belonging to S we can associate many vectors $\lambda\mathbf{x}$ that do not belong to S. For this purpose it suffices to take complex numbers λ such that $|\lambda|^2\, \mathbf{x}^H\mathbf{x} > 1$.

(b) Linear subspaces. Let S be an arbitrary linear space and S_1 a subset of S. If S_1 is still a linear space, S_1 is a linear subspace of S. However, it is worth noticing that any subset of S is not necessarily a linear subspace. To explain this point, we suppose that S is the space $\mathbb{C}^n$ and let $\mathbf{v}$ be an arbitrary vector of S. The set of vectors in the form $\lambda\mathbf{v}$ where λ is an arbitrary complex number is clearly a linear subspace of S. On the other hand, the set of vectors $\lambda\mathbf{v}$ such that $|\lambda| < 1$ is a subset of S but not a linear subspace.

Consider now n vectors $x_1, x_2, \ldots, x_n$ belonging to S. The linear subspace of S spanned by these vectors is the space whose elements x are in the form

$$x = \sum_{i=1}^{n} \lambda_i x_i$$

It is not difficult to verify that it is indeed a subspace of S.

(c) Scalar product and orthogonality in a linear space. Let us suppose that S is a linear space and let x and y be two vectors of S. A scalar product, or inner product, of x and y is an operation which introduces a complex number noted (x, y) and must satisfy the following properties

$$(x, y) = (y, x)^* \tag{3.123}$$

$$(z, \lambda x + \mu y) = \lambda(z, x) + \mu(z, y) \tag{3.124}$$

$$(x, x) \geq 0 \tag{3.125}$$

where the equality holds only for the vector $x = 0$. From (3.123) and (3.124) we deduce that

$$(\lambda x, y) = \lambda^*(x, y) \tag{3.126}$$

The Schwarz inequality follows from this relation and can be written as

$$|(x, y)|^2 \leq (x, x)\,(y, y) \tag{3.127}$$

In order to prove this inequality, we generalize the method that we introduced on p. 52. For this purpose, let us write

$$(x, y) = |(x, y)|\exp(j\phi) = m\exp(j\phi) \tag{3.128}$$

where m is the modulus of the complex number (x, y). Taking an arbitrary complex number λ, let us express (3.125) for the vectors $\lambda x + y$. Using all the properties of the scalar product this gives

$$(\lambda x + y, \lambda x + y) = |\lambda|^2(x, x) + \lambda^*(x, y) + \lambda(x, y)^* + (y, y) \geq 0$$

Suppose now that we apply this inequality for the complex numbers $\lambda = a\exp(j\phi)$ where ϕ is the same as in (3.128). This gives

$$a^2(x, x) + 2am + (y, y) \geq 0$$

As this quantity must be non-negative whatever the value of a, we deduce, as after (3.28), the Schwarz inequality (3.127).

The concept of orthogonality is a direct consequence of the definition of a scalar product. Two vectors x and y of a linear space are *orthogonal* if

$$(x, y) = 0 \tag{3.129}$$

It is also possible to define orthogonality to a subspace. A vector x is said to be orthogonal to a linear subspace S_1 of a linear space S if x is orthogonal to all the vectors of S_1.

It is important to note that if x is orthogonal to S itself, it is the null vector. In fact, if x is orthogonal to all the vectors of S, it is orthogonal to itself because x belongs to S and it verifies $(x, x) = 0$, which implies that $x = 0$.

Finally, it is possible to introduce the concept of orthogonal linear subspaces. Two subspaces S_1 and S_2 of S are orthogonal if any vector x_1 of S_1 is orthogonal to any vector x_2 of S_2. It is interesting to note that the intersection of two orthogonal subspaces is reduced to the vector $x = 0$. In fact, if x belongs to S_1 and S_2 that are orthogonal, then $(x, x) = 0$, which gives $x = 0$.

Multiplicity of the scalar products. It is important to note that many different scalar products can be constructed on a linear space. As a consequence, the concept of orthogonality is not unique. In order to explain this point, let us consider the linear space $\mathbb{C}^n$. The standard scalar product in this space is defined by

$$(\mathbf{u}, \mathbf{v}) = \mathbf{u}^H\mathbf{v} \tag{3.130}$$

which has the analytical form (3.68) where $\mathbb{M}$ is the identity matrix. However, let us consider this equation especially in the case where $\mathbb{M}$ is a PD matrix. It is clear that, with this assumption, the quantity

$$(\mathbf{u}, \mathbf{v})_{\mathbb{M}} \triangleq \mathbf{u}^H\mathbb{M}\mathbf{v} \tag{3.131}$$

is still a scalar product. In fact, as $\mathbb{M}$ is PD it is Hermitian, which gives (3.123). Furthermore, (3.124) is a direct consequence of (3.68) and (3.125) is the definition of a PD matrix. Therefore, there are as many scalar products as there are positive definite matrices. As the concept of orthogonality depends directly on the scalar product, the orthogonality of two vectors of $\mathbb{C}^n$ is not an intrinsic property, but a property that is relative to a specific scalar product.

As this point often seems surprising, we will give some examples. One way to introduce various scalar products is to consider the matrices $\mathbb{M}$ in (3.131) that are diagonal with positive diagonal elements. In this case (3.69) becomes

$$(\mathbf{u}, \mathbf{v})_{\mathbb{M}} = \sum_{i=1}^{n} \lambda_i u_i^* v_i \tag{3.132}$$

and the standard scalar product appears when all the λ_is are equal to one. It is clear in (3.132) that if this expression is null when the λ_is are equal, which means that **u** and **v** are orthogonal with the standard scalar product, the scalar product is no longer null when the λ_is are different. Therefore, orthogonal vectors are no longer orthogonal when the scalar product is transformed.

There is a general way of writing (3.131). This makes use of the spectral decomposition of any PD matrix given by (3.85). Using this expression, the scalar product of two vectors **a** and **b** defined by (3.131) becomes

$$(\mathbf{a}, \mathbf{b})_{\mathbb{M}} = \sum_{i=1}^{n} \lambda_i (\mathbf{a}^H \mathbf{u}_i)(\mathbf{b}^H \mathbf{u}_i)^* = \sum_{i=1}^{n} \lambda_i a_i^* b_i \tag{3.133}$$

where a_i and b_i are the components of the vectors **a** and **b** in the eigenbasis defined by the orthonormal (in the standard sense) eigenvectors $\mathbf{u}_i$.

(d) Normed spaces. A space is said to be normed if it is possible to introduce a distance between vectors or points of this space or a norm of a vector. The norm of x is noted $\|x\|$ and is a non-negative quantity. If $\|x\| = 0$, the vector x is the null vector. The distance between two points x and y is the norm of $x - y$. This distance must satisfy the triangle inequality

$$\|x + y\| \le \|x\| + \|y\| \tag{3.134}$$

which is pictured in Figure 3.3 corresponding to $\mathbb{R}^2$. This also gives

$$\|x - y\| = \|x - z + z - y\| \le \|x - z\| + \|z - y\| \tag{3.135}$$

(e) Pre-Hilbert spaces. Let S be a linear space in which a scalar product is introduced. This scalar product allows us to introduce a norm defined by

$$\|x\| = (x, x)^{1/2} \tag{3.136}$$

In fact, to be a norm, this quantity must satisfy the triangle inequality (3.134). This is a consequence of the Schwarz inequality (3.127). In fact, using (3.136) we obtain

$$\|x + y\|^2 = [\|x\| + \|y\|]^2 + P$$

where

$$P = (x, y) + (y, x) - 2[(x, x)(y, y)]^{1/2}$$
$$= 2\{|(x, y)|\cos\phi - [(x, x)(y, y)]^{1/2}\}$$

where ϕ is the phase of (x, y). However, the Schwarz inequality gives

$$|(x, y)| \leq [(x, x)\ (y, y)]^{1/2}$$

which implies that $P \leq 0$ and gives the triangle inequality. From the norm it is possible to define the distance between two points, or vectors, of S by

$$d(x, y) = \|x - y\| \tag{3.137}$$

A linear space S with a scalar product (x, y) and with a norm defined by (3.136) is said to be a pre-Hilbert space H. Note that, as the scalar product is generally not unique, there are several different pre-Hilbert spaces which can be constructed from a given linear space S.

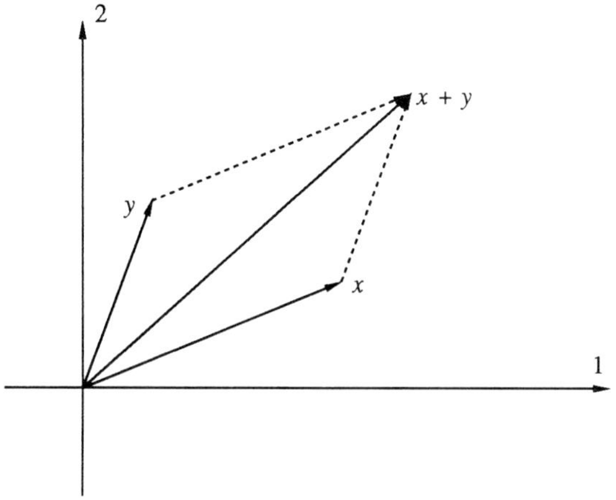

Figure 3.3 Triangle inequality in the plane.

(f) Hilbert spaces. A Hilbert space H is a pre-Hilbert space which is *complete*. This terminology means that any Cauchy sequence of vectors of H converges to a vector of H. A Cauchy sequence of vectors x_n is a sequence such that $\|x_m - x_n\|$ converges to zero when m and n independently tend to infinity.

It is obvious that any pre-Hilbert space of finite dimension, or spanned by a finite number of vectors u_i is complete. The problem thus only exists for pre-Hilbert spaces of infinite dimension.

Let us explain this point with an example. Consider the space S of real functions $x(t)$ defined for $0 \leq t \leq 1$ and that are continuous. This space is clearly a real linear space. In this space we can introduce a product defined by

$$(x, y) \triangleq \int_0^1 x(t)\, y(t)\, dt \tag{3.138}$$

and it is easy to verify that it satisfies all the properties of a *scalar product*. With this scalar product S becomes a pre-Hilbert space H. Consider the function $f(t)$ equal to 0 if

$t < \alpha < 1$ and to 1 if $t > \alpha$, where α is given. This function does not belong to H, because of the discontinuity for $t = \alpha$. However, it is possible to introduce many sequences of functions $x(t)$ belonging to H and which converge to $f(t)$, which means that $\|x_n(t) - f(t)\| \to 0$ when $n \to \infty$. Such sequences are of course Cauchy sequences, but the limit does not belong to H. Consequently, H is a pre-Hilbert space but not a Hilbert space.

(g) Projection theorem. Consider a subspace S of a given Hilbert space H and a vector x of H. Then x can be uniquely represented by a sum of two vectors of H in the form

$$x = \hat{x} + \tilde{x} \qquad (3.139)$$

where $\hat{x}$ belongs to S while $\tilde{x}$ is orthogonal to S.

Further, to any vector w of S we have

$$d(x, w) \geq d(x, \hat{x}) \qquad (3.140)$$

The proof of this theorem can be found in more specialized works (see, for example, [Cramér], p. 100).

The projection theorem extends a concept well known in classical geometry to arbitrary Hilbert spaces and is presented in Figure 3.4. The vector $\hat{x}$ appearing in (3.139) is called the *projection* of x onto the subspace S and is written

$$\hat{x} = \text{Proj}[x \mid S] \qquad (3.141)$$

This projection satisfies the *orthogonality principle*, written as

$$x - \hat{x} \perp S \qquad (3.142)$$

which means that $x - \hat{x}$, equal to $\tilde{x}$, is orthogonal to S. The projection theorem can be expressed in another form: to any vector x of H it is possible to associate its projection (3.141) and that is defined by the orthogonality principle (3.142).

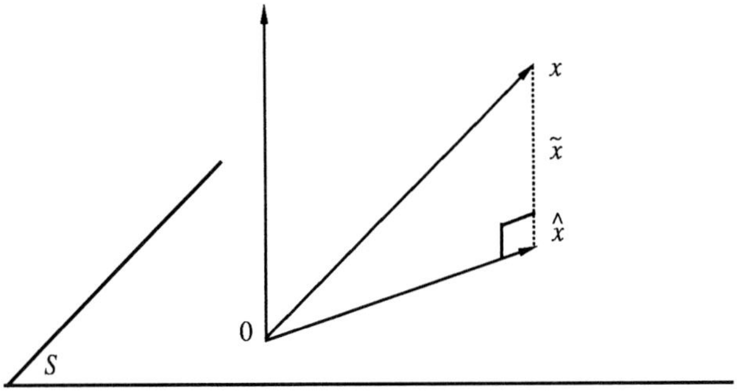

Figure 3.4 Representation of the projection theorem.

Finally, the *distance* between x and S is specified by

$$d(x,\ S) = \|\tilde{x}\| \tag{3.143}$$

From the triangle inequality we obtain

$$0 \leq \|\tilde{x}\| \leq \|x\| \tag{3.144}$$

If $\|\tilde{x}\| = 0$, then x belongs to S and is equal to its projection. On the other hand, if $\|\tilde{x}\| = \|x\|$, we then have $\hat{x} = 0$, which means that x is orthogonal to S.

(h) Projection theorem and complete subspaces. The proof, not explicitly given here, of the projection theorem needs a subspace S which should be a Hilbert subspace of H and therefore S should be complete. If this property is not satisfied, S is only a *pre-Hilbert space* and the projection theorem can be non-valid. This means that the vector x can have no projection onto S and an example of this situation is given in Problem 3.18. However, in many cases it is not necessary to prove first that S is complete. In fact, if an element $\hat{x}$ of S satisfying the orthogonality principle (3.142) exists, it satisfies (3.140) for any element w of S and then gives the minimum distance. This is also discussed in Problem 3.18.

(i) Vector sum of orthogonal subspace and projection. Let S_1 and S_2 be two orthogonal subspaces of a Hilbert space H. To any pair of vectors x_1 and x_2 belonging to S_1 and S_2 respectively, and any pair of complex numbers λ_1 and λ_2, it is possible to associate a set of vectors x in the form

$$x = \lambda_1 x_1 + \lambda_2 x_2 \tag{3.145}$$

By this operation we generate a subspace S called the vector sum of S_1 and S_2 and noted

$$S = S_1 \oplus S_2 \tag{3.146}$$

This is especially interesting when S_1 and S_2 are orthogonal subspaces. For example, the complementary space of S, $S^\perp$, is the space which is orthogonal to S and such that

$$S \oplus S^\perp = H \tag{3.147}$$

The relation (3.139) is a particular case of (3.145) and it results from the orthogonality principle that

$$\tilde{x} = \text{Proj}[x|S^\perp] \tag{3.148}$$

because $x - \tilde{x}$ is orthogonal to $S^\perp$.

More generally, let us show that if S is given by (3.146) where S_1 and S_2 are orthogonal, then

$$\text{Proj}[x|S] = \text{Proj}[x|S_1] + \text{Proj}[x|S_2] \tag{3.149}$$

This equation can also be written as

$$\hat{x} = \hat{x}_1 + \hat{x}_2 \tag{3.150}$$

and the result is represented in Figure 3.5.

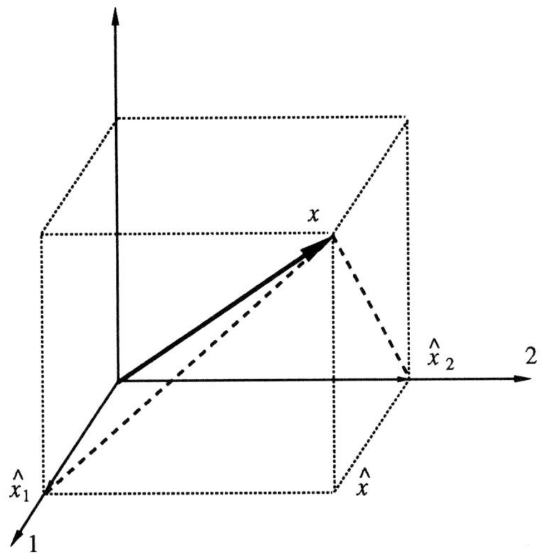

Figure 3.5 Projection onto the sum of two orthogonal subspaces.

In order to prove (3.150) we note that, because of the orthogonality principle, we have to prove that whatever u belonging to S we must have

$$(x, u) = (\hat{x}, u) = (\hat{x}_1 + \hat{x}_2, u) \tag{3.151}$$

where u has the general form $\lambda_1 u_1 + \lambda_2 u_2$ and u_1 and u_2 belong to S_1 and S_2 respectively. The first term of (3.151) can be written as

$$(x, u) = (x, \lambda_1 u_1 + \lambda_2 u_2) = \lambda_1 (x, u_1) + \lambda_2 (x, u_2) \tag{3.152}$$

But as $x - \hat{x}_1$ is orthogonal to S_1 we have

$$(x, u_1) = (\hat{x}_1, u_1) \tag{3.153}$$

and by the same reasoning

$$(x, u_2) = (\hat{x}_2, u_2) \tag{3.154}$$

As a result we obtain

$$(x, u) = \lambda(\hat{x}_1, u_1) + \lambda_2(\hat{x}_2, u_2) \tag{3.155}$$

The last term of (3.151) is

$$(\hat{x}_1 + \hat{x}_2, \lambda_1 u_1 + \lambda_2 u_2) = \lambda_1(\hat{x}_1, u_1) + \lambda_2(\hat{x}_2, u_2) \tag{3.156}$$

because $(\hat{x}_1, u_2) = 0$ as the vectors belong to orthogonal subspaces. Similarly $(\hat{x}_2, u_1) = 0$. Comparing (3.155) and (3.156) we obtain (3.151) that is the proof of (3.149) or (3.150).

This can obviously be repeated for a number of orthogonal subspaces larger than two and if

$$S = S_1 \oplus S_2 \oplus \ldots \oplus S_n \tag{3.157}$$

we have

$$\hat{x} = \hat{x}_1 + \hat{x}_2 + \ldots + \hat{x}_n \tag{3.158}$$

This shows the interest of decomposing an arbitrary subspace as a vector sum of orthogonal subspaces.

(j) Separable Hilbert spaces. The final concept which is useful in some questions is that of the separable Hilbert space. The Hilbert space is said to be *separable* if it contains a sequence of vectors x_i such that the subspace spanned by all these vectors is identical to H. If the vectors are orthogonal—and we will see later that it is always possible to realize this situation by an orthogonalization procedure—we can say the H is generated by a complete orthonormal sequence of vectors. This is a classical procedure in ordinary geometry.

3.7.2 The Hilbert Space of Scalar Random Variables

Consider a probability space $(\Omega, \mathcal{F}, P)$ and the set of all the second order scalar RVs $X(\omega)$. This set is obviously a linear space the vectors, or points, of which are the RVs. In order to arrive at a Hilbert space, we need a scalar product. It is easy to show that the quantity $E(X^*Y)$ satisfies all the requirements needed to be a scalar product, and we then define (X, Y) by

$$(X, Y) = E(X^*Y) \tag{3.159}$$

Note that the relation $(X, X) = 0$ does not mean that $X(\omega) = 0$ for every ω, but only almost everywhere in Ω. We thus make no distinction between two RVs X and Y, which are different but are such that $d(X, Y) = 0$. They are considered equivalent and are almost surely equal or equal with probability one.

It is important to note that the scalar product (3.159) is not associated to a particular realization of the RVs $X(\omega)$ and $Y(\omega)$, but to all the functions $X(\omega)$ and $Y(\omega)$. To understand this clearly, let us return to the case of the functions $x(t)$ defined for $0 \le t \le 1$. The scalar product of such functions is defined by

$$(x, y) = \int_0^1 x^*(t)y(t)dt \tag{3.160}$$

and cannot be deduced from the values of $x(t)$ and $y(t)$ at an arbitrary time instant. Furthermore, it is clear in (3.160) that (x, x) can be null without having $x(t) = 0$ for every t.

Finally, if we use the least square convergence defined by (3.116), we find that the Hilbert space constructed with (3.159) is complete.

These concepts have a direct application in *estimation problems*, briefly introduced here and considered in more detail later. Consider a set of n arbitrary second order RVs X_i, $1 \le i \le n$ defined on the same probability space. Any linear combination of such RVs is also second order, and the space of all these combinations is a subspace S of H, space of the scalar second order RVs defined on $(\Omega, \mathcal{F}, P)$. To any RV X it is possible to associate, using (3.141), the projection $\hat{X}$. This projection is called the *linear mean square estimation* of X in terms of X_i because of (3.140). This means that for any linear combination Y of X_i the error $E(|X - Y|^2)$ satisfies

$$E(|X - Y|^2) \ge E(|X - \hat{X}|^2) \tag{3.161}$$

It follows that $\hat{X}$ is the best linear mean square estimation of X in terms of the X_is. The last term of (3.161) is the minimum mean square error in the problem.

To close this section, note that all these concepts can be extended to vector RVs, as shown in the chapter dealing with estimation problems.

PROBLEMS

3.1 Consider the matrix

$$\mathbb{M} = \begin{bmatrix} \lambda_1 & a \\ 0 & \lambda_2 \end{bmatrix}$$

where λ_1 and λ_2 are two distinct positive numbers and a is real.

(a) Calculate the eigenvalues of $\mathbb{M}$.

(b) Determine the corresponding eigenvectors.

(c) Determine the condition on a ensuring that these vectors are orthogonal.

(d) Determine the condition on a in terms of λ_1 and λ_2 ensuring that $\mathbb{M}$ is PD for real vectors.

(e) Deduce from the previous results that a matrix can be PD without having orthogonal eigenvectors and that a matrix with positive eigenvalues is not necessarily PD.

3.2 Consider the matrix

$$\mathbb{M} = \begin{bmatrix} a & b \\ -b & a \end{bmatrix}$$

where a and b are two real numbers.

(a) Find the condition on these numbers ensuring that $\mathbb{M}$ is NND for real vectors.

(b) Calculate the eigenvalues of $\mathbb{M}$ and find the condition on these eigenvalues ensuring that $\mathbb{M}$ is NND for real vectors.

(c) Calculate the normalized eigenvectors $\mathbf{u}_1$ and $\mathbf{u}_2$ of $\mathbb{M}$. Indicate how $\mathbf{u}_2$ can be obtained when $\mathbf{u}_1$ is known.

(d) Calculate $\mathbf{u}_1^H\mathbf{u}_2$, $\mathbf{u}_1^T\mathbf{u}_1$ and $\mathbf{u}_2^T\mathbf{u}_2$.

(e) Write the matrix $\mathbb{M}$ in terms of λ_1 and $\mathbf{u}_1$.

3.3 Consider a real square matrix $\mathbb{M}$. What is the consequence of the fact that $\mathbb{M}$ is NND for real vectors on the eigenvalues of $\mathbb{M}$? Consider the cases of real and non real eigenvalues separately.

3.4 Consider the matrix $\mathbb{M} = 2\,\mathrm{Re}[\lambda \mathbf{u}\mathbf{u}^H]$ where λ is a complex number with a non-null imaginary part and $\mathbf{u}$ a normalized complex vector of $\mathbb{C}^N$.

(a) Find the condition on $\mathbf{u}$ ensuring that $\mathbf{u}$ is an eigenvector of $\mathbb{M}$. What is the corresponding eigenvalue?

(b) Is it possible for $\mathbf{u}$ to be real when the condition of (a) is fulfilled?

(c) Calculate $\mathbb{M}\mathbf{u}^*$ and deduce the property of the vector $\mathbf{u}^*$.

(d) Calculate $\mathbb{M}$ in terms of α, β, $\mathbf{a}$ and $\mathbf{b}$ when $\lambda = \alpha + j\beta$ and $\mathbf{u} = \mathbf{a} + j\mathbf{b}$.

(e) Deduce the condition on λ ensuring that $\mathbb{M}$ is NND for real vectors.

(f) Show that $\mathbb{M}$ cannot be symmetric and find the condition ensuring that $\mathbb{M}$ is either antisymmetric ($\mathbb{M} = -\mathbb{M}^T$) or of the form $\mathbb{D} + \bar{\mathbb{M}}$ where $\mathbb{D}$ is diagonal and $\bar{\mathbb{M}}$ antisymmetric.

3.5 Consider two complex square matrices $\mathbb{A}$ and $\mathbb{B}$ and suppose that they are NND for complex vectors.

(a) Find the condition on $\mathbb{A}$ and $\mathbb{B}$ ensuring that the product $\mathbb{A}\mathbb{B}$ is Hermitian.

(b) Suppose that this condition is not satisfied. Is the matrix $\mathbb{A}\mathbb{B}$ NND?

(c) Assuming that $\mathbb{A}$ is PD, show that the eigenvalues of $\mathbb{A}\mathbb{B}$ are non-negative.

3.6 Consider the square matrix $(\mathbb{A}\mathbb{B})^k$, k integer, where $\mathbb{A}$ and $\mathbb{B}$ are square real matrices which are also PD for complex vectors. Show that $\mathrm{Tr}[(\mathbb{A}\mathbb{B})^k] \geq 0$, where $\mathrm{Tr}[\mathbb{M}]$ means the trace of $\mathbb{M}$.

3.7 Suppose that the matrix $\mathbb{M}$ defined by (3.73) is NND and purely imaginary ($\mathbb{A} = \mathbb{O}$). What is the consequence on matrix $\mathbb{B}$?

3.8 **Structure of the Cholesky factorization**

(a) Consider a zeromean N dimensional random vector $\mathbf{V}$ with a diagonal covariance matrix Γ_V or $\Gamma_V = \mathrm{diag}\,[\{d_i\}]$, $d_i \geq 0$. Let $\mathbf{X}$ be another vector deduced from $\mathbf{V}$ by the transformation $\mathbf{X} = \mathbb{A}\mathbf{V}$, where $\mathbb{A}$ is an $N \times N$ matrix. Calculate the covariance matrix Γ_X of $\mathbf{X}$.

(b) Let us now consider an arbitrary covariance matrix Γ and its Cholesky factorization (3.105), where $\mathbb{A}$ is a triangular matrix with diagonal elements equal to one and $\mathbb{D}$ is a diagonal matrix. Show that there is the same number of independent parameters in Γ and in its Cholesky factorization.

(c) Suppose that the Cholesky factorization has been obtained. Calculate the determinant of Γ and find the condition on the d_is ensuring that Γ is full rank.

3.9 **Calculation of the Cholesky factorization.** Let $\mathbb{R}$ be an $N \times N$ Hermitian matrix and its Cholesky factorization $\mathbb{A}\mathbb{D}\mathbb{A}^H$, where $\mathbb{A}$ is a lower $N \times N$ triangular matrix with diagonal elements equal to one and $\mathbb{D}$ a diagonal matrix.

(a) Let us suppose that $\mathbb{R}$ is invertible and call $\mathbb{R}_n$ the square matrix deduced from $\mathbb{R}$ by taking the first n rows and columns. Is $\mathbb{R}_n$ Hermitian or not?

(b) Let us call $\mathbb{A}_n$ and $\mathbb{D}_n$ the matrices appearing in the Cholesky factorization of the matrix $\mathbb{R}_n$ and call $\mathbb{B}_n$ the matrix $\mathbb{A}_n^{-1}$. We also suppose that the matrices $\mathbb{R}_{n+1}$, $\mathbb{A}_{n+1}$, $\mathbb{B}_{n+1}$ and $\mathbb{D}_{n+1}$ are partitioned in the forms

$$\mathbb{R}_{n+1} = \begin{bmatrix} \mathbb{R}_n & \mathbf{u}_n \\ \mathbf{u}_n^H & r_{n+1} \end{bmatrix} \qquad \mathbb{A}_{n+1} = \begin{bmatrix} \mathbb{A}_n & \mathbf{0} \\ \mathbf{a}_n^H & 1 \end{bmatrix}$$

$$\mathbb{B}_{n+1} = \begin{bmatrix} \mathbb{B}_n & \mathbf{0} \\ \mathbf{b}_n^H & 1 \end{bmatrix} \qquad \mathbb{D}_{n+1} = \begin{bmatrix} \mathbb{D}_n & \mathbf{0} \\ \mathbf{0}^H & d_{n+1} \end{bmatrix}$$

Give the recursive algorithm allowing the calculation of the matrices $\mathbb{A}_{n+1}$, $\mathbb{B}_{n+1}$ and $\mathbb{D}_{n+1}$ in terms of the matrix elements of $\mathbb{R}_{n+1}$, $\mathbb{A}_n$, $\mathbb{B}_n$ and $\mathbb{D}_n$. Also give the initial elements of the recursion.

3.10 Another characterization of a PD Hermitian matrix. As in Problem 3.9, consider a Hermitian square matrix $\mathbb{R}$ and the submatrices $\mathbb{R}_n$, $1 \le n \le N$.

(a) Show that, if $\mathbb{R}$ is PD, $\det(\mathbb{R}_n) > 0$, $1 \le n \le N$.

(b) Show that the condition $\det(\mathbb{R}) > 0$ is not sufficient to ensure that $\mathbb{R}$ is PD.

(c) Writing the Cholesky factorization of the submatrices $\mathbb{R}_n$ as $\mathbb{R}_n = \mathbb{A}_n \mathbb{D}_n \mathbb{A}_n^H$, deduce that if $\det(\mathbb{R}_n) > 0$, $1 \le n \le N$, then $\mathbb{R}$ is PD.

3.11 Brownian matrices. Consider a set of N uncorrelated zeromean RVs U_i with the variances v_i. We define another set of N RVs X_n by

$$X_n = \sum_{i=1}^{n} U_i \quad , \qquad 1 \le n \le N$$

(a) Calculate the variance γ_n of the RVs X_n.

(b) Introducing the random vector $\mathbf{X}$ with components X_n, calculate its covariance matrix Γ and show that it can only be expressed in terms of the γ_ns. This matrix is sometimes called a Brownian matrix.

(c) Find the Cholesky factorization of Γ.

(d) Find the condition on the elements γ_n ensuring that a Brownian matrix is PD.

(e) Introducing the vector $\mathbf{U}$ with components U_i, show that $\mathbf{X} = \mathbb{S}\mathbf{U}$ and determine the matrix $\mathbb{S}$. Deduce the Cholesky factorization of Γ directly.

3.12 Using the same vector $\mathbf{U}$ as in the previous problem, we construct the vector $\mathbf{X}$ with components X_i defined by $X_1 = U_1$ and $X_i = U_i - U_{i-1}$, $2 \le i \le N$.

(a) Show that $\mathbf{X} = \mathbb{A}\mathbf{U}$ and determine the matrix $\mathbb{A}$.

(b) Calculate the covariance matrix Γ of $\mathbf{X}$ in terms of the variances of the RVs U_i.

(c) Deduce the Cholesky factorization of Γ.

(d) Find the condition on the matrix elements of Γ ensuring that Γ is PD.

3.13 Let $\mathbf{Y}$ be the vector the components of which are $Y_n = m_n X_n$, $1 \le n \le N$, where m_n is deterministic and X_n is defined in Problem 3.11.

(a) Using the same notations as in point (e) of this problem, we write $\mathbf{Y} = \mathbb{A}\mathbf{U}$. Show that $\mathbb{A} = \mathbb{M}\mathbb{S}$ where $\mathbb{S}$ is the matrix used for writing $\mathbf{X}$ in terms of $\mathbf{U}$.

(b) Deduce the covariance matrix $\mathbb{R}$ of $\mathbf{Y}$ in terms of $\mathbb{M}$, $\mathbb{S}$ and of a matrix $\mathbb{D}$ defined by the v_is, variances of the RV U_i.
(c) Calculate the matrix elements r_{ij} of $\mathbb{R}$.
(d) Introducing the quantities $\alpha_k = m_k$ and $\beta_k = m_k \gamma_k$, where γ_k is the variance of X_k calculated in (a) of Problem 3.11, express r_{ij} in terms of α and β.
(e) Using $\mathbb{R}$ defined by its matrix elements r_{ij}, find the matrices $\mathbb{M}$, $\mathbb{S}$ and $\mathbb{D}$ appearing in point (b). In order to do so, calculate first α_k and β_k, assuming $\beta_1 = 1$.
(f) Deduce a condition for the matrix elements r_{ij} ensuring that $\mathbb{R}$ is NND.

3.14 We want to extend the results of Example 3.2 to the case of multiple signals. To do this, we introduce the vector $\mathbf{X}$ of $\mathbb{C}^n$ defined by

$$\mathbf{X} = \mathbf{N} + \sum_{i=1}^{k} A_i \, \mathbf{s}_i \ , \qquad k < n$$

where $\mathbf{N}$ is a random vector, the $\mathbf{s}_i$s are deterministic vectors and the A_is are k uncorrelated RVs with variance σ_i^2 and uncorrelated with $\mathbf{N}$.
(a) Calculate the covariance matrix Γ_X of $\mathbf{X}$.
(b) Assuming that the covariance matrix Γ_N of $\mathbf{N}$ is $\sigma_N^2 \, \mathbb{I}$ and that the k vectors $\mathbf{s}_i$ are orthonormal, calculate the eigenvectors and eigenvalues of Γ_X.
(c) What is the property of the eigenvectors associated with the smallest eigenvalue?

3.15 **On the rank of some covariance matrices.** Consider two real and zeromean RVs X and Y such that $E(X^2) = E(Y^2) = 1$ and $E(XY) = c$. We introduce a tri-dimensional random vector $\mathbf{X}$ with components $X_1 = X$, $X_2 = X$ and $X_3 = Y$.
(a) Calculate the covariance matrix Γ_3 of $\mathbf{X}$.
(b) Calculate the rank and the determinant of Γ_3 and discuss the problem in terms of the values of c.
(c) Now consider an N x N NND matrix Γ_N and the matrices Γ_n obtained by the same procedure as in Problem 3.9 (a). Suppose that $\det(\Gamma_n) \neq 0$ for $1 \leq n \leq s$ and also that $\det(\Gamma_{s+1}) = 0$. Deduce from the previous question that the matrix Γ_n can be of a rank larger than s.
(d) Show that the same kind of situation can appear for a diagonal covariance matrix.

3.16 Prove the Schwarz inequality (3.49).

3.17 From the random vector $\mathbf{X}$ with mean value $\mathbf{m}$ and covariance matrix Γ we deduce the scalar RV defined by $Y = \mathbf{h}^T \mathbf{x}$.
(a) Calculate the mean value of Y.
(b) Calculate $E(Y^2)$ and deduce from (a) the variance of Y.

3.18 **Projection on incomplete spaces.** Consider a random point ω uniformly distributed on the interval $[-1, +1]$ and the real RVs $X(\omega)$.
(a) Write the condition ensuring that $X(\omega)$ is second order.
(b) Show that the set of all the RVs for which $X(\omega)$ is a continuous function of ω is a pre-Hilbert space P.
(c) Show that P is not complete. To do this, consider the RV $R(\omega)$ defined by $R(\omega) = 0$ if $0 < a < |\omega| \leq 1$ and $R(\omega) = 1$ if $|\omega| \leq a$ and show that $R(\omega)$

does not belong to P, but can be considered as the limit in the mean square sense of many sequences of RVs belonging to P. Give an example of such a sequence.

(d) Consider the subset S of P of RVs satisfying $X(\omega) = 0$ for $\omega < 0$. Show that S is a pre-Hilbert subspace of P and is not complete.

(e) Consider an RV $X(\omega)$ belonging to P but not to S. Applying the orthogonality principle, show that the RV $\hat{X}(\omega)$ obtained does not necessarily belong to S. Whenever this is the case, show that there is no element of S with a minimum distance to $X(\omega)$.

(f) Now suppose that $X(\omega)$ belongs to P and also satisfies $X(0) = 0$. Show that, even though S is not a complete subspace, $X(\omega)$ has a projection $\hat{X}(\omega)$ onto S. Calculate this projection by using the orthogonality principle.

3.19 Consider a zeromean real RV and the random vector with components 1, X and X^2 introducing the second order matrix (3.109').

(a) What happens to the RVs X and X^2 when the inequality (3.111) becomes an equality?

(b) Show that X almost surely satisfies the relation

$$X^2 - \frac{m_3}{m_2} X - m_2 = 0.$$

(c) Deduce from this relation the possible values of X and their probabilities.

(d) Discuss the problem when $m_3 = 0$.

(e) Conversely, show that any RV X taking two values only and such that $m_1 = 0$, satisfies (3.111) where the inequality is an equality.

3.20 **Polynomial expansion of an RV.** Let X and Y be two scalar RVs defined on the same probability space and suppose that all the moments of Y are finite. We want to find the best linear mean square estimation of X in terms of Y^0, Y^1, ... Y^n written as

$$\hat{X} = \alpha_0 + \alpha_1 Y + \alpha_2 Y^2 + \ldots + \alpha_n Y^n$$

(a) As the subspace S introduced in the projection theorem is spanned by the set of RVs Y^k, $0 \leq k \leq n$, write the $n + 1$ equations characterizing the orthogonality principle using the notations $m_k = E(Y^k)$ and $r_k = E(XY^k)$.

(b) Supposing that $n = 2$ and assuming that the PDF $p(x, y)$ of the pair of RVs X and Y is symmetric, calculate α_0, α_1 and α_2. What is the particular property of $\hat{X}$?

(c) Carry out the same calculation as in (**b**) assuming that only the PDF $p(y)$ is symmetric.

(d) Find the relation between this problem and Problem 3.19.

Chapter 4

Multidimensional Random Variables

4.1 INTRODUCTION

The concept of random vectors or of multidimensional random variables was introduced in the previous chapter, especially in Section 3.4, where the distribution function was defined. However, most of the chapter was devoted to the study of second-order properties, thereby justifying the title. This means that knowledge about the random vector was limited to second-order moments and especially to the mean value and the covariance matrix.

However, there are many problems where this knowledge is insufficient and the purpose of this chapter is to enlarge this knowledge. In particular, the concepts of conditional probabilities introduced in Chapter 1 and of multidimensional probability distribution will play an important role. More generally, we will study those problems concerning random vectors applicable to the study of random signals that cannot be solved with second-order methods only.

4.2 CONDITIONAL DISTRIBUTIONS

Conditional distributions are one of the newest topics that appear when our study is not limited to second-order properties. As this topic is generally considered difficult, we shall present the matter very progressively. In this entire section the basic expression is, of course, the definition of the conditional probability given by (1.24) or

$$P(B|A) = \frac{P(A.B)}{P(A)} \tag{4.1}$$

and the definition of independent events characterized by

$$P(A.B) = P(A)\ P(B) \tag{4.2}$$

4.2.1 Conditional Distribution with Scalar Random Variables

Let $X(\omega)$ be a scalar RV, as presented in Chapter 2, and let A be an event concerning this RV. For example A can be the event corresponding to $a < X \leq b$, as presented in Section 2.2.2, the probability of which is $F(b) - F(a)$. Similarly, A can be the event A_+ considered in Example 2.7, which introduces the probability P_+ given by (2.51). Clearly many other examples can be given.

The *conditional distribution function* (DF) of X with respect to A is given by the general expression (4.1) which becomes

$$F(x|A) = \frac{P[(X \leq x).A]}{P(A)} \tag{4.3}$$

This function has all the properties of a distribution function, as introduced in Section 2.2.2. In particular, it is obvious in (4.3) that $F(+\infty|A) = 1$, exactly as in (2.7).

If $F(x|A)$ is everywhere differentiable, we can introduce a conditional probability distribution function (PDF) defined as in (2.10) by

$$p(x|A) = \frac{dF(x|A)}{dx} \tag{4.4}$$

Similarly *conditional expectations* can be defined as in (2.18) by

$$E[h(X)|A] = \int h(x)\, dF(x|A) \tag{4.5}$$

which takes the forms of (2.18') or (2.18") depending on whether the RV is discrete or continuous. Of particular interest are the conditional moments, and especially the conditional mean. They are obtained by replacing $h(x)$ by x^k in (4.5).

Finally it is possible to introduce the *conditional characteristic function* defined by (4.5), where $h(x) = \exp(jux)$. This characteristic function enjoys all the properties presented in Section 2.5, and notably its expansion can generate the conditional moments, provided they are finite.

Let us now suppose that a set of events A_i constitutes a *partition* of Ω. For each event A_i it is possible to construct a conditional DF such as (4.3), and it results from (1.38) that the *a priori* DF $F(x)$ can be written as

$$F(x) = \sum_{i=1}^{n} F(x|A_i)\, P(A_i) \tag{4.6}$$

Applying this expression to the calculation of the expectation, we deduce from (2.18), (4.5) and (4.6) that

$$E[h(X)] = \sum_{i=1}^{n} E[h(X)|A_i]\, P(A_i) \tag{4.7}$$

This is a very important relation between *a priori* and conditional expectation values.

Example 4.1 Conditional probabilities with Poisson distribution. Suppose that $X(\omega)$ is a discrete RV with a Poisson distribution of mean value m, defined by (2.42). As in (2.48) let us call A_+ and A_- the events to obtain an even or an odd number respectively. The probabilities of A_+ and A_- are given by (2.51) and (2.52) respectively.

Let us now calculate the conditional probability of obtaining the even number $2k$ conditionally to A_+. It is clear that

$$P[(X = 2k).\ A_+] = P[X = 2k] = e^{-m}\frac{m^{2k}}{(2k)!}$$

Applying (4.1) and (2.51) we obtain

$$P[(X = 2k)|A_+] \stackrel{\Delta}{=} p(2k|+) = \frac{1}{\text{ch}m}\ \frac{m^{2k}}{(2k)!} \tag{4.8}$$

Using exactly the same procedure for odd numbers, we obtain

$$p(2k + 1|-) = \frac{1}{\text{sh}m}\ \frac{m^{2k+1}}{(2k + 1)!} \tag{4.8'}$$

Furthermore it is obvious that

$$p(2k|-) = p(2k + 1|+) = 0$$

As A_+ and A_- are a partition of Ω, we deduce that

$$p(2k) = p(2k|+)P_+ + p(2k|-)P_- = p(2k|+)P_+$$

which can be verified with (2.51) and (4.8).

In order to obtain the conditional moments, the simplest way is to calculate the conditional characteristic function. Using (4.8) and (2.55), we deduce that

$$\phi(u|+) = \sum_{k=0}^{\infty} \frac{e^{ju2k}\ m^{2k}}{(\text{ch}\,m)(2k)!} = \frac{\text{ch}(m\,e^{ju})}{\text{ch}\,m} \tag{4.9}$$

Similarly, we have

$$\phi(u|-) = [1/\text{sh}\,m]\ \text{sh}[m\exp(ju)]$$

It is easy to find that the characteristic function given by (2.91) satisfies

$$\phi(u) = \phi(u|+)\ P_+ + \phi(u|-)\ P_-$$

Let us now calculate the conditional moments. For this purpose, there is a difference between (2.91) and (4.9). In (2.91) it is easy to make a limited expansion in powers of u, as in Problem 2.3, because the term $e^{ju} - 1$ appearing in the expression of the characteristic function tends to zero when $u \to 0$. This is no longer the case for the term $m\ e^{ju}$ in (4.9). Thus the simplest way is to calculate the derivative of (4.9) and to apply (2.66). To do this, we have to calculate the first two derivatives of $\phi(u|+)$ given by (4.9). From this calculation we obtain

$$\phi'(u|+) = (\text{ch}m)^{-1}\; jm\,e^{ju}\,\text{sh}(m\,e^{ju})$$

$$\phi''(u|+) = -\,m\,(\text{ch}m)^{-1}\,[e^{ju}\,\text{sh}(m\,e^{ju}) + m\,e^{2ju}\,\text{ch}(m\,e^{ju})]$$

and applying (2.66) we obtain

$$m_1(+) = m\,\text{th}m$$

$$m_2(+) = m\,\text{th}m + m^2$$

$$\sigma^2(+) = m\,\text{th}m + m^2(1 - \text{th}^2 m)$$

The same calculation can be realized for $\phi(u|-)$ and the result is

$$m_1(-) = m\,\text{cth}m$$

$$m_1(-) = m\,\text{cth}m + m^2$$

$$\sigma^2(-) = m\,\text{cth}m + m^2(1 - \text{cth}^2 m)$$

It is interesting to discuss the behavior of these quantities when $m \to 0$. We easily find that $m_1(+)$, $m_2(+)$, $\sigma^2(+)$, and $\sigma^2(-)$ tend to zero. On the other hand $m_1(-)$ and $m_2(-)$ tend to one. This can be understood from the probabilities (4.8) and (4.8'). These expressions show that $p(0|+)$ and $p(1|-)$ tend to one when $m \to 0$. Conditionally to A_+ or A_- this means that the RV X only takes the values 0 or 1 respectively.

Example 4.2 Conditional probabilities with exponential distributions. Suppose that $X(\omega)$ is a continuous RV with a two-sided exponential distribution as in (2.33). Let A_+ and A_- be the events that $X(\omega)$ is positive or negative respectively. It results from (2.33) that

$$P(A_+) = P(A_-) = 1/2 \tag{4.10}$$

Applying (4.3) and (4.4) we find that the conditional probabilities of X are

$$p(x|+) = u(x)\;a\exp(-\,ax)$$

$$p(x|-) = u(-\,x)\;a\exp(ax)$$

The first PDF is the one-sided exponential distribution (2.30). We deduce from (2.75) that the first two conditional moments are

$$m_1(+) = 1/a \;;\; m_2(+) = 2/a^2 \;;\; \sigma^2(+) = 1/a^2$$

$$m_1(-) = -\,1/a \;;\; m_2(-) = 2/a^2 \;;\; \sigma^2(-) = 1/a^2$$

The application of (4.7) with (4.10) gives the *a priori* moments of X in (2.78). However, if we apply (4.7) to the conditional variance, we obtain $1/a^2$ instead of $2/a^2$ given by (2.34). This means that the formula (4.7) cannot be applied with the conditional variance. This is due to the fact that the variance is not an expectation value like (4.5) with a function $h(x)$ independent of A. In fact, the variance can be written as in (2.22), but in this expression the mean value depends on A, because it is the conditional first moment. It is thus necessary to take some care when using (4.7).

Example 4.3 Conditional probabilities with normal distribution. Let us consider the same problem as in the previous example when $X(\omega)$ is an $N(0, \sigma^2)$ RV. The conditional PDF is deduced from (2.35) which gives

$$p(x|+) = u(x)2(2\pi\sigma^2)^{-1/2}\exp(-x^2/2\sigma^2) \tag{4.11}$$

After simple integration by parts we obtain

$$m_1(+) = -m_1(-) = 2\sigma/\sqrt{2\pi}$$

$$m_2(+) = m_2(-) = \sigma^2$$

As a result the conditional variances are

$$\sigma^2(+) = \sigma^2(-) = (1 - 2/\pi)\,\sigma^2$$

We can present the same conclusion concerning the variance as in the previous example.

4.2.2 General Conditioning for Two Random Variables

Consider a two-dimensional RV (X, Y) defined by its DF (3.2) and an arbitrary event A related to $(X\,Y)$. It is possible to introduce a *conditional distribution* function by

$$F(x, y \mid A) = P\{(X \le x) \,.\, (Y \le y) \mid A\} \tag{4.12}$$

and by applying (4.1) we obtain

$$F(x, y \mid A) = \frac{P\{(X \le x) \,.\, (Y \le y) \,.\, A\}}{P(A)} \tag{4.13}$$

In many cases the event A is associated with a domain $D(A)$ of the plane (x, y) and $P(A)$ is then given by (3.3).

Example 4.4 Distribution function with an energy condition. Suppose that we impose the RV (X, Y) to satisfy

$$X^2 + Y^2 \le E \tag{4.14}$$

which defines the event A. As we will see later $X^2 + Y^2$ can be interpreted as an energy of the RV, and condition (4.14) means that this energy is bounded by E. The domain $D(A)$ associated with this condition is defined by a circle of radius $E^{1/2}$, and the event $(X \le x) \,.\, (Y \le y) \,.\, A$ appearing in the numerator of (4.13) is associated with the domain $D(x, y, A)$ represented in Fig. 4.1.

By applying (3.3) we can then calculate $P\{D(x, y, A)\}$ and $P\{D(A)\}$ and we obtain the DF by

$$F(x, y \mid A) = \frac{P\{D(x, y, A)\}}{P\{D(A)\}} \tag{4.15}$$

The final result depends of course on the DF $F(x, y)$ of the RV (X, Y).

As previously, we can also introduce *conditional marginal distributions* as, for example,

$$F_X(x \mid A) = P\{(X \leq x) \mid A\} \tag{4.16}$$

If $F(x,\ y \mid A)$ has already been calculated, this DF is given by the same rule as in (3.8), or

$$F_X(x \mid A) = F(x,\ +\infty \mid A) \tag{4.17}$$

For example, for the energy condition represented in Fig. 4.1, the domain $D(x, A)$ is obtained by taking $y \to \infty$, which removes the straight line parallel to the x-axis.

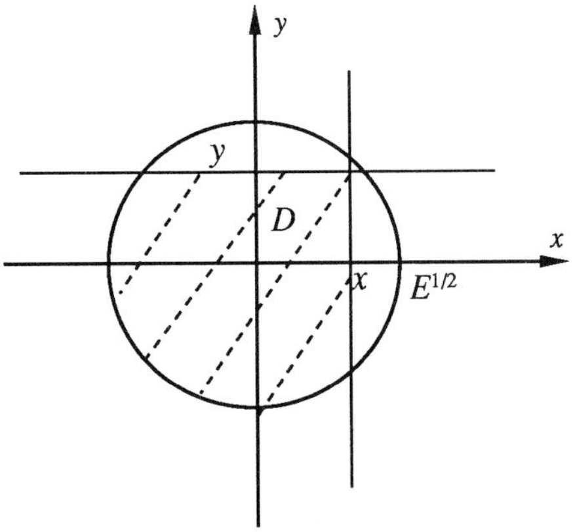

Figure 4.1 Domain $D(x, y, A)$ associated with energy condition.

It is clear that the previous discussion, also using only the DF $F(x,\ y)$, is likewise valid for continuous and discrete RVs. In the first case it is possible to use a relation such as (3.4) to introduce conditional probability densities. For example, if (4.15) is twice differentiable, we can apply the same expression as in (3.4) which defines the conditional PDF as

$$p(x,\ y|A) = \frac{\partial^2 F(x,\ y|A)}{\partial x\, \partial y}$$

Similarly the marginal conditional PDF is deduced by a derivation with respect to x of (4.17) and can also be expressed as in (3.11) by

$$p_X(x|A) = \int p(x,\ y|A)\ dy$$

Example 4.5 Energy condition with normal vectors. Let XY be a two-dimensional RV such that X and Y are two independent RVs with the same distribution $N(0,\ \sigma^2)$, as defined in (2.35). The PDF of this RV can then be written as

$$p(x,\ y) = \frac{1}{2\pi\sigma^2}\ \exp\left(-\ \frac{x^2\ +\ y^2}{2\sigma^2}\right)$$

Suppose that A is the event discussed in Example 4.4. The probability $P(A)$ can be calculated by using (3.3) or

$$P(A) = \int\int_{D(A)} p(x, y)\, dx\, dy$$

where $D(A)$ is the domain corresponding to the points located inside the circle of radius $a = E^{1/2}$ and centered at the origin. Using polar coordinates r, θ we easily find that $P(A)$ becomes

$$P(A) = \int_0^{2\pi} d\theta \int_0^a f(r)\, rdr = 2\pi \int_0^a f(r)\, rdr$$

where $f(r)$ is

$$f(r) = (2\pi\sigma^2)^{-1} \exp(-r^2/2\sigma^2)$$

Using the variable $u = r^2/2\sigma^2$ we easily obtain

$$P(A) = 1 - \exp(-a^2/2\sigma^2)$$

As a result, the conditional PDF $p_X(x|A)$ is equal to 0 if $x^2 + y^2 > a^2$ and if $x^2 + y^2 < a^2$ its value is

$$f(x, y; a) = (2\pi\sigma^2)^{-1}[1 - \exp(-a^2/2\sigma^2)]^{-1} \exp[-(x^2 + y^2)/2\sigma^2]$$

In order to calculate the marginal PDF $p_X(x|A)$ we have to make an integration with respect to the variable y. As $p(x, y|A) = 0$ if $x^2 + y^2 > a^2$, we obtain

$$p(x|A) = \int_{-\alpha(x)}^{+\alpha(x)} f(x, y)\, dy$$

where $\alpha^2(x) = a^2 - x^2$. This last integration can be realized by using the error function $\mathrm{erf}(x)$ defined in Problem 2.9.

4.2.3 Conditional Distribution between *X* and *Y*

In many cases we are interested in the distribution of the RV X with a condition only on the RV Y. The simplest case is, of course, when the two events are $X \le x$ and $Y \le y$. Using (4.1) and (3.2) we immediately obtain

$$F(x \mid Y \le y) \triangleq P\{(X \le x) \mid (Y \le y)\} = \frac{F(x, y)}{F(+\infty, y)} \tag{4.18}$$

The denominator is the marginal DF of Y, as given by (3.8).

But for many situations we are more interested in the event $Y = y$ than $Y \le y$. As this event has a null probability for continuous RVs, it is necessary to treat the continuous and the discrete cases separately.

(a) Discrete case. Starting from (3.6) and applying (4.1) we immediately obtain

$$p_{i|j} \triangleq P\{(X = x_i) \mid (Y = y_j)\} = \frac{p_{i,j}}{q_j} \tag{4.19}$$

where

$$q_j \triangleq P(Y = y_j) = \sum_i p_{i,j} \tag{4.20}$$

the last equation being the same as (3.12).

(b) Continuous case. We suppose that $F(x, y)$ is everywhere differentiable, which introduces a PDF $p(x, y)$ by (3.4). From this PDF we can calculate two marginal PDFs $p_X(x)$ and $p_Y(y)$.

By definition the *conditional probability density* of X given $Y = y$ is

$$p_X(x \mid y) \triangleq \frac{p(x, y)}{p_Y(y)} \tag{4.21}$$

It is important to note that this is a *definition* and not a consequence of (4.1). In fact, (4.21) gives

$$p(x, y) = p_Y(y)\, p_X(x \mid y) \tag{4.22}$$

which is a relation between *densities* and not probabilities.

The conditional probability density satisfies the following relations, deduced immediately from its definition

$$\int p_X(x \mid y)\, dx = 1 \tag{4.23}$$

$$\int p_X(x \mid y)\, p_Y(y)\, dy = p_X(x) \tag{4.24}$$

which is similar to (1.38).

Let us now explain the relation (4.22). Multiplying it by $\Delta x\, \Delta y$ we obtain, by an obvious extension of (2.11) or (3.5), the following terms

$$p(x, y)\, \Delta x\, \Delta y = P\{(x < X \le x + \Delta x) \,.\, (y < Y \le y + \Delta y)\} \tag{4.25}$$

$$p_Y(y)\, \Delta y = P\{(y < Y \le y + \Delta y)\} \tag{4.26}$$

which give, with (4.21) and (4.1)

$$p_X(x \mid y)\Delta x = P\{(x < X \le x + \Delta x) \mid (y < Y \le y + \Delta y)\} \tag{4.27}$$

This is the interpretation of the meaning of the conditional PDF.

It is obvious that we can also introduce a PDF $p_Y(y \mid x)$ by a relation quite similar to (4.21).

Finally, it is important to point out that conditional distributions are sometimes called *a posteriori* distributions, in contrast to *a priori* distributions related to marginal distributions.

It is clear that the knowledge of $p_Y(y)$ and $p_X(x|y)$ is equivalent to that of $F(x|Y \le y)$ and $F_Y(y)$. In fact, applying (4.22), we obtain $p(x, y)$ which by integration gives $F(x, y)$ and then $F_Y(y)$ which is equal to $F(+\infty, y)$. Conversely, starting from $F_Y(y)$ and $F(x|Y \le y)$, we deduce $F(x, y)$ from (4.18) and by derivation we obtain $p(x, y)$. From this PDF we calculate $p_X(y)$ by integration with respect to the variable x, which finally gives $p_X(x|y)$ by using (4.21).

4.2.4 Extension to Vector Random Variables

All the previous results can easily be extended if X and Y are vector RVs, or random vectors. The DF of a pair of random vectors $\mathbf{X}$ and $\mathbf{Y}$ is defined by (3.64). By using a derivation procedure like (3.54) we can deduce the PDF $p(\mathbf{x}, \mathbf{y})$, which means a function of x_i, $1 \le i \le m$ and y_j, $1 \le j \le n$.

The conditional PDF is defined by the same expression as (4.21) which now takes the form

$$p_{\mathbf{X}}(\mathbf{x} \mid \mathbf{y}) \triangleq \frac{p(\mathbf{x}, \mathbf{y})}{p_{\mathbf{Y}}(\mathbf{y})} \tag{4.28}$$

The marginal PDF appearing in the denominator is, of course, defined as in (3.11) by integration of $p(\mathbf{x}, \mathbf{y})$ over the whole space $\mathbb{R}^m$ where $\mathbf{x}$ is defined.

4.2.5 Independent Random Vectors

The independence between two vectors $\mathbf{X}$ and $\mathbf{Y}$ is characterized by (3.18) which now becomes $F_{\mathbf{X}}(\mathbf{x}, \mathbf{y}) = F_{\mathbf{X}}(\mathbf{x})\, F_{\mathbf{Y}}(\mathbf{y})$. For continuous RVs the application of (3.54) gives

$$p(\mathbf{x}, \mathbf{y}) = p_{\mathbf{X}}(\mathbf{x})\, p_{\mathbf{Y}}(\mathbf{y}) \tag{4.29}$$

which also characterizes the independence. For discrete scalar RVs, the independence between X and Y is characterized from (3.6) by

$$p(i, j) = p_X(i)\, p_Y(j) \tag{4.30}$$

4.3 CONDITIONAL EXPECTATION

In this section we assume that the RV (X, Y) is continuous, leaving the reader to transpose the results to the discrete case as an exercise.

The conditional PDF $p_X(x \mid y)$ allows us to introduce conditional expected values. As in (2.18), the conditional expected value of $h(X)$ is defined by

$$E_X\{h(X) \mid y\} = \int h(x)\, p_X(x \mid y)\, dx \tag{4.31}$$

The subscript X appears to specify that this expected value corresponds to the RV X. Where no confusion is possible, this subscript will be omitted.

It is very important to keep clearly in mind that the expected value given by (4.31) is still a function of y. It is then possible to calculate its expected value with respect to y, which is

$$E_Y[E_X\{h(X) \mid Y\}] = \int E_X\{h(X) \mid y\}\, p_Y(y)\, dy \tag{4.32}$$

By replacing $E_X\{h(X) \mid y\}$ with its value of (4.31), and by using (4.22) and comparing this to (3.15), we obtain

$$E_Y[E_X\{h(X) \mid Y\}] = E\{h(X)\} \tag{4.33}$$

This last term can also be written as $E_X\{h(X)\}$ and is the expected value in the marginal distribution. In conclusion, by taking the expected value with respect to the variable of the conditioning, we obtain the marginal expected value.

Among all the possible functions, the powers such as x^k are of particular importance, and if $h(x) = x^k$ we obtain the conditional moments in (4.31).

The first conditional moment will play an important role later and is sometimes called the *regression* $r(y)$ defined by

$$r(y) = E_X(X \mid y) = E(X \mid y) \tag{4.34}$$

Note immediately that, as we shall often see later, an RV $(X\,Y)$ can be zeromean without having a null regression.

Let us show that the regression enjoys a very interesting property, studied more carefully in a more general context. Consider an arbitrary function $g(y)$ provided that $g(Y)$ is a second order RV. The random variables $X - r(Y)$ and $g(Y)$ are then *uncorrelated* and also orthogonal.

To prove this, note first that $X - r(Y)$ is a zeromean RV, precisely because of (4.33). We then have to prove that

$$E\{[X - r(Y)]\, g(Y)\} = 0 \tag{4.35}$$

which characterizes orthogonality and, for zeromean RVs, uncorrelation as seen p. 58. This expression can be written as

$$E\{\ \} = \iint [x - r(y)]\, g(y)\, p(x, y)\, dx\, dy \tag{4.36}$$

Using (4.22) we obtain

$$E\{\ \} = \iint [x - r(y)]\, p_X(x \mid y) g(y)\, p_Y(y)\, dx\, dy \tag{4.37}$$

However, the integration with respect to x gives $r(y) - r(y) = 0$ from (4.31) and (4.34). This gives (4.35).

This property has an interesting interpretation within the framework of the concepts of Hilbert spaces studied in Section 3.7. In fact, (4.35) is an equation of orthogonality stating that $X - r(Y)$ is orthogonal to $g(Y)$, whatever the function $g(.)$, provided that $g(Y)$ is second order. This exhibits the orthogonality principle, and we can conclude that $r(Y)$ is *the projection of* X onto the subspace spanned by all the RVs $g(Y)$, which is of course a subspace of the Hilbert space of second order RVs defined on $(\Omega, \mathcal{F}, P)$. This result will have a very important application in estimation theory studied later. It is represented in Figure 4.2. In this figure $H(y)$ indicates the subspace generated by the RVs

$g(Y)$. It is important to note that X and $r(Y)$ have the same mean value, which implies that $X - r(Y)$ is a zeromean RV.

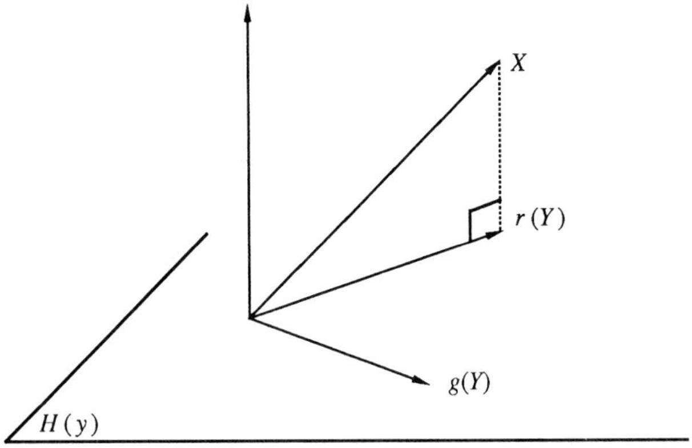

Figure 4.2 Regression and projection.

Before leaving this section, let us note that the concept of conditional expectation value can be used in a more general sense. In fact, to any DF such as $F(x, y \mid A)$ defined by (4.12), it is possible to associate an expected value such as $E\{h(X, Y) \mid A\}$. As A is quite arbitrary, we see that we have a large degree of freedom in this operation.

All these results can be transposed without difficulty to *random vectors*. The principles are exactly the same, although the notation can become a little cumbersome because of the multiplicity of the indices. We encourage the reader to practice by working out some of the problems presented at the end of this chapter.

As an example we shall return to the regression and its fundamental property. This regression can now be written as

$$\mathbf{r}(\mathbf{y}) = E(\mathbf{X} \mid \mathbf{y}) \tag{4.38}$$

Note that $\mathbf{r}(\mathbf{y})$ is a vector-valued function of the vector $\mathbf{y}$, and $\mathbf{r}(\mathbf{y})$ belongs to $\mathbb{R}^m$. Consider now a vector-valued function $\mathbf{g}(\mathbf{y})$ taking its values in $\mathbb{R}^s$, where s is arbitrary. Using exactly the same calculation as in (4.35), we obtain

$$E\{[\mathbf{X} - \mathbf{r}(\mathbf{Y})]\, \mathbf{g}^T(\mathbf{Y})\} = \mathbb{O} \tag{4.39}$$

which means that the vector $\mathbf{X} - \mathbf{r}(\mathbf{Y})$, which is zeromean, is uncorrelated to any vector $\mathbf{g}(\mathbf{Y})$ depending only on the random vector $\mathbf{Y}$. By extension of (4.35) the relation (4.39) is still a characterization of the *orthogonality principle*, a point that will be discussed in more detail later. This means that $\mathbf{r}(\mathbf{Y})$ is still the projection of the random vector $\mathbf{X}$ onto the Hilbert subspace spanned by all the functions $\mathbf{g}(\mathbf{Y})$. This will be applied extensively in estimation problems.

4.4 FUNCTIONS OF RANDOM VECTORS

To a given vector $\mathbf{x}$ of $\mathbb{R}^m$ it is possible to associate a vector $\mathbf{y}$ of $\mathbb{R}^n$ by a deterministic transformation, or

$$\mathbf{y} = \mathbf{h}(\mathbf{x}) \tag{4.40}$$

If $\mathbf{X}$ is now random, $\mathbf{Y} = \mathbf{h}(\mathbf{X})$ is also a random vector and it is interesting to calculate its DF. The principle is exactly the same as for scalar RVs, although calculations often become rather complex. Let us call $D(\mathbf{y})$ the region of $\mathbb{R}^m$ defined by

$$D(\mathbf{y}) \triangleq \left\{ \mathbf{x} \mid \prod_{i=1}^{n} [h_i(\mathbf{x}) \le y_i] \right\} \tag{4.41}$$

where the y_is are the n components of $\mathbf{y}$. It is clear that the DF $F_Y(\mathbf{y})$ of the random vector $\mathbf{Y}$ is

$$F_Y(\mathbf{y}) = \int_{D(\mathbf{y})} d^m F_X(\mathbf{x}) \tag{4.42}$$

where F_X is the DF of $\mathbf{X}$. The complexity of the calculation now depends only on the structures of F_X and of $\mathbf{h}(\mathbf{x})$. For example, we may be interested in energy problems, which gives $h(\mathbf{x}) = \mathbf{x}^T\mathbf{x}$, as in (3.59). As a result the domain $D(y)$ becomes

$$D(y) = \left\{ \mathbf{x} \mid \sum_{i=1}^{m} x_i^2 \le y \right\} \tag{4.43}$$

We leave it to the reader to study some of the problems presented at the end of the chapter in order to become familiar with the problem.

One case is of particular interest: it corresponds to the situation where the transformation (4.40) is invertible. This of course requires that $m = n$, and we have $\mathbf{x} = \mathbf{h}^{-1}(\mathbf{y})$.

This assumption is the same as that introduced in the case of scalar RVs. In fact, in order to obtain (2.99) it was necessary to assume that the transformation $y = h(x)$ was strictly monotone, in such a way that it was possible to write $x = h^{-1}(y)$. The procedure used to obtain the PDF was to calculate the DF first and then its derivative. This procedure is not directly applicable in the multivariable case, and we will use another approach.

We shall first recall a result concerning the calculus of multiple integrals. Suppose that we want to calculate the integral

$$I = \int_D f(\mathbf{x})\, d\mathbf{x} \tag{4.44}$$

which is a notation of an nth-order integral, already explained in (3.56'). It is sometimes useful to make a change of variables, or to pass from the variables x_i to other variables

y_j. It is shown in standard theory of multiple integral calculus that the change of variables needs the introduction of the so-called *Jacobian determinant.* If the transformation $\mathbf{y} = \mathbf{h}(\mathbf{x})$ is a one to one correspondence, or if we can write $\mathbf{x} = \mathbf{h}^{-1}(\mathbf{y})$, we can express the components x_k of the vector $\mathbf{x}$ in terms of the components y_l of the vector $\mathbf{y}$. Let us introduce the matrix $\mathbb{M}$ of which the elements are

$$M_{kl} = \frac{\partial x_k}{\partial y_l} \tag{4.45}$$

This matrix has the form

$$\mathbb{M} = \begin{bmatrix} \dfrac{\partial x_1}{\partial y_1} & \dfrac{\partial x_1}{\partial y_2} & \cdots\cdots & \dfrac{\partial x_1}{\partial y_n} \\ \dfrac{\partial x_2}{\partial y_1} & \dfrac{\partial x_2}{\partial y_2} & \cdots\cdots & \dfrac{\partial x_2}{\partial y_n} \\ \cdot & \cdot & \cdot\ \cdot\ \cdot & \\ \dfrac{\partial x_n}{\partial y_1} & \dfrac{\partial x_n}{\partial y_2} & \cdots\cdots & \dfrac{\partial x_n}{\partial y_n} \end{bmatrix}$$

The Jacobian determinant is the determinant of this matrix, or

$$J = \det(\mathbb{M})$$

Note that this Jacobian determinant is a function of the vector $\mathbf{y}$, because the x_ks are expressed in terms of the y_ls.

In the case where the vectors $\mathbf{x}$ and $\mathbf{y}$ belong to $\mathbb{R}^1$, or are scalars, the matrix $\mathbb{M}$ also becomes scalar and is equal to its determinant, or

$$J = \frac{dx}{dy}$$

The fact that the transformation $\mathbf{y} = \mathbf{h}(\mathbf{x})$ is invertible implies that the Jacobian determinant cannot be null, and therefore has a constant sign. The same is true for the scalar case, where positive or negative derivatives imply strictly increasing or decreasing functions. More exactly, the Jacobian determinant, or the derivative in the scalar case, can only be null on isolated points, which play no role in integral calculations.

Let us now suppose that the function $f(\mathbf{x})$ in (4.44) is non-negative. In this case it can be shown that we have

$$I = \int_{D'} f[h^{-1}(\mathbf{y})]\ |J(\mathbf{y})|\ d\mathbf{y} \tag{4.46}$$

where D' is the domain deduced from D by the transformation $\mathbf{y} = \mathbf{h}(\mathbf{x})$. In the scalar case this relation becomes

$$I = \int_{D'} f[x(y)]\ \frac{dx}{dy}\ dy$$

which is the standard rule for the change of variables in single integrals.

Let us give two examples of this kind of calculation. The first was partially presented in Example 4.5. Let $f(\mathbf{x})$ be the function appearing in the PDF corresponding to two normal independent RVs, or

$$f(\mathbf{x}) = \exp[-(x_1^2 + x_1^2)/2\sigma^2]$$

and suppose that D is the set of points located inside the circle of center 0 and of radius a. In order to calculate the integral, it is convenient to use polar coordinates or to introduce the variables

$$y_1 = r, \quad r > 0 \quad ; \quad y_2 = \theta, \quad 0 \le \theta \le 2\pi$$

The transformation is defined by the equations

$$x_1 = y_1 \cos y_2 \qquad x_2 = y_1 \sin y_2$$

and it is obviously invertible. We then have to calculate the domain D' and the Jacobian determinant. The domain D' is, of course, the set of points such that $0 \le y_1 \le a$ and $0 \le y_2 < 2\pi$, and the relation between the two domains is represented in Figure 4.3.

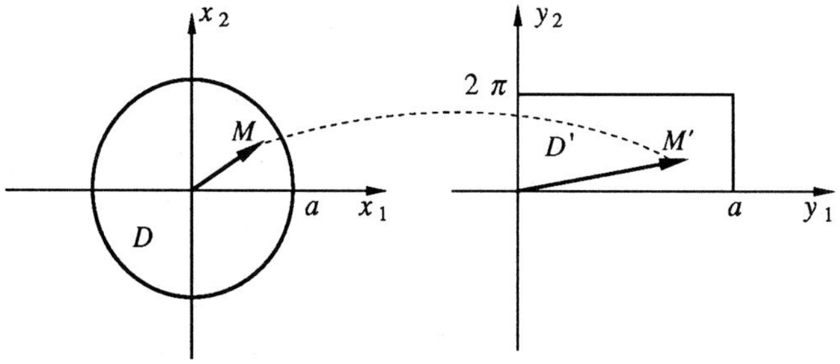

Figure 4.3 Polar coordinates transformation.

The matrix $\mathbb{M}$ defined by (4.45) is

$$\mathbb{M} = \begin{bmatrix} \cos y_2 & -y_1 \sin y_2 \\ \sin y_2 & y_1 \cos y_2 \end{bmatrix}$$

and its determinant is

$$J = y_1(\cos^2 y_2 + \sin^2 y_2) = y_1$$

and is always positive because $y_1 > 0$.

Finally $x_1^2 + x_2^2 = y_1^2$ and the integral I becomes

$$I = \iint_{D'} \exp(-y_1^2/2\sigma^2)\, y_1 \, dy_1 \, dy_2$$

$$= \int_0^{2\pi} dy_2 \int_0^a \exp(-y_1^2/2\sigma^2)\, y_1\, dy_1$$

$$= 2\pi \int_0^a \exp(-y_1^2/2\sigma^2)\, y_1\, dy_1$$

This last integral can be obtained by the change of variables $u = y_1^2/2\sigma^2$, which gives

$$I = 2\pi\sigma^2 \int_0^{a^2/2\sigma^2} \exp(-u)\, du$$

$$= 2\pi\sigma^2 \,[1 - \exp(-a^2/2\sigma^2]$$

Finally, if $a \to \infty$, $I = 2\pi\sigma^2$. This result justifies the value of the normalization coefficient appearing in the mathematical expression of the normal distribution.

The same calculation can be presented in the *tridimensional case*, where we want to calculate the integral of the function

$$f(\mathbf{x}) = \exp[-(x_1^2 + x_2^2 + x_3^2)/2\sigma^2]$$

inside domain D which is now a sphere of center 0 and radius a. For this calculation we can use spherical coordinates represented in Figure 4.4.

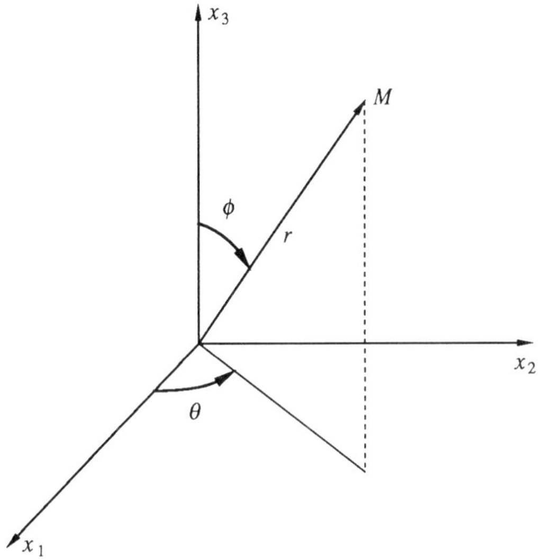

Figure 4.4 Spherical coordinates.

The new coordinates of M are

$$y_1 = r_1, \qquad r > 0$$

$$y_2 = \phi, \qquad 0 \le \phi \le \pi$$

$$y_3 = \theta, \qquad 0 \le \theta \le 2\pi$$

and the relation between the two systems of coordinates is

$$x_1 = r \sin\phi \cos\theta = y_1 \sin y_2 \cos y_3$$

$$x_2 = r \sin\phi \sin\theta = y_1 \sin y_2 \sin y_3$$

$$x_3 = r \cos\phi = y_1 \cos y_2$$

The domain of integration corresponding to the sphere in the new coordinates is represented in Figure 4.5.

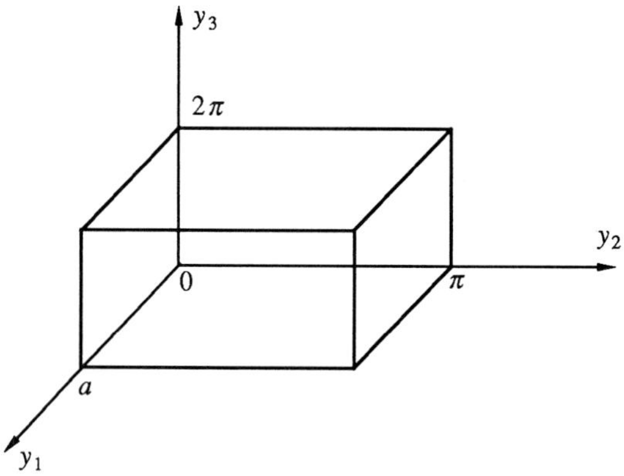

Figure 4.5 Domain D' corresponding to the sphere.

The matrix $\mathbb{M}$ defined by (4.45) is

$$\mathbb{M} = \begin{bmatrix} \sin y_2 \cos y_3 & y_1 \cos y_2 \cos y_3 & -y_1 \sin y_2 \sin y_3 \\ \sin y_2 \sin y_3 & y_1 \cos y_2 \sin y_3 & y_1 \sin y_2 \cos y_3 \\ \cos y_2 & -y_1 \sin y_2 & 0 \end{bmatrix}$$

and its determinant is

$$J = y_1^2 \sin y_2$$

which has a constant sign and is null only at the points $y_1 = 0$ and $y_2 = 0$ or $y_2 = \pi$.

Combining all these results, we obtain

$$I = \int_0^{2\pi} dy_3 \int_0^{\pi} \sin y_2 \, dy_2 \int_0^{a} \exp(-y_1^2/2\sigma^2)\, y_1^2 \, dy_1$$

which, after integration with respect to y_2 and y_3, gives

$$I = 4\pi \int_0^{a} y_1^2 \exp(-y_1^2/2\sigma^2)\, dy_1$$

This last integral can be calculated by parts and finally expressed in terms of the error function introduced in Problem 2.9. If $a \to \infty$ we easily obtain $I = [(2\pi)^{1/2}\sigma]^3$, which is still related to the normalization factor appearing in (2.35).

Let us now go back to the initial problem which is the calculation of the PDF $q_Y(\mathbf{y})$ of the vector RV $\mathbf{Y}$ in terms of the PDF $q_X(\mathbf{x})$ of $\mathbf{X}$ where $\mathbf{Y}$ is deduced from $\mathbf{X}$ by the invertible transformation

$$\mathbf{y} = \mathbf{h}(\mathbf{x}) \quad ; \quad \mathbf{x} = \mathbf{h}^{-1}(\mathbf{y})$$

Let D be an arbitrary domain in the space of the variable $\mathbf{x}$ and let D' be the domain deduced from D by the transformation $\mathbf{y} = \mathbf{h}(\mathbf{x})$. The basic relation is the conservation of the probability measure, which means that whatever D, we must have

$$\int_D p_X(\mathbf{x})\, d\mathbf{x} = \int_{D'} q_Y(\mathbf{y})\, d\mathbf{y}$$

Comparing (4.44) and (4.46) we then deduce that

$$q_Y(\mathbf{y}) = p_X[\mathbf{h}^{-1}(\mathbf{y})]\,|J(\mathbf{y})| \tag{4.47}$$

where $J(\mathbf{y})$ is the Jacobian determinant, or the determinant of the matrix $\mathbb{M}$ defined by (4.45). It is clear that in the scalar case (4.47) is equivalent to (2.99).

This general method can be extended for transformations which are not invertible, by decomposing such transformations into sums of other invertible transformations. However, the method rapidly becomes rather complex, and it is equivalent to use (4.42) and to calculate the derivatives.

This method can also be extended to the case $m \neq n$, if some conditions are fulfilled. Suppose that $n < m$ and call $\mathbf{h}_n(\mathbf{x})$ and $\mathbf{h}_p(\mathbf{x})$, $p = m - n$, two transformations such that by concatenation of $\mathbf{h}_n$ and $\mathbf{h}_p$ we get a transformation of $\mathbb{R}^m$ into itself, which is invertible. Applying the previous method, we obtain a PDF of a vector $\mathbf{y}$ of $\mathbb{R}^m$, and by integration over the $m - n$ variables, artificially introduced, we obtain the PDF of $\mathbf{Y}$, vector of $\mathbb{R}^n$.

Example 4.6 Linear transformation of a spherically invariant random vector. Consider a random vector of $\mathbb{R}^n$ with a PDF defined by

$$p(\mathbf{x}) = f(\mathbf{x}^T\mathbf{x}) \tag{4.48}$$

which is an extension to $\mathbb{R}^n$ of the PDF given by (3.20). This PDF is sometimes called spherically invariant, or spherically symmetrical. The scalar function $f(.)$ must, of course, be such that (3.56) holds.

Because of the symmetry of the term $\mathbf{x}^T\mathbf{x}$, we easily obtain $\mathbf{m} = E(\mathbf{X}) = \mathbf{0}$ and

$$\mathbb{\Gamma} = E(\mathbf{X}\mathbf{X}^T) = \sigma^2\mathbb{I} \tag{4.49}$$

where σ^2 is the variance of the RVs X_i, components of $\mathbf{X}$.

Consider now the random vector

$$\mathbf{Y} = \mathbb{A}\mathbf{X} \tag{4.50}$$

where $\mathbb{A}$ is a square and regular matrix with the inverse $\mathbb{A}^{-1}$. As $\mathbf{x} = \mathbb{A}^{-1}\mathbf{y}$, the term J appearing in (4.47) can be written

$$J = \det(\mathbb{A}^{-1}) = [\det(\mathbb{A})]^{-1} \tag{4.51}$$

where $\det(\mathbb{A})$ means the determinant of $\mathbb{A}$. To obtain this last result, we have used the well-known result concerning the determinant of a product of matrices $\mathbb{A}$ and $\mathbb{B}$ which shows that $\det(\mathbb{A}\mathbb{B}) = \det(\mathbb{A})\det(\mathbb{B})$. Furthermore, $\mathbf{h}^{-1}(\mathbf{y})$ appearing in (4.47) becomes $\mathbb{A}^{-1}\mathbf{y}$. Replacing this value in (4.48), we obtain

$$q_Y(\mathbf{y}) = |\det(\mathbb{A})|^{-1}\, f(\mathbf{y}^T\mathbb{A}^{-T}\mathbb{A}^{-1}\mathbf{y}) \tag{4.52}$$

Note that $\mathbb{A}^{-T}$ means $(\mathbb{A}^{-1})^T$. Let us now write this expression in a more interesting form. We deduce from (4.50) that $E(\mathbf{Y}) = \mathbf{0}$; the covariance matrix of $\mathbf{Y}$ is then

$$\Gamma_Y = E(\mathbf{Y}\mathbf{Y}^T) = \mathbb{A}\,\Gamma_X\,\mathbb{A}^T = \sigma^2\,\mathbb{A}\mathbb{A}^T \tag{4.53}$$

As the inverse of a product of matrices $\mathbb{A}\,\mathbb{B}$ is $\mathbb{B}^{-1}\,\mathbb{A}^{-1}$, we obtain

$$\Gamma_Y^{-1} = \sigma^{-2}\,\mathbb{A}^{-T}\,\mathbb{A}^{-1} \tag{4.54}$$

and using the fact that $\det(\mathbb{A}) = \det(\mathbb{A}^T)$ we have

$$\det(\Gamma_Y) = \sigma^{2n}\,\{\det(\mathbb{A})\}^2 \tag{4.55}$$

Combining all these results, we can write (4.52) as

$$q_Y(\mathbf{y}) = \frac{\sigma^n}{\{\det(\Gamma_Y)\}^{1/2}}\, f(\sigma^2\,\mathbf{y}^T\Gamma_Y^{-1}\mathbf{y}) \tag{4.56}$$

Let us apply this result to the very important case of *normal distribution.* Suppose that all the components of $\mathbf{X}$ are independent and $N(0,\ 1)$. In this case $p(\mathbf{x})$ becomes, from (2.35),

$$p(\mathbf{x}) = (2\pi)^{-n/2}\exp(-\frac{1}{2}\,\mathbf{x}^T\mathbf{x}) \tag{4.57}$$

The PDF of $\mathbf{Y}$ given by (4.50) is then

$$p(\mathbf{y}) = (2\pi)^{-n/2}\,\{\det(\Gamma_Y)\}^{-1/2}\exp(-\frac{1}{2}\,\mathbf{y}^T\,\Gamma_Y^{-1}\mathbf{y}) \tag{4.58}$$

which is the PDF of a normal zeromean random vector with covariance matrix Γ_Y. This PDF is also written $N(\mathbf{0},\ \Gamma_Y)$. This distribution will be studied extensively in Section 4.6.

4.5 CHARACTERISTIC FUNCTIONS. MOMENTS AND CUMULANTS

The characteristic function $\phi(\mathbf{u})$ is defined as in (2.53). For this, consider a real vector $\mathbf{u}$ with components $u_1,\ u_2,\ \ldots,\ u_n$ and a real random vector with components $X_1,\ X_2,\ \ldots,\ X_n$. The function $\phi(\mathbf{u})$ is defined, as in (3.57), by

$$\phi(\mathbf{u}) \triangleq E[\exp(j\mathbf{u}^T\mathbf{X})] = \int_{\mathbb{R}^n}\exp(j\mathbf{u}^T\mathbf{x})\; d^nF(\mathbf{x}) \tag{4.59}$$

It is clear that the scalar product $\mathbf{u}^T\mathbf{x}$ is given by

$$\mathbf{u}^T\mathbf{x} = u_1x_1 + u_2x_2 + \ldots + u_nx_n$$

Similarly, the second characteristic function is defined by

$$\psi(\mathbf{u}) = \ln\{\phi(\mathbf{u})\} \tag{4.60}$$

and the same precaution is necessary as after (2.92).

All the properties indicated in Section 2.5 can be transposed to $\phi(\mathbf{u})$ without difficulty. In particular, in the case of a continuous random vector, we can transpose (2.60) which becomes

$$p(\mathbf{x}) = (2\pi)^{-n} \int_{\mathbb{R}^n} \exp(-j\mathbf{u}^T\mathbf{x})\ \phi(\mathbf{u})\ d\mathbf{u} \tag{4.61}$$

The use of random vectors or of multivariate RVs introduces the concept of dependence that does not appear with scalar RVs. To simplify the discussion, let us assume that $n = 2$, as the generalization to arbitrary n introduces only difficulties in notation. In this case (4.59) can be written as

$$\phi(u, v) = \iint \exp\{j(ux + vy)\}\ d^2F(x, y) \tag{4.62}$$

At first we note that if X and Y are *independent*, $F(x, y)$ is given by (3.18) that gives

$$\phi(u, v) = \phi_X(u)\ \phi_Y(v) \tag{4.63}$$

where $\phi_X(u)$ and $\phi_Y(v)$ are the marginal characteristic functions of X and Y.

Let us now calculate $\phi(u, 0)$. The integral in y introduces the marginal distribution given by (3.8), and as in (3.14) we obtain

$$\phi_X(u) = \phi(u, 0) \tag{4.64}$$

which means that the marginal characteristic function is deduced from $\phi(u, v)$ by replacing the variable v by 0.

Let us now look at the *conditional characteristic function* defined by

$$\phi_X(u \mid y) = E\{\exp(juX) \mid y\} \tag{4.65}$$

In the continuous case this gives, with (4.21)

$$\phi_X(u \mid y) = \int \exp(jux)\ \frac{p(x, y)}{p_Y(y)}\ dx \tag{4.66}$$

As $p(x, y)$, and then $p_Y(y)$, can be deduced from $\phi(u, v)$, it is possible to express $\phi_X(u \mid y)$ in terms of $\phi(u, v)$ by an expression calculated in Problem 4.23.

Let us now return to a random vector $\mathbf{X}$ of arbitrary dimension and calculate the characteristic function of the vector $\mathbf{Y}$ defined by a linear transformation $\mathbf{y} = \mathbb{A}\mathbf{x}$, where $\mathbb{A}$ is a rectangular matrix. We have

$$\phi_Y(\mathbf{v}) \triangleq E\{\exp(j\mathbf{v}^T\mathbf{Y})\} = E\{\exp(j\mathbf{v}^T\ \mathbb{A}\ \mathbf{X})\} \tag{4.67}$$

Noting that (4.59) gives the value of $E\{\exp(j\mathbf{u}^T X)\}$ and that $\mathbf{v}^T \mathbb{A} = (\mathbb{A}^T \mathbf{v})^T$, we obtain

$$\phi_Y(\mathbf{v}) = \phi_X(\mathbb{A}^T \mathbf{v}) \tag{4.68}$$

A particular example appears when $\mathbb{A} = \mathbf{c}^T$, where $\mathbf{c}$ is a real column vector. In this case Y is a scalar RV and v is also scalar, which gives

$$\phi_Y(v) = \phi_X(v\,\mathbf{c}) \tag{4.69}$$

This means that the characteristic function of the scalar RV

$$Y = c_1 X_1 + c_2 X_2 + \dots + c_n X_n$$

can be expressed in terms of the characteristic function of the random vector **X** defined by (4.59) in the form

$$\phi_Y(v) = \phi_X(c_1 v,\ c_2 v,\ \dots,\ c_n v)$$

This expression is especially interesting when the RVs X_i are independent, because, as in (4.63), we have

$$\phi_X(\mathbf{u}) = \phi_1(u_1)\,\phi_2(u_2)\dots\ \phi_n(u_n) \tag{4.70}$$

where $\phi_i(u_i)$ is the characteristic function of the component X_i. In this case $\phi_Y(v)$ takes the form

$$\phi_Y(v) = \prod_{i=1}^{n} \phi_i(c_i v) \tag{4.71}$$

If all the c_is are equal to 1, we obtain the characteristic function of a *sum of independent* RVs. Finally the characteristic function of a sum of n RVs X_i which are IID, meaning that they are independent and with the same distribution, is

$$\phi_Y(v) = [\phi_X(v)]^n \tag{4.71'}$$

which has many applications in what follows.

Let us present some interesting applications of this relation. First, consider the case where the X_is are $N(m,\ \sigma^2)$. Their common characteristic function is given by (2.83). Using (4.71) we deduce that

$$\phi_Y(v) = \exp[j\,n\,m\,v - (1/2)\,n\,\sigma^2 v^2]$$

which is the characteristic function of a distribution $N(nm,\ n\sigma^2)$. We can apply the same method when the X_is have a Poisson distribution characterized by (2.91). In this case we obtain

$$\phi_Y(v) = \exp[nm(e^{jv} - 1)]$$

which is still a Poisson distribution but with the mean value nm. It is sometimes said that the normal and Poisson distributions are *stable* with respect to the addition.

Let us now compare (2.87) and (2.89). We deduce that a binomial RV can be considered as the sum of n IID random variables that are of the type "heads and tails." This explains why the mean and the variance are given by (2.40) and (2.41), or deduced from (2.36) and (2.37) by multiplication by n.

Let us now look at the *calculation of the moments* by the characteristic function. As $\phi(\mathbf{u})$ defined by (4.59) can be written

$$\phi(\mathbf{u}) = E\{\exp(j\mathbf{u}^T\mathbf{X})\}$$

the derivatives at the origin give the moments by an expression which is obviously more complex than in the scalar case and is

$$E(X_1^{k_1}X_2^{k_2}...X_n^{k_n}) = j^{-k}\left(\frac{\partial}{\partial u_1}\right)^{k_1}\left(\frac{\partial}{\partial u_2}\right)^{k_2}...\left(\frac{\partial}{\partial u_n}\right)^{k_n}\phi(\mathbf{u})\Big|_{\mathbf{u}=\mathbf{0}} \quad (4.72)$$

with $k = k_1 + k_2 + \ldots + k_n$. This expression means that the partial derivative of $\phi(\mathbf{u})$ is calculated for $\mathbf{u} = \mathbf{0}$. Instead of calculating the derivatives in (4.72), it is sometimes easier to identify the expansions of $\phi(\mathbf{u})$ and of $E[\exp(j\mathbf{u}^T\mathbf{X})]$, exactly as in the scalar case. This was used in most of the cases presented in Example 2.8. It is clear that the expansion of the characteristic function defined by (4.59) is

$$\phi(\mathbf{u}) = \sum_{k=0}^{\infty}\frac{j^k}{k!}E[(\mathbf{u}^T\mathbf{X})^k]$$

and the problem is to calculate the expectation, which is not very tedious for low values of k.

Applying exactly the same procedure to the second characteristic function $\psi(\mathbf{u})$ defined from $\phi(\mathbf{u})$ by (4.60), we obtain the *cumulants* which are then expressed by

$$c(X_1^{k_1}, X_2^{k_2},\ldots, X_n^{k_n}) = j^{-k}\left(\frac{\partial}{\partial u_1}\right)^{k_1}\left(\frac{\partial}{\partial u_2}\right)^{k_2}...\left(\frac{\partial}{\partial u_n}\right)^{k_n}\psi(\mathbf{u})\Big|_{\mathbf{u}=\mathbf{0}} \quad (4.73)$$

It is important to know that, whereas the moments can be defined either by (4.72) or by an expression like (3.57), the cumulants are not directly expected values of functions of the random vector $\mathbf{X}$. This justifies the fact that the argument appearing in c of (4.73) is not a product, as in the expected value (4.72). Cumulants, although less known than moments, play an important role in some questions of signal processing. Let us investigate some of their *most important properties*.

(a) As it is possible to permute the order of the partial derivatives in (4.73) without changing the result, we find that the cumulants, as well as the moments, are symmetric functions.

(b) Noting that the characteristic function of the random vector with components $a_1X_1, a_2X_2, \ldots, a_nX_n$ is $\phi(a_1u_1, a_2u_2, \ldots, a_nu_n)$, we obtain from (4.73) that

$$c(a_1 X_1, a_2 X_2, \ldots, a_n X_n) = a_1 a_2 \ldots a_n \, c(X_1, X_2, \ldots, X_n) \tag{4.74}$$

an expression which is also valid for the moments.

(c) Using the properties of the partial derivatives and the definition of the characteristic functions, we obtain

$$c(X + Y, Z_1, \ldots, Z_n) = c(X, Z_1, \ldots, Z_n) + c(Y, Z_1, \ldots, Z_n) \tag{4.75}$$

an expression obviously valid for the moments.

(d) If the RVs X_i appearing in (4.73) are *independent*, the characteristic function can be factorized as in (4.70) and the second characteristic function becomes a sum of functions $\psi(u_i)$. As a result, the partial derivative vanishes and we obtain

$$c(X_1^{k_1}, X_2^{k_2}, \ldots, X_n^{k_n}) = 0 \tag{4.76}$$

which is one of the most important properties of the cumulants.

(e) As in the case of scalar RVs, there are some relations between moments and cumulants (see Problem 2.3). These relations are, of course, more complex than in the scalar case, but from an expansion limited to the second order in u and v we easily obtain the relations

$$c(X^2) = \sigma_x^2 \;;\; c(Y^2) = \sigma_y^2 \;;\; c(X, Y) = \gamma_{xy} \tag{4.77}$$

where σ_x^2 and γ_{xy} are defined by (3.27) and (3.26) respectively. The cumulant $c(X, Y)$ is then nothing other than the covariance between X and Y.

Using the same procedure of expansion limited to the fourth order, we obtain for zeromean RVs the cumulants of order three and four by

$$c(X_i, X_j, X_k) = E(X_i X_j X_k) \tag{4.78}$$

$$c(X_i, X_j, X_k, X_l) = E(X_i X_j X_k X_l) - E(X_i X_j)\, E(X_k X_l)$$
$$- E(X_i X_k)\, E(X_j X_l) - E(X_i X_l)\, E(X_j X_k) \tag{4.79}$$

These formulae are absolutely general provided the RVs have a zero mean value. But they can be equal, which gives

$$c(X^4) = E(X^4) - 3\{E(X^2)\}^2 \tag{4.80}$$

as obtained in Problem 2.3. Furthermore, if the RVs X_i are independent but not necessarily with zero mean values, we deduce from point **(d)** that

$$c(X_i, X_j, X_k, X_l) = c(X_i^4)\, \delta[i, j, k, l] \tag{4.81}$$

where $\delta[i, j, k, l]$ is the extension of the Kronecker delta coefficient, taking only the values 0 or 1 and the last one only if $i = j = k = l$. At this step we can note the advantage of the cumulants over the moments. Using (4.79) we see that the expectation value $E(X_i X_j X_k X_l)$ for independent RVs is much more complex. In fact, if the RVs X_i are zeromean-valued and independent, we have

$$E(X_i X_j X_k X_l) = E(X_i X_j)\, E(X_k X_l) + E(X_i X_k)\, E(X_j X_l) + E(X_i X_l)\, E(X_j X_k) + \left[E(X_i^4) - 3\{E(X_i^2)\}^2 \right] \delta[i, j, k, l]$$

and it is easier to use (4.81) than this expression. Note that the last term of this equation or of (4.81) can be expressed with the kurtosis defined by (2.25).

4.6 NORMAL RANDOM VECTORS

4.6.1 Probability Density Function

The PDF of a normal or Gaussian zeromean random vector has been given in (4.58). Suppose now that the mean value is non-vanishing. By replacing $\mathbf{y}$ by $\mathbf{x} - \mathbf{m}$, we obtain the general expression of the PDF of a vector $\mathbf{X}$ of mean value $\mathbf{m}$ and covariance matrix $\mathbb{\Gamma}$ by

$$\boxed{p(\mathbf{x}) = (2\pi)^{-n/2} \{\det(\mathbb{\Gamma})\}^{-1/2} \exp\{-\frac{1}{2} (\mathbf{x} - \mathbf{m})^T \mathbb{\Gamma}^{-1} (\mathbf{x} - \mathbf{m})\}} \tag{4.82}$$

This PDF is completely defined by $\mathbf{m}$ and $\mathbb{\Gamma}$ and noted $N(\mathbf{m}, \mathbb{\Gamma})$.

If $\mathbb{\Gamma}$ is a *diagonal matrix* we immediately see in (4.82) that $p(\mathbf{x})$ can be factorized as a product of PDF of the type $N(m_i, \sigma_i^2)$. This means that uncorrelated normal RVs are independent. This is a very important result, widely used in applications.

Another aspect of normal random vectors which become apparent is that all their statistical properties are *completely specified* as soon as the first and second order properties are known. This greatly simplifies calculations that become almost impossible without the Gaussian assumption. This mathematical property would be without interest if normality were not often met in practical applications. But, fortunately, physical arguments show that many phenomena can be well represented by a normal distribution, which is a consequence of the central limit theorem discussed in section 4.7.2. These arguments will be expanded later within the framework of Gaussian random signals.

As a normal random vector is completely defined by its mean value and its covariance matrix, it is clear that the results of Section 3.5 on second order properties of random vectors will be of special interest. Note immediately that we only defined real normal vectors, leaving the study of the complex case for later.

Before leaving this topic, let us remark that in order to write the PDF in the form of (4.82), it is necessary for the matrix $\mathbb{\Gamma}$ to have an inverse, which means that it is positive definite. However, it is possible to use the expression $N(\mathbf{m}, \mathbb{\Gamma})$ even when $\mathbb{\Gamma}$ has some null eigenvalues. We need to be careful here, as explained on p. 67. Physically this means that if $\mathbf{X}$ is a random vector $N(\mathbf{m}, \mathbb{\Gamma})$ where $\mathbb{\Gamma}$ has some null eigenvalues, the components of $\mathbf{X}$ are related by some linear equations. In fact, if the rank of $\mathbb{\Gamma}$ is p, the vector is normal in a space with p dimensions. The simplest example to illustrate this situation corresponds to the case of a two-dimensional vector (X, Y), such that $Y = \alpha X$. It is clear that if X is $N(m, \sigma^2)$, Y is $N(\alpha m, \alpha^2\sigma^2)$ and the pair X, Y is normal with a covariance matrix of rank one. Other more interesting examples will be discussed later.

4.6.2 Characteristic Function, Moments, and Cumulants

The characteristic function of a distribution $N(\mathbf{m}, \mathbb{\Gamma})$ can be directly calculated from the definition (4.59) where $d^nF(\mathbf{x})$ is replaced by $p(\mathbf{x})dx_1\ dx_2\ \ldots\ dx_n$, symbolically written as $p(\mathbf{x})d\mathbf{x}$. The calculation is an extension to $\mathbb{R}^n$ of that realized for the scalar case. But there is a more elegant way of arriving at the result, by using some results obtained in previous sections.

Let us first consider a random vector $\mathbf{X}$ with the distribution $N(\mathbf{0}, \mathbb{I})$, where $\mathbb{I}$ is the identity matrix of $\mathbb{R}^n$. As noticed above, the components of this vector are scalar independent Gaussian RVs, and using (4.70) and (2.84), we obtain its characteristic function by

$$\phi(\mathbf{u}) = \exp\left(-\frac{1}{2}\ \mathbf{u}^T\mathbf{u}\right) \tag{4.83}$$

Now consider the random vector $\mathbf{Y}$ defined by (4.50). Its PDF is given by (4.58), which is a distribution $N(\mathbf{0}, \mathbb{A}\mathbb{A}^T)$, because of (4.53). The characteristic function of $\mathbf{Y}$ is given by (4.68), where $\phi_X(\mathbf{u})$ is given by (4.83). We then have

$$\phi_Y(\mathbf{v}) = \exp\left(-\frac{1}{2}\ \mathbf{v}^T\mathbb{A}\mathbb{A}^T\mathbf{v}\right) \tag{4.84}$$

As $\mathbb{\Gamma}_Y = \mathbb{A}\mathbb{A}^T$, and by omitting the subscript Y, we find that the characteristic function of the distribution $N(\mathbf{0}, \mathbb{\Gamma})$ is

$$\phi(\mathbf{u}) = \exp\left(-\frac{1}{2}\ \mathbf{u}^T\mathbb{\Gamma}\mathbf{u}\right) \tag{4.85}$$

Finally, if the mean value is $\mathbf{m}$ instead of $\mathbf{0}$, we obtain

$$\boxed{\phi(\mathbf{u}) = \exp(j\mathbf{m}^T\mathbf{u})\ \exp\left(-\frac{1}{2}\ \mathbf{u}^T\mathbb{\Gamma}\mathbf{u}\right)} \tag{4.86}$$

which is the characteristic function associated with the distribution $N(\mathbf{m}, \mathbb{\Gamma})$.

Let us now present some comments on this result. We first note that it extends the result obtained in the scalar case to the multi-dimensional case, which means that the Fourier transform of a Gaussian shape is still Gaussian.

Second, even though the expressions (4.82) and (4.86) are similar, the latter is much simpler. In particular, there is no constant coefficient before the exponential functions, because $\phi(\mathbf{0}) = 1$.

Third, if we consider (4.86) as a definition of a normal distribution $N(\mathbf{m}, \mathbb{\Gamma})$, it is no longer necessary for $\mathbb{\Gamma}$ to be a full rank matrix because we do not use the inverse of the matrix $\mathbb{\Gamma}$. If that is not so, we cannot calculate $p(\mathbf{x})$ by (4.82) and before that we must reduce the dimension of the vectors $\mathbf{x}$ and $\mathbf{u}$ to arrive at a positive definite matrix. This operation eliminates the subspace of null eigenvalues of $\mathbb{\Gamma}$.

Finally, the use of (4.54) shows that all the components of $\mathbf{X}$ have normal marginal distributions.

The second characteristic function defined by (4.60) is then

$$\psi(\mathbf{u}) = j\ \mathbf{m}^T\mathbf{u} - \frac{1}{2}\ \mathbf{u}^T\mathbb{\Gamma}\mathbf{u} \tag{4.87}$$

From these expressions it is possible to calculate the moments and cumulants with (4.72) and (4.73). This last expression combined with (4.87) shows that *all the cumulants of an order greater than two are null*, which is a fundamental property of Gaussian random vectors. In other words, the only non-vanishing cumulants of a normal vector $\mathbf{X}$ with components X_i are

$$c(X_i) = m_i \quad ; \quad c(X_i, X_j) = \gamma_{ij} \tag{4.88}$$

where m_i is the mean value of X_i and γ_{ij} is the covariance of X_i and X_j.

The calculation of the moments is a little more tedious. By using (4.72) with $k = 1$ or 2, we easily obtain

$$E(\mathbf{X}) = \mathbf{m} \quad ; \quad E(\mathbf{X}\mathbf{X}^T) = \mathbb{\Gamma} + \mathbf{m}\mathbf{m}^T \tag{4.89}$$

In reality, the first equation is a direct consequence of the symmetry of (4.82).

For higher-order moments we will consider only the case of zeromean normal RVs with the characteristic function given by (4.85). Instead of using the general expression (4.72), which needs the calculation of a lot of derivatives, we use McLaurin expansions of $E\{\exp(j\ \mathbf{u}^T\mathbf{X})\}$ and of (4.85) and identify the coefficients of these expansions.

By this method we see that it is impossible to obtain terms with an odd power in u_i from (4.85). In consequence, all the odd moments of a zeromean normal vector are null, and are written as

$$E(X_{i_1} X_{i_2} \dots X_{i_{2k+1}}) = 0 \tag{4.90}$$

where the coefficients i_j are $2k + 1$ integers satisfying $1 \le i_j \le n$, n being the dimension of the vector $\mathbf{X}$.

Consider now the even moments. The identification of the two McLaurin expansions immediately gives

$$\frac{1}{(2k)!}\ E\{(\mathbf{u}^T\mathbf{X})^{2k}\} = \frac{1}{2^k\ k!}\ (\mathbf{u}^T\mathbb{\Gamma}\mathbf{u})^k \tag{4.91}$$

from which all the moments of order $2k$ can be deduced. If, for example, all the u_is except u_1 are vanishing, we find

$$E(X_1^{2k}) = (2k - 1)!!\ \gamma_{11}^k \tag{4.92}$$

where the γ_{ij}s are the matrix elements of $\mathbb{\Gamma}$. We then find (2.86) again. Let us now calculate $E(X_1 X_2 X_3 X_4)$. In order to do so, we take $k = 2$ and $u_i = 0$ for $i > 4$. On the l.h.s. of (4.91) there are 4! terms $u_1 u_2 u_3 u_4 E(X_1 X_2 X_3 X_4)$, as a result of all the possible permutations of the u_is. On the r.h.s. of (4.91) we see that the coefficient of $u_1 u_2 u_3 u_4$ is $(\gamma_{12} \gamma_{34} + \gamma_{13} \gamma_{24} + \gamma_{14} \gamma_{23})$ and there are 2^2 2! such terms as a result of pairs of γ_{ij} and permutations of the indices i, j for each pair. In conclusion we obtain

$$E(X_1 X_2 X_3 X_4) = \gamma_{12} \gamma_{34} + \gamma_{13} \gamma_{24} + \gamma_{14} \gamma_{23} \tag{4.93}$$

This procedure is quite general and we have

$$E(X_i X_j X_k X_l) = \gamma_{ij} \gamma_{kl} + \gamma_{ik} \gamma_{jl} + \gamma_{il} \gamma_{jk} \tag{4.94}$$

whatever the values of the indices. If they are equal to 1, we again find (4.92) for $k = 2$. The procedure can be extended to arbitrary values of k and the result is

$$E(X_{i_1} X_{i_2} \ldots X_{i_{2k}}) = \sum_G \gamma_{i_1 i_2} \gamma_{i_3 i_4} \cdots \gamma_{i_{2k-1} i_{2k}} \tag{4.95}$$

In this expression G means that the sum is extended to all the permutations of the i_js giving distinct terms. These are called *Gaussian permutations* and there are $(2k - 1)!!$ of them. In fact, there are $(2k)!$ permutations of the $2k$ integers i_j. But because of the symmetry $\gamma_{ij} = \gamma_{ji}$ and of the permutations between the terms γ, there are $2^k k!$ equal terms. Dividing $(2k)!$ by $2^k k!$ we get $(2k - 1)!!$. In (4.95) the i_js are not necessarily distinct. If they are all equal, we again find (4.92). Finally, we note that if we use (4.93) in (4.79) we obtain, as expected, a cumulant equal to zero.

4.6.3 Linear Transformations of Normal Random Vectors

One of the major properties of normal distribution is that it remains normal in any linear transformation of the random vector. This means that if the random vector $\mathbf{X}$ is $N(\mathbf{m}, \mathbb{\Gamma})$, the vector $\mathbf{Y} = \mathbb{A}\mathbf{X}$ is $N(\mathbb{A}\mathbf{m}, \mathbb{A}\mathbb{\Gamma}\mathbb{A}^T)$. In order to verify this property, let us calculate the characteristic function $\phi_Y(\mathbf{v})$ of $\mathbf{Y}$. This function is given by (4.68), and by replacing $\phi_X(\mathbf{u})$ in this expression by (4.86) we obtain

$$\phi_Y(\mathbf{v}) = \exp(j\mathbf{m}^T \mathbb{A}^T \mathbf{v}) \exp\left(-\frac{1}{2} \mathbf{v}^T \mathbb{A}\mathbb{\Gamma}\mathbb{A}^T \mathbf{v}\right) \tag{4.96}$$

which shows that $\mathbf{Y}$ is $N(\mathbb{A}\mathbf{m}, \mathbb{A}\mathbb{\Gamma}\mathbb{A}^T)$.

This result is fundamental, because of the importance of *linear systems* in signal processing problems. It means that the Gaussian character of a random vector, or of a random signal, is preserved in any linear transformation. This has many and various consequences studied later.

It is of some interest to note that this result does not mean that the sum of two normal RVs is necessarily normal. In fact, it is perfectly possible to construct two RVs X and Y that are separately normal, but such that the pair (X, Y) is not a normal random vector. In this case the sum $X + Y$ has no reason to be normal. Let us examine now an example of such a situation.

Example 4.7 The sum of two normal RVs can be non normal. Consider an RV X which is $N(0, 1)$ and another RV Y defined as follows: Y is equal to X with the probability p and to $-X$ with the probability $q = 1 - p$. Let us call $F_X(x)$ the DF of X defined by (2.3). The DF of Y can be written as

$$F_Y(y) \stackrel{\Delta}{=} P(Y \leq y) = p\, P(X \leq y) + q\, P(-X \leq y) \tag{4.97}$$

As $P(-X \leq y) = P(X \geq -y) = 1 - F_X(-y)$ because $F_X(y)$ is a continuous function, and as $F_X(-y) = 1 - F_X(y)$ because of the symmetry of the normal PDF, we obtain

$$F_Y(y) = (p + q)\, F_X(y) = F_X(y) \tag{4.98}$$

It follows that Y is still an $N(0, 1)$ RV. Consider now the sum $Z = X + Y$. This RV is equal to $2X$ or to 0 with the probabilities p and q respectively. Thus its distribution function is $p\, F_X(z/2) + q$ if $z > 0$ or $p\, F_X(z/2)$ if $z < 0$, and it follows that Z is no longer a normal RV. This result does not contradict the conservation of the normality in addition. In fact, this general result is valid if the random vector, or the pair X, Y, is normal and has a characteristic function given by (4.86). It is sometimes said that X and Y are *jointly normal*. But in our example X, Y is not a normal random vector, even though the marginal distribution of X and Y are normal. To verify the point, let us calculate the characteristic function $\phi(u, v)$ of X, Y. We deduce from (4.62) that $\phi(u, v)$ is

$$\phi(u, v) = p\, E[\exp\{j(u + v)\, X\}] + q\, E[\exp\{j(u - v)\, X\}] \tag{4.99}$$

and as X is $N(0, 1)$, this gives

$$\phi(u, v) = p \exp\left[-(1/2)(u + v)^2\right] + q \exp\left[-(1/2)(u - v)^2\right] \tag{4.100}$$

We verify that $\phi(u, 0) = \phi(0, u) = \exp\{-(1/2)u^2\}$ but the characteristic function of Z is

$$\phi_Z(u) = \phi(u, u) = p \exp(-2u^2) + q \tag{4.101}$$

which, of course, does not have the form of (2.84). It is possible to construct many other examples showing the same kind of result.

4.6.4 Conditional Distributions, Regression

Consider two random vectors $\mathbf{X}$ and $\mathbf{Y}$ of arbitrary dimension, zeromean, and jointly normal. This means that the PDF of the pair $\mathbf{X}$, $\mathbf{Y}$ has the form deduced from (4.82) and written

$$p(\mathbf{x}, \mathbf{y}) = (2\pi)^{-n/2}\, [\det(\mathbb{\Gamma})]^{-1/2} \exp[-(1/2)\, q(\mathbf{x}, \mathbf{y})] \tag{4.102}$$

where n is the sum of the dimension of $\mathbf{x}$ and $\mathbf{y}$ and $q(\mathbf{x}, \mathbf{y})$ is the quadratic form $(\mathbf{x}, \mathbf{y})^T$ $\mathbb{\Gamma}^{-1}(\mathbf{x}, \mathbf{y})$ written as

$$q(\mathbf{x}, \mathbf{y}) = [\mathbf{x}^T \mathbb{M}_x \mathbf{x} + 2\mathbf{x}^T \mathbb{M}_{xy} \mathbf{y} + \mathbf{y}^T \mathbb{M}_y \mathbf{y}] \tag{4.103}$$

To arrive at this expression we have assumed that, like $\mathbb{\Gamma}$, $\mathbb{\Gamma}^{-1}$ is symmetric. This is a direct consequence of the property $\mathbb{\Gamma}\mathbb{\Gamma}^{-1} = \mathbb{I}$. The matrices $\mathbb{M}$ will be calculated later, but now their values do not play any role in the following discussion. The $n \times n$ matrix $\mathbb{\Gamma}$ can be written as

$$\mathbb{\Gamma} = \begin{bmatrix} \mathbb{\Gamma}_x & \mathbb{\Gamma}_{xy} \\ \\ \mathbb{\Gamma}_{yx} & \mathbb{\Gamma}_y \end{bmatrix} \tag{4.104}$$

The conditional PDF of $\mathbf{X}$ given $\mathbf{y}$ is given by (4.28) where $p_Y(\mathbf{y})$ is the marginal distribution of $\mathbf{Y}$. But as the pair $\mathbf{X}$, $\mathbf{Y}$ is normal, the random vector $\mathbf{Y}$ is also $N(\mathbf{0}, \mathbb{\Gamma}_y)$ and its PDF can be written as

$$p_Y(\mathbf{y}) = k \exp\left[-(1/2)q'(\mathbf{y})\right] \tag{4.105}$$

where $q'(\mathbf{y})$ is simply $\mathbf{y}^T \mathbb{\Gamma}_y^{-1} \mathbf{y}$ and k is given by (4.82) in which $\mathbb{\Gamma}$ is replaced by $\mathbb{\Gamma}_y$. Using (4.28) we see that $p_X(\mathbf{x} \mid \mathbf{y})$ still has the general structure of a normal PDF because the difference of two quadratic forms is still a quadratic form. As a conclusion we deduce that *the conditional distribution of* $\mathbf{X}$ *for a given value* $\mathbf{y}$ *remains normal.* It can then be written as $N[\mathbf{m}(\mathbf{y}), \mathbb{V}(\mathbf{y})]$, and, in order to specify completely the conditional PDF, the conditional mean and covariance matrix remain to be calculated.

Conditional mean. This quantity is the *regression* $\mathbf{r}(\mathbf{y})$ defined by (4.38). Replacing the mean value $\mathbf{m}$ in (4.82) by $\mathbf{r}(\mathbf{y})$, we must find the exponential of a quadratic form in $\mathbf{x}$ and $\mathbf{y}$ as in (4.102). It follows that $\mathbf{r}(\mathbf{y})$ is linear in $\mathbf{y}$ and can be written as

$$\mathbf{r}(\mathbf{y}) = \mathbb{A}\,\mathbf{y} + \mathbf{c} \tag{4.106}$$

where $\mathbb{A}$ is a rectangular matrix and $\mathbf{c}$ a constant vector. Taking the expected value and using (4.33), we find that $\mathbf{c} = \mathbf{0}$. In order to determine $\mathbb{A}$ we can use the general property (4.39) with $\mathbf{g}(\mathbf{Y}) = \mathbf{Y}$. This takes the form

$$E\{[\mathbf{X} - \mathbb{A}\,\mathbf{Y}]\,\mathbf{Y}^T\} = \mathbb{O} \tag{4.107}$$

which gives

$$\mathbb{A} = \mathbb{\Gamma}_{xy}\,\mathbb{\Gamma}_y^{-1} \tag{4.108}$$

In conclusion, we arrive at this fundamental result deduced from the normal distribution: the *conditional expected value is linear* and can be written

$$\mathbf{r}(\mathbf{y}) = E(\mathbf{X} \mid \mathbf{y}) = \mathbb{\Gamma}_{xy}\,\mathbb{\Gamma}_y^{-1}\,\mathbf{y} \tag{4.109}$$

This has several applications as seen below.

Conditional covariance matrix. This matrix $\mathbb{V}(\mathbf{y})$ must be used in (4.82), replacing $\mathbb{\Gamma}$, in order to arrive at a quadratic form in $\mathbf{x}$ and $\mathbf{y}$ when $\mathbf{m}$ is replaced by $\mathbf{r}(\mathbf{y})$ given by (4.109). As a result we deduce that $\mathbb{V}(\mathbf{y})$ does not depend on $\mathbf{y}$. We can verify this fact by noting that (4.107) shows that $\mathbf{X} - \mathbb{A}\,\mathbf{Y}$ and $\mathbf{Y}$ are uncorrelated. But as they are normal they are *independent*, which means that

$$\mathbb{V}(\mathbf{y}) \triangleq E\{(\mathbf{X} - \mathbb{A}\,\mathbf{Y})\,(\mathbf{X} - \mathbb{A}\,\mathbf{Y})^T \mid \mathbf{y}\}$$
$$= E\{(\mathbf{X} - \mathbb{A}\,\mathbf{Y})\,(\mathbf{X} - \mathbb{A}\,\mathbf{Y})^T\} = \mathbb{V} \tag{4.110}$$

Thus, using (4.108), we immediately obtain

$$\mathbb{V} = \Gamma_x - \Gamma_{xy}\,\Gamma_y^{-1}\,\Gamma_{yx} \tag{4.111}$$

To summarize these results, we can say that when $\mathbf{X}$ and $\mathbf{Y}$ are jointly zeromean normal the conditional PDF (4.28) is also normal and takes the form $N(\Gamma_{xy}\,\Gamma_y^{-1}\,\mathbf{y}, \Gamma_x - \Gamma_{xy}\,\Gamma_y^{-1}\Gamma_{yx})$.

4.6.5 Two-dimensional Normal Random Variables

We shall now apply the general results presented above to the particular case of two-dimensional RVs. The second order properties were presented in Section 3.3, and we shall investigate the consequences of the assumption of normality. In other words, we leave the vector and matrix notations, although the mathematical expressions may appear somehow more tedious than in the general case with vectors and matrices.

As the normal distribution is completely defined by using only second order properties, we will use the notations introduced in Section 3.3. It was shown there that a two-dimensional RV (X, Y) has first order moments given by (3.24) while the second order moments are completely specified by σ_x, σ_y, and c given by (3.33). Using this relation, the covariance matrix in (4.82) can be written as in (3.51) and thus takes the form

$$\Gamma = \begin{bmatrix} \sigma_x^2 & c\,\sigma_x\sigma_y \\ c\,\sigma_x\sigma_y & \sigma_y^2 \end{bmatrix} \tag{4.112}$$

Its determinant is, of course,

$$\det(\Gamma) = (\sigma_x\sigma_y)^2\,(1 - c^2) \tag{4.113}$$

and its inverse is

$$\Gamma^{-1} = \frac{1}{1-c^2}\begin{bmatrix} \dfrac{1}{\sigma_x^2} & -\dfrac{c}{\sigma_x\sigma_y} \\ -\dfrac{c}{\sigma_x\sigma_y} & \dfrac{1}{\sigma_y^2} \end{bmatrix} = \frac{(\sigma_x^2\,\sigma_y^2)^{-1}}{1-c^2}\begin{bmatrix} \sigma_y^2 & c\,\sigma_x\sigma_y \\ c\,\sigma_x\sigma_y & \sigma_x^2 \end{bmatrix} \tag{4.114}$$

As a result, the quadratic form in (4.82) takes the form

$$Q(x, y) = \frac{1}{1-c^2}\left[\frac{(x-m_x)^2}{\sigma_x^2} - 2c\,\frac{(x-m_x)(y-m_y)}{\sigma_x\sigma_y} + \frac{(y-m_y)^2}{\sigma_y^2}\right] \tag{4.115}$$

and the PDF (4.82) can be written as

$$p(x,\ y) = \frac{1}{2\pi\,\sigma_x\sigma_y(1-c^2)^{1/2}}\exp\left[-\frac{1}{2}\,Q(x,\ y)\right] \tag{4.116}$$

The corresponding characteristic function is simpler to write out and (4.86) can be written in the form

$$\phi(u,\ v) = \exp[j(um_x + vm_y)]\exp[-(1/2)(\sigma_x^2\,u^2 + 2c\sigma_x\sigma_y\ uv + \sigma_y^2\,v^2)] \tag{4.117}$$

The PDF (4.116) has a maximum for $x = m_x$ and $y = m_y$, and if we consider the RVs $X - m_x$ and $Y - m_y$, which are zeromean, we obtain a translation of the coordinates and the maximum is obtained for $x = y = 0$. To simplify the expression we can use reduced coordinates replacing x and y by x/σ_x and y/σ_y respectively, which is equivalent to considering the case where $\sigma_x = \sigma_y = 1$. In other words, the PDF of a two-dimensional normal RV corresponding to the law $N(0,\ 0;\ 1,\ 1,\ c)$ is given by (4.115) and (4.116) with $m_x = m_y = 0$ and $\sigma_x = \sigma_y = 1$. This PDF depends only on the correlation coefficient c, and $Q(x,\ y)$ can now be written as

$$Q(x,\ y) = \frac{1}{1-c^2}(x^2 - 2cxy + y^2) \tag{4.118}$$

The general shape of this PDF is presented in Figure 4.6.

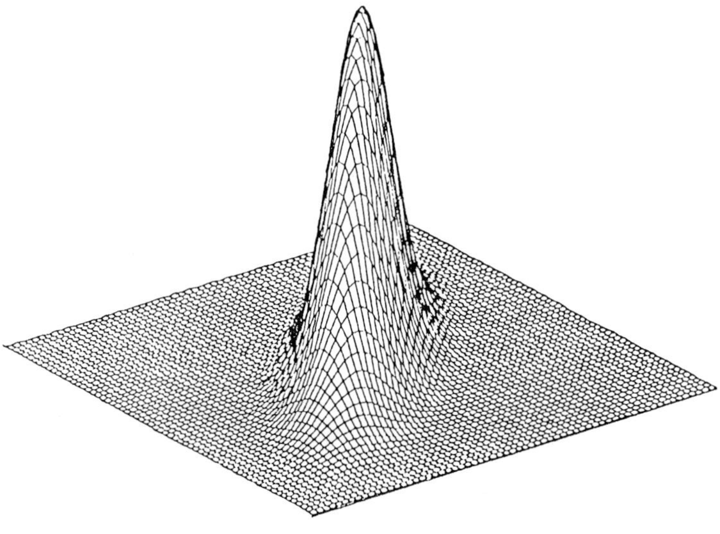

Figure 4.6 Density function of the distribution $N(0,\ 0;\ 1,\ 1,\ 0.8)$.

The curves of constant PDF are given by the equation

$$x^2 - 2\,c\,x\,y + y^2 = k^2 \tag{4.119}$$

These curves are ellipses, one example of which appears in Figure 4.7.

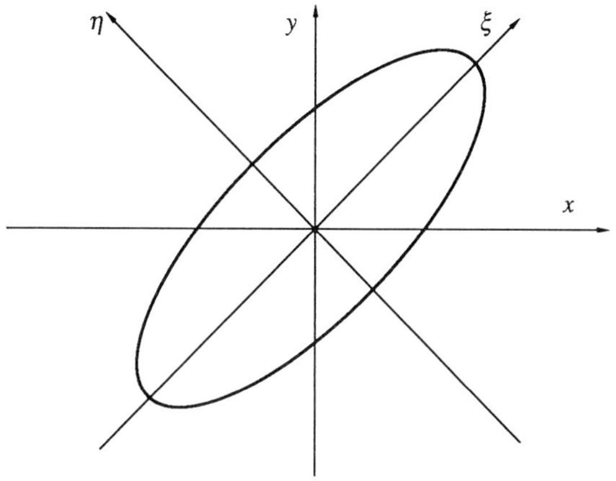

Figure 4.7 Ellipse of constant PDF.

Whatever the values of c, the axes of these ellipses are obtained from a rotation of $\pi/4$ of the axes Ox and Oy. The simplest way to verify this point is to take new coordinates ξ and η defined by

$$\xi = \frac{x + y}{\sqrt{2}} \quad ; \quad \eta = \frac{-x + y}{\sqrt{2}} \tag{4.120}$$

which gives

$$x = \frac{\xi - \eta}{\sqrt{2}} \quad ; \quad y = \frac{\xi + \eta}{\sqrt{2}} \tag{4.121}$$

Using these expressions in (4.119), we obtain

$$(1 - c)\xi^2 + (1 + c)\eta^2 = k^2 \tag{4.122}$$

Let us now make some comments related to the values of the correlation coefficient c. If $c = 0$, the ellipse becomes a circle and the PDF can be factorized in a product of two distributions $N(0, 1)$, as appears in Fig. 2.4. In fact, if $c = 0$, the covariance matrix (4.112) is diagonal and whatever the values of their mean values and variances, the two RVs X and Y are independent, as seen in (4.115) and (4.116).

Consider now the opposite case, which is a little more complex. If $c \to +1$, the ellipse of constant PDF tends to be confused with the straight line $c \to +1$, This is not at all surprising; in fact, we saw on p. 52 that if $|c| = 1$, X and Y are proportional. In our case, where X and Y have the same variance, we deduce that if $c = 1$ we have $Y = X$ and if $c = -1$ we have $Y = -X$. In the first case the PDF of X, Y is concentrated on the straight line $y = x$, and in the second case on the straight line $y = -x$. These two lines can be considered as degenerated ellipses.

Let us now examine the structures of the *a priori* and *a posteriori* distributions. To obtain the marginal distribution of X it is sufficient to take $v = 0$ in (4.117), as seen in (4.64). It follows that X is simply $N(m_x, \sigma_x^2)$. Supposing now that $m_x = m_y = 0$, the conditional distribution of X for a given value of y is still normal, with mean m and variance v given by (4.109) and (4.111), which gives, with (4.112)

$$m = c(\sigma_x/\sigma_y)y \quad ; \quad v = \sigma_x^2\,(1 - c^2) \tag{4.123}$$

4.6.6 Complex Normal Random Vectors

We have already noticed several times that zeromean *real* normal random vectors have a probability distribution completely defined by the covariance matrices, as in (4.82) and in (4.86). As the concept of the covariance matrix is also defined for *complex* random vectors, as seen in (3.90), it is natural to ask if a complex random vector is also completely defined as soon as its covariance matrix is known. For this purpose we must first define the concept of a complex normal random vector precisely.

A *complex* random vector **Z** is said to be *normal* if its real and imaginary parts, noted **X** and **Y** respectively, are jointly normal. If we assume that they are zeromean (which is not a restriction of the generality), the PDF or the characteristic function of this pair of vectors is completely defined from the covariance matrix given by (4.104). If **X** and **Y** are vectors of $\mathbb{R}^n$, Γ is a $2n \times 2n$ matrix and (4.104) presents a partition in four blocks which are $n \times n$ real matrices.

On the other hand, the covariance matrix Γ_z of the random vector **Z** is an $n \times n$ matrix equal to $E(\mathbf{Z}\,\mathbf{Z}^H)$, and using $\mathbf{Z} = \mathbf{X} + j\,\mathbf{Y}$, this gives

$$\Gamma_z = \mathbb{A} + j\mathbb{B} = \Gamma_x + \Gamma_y + j(-\Gamma_{xy} + \Gamma_{yx}) \tag{4.124}$$

which exhibits the real and imaginary parts $\mathbb{A}$ and $\mathbb{B}$. Note that these matrices satisfy the basic relation (3.78) because $\Gamma_{xy}^T = \Gamma_{yx}$. It is obvious that without other assumptions, the knowledge of Γ_z, i.e. of $\mathbb{A}$ and $\mathbb{B}$, is not sufficient to determine the three matrices Γ_x, Γ_y and Γ_{xy} appearing in (4.124).

The conclusion is thus clear: a complex random vector **Z** is *not* completely described by its covariance matrix. This is due to the fact that there is another second order moment of **Z** which must be introduced. This moment is simply the matrix $\mathbb{C}$ defined by

$$\mathbb{C} = E(\mathbf{Z}\mathbf{Z}^T) = \mathbb{P} + j\mathbb{Q} \tag{4.125}$$

Replacing **Z** by $\mathbf{X} + j\mathbf{Y}$, we easily find

$$\mathbb{P} = \Gamma_x - \Gamma_y \quad ; \quad \mathbb{Q} = \Gamma_{xy} + \Gamma_{yx} \tag{4.126}$$

Note that $\mathbb{P}$ and $\mathbb{Q}$ are symmetric. Using $\mathbb{A}$, $\mathbb{B}$, $\mathbb{P}$ and $\mathbb{Q}$, we can easily determine the matrices of (4.104), the result of which is

$$\mathbb{\Gamma}_x = (1/2)\,(\mathbb{A} + \mathbb{P}) \;\; ; \;\; \mathbb{\Gamma}_y = (1/2)\,(\mathbb{A} - \mathbb{P}) \tag{4.127}$$

$$\mathbb{\Gamma}_{xy} = (1/2)\,(\mathbb{Q} - \mathbb{B}) \tag{4.128}$$

In many problems we assume that $\mathbb{C} = \mathbb{0}$, an assumption which will be discussed later. In this case we deduce that

$$\mathbb{\Gamma}_x = \mathbb{\Gamma}_y = (1/2)\,\mathbb{A} \tag{4.129}$$

$$\mathbb{\Gamma}_{xy} = -\,\mathbb{\Gamma}_{yx} = -\,(1/2)\,\mathbb{B} \tag{4.130}$$

A normal complex random vector satisfying this assumption has a probability distribution which is completely defined by its covariance matrix $\mathbb{\Gamma}_z$. In fact, the knowledge of $\mathbb{\Gamma}_z$ gives $\mathbb{A}$ and $\mathbb{B}$ and then, with (4.129) and (4.130), completely specifies the matrix $\mathbb{\Gamma}$ of (4.104), which also determines the PDF by (4.82).

We will now approach the same problem from a different aspect, giving the structure of the PDF and of the characteristic function. The PDF of a complex normal random vector $\mathbf{Z} = \mathbf{X} + j\mathbf{Y}$ is a function $p(\mathbf{x}, \mathbf{y})$ which can also be written simply as $p(\mathbf{z})$, where $\mathbf{z} = \mathbf{x} + j\mathbf{y}$. This function has an analytical structure given by (4.82) in terms of $\mathbf{x}$ and $\mathbf{y}$ but has no reason to be expressed only in terms of $\mathbf{z}$. The same can be said of the characteristic function, which can be written as $\phi(\mathbf{u}, \mathbf{v}) = \exp\,[-(1/2)\,Q(\mathbf{u}, \mathbf{v})]$ with

$$Q(\mathbf{u}, \mathbf{v}) = \mathbf{u}^T\,\mathbb{\Gamma}_x\,\mathbf{u} + \mathbf{v}^T\,\mathbb{\Gamma}_y\,\mathbf{v} + 2\,\mathbf{u}^T\,\mathbb{\Gamma}_{xy}\,\mathbf{v} \tag{4.131}$$

which is a direct consequence of (4.86) and (4.104). It is clear that $Q(\mathbf{u}, \mathbf{v})$ is a function of the complex vector $\mathbf{w} = \mathbf{u} + j\,\mathbf{v}$ but cannot be expressed directly in terms of $\mathbf{w}$.

Let us now find the conditions on the matrices appearing in (4.104) which ensure that the quadratic form (4.131) can be written as

$$Q(\mathbf{u}, \mathbf{v}) = Q(\mathbf{w}) = \alpha\,\mathbf{w}^H\,\mathbb{\Gamma}_z\,\mathbf{w} \tag{4.132}$$

where α is real and $\mathbb{\Gamma}_z$ is the covariance matrix (4.124). Using the properties (3.78), we easily find that

$$Q(\mathbf{w}) = \alpha\,[\mathbf{u}^T\mathbb{A}\,\mathbf{u} + \mathbf{v}^T\,\mathbb{A}\,\mathbf{v} - 2\,\mathbf{u}^T\,\mathbb{B}\,\mathbf{v}] \tag{4.133}$$

Comparing (4.131) and (4.133) we see immediately that if (4.129) and (4.130) hold, then we have (4.132) with $\alpha = 1/2$. Conversely, if we assume that (4.132) is valid whatever the values of $\mathbf{u}$ and $\mathbf{v}$, we can find (see Problem 4.34) that $\alpha = 1/2$ and that (4.129) and (4.130) are satisfied. In conclusion, if $\mathbb{C} = E[\mathbf{Z}\mathbf{Z}^T] = \mathbb{0}$, which means that $\mathbb{\Gamma}_x = \mathbb{\Gamma}_y$ and $\mathbb{\Gamma}_{xy} = -\mathbb{\Gamma}_{yx}$, then the characteristic function of the corresponding normal vector can be written as

$$\boxed{\phi(\mathbf{u}, \mathbf{v}) = \phi(\mathbf{w}) = \exp\{-(1/4)\mathbf{w}^H\,\mathbb{\Gamma}_z\,\mathbf{w}\}} \tag{4.134}$$

Let us apply this relation for $n = 1$, where $\mathbf{u}$ and $\mathbf{v}$ are real numbers. The condition $\gamma_{xy} = -\gamma_{yx}$ and the property $\gamma_{xy} = \gamma_{yx}$ gives $\gamma_{xy} = 0$, and (4.134) becomes

$$\phi(u, v) = \phi(w) = \exp\{-(1/4)\sigma_z^2 \ |w|^2\} \tag{4.135}$$

where σ_z^2 is given by (3.44). The relation (4.129) gives $\sigma_x^2 = \sigma_y^2 = \sigma^2$, and (4.135) also becomes

$$\phi(u, v) = \exp\,[-(1/2)\sigma^2\,(u^2 + v^2)] = \exp\,[-(1/2)\sigma^2\,|w|^2] \tag{4.136}$$

In conclusion, in the scalar case the condition (4.129) and (4.130) means that the RVs X and Y are jointly normal, independent and with the same variance. In this scalar case the PDF deduced from (4.82) is

$$p(x, y) = \frac{1}{2\pi\sigma^2}\exp\left[-\frac{1}{2\sigma^2}\,(x^2 + y^2)\right] \tag{4.137}$$

which can also be written in terms of the complex number $z = x + jy$ and the variance $\sigma_z^2 = 2\,\sigma^2$ by

$$p(z) = \frac{1}{\pi\sigma_z^2}\exp\left(-\frac{1}{\sigma_z^2}\,|z|^2\right) \tag{4.138}$$

Considering now the *real two-dimensional* vector **Z** of components X and Y, we see that (4.137) takes the form of (4.48) characterizing a spherically invariant random vector. In this case it is said that the normal complex RV Z is *circular*, which is equivalent to spherically invariant or spherically symmetric. By an extension of this vocabulary, it is also said that a complex normal random vector such that $\mathbb{C}$ defined by (4.125) is null or satisfying (4.129) and (4.130) is also *circular*.

To terminate this discussion, it is necessary to evaluate the PDF of a normal complex circular random vector associated to the characteristic function (4.134), or to extend to the n-dimensional case the result (4.138) obtained for $n = 1$. For this we note that it is possible to express (4.134) in the form

$$\phi(\mathbf{u}, \mathbf{v}) = \phi(\mathbf{w}) = \exp\{-(1/2)Q(\mathbf{w})\} \tag{4.139}$$

where $Q(\mathbf{w})$ is given by (4.133) with $\alpha = 1/2$. The PDF associated with this characteristic function has the form (4.102) and the matrices $\mathbb{M}$ coming from a partition in four blocks of $\mathbb{\Gamma}^{-1}$ remain to be calculated. It results from the assumption of circularity that (4.104) takes the form

$$\mathbb{\Gamma} = (1/2)\begin{bmatrix} \mathbb{A} & -\mathbb{B} \\ \mathbb{B} & \mathbb{A} \end{bmatrix} \tag{4.140}$$

where $\mathbb{A}$ and $\mathbb{B}$ are defined by (4.124) or (4.129) and (4.130). By writing $\mathbb{\Gamma}\mathbb{\Gamma}^{-1} = \mathbb{I}$ we easily obtain

$$\mathbb{\Gamma}^{-1} = 2 \begin{bmatrix} \mathbb{V} & \mathbb{M} \\ \mathbb{M}^T & \mathbb{V} \end{bmatrix} \tag{4.141}$$

with

$$\mathbb{V} = (\mathbb{A} + \mathbb{B}\mathbb{A}^{-1}\mathbb{B})^{-1} \tag{4.142}$$

$$\mathbb{M} = \mathbb{A}^{-1}\mathbb{B}\mathbb{V} \tag{4.143}$$

which can be verified by identification. As a result we obtain in (4.102)

$$q(\mathbf{x}, \mathbf{y}) = 2(\mathbf{x}^T \mathbb{V}\mathbf{x} + \mathbf{y}^T \mathbb{V}\mathbf{y} + 2\, \mathbf{x}^T \mathbb{M}\mathbf{y}) \tag{4.144}$$

Let us now show that this quadratic form can be written in terms of the inverse of the covariance matrix $\mathbb{\Gamma}_z$ defined by (4.124). To do this we must calculate its inverse, which can be written as

$$\mathbb{\Gamma}_z^{-1} = \mathbb{R} + j\mathbb{S} \tag{4.145}$$

Writing $\mathbb{\Gamma}_z \mathbb{\Gamma}_z^{-1} = \mathbb{I}$ and equating the real and imaginary parts, we easily find

$$\mathbb{R} = \mathbb{V} \tag{4.146}$$

$$\mathbb{S} = -\mathbb{M} \tag{4.147}$$

Using these expressions and making the same calculation as that used to arrive at (3.77), we find that the quadratic form (4.144) can be written as

$$q(\mathbf{x}, \mathbf{y}) = 2\, \mathbf{z}^H \mathbb{\Gamma}_z^{-1} \mathbf{z} \tag{4.148}$$

We have now to express the coefficient before the exponential ensuring that $p(\mathbf{x}, \mathbf{y})$ is normalized or to find an expression of det ($\mathbb{\Gamma}$) appearing in (4.102) in terms of $\mathbb{\Gamma}_z$. Using (4.129) and (4.130) and applying the fundamental property which states that a determinant is unchanged when we add or subtract rows or columns, we can write

$$\det(\mathbb{\Gamma}) = \det\left\{ (1/2) \begin{bmatrix} \mathbb{A} & -\mathbb{B} \\ \mathbb{B} & \mathbb{A} \end{bmatrix} \right\} = \det\left\{ (1/2) \begin{bmatrix} \mathbb{A} + j\mathbb{B} & -\mathbb{B} \\ \mathbb{B} - j\mathbb{A} & \mathbb{A} \end{bmatrix} \right\} \tag{4.149}$$

which gives with (4.124)

$$\det(\mathbb{\Gamma}) = \det\left\{ (1/2) \begin{bmatrix} \mathbb{\Gamma}_z & -\mathbb{B} \\ -j\mathbb{\Gamma}_z & \mathbb{A} \end{bmatrix} \right\} = \det\left\{ (1/2) \begin{bmatrix} \mathbb{\Gamma}_z & -\mathbb{B} \\ \mathbb{0} & \mathbb{\Gamma}_z^* \end{bmatrix} \right\} \tag{4.150}$$

But as $\mathbb{\Gamma}_z$ is positive definite, det ($\mathbb{\Gamma}_z$) > 0, and it follows that

$$\det(\Gamma) = (1/2)^{2n} \; [\det(\Gamma_z)]^2 \tag{4.151}$$

Inserting (4.148) and (4.151) in (4.102) we find

$$\boxed{p(\mathbf{x}, \mathbf{y}) = p(\mathbf{z}) = \pi^{-n} \; [\det(\Gamma_z)]^{-1} \; \exp(-\,\mathbf{z}^H \, \Gamma_z^{-1} \, \mathbf{z})} \tag{4.152}$$

which is the general expression of the PDF of a *normal circular complex random vector*. Note that this expression is given explicitly only in terms of the matrix Γ_z and its inverse. To avoid any confusion with the real case and also remind that circularity is a particular property of normal vectors, this PDF is sometimes noted $N_c(\mathbf{0}, \Gamma_z)$. Note also that this expression is the Fourier transform of (4.134), and as previously, we see that the characteristic function has a much simpler expression than the PDF. For $n = 1$ we again find (4.138), noting that Γ_z is reduced to σ_z^2.

We shall conclude this section by considering the problem of the *moments* of a normal complex random vector. We wish to show that the general expressions given in the section 4.6.2 for the real case are still valid for the complex case. But, immediately, a new problem arises. We have seen that even for the second order there are two distinct moments which are $E(Z_1 \; Z_2)$ and $E(Z_1 \; Z_2^*)$. The same is valid for higher-order moments, and, starting from a vector $\mathbf{Z}$ with components Z_i, the general k-order moments must be written as

$$m_k[\{i_j\};\{\varepsilon_j\}] = E\left[Z_{i_1}^{\varepsilon_1} \; Z_{i_2}^{\varepsilon_2} \; \ldots \; Z_{i_k}^{\varepsilon_k}\right] \tag{4.153}$$

where the i_js are integers taken between 1 and n, while ε_j are symbols taking the values +1 or *, which introduces the choice between Z or Z*.

Without any further assumption the calculation of $m_k[\{i_j\};\{\varepsilon_j\}]$ is always possible by taking $Z = X + jY$ and using the results for real RVs which seems *a priori* very tedious. However, we will see that some order can be introduced.

First we note that if the random vector $\mathbf{Z}$ is zeromean, the real and imaginary parts of all its components satisfy (4.90). Then all the odd moments in the form m_{2p+1} are null. We must then consider only even moments for which we will use the general expressions (4.93), (4.94), and (4.95) valid in the real case. In order to simplify the discussion, consider the moment

$$m_4 = E(Z_1 \; Z_2 \, Z_3 \; Z_4) \tag{4.154}$$

In this expression we must replace each Z by $X + jY$, and then apply the general relation (4.93). It follows that m_4 can be written

$$m_4 = E\{(X_1 + j\,Y_1)\,(X_2 + j\,Y_2)\,(X_3 + j\,Y_3)\,(X_4 + j\,Y_4)\} \tag{4.155}$$

Here we have a number of fourth order moments mixing the X_is and the Y_js. But each term contains four RVs X_i or Y_j and as they are jointly normal we can apply (4.93). The formula also exhibits a pairing effect; for example, the first of the three terms of the right-hand side of (4.93) forms a grouping of the variables with the indices (1, 2) and (3, 4).

Limiting our attention to the group (1, 2), we obtain four terms which are $E(X_1 X_2) - E(Y_1 Y_2) + jE(X_1 Y_2) + jE(Y_1 X_2)$. This sum of terms is none other than $E(Z_1 Z_2)$, and repeating the argument, we find that

$$m_4 = E(Z_1 Z_2)\, E(Z_3 Z_4) + E(Z_1 Z_3)\, E(Z_2 Z_4) + E(Z_1 Z_4)\, E(Z_2 Z_3) \quad (4.156)$$

which is exactly (4.93) extended to the complex case. The same reasoning can be made in the general case, and as a result we find that (4.95) can be applied to (4.153), which gives

$$m_{2k}[\{i_j\};\{\varepsilon_j\}] = \sum_G E\left(Z_{i_1}^{\varepsilon_1} Z_{i_2}^{\varepsilon_2}\right)\ldots E\left(Z_{i_{2k-1}}^{\varepsilon_{2k-1}} Z_{i_{2k}}^{\varepsilon_{2k}}\right) \quad (4.157)$$

where G has the same meaning as after (4.95).

An interesting situation appears when the vector $\mathbf{Z}$ is *normal circular*. In this case $E(\mathbf{Z}\mathbf{Z}^T) = \mathbb{0}$, which means that $E(Z_i Z_j) = 0$. As a result many terms in (4.157) are null and m_{2k} is non-null only if there are k symbols ε_j equal to one and k equal to *. As an example, the only non-vanishing four order moments are in the form

$$E(Z_1 Z_2 Z_3^* Z_4^*) = E(Z_1 Z_3^*)\, E(Z_2 Z_4^*) + E(Z_1 Z_4^*)\, E(Z_2 Z_3^*) \quad (4.158)$$

In particular this gives

$$E(|Z|^4) = 2\,[E(|Z|^2)]^2 = 2\sigma_z^4 \quad (4.159)$$

which replaces the value $3\sigma^4$ deduced from (2.86).

4.7 CONVERGENCE OF RANDOM VARIABLES

The concept of mean square convergence was introduced in Section 3.6 and there we used only second order properties of RVs for its definition. There are, however, several other kinds of convergence, which will be discussed below. Although their relationships and some applications will also be studied, it is not our intention to develop in detail many problems related to the convergence of RVs. These problems rapidly become very abstract and are discussed extensively in specialized works on probability theory. Our aim here is to introduce the minimum of knowledge about convergence concepts that will be sufficient to understand the problems of signal processing in which they are used.

4.7.1 Definitions and Principal Properties of Convergence

The concept of convergence is very simply defined for deterministic numbers and is the basis of mathematical analysis. In particular the notions of continuity, derivation, and integration of functions use the convergence extensively. The same is true for random signals, but here a new difficulty arises generated by the multiple kinds of convergence possible. We will begin with the basic definitions of convergence modes. In all that follows we consider a sequence of RVs $X_k(\omega)$ and an RV $X(\omega)$ defined on the same probability space. The number k is integer, and we shall study the situation when k becomes infinite.

(a) Convergence in distribution. Let us call $F_k(x)$ and $F(x)$ the DFs of X_k and X respectively. We say that X_k converges in distribution to X if the functions $F_k(x)$ tend to $F(x)$ on every point of continuity of $F(x)$.

As stated by its denomination, this mode of convergence is uniquely defined in terms of the distribution functions $F_k(x)$ and $F(x)$. In particular nothing is said about the correlation between the RVs X_k.

In applications it is sometimes more interesting to use the characteristic functions related to the DF by (2.53). It is possible to show that the convergence in distribution is equivalent to the convergence of the characteristic function $\phi_k(u)$ to a characteristic function if this last convergence is uniform in the vicinity of $u = 0$.

The convergence in distribution can be extended without difficulty to vector RVs, the relationship between the DF and the characteristic function now being defined by (4.59).

(b) Convergence in probability. It is said that X_k tends in probability to X if for every $\varepsilon > 0$ we have $P(|X_k - X| > \varepsilon) \to 0$ when $k \to \infty$. A convergence criterion exists to ensure this convergence without any knowledge of the limit that is analogous to a Cauchy criterion. This criterion states that X_k is convergent in probability if and only if to any pair of positive numbers ε and η we can associate a number N such that when k and k' are greater than N we have $P(|X_k - X_{k'}| > \eta) < \varepsilon$.

(c) Convergence in quadratic mean. This mode of convergence is defined on p. 71 and a convergence criterion is also given.

(d) Convergence with probability one. It is said that X_k converges with probability one, or almost everywhere, or almost surely, to X if $P[X_k \to X] = 1$. This means that the set of points ω for which $X_k(\omega)$ is not convergent has a zero probability measure. This mode of convergence is probably the most important in applications. The implication is that for any experiment defined by a point ω, the result of this experiment $X_k(\omega)$ converges to a value $X(\omega)$, except for events of null probability.

It is now of great interest to know the relations between these modes of convergence. This is summarized in Figure 4.8.

It is important to note that all the situations that are not specified by an arrow are possible. For example, the convergence in probability does not imply the quadratic mean convergence. Note also that there is in general no specific relation between the strongest kinds of convergence, i.e., in quadratic mean and with probability one.

4.7.2 The Central Limit Theorem

Consider a sequence of independent and identically distributed (IID) random variables X_k. To simplify, let us suppose that they are zeromean and with the same variance σ^2 and let Y_n be the RV

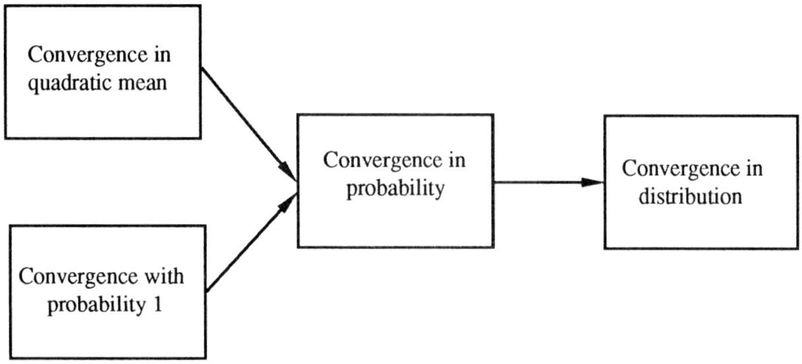

Fig. 4.8 Relations between modes of convergence.

$$Y_n = \frac{1}{\sqrt{n}} \sum_{k=1}^{n} X_k \tag{4.160}$$

It is clear that Y_n is zeromean, as are the X_ks, and that the variance of Y_n and of X_k are the same, because of the normalization factor before the sum. Let us call $\phi_X(u)$ the characteristic function of the RVs X_k and $\phi_Y(u)$ that of Y_n. Using (4.71), we obtain

$$\phi_Y(u) = \left[\phi_X\left(\frac{u}{\sqrt{n}}\right)\right]^n \tag{4.161}$$

and expressing this in terms of the second characteristic functions defined by (2.92), we obtain

$$\psi_Y(u) = n\ \psi_X\left(\frac{u}{\sqrt{n}}\right) \tag{4.162}$$

As indicated after (2.94), the McLaurin expansion of the second characteristic function gives the first cumulants, and using (2.92) and (2.64) with $m = 0$, we obtain

$$\psi_Y(u) = n\left[-\frac{1}{2n}\sigma^2 u^2 + \varepsilon(n)\right] \tag{4.163}$$

where $\varepsilon(n)$ is a sum of terms in the form $d_k\ u^k\ n^{-k/2},\ k > 2$. If all the cumulants are finite, the coefficients d_k are also finite. In conclusion,

$$\lim_{n\to\infty} \psi_Y(u) = -(1/2)\,\sigma^2 u^2 \tag{4.164}$$

But the r.h.s. of this equation is the second characteristic function of a normal distribution $N(0,\ \sigma^2)$, which means that Y_n converges in distribution to this normal distribution.

Let us comment this fundamental result. It essentially indicates that the sum conveniently normalized of independent RVs tends to become normal. This is of fundamental physical interest. In fact, a large number of physical phenomena can be considered as a sum of independent contributions, sometimes called microscopic phenomena.

If this situation arises the result can be considered as normal, or at least approximately normal. For example, if you enter a large room where hundreds of people are talking all at once, you hear a noise although you cannot distinguish any particular conversation. This noise is normal, as it can easily be verified experimentally.

Because of its physical importance, the central limit theorem can be extended in various ways which cannot be presented in detail here. For example, the assumption of independence can be replaced by less strong hypotheses, and the same can be said for the assumption of identical distributions. Roughly speaking, the most important aspect is to have a sum of contributions of the same order. Taking again our example of the noise in a room, if there is a source (a speaker, or a musical instrument) that is much louder than all the other sources, the impression of complete disorder disappears and this particular source can be recognized, even though it may be with difficulty. In this case the normal distribution usually disappears.

4.7.3 Convergence to Poisson Distribution

Consider a set of head and tail independent RVs X_k and, using the same notation as in Example 2.5, let Y_n be the sum

$$Y_n = \sum_{k=1}^{n} X_k \tag{4.165}$$

The characteristic function of the X_ks is given by (2.87) and that of Y_n by (4.71') which gives (2.89). It follows that the mean and variance of Y_n are given by (2.40) and (2.41). If $n \to \infty$ these quantities become infinite, and it is then necessary to introduce a normalization factor. Suppose that $p = m/n$, which gives

$$E(Y_n) = m \; ; \; \sigma_n^2 = m\{1 - (m/n)\} \tag{4.166}$$

As a result, when $n \to \infty$ we have $E(Y) = \sigma_Y^2 = m$. The characteristic function of Y_n is deduced from (2.89), which gives

$$\phi_n(u) = \left[1 + (m/n)\{\exp(ju) - 1\}\right]^n \tag{4.167}$$

and the second characteristic function is

$$\psi_n(u) = n \ln \left[1 + (m/n)\{\exp(ju) - 1\}\right] \tag{4.168}$$

By developing the logarithm, we find that

$$\lim_{n\to\infty} \psi_n(u) = m\,\{\exp(ju) - 1\} \tag{4.169}$$

and comparing this with (2.94), we see that the r.h.s. of (4.169) is the second characteristic function of a Poisson distribution of mean value m. This means that the Poisson distribution can be considered as a limit form of the binomial distribution, with the condition $p = m/n$. As this probability tends to zero, the term *small probabilities* is sometimes used.

4.7.4 Some Examples of Convergence

(a) Determination of the mean value. Let us take the same assumptions as in 4.7.2 and consider the RV Y_n defined by

$$Y_n = \frac{1}{n} \sum_{k=1}^{n} X_k \tag{4.170}$$

which differs uniquely from (4.160) by the normalization coefficient. As the X_ks are IID, we obtain

$$m_Y = m_X \quad ; \quad \sigma_Y^2 = (1/n)\, \sigma_X^2 \tag{4.171}$$

Physically Y_n is the *sample mean* of X_k and as $\sigma_Y^2 = E[(Y_n - m_Y)^2]$ we deduce from (4.171) and from the definition (3.112) that Y_n converges to the mean value of X in the mean square sense. Using the Tchebyshev inequality (2.23), we also deduce that $P(|Y_n - m_X| \geq \varepsilon)$ tends to zero when $n \rightarrow \infty$. This is also a consequence of the point made previously that mean square convergence leads to convergence in probability.

This gives a physical interpretation of the result. The mean values of X_k and Y_n are the same, and when n is sufficiently large the probability that the sample mean and the mean value differ by a quantity greater than ε is arbitrarily small.

Note that the proof is entirely deduced from (4.171). But the assumption of independence is not necessary to obtain this relation. In fact, (4.171) is still valid once the X_ks are uncorrelated. It is even possible to remove this condition, and as seen later within the framework of ergodic properties, the same result can be obtained for some cases of correlated RVs X_k.

This result is sometimes called the *weak law* of large numbers, while the *strong law* of large numbers is devoted to the convergence with probability one. By using methods outside the scope of this book it is possible to show the following result. If the RVs X_k are identically distributed, second order, independent, and with the mean value m, then Y_n converges with probability one to m.

(b) Relative frequency and probability. Let us return to the problem indicated on p. 16 and summarized by (1.56). We introduce the RV X, only taking the values 0 and 1. Suppose that $X = 1$ if A is realized and $X = 0$ otherwise. It is clear that $P(X = 1) = P(A)$ and also that $E(X) = P(A)$. The number $n(A)$ is the number of experiments, giving the result A when using repeated trials. This number is equal to the sum $X_1 + X_2 + \ldots + X_n$, because of the structure of the RVs X_i. Consequently the quantity $\pi_n(A)$ appearing in (1.56) has the same meaning as Y_n in (4.170). As thc RVs X_i are second order and independent and with the same expected value $E(A)$, we deduce from the discussion above that the relative frequency $\pi_n(A)$ converges to $P(A)$ in the quadratic mean sense and with probability one, when n tends to infinity. This is the basic relation between probability and relative frequency in independent and repeated trials.

(c) Convergence of the convolution. Let us return to the problem discussed on p. 72 relating to the output of a discrete-time filter given by (3.119). For simplification, consider the quantity S given by (3.120). The question is to find a condition ensuring the convergence with probability one of the series connected to the discrete-time convolution. Supposing that the RVs X_l satisfy $E[|X_l|] < M$ and applying the Fubini theorem allowing the permutation of expectation value and summation, we obtain

$$E\left[\sum_{l=0}^{\infty} |h_l||X_l|\right] = \sum_{l=0}^{\infty} |h_l| \; E[|X_l|] \le M \sum_{l=0}^{\infty} |h_l| \tag{4.172}$$

If the last series of this equation is finite, we have with probability one

$$\sum_{l=0}^{\infty} |h_l||X_l| < +\infty \tag{4.173}$$

which shows that the series (3.120) is convergent with probability one.

This has a very simple physical meaning. The condition $\Sigma\, |h_l| < +\infty$ is characteristic of a strictly stable filter, which means that any bounded input gives a bounded output. We then see that if the RVs X_k of the sequence satisfy in the mean a kind of stability condition, the random variable S is almost surely finite. It is clear that this stability condition also leads to S being convergent in the quadratic mean sense.

4.8 FURTHER RESULTS ON RANDOM VECTORS

The purpose of this section is to study various topics illustrating results presented in previous sections. Some points are not only illustrations of the same nature as the problems at the end of the chapter, but also introduce new ideas which are applicable to signal processing problems.

4.8.1 Entropy of Random Vectors

The concept of entropy is widely used in statistical mechanics, and we recommend that readers interested in this field of physics consult the large selection of books devoted to this question. Roughly speaking, entropy is a measure of disorder or uncertainty. It is defined for *discrete* RVs by

$$H[\{p_i\}] = -\sum p_i \ln(p_i) \tag{4.174}$$

where p_i is the probability associated to the value x_i, exactly as in (2.8). It is important to note that H is completely independent of the values x_i of the RV. Another RV Y, taking other values y_i, but with the same probabilities has the same entropy. In terms of DF the entropy is only a function of the steps of this function but does not depend on the points x_i where these steps appear. As a consequence there is no difference between random variables and random vectors, and the space where the RV is defined plays no role in entropy problems.

The concept of entropy is at the foundation of *information theory*, and the function (4.174) has many interesting properties discussed in works that deal with this field (see [Gallagher]). For example, it is very easy to show that if there are N terms in (4.174), the entropy is maximum for $p_i = 1/N$, which corresponds to equiprobability of the N possible values.

By analogy with (4.174) the entropy of a *continuous* random vector is defined by

$$H[p(.)] \triangleq - \int p(\mathbf{x}) \ln[p(\mathbf{x})]\, d\mathbf{x} \tag{4.175}$$

but this quantity has quite different properties from (4.174). The most fundamental point is that whereas $H[\{p_i\}]$ is always nonnegative, $H[p(.)]$ can *a priori* have negative values. This is due to the fact, already indicated, that $p(\mathbf{x})$ is not a probability but a density which can have values greater than one. Furthermore, the entropy (4.175) does not hold the invariance property discussed for (4.174). In fact, if we transform the random vector $\mathbf{X}$ into another $\mathbf{Y}$ by an inversible transformation $\mathbf{Y} = \mathbf{h}(\mathbf{X})$, there is no reason for the quantity $H[p(.)]$ to remain the same, as it does for discrete RVs. Even with these limitations, this quantity plays an important role in aspects of signal processing, particularly in the so-called *maximum entropy method.* This relates to entropy of a normal distribution which will now be calculated.

Example 4.8 Entropy of normal distribution. Using (4.82) in (4.175) we have to calculate the logarithm of $p(\mathbf{x})$ which is

$$\ln[p(\mathbf{x})] = -(n/2)\ln(2\pi) - (1/2)\ln[\det(\Gamma)] - (1/2)(\mathbf{x}-\mathbf{m})^T\,\Gamma^{-1}\,(\mathbf{x}-\mathbf{m}) \tag{4.176}$$

Only the last term is a function of $\mathbf{x}$, and it is easy to verify that it can take the form

$$(\mathbf{x}-\mathbf{m})^T\,\Gamma^{-1}\,(\mathbf{x}-\mathbf{m}) = \mathrm{Tr}\{(\mathbf{x}-\mathbf{m})(\mathbf{x}-\mathbf{m})^T\,\Gamma^{-1}\} \tag{4.177}$$

As the trace is a linear operation, its expected value can be calculated by using (3.63). We then find Γ and as $\mathrm{Tr}(\Gamma\Gamma^{-1}) = \mathrm{Tr}(\mathbb{I}) = n$, we finally obtain

$$H_N(\Gamma) = (n/2)\ln(2\pi e) + (1/2)\ln(\det\Gamma) \tag{4.178}$$

We can see that this entropy is only a function of the covariance matrix Γ and is independent of the mean value $\mathbf{m}$. The subscript N is related to the normal distribution. We can use exactly the same procedure for normal complex circular distribution. Using (4.152), we obtain

$$H_{N_C}(\Gamma) = n\ln(\pi e) + \ln(\det\Gamma) \tag{4.179}$$

In both cases the entropy is only a function of $\ln(\det\Gamma)$. As the determinant is the product of the eigenvalues, it is perfectly possible to obtain negative values for H, as pointed out above. This can be verified when $n = 1$ where $\det\Gamma = \sigma^2$. If σ^2 is sufficiently small the entropy becomes negative.

We can also verify that the entropy is not invariant in linear transformations. This is obvious because a normal random vector remains normal in a linear transformation but its covariance matrix is changed, as seen in (4.96). More precisely, starting from a real vector **X** which is $N(\mathbf{0}, \mathbb{\Gamma})$ we can deduce a vector **Y** by the real transformation $\mathbb{Y} = \mathbb{A}\ \mathbb{X}$. It results from (4.96) that this vector is $N(\mathbf{0}, \mathbb{A}\mathbb{\Gamma}\mathbb{A}^T)$. The last term of (4.178) must then be replaced by $\ln\{\det(\mathbb{A}\mathbb{\Gamma}\mathbb{A}^T)\}$. If $\mathbb{A}$ is a square matrix the determinant can be calculated by a product of determinants, and we obtain

$$H_N(\mathbb{A}\mathbb{\Gamma}\mathbb{A}^T) = H_N(\mathbb{\Gamma}) + \ln|\det\mathbb{A}| \tag{4.180}$$

Nevertheless the entropy remains the same in all these linear transformations such that $\det\mathbb{A} = 1$. This is especially the case if $\mathbb{A}$ is a triangular matrix with diagonal elements equal to one.

The *maximum entropy* method is based on the following fundamental result.

Proposition. If a vector **X** is zeromean and with the covariance matrix $\mathbb{\Gamma}$, its entropy satisfies $H_X \leq H_N(\mathbb{\Gamma})$ and the equality holds if and only if **X** is $N(\mathbf{0}, \mathbb{\Gamma})$.

Proof. Let us call $q(\mathbf{x})$ the PDF of **X** and $p_N(\mathbf{x})$ the PDF associated with the distribution $N(\mathbf{0}, \mathbb{\Gamma})$ and given by (4.82). As $\ln[p_N(\mathbf{x})]$ is given by (4.176) and as $q(\mathbf{x})$ and $p_N(\mathbf{x})$ give the same covariance matrix $\mathbb{\Gamma}$, we can write

$$\int q(\mathbf{x}) \ln[p_N(\mathbf{x})]\, d\mathbf{x} = \int p_N(\mathbf{x}) \ln[p_N(\mathbf{x})]\, d\mathbf{x} = -H_N(\mathbb{\Gamma}) \tag{4.181}$$

We deduce that the entropy H_X of the PDF $q(\mathbf{x})$ satisfies

$$H_X - H_N(\mathbb{\Gamma}) = \int q(\mathbf{x}) \ln\{p_N(\mathbf{x})/q(\mathbf{x})\}\, d\mathbf{x} \tag{4.182}$$

But it is well known that $\ln x \leq x - 1$ and the equality holds if and only if $x = 1$. It follows that

$$H_X - H_N(\mathbb{\Gamma}) \leq 0 \tag{4.183}$$

and the equality holds if and only if $p(\mathbf{x}) = p_N(\mathbf{x})$.

This means that in all the problems in which we are interested in maximizing the entropy for given second order properties, we can introduce the normal distribution immediatly. This will be applied later when considering prediction problems.

4.8.2 Other Properties of the Regression

The regression $\mathbf{r}(\mathbf{y})$ defined by (4.38) satisfies the orthogonality equation (4.39). Furthermore, if **X** and **Y** are two independent zeromean random vectors, it results from (4.29) that $E(\mathbf{X}|\mathbf{y}) = E(\mathbf{X}) = \mathbf{0}$, and the regression is then null. This is also the case with the cross-covariance matrix defined by (3.65). Conversely, if this matrix is null, we can only say that **X** and **Y** are uncorrelated. Let us now investigate the consequences of the property $\mathbf{r}(\mathbf{y}) = \mathbf{0}$. A result of (4.39) is that **X** is uncorrelated with any function $\mathbf{g}(\mathbf{Y})$, provided that it is a second order random vector. Of course, by taking $\mathbf{g}(\mathbf{Y}) = \mathbf{Y}$, we find that **X** and **Y** are uncorrelated. But the nullity of the regression does not mean that **X** and **Y** are independent. To verify this point, let us take another look at spherically invariant RVs and suppose, as in (4.48), that

$$p(\mathbf{x}, \mathbf{y}) = f(\mathbf{x}^T\mathbf{x} + \mathbf{y}^T\mathbf{y}) \tag{4.184}$$

where $f(.)$ is still such that (3.56) holds. This function introduces independent random vectors **X** or **Y** only if $f(.)$ takes an exponential form, which characterizes a normal distribution for which we know that independence and noncorrelation are synonymous. Let us then suppose that $f(.)$ is not an exponential function. It results from (4.28) that $\mathbf{r}(\mathbf{y}) = \mathbf{0}$ if and only if

$$\int \mathbf{x}\, p(\mathbf{x}, \mathbf{y})\, d\mathbf{x} = \mathbf{0} \tag{4.185}$$

which is obviously the case for $p(\mathbf{x}, \mathbf{y})$ given by (4.184).

Let us now investigate the case where the regression is linear, as in (4.109). This property is not a characteristic of the normal law and can happen to many other distributions, of which we shall only give one case here. For this we start from a pair of vectors **X** and **Y** with a PDF of the form

$$p(\mathbf{x}, \mathbf{y}) = f\{q(\mathbf{x}, \mathbf{y})\} \tag{4.186}$$

where $q(\mathbf{x}, \mathbf{y})$ is a quadratic form as in (4.103). This is a generalization of the structure (4.184), and the normal PDF appears as a particular case. We assume that the matrices $\mathbb{M}_x$ and $\mathbb{M}_y$ are positive definite, while $\mathbb{M}_{xy}$ is in general rectangular. Let us write $q(\mathbf{x}, \mathbf{y})$ as

$$q(\mathbf{x}, \mathbf{y}) = (\mathbf{x} - \mathbb{A}\mathbf{y})^T\, \mathbb{M}_x (\mathbf{x} - \mathbb{A}\mathbf{y}) + \mathbf{y}^T \mathbb{B}\, \mathbf{y} \tag{4.187}$$

The identification between (4.103) and (4.187) gives

$$\mathbb{A} = -\,\mathbb{M}_x^{-1}\, \mathbb{M}_{xy}\ ;\ \mathbb{B} = \mathbb{M}_y - \mathbb{M}_{yx}\, \mathbb{M}_x^{-1}\, \mathbb{M}_{xy} \tag{4.188}$$

As a result of the form in (4.187) we deduce that

$$\int (\mathbf{x} - \mathbb{A}\mathbf{y})\, p(\mathbf{x}, \mathbf{y})\, d\mathbf{x} = 0 \tag{4.189}$$

which generalizes (4.185) and shows that

$$r(\mathbf{y}) = E(\mathbf{X}\,|\,\mathbf{y}) = \mathbb{A}\mathbf{y} \tag{4.190}$$

The matrix $\mathbb{A}$ has the form given by (4.188) in terms of the matrices $\mathbb{M}$, but also the form (4.108) in terms of the matrices $\mathbb{\Gamma}$. This is in fact deduced from (4.107) which uses the orthogonality property of the regression and does not depend on the normal law. On the other hand the conditional covariance matrix $\mathbb{V}(\mathbf{y})$ has no reason to be independent of **y**, as in the normal case.

4.8.3 Conditional Distributions and Functions of Random Vectors

Functions of random vectors were discussed in Section 4.4. Starting from a random vector **X**, (4.40) introduces another random vector **Y**, and our purpose is to study the pair (**X**, **Y**) and deduce some interesting properties. The situation is quite different for the discrete and continuous cases and we will begin with the former.

Suppose that the vector **X** takes the values $\mathbf{x}_i$ with the probabilities p_i. The transformation (4.40) associates to each vector $\mathbf{x}_i$ a vector $\mathbf{y}_i$, but several vectors $\mathbf{y}_j$ can come from the same vector $\mathbf{x}_i$. We introduce the set S_k of vectors $\mathbf{x}_i$ defined by

$$S_k = \{\mathbf{x}_i \mid \mathbf{h}(\mathbf{x}_i) = \mathbf{y}_k\} \tag{4.191}$$

It follows that we obtain

$$q_k = P(\mathbf{Y} = \mathbf{y}_k) = \sum_i p_i \tag{4.192}$$

where the sum is extended to all those vectors $\mathbf{x}_i$ belonging to S_k. Let us now consider the pair (**X**, **Y**). The corresponding probability is given by

$$P\{(\mathbf{X} = \mathbf{x}_i).(\mathbf{Y} = \mathbf{y}_k)\} = p_i\, \delta[\mathbf{y}_k, \mathbf{h}(\mathbf{x}_i)] \tag{4.193}$$

where $\delta[\ ]$ has the same meaning as in (4.81). This relation means that when the event $\mathbf{x}_i$ is realized, the value of **Y** is specified. We shall now calculate the marginal distributions. By summation over all the values of $\mathbf{y}_k$ we obtain p_i, because there is only one $\mathbf{y}_k$ associated with a given $\mathbf{x}_i$. Conversely, the summation over the $\mathbf{x}_i$ for a given value of $\mathbf{y}_k$ gives q_k appearing in (4.192). The conditional probability $P\{(\mathbf{Y} = \mathbf{y}_k) \mid (\mathbf{X} = \mathbf{x}_i)\}$ is obviously $\delta[\mathbf{y}_k, \mathbf{h}(\mathbf{x}_i)]$, which is characteristic of a deterministic transformation such as (4.40). Finally, the last conditional probability is

$$P\{(\mathbf{X} = \mathbf{x}_i) \mid (\mathbf{Y} = \mathbf{y}_k)\} = (1/q_k)\, p_i\, \delta[\mathbf{y}_k, \mathbf{h}(\mathbf{x}_i)] \tag{4.194}$$

where q_k is given by (4.192). An example in which these ideas are applied follows.

Example 4.9 From Poisson to multinomial distribution. The probabilities associated with a Poisson distribution of mean value m are given by (2.42). Suppose now that we are given k independent RVs X_i with the same Poisson distribution, and that these RVs are the components of the random vector **X**. This vector takes only the values $\mathbf{X} = \mathbf{n}$, where $\mathbf{n}$ is a vector with components $n_1, n_2, \ldots, n_k$, which are all non-negative integers. Because of the independence, using (2.42), we can write the probability $p_\mathbf{n}$ that $\mathbf{X} = \mathbf{n}$ as

$$p_\mathbf{n} = \prod_{j=1}^{k} (e^{-m}\, m^{n_j})/(n_j!) = \frac{e^{-km}\, m^{\Sigma n_j}}{\Pi(n_j!)} \tag{4.195}$$

Let us now introduce the RV Y defined by

$$Y = \sum_{j=1}^{k} X_j \tag{4.196}$$

It is clear that this sum also takes only integer values, as do the RVs X_j, components of **X**. Furthermore, the event $\mathbf{X} = \mathbf{n}$ implies that $X_j = n_j$ and the corresponding value of Y is $y = \Sigma\, n_j$. Applying (4.193) and (4.195), we obtain

$$P\{(\mathbf{X} = \mathbf{n}).(Y = y)\} = \frac{e^{-km}\, m^y}{\prod_j (n_j!)}\; \delta[y, \Sigma n_j] \tag{4.197}$$

which can be written as

$$P\{(\mathbf{X} = \mathbf{n}).(Y = y)\} = \frac{e^{-km}\, (km)^y}{y!} \cdot \frac{y!}{n_1!\, n_2! \ldots n_k!} \cdot \frac{1}{k^y}\, \delta[y, \Sigma n_j] \tag{4.198}$$

We now introduce the marginal distribution of Y. This is a sum of independent Poisson RVs. It follows that Y is still a Poisson RV with mean value km. To show this result it is sufficient to apply (4.71) where $\phi_X(u)$ is replaced by (2.91). As the first term of the r.h.s. of (4.198) is precisely the probability that $Y = y$, we finally obtain, with the introduction of $\pi = 1/k$,

$$P\{(\mathbf{X} = \mathbf{n})|(Y = y)\} = \frac{(\Sigma n_j)!}{n_1!\, n_2! \ldots n_k!} \cdot \pi^{n_1}\, \pi^{n_2} \ldots \pi^{n_k}\, \delta[y, \Sigma n_j] \tag{4.199}$$

This last function depends of course on y because $y = \Sigma n_j$. The probability (4.199) is a particular case of a *multinomial distribution.* We shall now explore some of its properties.

A general multinomial distribution is defined by

$$P\{(\mathbf{X} = \mathbf{n})\} = \frac{N!}{n_1!\, n_2! \ldots n_k!} \cdot p_1^{n_1}\, p_2^{n_2} \ldots\, p_k^{n_k}\, \delta[N, \Sigma n_i] \tag{4.200}$$

where $\Sigma p_i = 1$. The binomial distribution defined by (2.38) is a particular case obtained for $k = 2$. The simplest way to study its property is to calculate its characteristic function defined by (4.59), or

$$\phi(\mathbf{u}) = \sum P(\mathbf{X} = \mathbf{n}) \exp(j\mathbf{u}^T\mathbf{n}) \tag{4.201}$$

As $\exp(j\mathbf{u}^T\mathbf{n}) = \exp(j\Sigma u_i n_i) = \Pi \exp(j u_i n_i)$, we deduce that

$$\phi(\mathbf{u}) = \left\{ \sum_{i=1}^{k} p_i \exp(ju_i) \right\}^N \tag{4.202}$$

It is more convenient to use the property $\Sigma p_i = 1$, and to write this function as

$$\phi(\mathbf{u}) = \{1 + p_1(e^{ju_1} - 1) + p_2(e^{ju_2} - 1) + \ldots + p_k(e^{ju_k} - 1)\}^N \tag{4.203}$$

From this characteristic function we deduce some interesting points. If we are interested in the marginal distribution of the component X_1 we have, after (4.64), only to replace u_2, u_3, ..., u_k by 0, which results in a binomial distribution characterized by (2.89). Thus any component of the vector $\mathbf{X}$ has a marginal distribution which is binomial. In order to calculate the covariance matrix of the vector $\mathbf{X}$, we have to know moments such as $E(X_m X_n)$. For this we make all the u_is equal to zero except u_m and u_n, thus arriving at a trinomial distribution. Finally, it is important to note that the conditional probability (4.199) no longer depends on the parameter m of the Poisson distribution which was the starting point of the calculation.

Let us now investigate the *continuous* case by assuming that **X** is a continuous random vector and that the transformation (4.40) is such that **Y** is also a continuous random vector. As the only PDF is $p(\mathbf{x})$ related to **X**, we can write, as in (4.193),

$$p(\mathbf{x}, \mathbf{y}) = p(\mathbf{x})\ \delta\{\mathbf{y} - \mathbf{h}(\mathbf{x})\} \tag{4.204}$$

where $\delta\{\ \}$ is the multivariable Dirac distribution. In order to simplify the discussion, we assume that $h(\mathbf{x})$ and y are scalars, which gives

$$p(\mathbf{x}, y) = p(\mathbf{x})\ \delta\{ y - h(\mathbf{x})\} \tag{4.205}$$

It is obvious that the integral in **x** and y is equal to one and that the conditional PDF $p(y \mid \mathbf{x})$ is, from (4.37),

$$p(y \mid \mathbf{x}) = \delta\{ y - h(\mathbf{x})\} \tag{4.206}$$

This simply means that if **x** is given, y is no longer random. In fact, as a result of (2.13), $\delta(x - x_i)$ is the PDF of the RV equal to x_i *a priori* given. On the other hand the marginal PDF of y is obtained by integration of $p(\mathbf{x}, y)$ over the space where **x** is defined, and there is no simple expression such as (4.192) for the discrete case. We are already aware of this difficulty because $p(y)$ is nothing other than the derivative of the distribution function given by (4.42). The above discussion has an important field of application in signal processing problems when the concept of *sufficient statistic* appears.

The concept of sufficient statistic. In many estimation problems we work with probability distributions which depend on one or several parameters. For example the Poisson distribution depends only on the parameters m, as seen in (2.42). The univariate normal distribution $N(m, \sigma^2)$ depends only on two parameters, the mean m and the variance σ^2, as seen in (2.31). It is very interesting to estimate these parameters from the observation of a sequence of independent realizations of RVs with this distribution. For example, it can be important to estimate the mean and the variance from the observation of a sequence of N independent RVs assumed to be normal. Let us call $X_1, X_2, \ldots, X_N$ these obsevations, and **X** the vector with components X_i. The PDF of **X** is then a function of a parameter θ, assumed for simplification to be scalar, and can be written as $p(\mathbf{x}\ ;\ \theta)$. In order to estimate θ we construct a scalar function $y = h(\mathbf{x})$, as in (4.40), which is called a *statistic*.

A statistic $y = h(\mathbf{x})$ is said to be *sufficient* for the parameter θ if the conditional PDF $p(\mathbf{x} \mid y)$ no longer depends on the parameter θ. This means that $y = h(\mathbf{x})$ contains all the information which can be extracted from the observation concerning the unknown parameter θ. In other words, if we are only interested in the estimation of θ, the random vector **X** can be reduced to its sufficient statistic.

Example 4.10 Some examples of sufficient statistics. The first example is a direct consequence of the discussion accompanying Example 4.9. Suppose that we want to estimate the mean value of a Poisson distribution from the observations of N random variables that are independent and that have the same

Poisson distribution. Taking again the terms at the end of the example, we can say that sum Y defined by (4.196) is a sufficient statistic for the estimation of the parameter m appearing in (4.195) because of (4.199) where m disappears.

Let us now construct a similar example in the case of a normal distribution. Suppose that in Example 4.9 the k independent and identically distributed (IID) RVs X_i are $N(m, \sigma^2)$. If we reconsider their sum Y given by (4.196), we can show that Y is still a sufficient statistic for the parameter m. The first point to note is that Y is still normal and, more precisely, is $N(km, k\sigma^2)$ as seen in section 4.6.3. The PDF of $\mathbf{X}$ is given by (4.82), where $\Gamma = \sigma^2\mathbb{I}$ because of the independence assumption, and can then be written as

$$p(\mathbf{x}) = \alpha \exp\{-(1/2\sigma^2)(\mathbf{x}^T\mathbf{x} - 2m\mathbf{c}^T\mathbf{x} + km^2)\} \tag{4.207}$$

where α is a constant independent of m and $\mathbf{c}$ the vector the components of which are all equal to one. From this definition (4.40) can be written as $Y = \mathbf{c}^T\mathbf{X}$, and $p(\mathbf{x}, y)$ given by (4.205) becomes

$$p(\mathbf{x}, y) = p(\mathbf{x})\ \delta(y - \mathbf{c}^T\mathbf{x}) \tag{4.208}$$

As Y is $N(km, k\sigma^2)$, its PDF is

$$p(y) = \alpha' \exp\{-(1/2\ k\sigma^2)\ (y^2 - 2\ kmy + k^2m^2)\} \tag{4.209}$$

and $p(\mathbf{x}|y)$ is given by $p(\mathbf{x}, y)/p(y)$. Because of the Dirac distribution in (4.208) we can replace in the denominator y by $\mathbf{c}^T\mathbf{x}$. When we do this we find that

$$(1/k)\ (y^2 - 2\ kmy + k^2m^2) = (1/k)\ (\mathbf{c}^T\mathbf{x})^2 - 2\ m\mathbf{c}^T\mathbf{x} + km^2 \tag{4.210}$$

Comparing this with (4.207), we conclude that the terms depending on m disappear in $p(\mathbf{x}|y)$ because of the division of $p(\mathbf{x},y)$ by $p(\mathbf{y})$. The conditional PDF $p(\mathbf{x}|y)$ is then independent of m and Y defined by (4.196) is a sufficient statistic for m. Note that it is not a sufficient statistic for σ^2 because this parameter remains present in $p(\mathbf{x}|y)$.

Let us now consider the *basic detection problem* which is a choice between two hypotheses from an observation $\mathbf{X}$. In practical application this means that we have to decide between two hypotheses H_0 and H_1 from an observation $\mathbf{X}$. For example, H_0 can mean that the observation is only due to noise, while H_1 means that a signal is present. The two hypotheses are said to be simple if the PDF of $\mathbf{X}$ is known in the two cases, and are noted $p_0(\mathbf{x})$ and $p_1(\mathbf{x})$. In this situation it can be shown that the *likelihood ratio* defined by

$$L(\mathbf{x}) \triangleq \frac{p_1(\mathbf{x})}{p_0(\mathbf{x})} \tag{4.211}$$

and studied in Problem 2.5 is a sufficient statistic. As previously, let us introduce the RV Y defined by the function $y = L(\mathbf{x})$. In detection problems a statistic $y = h(\mathbf{x})$ is said to be sufficient if conditionally to y the two hypotheses cannot be discerned or if

$$p_0(\mathbf{x}|y) = p_1(\mathbf{x}|y) \tag{4.212}$$

This means that $L(\mathbf{x})$ contains all the information necessary for a choice to be made between them. The method is quite similar and starts from (4.205), which allows us to write (4.212) in the form

$$\frac{p_1(\mathbf{x})\ \delta\{y - L(\mathbf{x})\}}{p_1(y)} = \frac{p_0(\mathbf{x})\ \delta\{y - L(\mathbf{x})\}}{p_0(y)} \tag{4.213}$$

By using the method given in Problem 2.5, we can show that $p_1(y) = y\ p_0(y)$. Furthermore, because of the Dirac distribution, we have $y = L(\mathbf{x})$ or $p_1(\mathbf{x}) = y\ p_0(\mathbf{x})$. Replacing p_1 with these values in the l.h.s. of (4.213), we obtain the r.h.s., which shows the result. It is for this reason that the likelihood ratio is so important in all detection problems.

4.8.4 More on Circularity

Normal circular complex random vectors were defined by (4.152) and (4.134). Although the assumption of circularity introduces a particular class of normal complex random vectors as defined on p. 120, it is so important that it is sometimes included in the definition (see [Brillinger] p. 89). Let us investigate some of its properties.

Consider first a scalar normal circular RV. Its PDF is given by (4.138). If we introduce the real and imaginary parts of z, writing $z = x + j\,y$, we obtain a PDF $p(x, y)$ given by (4.137) in which $\sigma^2 = (1/2)\sigma_z^2$, which is a particular case of (4.129). This shows that the real and imaginary parts X and Y of Z are two IID normal RVs with the distribution $N(0, \sigma^2)$. But it is well known that when we use complex numbers it is also interesting to work with amplitude and phase, writing z as $a\exp(j\phi)$. Similarly, two other RVs A and Φ can be associated to a complex RV Z such that

$$Z = A\exp(j\Phi) \tag{4.214}$$

The problem is to calculate the PDF $q(a, \phi)$ of the pair A, Φ. If we assume that $A \geq 0$ and that $0 \leq \Phi < 2\pi$, the transformation between (X, Y) and (A, Φ) is that presented on p. 100. As a result (4.47) can be written in the form

$$q(a, \phi) = p(a\cos\phi,\ a\sin\phi)\ a \tag{4.215}$$

Using (4.137) we obtain

$$q(a, \phi) = (1/2\pi\sigma^2)\, a\exp(-a^2/2\sigma^2) \tag{4.216}$$

As this function is independent of ϕ, we deduce that the two RVs A and Φ are independent and that Φ is uniformly distributed between 0 and 2π, while A is defined by the PDF

$$p(a) = (a/\sigma^2)\exp(-a^2/2\sigma^2),\ a \geq 0 \tag{4.217}$$

called a *Rayleigh distribution.* The moments of this distribution are very easy to calculate, as proposed in Problem 4.43.

Suppose now that the complex RV is no longer zeromean-valued, and that its mean value is m_z, which is also a complex number $m_x + j\,m_y$. Its PDF is now defined by (4.138) where z must be replaced by $z - m_z$. If we use the real and imaginary parts X and Y, the corresponding PDF is given by (4.137) where x and y are replaced by $(x - m_x)$ and $(y - m_y)$ respectively. In order to calculate the PDF of the RVs A and Φ, we again use (4.47). If we write the complex number m_z as $m\exp(j\theta)$, we obtain

$$p(a, \phi) = \frac{a}{2\pi\sigma^2} \cdot \exp\left\{\frac{-1}{2\sigma^2}\,(a^2 - 2\,a\,m\cos(\phi - \theta) + m^2)\right\} \qquad (4.218)$$

We immediately see that there is no longer independence between A and Φ. The marginal PDF of A can be obtained by integration on the variable ϕ, and using the property

$$(1/2\pi)\int_0^{2\pi} \exp(x\cos\alpha)\,d\alpha = I_0(x) \qquad (4.219)$$

where $I_0(x)$ is the modified Bessel function, we obtain

$$p(a) = \frac{a}{\sigma^2}\exp\left(-\frac{a^2+m^2}{2\sigma^2}\right) I_0\left(\frac{am}{\sigma^2}\right) \qquad (4.220)$$

which is sometimes called a Rayleigh-Rice distribution. It is represented in Figure 4.9 for $\sigma = 1$ and various values of m. Again we find (4.217) for $m = 0$, and the Rayleigh distribution is a particular case of Rayleigh-Rice distribution. It appears that for $m \geq 4$ the curves are quite similar. This is a result of the structure of the Bessel function. In fact for large values of x we can approximate $I_0(ma)$ by $(2\pi ma)^{1/2}\exp(ma)$. The points appearing in the curve corresponding to $m = 4$ of Figure 4.9 are calculated with this approximation which appears quite sufficient. The approximation shows that for $m \geq 4$ the Rayleigh-Rice distribution can be approximated by a normal distribution $N(m, 1)$, which appears for the curves of the figure corresponding to $m = 4$, 5 and 6.

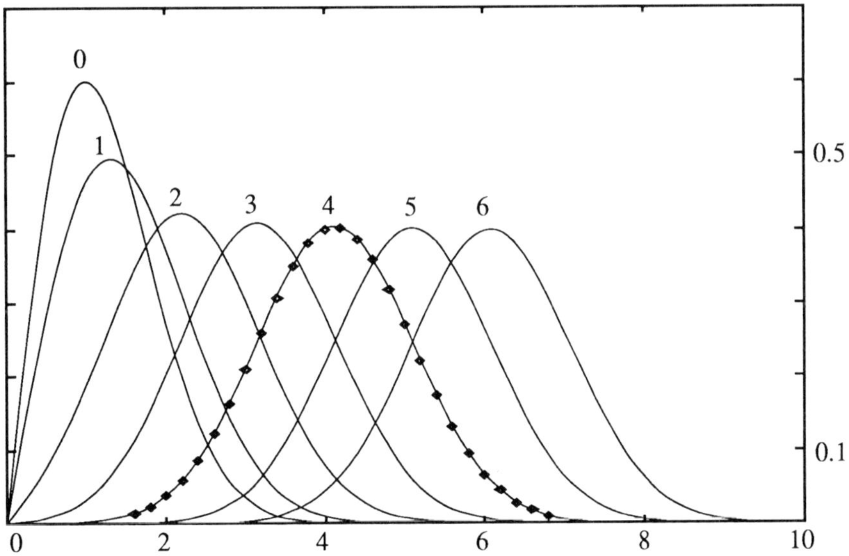

Figure 4.9 Rayleigh-Rice probability density function for several values of m; $\sigma = 1$.

Let us now consider two circular normal RVs Z_1 and Z_2 that are the two components of a vector $\mathbf{Z}$ with an arbitrary covariance matrix Γ_z. The corresponding PDF is given by (4.152) with $n = 2$. To simplify the calculation let us assume that $E(|Z_1|^2) = E(|Z_2|^2) = \sigma_z^2$. We can then write $E(Z_1 Z_2^*) = c_z \sigma_z^2$. But note that the correlation coefficient c_z can be complex. It follows that the two rows of the matrix Γ_z are $\sigma_z^2 [1, c_z]$ and $\sigma_z^2 [c_z^*, 1]$. We now have a problem similar to that seen with two-dimensional real normal RVs. After some simple algebra we can write (4.152) in the form

$$p(\mathbf{z}) = p(z_1, z_2) = (\pi\sigma_z^2)^{-2} (1 - |c_z|^2)^{-1} \exp\{-Q(\mathbf{z})\} \tag{4.221}$$

with

$$Q(\mathbf{z}) = [\sigma_z^2 (1 - |c_z|^2)]^{-1} [|z_1|^2 + |z_2|^2 - 2 \operatorname{Re}(c_z z_1^* z_2)] \tag{4.222}$$

As in (4.137), it is interesting to write this expression in terms of x_1, x_2 and y_1, y_2, real and imaginary parts of z_1 and z_2. For this, we note that as Z_1 and Z_2 are separately normal circular, we have $\sigma_z^2 = 2\sigma^2$, where σ^2 is the variance common to X_1, X_2, Y_1, and Y_2. Furthermore, as the correlation coefficient c_z must satisfy $|c_z| \leq 1$, as a result of (3.50), it can be written as $c_z = c \exp(j\theta)$. With these notations, the PDF can be written as

$$p(x_1, y_1, x_2, y_2) = k \exp(-\alpha Q) \tag{4.223}$$

$$k = (2\pi\sigma^2)^{-2} (1 - c^2)^{-1} \tag{4.224}$$

$$\alpha = [2\sigma^2(1 - c^2)]^{-1} \tag{4.225}$$

$$Q = x_1^2 + y_1^2 + x_2^2 + y_2^2 - 2c \cos\theta(x_1 x_2 + y_1 y_2) - 2c \sin\theta(x_2 y_1 - x_1 y_2) \tag{4.226}$$

which is the most general expression of the PDF of two normal complex circular RVs with the same variance.

As previously, it is interesting to write out this expression in terms of amplitude and phase. Writing $x = a\cos\phi$ and $y = a\sin\phi$ for z_1 and z_2, we easily obtain

$$p(a_1, a_2, \phi_1, \phi_2) = k\, a_1 a_2 \exp(-\alpha Q) \tag{4.227}$$

where Q now has the form

$$Q = a_1^2 + a_2^2 - 2a_1 a_2\, c \cos(\phi_1 - \phi_2 - \theta) \tag{4.228}$$

For $c = 0$ we obtain the product of two Rayleigh distributions. We will see later that there are some particular signals for which c_z is real, which means that $\theta = 0$ or π. Coming back to (4.226), this means that $p(.)$ can be expressed as the product of a function of x_1 and x_2 and another of y_1 and y_2. Thus the real parts of Z_1 and Z_2 and their imaginary parts are independent, as in the case of one normal circular RV. Finally, marginal distribution can be obtained by integrations over some variables, as seen in several problems later.

4.8.5 Second Order Properties after Nonlinear Transformations

If the transformation (4.40) is linear it is easy to calculate the first and second order moments of $\mathbf{Y}$ in terms of those of $\mathbf{X}$ (see the example in Section 4.6.3). Although linear transformations are practically the most important, we have also discussed examples of nonlinear transformations which are of great interest. Unfortunately, there is no general way of calculating the moments of **Y** in terms of those of **X**, but for some special classes of transformations this is possible. To simplify, suppose that the transformation is "diagonal", which means that **X** and **Y** have the same dimension and that each component of **Y** depends only on one component of **X**. We consider here only two components, and write

$$Y_1 = h_1(X_1) \quad ; \quad Y_2 = h_2(X_2) \tag{4.229}$$

The class of transformations considered here is such that the functions $h(.)$ have a Fourier transform, even in the distribution sense. This allows us to write

$$h(x) = \int H(u) \exp(jux)\, du \tag{4.230}$$

It follows that we have

$$E(Y) = E\{h(X)\} = \int H(u)\, \phi(u)\, du \tag{4.231}$$

where $\phi(u)$ is the characteristic function of X, and

$$E(Y_1 Y_2) = \int\int H_1(u)\, H_2(v)\, \phi(u, v)\, du\, dv \tag{4.232}$$

where $\phi(u, v)$ is the characteristic function of the pair (X_1, X_2).

Suppose now that the RVs X_1 and X_2 or X are zeromean and normal. This means that $\phi(u)$ in (4.231) can be written as $\exp\{-(1/2)\, \gamma_x\, u^2\}$. Using this expression in (4.231), and taking the derivative with respect to γ_x, we obtain

$$\frac{d}{d\gamma_x} E(Y) = -\frac{1}{2} \int H(u)\, u^2\, \phi(u)\, du \tag{4.233}$$

Returning to (4.230), we see that the second derivative of $h(x)$ appears, which gives

$$\frac{d}{d\gamma_x} E(Y) = \frac{1}{2}\, E\{h''(X)\} \tag{4.234}$$

This introduces a procedure allowing a direct calculation of the mean value of $h(X)$, which is of course interesting when the transformation $h''(x)$ is much simpler than $h(x)$.

Example 4.11 Calculation of mean values after nonlinear transformations. Consider first the very simple example of a quadratic transformation defined by $y = x^2$. As $h''(x) = 2$, we obtain

$$\frac{d}{d\gamma_x} E(Y) = 1 \tag{4.235}$$

or

$$E(Y) = \gamma_x + k \tag{4.236}$$

But for $\gamma_x = 0$ we know that $X = 0$ with probability one and it follows that $k = 0$. The result $E(Y) = \gamma_x$ can of course be obtained directly, because it is the definition of the variance.

Suppose now that $y = |x|$, which corresponds to the input-output relationship of a device called a *linear rectifier.* We obviously have $y'(x) = Sg(x)$, where $Sg(x) = x/|x|$, and $y''(x) = 2\delta(x)$. Using the normal distribution $N(0, \gamma_x)$ given by (2.35), we obtain

$$\frac{d}{d\gamma_x} E(X) = (2\pi\gamma_x)^{-1/2} \tag{4.237}$$

which, after integration, gives

$$E(|X|) = 2(2\pi)^{-1/2} (\gamma_x)^{1/2} + k \tag{4.238}$$

As previously, we obtain $k = 0$ by taking the case where $\gamma_x = 0$. This finally gives

$$E(|X|) = (2/\pi)^{1/2} \sigma_x \tag{4.239}$$

where $\sigma_x^2 = \gamma_x$.

Let us now consider the crosscorrelation (4.232) between the outputs Y_1 and Y_2 defined by (4.229). Suppose again that X_1 and X_2 are jointly normal with the characteristic function $\phi(u, v)$ defined by (4.117) where $m_x = m_y = 0$. Replacing σ_x^2, σ_y^2 and $c\,\sigma_x\,\sigma_y$ by σ_1^2, σ_2^2 and γ respectively, we obtain

$$\frac{\partial\phi}{\partial\gamma} = -uv\,\phi(u, v) \tag{4.240}$$

If we insert this expression into (4.232) and repeat the same procedure, we obtain

$$\frac{\partial E(Y_1Y_2)}{\partial\gamma} = E[h_1'(X_1)\, h_2'(X_2)] \tag{4.241}$$

This operation can be repeated, which gives

$$\left(\frac{\partial}{\partial\gamma}\right)^k E(Y_1Y_2) = E[h_1^{(k)}(X_1)\, h_2^{(k)}(X_2)] \tag{4.242}$$

Example 4.12 Calculation of crosscorrelations after nonlinear transformations. As in the last example, consider first the quadratic transformation $y = x^2$ and suppose that $h_1 = h_2$. Using (4.241), we obtain

$$\frac{\partial E(Y_1Y_2)}{\partial\gamma} = 4\, E(X_1X_2) = 4\,\gamma \tag{4.243}$$

As a result, we have $E(Y_1Y_2) = 2\,\gamma^2 + k$, and by taking $\gamma = 0$, we obtain

$$E(Y_1Y_2) = 2\,\gamma^2 + \sigma_1^2\,\sigma_2^2 \tag{4.244}$$

which can also be deduced directly from (4.93).

Let us now consider the case of the system called *perfect clipping*, such that the input-output relationship is $y = Sg(x)$. This system is very interesting from a practical point of view because it transforms an arbitrary signal into a signal with only two values, which makes calculations much simpler. As the derivative of $h(x) = Sg(x)$ is $2\delta(x)$, we deduce from (4.242) that

$$\left(\frac{\partial}{\partial \gamma}\right)\{E(Y_1Y_2)\} = 4\, E[\delta(X_1)\, \delta(X_2)] \tag{4.245}$$

Assuming that in (4.116) $\sigma_x = \sigma_y = \sigma$ and $\gamma = c\sigma^2$, we obtain

$$\frac{\partial}{\partial c}\{E(Y_1Y_2)\} = \frac{2}{\pi} \cdot \frac{1}{\sqrt{1 - c^2}} \tag{4.246}$$

By integration and elimination of the constant of integration with the case where $c = 0$, we obtain

$$E(Y_1Y_2) = (2/\pi) \text{ Arc sin } c \tag{4.247}$$

This result can also be obtained by a direct calculation proposed in Problem 4.44. Note that the result depends only on c and not on γ. This is natural because a variation of σ realizes a scaling effect on the pair X,Y without changing their correlation coefficient, and, since only the signs of X and Y appear in the transformation, the clipping operation eliminates any scaling effect.

Finally, let us consider a case where the two transformations h_1 and h_2 are different, and suppose that $h_2(X_2) = X_2$. Using (4.241), we obtain

$$\left(\frac{\partial}{\partial \gamma}\right)\{E(Y_1\, Y_2)\} = E\{h'(X_1)\} \tag{4.248}$$

But as the last term does not depend on γ, we obtain

$$E\{h(X_1)\, X_2\} = E\{h'(X_1)\}\; E(X_1\, X_2) \tag{4.249}$$

This is an interesting relation used in some correlation calculations, as $E\{h(X_1)\, X_2\}$ can be easier to compute than $E(X_1X_2)$. This is especially the case when $h(x) = \text{Sg}(x)$, which gives

$$E\{\text{Sg}(X_1)\, X_2\} = (2/\pi)^{1/2}\, \sigma^{-1} E(X_1\, X_2) \tag{4.250}$$

4.8.6 Some Statistical Properties of Quadratic Forms

Many systems of practical interest are quadratic, and in the framework of nonlinear systems the quadratic ones are the simplest. Their study is almost reduced to that of quadratic forms of random vectors. Suppose that $\mathbf{X}$ is a real zeromean random vector of $\mathbb{R}^n$ and let Y be the RV defined by

$$Y = \mathbf{X}^T \mathbb{M} \mathbf{X} \tag{4.251}$$

where $\mathbb{M}$ is a real matrix. This relation can be written as

$$Y = (1/2)\, \mathbf{X}^T(\mathbb{M} + \mathbb{M}^T)\mathbf{X} \tag{4.252}$$

and as $\mathbb{M} + \mathbb{M}^T$ is symmetric, there is no loss of generality in assuming that $\mathbb{M}$ in (4.251) is symmetric. Using the same procedure as in (4.177), we can write (4.251) in the form

$$Y = \text{Tr}(\mathbb{M}\mathbf{X}\mathbf{X}^T) = \text{Tr}(\mathbf{X}\mathbf{X}^T\mathbb{M}) \tag{4.253}$$

As the trace is a linear operation, we immediately deduce that

$$E(Y) = \text{Tr}(\mathbb{\Gamma}\mathbb{M}) = \text{Tr}(\mathbb{M}\mathbb{\Gamma}) \tag{4.254}$$

where $\mathbb{\Gamma}$ is the covariance matrix of **X**.

On the other hand, the calculation of the variance of Y needs the knowledge of fourth order moments of **X**. A general calculation will be presented in another context in a later chapter and we will restrict our study here to the case where **X** is $N(0, \mathbb{\Gamma})$. With this assumption we can use (4.93) to calculate the fourth order moments of **X** in terms of the matrix elements of $\mathbb{\Gamma}$. Developing (4.251) as in (3.68), we obtain

$$E(Y^2) = \sum_{i,\,j,\,k,\,l} M_{ij} M_{kl} E(X_i X_j X_k X_l) \tag{4.255}$$

As the expected value given by (4.93) is a sum of three terms, we also have a sum of three terms to obtain $E(Y^2)$. The first is

$$T_1 = \sum_{i,\,j,\,k,\,l} M_{ij}\, M_{kl}\, \gamma_{ij}\, \gamma_{kl} \tag{4.256}$$

Grouping the terms i, j and k, l, we obtain

$$T_1 = \{E(Y)\}^2 \tag{4.257}$$

When we use the fact that both $\mathbb{M}$ and $\mathbb{\Gamma}$ are symmetric, the second term can easily be written as

$$T_2 = \sum_{i,\,j,\,k,\,l} M_{ij}\, \gamma_{jk}\, M_{kl}\, \gamma_{li} \tag{4.258}$$

which is the sum of the diagonal elements of the matrix $\mathbb{M}\mathbb{\Gamma}\mathbb{M}\mathbb{\Gamma} = (\mathbb{M}\mathbb{\Gamma})^2$. This gives

$$T_2 = \text{Tr}\{(\mathbb{M}\mathbb{\Gamma})^2\} = \text{Tr}\{(\mathbb{\Gamma}\mathbb{M})^2\} \tag{4.259}$$

The third term can be calculated in the same way so that

$$\text{Var}(Y) = E(Y) - \{E(Y)\}^2 = 2\text{Tr}\{(\mathbb{\Gamma}\mathbb{M})^2\} \tag{4.260}$$

We have thus obtained the mean value and the variance of Y by (4.254) and (4.260) respectively. Incidentally, this calculation shows that if $\mathbb{M}$ is symmetric and $\mathbb{\Gamma}$ symmetric and NND, $\text{Tr}\{(\mathbb{M}\mathbb{\Gamma})^2\}$ is non-negative, a result already obtained in Problem 3.6. Similar results can be obtained for complex normal vectors, as indicated in Problem 4.49.

Example 4.13 Calculation of the mean and variance of Y when X is $N(\mathbf{m}, \mathbb{\Gamma})$. This is the same problem as above, with the only difference that **X** is no longer zeromean. The term $E(\mathbf{X}\mathbf{X}^T)$ appearing in the mean value of Y given by (4.253) is deduced from (3.61), which gives

$$E(Y) = \text{Tr}(\mathbb{\Gamma}\mathbb{M}) + \text{Tr}(\mathbb{M}\mathbf{m}\mathbf{m}^T) = \text{Tr}(\mathbb{\Gamma}\mathbb{M}) + \mathbf{m}^T\mathbb{M}\mathbf{m} \tag{4.261}$$

This expression replaces (4.254) when the mean value **m** is non-null.

The calculation of the variance of Y is more complicated, as we now have to calculate a fourth order moment as in (4.93) for non-zeromean normal RVs. For this it is interesting to replace $\mathbf{X}$ by $\mathbf{Z} + \mathbf{m}$, where $\mathbf{Z}$ is now $N(\mathbf{0}, \mathbb{\Gamma})$. It follows that Y of (4.251) takes the form

$$Y = (\mathbf{Z} + \mathbf{m})^T \mathbb{M}(\mathbf{Z} + \mathbf{m}) = \mathbf{Z}^T\mathbb{M}\mathbf{Z} + 2\mathbf{m}^T\mathbb{M}\mathbf{Z} + \mathbf{m}^T\mathbb{M}\mathbf{m} \tag{4.262}$$

after using the point that $\mathbb{M}$ is symmetric. The calculation of Y^2 introduces various terms, but all those introducing odd powers in $\mathbf{Z}$ disappear when taking the expected value because of (4.90). Thus $E(Y^2)$ takes the form

$$E(Y^2) = T_4 + T_2 + T_0 \tag{4.263}$$

with

$$T_4 = E\{(\mathbf{Z}^T\mathbb{M}\mathbf{Z})^2\} \tag{4.264}$$

$$T_2 = 2\mathbf{m}^T\mathbb{M}\mathbf{m}E(\mathbf{Z}^T\mathbb{M}\mathbf{Z}) + 4\mathbf{m}^T\mathbb{M}E(\mathbf{Z}\mathbf{Z}^T)\,\mathbb{M}\mathbf{m} \tag{4.265}$$

$$T_0 = (\mathbf{m}^T\mathbb{M}\mathbf{m})^2 \tag{4.266}$$

The term T_4 is exactly the same as $E(Y^2)$ calculated previously when $\mathbf{Z}$ was zeromean and equal to $\mathbf{X}$. Its value is then

$$T_4 = \mathrm{Tr}^2(\mathbb{\Gamma}\mathbb{M}) + 2\mathrm{Tr}\{(\mathbb{\Gamma}\mathbb{M})^2\} \tag{4.267}$$

The term T_2 can be written as

$$T_2 = 2\mathbf{m}^T\mathbb{M}\mathbf{m}\ \mathrm{Tr}(\mathbb{\Gamma}\mathbb{M}) + 4\mathbf{m}^T\mathbb{M}\mathbb{\Gamma}\mathbb{M}\mathbf{m} \tag{4.268}$$

Combining all these terms and subtracting $E^2(Y)$ given by (4.261), we finally obtain

$$v_y = \mathrm{Var}(Y) = 2\mathrm{Tr}\{(\mathbb{\Gamma}\mathbb{M})^2\} + 4\mathbf{m}^T\mathbb{M}\mathbb{\Gamma}\mathbb{M}\mathbf{m} \tag{4.269}$$

which replaces (4.260) when $\mathbf{X}$ is no longer zeromean but still normal. When $\mathbf{m} = \mathbf{0}$ we obviously return to (4.260).

Apart from the mathematical calculations it is interesting to interpret the last term of (4.269). It is well known that in linear systems the output variance is invariant when we change the input mean value. This disappears for quadratic systems, and more generally for nonlinear systems, and is due to an interference effect between the components of the input signal. This phenomenon will be examined later in some examples.

4.8.7 Optimum Mean and Variance Estimators

Consider a real random vector $\mathbf{X}$ with mean value $m\mathbf{s}$ and covariance matrix $\mathbb{\Gamma}$. Suppose that we want to estimate m by a *linear procedure* giving a minimum variance. To do this we introduce the RV $Y = \mathbf{h}^T\mathbf{X}$ and calculate $\mathbf{h}$ such that $E(Y) = m$ and the variance of Y is minimum.

There are various physical situations corresponding to these statistical assumptions. The most common appears when $\mathbf{s} = \mathbf{r}$, vector with the components $r_i = 1$. This corresponds to a vector the components of which are RVs with the same mean value m. On the other hand, there are many cases, as on p. 68, where the vector $\mathbf{X}$ can be considered as a sum of a disturbing noise $\mathbf{N}$ and a vector $m\mathbf{s}$ containing information. Usually the noise $\mathbf{N}$ has a zeromean value and

$$\mathbf{X} = \mathbf{N} + m\mathbf{s} \tag{4.270}$$

is a vector of mean value $m\,\mathbf{s}$. In amplitude modulation methods the vector $\mathbf{s}$ is known and the information is contained in the parameter m which must be estimated.

As the mean value of Y is $m\ \mathbf{h}^T\mathbf{s}$, the condition that $E(Y) = m$ can be written as

$$\mathbf{h}^T\mathbf{s} = 1 \tag{4.271}$$

Furthermore, the variance of Y can be deduced from the results of Chapter 3, and is calculated in Problem 3.17. Its value is

$$v_y = \mathbf{h}^T \mathbb{\Gamma} \mathbf{h} \tag{4.272}$$

and we assume hereafter that $\mathbb{\Gamma}$ is positive definite. The mathematical problem is now to find the vector $\mathbf{h}$ satisfying (4.271) such that (4.272) is minimum, called a minimum under constraint.

There are many ways to solve this problem, but the simplest is to use the statement on p. 75 concerning the scalar product associated with any PD matrix. As in (3.131), we can introduce the scalar product between two vectors $\mathbf{a}$ and $\mathbf{b}$ by

$$(\mathbf{a},\ \mathbf{b})_{\mathbb{\Gamma}} = (\mathbf{a},\ \mathbf{b}) \triangleq \mathbf{a}^T \mathbb{\Gamma} \mathbf{b} \tag{4.273}$$

With this product (4.271) can be written as

$$(\mathbf{h},\ \mathbb{\Gamma}^{-1}\mathbf{s}) = 1 \tag{4.274}$$

As any scalar product satisfies the Schwarz inequality, we deduce that

$$(\mathbf{h},\ \mathbb{\Gamma}^{-1}\mathbf{s})^2 \le (\mathbf{h},\ \mathbf{h})\ (\mathbb{\Gamma}^{-1}\mathbf{s},\ \mathbb{\Gamma}^{-1}\mathbf{s}) \tag{4.275}$$

Using (4.273), this can be written as

$$(\mathbf{h}^T\mathbf{s})^2 \le (\mathbf{h}^T\mathbb{\Gamma}\mathbf{h})\ (\mathbf{s}^T\ \mathbb{\Gamma}^{-1}\mathbf{s}) \tag{4.276}$$

which shows with (4.271) and (4.272) that the variance v_y satisfies

$$v_y \ge (\mathbf{s}^T\ \mathbb{\Gamma}^{-1}\mathbf{s})^{-1} \tag{4.277}$$

The minimum is obtained when the vectors $\mathbf{h}$ and $\mathbb{\Gamma}^{-1}\mathbf{s}$ are proportional, which gives the optimum vector $\mathbf{h}_o$ by

$$\mathbf{h}_o = (\mathbf{s}^T\ \mathbb{\Gamma}^{-1}\mathbf{s})^{-1}\ \mathbb{\Gamma}^{-1}\mathbf{s} \tag{4.278}$$

Let us now apply this result to the particular case of the measurement of the mean value from a sequence of N correlated RVs with the same mean value. In this case the vector $\mathbf{s}$ must be replaced by the vector $\mathbf{r}$ introduced previously. The situation is particularly simple when the RVs become uncorrelated, in which case $\mathbb{\Gamma} = \sigma^2\ \mathbb{I}$. As a result we again find the estimator (4.170) introduced *a priori* and sometimes called the *sample mean.* We then know that in the class of linear estimators this is what gives the minimum variance.

Consider now the more difficult problem of *variance estimation*, for which results given in Section 4.8.6 will be used. We will first introduce the preliminary assumptions of the problem. We start from a zeromean random vector $\mathbf{X}$ with a covariance matrix $v\,\mathbb{R}$ where $\mathbb{R}$ is a normalized PD matrix. This means that all the diagonal elements r_{ii} satisfy $r_{ii} = 1$, and that all the components of the vector $\mathbf{X}$ have the same variance v. In other words, we know the matrix $\mathbb{R}$ and we want to estimate the variance v of the components X_i of $\mathbf{X}$. In many cases these components are assumed to be uncorrelated, in which case $\mathbb{R}$ is simply the identity matrix.

As the mean value can be estimated by a linear filter, it is reasonable to suppose that the variance can be estimated by a *quadratic* filter, and we introduce the RV Y defined by (4.251) where $\mathbb{M}$ is symmetric. The problem is then to find $\mathbb{M}$ such that the estimator is *unbiased*, which means that $E(Y) = v$, and that its variance is minimum. Using (4.254), the first condition gives

$$\mathrm{Tr}(\mathbb{M}\mathbb{R}) = 1 \tag{4.279}$$

For the calculation of the variance of Y we introduce, as in 4.8.6, the assumption that $\mathbf{X}$ is $N(\mathbf{0},\ v\,\mathbb{R})$, which gives (4.260). In this expression Γ must be replaced by $v\,\mathbb{R}$ where v is unknown. But whatever its value, the variance of Y is minimum if $\mathrm{Tr}[(\mathbb{M}\mathbb{R})^2]$ is minimum. In conclusion, we encounter the same kind of problem as we had for the estimation of the mean: find the minimum of $\mathrm{Tr}[(\mathbb{M}\mathbb{R})^2]$ with the constraint (4.279). The analogy of the problem leads us to seek the solution in a similar way.

First note that $\mathbb{M}$ and $\mathbb{R}$ are square symmetric matrices, for example with n rows and columns. The space the elements of which are symmetric matrices is a linear space. Consider a given PD matrix $\mathbb{R}$ and two arbitrary symmetric matrices $\mathbb{P}$ and $\mathbb{Q}$. The quantity

$$< \mathbb{P},\ \mathbb{Q} > \triangleq \mathrm{Tr}[\mathbb{P}\mathbb{R}\mathbb{Q}\mathbb{R}] \tag{4.280}$$

has all the properties required to be a scalar product of elements of the space considered. In fact $< \mathbb{P}, \mathbb{Q} >$ is linear both in terms of $\mathbb{P}$ and $\mathbb{Q}$, and $< \mathbb{P}, \mathbb{P} >$ is positive, which results from (4.260). Furthermore, $< \mathbb{P}, \mathbb{P} > = 0$ implies that $\mathbb{P} = \mathbb{0}$.

This last point is a little more difficult to demonstrate. If $< \mathbb{P},\ \mathbb{P} > = 0$, this means that the RV $Y = \mathbf{X}^T\mathbb{P}\mathbf{X}$ has a null variance when $\mathbf{X}$ is a normal random vector $N(\mathbf{0},\ \mathbb{R})$. As seen on p. 29, this means that the RV Y is equal with probability one to its mean value given by (4.254). Using (4.253), we can then write

$$\mathrm{Tr}[\mathbb{P}(\mathbf{X}\mathbf{X}^T - \mathbb{R})] = 0 \tag{4.281}$$

with probability one. As $\mathbf{X}$ is normal, there is no limitation on its possible values, and it follows that we must have whatever the value of the vector $\mathbf{x}$

$$\mathrm{Tr}[\mathbb{P}(\mathbf{x}\mathbf{x}^T - \mathbb{R})] = 0 \tag{4.282}$$

or

$$\sum_{i,\,j} P_{ij}\,(x_i\,x_j - R_{ij}) = 0 \tag{4.283}$$

whatever the values of x_i and x_j. By taking all the x_is equal to zero except one, we deduce immediately that $P_{ii} = 0$. Then, using two non-null x_i and x_j and remembering that $P_{ij} = P_{ji}$, we deduce $P_{ij} = 0$, which completes the proof that $\mathbb{P} = \mathbb{0}$.

Having shown that (4.280) is effectively a scalar product, we can return to our initial problem. Noting that (4.279) can be written $< \mathbb{M}, \mathbb{R}^{-1} > = 1$, this problem can be stated as follows: find a matrix $\mathbb{M}$ such that $< \mathbb{M}, \mathbb{R}^{-1} > = 1$ and $< \mathbb{M}, \mathbb{M} >$ is minimum. This is exactly the same problem as that deduced from (4.271) and (4.272). The solution is then the same. From the Schwarz inequality we deduce that

$$< \mathbb{M}, \mathbb{R}^{-1} >^2 \leq < \mathbb{M}, \mathbb{M} > < \mathbb{R}^{-1}, \mathbb{R}^{-1} > \tag{4.284}$$

Noting that $2\,v^2 < \mathbb{M}, \mathbb{M} >$ is the variance v_y of $\mathbf{X}^T \mathbb{M}\, \mathbf{X}$, we see that it satisfies

$$(v_y / v^2) \geq 2 < \mathbb{R}^{-1}, \mathbb{R}^{-1} >^{-1} = 2[\mathrm{Tr}(\mathbb{I})]^{-1} = 2n^{-1} \tag{4.285}$$

The minimum is obtained when $\mathbb{M}$ is proportional to $\mathbb{R}^{-1}$ and the condition (4.279) gives $\mathbb{M}_o = n^{-1}\,\mathbb{R}^{-1}$. As a consequence, the optimum quadratic estimator of the variance can be written

$$Y = (1/n)\,\mathbf{X}^T\,\mathbb{R}^{-1}\,\mathbf{X} \tag{4.286}$$

We verify immediately that it is unbiased and that its variance when $\mathbf{X}$ is $N(\mathbf{0}, \mathbb{R})$ is $2\,n^{-1}$. A particularly interesting case appears when the components of $\mathbf{X}$ are independent, in which case $\mathbb{R} = \mathbb{I}$, giving

$$Y = \frac{1}{n} \sum_{i=1}^{n} X_i^2 \tag{4.287}$$

This quantity is sometimes called the *sample variance* of Y. Note that it was obtained on the assumption that the vector $\mathbf{X}$ is zeromean, and can therefore be used only when this assumption is fulfilled. We shall now look at the probability distribution of Y.

Example 4.14 Chi-squared distribution (χ^2). Writing $Y = n^{-1}\,V$ we see that V is a sum of the square of n independent $N(0, 1)$ RVs. This RV is sometimes called χ^2 RV and its distribution a χ^2 *distribution with n degrees of freedom*. It is very easy to calculate the corresponding characteristic function. The first step consists of calculating the characteristic function of the square of an RV $N(0, 1)$. This is defined by

$$\phi_1(u) = E(e^{juX^2}) = (2\pi)^{-1/2} \int_{-\infty}^{+\infty} \exp\{-(1/2)\,x^2\,(1 - 2ju)\}\,dx \tag{4.288}$$

As the integral of $\exp(-x^2/2a^2)$ is equal to $(2\pi)^{1/2}\,a$ if $\mathrm{Re}(a) > 0$, we obtain

$$\phi_1(u) = (1 - 2ju)^{-1/2} \tag{4.289}$$

From this expression and from the independence of the RVs appearing in (4.287) we find that the characteristic function of V, or of a χ^2 distribution with n degrees of freedom, is

$$\phi_n(u) = (1 - 2ju)^{-n/2} \tag{4.290}$$

The PDF associated with this characteristic function is calculated in Problem 4.50, and the final result is

$$p_n(x) = 2^{-n/2} \left[\Gamma(n/2)\right]^{-1} x^{(n-2)/2} e^{-x/2} u(x) \tag{4.291}$$

where $u(x)$ is the unit step function and $\Gamma(x)$ the gamma function defined for $x > 0$ by

$$\Gamma(x) = \int_0^\infty u^{x-1} e^{-u} du \tag{4.292}$$

This function satisfies the recursion $\Gamma(x+1) = x\,\Gamma(x)$ and $\Gamma(1/2) = \pi^{1/2}$, which allows the calculation of $\Gamma(n/2)$.

We are now at the final step of our discussion. Suppose that we have a vector $\mathbf{X}$ which is $N(m\,\mathbf{s}, v\,\mathbb{R})$ and that we want to estimate from $\mathbf{X}$ the parameters m and v. To do this we make a linear estimation for m and a quadratic estimation for v. We wish to find those estimators which are unbiased and have a minimum variance.

In order to estimate m with a linear operation such as $\mathbf{h}^T \mathbf{X}$ with the minimum variance, we must take $\mathbf{h} = \mathbf{h}_o$ which is given by (4.278). In this expression $\mathbb{\Gamma}$ must be replaced by $v\,\mathbb{R}$, and the structure of (4.278) shows that v disappears in this equation. Thus the optimum estimate of m can be written as

$$\hat{m} = (\mathbf{s}^T \mathbb{R}^{-1}\mathbf{s})^{-1}\, \mathbf{s}^T \mathbb{R}^{-1}\mathbf{X} \tag{4.293}$$

Let us now consider the quadratic estimate of the variance given by (4.251). The problem is to find the matrix $\mathbb{M}$ such that $E(Y) = v$ with a minimum variance. As previously, we assume, without loss of generality, that $\mathbb{M}$ is symmetric and we add the new assumption that $\mathbb{M}$ is NND. In reality this assumption is not restrictive as it means that Y given by (4.251) is positive, which is reasonable as an estimate of the variance. The mean value of Y is given by (4.261), which gives

$$E(Y) = v\ \mathrm{Tr}(\mathbb{R}\mathbb{M}) + m^2\mathbf{s}^T\mathbb{M}\mathbf{s} \tag{4.294}$$

and the variance is given by (4.269), which becomes

$$v_y = 2v^2\ \mathrm{Tr}\{(\mathbb{M}\mathbb{R})^2\} + 4m^2\ v\ \mathbf{s}^T\mathbb{M}\mathbb{R}\mathbb{M}\mathbf{s} \tag{4.295}$$

The assumption of the unbiased estimator, whatever v and m, gives, from (4.294),

$$\mathrm{Tr}(\mathbb{R}\mathbb{M}) = 1 \qquad ; \qquad \mathbf{s}^T\mathbb{M}\mathbf{s} = 0 \tag{4.296}$$

The second equation implies that $\mathbb{M}\,\mathbf{s} = \mathbf{0}$, because $\mathbb{M}$ is NND. This can be obtained by factoring $\mathbb{M}$ as $\mathbb{A}\,\mathbb{A}^T$, as seen on p. 67, and deducing that $\mathbb{A}^T\mathbf{s} = \mathbf{0}$ or $\mathbb{M}\mathbf{s} = \mathbf{0}$. In consequence, the last term of (4.295) vanishes. Finally, the problem is reduced to finding the matrix $\mathbb{M}$ symmetric and NND such that $\mathrm{Tr}\{(\mathbb{M}\mathbb{R})^2\}$ is minimum with the double constraint (4.296). In order to solve this problem we will use a *geometric approach* making use of the scalar product (4.280). In this context the problem is to find the matrix $\mathbb{M}$ such that $<\mathbb{M},\mathbb{M}>$ is minimum with the constraints

$$(a)\ <\mathbb{M},\mathbb{A}> = 1 \qquad ; \qquad (b)\ <\mathbb{M},\mathbb{B}> = 0 \tag{4.297}$$

where

$$\mathbb{A} = \mathbb{R}^{-1} \quad ; \quad \mathbb{B} = \mathbb{R}^{-1}\mathbf{s}\,\mathbf{s}^T\mathbb{R}^{-1} \tag{4.298}$$

When $\mathbb{B} = \mathbb{O}$ the constraint (b) disappears and we return to the problem solved previously, giving (4.286). To take into account the new constraint (b) we will use the *projection theorem* introduced on p. 78. As the space of square symmetric matrices is linear and of finite dimension, the scalar product (4.280) allows us to write any matrix $\mathbb{Q}$ of this space in the form

$$\mathbb{Q} = \mathbb{Q}^{\perp} + k_Q\,\mathbb{B} \tag{4.299}$$

where $\mathbb{Q}^{\perp}$ and $\mathbb{B}$ are orthogonal. As a result of this orthogonality property,

$$k_Q = <\mathbb{Q}, \mathbb{B}> / <\mathbb{B}, \mathbb{B}> \tag{4.300}$$

Furthermore, the constraint (b) means that $\mathbb{M}$ and $\mathbb{B}$ are orthogonal, which means that $k_M = 0$ and that $\mathbb{M} = \mathbb{M}^{\perp}$. Applying the decomposition (4.299) to the matrix $\mathbb{A}$, we deduce that $<\mathbb{M}, \mathbb{A}> = <\mathbb{M}, \mathbb{A}^{\perp}>$ where $\mathbb{A}^{\perp}$ is deduced from (4.299) and (4.300) and is

$$\mathbb{A}^{\perp} = \mathbb{A} - \{<\mathbb{A}, \mathbb{B}> / <\mathbb{B}, \mathbb{B}>\}\,\mathbb{B} \tag{4.301}$$

At this step we can again apply the Schwarz inequality, as in (4.284), which, with (a), gives

$$1 = <\mathbb{M}, \mathbb{A}> = <\mathbb{M}, \mathbb{A}^{\perp}>^2 \leq <\mathbb{M}, \mathbb{M}><\mathbb{A}^{\perp}, \mathbb{A}^{\perp}> \tag{4.302}$$

and the optimum matrix $\mathbb{M}_o$, giving the minimum value of $<\mathbb{M}, \mathbb{M}>$, is proportional to $\mathbb{A}^{\perp}$. The constraint (a) finally gives

$$\mathbb{M}_o = <\mathbb{A}^{\perp}, \mathbb{A}^{\perp}>^{-1}\,\mathbb{A}^{\perp} \tag{4.303}$$

Noting that $<\mathbb{A}^{\perp}, \mathbb{A}^{\perp}> = <\mathbb{A}, \mathbb{A}^{\perp}>$, and using (4.280), (4.298) and (4.301) we find

$$<\mathbb{A}, \mathbb{A}^{\perp}> = n - 1 \tag{4.304}$$

It follows that $\mathbb{M}_o$ takes the form

$$\mathbb{M}_o = (n-1)^{-1}\,\{\mathbb{R}^{-1} - (\mathbf{s}^T\mathbb{R}^{-1}\mathbf{s})^{-1}\,\mathbb{R}^{-1}\mathbf{s}\,\mathbf{s}^T\mathbb{R}^{-1}\} \tag{4.305}$$

Inserting this value in (4.295), we find that the minimum variance is

$$(v_y/v^2) = 2(n-1)^{-1} \tag{4.306}$$

which must be compared to (4.285).

Finally, let us apply these results to the problem of the estimation of the variance of a sequence of n IID RVs of unknown mean value. As all the components of $\mathbf{s}$ now have the same value $s_i = 1$, we deduce that

$$\mathbf{X}^T\,\mathbb{M}_o\mathbf{X} = (n-1)^{-1}\left[\sum_{i=1}^{n} X_i^2 - n^{-1}\left(\sum_{i=1}^{n} X_i\right)^2\right] \tag{4.307}$$

But with the same assumptions (4.293) becomes

$$\hat{m} = n^{-1} \sum_{i=1}^{n} X_i \tag{4.308}$$

which allows us to write (4.307) in the form

$$\hat{v} = \mathbf{X}^T \mathbb{M}_o \mathbf{X} = (n-1)^{-1} \left\{ \sum_{i=1}^{n} (X_i - \hat{m})^2 \right\} \tag{4.309}$$

These expressions of $\hat{m}$ and $\hat{v}$ are called the *sample mean* and the *sample variance* respectively. It results from our discussion that they are unbiased and with minimum variance estimators of the unknown mean m and variance v.

It is interesting to discuss their statistical properties in more detail. Suppose that the random vector $\mathbf{X}$ is $N(m\mathbf{s}, v\,\mathbb{I})$. In this case $\hat{m}$ and $\hat{v}$ can be written as

$$\hat{m} = (\mathbf{s}^T\mathbf{s})^{-1}\mathbf{s}^T\mathbf{X} \tag{4.310}$$

and

$$\hat{v} = (n-1)^{-1} \{\mathbf{X}^T\mathbf{X} - (\mathbf{s}^T\mathbf{s})^{-1}(\mathbf{s}^T\mathbf{X})^2\} \tag{4.311}$$

These expressions have a simple geometric interpretation, which greatly simplifies the study of their statistical properties. Consider the space $\mathbb{R}^n$ with the standard scalar product $\mathbf{u}^T\mathbf{v}$ between two vectors $\mathbf{u}$ and $\mathbf{v}$. In this space the matrix $\mathbb{P}_s = (\mathbf{s}^T\mathbf{s})^{-1}\mathbf{s}\mathbf{s}^T$ corresponds to the orthogonal projection onto the subspace of vectors proportional to $\mathbf{s}$. In fact, we can verify that $\mathbb{P}_s^2 = \mathbb{P}_s$ and that $\mathbb{P}_s\mathbf{s} = \mathbf{s}$. On the other hand the matrix $\mathbb{P}^\perp = \mathbb{I} - \mathbb{P}_s$ corresponds to the projection onto the subspace orthogonal to $\mathbf{s}$. Thus $(\mathbb{P}^\perp)^2 = \mathbb{P}^\perp$ and $\mathbb{P}^\perp\mathbf{s} = \mathbf{0}$. Using these matrices, $\hat{v}$ defined by (4.311) can be written as

$$\hat{v} = (n-1)^{-1}\mathbf{X}^T\mathbb{P}^\perp\mathbf{X} \tag{4.312}$$

Now consider an arbitrary set of orthonormal vectors $\mathbf{u}_i$ of $\mathbb{R}^n$ with the only constraint that $\mathbf{u}_1 = (\mathbf{s}^T\mathbf{s})^{-1/2}\mathbf{s}$. As the covariance matrix of $\mathbf{X}$ is $v\,\mathbb{I}$, the vectors $\mathbf{u}_i$ are a set of eigenvectors. We can decompose the vector $\mathbf{X}$ in the form (3.97), where the components X_i are uncorrelated because of (3.98). The first component is given by (3.97) or $X_1 = (\mathbf{s}^T\mathbf{s})^{-1/2}\mathbf{s}^T\mathbf{X}$, which is proportional to $\hat{m}$ given by (4.310). Furthermore, $P^\perp\mathbf{X}$ can be written

$$P^\perp\mathbf{X} = \sum_{i=2}^{n} X_i\,\mathbf{u}_i \tag{4.313}$$

As the X_is are uncorrelated and thus independent because of the normal assumption, $\mathbb{P}^\perp\mathbf{X}$ and $\hat{m}$ are independent. Furthermore, from (3.98) we see that all the components X_i have the same variance because $\mathbb{R} = v\,\mathbb{I}$. Finally, as $\mathbb{P}^\perp$ is symmetrical, we have $\mathbb{P}^\perp = (\mathbb{P}^\perp)^T\mathbb{P}^\perp$, and as a result of the orthonormality of the vectors $\mathbf{u}_i$, (4.312) becomes

$$\hat{v} = (n-1)^{-1}\sum_{i=2}^{n} X_i^2 \tag{4.314}$$

The conclusion is that $\hat{m}$ and $\hat{v}$ are independent RVs, that $\hat{m}$ is normal $N(m, v)$, and that $\hat{v}$ is an RV with a χ^2 distribution with $(n-1)$ degrees of freedom. This completely specifies the statistical properties of the pair of RVs $\hat{m}$ and $\hat{v}$.

4.8.8 Random Sums and Related Problems

Sums of random variables were often discussed above. The randomness of the results stemmed from the fact that each term of the sum was random, but not because the number of terms appearing in the sum could also be random. In studying this aspect here, we will show that it is not a conceptual problem but can appear in real physical applications. Consider a switchboard receiving various telephone calls of different durations, and suppose that we want to know the total duration during which the switchboard is busy in one day. In a first approximation we can consider that the number of calls processed during one day is random and that the duration of each call is also random. In consequence the total duration is the sum of a random number of random variables, or a *random sum.* The same type of situation arises when analyzing the output of a photomultiplier mentioned on p. 4. Because of the random aspect of the secondary emission, we can associate to each initial photoelectron a random number of secondary electrons, and as the number of initial photoelectrons is random, the total number of secondary electrons appears as a random sum.

The tool most adapted to the discussion of the statistical properties of random sums is the characteristic function. We shall first establish the fundamental result before discussing more sophisticated situations. Consider a sum of independent and identically distributed (IID) RVs X_i and let Y be defined by

$$Y = \sum_{i=1}^{N} X_i \tag{4.315}$$

where N is an RV independent of the X_is. The conditional characteristic function of Y for a given value n of N is defined as in (4.65), and as in (4.71') it can be expressed in terms of the characteristic function $\phi(u)$ of the X_is, or

$$\phi_Y(u|n) = [\phi(u)]^n \tag{4.316}$$

Using the general property (4.32) we deduce that the marginal characteristic function of Y becomes

$$\phi_Y(u) = E\{[\phi(u)]^N\} \tag{4.317}$$

If we then use the generating function of N defined by $g(s) = E(s^N)$ and discussed in Problem 2.4, we obtain

$$\phi_Y(u) = g[\phi(u)] \tag{4.318}$$

It is also possible to use the characteristic function of N, which gives

$$\phi_Y(u) = \phi_N[-j\ \ln\phi(u)] = \phi_N[-j\psi(u)] \tag{4.319}$$

where $\psi(u)$ is the second characteristic function defined by (2.92). As the characteristic function completely specifies the statistical properties of an RV, we conclude that the random sum (4.315) is completely defined either by (4.318) or by (4.319).

Example 4.15 Some examples of random sums. Suppose that all the RVs X_i are equal to one with probability one. Their characteristic function $\phi(u)$ is then equal to $\exp(ju)$ and, as a result of (4.319), $\phi_Y(u)$ becomes $\phi_N(u)$. This is to be expected as Y of (4.315) is in this case simply equal to the number N of RVs which appear in the sum.

We have seen previously that many characteristic functions of discrete non-negative RVs depend on u through the term $\exp(ju) - 1$. This is particularly the case for the Poisson distribution, as seen in (2.91). In such situations $\phi_Y(u)$ defined by (4.319) is a function of u through the term $\exp[\psi(u)] - 1$. For example, the characteristic function of a sum of a Poisson distributed random number with the mean value μ of normal $N(m, \sigma^2)$ RVs can be written

$$\phi_Y(u) = \exp[\mu(e^{jmu - (1/2)\sigma^2 u^2} - 1)] \tag{4.320}$$

From this characteristic function it is possible to deduce all the moments of Y. By an expansion of $\phi_Y(u)$ limited to terms in u^2 we obtain

$$m_Y = \mu m \quad ; \quad \sigma_Y^2 = \mu\sigma^2 + \mu m^2 \tag{4.321}$$

This can be interpreted as follows. For the calculation of the mean value of Y we can use the expression valid for non-random sums by replacing the number of terms by the mean value $E(N) = \mu$. For the variance, the first term has the same interpretation but there is an additional contribution due to the randomness of the number of terms in the sum. This last term disappears if $m = 0$ or if the RVs X_i are zeromean.

The central limit theorem for random sums. This theorem was discussed in Section 4.7.2, and we wish to investigate the consequence of introducing a random number n in (4.160). As in that equation, we will assume that the RVs X_k are zeromean. On the other hand, we will write the number of terms in the form $N = k\,M$ where k is a given positive integer and M a discrete RV taking only non negative values. The normalization factor $n^{-1/2}$ in (4.160) is replaced by $\{E(N)\}^{-1/2}$ in order to divide by a non-random number. Noting as m the mean value $E(M)$, the normalization factor becomes $(k\,m)^{-1/2}$. When we take into account this factor, and note that N in (4.317) is now kM, we obtain the characteristic function of Y which is

$$\phi_Y(u) = \phi_M[-jk\,\psi(u/\sqrt{km})] \tag{4.322}$$

Suppose now that k goes to infinity. Using exactly the same procedure as that after (4.162), we find that $\phi_Y(u)$ converges to the function

$$\phi_1(u) = \phi_M(j\sigma^2 u^2/2m) \tag{4.323}$$

When compared with (4.319), we see that $\phi_1(u)$ is the characteristic function of a sum of a random number of terms that are IID $N(0, \sigma^2/m)$ RVs and the random number has the characteristic function $\phi_M(u)$. In other words, it can be said that the central limit theorem appears in the argument of the function $\phi_M(.)$. Finally, it is clear that if M is non-random, its characteristic function is $\exp(jum)$, and (4.323) becomes (4.164).

Let us now present an interpretation of this result. The factor k introduced to define the random number N is simply a scaling factor applied to a given random number M. The result is then not surprising: for each realization $M(\omega_0)$ of the RV $M(\omega)$ we can apply the classical central limit theorem that introduces the normal distribution, resulting in a random number of normal RVs.

Let us look at another form of the central limit theorem. Suppose that $k = 1$ in (4.322), which means that $M = N$, and that m goes to infinity. Taking some examples before the more general situation, suppose that $N = M$ is a Poisson RV with mean value m. Its characteristic function $\phi_N(u)$ takes the form (2.91) and, inserting this in (4.322), we obtain

$$\phi_Y(u) = \exp\{m[\exp\psi(u/\sqrt{m}) - 1]\} \tag{4.324}$$

Expanding ψ in the power of $u/\sqrt{m}$ we see that when m goes to infinity $\phi_Y(u)$ converges to the function

$$\phi_2(u) = \exp(-\sigma^2 u^2/2) \tag{4.325}$$

characterizing an $N(0, \sigma^2)$ distribution. In other words, we find again the classical central limit theorem in which σ^2 is the variance of the RVs X_i, elements of the sum. It is easy to verify that we obtain the same result when, instead of a Poisson distribution, we take a binomial distribution with the characteristic function (2.89). In this case the mean value is $m = n\,p$, and as p is smaller than one, the condition $m \to \infty$ corresponds to $n \to \infty$.

But this result is not at all general, as is shown in the following example. Suppose that the RV N has a discrete exponential distribution, or that

$$p_k = (m+1)^{-1}\,[m/(m+1)]^k \tag{4.326}$$

The corresponding characteristic function is

$$\phi_N(u) = [1 - m(e^{ju} - 1)]^{-1} \tag{4.327}$$

This distribution is sometimes called a boson distribution, as it appears in statistical mechanics when discussing the statistical properties of bosons. The mean value is of course m. Inserting this function in (4.322) with $k = 1$, we see that for $m \to \infty$ we obtain the limit

$$\phi_3(u) = [1 + \sigma^2 u^2/2]^{-1} \tag{4.328}$$

which is (2.76) with $a^2 = 2/\sigma^2$. The PDF is then (2.33) with $a = \sqrt{2}/\sigma$, or a two-sided exponential distribution. In other words, we obtain a central limit theorem introducing a limit distribution which is no longer normal. The question is analyzed in more detail in

Problem 4.55, where it is shown that a necessary condition for obtaining a normal limit distribution is that (σ/m) tends to zero when m goes to infinity, where m and σ^2 are the mean and variance of the number of terms N of the sum. This condition is fulfilled for a Poisson distribution that gives a variance $\sigma^2 = m$, but not for the boson distribution which introduces the variance $\sigma^2 = m(m+1)$.

4.8.9 Linear Forms and Normality

Suppose that we are given a set of N IID zeromean RVs U_i, and let us introduce the two RVs X and Y defined by

$$X = \sum_{i=1}^{N} a_i U_i \quad ; \quad Y = \sum_{i=1}^{N} b_i U_i \tag{4.329}$$

By applying the properties given on p. 108 we can compute the cumulants $c(X^k, X^l)$, and from the independence of the U_is we find that

$$c(X^k, X^l) = c_{k+l} \sum_{i=1}^{N} a_i^k b_i^l \tag{4.330}$$

The two RVs X and Y are uncorrelated if $c(X, Y) = 0$, which implies

$$\sum_{i=1}^{N} a_i b_i = 0 \tag{4.331}$$

assumed to be valid in what follows.

We shall suppose that the PDF of the U_is is symmetric, which implies that U_i and $-U_i$ have the same distribution. As a consequence the cumulants of U_i and of $-U_i$ are the same, and from (4.74), $c_{2k+1} = 0$. From (4.330), we deduce that

$$c(X^2, Y^{2k}) = c_{2(k+1)} \sum_{i=1}^{N} a_i^2 b_i^{2k} \tag{4.332}$$

If all the coefficients a_i and b_i are non-null, the last term is positive. This shows that if X and Y are *independent*, which means $c(X^2, Y^{2k}) = 0$, then $c_{2k} = 0$ for $k > 1$. This characterizes the normal distribution. In conclusion, if the RVs X and Y defined by (4.329) are not only uncorrelated but also independent, the U_is are normal RVs with the same distribution. This proof also shows that if X and Y are only independent until an order q, meaning that all the cumulants (4.330) are null for $k + l \leq q$, the U_is are also normal until the same order. This is a very interesting and progressive definition of normality.

We now suppose that all the a_is are positive and the b_is non-null, relaxing the assumption of the symmetry of the PDF of the U_is. We will still show that if the RVs are independent, the U_is are also normal. To do this, we apply (4.330) to calculate $c(X, Y^{2p})$ and $c(X^2, Y^{2p})$ and, as these expressions must be null, we deduce that $c_{2p+1} = c_{2p+2} = 0$. The only non-null cumulant is then c_2, because of (4.331), which characterizes the normal distribution. This can be especially applied to the sample mean (4.308). We saw in the discussion of (4.312) that $\hat{m}$ is uncorrelated with the components of $\mathbb{P}^{\perp}\mathbf{X}$. If we now introduce independence, it follows that the vector **X** is normal. There are other characterizations of the normal law which are discussed in problems.

PROBLEMS

4.1 In many alarm systems the alarm is triggered by comparing some signals to a threshold. Suppose that the observation used for the decision is a set of n independent RVs X_i with the DFs $F_i(x)$.

(a) Calculate the probability of alarm $\alpha_1(t)$ if the alarm is decided when all the X_is are greater than a given threshold t. Consider the particular case where $F_i(x) = F(x)$.

(b) Assuming that $F_i(x) = F(x)$, calculate the probability of alarm $\alpha_2(t)$ if the alarm is triggered when at least one of the X_is is greater than a threshold t.

(c) Compare $\alpha_1(t)$ and $\alpha_2(t)$.

4.2 Show that the conditional distribution function (4.12) satisfies all the requirements introduced in Section 3.2.1 for any distribution function.

4.3 **Order statistics.** Let $X_1, X_2, ..., X_k$ be k IID RVs characterized by the DF $F(x)$. To each outcome ω of these RVs we associate the RVs $Y_i(\omega)$ defined as follows: the $Y_i(\omega)$s are deduced from the $X_i(\omega)$s by the permutation such that $Y_1(\omega) \leq Y_2(\omega) \leq ... \leq Y_k(\omega)$. Show that the PDF of $Y_q(\omega)$, $1 \leq q \leq k$, is

$$p_q(y) = \frac{k!}{(q-1)!(k-q)!} F^{q-1}(y)[1-F(y)]^{k-q} p(y)$$

where $p(x)$ is the PDF deduced from $F(x)$.

4.4 Suppose that (X, Y) is a continuous RV with a PDF $p(x, y)$. Show that the conditional PDF $p(x, y|A)$ deduced from (4.13) is $[p(A)]^{-1}p(x, y)$ if $(x, y) \in D(A)$ and 0 otherwise.

4.5 Apply the result of the previous problem to the calculation of the PDF $p(x, y)$ deduced from (4.15) when A corresponds to the energy condition (4.14) and when X and Y are independent and $N(0, \sigma^2)$.

4.6 **Sum of two RVs**

(a) Using (4.204) calculate the PDF $p(x, y, z)$ of the triplet of RVs X, Y, and Z where (X, Y) is a pair of continuous RVs with the PDF $p(x, y)$ and $Z = X + Y$.

(b) Deduce the PDF $p_Z(z)$ of Z.

(c) Calculate the characteristic function $\phi_Z(w)$ of Z in terms of $\phi(u, v)$, characteristic function of X, Y.

(d) Write all the previous results when X and Y are independent and find again (4.70)

4.7 **Sum of two RVs** (continued)

The notations are the same as in the previous problem.

(a) Calculate the conditional PDFs $p(x, y|z)$ and $p_X(x|z)$.
(b) Write the conditional PDF $p(y, z|x)$ in terms of the conditional PDF $p(x|z)$.
(c) Calculate the moments $E(X|z)$, $E(Y|z)$, and $E(XY|z)$.
(d) Calculate $E(X|z) + E(Y|z)$ and interpret the result.
(e) Assuming that X and Y are two normal independent $N(0, \sigma^2)$ RVs, show that $p_X(x|z)$ is a normal PDF $N(z/2, \sigma^2/2)$.

4.8 Sum of two RVs (continued)

(a) Using still the same notations as in the previous two problems, calculate the conditional characteristic function $\phi(u, v|z)$ of X, Y in terms of the caracteristic function $\phi(u, v)$.
(b) Using this result, find again the result of point (e) of the previous problem.

4.9 Let X, Y be a pair of RVs with the PDF $p(x, y) = \alpha\,(a^2 - x^2 - y^2)$ for $0 < x^2 + y^2 < a^2$ and $p(x, y) = 0$ otherwise.

(a) Calculate the constant α and the marginal distribution of X.
(b) Calculate the covariance matrix of (X, Y).
(c) Give the value of the regression $r(y)$ defined by (4.34).

4.10 Suppose that X and Y are two discrete IID RVs with a Poisson distribution of mean value m. Show that $P[(X = k)|X + Y = p] = 2^{-p}\binom{p}{k}$.

4.11 (a) Calculate the DF of the RV $Z = \text{Max}(X_1, X_2, \ldots, X_n)$ where the X_is are the components of a random vector $\mathbf{X}$ with the DF $F(\mathbf{x})$. (Hint : interpret the event $Z \le z$ in terms of X_i).
(b) Calculate the PDF of Z in terms of $p(\mathbf{x})$, PDF of $\mathbf{X}$.
(c) Consider the case where the X_is are independent.

4.12 Consider n independent RVs X_i defined by their DF $F_i(x)$. Calculate the DF of the RV $S = \text{Min}[X_1, X_2, \ldots, X_n]$.

4.13 Suppose that the three RVs $X(\omega)$, $Y(\omega)$, and $Z(\omega)$ are the coordinates of a random point of $\mathbb{R}^3$.

(a) What is the meaning of the fact that $M(\omega)$ is uniformly distributed on the sphere $x^2 + y^2 + z^2 = a^2$?
(b) Assuming this distribution in all what follows, calculate the marginal PDF $p(x)$ of $X(\omega)$.
(c) Calculate the conditional PDF $p(x, y|z)$.
(d) Give the conditional PDF $p(z|x, y)$.
(e) Calculate the characteristic function $\phi(u, v, w)$ of X, Y, Z and express this function in terms of $r = (u^2 + v^2 + w^2)^{1/2}$.

4.14 Consider two random vectors $\mathbf{X}$ and $\mathbf{Y}$ and let us write $\mathbf{m}_X$ and $\mathbb{\Gamma}_X$ the mean value and the covariance matrix of $\mathbf{X}$. Using the regression $\mathbf{r}(\mathbf{y})$ defined by (4.38) we define the matrix $\mathbb{R}(\mathbf{y})$ by $E\{[\mathbf{X} - \mathbf{r}(\mathbf{y})]\,[\mathbf{X} - \mathbf{r}(\mathbf{y})]^T|\mathbf{y}\}$. By taking the expected value with respect to $\mathbf{Y}$ we define $\mathbf{m} = E[\mathbf{r}(\mathbf{Y})]$, $\mathbb{P} = E[\mathbb{R}(\mathbf{Y})]$ and $\mathbb{C} = E\{[\mathbf{r}(\mathbf{Y}) - \mathbf{m}]\,[\mathbf{r}(\mathbf{Y}) - \mathbf{m}]^T\}$.

(a) Express $\mathbf{m}$ in terms $\mathbf{m}_X$.
(b) Express $\mathbb{P}$ in terms of $\mathbb{\Gamma}_X$ and $\mathbb{C}$.

(c) Show and interpret the relation $\Gamma_X > \mathbb{C}$.

4.15 Suppose that the vectors $\mathbf{X}$ and $\mathbf{Y}$ are zeromean and jointly normal. The vector $\mathbf{X}$ is transformed into $\mathbf{h}(\mathbf{X})$ by a deterministic transformation. Show that $E[\mathbf{h}(\mathbf{X})\mathbf{Y}^T] = \mathbb{K}\ E[\mathbf{X}\mathbf{Y}^T]$ where $\mathbb{K}$ is a matrix independent of the distribution of $\mathbf{Y}$. (Hint : use the orthogonality principle (4.39) and the structure of the regression given in Section 4.6.4).

4.16 **Other characterization of independence.** It results from (4.39) that if $\mathbf{r}(\mathbf{Y}) = \mathbf{0}$, $\mathbf{X}$ and $\mathbf{g}(\mathbf{Y})$ are uncorrelated.
(a) Show the converse property.
(b) Show now that $\mathbf{X}$ and $\mathbf{Y}$ are independent if and only if $\mathbf{h}(\mathbf{X})$ and $\mathbf{g}(\mathbf{Y})$ are uncorrelated whatever the functions $\mathbf{h}(.)$ and $\mathbf{g}(.)$.

4.17 **From uncorrelation to independence.** We saw in Example 3.1 some cases of RVs which are uncorrelated but not independent. We will study other kinds of such a situation. For this purpose consider two zeromean RVs X and Y and let us call $r(y)$ the regression defined by (4.34) and c the correlation coefficient defined by (3.33).
(a) Show that if X and Y are independent, then $r(y) = 0$ and $c = 0$.
(b) Show that if $r(y) = 0$, then $c = 0$.
(c) Suppose that Y is zeromean and satisfies $E(Y^2) = 1$ and $E(Y^3) = 0$ and let X be equal to $Y^2 - 1$. Show that $c = 0$.
(d) Using the same assumptions as in (c), calculate $r(y)$ and show that if $y^2 \neq 1$, $r(\mathbf{y}) \neq 0$.
(e) Deduce from the previous results that (X, Y) is a pair of two uncorrelated RVs, but that X and $g(Y)$ are in general correlated. Calculate the PDF of (XY) and show that X and Y are not independent.
(f) Suppose now that $X = AU$ and $Y = AV$ where $A > 0$ and A, U, and V are three independent RVs such that $E(U) = E(V) = 0$. Show that X and Y are uncorrelated.
(g) Using the same assumptions as in (f) calculate the PDF of (X, Y) in terms of the PDFs of A, U, and V and show that X and Y are not independent. Deduce also that $r(y) = 0$.
(h) Deduce from all these results a hierarchy between uncorrelation, null regression and independence.

4.18 Let ϕ an RV uniformly distributed in $[0, 2\pi)$ and X, Y the RVs (3.22).
(a) Show that the regression $r(y)$ defined by (4.34) is null.
(b) Calculate the PDF of the pair (X, Y) and deduce that X and Y are not independent.

4.19 Suppose that X, Y, and Z are three $N(0, \sigma^2)$ independent RVs and consider the RVs S and T defined by $S = X/Z$ and $T = Y/Z$. By introducing the RV $R = Z$, calculate the PDF of the triplet (S, T, R). Deduce from this result the PDF $p(s, t)$ of the pair (S, T).

4.20 Let A and Φ be a two-dimensional RV with the condition $A \geq 0$ and $0 \leq \Phi < 2\pi$. Suppose that the PDF of this RV is $q(a, \phi)$. We want to study the two-dimensional RV X, Y deduced from A, Φ by the transformation $X = A\cos\Phi$, $Y = A\sin\Phi$.

(a) Calculate the PDF $p(x, y)$ of (X, Y) in terms of $q(a, \phi)$.
(b) Give the expression of $q(a, \phi)$ when A and Φ are independent and Φ is uniformly distributed in $[0, 2\pi)$.
(c) Introducing the RV $U = A^2$, express the PDF $p_U(u)$ in terms of the PDF $p_A(a)$. Deduce the expression of $p(x, y)$ in terms $p_U(u)$.
(d) Calculate the marginal PDF $p_X(x)$ when $p_U(u)$ is a one-sided exponential PDF.
(e) Calculate $p_X(x)$ when $U = 1$.
(f) Make a direct calculation of $p_X(x)$ when $X = \cos\Phi$ and Φ is uniformly distributed in $[0, 2\pi)$.

4.21 (a) Starting from a two-dimensional RV (X, Y) defined by a PDF $p(x, y)$, calculate the DF $F(z)$ of the RV $Z = X^2 + Y^2$ in terms of $p(x, y)$.
(b) Find explicitly $F(z)$ when X and Y are two independent $N(0, \sigma^2)$ RVs.
(c) Using the same assumption as in (b), find the PDF $p(z)$ of Z and the moments $m_k = E(Z^k)$.
(d) Using the formula giving the higher-order moments of a normal RV, calculate directly m_1, m_2, m_3, and m_4.

4.22 In many communication systems a signal X is modulated and transformed into an other Y. In all this problem we study the linear modulation characterized by $Y = aX + b$, where X is a continuous RV with a PDF $p_X(x)$. Similarly the PDF of Y is called $p_Y(y)$.
(a) Assuming that $a = 1$, calculate $p_Y(y)$ in terms of $p_X(x)$.
(b) Make the same calculation when $b = 0$ and $a > 0$.
(c) Make the same calculation when $b = 0$ and $a < 0$.
(d) Give $p_Y(y)$ in terms of $p_X(x)$ with only the condition $b = 0$.
(e) We assume now that the modulation is random which means that a becomes an RV A. We suppose that A is continuous-valued and call $p(a, x)$ the PDF of the pair (A, X). Calculate the conditional PDF $p_Y(y|a)$ and deduce the PDF $p_Y(y)$ of Y.
(f) Discuss the case when A and X are independent.
(g) Assuming still that A and X are independent, calculate the characteristic function $\phi_Y(u)$ in terms of $p(a)$ and of $\phi_X(u)$, characteristic function of X.
(h) Complete the calculation of $\phi_Y(u)$ when X is $N(0, \sigma^2)$ and when $S = A^2$ is an exponential RV with the PDF $p(s) = \alpha \exp(-\alpha s)$. Deduce the corresponding PDF of the modulated signal Y.

4.23 Calculate the conditional characteristic function $\phi_X(u|y)$ given by (4.66) in terms of $\phi(u, v)$ defined by (4.62) assuming that the RVs X and Y are continuous-valued.

4.24 We want to extend the result obtained in the previous problem to the vector case and deduce some interesting results concerning the conditional mean. To achieve this consider two random vectors $\mathbf{X}$ and $\mathbf{Y}$ and let us call $\phi(\mathbf{u}, \mathbf{v})$ their characteristic function.
(a) Express the conditional characteristic function $\phi_X(\mathbf{u}|\mathbf{y})$ in terms of $\phi(\mathbf{u}, \mathbf{v})$. Show that the result is the ratio of two functions $N(\mathbf{u}, \mathbf{y})$ and $D(\mathbf{y}) = N(\mathbf{0}, \mathbf{y})$.
(b) Deduce the regression $\mathbf{r}(\mathbf{y})$ defined by (4.38) from the previous result.

(c) Assuming that $\mathbf{X}$ and $\mathbf{Y}$ are jointly normal and zeromean, find again (4.109).

(d) Find again (4.190) if $\phi(\mathbf{u}, \mathbf{v}) = f\{Q(\mathbf{u}, \mathbf{v})\}$ where $Q(\mathbf{u}, \mathbf{v})$ is a quadratic form in $\mathbf{u}$ and $\mathbf{v}$ as in (4.131).

4.25 Calculate the characteristic function of the RV $Y = X_1 + X_2 + ... + X_n$, where the X_is are n IID RVs with an one-sided exponential distribution.

4.26 Distributions stable by addition. Suppose that $Y = X_1 + X_2 + ... + X_N$ where the X_is are N independent RVs with characteristic functions $\phi_i(\mathbf{u})$.

(a) Suppose that the X_is are $N(0, \sigma_i^2)$. Show that Y remains normal and calculate its variance.

(b) Suppose that the X_is are discrete valued with Poisson distribution of mean values m_i. Show that Y is still Poisson distributed and calculate its mean value.

(c) Suppose that the X_is have binomial distributions defined by (2.38) with the same p but not necessarily the same n_i. Show that Y has still a binomial distribution and give the parameters n and p of this distribution.

4.27 (a) Show that the function $\phi(u, v) = \exp[-(1/2)(u^2 + v^2)]\ \mathrm{ch}(c\,u\,v)$, where $|c| < 1$, is a characteristic function.

(b) Calculate the corresponding PDF and explain the structure of the RV (X, Y) described by this PDF.

(c) Calculate the marginal PDF of X and Y and show that we met a situation similar to that presented in Example 4.7.

4.28 Reproducing property of the normal kernel. Let us write $g(\mathbf{x}; \mathbf{m}, \Gamma)$ the PDF of an RV $N(\mathbf{m}, \Gamma)$ given by (4.82). We want to establish the reproducing property expressed as

$$\int g(\mathbf{x}; \mathbf{m}_x, \Gamma_x)\, g(\mathbf{x}; \mathbf{m}_y, \Gamma_y)\, d\mathbf{x} = g(\mathbf{m}_x; \mathbf{m}_y, \Gamma_x + \Gamma_y)$$

(a) Show that $g(\mathbf{m}_x; \mathbf{m}_y, \Gamma_x + \Gamma_y) = g(\mathbf{m}_y; \mathbf{m}_x, \Gamma_x + \Gamma_y)$.

(b) Show that $g(2\mathbf{m}_y - \mathbf{x}; \mathbf{m}_y, \Gamma_y) = g(\mathbf{x}; \mathbf{m}_y, \Gamma_y)$.

(c) Consider two independent random vectors $\mathbf{X}$ and $\mathbf{Y}$ with the PDFs $g(\mathbf{x}; \mathbf{m}_x, \Gamma_y)$ and $g(\mathbf{y}; \mathbf{m}_y, \Gamma_y)$ and let $\mathbf{Z} = \mathbf{X} + \mathbf{Y}$. Give the mean value and the covariance matrix of $\mathbf{Z}$.

(d) Calculate the PDF of $\mathbf{Z}$ in terms of the PDFs of $\mathbf{X}$ and $\mathbf{Y}$.

(e) Using all these results, deduce the reproducing property.

4.29 Let X and Y be two normal zeromean RVs characterized by the PDF (4.116) and consider the RV Z defined by $Z = \cos(X - Y)$.

(a) Calculate $E(Z)$. Explain the result obtained when $\sigma_x = \sigma_y$ and $c = 1$.

(b) Calculate $E(Z^2)$ and the variance σ_z^2. Discuss also the case when $\sigma_x = \sigma_y$ and $c = 1$.

(c) Give the principle of the calculation of $E(Z^n)$.

4.30 Let $\mathbf{X}$ be a zeromean random vector with the covariance matrix Γ. Show that if for any vector $\mathbf{a}$ the scalar RV $Y = \mathbf{a}^T\mathbf{X}$ is normal, then $\mathbf{X}$ is $N(\mathbf{0}, \Gamma)$.

4.31 Let $\mathbf{X}$ be a zeromean two-dimensional real normal vector $N(\mathbf{0}, \mathbb{\Gamma})$ and $\mathbf{Y}$ the vector deduced from $\mathbf{X}$ by the linear transformation $\mathbf{Y} = \mathbb{A}\mathbf{X}$ where $\mathbb{A}$ is the matrix

$$\mathbb{A} = \begin{bmatrix} \cos\theta & \sin\theta \\ -\sin\theta & \cos\theta \end{bmatrix}$$

(**a**) Give a geometric interpretation of the action of the matrix $\mathbb{A}$ on the vector $\mathbf{X}$.
(**b**) Assuming that $\mathbb{\Gamma}$ is written as in (4.112), give the expression of the covariance matrix of $\mathbf{Y}$.
(**c**) Find the values of θ such that the components of $\mathbf{Y}$ are independent.
(**d**) Discuss the particular cases where $\sigma_x = \sigma_y$ or $c = 0$ or $|c| = 1$.

4.32 Calculate the moments $E(X_iX_j)$, $E(X_iX_jX_k)$, and $E(X_iX_jX_kX_l)$ for a normal vector $\mathbf{X}$ with the PDF $N(\mathbf{m}, \mathbb{\Gamma})$. Find again the known results when $\mathbf{m} = \mathbf{0}$.

4.33 **Other characterization of normal circular vectors.** Let $\mathbf{Z}$ be a complex vector and $\mathbf{S}$ the vector defined by $\mathbf{S} = \exp(j\alpha)\mathbf{Z}$.
(**a**) Calculate the matrices $\mathbb{\Gamma}_s$ and $\mathbb{C}_s$ in terms of $\mathbb{\Gamma}_z$ and $\mathbb{C}_z$, where the matrices $\mathbb{\Gamma}$ and $\mathbb{C}$ are defined by (4.124) and (4.125) respectively.
(**b**) Show that if $\mathbf{Z}$ is normal circular, $\mathbf{S}$ has the same distribution as $\mathbf{Z}$. Give an interpretation of this result.
(**c**) Conversely show that if $\mathbf{Z}$ and $\mathbf{S}$ are normal and have the same PDF whatever the value of α, this distribution is normal circular.

4.34 Using the notation introduced on p. 119 show that if (4.132) is valid whatever the values of $\mathbf{u}$ and $\mathbf{v}$, then we must have $\alpha = 1/2$ and the relations (4.129) and (4.130) must be valid.

4.35 Show that if a complex random vector $\mathbf{Z}$ is $N_c(\mathbf{m}, \mathbb{\Gamma})$, then the complex vector $\mathbb{A}\mathbf{Z}$ is normal circular and specify its mean value and covariance matrix.

4.36 Show that if $\mathbf{Z}$ is $N_c(\mathbf{0}, \mathbb{\Gamma})$ and if $\mathbb{\Gamma}$ is a real matrix, the real and imaginary parts of $\mathbf{Z}$ are independent.

4.37 (**a**) Show that if a real vector $\mathbf{X}$ is $N(\mathbf{0}, \mathbb{\Gamma})$ and if $\mathbb{\Gamma}$ is diagonal, the components X_i of $\mathbf{X}$ are independent RVs.
(**b**) Consider now a complex vector $\mathbf{Z} = \mathbf{X} + j\mathbf{Y}$ and the real vector $\mathbf{R}$ with components X_1, Y_1, X_2, Y_2, ..., X_n, Y_n. Show that if $\mathbf{Z}$ is $N_c(\mathbf{0}, \mathbb{\Gamma})$ and if $\mathbb{\Gamma}$ is diagonal, the components of $\mathbf{R}$ are independent RVs.
(**c**) The property (**b**) is not necessarily true if $\mathbf{Z}$ is no longer normal circular. To show this, consider two real vectors of the same dimension $\mathbf{X} = \mathbf{S} + \mathbf{M}$ and $\mathbf{Y} = \mathbf{S} + \mathbf{N}$, where $\mathbf{S}$, $\mathbf{M}$, $\mathbf{N}$ are zeromean, independent, and normal, and suppose that $\mathbb{\Gamma}_M$, $\mathbb{\Gamma}_N$ and $\mathbb{\Gamma}_S$ are diagonal. Show that the covariance matrix of the vector $\mathbf{Z} = \mathbf{X} + j\mathbf{Y}$ is diagonal, that $\mathbf{Z}$ is no longer circular and that the property (**b**) does not hold.

4.38 There is another way to arrive at (4.151) by comparing the eigenelements of $\mathbb{\Gamma}$ and $\mathbb{\Gamma}_z$.

(a) Show that if $\mathbf{w} = \mathbf{u} + j\mathbf{v}$ is an eigenvector of Γ_z with the eigenvalue λ, then the vector $[\mathbf{u}^T, \mathbf{v}^T]^T$ is an eigenvector of Γ with the eigenvalue $\lambda/2$.
(b) Deduce that Γ cannot have more than n distinct eigenvalues.
(c) As the determinant is the product of the eigenvalues, deduce (4.151).
(d) Show that $[\mathbf{v}^T, -\mathbf{u}^T]^T$ is an eigenvector of Γ with the same eigenvalue as in (a).
(e) Deduce and compare the spectral decompositions of Γ and Γ_z.

4.39 We want to extend (4.92) to normal complex circular RVs.
(a) Using the general expression of the moments, calculate $E(|Z|^{2k})$ and express the result in terms of σ_z^2 or σ^2 appearing in (4.137) or (4.138).
(b) Deduce the characteristic function $\phi(u)$ of $|Z|^2$ from these moments.
(c) Deduce the PDF of $|Z|^2$ from this characteristic function.

4.40 (a) Let **A**, **B**, **C**, **D** be four zeromean real jointly normal random vectors. Show that $E(\mathbf{A}^T\,\mathbf{B}\;\mathbf{C}^T\,\mathbf{D}) = \mathrm{Tr}(\Gamma_{AB})\,\mathrm{Tr}(\Gamma_{CD}) + \mathrm{Tr}(\Gamma_{AC}\Gamma_{BD}) + \mathrm{Tr}(\Gamma_{AD}\,\Gamma_{BC})$.
(b) Assuming that the vectors are complex normal circular, calculate $E(\mathbf{A}^H\,\mathbf{B}\;\mathbf{C}^H\,\mathbf{D})$.

4.41 Some relations between modes of convergence
(a) Show that if X_k converges to X in quadratic mean, it converges also in probability.
(b) Show that if X_k converges to X with probability one, it converges also in probability.

4.42 Entropy of two normal random vectors. The entropy of a zeromean normal random vector **X** is given by (4.178). Consider now two zeromean jointly normal random vectors **X** and **Y** with a PDF as in (4.102).
(a) Calculate the conditional entropy $H(\mathbf{x}|\mathbf{y})$ defined as $E[\ln\{p(\mathbf{X}|\mathbf{Y})\}]$.
(b) Compare this entropy with that of the random vector $\mathbf{X} - \mathbf{r}(\mathbf{y})$, where $\mathbf{r}(\mathbf{y})$ is the regression defined by (4.38).
(c) The mutual information between **X** and **Y** is defined by $I(\mathbf{x}, \mathbf{y}) = H(\mathbf{x}) - H(\mathbf{x}|\mathbf{y})$, where $H(\mathbf{x})$ is the entropy defined by (4.175). Calculate $I(\mathbf{x}, \mathbf{y})$ in terms of the matrices Γs appearing in (4.104).
(d) Show that $H(\mathbf{x}, \mathbf{y}) = H(\mathbf{y}) + H(\mathbf{x}|\mathbf{y})$.

4.43 Moments of the Rayleigh distribution. This PDF is defined by (4.217).
(a) Using the results concerning the moments of the distribution $N(0, \sigma^2)$, calculate the odd moments m_{2k-1} of the Rayleigh distribution.
(b) In order to calculate the even moments m_{2k} find first a recursion between m_{2k} and m_{2k-2} and deduce the values m_{2k}.
(c) Give the value of the variance.

4.44 Some properties of perfect clipping
(a) The PDF of a two dimensional normal RV (X, Y) is given by (4.115) and (4.116). Assuming that $m_x = m_y = 0$ and that $\sigma_x^2 = \sigma_y^2 = \sigma^2$, show that the probability P defined by $P = P[(X > 0).(Y > 0)]$ is equal to $1/4 + (1/2\pi)$ Arc sin$c$.
(b) Deduce the value of $E[\mathrm{Sg}(X)\;\mathrm{Sg}(Y)]$ and find again (4.237).
(c) Introducing the two RVs S and T defined by $S = \mathrm{Sg}(X)$ and $T = \mathrm{Sg}(Y)$, calculate the moment $E(S^k\,T^l)$.

(d) Give the PDF $p(s, t)$ and the characteristic function $\phi(\alpha, \beta)$ of the RV (ST).

4.45 Consider a pair of normal zeromean RVs (X, Y) characterized by the covariance matrix (4.112). A linear rectifier is a system that transforms the pair (X, Y) into another (S, T) such that $S = |X|$ and $T = |Y|$.

(a) Calculate the moments m_k of S.

(b) Show that $E(ST) = 2\sigma_x \sigma_y \pi^{-1}(\cos\alpha + \alpha\sin\alpha)$ where $\sin\alpha = c$. Give an interpretation of the result obtained for $c = 0$ or $|c| = 1$.

(c) Deduce the covariance matrix of the pair of RVs (S, T).

4.46 Intensity of a random field. A random field can often be represented by a complex normal circular $N_c(\mathbf{0}, \mathbb{\Gamma})$ RV $\mathbf{Z}$. The intensity of the field corresponding to the component Z_k is defined by $I_k = |Z_k|^2$.

(a) Calculate the PDF $p_k(i)$ of I_k and its moments.

(b) Calculate the covariance matrix of the vector $\mathbf{I}$ with components I_k in terms of the covariance matrix $\mathbb{\Gamma}$.

(c) Calculate the higher order moments of I_k.

4.47 Correlation between some linear and quadratic forms. Let $\mathbf{X}$ be a random vector $N(0, \sigma^2\mathbb{I})$ and consider the linear form $y = \mathbf{a}^T\mathbf{X}$ and the quadratic form $z = \mathbf{X}^T\mathbb{M}\mathbf{X}$ where $\mathbb{M}$ is symmetric. We want to study the consequences of the condition $\mathbb{M}\mathbf{a} = \mathbf{0}$.

(a) Show that the eigenvectors $\mathbf{u}_i$ of $\mathbb{M}$ are orthogonal and write the spectral decomposition of $\mathbb{M}$.

(b) Show that $\mathbf{a}$ is eigenvector of $\mathbb{M}$ and give the corresponding eigenvalue.

(c) Write the random vector $\mathbf{X}$ as in (3.97) and specify the statistical properties of the components X_i.

(d) Show that z is a sum of independent RVs.

(e) Deduce that y an z are also independent.

(f) Deduce that $\hat{m}$ and $\hat{v}$ defined by (4.310) and (4.311) are independent as shown after (4.313).

4.48 Linear and quadratic forms (continued). We want to extend the results of the previous problem to the case of a vector $\mathbf{X}$ with correlated components. Suppose that $\mathbf{X}$ is $N(\mathbf{0}, \mathbb{\Gamma})$ and consider the same linear and quadratic forms $y = \mathbf{a}^T\mathbf{X}$ and $z = \mathbf{X}^T\mathbb{M}\mathbf{X}$.

(a) Introducing a factorization of $\mathbb{\Gamma}$ as in (3.99) and the vector $\mathbf{Y} = \mathbb{A}^{-1}\mathbf{X}$, show that $\mathbf{Y}$ is $N(\mathbf{0}, \mathbb{I})$.

(b) Write y and z in terms of $\mathbf{Y}$ and deduce that if $\mathbb{M}\mathbb{\Gamma}\mathbf{a} = \mathbf{0}$ the RVs y and z are independent.

(d) Deduce that $\hat{m}$ defined by (4.293) and $\hat{v}$ defined by (4.251) where $\mathbb{M}$ is the matrix defined by (4.305) are independent.

4.49 Assuming that $\mathbf{X}$ is a complex random vector with a distribution $N_c(\mathbf{0}, \mathbb{\Gamma})$ defined by (4.152), calculate the expectation and the variance of $Y = \mathbf{Z}^H\mathbb{M}\mathbf{Z}$, where $\mathbb{M}$ is Hermitian.

4.50 Probability density function of a χ^2 distribution

(a) Applying the results of Problem 2.10, calculate the PDF $p_1(x)$ corresponding to the characteristic function (4.289).

(b) Using (2.73) calculate the PDF $p_2(x)$ corresponding to the characteristic function $\phi_2(u)$ defined by (4.290).

(c) Find the relation between the first derivatives of $\phi_n(u)$ and $\phi_{n+2}(u)$. Deduce by Fourier transformation the relation between $p_{n+2}(x)$ and $p_n(x)$.

(d) From all these results deduce (4.291).

4.51 Show that the characteristic function of Y defined by (4.251) where $\mathbf{X}$ is $N(0, \mathbb{\Gamma})$ can be expressed as $\phi_Y(u) = [\det(\mathbb{I} - 2\,ju\mathbb{\Gamma}\mathbb{M})]^{-1/2}$. Express this function when $\mathbb{\Gamma} = \mathbb{M} = \mathbb{I}$ and compare with (4.290).

4.52 Let X_1, X_2, X_3 be three IID RVs with the PDF $N(0, 1)$. Consider the three other RVs Y_1, Y_2, Y_3 defined by $Y_1 = X_1 - X_2$, $Y_2 = X_2 - X_3$, and $Y_3 = X_3 - X_1$.

(a) Calculate the covariance matrix of the random vector $\mathbf{Y} = [Y_1, Y_2, Y_3]^T$.

(b) Determine the rank of this covariance matrix and explain the consequence of the result on the Y_is.

(c) Calculate the characteristic function of the RV S defined by $S = Y_1^2 + Y_2^2 + Y_3^2$.

(d) Show that S and $T = X_1 + X_2 + X_3$ are independent and deduce the characteristic function $\phi(u, v)$ of the pair (S, T).

4.53 On the Cramér-Rao bound for the mean estimator. Suppose that $\mathbf{X}$ is a random vector with the distribution $N(m\mathbf{s}, \mathbb{\Gamma})$. The optimum estimator of m calculated in Section 4.8.7 is $\mathbf{h}_o^T\mathbf{X}$ where $\mathbf{h}_o$ is given by (4.278) and its variance is given by (4.277). It is possible to show that the variance of any unbiased estimator is greater or equal to $1/I_m$ (Cramér-Rao bound), where I_m is the Fisher information. This quantity is deduced from the PDF function $p(\mathbf{x}; m)$ of $\mathbf{X}$ by $I_m = E\{[\partial \ln p(\mathbf{x}; m)/\partial m]^2\}$. Calculate this information and deduce that the optimum estimator is efficient, which means that its variance reaches the Cramér-Rao bound.

4.54 On the Cramér-Rao bound for the variance estimator. Assuming that $\mathbf{X}$ is $N(0, v\mathbb{R})$, show that the estimator (4.286) is efficient by using the same arguments as in the previous problem (Hint : use the results of Section 4.8.6 systematically).

4.55 Suppose that in (4.322) $k = 1$ and $M = N$.

(a) Expand the second characteristic function $\Psi_Y(u)$ in u^k for $k \le 4$ and express the result in terms of the cumulants c_k of X, of the mean value m and variance σ^2 of N.

(b) Deduce that a necessary condition to obtain a normal distribution for Y when $m \to \infty$ is that $(\sigma/m) \to 0$ for $m \to \infty$.

(c) Show that this condition is satisfied if N is a Poisson or a binomial RV, but not if N has the boson distribution (4.326).

4.56 Using an expansion of (4.319) limited to the term u^2, calculate the mean value m_Y and the variance σ_Y^2 in terms of m_N, m_X, σ_N^2, σ_X^2, mean values and variances of the RVs N and X. Give an interpretation of the result and explain the difference with the interpretation of (4.321).

Chapter 5

Statistical Description of Random Signals

The intuitive concept of a random signal was introduced in Chapter 1, and a review of random variables was given in Chapters 2, 3, and 4. We will now combine the results of those chapters to present a general description of random signals. The purpose of this chapter is to introduce methods allowing the description of signals such as those in Figures 1.2, 1.3, and 1.4. These signals seem quite different, but can be described by similar methods. It was mentioned that these signals are particular trajectories of random signals, which will now be described more precisely.

5.1 THE FAMILY OF FINITE-DIMENSIONAL DISTRIBUTIONS

Let $(\mathscr{S}, \mathscr{F}, P)$ be an arbitrary probability space and a parameter set T. The elements of T are usually time instants. A real random signal $X(t; s)$ is a real valued function such that for every $t \in T$, $X(t; s)$ is an RV defined on $(\mathscr{S}, \mathscr{F}, P)$. Note that ω used in the previous chapters is replaced by s, in order to avoid any confusion with angular frequency. For a given value s_0 of s, $X(t; s_0)$ is a deterministic function which can, for example, be represented by a curve. This is called a sample function, or a trajectory of the random signal; examples of trajectories were presented in Chapter 1. Consequently, a random signal is a family of trajectories which depend on a parameter s referring to the randomness. In most applications the parameter set T is $(-\infty, +\infty)$. If this set is continuous, we have a continuous-time (CT) random signal, and if it has only a countable number of time instants t_i, we have a discrete-time (DT) random signal.

It is important to note that a time point process, or a random distribution of points on time, can also be described by a function such as $X(t; s)$. For this it is sufficient to assume that it is a step-wise function with constant increments at each random point of the point process. A realization of the point process, or a set of random points $\{t_i\}$, is then equivalent to a trajectory of $X(t; s)$. As there is often confusion on this, we recall that a point process is a CT signal and not a DT signal, even though the events are discrete. This is due to the fact that the possible positions of the time instants $\{t_i\}$ belong to a continuous set, namely $(-\infty, +\infty)$, as seen in Figure 1.3. It is sometimes more convenient to work with the *increments* of $X(t; s)$ rather than with $X(t; s)$ itself. It is clear that $\Delta X(t, \tau; s) = X(t; s) - X(t-\tau; s)$ is the random number of points of the process in the interval $(t-\tau, t)$, which is easier to measure than the signal $X(t; s)$.

The fundamental idea of a family of distributions is to try to describe a random signal as a collection of a finite number of random variables. For this purpose, suppose that n time instants t_i, $1 \le i \le n$, are arbitrarily selected. The n values of the signal $X(t_i; s)$ can be considered as the components of an n-dimensional random vector $\mathbf{X}$. We know that such a vector is completely described by its distribution defined by

$$F_n(x_1, x_2, \ldots, x_n; t_1, t_2, \ldots, t_n) =$$
$$P\left[\{X(t_1) \le x_1\} \cdot \{X(t_2) \le x_2\} \ldots \{X(t_n) \le x_n\}\right] \tag{5.1}$$

It is important to note that this function, just like any distribution function, depends not only on $x_1, x_2, \ldots, x_n$, but also on $t_1, t_2, \ldots, t_n$ which are the time instants arbitrarily chosen. The family of finite DFs is the set of all functions as in (5.1) whatever the number n, and for each n whatever the values of t_i.

In the following we admit that a CT random signal is completely described by this family. Note that this procedure corresponds to an arbitrary sampling of the CT signal $X(t; s)$ and moreover reduces the study of random signals, or random functions, to that of RVs. It is clear that from a strictly mathematical viewpoint this definition is not completely satisfactory because it reduces the study of the continuous set T to that of a countable version, which is not necessarily possible. However, for all the applications discussed later, the description of a random signal by its family of DFs is quite sufficient. In the DT case these problems disappear, and as a DT random signal is a sequence of RVs, it is clear that the DF of the ensemble of these RVs completely describes the signal.

The DF (5.1), considered only as a function of the x_is, must satisfy all the properties of DFs given in the previous chapters. But there are additional properties due to the variable t_is. The *symmetry condition* requires that F be invariant if we permute the pair (x_i, t_i) and the pair (x_j, t_j). The *consistency condition* is expressed by

$$\lim_{x_n \to \infty} F_n(x_1, x_2, \ldots, x_n; t_1, t_2, \ldots, t_n) =$$
$$F_{n-1}(x_1, x_2, \ldots, x_{n-1}; t_1, t_2, \ldots, t_{n-1}) \tag{5.2}$$

Even though from a strictly mathematical viewpoint the family of finite dimensional DFs is insufficient to characterize a random signal completely, this family is very complex, and without specific details it is impossible to know it exactly. It is for this reason that in many cases only a much simpler statistical description of the signal is possible.

One method is to limit the study of the family to an arbitrary value of n. If $n = 1$, we have the function

$$F_1(x; t) = P\{X(t) \le x\} \tag{5.3}$$

This function is the probability distribution of the signal at one instant t, whatever its values. In many physical situations the signal exhibits some *stationarity* properties, which are studied in more detail later. In the case of (5.3) this leads to

$$F_1(x;\ t) = F(x) \tag{5.4}$$

which means that the DF of $X(t;\ s)$ is independent of t. If $n = 2$, we obtain a function $F(x_1, x_2; t_1, t_2)$ describing a pair of RVs. The stationarity property in this case gives

$$F_2(x_1,\ x_2;\ t_1,\ t_2) = F(x_1,\ x_2;\ t_1 - t_2) \tag{5.5}$$

which means that only the time interval $t_1 - t_2$ appears. When even the distribution functions are unknown, a less precise description of the random signal can be reached by using some expected values.

Before closing this section let us make some final comments and some extensions. We saw in Chapter 2 that the DF characterizes the properties of an RV regardless of whether this RV is continuous- or discrete-valued. The same is of course valid for random signals. But if we know, for example, that the signal $X(t;\ s)$ is continuous-valued (which has nothing to do with continuity in time), we can use the probability density instead of the DF. On the other hand, if the signal is discrete-valued, we can use the probabilities of these values instead of the DF.

The following section will deal with two extensions of the previous material. The first is relative to the complex-valued random signal. Without further indication, a complex random signal

$$X(t;\ s) = A(t;\ s) + j\,B(t;\ s) \tag{5.6}$$

is simply a collection of two real signals $A(t;\ s)$ and $B(t;\ s)$, real and imaginary parts respectively. This immediately leads to the case of the vector random signal $\mathbf{X}(t;\ s)$. Such a signal is described by its k components $X_1(t;\ s)$, $X_2(t;\ s)$, ..., $X_k(t;\ s)$, which can be either real- or complex-valued.

5.2 EXPECTATIONS

5.2.1 Mean and Variance

The mean value of a random signal is defined by

$$m(t) \triangleq E\{X(t;\ s)\} \tag{5.7}$$

The calculation of $m(t)$ needs only the knowledge of the DF $F_1(x\,;\,t)$ given by (5.3). Conversely, if we know $m(t)$ we cannot deduce $F_1(x\,;\,t)$ and the knowledge of the mean value is much lower than that of the DF $F_1(x\,;\,t)$.

If $m(t) = 0$, the random signal is said to be zeromean. If this is not the case, we immediately see that the signal

$$X_0(t) \triangleq X(t) - m(t) \tag{5.8}$$

is zeromean-valued. It is often an advantage to work with zeromean signals and to use (5.8) systematically. We will assume below that the signals are zeromean-valued, even when this is not explicitly specified.

If $F_1(x\,;\,t)$ does not depend on time, then neither does $m(t)$. This is especially the case when the random signal is stationary, but conversely $m(t) = 0$ does not mean this stationarity, as will be seen later. Finally, let us note that $m(t)$ and $X(t;\ s)$ have the same geometric nature: real or complex scalars or real or complex vectors.

The variance of a scalar signal is given in the real case by

$$\sigma^2(t) = E\{X^2(t;\ s)\} - m^2(t) \tag{5.9}$$

or in the complex case by

$$\sigma^2(t) = E\{|X(t;\ s)|^2\} - |m(t)|^2 \tag{5.10}$$

As the mean value, the variance does not depend on t when the signal is stationary. In the vector case we can introduce the instantaneous covariance matrix, defined as in (3.90) by

$$\mathbb{\Gamma}(t) = E\{\mathbf{X}(t;\ s)\ \mathbf{X}^H(t;\ s)\} - \mathbf{m}(t)\ \mathbf{m}^H(t) \tag{5.11}$$

In the real case the Hermitian operation is replaced by the transposition, as in (3.63).

The mean and variance of a scalar real random signal $X(t;\ s)$ have an important physical meaning. Let us return to the example of the noise delivered by a resistor due to the thermal motion of electrons discussed in Chapter 1. If $X(t;\ s)$ represents the random voltage observed, it is clear that $m(t) = 0$, which describes the Ohm law. On the other hand, $\sigma^2(t)$ describes the mean power due to the thermal noise, and can be physically measured, as will be described later.

5.2.2 Covariance and Correlation Functions

All the previous expected values are calculated at one time instant t and then deduced from the DF $F_1(x;\,t)$. Let us now consider two time instants t_1 and t_2. From the DF $F_2(x_1, x_2\,;\ t_1, t_2)$ we can deduce the *covariance function* of the signal defined by

$$\gamma(t_1,\ t_2) = E\{X(t_1)\ X(t_2)\} - m(t_1)\ m(t_2) \tag{5.12}$$

In the complex case $X(t_2)$ and $m(t_2)$ are replaced by their complex conjugates. This covariance function will play a very important role in what follows. Its knowledge includes that of the variance $\sigma^2(t)$ because (5.9) leads to

$$\sigma^2(t) = \gamma(t,\ t) \tag{5.13}$$

In the case of complex vector signals, the covariance function is the matrix

$$\mathbb{\gamma}(t_1,\ t_2) = E\{\mathbf{X}(t_1)\ \mathbf{X}^H(t_2)\} - \mathbf{m}(t_1)\ \mathbf{m}^H(t_2) \tag{5.14}$$

As mentioned above, we assume that the signals are zeromean-valued unless otherwise specified. Thus, in most of the following calculations the term $m(t)$ will not appear in expressions of the covariance function.

In many situations, and particularly when some stationarity appears, the DF F_2 satisfies (5.4) and (5.5). In this case the covariance function also satisfies

$$\gamma(t_1,\ t_2) = \gamma(t_1 - t_2) \tag{5.15}$$

In order to avoid any confusion between $\gamma(\tau)$ and $\gamma(t_1, t_2)$, the former is called the *correlation function* of the signal. The correlation function is related to the covariance function by

$$\gamma(\tau) = \gamma(t, t - \tau) \tag{5.16}$$

and for a *stationary complex random signal* with mean value m it is then defined by

$$\boxed{\gamma(\tau) = E\{X(t)\ X^*(t - \tau)\} - |m|^2} \tag{5.17}$$

Let us present some comments on this very important function.

(a) Real signals. In the case of real signals the correlation function takes the form

$$\gamma(\tau) = E[X(t)\ X(t - \tau)] - m^2 \tag{5.17'}$$

(b) Consequences of the stationarity. It is clear that the correlation function does not depend on t, but only on the delay τ, because of the stationarity. It is therefore possible to replace t by $t + \tau$ in (5.17) or (5.17'). This gives

$$\gamma(\tau) = E[X(t + \tau)\ X^*(t)] - |m|^2$$

or

$$\gamma(\tau) = E[X(t + \tau)\ X(t)] - m^2$$

Compared to (5.17'), this expression shows that the correlation function of a *real signal* is an even function of τ. Furthermore it is important to note that the variance and the correlation function are related through

$$\sigma^2 = \gamma(0) \tag{5.18}$$

(c) Vocabulary. Let us assume that $m \neq 0$. In this case, the function

$$m_2(\tau) \triangleq E[X(t)\ X^*(t - \tau)]$$

and the correlation function $\gamma(\tau)$ are different. It is a rather common habit to call $\gamma(\tau)$ the "covariance function" and $m_2(\tau)$ the "correlation function." We indicated on p. 58 why this habit is not satisfactory, and we will repeat the arguments. Let τ_1 be a delay for which $\gamma(\tau_1) = 0$. For this delay the RVs $X(t)$ and $X(t - \tau_1)$ satisfy

$$E[X(t)\, X^*(t - \tau_1)] = E[X(t)]\, E[X^*(t - \tau_1)] = |m|^2$$

which characterizes uncorrelated RVs. Thus the zeros of the "covariance function" characterize the delays for which $X(t)$ and $X(t - \tau)$ are *uncorrelated.* For these delays $m_2(\tau) = |m|^2$, and the "correlation function" is non-null. Conversely, the zeros of the "correlation function" do not characterize uncorrelated values of the signal. That is why we prefer to use the term correlation function for $\gamma(\tau)$, which implies that $\gamma(\tau) = 0$ means uncorrelation, while $m_2(\tau)$ is called the stationary *second order moment* of $X(t)$. This distinction disappears, of course, for zeromean signals.

In conclusion, in all that follows the distinction between correlation and covariance functions is only a result of the stationarity of the signal, as seen in (5.16).

(d) Extensions to the discrete-time case. All the previous definitions can be extended without difficulty to the case of DT signals. Let us note $X[k]$ such a random signal, where k is integer. The covariance function of a complex signal is given by

$$\gamma[k_1, k_2] = E\{X[k_1]\, X^*[k_2]\} - m[k_1]\, m^*[k_2]$$

and if some stationarity is introduced, the correlation function (5.17) becomes

$$\gamma[p] = E\{X[k]\, X^*[k - p]\} - |m|^2 \tag{5.17''}$$

where p is also integer.

(e) Physical meaning. The covariance function is a numerical means of evaluating the randomness of a signal. This is especially clear in the stationary case, or with the correlation function. If the correlation function is null, this means that $X(t)$ and $X(t - \tau)$ are uncorrelated whatever the value of t. We will see later that this means that the signal at time t has lost the memory of its value at $t - \tau$.

(f) Extensions to vector signals. For vector signals, $\boldsymbol{\gamma}(t_1, t_2)$ and $\boldsymbol{\gamma}(\tau)$ are square matrices. Note that as $\boldsymbol{\gamma}(0) = E\{\mathbf{X}(t)\, \mathbf{X}^H(t)\}$ it is a non-negative definite matrix. This leads to a new terminology problem. The k components $X_i(t)$ of a vector random signal $\mathbf{X}(t)$ are scalar random signals. The matrix elements of the square matrix $\boldsymbol{\gamma}(t_1, t_2)$ are

$$\gamma_{i,j}(t_1, t_2) = E\{X_i(t_1)\, X_j^*(t_2)\} \tag{5.19}$$

and in the stationary case this becomes

$$\gamma_{i,j}(\tau) = E\{X_i(t)\, X_j^*(t - \tau)\} \tag{5.20}$$

Comparing this with (5.12) or (5.17), we call the diagonal elements $\gamma_{i,i}$ autocovariance or *autocorrelation functions* and the off-diagonal elements $\gamma_{i,j}$ crosscovariance or *crosscorrelation functions.* The term intercorrelation is sometimes used instead of crosscorrelation.

5.2.3 Properties of Covariance Functions

The covariance function of a signal $X(t)$ is not arbitrary and exhibits some very important properties, among which only the most significant will be presented.

(a) Existence. From the Schwarz inequality (3.49) for RVs we deduce that

$$|\gamma(t_1, t_2)|^2 \leq \gamma(t_1, t_1)\,\gamma(t_2, t_2) \tag{5.21}$$

As a result, if

$$E\{|X(t)|^2\} = \gamma(t, t) < +\infty \tag{5.22}$$

for any t, the covariance does exist. Random signals satisfying (5.22) are called *second order random signals*. As seen in (3.31) a new application of the Schwarz inequality shows that the mean value is also bounded, because

$$|E\{X(t).1\}|^2 \leq E\{|X(t)|^2\}.1 < +\infty \tag{5.23}$$

(b) Second order stationary random signals. A signal is second order stationary if it satisfies (5.22) and if its mean value and covariance function satisfy

$$m(t) = m \quad ; \quad \gamma(t_1, t_2) = \gamma(t_1 - t_2) \tag{5.24}$$

Stronger kinds of stationarity will appear later.

(c) Symmetry. We see from (5.12) that for a real signal $X(t)$

$$\gamma(t_1, t_2) = \gamma(t_2, t_1) \tag{5.25}$$

and (5.14) gives

$$\boldsymbol{\gamma}(t_1, t_2) = \boldsymbol{\gamma}^H(t_2, t_1) \tag{5.26}$$

In the stationary case this gives symmetry relations for the correlation function. For complex signals we have

$$\gamma(\tau) = \gamma^*(-\tau) \tag{5.27}$$

and for real signals the correlation function is an even function or

$$\gamma(\tau) = \gamma(-\tau) \tag{5.28}$$

In the case of complex vector signals the matrix correlation function satisfies

$$\boldsymbol{\gamma}(\tau) = \boldsymbol{\gamma}^H(-\tau) \tag{5.29}$$

and we verify that, for $\tau = 0$, this matrix is Hermitian because it is the covariance matrix of the vector $\mathbf{X}(t)$, as seen in (5.14) where $t_1 = t_2 = t$.

(d) Extreme values. In the stationary case (5.21) gives

$$|\gamma(\tau)| \leq \gamma(0) \tag{5.30}$$

which means that a correlation function has its absolute maximum for $\tau = 0$. This allows the introduction of the normalized correlation function defined by

$$\tilde{\gamma}(\tau) = \gamma(\tau)/\gamma(0)$$

This function satisfies $|\tilde{\gamma}(\tau)| \leq 1$ and $\tilde{\gamma}(0) = 1$. Note that these relations are the same as (2.56) valid for characteristic functions. On the other hand, if $\tau \to \infty$ the RVs $X(t)$ and $X(t-\tau)$ tend in general to become uncorrelated. If this is the case, we have

$$\lim_{\tau \to \infty} \gamma(\tau) = 0 \tag{5.31}$$

This result is not absolutely general, and we give here an example where it is not true. Suppose that $X(t)$ is a stationary zeromean random signal which satisfies (5.31), and consider

$$Y(t) = X(t) + Z \tag{5.32}$$

where Z is a zeromean RV uncorrelated with $X(t)$. As $E[ZX(t)] = 0$, we obtain

$$\gamma_Y(\tau) = \gamma_X(\tau) + \sigma_Z^2 \tag{5.33}$$

and if $\gamma_X(\tau)$ satisfies (5.31), $\gamma_Y(\tau)$ does not. This is due to the long memory effect introduced by the random variable Z which is time-independent.

(e) Addition of covariances. Consider two uncorrelated zeromean random signals $X_1(t)$ and $X_2(t)$. This means that

$$E\{X_1(t_i)\, X_2^*(t_j)\} = 0 \tag{5.34}$$

Let γ_1 and γ_2 be the covariances of X_1 and X_2 respectively, while γ is the covariance of the sum $X_1 + X_2$. By direct calculation we deduce from (5.34) that

$$\gamma(t_i, t_j) = \gamma_1(t_i, t_j) + \gamma_2(t_i, t_j) \tag{5.34'}$$

This can of course be applied to a sum of arbitrary numbers of uncorrelated signals. This result means that the class of covariance functions is closed by addition.

(f) Multiplication of covariances. Let us now suppose that the random signals $X_1(t)$ and $X_2(t)$ are zeromean and *independent*, and consider their product $Y(t) = X_1(t)\, X_2(t)$. As $Y(t)$ is also zeromean, the covariance function of $Y(t)$ is

$$\gamma_Y(t_i, t_j) = E\{X_1(t_i)\, X_2(t_i)\, X_1(t_j)\, X_2(t_j)\}$$

Because of the independence assumption, we obtain

$$\gamma_Y(t_i, t_j) = \gamma_1(t_i, t_j)\, \gamma_2(t_i, t_j) \tag{5.35}$$

which now means that the class of covariance functions is closed by multiplication. In other words the sum or product of covariances remain covariance functions.

(g) Non-negative property. The covariance functions are not arbitrary and satisfy an essential property of non-negativeness. As the same is true for *correlation functions*, we will show this property in the latter case. Consider first the complex *scalar* case. Let λ_i be n arbitrary complex numbers and t_i be n arbitrary time instants. Consider the RV Z defined by

$$Z \triangleq \sum_{i=1}^{n} \lambda_i X(t_i)$$

By saying that $E(|Z|^2) \geq 0$, we obtain

$$E(|Z|^2) = \sum_{i=1}^{n} \sum_{j=1}^{n} \lambda_i \lambda_j^* \gamma(t_i - t_j) \geq 0 \tag{5.36}$$

This inequality characterizes a function $\gamma(\tau)$ which is non-negative definite. Such functions exhibit very nice properties that will be studied in detail in a later chapter. The same relation is valid for covariance functions and for this it suffices to replace $\gamma(t_i - t_j)$ by $\gamma(t_i, t_j)$ in (5.36). Note the analogy with (3.91) valid for vectors.

For those not familiar with *non-negative definite functions*, it is important to point out that "non-negative definite" does not mean "non-negative." To show this convincingly, let us recall that $\gamma(\tau)$ is in general complex, at least for a complex-valued random signal. Finally, note that we have already seen this kind of property in the study of characteristic functions in Section 2.5. In fact (5.38) is equivalent to (2.57) valid for characteristic functions.

Consider now a *vector* complex random signal. Instead of arbitrary numbers λ_i, consider arbitrary vectors $\mathbf{u}_i$ and the RV

$$Z \triangleq \sum_{i=1}^{n} \mathbf{u}_i^H \mathbf{X}(t_i)$$

Again, by saying that $E(|Z|^2) \geq 0$, we obtain

$$\sum_{i=1}^{n} \sum_{j=1}^{n} \mathbf{u}_i^H \boldsymbol{\gamma}(t_i - t_j)\, \mathbf{u}_j \geq 0 \tag{5.37}$$

where $\boldsymbol{\gamma}(\tau)$ is the matrix-valued correlation function of the random signal $\mathbf{X}(t)$ defined by (5.14). Note that to arrive at (5.37) we used the well-known property $\mathbf{u}^H \mathbf{X} = (\mathbf{X}^H \mathbf{u})^*$.

It is possible to show that these properties of non-negativeness are characteristic of the correlation functions. This means that to any function $\gamma(\tau)$ satisfying (5.36) it is possible to associate at least one random signal such that its correlation function is precisely $\gamma(\tau)$. We will not give the proof of this result here, as it is similar to that obtained for random vectors in Section 3.5.2, p. 65.

(h) Correlation time. As indicated by (5.31), the correlation function of a zeromean random signal generally decreases to zero when $|\tau|$ increases. But the decrease can be fast or slow, and the correlation time t_c is a parameter which describes this phenomenon. More precisely if $\tau > t_c$ the RVs $X(t)$ and $X(t-\tau)$ become practically uncorrelated. The situation is similar to that found when defining the time constant of a filter (see [Picinbono], p. 138). There are several analytical definitions for t_c which will be presented later.

5.2.4 Examples of Correlation Functions

The correlation function of a random signal is one of its most important characteristics and is used in many areas of signal processing problems. It is therefore worth gaining some familiarity with such functions, although we cannot present all the possible correlation functions. In reality, as indicated just above, any non-negative definite function of τ can be the correlation function of a random signal, and the number of such functions is not finite. Our purpose will be to present only those which appear most frequently in applications.

First, it is important to introduce a characterization of a non-negative definite function other than the one expressed by (5.36). In fact, it is very difficult to verify that it is satisfied by a given function $\gamma(\tau)$. However, we saw in Chapter 2 that the characteristic functions, which have the same property as the correlation functions, can be specified as the Fourier transform of PDF, which are non-negative functions. Although this result is the main subject of the next chapter, we will now admit that any function $\gamma(\tau)$ the Fourier transform of which is non-negative is a correlation function. This Fourier transform, noted $\Gamma(\nu)$ and extensively studied in the next chapter, is called the *power spectrum* or the *spectral density* of the random signal $X(t)$. In other words, the class of correlation functions is equivalent to the class of power spectra, and we will restrict our presentation to the most important elements of this class.

(a) Exponential correlation function. Let us consider the function defined by

$$\gamma(\tau) = \exp(-a|\tau|) \tag{5.38}$$

It is a normalized exponential correlation function and the function $\exp(-|x|)$ is represented in Figure 5.1.

This satisfies (5.31) and the correlation time t_c is often considered as equal to $1/a$. The corresponding power spectrum defined by

$$\Gamma(\nu) = \int \gamma(\tau) \exp(-2\pi j \nu \tau)\ d\tau$$

is equal to

$$\Gamma(\nu) = \frac{2a}{a^2 + 4\pi^2 \nu^2} \tag{5.39}$$

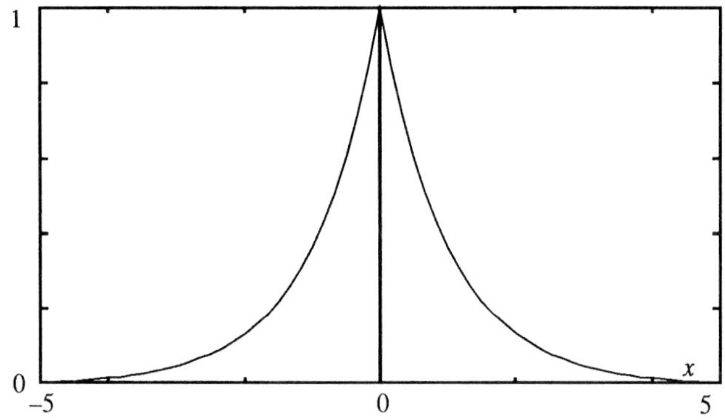

Figure 5.1 Exponential correlation function $\exp(-|x|)$.

and it is sometimes said that it has a Lorentzian shape. This spectrum is proportional to the function $(1+x^2)^{-1}$ represented in Figure 5.2.

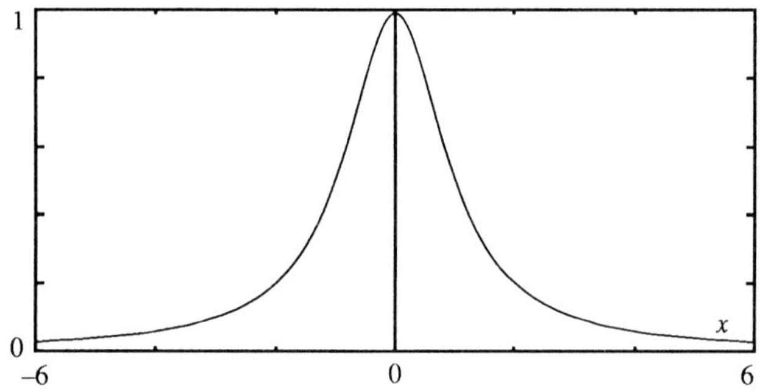

Figure 5.2 Power spectrum of an exponential correlation function.

The discrete-time version of an exponential correlation function is given by

$$\gamma[k] = a^{|k|} \quad , \quad |a| < 1$$

and will play an important role in Chapter 9 in the framework of autoregressive signals.

(b) High-frequency exponential correlation function. This is defined by

$$\gamma(\tau) = e^{-a|\tau|}\cos(\omega_0 \tau) \tag{5.40}$$

where ω_0 is an arbitrary angular frequency. If $a = 0$, we simply obtain the function $\cos(\omega_0 \tau)$ and the power spectrum is limited to two spectral lines at the frequencies ω_0 and $-\omega_0$. If $\omega_0 = 0$ we return to (5.38).

(c) Band-limited correlation function. Suppose that the power spectrum is constant in the frequency band $[-B, +B]$, and null otherwise. Such a spectrum is presented in Figure 5.3.

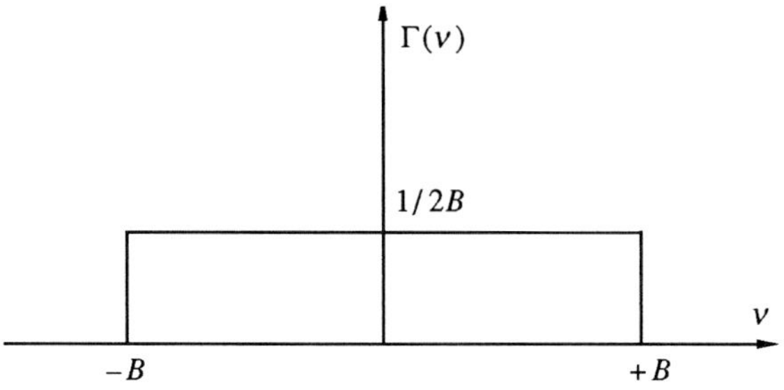

Figure 5.3 Constant band-limited spectrum.

The corresponding correlation function is

$$\gamma(\tau) = \text{sinc}(2B\,\tau) \tag{5.41}$$

where the function sinc x is defined by

$$\text{sinc}\, x = \frac{\sin(\pi x)}{\pi x}$$

and appears in Figure 5.4. The correlation time can be taken as equal to $1/B$. However, it is clear that the function sinc x decreases more slowly that $\exp(-x)$.

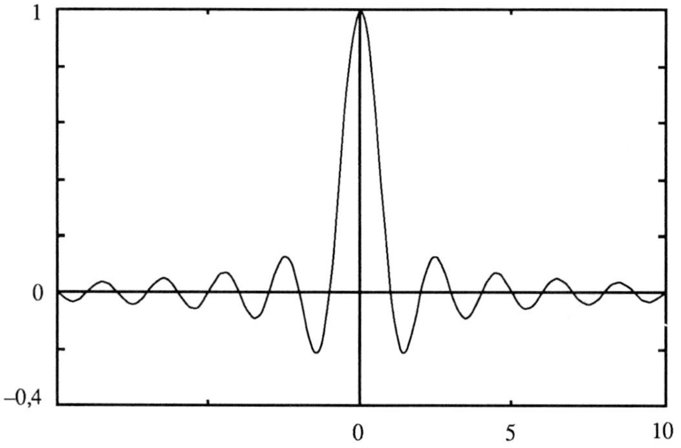

Figure 5.4 The function sinc x.

(d) High-frequency band-limited correlation function. The procedure is the same as in (**b**), which gives

$$\gamma(\tau) = \operatorname{sinc}(2B\tau)\cos(\omega_0\tau) \tag{5.41'}$$

In most applications it is assumed that $(B/\omega_0) << 1$, which is called the narrow-band situation. This means that the function $\operatorname{sinc}(2B\tau)$ is almost constant during one period of the periodic function $\cos(\omega_0\tau)$. It can be said that there is an amplitude modulation of the cosine signal. The same approximation can be introduced for the case of (**b**), and is characterized by $(a/\omega_0) << 1$.

(e) Gaussian correlation function. This has the form

$$\gamma(\tau) = \exp(-a^2\tau^2)$$

represented in Figure 2.4. The power spectrum has the same shape, because we saw in Chapter 2 that the Fourier transform of a Gaussian shape remains Gaussian.

5.2.5 Applications of Correlation Functions

There are so many problems in which the correlation function is used that a complete chapter would be necessary to describe them all. We shall limit ourselves to the presentation of only a few, because many others will appear in the following chapters.

(a) Interference experiments. Let $X(t)$ be an arbitrary random signal and consider the signal deduced from $X(t)$ by

$$Y(t) = X(t) + X(t-d)$$

This is typical for what is realized in an interference experiment, where two coherent sources deliver the same signal and the delay d is a result of the propagation effect. However, the superposition of a signal and the same delayed by d appears in many cases other than optical interferences.

Suppose that $X(t)$ is real, stationary and zeromean and let $\gamma(\tau)$ be its correlation function. We are first interested in the statistical properties of $Y(t)$ at time t. It is clear that the mean value of $Y(t)$ is null. Let us calculate its variance equal to $E[Y^2(t)]$. Starting from the definition of $Y(t)$ we obtain

$$\sigma_Y^2 = E\left\{[X(t) + X(t-d)]^2\right\} = 2[\gamma(0) + \gamma(d)]$$

This variance depends on d and its maximum is obtained for $d = 0$ which, because of (5.18), gives

$$\sigma_Y^2 = 4\sigma_X^2$$

On the other hand the minimum possible value of $\gamma(d)$ is $-\gamma(0) = -\sigma_X^2$, because of (5.30), in which case we have $\sigma_Y^2 = 0$. In the first case we have a coherent addition and in the second a coherent subtraction. This especially appears when we use the correlation function (5.40) with $a = 0$, which gives

$$\sigma_Y^2 = 2[1 + \cos(\omega_0 d)]$$

This describes the classical interference pattern with alternation of maxima and minima. Suppose now that $\gamma(d)$ satisfies (5.31). We deduce that when $d \to \infty$, $\sigma_Y^2 \to 2\sigma_X^2$. In this case the RVs $X(t)$ and $X(t-d)$ become uncorrelated and we again find the rule (3.37) concerning the incoherent sum. So, the interference phenomenon disappears for large values of the delay d, which is a classical result consequent to the properties of the correlation function.

Let us now calculate the correlation function of $Y(t)$. It is defined by (5.17') with $m = 0$, which gives

$$\gamma_Y(\tau) = E\{[X(t) + X(t-d)][X(t-\tau) + X(t-\tau-d)]\}$$

$$= 2\gamma(\tau) + \gamma(\tau+d) + \gamma(\tau-d)$$

resulting in σ_Y^2 for $\tau = 0$. It is a function of τ and d and its behavior depends on the structure of $\gamma(\tau)$, correlation function of $X(t)$.

(b) Localization of a source by correlation. Suppose that a radiation source is located at a great distance and that it generates plane waves. The field of waves is measured by two sensors as indicated in Figure 5.5. Let d be the distance between the two sensors. Suppose that the signal registered by the sensor S_1 is $X(t)$. Because of the propagation in the medium with a velocity v, the signal registered by the sensor S_2 is $X(t + \theta)$, where θ is the delay due to the propagation. The value of this delay is

$$\theta = (1/v)\, d \sin\phi$$

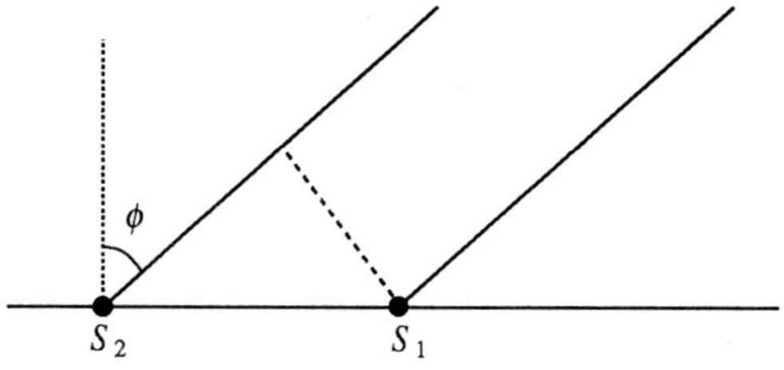

Figure 5.5 Localization of sources with two sensors.

and this is a function of the angle ϕ defining the position of the source. By using a correlator we can measure the cross-correlation between the two signals, which is

$$\gamma_c(\tau) = E[X(t)\, X(t + \theta - \tau)] = \gamma(\tau - \theta)$$

where $\gamma(\tau)$ is the correlation function of $X(t)$. Now, suppose that this function has only one maximum, necessarily for $\tau = 0$, such as all the correlation functions presented in the previous section. In this case we deduce that $\gamma_c(\tau)$ has a maximum for $\tau = \theta$, and the measurement of this maximum allows us to find the delay θ and thus the angle ϕ defining the angular position of the source, or the direction of the plane wave.

5.2.6 Higher-order Moments and Cumulants

The general kth order moment of a signal $X(t)$ is defined by

$$m_k(t_1, t_2, \ldots, t_k) = E\{X(t_1)\, X(t_2) \ldots X(t_k)\} \tag{5.42}$$

For complex signals there is not only one such moment but 2^k because each $X(t_i)$ can be replaced by its complex conjugate $X^*(t_i)$. In reality this number can be reduced because some moments are simply the complex conjugate of some others. For example for $k = 2$, there are only two moments which are

$$m_2(t_1, t_2) = E\{X(t_1)\, X(t_2)\} \; ; \; n_2(t_1, t_2) = E\{X(t_1)\, X^*(t_2)\} \tag{5.43}$$

The second one is much more important and appears in the definition of the covariance function $\gamma(t_1, t_2)$ defined as in (5.14) or (5.17).

Supposing now that the signal is real, let us introduce its *cumulants*. As for RVs, the first two cumulants are equal to the mean and the covariance [see (4.77)]. The higher-order cumulants are defined by

$$c_k(t_1, t_2, \ldots, t_k) = c\{X(t_1), X(t_2), \ldots, X(t_k)\} \tag{5.44}$$

where the r.h.s. means the cumulant of the RV $X(t_1)$, $X(t_2)$,..., $X(t_k)$, as defined by (4.73). Higher-order moments are certainly less used than second-order moments, but nevertheless appear in some practical situations. Let us give a simple example. In many signal processing systems we have to use the square of a given signal $X(t)$, or $Y(t) = X^2(t)$. The transformation from $X(t)$ to $Y(t)$ is called a quadratic transformation. If $X(t)$ is random and stationary, $Y(t)$ has the same properties and its correlation function is

$$\gamma_Y(\tau) = E[Y(t)\, Y(t - \tau)] - E\left\{[Y(t)]^2\right\} = m_4(t, t, t - \tau, t - \tau) - \gamma^2(0)$$

This shows that the fourth-order moment of $X(t)$ is necessary in order to calculate the correlation function of $Y(t)$.

5.3 STATIONARY RANDOM SIGNALS

The concept of stationarity was referred to several times above, and we shall now give its precise definition. The discussion is presented for CT signals and can be transposed without difficulty to the DT case.

5.3.1 Strict-Sense Stationarity

A random signal $X(t)$ is said to be strictly stationary if all its statistical properties are invariant by any translation of the time origin. This means that the two signals $X(t)$ and $Y(t) = X(t + \tau)$ have the same statistical properties, characterized in this book by the family of finite-dimensional distribution. The same is valid for the concept of stationarity. Thus the DF defined by (5.1) is invariant when all the t_is are changed to $t_i + \tau$, or

$$F_n(x_1, x_2, \ldots, x_n; t_1, t_2, \ldots, t_n) = F_n(x_1, x_2, \ldots, x_n; t_1 + \tau, t_2 + \tau, \ldots, t_n + \tau) \quad (5.45)$$

As τ is arbitrary, we immediately deduce that a stationary random signal cannot be a time-limited signal, as defined in [Picinbono] 2.2.1. This leads to the conclusion that stationarity is a concept which can be very useful, as will be seen later, but impossible to verify because any experiment in our real world has a beginning and an end. We will discuss this question at the end of this section. Note that strict-sense stationarity in the case of a complex random signal requires the use of its real and imaginary parts, and the DFs F_n are those of the ensemble of these real and imaginary parts. It is clear that if (5.45) is valid, then the function F_n appearing in this equation is only dependent on $n-1$ variables t_i. This was already noted for $n = 1$ or $n = 2$ in (5.4) and (5.5) respectively. The same property appears for the moment (5.42) and the cumulant (5.44).

5.3.2 Second-Order Stationarity

Instead of being defined by a DF, this stationarity is defined by moments up to the second order. A random signal is then second-order (or wide-sense, or weakly) stationary if it is second-order and if its mean value is constant and its covariance reduced to a correlation function.

It is obvious that second-order stationarity does not imply strict-sense stationarity. The inverse is also true, which is more surprising. This is due to the fact that a strictly stationary random signal is not necessarily second-order, if the distribution does not generate finite second-order moments. Note that second-order stationarity is also defined for complex signals; this is because mean and covariance functions are defined without change for real and complex signals.

Example 5.1 Constant signal. Consider the random signal defined by

$$X(t; s) = X(s) \quad (5.46)$$

where $X(s)$ is an RV defined on some probability space. As $X(t; s)$ does not depend on t, it is obvious that all the DFs F_n are independent of t, and the signal is strictly stationary. Furthermore, if $X(s)$ is second-order, then $X(t)$ is second-order stationary.

Example 5.2 Sinusoidal signal with random phase. Consider now the signal defined by

$$X(t; s) = \cos\{\omega t + \phi(s)\} \quad (5.47)$$

where $\phi(s)$ is a random phase. Suppose further that the RV $\phi(s)$ is uniformly distributed in $[0, 2\pi)$. As a result of this assumption we obtain $E\{X(t)\} = 0$. As $X(t_1)X(t_2) = (1/2)\cos\{\omega(t_1 + t_2) + 2\phi(s)\} + (1/2)\cos\{\omega(t_1 - t_2)\}$, we obtain

$$\gamma(t_1, t_2) = (1/2)\cos\{\omega(t_1 - t_2)\} \tag{5.48}$$

which shows that $X(t)$ is second-order stationary.

We shall now see that this signal is also strict-sense stationary. If we again take the signal $Y(t) = X(t + \tau)$ used in (5.45), we deduce from (5.47) that it can be written as $\cos\{\omega t + \psi\}$ where $\psi = \phi(s) + \omega\tau$. The phase $\phi(s)$ is then replaced by $\psi(s)$ in (5.47). But as $\phi(s)$ is uniformly distributed in $[0, 2\pi)$, the same arises for ψ and as a result $X(t)$ and $Y(t)$ have the same statistical definition.

Example 5.3 Complex sinusoidal signal. Consider the random signal

$$Z(t; s) = A(s)\exp(j\omega t) \tag{5.49}$$

where $A(s)$ is a zeromean second-order RV. It is clear that $E(Z) = 0$ and that $\gamma(\tau) = \sigma^2 \exp(j\omega\tau)$, where σ^2 is the variance of A. Then $Z(t)$ is second-order stationary. On the other hand, it is not strict-sense stationary. It is obvious that the DF of $Z(t)$ is not the same as that of $Z(t + \tau)$.

Example 5.4 Difference between auto and crosscorrelation. Consider two signals $X_1(t)$ and $X_2(t)$ of the type discussed in Example 5.2 with two angular frequencies ω_1 and ω_2 and the same random phase $\phi(s)$. These two signals, considered singly, are stationary. Let us now calculate their crosscorrelation $\gamma_{1,2}(t_i, t_j)$. Using the same procedure as before, we easily find

$$\gamma_{1,2}(t_i, t_j) = (1/2)\cos(\omega_1 t_i - \omega_2 t_j)$$

This function is clearly not a function of $(t_i - t_j)$, which means that the crosscovariance is not stationary. We then have a very simple example of two signals with a stationary autocovariance, or an autocorrelation function, but not a stationary crosscovariance. This is similar to the situation discussed in Example 4.7 concerning two RVs which are individually Gaussian but not jointly Gaussian.

Example 5.5 Random frequency. Consider the complex signal

$$Z(t; s) = \exp[j\{\Omega(s)t + \phi(s)\}] \tag{5.50}$$

where $\Omega(s)$ is a random angular frequency and $\phi(s)$ an arbitrary phase. Suppose that $\Omega(s)$ and $\phi(s)$ are two independent RVs. We have

$$E\{Z(t)\} = E[\exp\{j\Omega(s)t\}]\, E[\exp\{j\phi(s)\}] \tag{5.51}$$

In order to obtain a zero mean-value, the most usual procedure is to assume a phase ϕ uniformly distributed. With this assumption we deduce that

$$\gamma(\tau) \stackrel{\Delta}{=} E\{Z(t)Z^*(t - \tau)\} = E\{\exp(j\Omega\tau)\} \tag{5.52}$$

Comparing this with (2.53) we observe that the correlation function $\gamma(\tau)$ is none other than the characteristic function of the random frequency $\Omega(s)$.

5.3.3 Further Points on Stationarity

Even though the concept of stationarity plays a very important role in what follows, it remains essentially theoretical. In fact, it is impossible to verify stationarity experimentally because of the finite duration of any experiment in the real world. Physical situations are more accurately described by other concepts of stationarity, but with the trade-off of greater mathematical complexity.

(a) Semistationarity. Consider a stationary signal $X(t)$ and the signal $Y(t)$ equal to $X(t)$ if $t_1 \leq t \leq t_2$ and to zero if t does not satisfy the previous inequalities. It is clear that $Y(t)$ is not stationary, as is any signal of finite duration. Nevertheless, if t and $t + \tau$ are such that $Y(t)$ and $Y(t+\tau)$ are not null, the covariance function of $Y(t)$ satisfies (5.15), which means a stationary property. Roughly speaking, we can say that $Y(t)$ has all the stationarity properties of $X(t)$ for all those time instants which belong to the interval $[t_1, t_2]$. From a practical point of view these instants can be considered as being the beginning and the end of the observation. Unfortunately there is no simple mathematical treatment of such signals because their properties depend on the arbitrary times t_1 and t_2. However, if $t_2 - t_1 >> t_c$, where t_c is the correlation time introduced on p. 172, $Y(t)$ has almost the same properties as $X(t)$, as will be shown later.

(b) Stationary and periodic signals. Instead of suppressing completely the signal outside the interval $[t_1, t_2]$, it is possible to repeat $X(t)$ by periodicity for any value of t. In other words, $Y(t) = X(t)$ for $t_1 \leq t < t_2$ and $Y(t) = Y(t + kT)$ with $T = t_2 - t_1$. Thus $Y(t)$ is periodic but still has a memory of some stationarity. In fact, it is clear that the covariance function $\gamma_Y(t; \tau) = E\{Y(t)\, Y(t-\tau)\}$ does not depend on t and can then be written as in (5.16). But, due to the periodicity, we also have $\gamma_Y(\tau) = \gamma_Y(\tau + kT)$, which means that the correlation function is also periodic. In conclusion, the signal $Y(t)$ is stationary and periodic and its correlation function is also periodic.

(c) Locally stationary signals. Consider a stationary signal $X(t)$ and a deterministic function $a(t)$. The signal $Y(t) = a(t)\, X(t)$ is no longer stationary because of the term $a(t)$, which can be considered as an instantaneous amplitude of $Y(t)$. The latter is thus obtained from an amplitude modulation of $X(t)$ by $a(t)$. The covariance function of $Y(t)$ is

$$\gamma_Y(t_1, t_2) = a(t_1)\, a(t_2)\, \gamma_X(t_1 - t_2) \tag{5.53}$$

Suppose now that the variations of $a(t)$ are very slow compared with the correlation time t_c of $X(t)$. This means that for time intervals such that $|t_1 - t_2| < t_c$ we have $a(t_1) \cong a(t_2)$, and if $|t_1 - t_2| > t_c$, $\gamma_X(t_1 - t_2)$ is approximately null. In this case we can write

$$\gamma_Y(t_1, t_2) \cong a^2\, \{(t_1 + t_2)/2\}\, \gamma_X(t_1 - t_2) \tag{5.54}$$

This means that, locally, $Y(t)$ is stationary but the correlation function is multiplied by an amplitude depending on time. It is easy to verify that as $\gamma_X(\tau)$ is positive definite, $\gamma_Y(t_1, t_2)$ also has this property.

5.4 LINEAR FILTERING OF RANDOM SIGNALS

A linear filter is a system the input-output relationship of which is defined by a convolution. For DT signals this convolution is written as in (3.119), and for CT signals the series is replaced by an integral. Let us now suppose that the input is random, as is the output, which can be written, for DT and CT respectively, as follows

$$Y[k] = \sum h[l]\, X[k-l] \tag{5.55}$$

$$Y(t) = \int h(\theta)\, X(t-\theta)\, d\theta \tag{5.56}$$

As usual when no limitation is specified on the series or on the integral, they are assumed to be extended from $-\infty$ to $+\infty$.

The first question which arises is the convergence of this series or integral, studied in a particular case on p. 72 for mean square convergence, and on p. 128 for convergence with probability one. We have seen that this last mode of convergence is ensured once the filters are strictly stable (see [Picinbono] p. 138) and the signals satisfy $E(|X|) < M$. The condition of convergence in quadratic mean will be studied in greater detail below and in this section we admit that the series (5.55) and the integral (5.56) are convergent in quadratic mean.

As the randomness of the input is transferred to the output, it is important to determine the statistical properties of the output from those of the input and from the impulse response of the filter.

It is in general extremely difficult, even impossible, to determine the family of finite-dimensional distributions of the output of a filter. Even the function $F_1(y\,;\,t)$ as defined in (5.4) is very difficult to calculate. To see where the difficulty lies, consider the DT case. If the filter has a finite impulse response, or is FIR, the series (5.55) becomes a sum of a finite number of random terms. The distribution function of a sum of RVs was studied in Chapter 4. For example, using the result (4.69) we can calculate the characteristic function of the RV $Y = \mathbf{h}^T\mathbf{X}$ which is (5.55) written for FIR filters. This can be written as $\phi_Y(u) = \phi_X(u\mathbf{h})$, where $\phi_X(\mathbf{u})$ is the characteristic function of the vector $\mathbf{X}$. To return to the DF, we must make an inverse multidimensional Fourier transform, not an easy task. It becomes still harder when the filter is no longer FIR, giving an infinite dimensional Fourier transform.

On the other hand, the expected values studied in Section 5.2 are much easier to obtain. For example, the mean value of $Y[k]$ defined by (5.55) is

$$m_Y[k] = \sum h[l]\, m_X[k-l] \tag{5.57}$$

which means that the convolution relation is still valid for the mean value. This is the same for CT, as seen from (5.56). Thus, if we pass the random signal $X[k]$ in a linear filter which gives the random signal $Y[k]$, the convolution relation is valid for the random elements as well as for their mean values. In consequence, if the input is zeromean valued, the output also has this property.

Consider now the second order properties in the case of a CT signal. Assuming that the input is zeromean valued, we deduce from the definition (5.14) applied in the scalar case and from (5.56) that

$$\gamma_Y(t_1, t_2) = \iint h(\theta_1)\, h^*(\theta_2)\, \gamma_X(t_1 - \theta_1, t_2 - \theta_2)\, d\theta_1\, d\theta_2 \tag{5.58}$$

The same type of expression is valid for DT signals. If we assume that the signals and the filter are real, we deduce from (5.42) and (5.56) that

$$m_Y(t_1, \ldots, t_n) = \int \ldots \int h(\theta_1) \ldots h(\theta_n)\, m_X(t_1 - \theta_1, \ldots, t_n - \theta_n)\, d\theta_1 \ldots d\theta_n \tag{5.59}$$

We come now to the calculation of the output cumulants in terms of the input cumulants. Using the property (4.74) we deduce that the transformation of the cumulants is described by the same rules as the moments, which means that we can write the same equation as (5.59), now written for the DT case, as

$$c_Y[k_1, \ldots, k_n] = \sum_{l_1} \ldots \sum_{l_n} h[l_1] \ldots h[l_n] c_X[k_1 - l_1, \ldots, k_n - l_n] \tag{5.60}$$

Let us now turn to the problem of stationarity. The basic property of a linear filter is that it is time invariant (see [Picinbono] p. 5). It is then obvious that if the input of a linear filter is stationary in any sense, the output has the same property. In particular, if the mean and the covariance are stationary, the relations (5.57) and (5.58) become

$$m_Y = m_X \sum h[k] \tag{5.61}$$

$$\gamma_Y(\tau) = \iint h(\theta_1)\, h^*(\theta_2)\, \gamma_X(\tau + \theta_2 - \theta_1)\, d\theta_1\, d\theta_2 \tag{5.62}$$

It is obvious to write the similar expresions valid for the DT case.

5.5 ERGODICITY

Consider a CT signal $X(t; s)$ where the notation is the same as in Section 5.1. For a given value of s, say s_0, we have a function of time $X(t; s_0)$ which is called a *trajectory* of the signal (see p. 4). In physical problems it is precisely this trajectory which is observed. A practical method of determining the mean value of this signal consists in calculating the time average, which can be written

$$m(T; s_0) = \frac{1}{T} \int_0^T X(t; s_0)\, dt \tag{5.63}$$

The question is then to know if this quantity has any relation to the ensemble average or expected value defined by (5.7), particularly when T becomes very large. Ergodicity problems are related to the question of equivalence between time and statistical averages, and there are numerous ergodic theorems which will not be dealt with here because the matter becomes rapidly very abstract. We shall restrict ourselves to the discussion of the principal ideas used in practice, leaving the reader to consult more specialized books for a general theory.

Suppose that $X(t; s)$ has a constant mean value m. It is then possible to write this signal as $m + \{X(t; s) - m\}$ or $m + X_0(t; s)$, where $X_0(t; s)$ is zeromean valued, as in (5.8). Consider now the random variable $M(T; s)$ defined by replacing $X(t; s_0)$ in (5.63) by $X(t; s)$, which gives

$$M(T; s) = m + \frac{1}{T} \int_0^T X_0(t; s)\, dt \tag{5.64}$$

If the last term converges to zero when $T \to \infty$, we deduce an ergodic property for the measurement of the mean value. But as there are many types of convergence, there are also many types of ergodicity.

Consider first the quadratic mean ergodicity. For this we deduce from (5.64) that

$$E\{\, |M(T; s) - m|^2\} = \frac{1}{T^2} \int_0^T \int_0^T \gamma(t, \theta)\, dt\, d\theta \tag{5.65}$$

where $\gamma(t, \theta)$ is the covariance function of $X(t)$ defined by (5.12). In other words, a sufficient condition for ergodicity in quadratic mean is that the last term of (5.65) tends to zero when $T \to \infty$. The consequence of this type of convergence is the same as that discussed in 4.7.4 (**a**). Using again the Tchebyshev inequality, we can say that whatever the value of ε, the probability that $|M(T; s) - m|$ is greater than ε decreases to zero when $T \to \infty$.

It is possible to show that if $\gamma(t, \theta)$ satisfies the condition ensuring that (5.65) converges to zero for $T \to \infty$, then $M(T; s)$ also converges with probability one to m. This means that for any realization of the signal $X(t; s)$, say $X(t; s_0)$, the time average (5.63) converges to the mean value m, except in events of zero probability. This is the most significant ergodic result, because in practical experiments we almost always consider a particular trajectory $X(t; s_0)$ of the signal $X(t; s)$.

It is important to note that even though $X(t; s)$ has a constant expected value, we have not assumed that it is second order stationary. Thus ergodicity does note require stationarity. Of course, if $X(t)$ is second order stationary we have only to replace $\gamma(t, \theta)$ in (5.65) by $\gamma(t - \theta)$, as in (5.15). Conversely stationarity does not imply ergodicity, and this will appear in the next example.

Example 5.6 Ergodicity of some signals. Consider first the constant signal presented in Example 5.1. As it does not depend on t, (5.63) can be written as

$$m(T; s_0) = X(t; s_0) \tag{5.66}$$

and ergodicity does not hold. In fact, as $m(T;\ s_0)$ does not depend on T, its limit for $T \to \infty$ is always equal to $X(t;\ s_0)$ and not to the mean value of $X(t;\ s)$. This has a simple physical interpretation. Any trajectory of this signal is constant, and thus there is no relation between the evolution of $X(t;\ s)$ in time and its possible values at a given time instant. This is a very simple manifestation of a more general result. If a signal $X(t;\ s)$ is random only because it contains a finite number of parameters, it is in general non-ergodic because the randomness appears too simply and does not present any evolution in time. We will immediately show that this result is not absolutely general. Consider the signal defined by (5.47), and suppose that $\phi(s)$ is uniformly distributed. As indicated after (5.47) this leads to the result that $E\{X(t)\} = 0$. But the time average defined by (5.63) is also null when $T \to \infty$, which exhibits an ergodic property. However, this property is due to the particular distribution of the random phase $\phi(s)$. If this phase is no longer uniformly distributed, the time average remains null while the mean value has no reason to be null. The same is true for the signal defined by (5.49), which has a zero time average while its expectation value is $E\{A(s)\}\ \exp(j\omega t)$. This comment can also be made for the signal defined by (5.50).

Let us now consider the case of a DT signal. Here the integrals appearing in (5.64) and (5.65) are simply sums. In consequence, a condition ensuring the quadratic mean convergence of the time average to the expectation is

$$\lim_{N \to \infty} \frac{1}{N^2} \sum_{i=1}^{N} \sum_{j=1}^{N} \gamma[i,j] = 0 \tag{5.67}$$

where $\gamma[i,j]$ is the covariance function similar to (5.12) in the DT case. With a DT signal which is a sequence of uncorrelated RVs, as in 4.7.4 (**a**), we have $\gamma[i,j] = \sigma_X^2\ \delta[i,j]$, where $\delta[i,j]$ is the Kronecker-delta signal. Using this value in (5.67), we see immediately that the condition ensuring ergodicity is fulfilled. This is sometimes called the *weak law of large numbers*. Other examples will be studied in the next chapter.

5.6 CONTINUITY OF CONTINUOUS-TIME SIGNALS

Although the title of this section may appear surprising, it should be clear that a CT signal is a function which depends on a continuous parameter (here, time) but that this function can also be discontinuous.

Using exactly the same concepts as in classical analysis, a random CT signal is said to be continuous if $X(t + \Delta t)$ converges to $X(t)$ when $\Delta t \to 0$. Thus there are as many kinds of continuity concepts as there are types of convergence. As mean square continuity will be discussed in detail in a next section, we will look here at continuity with probability one. Thus $X(t)$ is continuous with probability one in t_0 if

$$\lim_{\Delta t \to 0} X(t_0 + \Delta t) = X(t_0) \tag{5.68}$$

with probability one. To clarify this concept we take the following example: consider a random point s uniformly distributed in the closed interval $[0, 1]$ and let the signal $X(t; s)$ be defined for $0 \le t \le 1$ by $X(t; s) = 0$ if $t < s$ and $X(t; s) = 1$ if $t \ge s$. For any value s_0 of the random parameter s, $X(t; s)$ is a discontinuous function with a step 1 at the time $t = s_0$. Thus any trajectory of $X(t; s)$ is discontinuous with probability one, which does not prevent the signal $X(t; s)$ from being continuous at any time instant t with probability one. In fact, the complementary event is that $X(t; s)$ is discontinuous in t_0 or that

$$\lim_{\Delta t \to 0} X(t_0 + \Delta t\ ;\ s) \ne X(t_0\ ;\ s) \tag{5.69}$$

This event appears only if $s = t_0$, which is an event of zero probability. Thus $X(t; s)$ is continuous almost everywhere. Compared with the previous discussion, this is not in contradiction with the fact that $X(t; s)$ is a discontinuous function of t with probability one.

We conclude by verifying that $X(t; s)$ is also continuous in quadratic mean at any time t. In fact, the RV $|X(t + \Delta t) - X(t)|$ takes the value 1 with the probability Δt and 0 with the probability $1 - \Delta t$. We then have $E\{[X(t + \Delta t) - X(t)]^2\} = \Delta t$, which tends to zero if $\Delta t \to 0$.

It is obvious that this discussion is not relevant to DT signals, because the concept of continuity in their case is meaningless.

5.7 POINT PROCESSES

Point processes were introduced on p. 4 where an example was given to illustrate the basic ideas. A point process is also sometimes called a stream of random events. We also noted that a point process is a CT signal, and that it can be easily described by a random signal $X(t; s)$ which is the number of points in the interval $[0, t)$ where the origin is arbitrary. Thus $X(t; s)$ is a step-wise function and presents steps of equal amplitude at all the points $\{t_i\}$ of the point process.

To the function $X(t; s)$ describing the point process we can apply all the results given in this chapter, but when we do so we lose the physical meaning of a point process. We give below the more interesting aspects of point processes for signal theory.

5.7.1 Description by Number of Points

As mentioned on p. 164, the increment of the function $X(t; s)$ previously introduced is the number of points of the process appearing in the interval $[t - \tau, t)$. More generally, let us call $N(t, T; s)$ the RV equal to the number of random points of the process in the interval $[t, t + T)$. It is of course a discrete RV taking only the values of non-negative integers. The concept of a family of distributions can be transposed to these RVs. More precisely, this is the distribution associated with the RVs $N_1, N_2, \ldots, N_k$ equal to the numbers of random points appearing in k arbitrary non-overlapping intervals. If this distribution is invariant under any change of the time origin, the point process is said to be strictly *stationary*. These numbers can be recorded physically by a specific device called a counter.

5.7.2 Description by Time Intervals

Instead of counting the number of points in an arbitrary time interval, it can also be interesting to measure the duration between successive events. This can be achieved by a device called a time-amplitude converter, which, as its name implies, converts the duration between two successive events into a pulse with an amplitude proportional to this duration. The concept of duration can even become more complex. For example instead of measuring the duration between successive events, we can measure the duration between one event and the kth successive event.

But the most fundamental difference can be explained in the following example. Consider an object subject to birth and death in the general sense. This could be the case of a living being, but also that of an object for which the birth is the time of its construction and the death the time when it is no longer used or useful. We call the lifetime the duration between birth and death, and common sense indicates that the lifetime of most objects is random. On the other hand, if we consider the duration between an arbitrary time (posterior to birth) and death, we introduce the concept of survival time. The same concepts can be applied to point processes: the duration between two successive events is called the *lifetime* while that between an arbitrary time and the first point of the process is the *survival time*. These quantities are of course continuous RVs, with different probability distributions. If they are the same, we have the very specific case of a point process which does not age.

5.7.3 Description as Distribution Process

In signal processing problems it is interesting to associate to a point process a signal to which many signal theory results can be applied. To explain this point, we return to the example of the photomultiplier discussed on p. 4. There we stated that the output signal appears as a random sequence of identical pulses. To describe pulse signals the appropriate means is the use of distribution, especially the Dirac distribution. (For further information on the properties of this, see [Oppenheim], p. 515 and [Picinbono], p. 41). For example, a signal pulse at time t_0 is described by the signal $\delta(t-t_0)$. Similarly, a periodic sequence of pulses at time instants $t_i = iT$ is called the Dirac comb signal (see [Picinbono], p. 45) and is represented by

$$x(t) = \delta_T(t) = \sum_i \delta(t - t_i) \tag{5.70}$$

With these ideas we can associate to any point process, i.e., a random distribution of points $\{t_i\}$, a random signal represented by

$$x(t) = \sum_i \delta(t - t_i) \tag{5.71}$$

It is important to note that the randomness of this signal is entirely due to the randomness of the distribution of the time instants t_i. Compared with (5.70), this signal could be called a random comb signal, but this expression is rarely used.

It is possible to generalize the situation a little more, again taking the example on p. 4. As its name implies, a photomultiplier contains a process of multiplication. Roughly speaking, the idea is as follows: as the charge of the electron is extremely small, each photoelectron emitted by the cathode generates, by secondary effect on other electrodes, a great number of other electrons emitted at about the same time. A multiplication process then takes place due to the secondary effect, but the number of electrons generated from one initial photoelectron is random. Consequently, not all the impulses in (5.71) have the same amplitude. A more convenient model for the signal generated is

$$x(t) = \sum_i A_i \ \delta(t - t_i) \tag{5.72}$$

where the A_is are random amplitudes. The signal $X(t)$ is then random both by reason of the randomness of the time instants of the point process and the amplitudes of the pulses.

Random signals such as (5.71) and (5.72) are sometimes called random distributions because of the term δ, which indicates that they should be manipulated with some care in view of the singular properties of distributions. In particular it is well known that the product of distributions has no meaning, and we cannot, without special treatment, calculate the covariance function as defined by (5.12). On the other hand, the output of a linear filter is very easy to calculate, since it is well known that the Dirac distribution is the neutral element of the convolution (see [Picinbono] p. 42).

5.7.4 Random Signals Generated by Point Processes

Point processes can be considered *per se* or as the source of certain random signals. Returning again to the example on p. 4, we can consider the statistical distribution of the points t_i where electrons are emitted (which is one of the most important topics in the field of statistical optics research), or we can consider signals generated by a random sequence of pulses. The most common method is to study the linear filtering of signals such as (5.71) and (5.72). If the latter signal is applied at the input of a linear filter described by (5.56), we obtain the output

$$X(t) = \sum_i A_i \ h(t - t_i) \tag{5.73}$$

To calculate the moments of this expression, it is interesting to change it into another form using the description of a point process presented in 5.7.1. The increment $dN(\theta)$ of the function $N(t)$ introduced there is the number of random points in the interval $[\theta, \theta + d\theta)$. When $d\theta$ tends to zero, $dN(t)$ can only take the values 0 or 1. This is a characteristic property of a point process without accumulation or without the possibility of presenting two time instants t_i and t_j such that $|t_i - t_j|$ is arbitrarily small. Point processes with the accumulation property are very specific and are not considered in what follows. Supposing first that $A_i = A$, which is a constant non-random amplitude, (5.73) can then be written as

$$X(t) = A \int h(t - \theta)\, dN(\theta) \tag{5.74}$$

which is certainly the most interesting expression for further calculations. In order to take into account the random amplitude, let us introduce a random function $M(t)$ such that $dM(\theta) = 0$ if $dN(\theta) = 0$ and $dM(\theta) = A(\theta)$ if $dN(\theta) = 1$. This means that the interval $[\theta, \theta + d\theta)$ contains one random point t_i and that the associated amplitude is $A(\theta)$. With this notation (5.73) becomes

$$X(t) = \int h(t - \theta)\, dM(\theta) \tag{5.75}$$

which gives (5.74) as a particular case.

With these expressions we can easily calculate the moments of $X(t)$. For example, the output mean value (5.57) becomes

$$m_X(t) = \int h(t - \theta)\, E[dM(\theta)] \tag{5.76}$$

In many situations we have $E[dM(\theta)] = \alpha\, d\theta$, which gives

$$m_X = \alpha \int h(t - \theta)\, d\theta = \alpha G(0) \tag{5.77}$$

where $G(\nu)$ is the frequency response of the filter. Let us explain the physical meaning of α. If the amplitude A is constant and equal to 1, we have $dM(\theta) = dN(\theta)$ and then

$$\alpha(\theta)\, d\theta = E[dM(\theta)] = E[dN(\theta)] = \lambda\, d\theta \tag{5.78}$$

Thus $\lambda\, d\theta$ is the mean number of points of the point process in the interval $[\theta, \theta + d\theta)$. In consequence λ is called the *density* of the point process. For stationary point processes this density is constant, as in (5.77) and (5.78). Otherwise λ can be a function of time $\lambda(\theta)$.

Similarly, the covariance function can be written, as in (5.58),

$$\gamma_X(t_1, t_2) = \iint h(t_1 - \theta_1)\, h(t_2 - \theta_2)\, E[dM(\theta_1)\, dM(\theta_2)] - m_X(t_1)\, m_X(t_2) \tag{5.79}$$

Here also stationarity conditions will introduce simplifications in the calculation, with the consequence that γ_X is only a function of $t_1 - t_2$. With more complexity kth order moments can be obtained by an obvious extension of (5.59).

5.8 SECOND ORDER RANDOM SIGNALS

The concept of second order random signals is defined on p. 169, and Chapter 3 is largely devoted to the properties of second order random vectors. We will now combine these results to discuss some additional properties. Note that the results presented below are quite general because the only assumption introduced in this section is that the signal $X(t)$ has a finite second order moment. This happens very often with signals encountered in most applications. We shall examine here problems of the analysis of second order random signals, systematically using the concept of mean square convergence. Let us recall that a CT signal $X(t; s)$ is a function of a continuous parameter about which we can ask questions concerning continuity, differentiation and integration. Throughout this section, unless otherwise indicated, we consider complex-valued scalar signals, and the transposition of results to real-valued signals poses no problem.

5.8.1 Mean Square Continuity

The concept of continuity was first introduced in Section 5.6, and we now restrict our attention to mean square continuity only.

Proposition. The signal $X(t; s)$ is mean square continuous at the time $t = t_0$ if and only if its covariance function $\gamma(t, \theta)$ is continuous at the point $t = \theta = t_0$.

Proof. Using the criterion (3.117), the mean square convergence is equivalent to

$$\lim E[X(t_0 + \Delta t)\, X^*(t_0 + \Delta \theta)] = c \tag{5.80}$$

when Δt and $\Delta\theta$ tend independently to zero. As the l.h.s. of (5.80) introduces the term $\gamma(t_0 + \Delta t, t_0 + \Delta\theta)$, we immediately obtain the result.

It is interesting to note that if $\gamma(t, \theta)$ is continuous at any point $t = \theta$, it is everywhere continuous. In fact, if $\gamma(t, \theta)$ is continuous in t_1, t_1, then $X(t_1 + \Delta t)$ converges to $X(t_1)$. The same can be said for the point t_2. As a result, when Δt and $\Delta\theta$ tend to zero, we obtain

$$\lim E[X(t_1 + \Delta t)\, X^*(t_2 + \Delta \theta)] = E[X(t_1)\, X^*(t_2)] \tag{5.81}$$

which is equivalent to

$$\lim \gamma(t_1 + \Delta t\, ;\, t_2 + \Delta \theta) = \gamma(t_1, t_2) \tag{5.82}$$

We shall give examples of mean square continuous and discontinuous signals later after having enlarged our knowledge of models of random signals.

5.8.2 Mean Square Differentiability

By definition the mean square derivative of the signal $X(t; s)$ is the mean square limit of $(1/\Delta t)[X(t + \Delta t) - X(t)]$ when $\Delta t \to 0$.

Proposition. The signal $X(t)$ has a mean square derivative at the time $t = t_0$ if and only if the covariance function $\gamma(t, \theta)$ is such that $\partial^2\gamma/\partial t\partial\theta$ exists and is finite at the point $t = \theta = t_0$.

Proof. Again using the criterion (3.117), we can say that the derivative at $t = t_0$ exists if and only if

$$\lim \frac{1}{\Delta t\ \Delta\theta} E\left\{[X(t_0 + \Delta t) - X(t_0)]\,[X^*(t_0 + \Delta\theta) - X^*(t_0)]\right\} = c \tag{5.83}$$

Using the definition of the covariance, this is equivalent to writing

$$\lim \frac{1}{\Delta t\ \Delta\theta}\,[\gamma(t_0 + \Delta t, t_0 + \Delta\theta) - \gamma(t_0 + \Delta t, t_0) - \gamma(t_0, t_0 + \Delta\theta) + \gamma(t_0, t_0)] = c \tag{5.84}$$

This is precisely the definition of the derivative $\partial^2\gamma(t, \theta)/\partial t\partial\theta$ at $t = \theta = t_0$. Using the same method as for mean square continuity, it is possible to show that if $\gamma(t, \theta)$ has a second derivative at $t = \theta$, this second derivative exists for any t and θ.

It is of some interest to study certain properties of the mean square derivative of $X(t)$ written $\dot{X}(t)$. Using the same procedure, we easily find the crosscovariance between a signal and its derivative as well as the covariance of this derivative. The results are

$$E\{X(t_1)\dot{X}^*(t_2)\} = \frac{\partial\gamma(t_1, t_2)}{\partial t_2} \tag{5.85}$$

$$E\{\dot{X}(t_1)X^*(t_2)\} = \frac{\partial\gamma(t_1, t_2)}{\partial t_1} \tag{5.86}$$

$$E\{\dot{X}(t_1)\dot{X}^*(t_2)\} = \frac{\partial^2\gamma(t_1, t_2)}{\partial t_1\partial t_2} \tag{5.87}$$

This procedure can obviously be extended to higher-order derivatives.

Finally, let us recall that, as explained in Section 5.6, mean square continuity or differentiability does not mean that any sample function $X(t; s_0)$ is a continuous or differentiable function.

5.8.3 Mean Square Series or Integrals

Although this problem has already been treated for series in a case of uncorrelated RVs, we shall reconsider the same problem assuming now that $X[k]$ is a DT random signal. Starting from (3.120), we see that the partial sum is

$$S_k = \sum_{l=1}^{k} h[l]\,X[l] \tag{5.88}$$

Using the covariance $\gamma[p, q]$ of $X[l]$ we deduce that the criterion (3.117) can be written as

$$\sum_{p=1}^{k} \sum_{q=1}^{l} h[p]\, h^*[q]\, \gamma[p, q] \rightarrow c \tag{5.89}$$

which means that the mean square convergence of S_k is equivalent to the convergence of series of the general term $h[p]\, h^*[q]\, \gamma[p, q]$.

The situation is the same for integrals. In other words, the integral

$$I = \int_a^b h(t)\, x(t)\, dt \tag{5.90}$$

where the interval $[a, b]$ is finite or infinite is convergent in the mean square sense if the covariance function $\gamma(t, \theta)$ is such that the integral

$$J = \int_a^b \int_a^b h(t)\, h^*(\theta)\, \gamma(t, \theta)\, dt\, d\theta \tag{5.91}$$

exists. The same result can be obtained for integrals of the type

$$K = \int_a^b h(t)\, dX(t) \tag{5.92}$$

used, for example, in (5.74) and (5.75). In this case it is sufficient to replace $\gamma(t, \theta)\, dt\, d\theta$ by $E[dX(t)\, dX(\theta)]$ in (5.91).

5.9 ENERGY AND POWER

Taking once again the example of resistors described on p. 4, let us call $X(t; s)$ the voltage observed accross a given resistor. Because of the Ohm law the mean value of $X(t; s)$ is null, but there are random fluctuations, especially due to the motion of the electrons at a non-zero temperature. The quantity $X^2(t; s)$ is proportional to the instantaneous power of the noise generated by the resistor. More generally, any quantity quadratic in X has an energetic interpretation, and we shall give the most important definitions used hereafter. We assume that $X(t; s)$ is zeromean and complex-valued, as for example when using complex notations in circuit theory.

The quantity

$$P(t; s) = |X(t; s)|^2 \tag{5.93}$$

is called the *random instantaneous power*. If $X(t; s)$ is second order, as assumed in the previous section, the signal $X(t; s)$ has a finite *instantaneous mean power* defined by (5.22), or

$$P(t) = E\{P(t; s)\} = \gamma(t, t) \tag{5.94}$$

If $X(t; s)$ is second order stationary, the instantaneous mean power is constant and becomes

$$P = \gamma(0) \tag{5.95}$$

where $\gamma(\tau)$ is the correlation function defined by (5.17). Comparing this with (5.18), we see that the mean power is also the variance of the signal, and we make no distinction between P and σ^2 for zeromean signals.

Noting that the energy is the time integral of the power, we can translate all the previous concepts to energy. For example, the *random energy* of a signal is defined by

$$J(s) = \int |X(t; s)|^2 \, dt \tag{5.96}$$

where the integral goes from $-\infty$ to $+\infty$. This random energy has no reason to be finite. If the integral is limited to a finite time interval, we obtain the energy of the signal in the corresponding interval. The *mean energy* is obtained by taking the expected value, which, if it is finite, can be written

$$J = \int P(t) \, dt = \int \gamma(t, t) \, dt \tag{5.97}$$

It is obvious that for stationary signals for which $P(t)$ is a constant given by (5.95), the mean energy is infinite. In the deterministic context this situation appears for periodic signals which have an infinite energy but can have a finite power.

For practical applications it is also interesting to introduce time averages, as considered in Section 5.5. By extending the integral (5.63) over the past we can introduce the *time-average power* defined by

$$P_T(s) = \frac{1}{2T} \int_{-T}^{+T} P(t; s) \, dt \tag{5.98}$$

and if some ergodic conditions are met we have, as is a case for the mean value considered in (5.63),

$$\lim_{T \to \infty} P_T(s) = P \tag{5.99}$$

where the limit can be considered in the mean square sense or with probability one. It is obvious that if the random energy $J(s)$ defined by (5.96) is finite, the limit of $P_T(s)$ when $T \to \infty$ is null. The same holds for deterministic signals: if they have a finite energy, the time average power is null. These definitions will be used extensively in the next chapter in the discussion of the harmonic decomposition of random signals.

5.10 ADDITIONAL REMARKS

Although most of the results in the previous sections were devoted to real scalar random signals, the extension to complex and *vector signals* was indicated at the end of Section 1. The concept of the covariance matrix of a complex vector signal is introduced by

(5.14). The linear filtering of these signals is given by a straightforward extension of (5.56). For example, suppose that $\mathbf{X}(t)$ is a complex vector of $\mathbb{C}^n$ and $\mathbf{Y}(t)$ a complex vector of $\mathbb{C}^m$. The filtering relation can be written as

$$\mathbf{Y}(t) = \int \mathbb{h}(\theta)\, \mathbf{X}(t - \theta)\, d\theta \tag{5.100}$$

This makes use of an impulse function $\mathbb{h}(t)$, which is of course an $m \times n$ matrix.

It is of more interest to consider signals which depend on a vector parameter. The most usual case appears when this parameter corresponds to the space, at 2 or 3 dimensions. For example, a random field can be described by a function such as $X(\mathbf{r}; s)$ where $\mathbf{r}$ means the space and s the probabilistic parameter. The concept of second order stationarity is unchanged. This means that the covariance function $\gamma(\mathbf{r}_1, \mathbf{r}_2)$, defined as in (5.12), is now only a function of $\mathbf{r}_1 - \mathbf{r}_2$. But there is another degree of freedom. For example, a field can be stationary and also homogeneous, in which case the covariance function is invariant under translation and rotation. This means that now $\gamma(\mathbf{r}_1, \mathbf{r}_2)$ is only a function of $|\mathbf{r}_1 - \mathbf{r}_2|$.

The concept of point processes can be easily extended to vector parameters. For example, a plane point process is simply a random distribution of points on a plane. In this case there is a more significant change in respect to the scalar case, where the concept of the number of points is still valid provided that time intervals are replaced by closed domains in the plane. On the other hand, the distance between successive points, called a duration for the time axis, no longer has any meaning on a plane, because there is no order in $\mathbb{R}^n$ when $n > 1$. This is exactly the same situation as that appearing in multidimensional filtering (see [Picinbono], p. 149). The concept of causality, related to an order in $\mathbb{R}$, cannot be defined for multidimensional filters because the concept of past and future has no clear meaning in a plane. Point processes in a plane have an obvious application when discussing noise in plane images, examples of which will be discussed below.

PROBLEMS

5.1 Calculate the DF defined by (5.1) for the signal $X(t; s)$ defined by (5.46) in terms of the DF $F(x)$ of the RV $X(s)$.

5.2 Calculate the DF defined by (5.1) for the signal $X(t; s) = X(s) f(t)$, where $X(s)$ is an RV defined by the DF $F(x)$ and $f(t)$ a positive deterministic signal. Express the result in terms of $F(x)$. Extend the result when $f(t)$ is no longer positive.

5.3 Consider the random signal $X(t) = f(At)$ where $f(.)$ is a strictly monotonic function and A an RV with the DF $F(a)$. Calculate the DF of $X(t)$ defined by (5.1) in terms of $F(a)$.

5.4 Consider the signal $X(t; s) = \exp[-A(s)t]$ where $A(s)$ is a continuous RV with the PDF $p(a)$.

(a) Using the results concerning two-sided Laplace transforms, show that the signal $X(t; s)$ is second order in an interval $[T_1, T_2]$.
(b) Express the mean value $m(t)$ and the variance $\sigma^2(t)$ of $X(t; s)$ in terms of the Laplace transform $P(\alpha)$ of $p(a)$.
(c) Make the same calculation for the covariance function $\gamma(t_1, t_2)$.
(d) Complete the calculation when $p(a)$ is the PDF of a uniform and of a normal RV.

5.5 A strict white noise $X[k]$ is a discrete-time signal such that the RVs $X[k]$ are IID with moments m_i. Suppose that $m_1 = 0$.
(a) Calculate the correlation function $\gamma[k]$ of $X[k]$ in terms of m_2.
(b) A quadratic device transforms $X[k]$ into $Y[k] = X^2[k]$. Calculate the mean value and the covariance function of $Y[k]$ in terms of m_2 and m_4. For this calculation use the results presented on p. 108.
(c) Is $Y[k]$ still a strict white noise?

5.6 The signals $X_1(t)$ and $X_2(t)$ observed at the outputs of two sensors can be written as $X_1(t) = S(t) + N_1(t)$ and $X_2(t) = S(t) + N_2(t)$, where the signals $S(t)$, $N_1(t)$, and $N_2(t)$ are zeromean, uncorrelated, and jointly stationary. Calculate the autocorrelation functions of $X_1(t)$ and $X_2(t)$ and the crosscorrelation function between $X_1(t)$ and $X_2(t)$. Calculate also the autocorrelation function of the sum $X_1(t) + X_2(t)$.

5.7 Introducing the random vector **X** with components $X(t_1)$, $X(t_2)$,..., $X(t_n)$, show that the condition (5.36) is equivalent to saying that the covariance matrix of **X** is non-negative definite.

5.8 Suppose that $X(t)$ is a second order stationary signal with the correlation function $\gamma(\tau)$. Write the covariance matrix of the vector $[X(t), X(t + \theta), X(t + 2\theta)]^T$ in terms of $\gamma(\tau)$. Using the results of Problem 3.10, show that the rectangular function $f(\tau)$ equal to 1 if $|\tau| \leq T$ and to 0 if $|\tau| > T$ cannot be a correlation function.

5.9 Consider the random signal $X(t; s) = M(s) \cos\omega t + N(s) \sin\omega t$ where $M(s)$ and $N(s)$ are two second order zeromean RVs. Let us call γ_M and γ_N the variances of M and N respectively and γ_{MN} the covariance between M and N.
(a) Calculate the mean value $m(t)$ of $X(t; s)$.
(b) Calculate the covariance function $\gamma(t_1, t_2)$ of $X(t; s)$ in terms of the quantities γ.
(c) Deduce the necessary and sufficient condition on the γs ensuring that the signal $X(t; s)$ is second order stationary.

5.10 Consider the signal $X(t; s) = A(s) \cos[\omega t + \Phi(s)]$ where $A(s)$ and $\Phi(s)$ are two independent RVs and $A(s) \geq 0$.
(a) Write the conditions on the RV Φ ensuring that $X(t; s)$ is zeromean and second order stationary.
(b) Show that these conditions are satisfied if Φ is uniformly distributed in $[0, 2\pi)$.
(c) Show the same if the PDF of Φ is $p(\phi) = (1/2\pi)(1 + a \cos 3\phi)$, $|a| \leq 1$.
(d) Find the necessary and sufficient condition on $p(\phi)$ ensuring that $X(t; s)$ is zeromean and second order stationary.

5.11 Combining (5.47) and (5.49) we consider the complex random signal $Z(t;\ s) = A(s)\exp j[\omega t + \Phi(s)]$.

(**a**) Show that $Z(t;\ s)$ is strict-sense stationary if and only if $A(s)$ and $\Phi(s)$ are independent and $\Phi(s)$ is uniformly distributed in $[0,\ 2\pi)$.

(**b**) Study the same problem for the signal of Problem 5.10.

(**c**) Deduce that the signal of Problem 5.9 is strictly stationary if and only if the RVs M and N are spherically invariant, as defined on p. 103.

5.12 **Stationary and periodic signals.** Assuming that $X(t)$ is a zeromean second order stationary signal with correlation function $\gamma(\tau)$, we define the signal $Y(t)$ as in 5.3.3 (**b**). Show that its covariance function $\gamma_Y(t_i,\ t_j)$ defined by (5.12) is only a function of $t_i - t_j$. Deduce that the correlation function $\gamma_Y(\tau)$ defined by (5.17) is periodic. Compare $\gamma_Y(\tau)$ and $\gamma_Y(T - \tau)$ and deduce that $\gamma_Y(\tau)$ can be constructed from the values of $\gamma(\tau)$ corresponding to $0 \leq \tau < T/2$.

5.13 Suppose that the correlation function $\gamma(\tau)$ of a zeromean stationary random signal $X(t;\ s)$ satisfies $\gamma(T) = \gamma(0)$.

(**a**) Deduce that $X(t)$ and $X(t + T)$ are equal with probability one.

(**b**) Deduce from the stationarity that $X(t)$ and $X(t + kT)$ are equal with probability one.

(**c**) Deduce that $\gamma(\tau)$ is a periodic function with the period T.

(**d**) Assuming that $\gamma(\tau)$ has a Fourier transform $\Gamma(\nu)$, it is possible to deduce from (5.36) that $\Gamma(\nu) \geq 0$. Use this property to show that if $\gamma(0) = \gamma(T)$, then $\gamma(\tau)$ is periodic with the period T.

5.14 Using the same notations as in the previous problem suppose now that $\gamma(T) = -\gamma(0)$.

(**a**) Deduce that $X(t) = (-1)^k X(t + kT)$ with probability one.

(**b**) Show that the correlation function is periodic with the period $2T$ and that $\gamma(T + \tau) = \gamma(T - \tau)$.

(**c**) Deduce the same results from the property $\Gamma(\nu) \geq 0$.

5.15 We want to show that the property $X(t;\ s) = X(t + kT;\ s)$ with probability one does not mean that almost all the trajectories $X(t;\ s_0)$ of $X(t;\ s)$ are periodic functions. For this purpose consider the deterministic signal $f(t)$ periodic with period 1 and let $Y(t;\ s)$ be the signal defined as follows: the variable s is random and uniformly distributed in the interval $[0,\ 1]$ and $Y(t;\ s) = 0$ if $t \neq s$ and 1 if $t = s$.

(**a**) Show that $X(t;\ s) = Y(t;\ s) + f(t)$ satisfies the property given at the beginning of this problem.

(**b**) Show that all the trajectories of $Y(t;\ s)$ are non-periodic functions.

(**c**) Explain the difference between the terms "periodic signals" and "signals with periodic trajectories".

5.16 A real random signal $X(t;\ s)$ is said to be second order *cyclostationary* if its mean value $m(t)$ and covariance functions $\gamma(t_1,\ t_2)$ satisfy $m(t + kT) = m(t)$ and $\gamma(t_1 + kT,\ t_2 + kT) = \gamma(t_1,\ t_2)$ for any integer k and a given period T.

(**a**) Show that a second order stationary signal is also second order cyclostationary with any value of T.

(**b**) Show that a deterministic and periodic signal is cyclostationary.

(c) A periodic function of period T is completely known if it is known in the interval $[0\ T]$. Introducing the plane $t_1 \times t_2$ show the same kind of property for the covariance function $\gamma(t_1, t_2)$ of a cyclostationary, of a stationary and of a stationary and periodic signal.

5.17 Amplitude modulation. Suppose that every T seconds a source of information sends a scalar message M_i and consider the signal $X(t)$ equal to $M_i\, f(t)$ if $(i-1)\,T < t \leq iT$, where $f(t)$ is a deterministic signal and i an integer. We assume that the M_is are a sequence of IID RVs with mean value μ and variance σ^2. This specifies the second order properties of the random signal $X(t; s)$.

(a) Calculate its mean value $m(t)$.

(b) Calculate its covariance function $\gamma(t_1, t_2)$.

(c) Show that $X(t; s)$ is cyclostationary as introduced in the previous problem.

(d) In the plane $t_1 \times t_2$ specify the domain where $\gamma(t_1, t_2)$ is non-null.

5.18 Random shift. Consider the random signal $X(t; s)$ and the RV $A(s)$ defined on the same probability space. We introduce the random shifted signal $Y(t; s) = X[t - A(s); s]$, where $A(s)$ means a random shift. Let us call $m(t)$ and $\gamma(t_1, t_2)$ the mean value and the covariance function of $X(t; s)$.

(a) Calculate the mean value and the covariance function $m_Y(t)$ and $\gamma_Y(t_1, t_2)$ of $Y(t; s)$.

(b) Complete the calculation in the case where $X(t; s)$ is second order stationary and interpret the result.

(c) Suppose now that $X(t; s)$ is cyclostationary, as defined in Problem 5.16. Show that if A is uniformly distributed in $[0, T]$, the shifted signal $Y(t)$ becomes stationary.

5.19 Using the results of the previous problem show that the covariance function of the amplitude modulation signal introduced in Problem 5.17 shifted with RV uniformly distributed in $[0, T)$ has a triangular shape if $f(t) = c$.

5.20 The impulse response of a discrete-time exponential filter is $h[k] = 0$ for $k < 0$ and $h[k] = a^k$, $|a| < 1$ for $k \geq 0$. Let $X[k]$ be the signal null for $k < 0$ and such that its values for $k \geq 0$ are IID RVs with mean value m and variance σ^2.

(a) Show that $X[k]$ is semistationary as defined on p. 180.

(b) Calculate the output $Y[k]$ of the exponential filter when the input is $X[k]$.

(c) Calculate the mean value of $Y[k]$ and indicate what happens when $k \to +\infty$.

(d) Calculate the covariance function of $Y[k]$ and show that $Y[k]$ becomes second order stationary when $k \to +\infty$.

5.21 Transpose all the results of Section 5.8 to second order stationary signals.

5.22 The correlation function of a real weak unit white noise $X(t)$ is $\delta(\tau)$, where $\delta(.)$ is the unit impulse function. Applying $X(t)$ at the input of a linear filter with a real impulse response $h(t)$ we obtain the signal $Y(t)$.

(a) Calculate the correlation function $\gamma_Y(\tau)$ of $Y(t)$ in terms of $h(t)$ and show that it is an even function.

(b) Give the condition on $h(t)$ ensuring that $Y(t)$ is a second order signal.

(c) Give the condition on $h(t)$ ensuring that $Y(t)$ is mean square continuous.

(d) Give the condition on $h(t)$ ensuring that $Y(t)$ has a mean square derivative and calculate the correlation function $\gamma_Y(\tau)$ of this derivative.

5.23 Apply the results of the two previous problems to the following filters.

(**a**) Exponential filter such that $h(t) = u(t)\exp(-at)$, where $u(t)$ is the unit step function.

(**b**) Rectangular filter such that $h(t) = 1/T$ if $0 \leq t \leq T$ and 0 otherwise.

(**c**) Gaussian filter such that $h(t) = \exp(-a^2 t^2)$.

5.24 Discuss the mean square ergodicity for mean value measurement of the three signals studied in the previous problem.

5.25 The correlation function of a real discrete-time, zeromean, second order, and unit white noise is $\delta[k]$, where $\delta[\ ]$ is the Kronecker-delta function. Applying this signal to a real linear filter with the impulse response $h[k]$ we obtain at the output a signal $X[k]$. Calculate the correlation function $\gamma_X[k]$ of this signal.

5.26 Calculate explicitly the correlation function $\gamma_X[k]$ of the previous problem for the following filters.

(**a**) $h[k] = a^k$ for $k \geq 0$ and $h[k] = 0$ otherwise.

(**b**) $h[0] = 1$; $h[1] = -a$ and $h[k] = 0$ for other values of k.

(**c**) $h[0] = a$; $h[1] = -1$ and $h[k] = 0$ for other values of k.

(**d**) Interpret the results obtained in (**b**) and (**c**).

5.27 **Stationary Poisson process.** A stationary Poisson process is a point process defined as follows: (**a**) the number of points t_i in the interval $[t, t + T)$ is a Poisson RV defined by the mean value λT, λ being called the density of the process ; (**b**) the RVs N_i introduced in Section 5.7.1 are independent.

(**a**) Let t be an arbitrary time instant and L the RV equal to the time interval between t and the first point of the process appearing after t. Calculate the PDF $p_L(l)$ of L.

(**b**) Let S be the RV equal to the lifetime of the process, as defined in Section 5.7.2. Calculate the PDF $p_S(s)$ of S. Deduce that a Poisson process does not age.

(**c**) Calculate the mean value and the variance of L.

5.28 **Poisson process** (continued). Suppose that the point process of time instants where photoelectrons are emitted is a Poisson process of density λ. Suppose that, by the secondary effect described in Section 5.7.3, a photoelectron appearing at t_i generates N_i other electrons. Assuming that the numbers N_i are IID random variables defined by the characteristic function $\phi_N(u)$, calculate the distribution of the total number of electrons appearing in the interval $[0, T)$. Assuming that $\phi_N(u)$ is the characteristic function of a Poisson RV of mean value m, calculate the mean value and the variance of the total number of electrons appearing in $[0, T)$.

5.29 **Bernouilli process.** Let $u[k]$ be a DT white noise which is a sequence of IID random variables. Suppose that these RVs take only the values 0 and 1 and let p be the probability that $u[k] = 1$. The signal appears as a random sequence of 0 and 1 and is called a *Bernouilli process*. Calculate the probability distribution of the random distance between two successive 1s. Calculate the mean value and the variance of this distribution.

5.30 Make the same calculation as in the previous problem for the distance between two successive 0s.

5.31 Considering the signal $u[k]$ of the previous two problems as a discrete-time point process similar to that represented in Figure 1.3, show that this point process does not age.

5.32 Let $u[k]$ be a DT signal (k integer) taking only integer values. This signal is processed by a parity operator system P. This system is characterized by the point that its output $v[k]$ at time k depends *only* on the input $u[k]$ at the same instant k (no memory effect), and furthermore $v[k] = 0$ if $u[k]$ is even and $v[k] = 1$ if $u[k]$ is odd.

(a) Suppose first that $u[k]$ takes only the values 0 or 1. Compare $u[k]$ and $v[k]$.

(b) Suppose now that $u[k]$ is a DT strict white noise, which means that the $u[k]$s are a sequence of IID random variables. Is it possible to say that $v[k]$ is a strict white noise?

(c) Suppose that $u[k]$ is a strict white noise characterized by the probabilities $p_n = P\{u[k] = n\}$, n being an integer which can be positive, negative or null. Calculate the mean value m_V and the correlation function $\gamma_V[q]$ of $v[k]$ in terms of the p_ns.

(d) Complete the calculation when p_n corresponds to a Poisson distribution of mean value m.

5.33 Let $u[k]$ be a Bernouilli process, as defined in Problem 5.29, and characterized by the probability $p = P\{u[k] = 1\}$.

(a) Let $x[k]$ be the signal defined by $x[k] = u[k] + u[k-1]$. Calculate the possible values of $x[k]$ and the corresponding probabilities.

(b) Calculate the mean value m_X and the variance v_X of $x[k]$.

(c) Calculate the correlation function $\gamma_X[q]$ of $x[k]$ and verify that $\gamma_X[0] = v_X$.

(d) The signal $x[k]$ is processed by the parity system described in the previous problem. Let $y[k]$ be the output of this system when its input is $x[k]$. Calculate the probability $\pi = P\{y[k] = 1\}$ in terms of p. Deduce the mean value and the variance m_Y and v_Y of $y[k]$. Compare the mean values and the variances of $x[k]$ and $y[k]$ when $p = 1/2$.

(e) Using the properties of the signal $u[k]$, calculate in terms of p the probabilities $p_Y[k, l] = P\{(y[k] = 1).(y[l] = 1)\}$, $k \neq l$.

(f) Deduce from the previous calculation the correlation function $\gamma_Y[q]$ of $y[k]$.

(g) Calculate the power spectra of $x[k]$ and $y[k]$ when $p = 1/2$. Compare the two signals $x[k]$ and $y[k]$.

Chapter 6

Spectral Properties of Random Signals

6.1 INTRODUCTION

In science and engineering literature the term "spectral" refers to the idea of frequency distribution. A spectral analyzer is a device measuring the power in a specified frequency band and the spectrum of a signal is related to its Fourier representation dealt with primarily for deterministic signals. We shall devote the present chapter to this fundamental topic, beginning with the reason for its importance. The spectral representation of a signal is, in principle, the decomposition of that signal as a sum of complex exponential signals in the form $\exp(j\omega t)$. We may wonder at the emphasis given to such signals, which are the basis of the Fourier analysis, or at the relevance of their mathematical analysis in practical applications. The answer to this, analyzed in [Picinbono], p. 12 and [Oppenheim] p. 167, is that complex exponential signals are the *eigensignals of linear filters*, introducing the concepts of transfer function or of frequency response. Because of the practical importance of linear filters, or linear and time-invariant systems, it is essential to be able to decompose any signal, regardless of whether it is random or deterministic, as a sum of complex exponential signals. This is the foundation of spectral representation.

In the case of random signals the question is complex. Roughly speaking, we may be interested in spectral representation of the *signal itself*, which introduces another random function in the frequency domain, or in certain mean values, especially the *covariance* or *correlation function.* These functions are no longer random, and knowledge acquired in the study of deterministic signals can be applied without restriction. But the specific properties of the covariance functions studied in the previous chapter will imply counterparts in the frequency domain, analyzed later.

After the question: representation of what?, comes the question, in which terms? Let us recall the expressions used for deterministic signals, which reappear in this context. To a continuous-time (CT) signal $x(t)$ it is possible to associate its Fourier or Laplace transforms defined by

$$X(\nu) = \int x(t)\, e^{-2\pi j \nu t}\, dt \tag{6.1}$$

$$X\{s\} = \int x(t)\, e^{-st} dt \tag{6.2}$$

For a discrete-time (DT) signal $x[k]$, k integer, we have similar relations for the Fourier or z-transform

$$X(\nu) = \sum x[k]\, e^{-2\pi j \nu k} \tag{6.3}$$

$$X\{z\} = \sum x[k]\, z^{-k} \tag{6.4}$$

It is well known that all these relations can be inverted, which means that $x(t)$ or $x[k]$ can be deduced from their transforms. The calculation of the Fourier transform often makes use of the discrete Fourier transform (DFT) which is a specific aspect of Fourier series. Thus the general problem studied here is to find out the consequences in these transforms when the signals are random, as described in the previous chapters.

A few comments on notation: until now all random elements were written with *capital letters* in order to avoid confusion with their possible values, as explained on p. 22 (note especially (2.3) and the discussion which follows). However, as we must be able to distinguish a pair of transforms, and as distribution functions are rarely used in this chapter, a random signal will now be written either $x(t; s)$ or, where no confusion is possible, $x(t)$. As in the previous chapter, the letter s refers to the probabilistic nature of $x(t; s)$.

Similarly, complex exponential signals can be written both in the form $\exp(2\pi j \nu t)$ or $\exp(j\omega t)$, using the normal frequency or the angular frequency $\omega = 2\pi\nu$. The Fourier transforms can be written with both notations and their relationship is well known (see [Picinbono], p. 29). But as the inversion formulae are much more symmetric when using the normal frequency, we shall use this in our calculations.

A final comment on the *organization* of this chapter. The first two sections deal with the spectral representation of the random signal itself. From a mathematical viewpoint these two sections are the most difficult, even though we do not attempt a complete analysis with more abstract arguments, which is outside the scope of this book. Readers more interested in applications can go through these sections fairly rapidly, looking essentially at the form of the equations, before arriving at the most important section devoted to the concept of the *power spectrum* or *spectral density*. This latter section can in fact be read almost independently, but logically it is better to understand clearly the difference between the concepts of the power spectrum of a signal and its representation. Readers or teachers with limited time may wish to go directly to Section 6.4 where the results are presented again in a simpler and more direct way.

6.2 SPECTRAL REPRESENTATIONS OF RANDOM SIGNALS

6.2.1 The Concept of Harmonizable Signals

A random CT signal $x(t; s)$, also written $x(t)$, is said to be harmonizable if it is possible to write it in the form

$$x(t;\, s) = \int d\overline{X}(\nu;\, s)\, e^{2\pi j \nu t} \tag{6.5}$$

which represents a sum of exponential signals with random complex amplitude $d\overline{X}(\nu;\, s)$, written simply $d\overline{X}(\nu)$. The same definition is valid for the DT case, and (6.5) becomes

$$x[k] \;=\; \int_{-1/2}^{+1/2} d\overline{X}(\nu)\, e^{2\pi j \nu k} \tag{6.6}$$

As there is a strong analogy between these two expressions, we will repeat the discussion for the two cases only when some specific differences appear. Instead of immediately discussing the conditions ensuring the decomposition of (6.5) and (6.6) and their precise mathematical meaning, we will follow an intuitive reasoning giving their interpretation and some of their main consequences.

First, if

$$d\overline{X}(\nu) = X(\nu)\, d\nu \tag{6.7}$$

we observe that (6.5) and (6.6) are the inversion formulae deduced from (6.1) and (6.3), which are well known for deterministic signals. Unfortunately the Fourier transform $X(\nu)$ of a random signal appearing in (6.1) and giving $x(t)$ in (6.7) and (6.5) is not always defined. The reason for this is as follows: many signals discussed above are assumed to be stationary, in which case, as seen in Section 5.9, their energy is infinite. But it is well known (see [Picinbono], p. 41), that the Fourier representation of signals with infinite energy, as for example the unit step signal, needs an extension of the classical Fourier representation using the distribution theory.

However, the representation (6.5) contains the case where only single frequencies are present, as for example that of the Fourier series. It is thus sufficient to assume that $\overline{X}(\nu)$ is a step-wise function, such as the distribution function of discrete RVs discussed in Section 2.2.3. If $\overline{X}(\nu)$ is a step-wise function which varies only at some given frequencies ν_k where it exhibits steps of random amplitudes X_k, we can write (6.5) as

$$x(t) = \sum_k X_k \exp(2\pi j \nu_k t) \tag{6.8}$$

Now if $\nu_k = kF$, where F is the fundamental frequency, we arrive at a Fourier series, and it is clear that $x(t)$ becomes periodic with the period $T = 1/F$. It is important to note that there is a marked difference between this situation and that seen for the function $M(t)$ of Section 5.7 on point processes. In fact, the frequencies ν_k where $\overline{X}(\nu)$ has its steps are deterministic and only the amplitude is random, while for the function $M(t)$ both the location of the steps and their amplitude are random. When the function $\overline{X}(t)$ has no step, involving the use of (6.5), we say that $x(t)$ has a *continuous spectral representation.* When, however, (6.5) can be written as (6.8) we say that $x(t)$ has a *discrete spectral representation* or exhibits spectral lines.

Let us now consider the problem of *linear filtering* which, as said before, is the physical motivation of the decomposition (6.5). If $x(t)$ is transformed into $y(t)$ by a linear filter, as in (5.56), we can generally write the same equation for $y(t)$ as (6.5) for $x(t)$, only replacing $d\overline{X}(\nu)$ by

$$d\overline{Y}(\nu) = G(\nu)\, d\overline{X}(\nu) \tag{6.9}$$

where $G(\nu)$ is the frequency response of the filter. As in the case of deterministic signals, this represents the great advantage of the spectral representation, where convolution is replaced by product.

This last property allows us to give a physical interpretation of the function $\overline{X}(\nu)$. Consider the perfect low pass filter F_{ν_0} with a frequency response given by

$$G_{\nu_0}(\nu) = 1 - u(\nu - \nu_0) \tag{6.10}$$

where $u(\nu)$ is the unit step function. This is of course an unrealizable filter, and its impulse function is a distribution. But this problem plays no role in our intuitive discussion. Applying $x(t)$ at its input and calculating the output $y_{\nu_0}(t)$ at $t = 0$, we deduce from (6.5) that

$$y_{\nu_0}(0) = \int_{-\infty}^{\nu_0} d\overline{X}(\nu) = \overline{X}(\nu_0) - \overline{X}(-\infty) \tag{6.11}$$

The result of this equation is that, except for the random variable $\overline{X}(-\infty)$, the random function $\overline{X}(\nu)$ is the value of the output at $t = 0$ of our filter when the input is $x(t)$. Similarly $\overline{X}(\nu_2) - \overline{X}(\nu_1)$, $\nu_1 < \nu_2$, is the output at $t = 0$ of a perfect band-pass filter receiving the input $x(t)$. Thus $\overline{X}(\nu)$ appears as a frequency random distribution function with a property similar to (2.6).

6.2.2 Existence of Harmonizable Signals

We shall now discuss in more detail (6.5), although we will not use all the mathematical ingredients necessary for a completely rigorous treatment. Consider a second order random signal with the covariance $\gamma(t_1, t_2)$ defined either by (5.12) with the complex conjugates or by (5.14) where the vectors are now scalars. This covariance function is said to be harmonizable if there exists a function $\overline{\Gamma}(\nu_1, \nu_2)$ with second order increments $d^2\overline{\Gamma}(\nu_1, \nu_2)$ such that

$$\gamma(t_1, t_2) = \iint d^2\overline{\Gamma}(\nu_1, \nu_2) \exp\{2\pi j(\nu_1 t_1 - \nu_2 t_2)\} \tag{6.12}$$

and

$$\iint |d^2\overline{\Gamma}(\nu_1, \nu_2)| < +\infty \tag{6.13}$$

Note that at present this is only a definition concerning the covariance and its generalized Fourier transform. But properties of the covariance can induce properties on the signal. Thus it is easy to deduce from (6.12) and (6.13) that the covariance is continuous at any point (t, t), which means, as seen in Section 5.8.1, that if the covariance is harmonizable, the signal $x(t)$ is mean square continuous. We now have the fundamental result.

If the covariance of $x(t)$ is harmonizable it is possible to find a random function $\overline{X}(\nu)$ such that the integral appearing in (6.5) is mean square convergent and equal with probability one to $x(t)$. Furthermore, we have

$$E\{\overline{X}(\nu_1)\,\overline{X}^*(\nu_2)\} = \overline{\Gamma}(\nu_1, \nu_2) \tag{6.14}$$

$$E\{d\overline{X}(\nu_1)\,d\overline{X}^*(\nu_2)\} = d^2\overline{\Gamma}(\nu_1, \nu_2) \tag{6.15}$$

The proof of this theorem, not given here, can be found in more specialized texts, such as [Blanc-Lapierre], p. 359, [Cramér], p. 129, [Loève, 1963], p. 474. It is of greater interest to discuss some of the consequences of this result.

The condition (6.13) ensuring that the covariance, and thus the signal $x(t)$, are harmonizable is very broad. To understand this clearly, let us study its meaning in the case of deterministic signals, which are a particular aspect of random signals. Let us call $x(t)$ a deterministic signal and $X(\nu)$ its Fourier transform. Its covariance is of course $\gamma(t_1, t_2) = x(t_1)\,x^*(t_2)$ and (6.12) can be written as

$$\gamma(t_1, t_2) = \iint X(\nu_1)\,X^*(\nu_2)\exp\{2\pi j(\nu_1 t_1 - \nu_2 t_2)\}d\nu_1\,d\nu_2 \tag{6.16}$$

which gives $d^2\overline{\Gamma}(\nu_1, \nu_2) = X(\nu_1)\,X^*(\nu_2)\,d\nu_1\,d\nu_2$. It follows that the condition (6.13) can be written

$$\int |X(\nu)|\,d\nu < +\infty \tag{6.17}$$

a condition met for a broad class of Fourier transforms. Nevertheless, there are some exceptions presented below.

Example 6.1. Harmonizable and non harmonizable signals. It is well known (see [Picinbono], p. 46) that the signal "sign of t", $v(t) = t/|t|$ has an FT equal to $1/j\pi\nu$. This FT does not satisfy (6.17) and then $v(t)$ is not harmonizable. One could object that this could be due to the fact that $v(t)$ is not random. In answer, consider the random signal $x(t; s) = A(s)\,v(t)$, where $A(s)$ is a second order RV with variance σ^2. The covariance of $x(t; s)$ is then $\sigma^2\,v(t_1)\,v(t_2)$, and we again find the same result. Similarly, the very classical signal $x(t) = u(t)\exp(-at)$, $a > 0$, is non-harmonizable because its FT $X(\nu) = (a + 2\pi j\nu)^{-1}$ has a non-integrable modulus. This is due to the fact that for high frequencies the behavior of $|X(\nu)|$ is in ν^{-1} which decreases to 0 slower to ensure (6.17). This is directly related to the fact that the two previous signals are not continuous at $t = 0$. On the other hand, the signal $x(t) = \exp(-a|t|)$ is harmonizable, because its FT is $2a(a^2 + 4\pi^2\nu^2)^{-1}$, but that is no longer the case for its derivative which again has a discontinuity at $t = 0$.

The harmonic decomposition of a random signal can sometimes be written by (6.7), which again gives the structure of a classical Fourier transform. This is the case of signals considered above which are the product of an RV $A(s)$, which is an amplitude, by a deterministic signal $x(t)$ having an FT $X(\nu)$. In this case (6.7) takes the form $d\overline{X}(\nu) = A(s)\,X(\nu)\,d\nu$ and (6.15) becomes

$$d^2\bar{\Gamma}(\nu_1, \nu_2) = \sigma_A^2 X(\nu_1) X^*(\nu_2)\, d\nu_1\, d\nu_2 \tag{6.18}$$

which means that $\bar{\Gamma}(\nu_1, \nu_2)$ has a derivative $\Gamma(\nu_1, \nu_2) = \sigma_A^2 X(\nu_1) X^*(\nu_2)$.

Conversely, suppose that we are given a signal $x(t)$ with a covariance $\gamma(t_1, t_2)$ having a bidimensional FT, which means that we have (6.12) with the relation

$$d^2\bar{\Gamma}(\nu_1, \nu_2) = \Gamma(\nu_1, \nu_2)\, d\nu_1\, d\nu_2 \tag{6.19}$$

and suppose that $|\Gamma(\nu_1, \nu_2)|$ is integrable in the whole plane $\nu_1 \times \nu_2$. In this case $x(t)$ is harmonizable and (6.15) can be written as

$$E\{d\bar{X}(\nu_1)\, d\bar{X}^*(\nu_2)\} = \Gamma(\nu_1, \nu_2)\, d\nu_1\, d\nu_2 \tag{6.20}$$

which allows us to use (6.7), and thus the signal $x(t)$ has a random FT $X(\nu)$.

6.2.3 Consequences of the Harmonic Decomposition

Suppose now that we have a signal $x(t)$ written as in (6.5). By taking the expected value, we obtain its mean value defined by (5.7), which can be written as

$$m(t) = \int E\{d\bar{X}(\nu)\} e^{2\pi j \nu t} \tag{6.21}$$

This corresponds to the Fourier transform of the mean value. In particular, if $x(t)$ has a zero mean value whatever t, we deduce that $E\{d\bar{X}(\nu)\} = 0$, which is very assumed.

Let us study the problem of the linear filtering of a harmonizable signal. The input-output relationship is a convolution which gives the relation (6.9) in the Fourier domain. As a result (6.15) can be written as

$$d^2\bar{\Gamma}_y(\nu_1, \nu_2) = G(\nu_1)\, G^*(\nu_2)\, d^2\bar{\Gamma}_x(\nu_1, \nu_2) \tag{6.22}$$

where $\bar{\Gamma}_x$ and $\bar{\Gamma}_y$ are the functions appearing in the Fourier decomposition (6.12) of the covariances of x and y respectively. Especially, if the frequency response $G(\nu)$ is bounded, or $|G(\nu)| < M$, the relation (6.13) also holds for y if it holds for x. Similarly, if the increments $d^2\bar{\Gamma}_x(\nu_1, \nu_2)$ take the form (6.19) this also appears for the increment of y, and we have

$$\Gamma_y(\nu_1, \nu_2) = G(\nu_1)\, G^*(\nu_2)\, \Gamma_x(\nu_1, \nu_2) \tag{6.23}$$

The two equations (6.22) and (6.23) will play an important role in what follows.

We shall now deal with some specific properties of the increments which appear in the harmonic decomposition of the signal or of its covariance. If $x(t)$ is a real signal we must have in (6.5)

$$d\bar{X}(\nu) = d\bar{X}^*(-\nu) \tag{6.24}$$

which is called a Hermitian symmetry, already encountered in the Fourier transform of real deterministic signals (see [Picinbono], p. 34). Furthermore, most of the classical properties of Fourier transforms can also be found for the increments $d\bar{X}(\nu)$. Let us look now at the increments appearing in (6.15). We obviously have

$$d^2\bar{\Gamma}(\nu_1, \nu_2) = d^2\bar{\Gamma}^*(\nu_2, \nu_1) \tag{6.25}$$

a relation which resembles the characteristic property of a Hermitian matrix (see p. 61). In reality the increments $d^2\bar{\Gamma}(\nu_1, \nu_2)$ enjoy all the properties met for elements of Hermitian matrices, if we treat them as particular matrices depending on continuous indices. As for the covariance matrices, they are also non-negative definite. In order to verify this point recall that the covariance function satisfies (5.13), or is non-negative for $t_1 = t_2 = t$. Applying that to $\gamma_y(0, 0)$, and using (6.22) and (6.12), we find that for any frequency response $G(\nu)$ we have

$$\iint G(\nu_2)\, d^2\bar{\Gamma}(\nu_1, \nu_2)\, G^*(\nu_2) \geq 0 \tag{6.26}$$

This is very much like saying that (3.69) is non-negative whatever the vector $\mathbf{u}$.

6.2.4 The Discrete-time Case

A discrete-time (DT) signal can be obtained either directly or by the sampling of a continuous-time (CT) signal. In the first case we can note such a signal, often called a time series, in the form $x[k]$ where k is an integer. It is now possible to reproduce all the earlier calculations as from (6.6) for $x[k]$, the only difference being that all the integrals in the calculations are limited to the interval $[-1/2, +1/2]$. This is because the term $\exp(2\pi j\nu k)$ which appears in (6.6) is periodic with a period equal to 1 for integers k. The integral (6.11) must therefore be extended over the interval $[-1/2, \nu_0]$ and the perfect low-pass filter used in (6.10) must also select the same frequency domain. As, apart from these differences, all the calculations are the same, they will not be repeated.

The situation changes completely when the DT signal is deduced from a CT signal by *sampling*. It is here that the problem of the relation between the harmonic decomposition of these two signals arises. This is discussed in detail for deterministic signals in [Picinbono], Chap. 3, and [Oppenheim], Chap. 8. Let us start from a CT signal $x(t)$ with the decomposition (6.5). We associate with this signal a DT signal $x[k]$ such that

$$x[k] = x(t_k) = x(kT) \tag{6.27}$$

where T is the sampling period. The problem now is to expand $x[k]$ as in (6.6) and to establish the relation between the random increments of the continuous and the discrete harmonic decomposition. This is exactly the same as for deterministic signals and the same notation will be used as in [Picinbono], p. 87. The output $x_s(t)$ of a perfect sampler of a CT signal $x(t)$ can be written

$$x_s(t) = \sum x(t_k)\, \delta(t - t_k) = x(t)\delta_T(t) \tag{6.28}$$

where $\delta_T(t)$ is the comb distribution. As the FT of this distribution is still a comb distribution, we can write

$$x_s(t) = \int d\bar{X}_s(\nu)\, \exp(2\pi j \nu t) \tag{6.29}$$

where

$$d\bar{X}_s(\nu) = (1/T) \sum_n d\bar{X}(\nu - n/T) \tag{6.30}$$

For those unfamiliar with the comb distribution, another method is possible: we calculate the signal $x[k]$ defined by (6.27) using (6.5). Due to the periodicity of the exponential term noted above, we can regroup all the terms associated with the frequency ν, which gives

$$x[k] = \int_{-1/2T}^{+1/2T} \sum_n d\bar{X}(\nu - n/T))\, \exp(2\pi j \nu k T) \tag{6.31}$$

This can be written more simply as

$$x[k] = T \int_{-1/2T}^{+1/2T} d\bar{X}_s(\nu)\, \exp(2\pi j \nu k T) \tag{6.32}$$

6.2.5 Other Harmonic Decompositions

The decomposition (6.5) is quite general and, as mentioned before, includes the case of Fourier series, which is a particular case of (6.8). It is well known that Fourier series are well adapted to time-limited or periodic signals. The random character of the signal only introduces the problem of convergence, which can be solved in the same way as for integrals. We can thus introduce Fourier coefficients of a signal $x(t)$ limited in the time interval $[0, T]$ by

$$X_k = \frac{1}{T} \int_0^T x(t)\, \exp\left(-2\pi j \frac{k}{T} t\right) dt \tag{6.33}$$

and the random signal is given by

$$x(t) = \sum_k X_k \exp\left(2\pi j \frac{k}{T} t\right) \tag{6.34}$$

From these expressions we can also deduce other similar ones concerning the Fourier series of the covariance function that is limited to the square ($T \times T$). The same procedure can be used for discrete Fourier transforms summarized by the classical expressions

$$x[m] = \sum_{n=0}^{N-1} X[n] w^{mn} \; ; \; X[n] = \frac{1}{N} \sum_{m=0}^{N-1} x[m] w^{-mn} \tag{6.35}$$

where $w = \exp(2\pi j/N)$. Remember the important fact that when we use the DFT starting from a CT signal we make several approximations: if the signal is not time-limited, there is first a windowing or a truncation of duration T and secondly a sampling of period θ, and the coefficient N appearing in (6.35) is such that $T = N\theta$. For a detailed analysis of these questions see [Picinbono], p. 94.

6.3 SPECTRAL REPRESENTATION AND STATIONARITY

In this section we shall investigate the consequences of stationarity on the spectral representation of signals. Although stationarity may be considered by some as a concept without meaning for the reasons given in Section 5.3.1, it has such important consequences that it cannot be ignored. In fact, it introduces the concept of *power spectrum*, which for practical applications may be the most significant in this book. Note that as we are working only with second order moments, the concept of second order stationarity is sufficient for our purposes. This form of stationarity is characterized by (5.24) and makes use of the correlation function (5.17).

We shall begin by treating the case of the first order moment seen in (5.24). Using $m(t) = m$ in (6.21), we deduce

$$E\{d\overline{X}(\nu)\} = m\,\delta(\nu)\,d\nu \tag{6.36}$$

where $\delta(\nu)$ is the Dirac distribution. This means that the function $E\{\overline{X}(\nu)\}$ is constant, except at the frequency $\nu = 0$, where a step of amplitude m is possible. Consider now the consequence of stationarity on (6.12). It is clear that $\gamma(t_1,\ t_2)$ is a function of $t_1 - t_2$ whatever the values t_1 and t_2 if and only if $d^2\overline{\Gamma}(\nu_1,\ \nu_2)$ is null outside the straight line $\nu_1 = \nu_2$, which can be written as

$$d^2\overline{\Gamma}(\nu_1,\ \nu_2) = d\overline{\Gamma}(\nu_1)\,\delta(\nu_1 - \nu_2)\,d\nu_2 \tag{6.37}$$

Furthermore, it results from (6.15) that for $\nu_1 = \nu_2$, $d^2\overline{\Gamma}(\nu_1,\ \nu_2) \geq 0$, and thus we also have $d\overline{\Gamma}(\nu) \geq 0$. Inserting (6.37) in (6.12), we obtain

$$\gamma(\tau) = \int \exp(2\pi j\nu\tau)\,d\overline{\Gamma}(\nu) \tag{6.38}$$

which gives the correlation function in terms of the increments of the function $\overline{\Gamma}(\nu)$. As these increments are non-negative, this function is non-decreasing, a property already met for the DF of RVs. The function $\overline{\Gamma}(\nu)$ is called the *spectral distribution function* of the stationary signal $x(t)$. If this distribution has a density, called *spectral density* and defined by

$$d\overline{\Gamma}(\nu) = \Gamma(\nu)\,d\nu \tag{6.39}$$

we have

$$\gamma(\tau) = \int \Gamma(\nu)\,e^{2\pi j\nu\tau}\,d\nu \tag{6.40}$$

This is the fundamental relation between the *correlation function* and the *spectral density* of a stationary signal $x(t)$. From now on we will use the function $\Gamma(\nu)$ in place of $\bar{\Gamma}(\nu)$. In fact, the only discontinuities appearing in practice are steps of $\bar{\Gamma}(\nu)$ that introduce Dirac distributions on $\Gamma(\nu)$ called *spectral lines* and the spectral density is much easier to manipulate than the spectral distribution function. Spectral lines appear when the signal contains constant or periodic components. For example, the signal (5.46) has a constant correlation function and its power spectrum is limited to a spectral line at the null frequency. Similarly the signal (5.47) has two spectral lines at angular frequencies ω and $-\omega$ and (5.49) has a spectral line at ω.

Some additional comments are in order here. It is clear that $\Gamma(\nu)$ *is non-negative*, because the increment (6.39) is non-negative. Furthermore, (6.40) appears as the FT of $\Gamma(\nu)$, and there is the same relation between $\Gamma(\nu)$ and $\gamma(\tau)$ as between $p(x)$ and $\phi(u)$ in (2.54). It follows that $\gamma(\tau)$ must be a non-negative definite function, a fact already shown in (5.36). Finally, let us interpret $\Gamma(\nu)$ in terms of power. As $\gamma(0) = E[|x(t)|^2]$ from (5.95) is the mean power P of the signal, we have from (6.40)

$$P = \int \Gamma(\nu)\, d\nu \tag{6.41}$$

This is why $\Gamma(\nu)$ is often called the *power spectrum*, or simply *the spectrum* of the random signal. Of course this power spectrum can be deduced from the correlation function by inverting (6.40), which gives

$$\Gamma(\nu) = \int \gamma(\tau)\, e^{-2\pi j \nu \tau}\, d\nu \tag{6.42}$$

It is clear on (6.40) and (6.42) that the correlation function $\gamma(\tau)$ and the power spectrum are related by a *Fourier transformation.* The same equations are valid for the DT case, which gives

$$\gamma[k] = \int_{-1/2}^{+1/2} \Gamma(\nu)\, e^{2\pi j \nu k}\, d\nu \tag{6.40'}$$

$$\Gamma(\nu) = \sum \gamma[k]\, e^{-2\pi j \nu k} \tag{6.42'}$$

Finally, if we use angular frequencies instead of normal frequencies, we obtain for the CT case

$$\gamma(\tau) = \frac{1}{2\pi} \int S(\omega)\, e^{j\omega\tau}\, d\omega \tag{6.40''}$$

$$S(\omega) = \int \gamma(\tau)\, e^{-j\omega\tau}\, d\tau \tag{6.42''}$$

which gives $S(\omega) = \Gamma(\omega/2\pi)$ and for the DT case

$$\gamma[k] = \frac{1}{2\pi} \int_{-\pi}^{+\pi} S(\omega)\, e^{j\omega k}\, d\omega \tag{6.40'''}$$

$$S(\omega) = \sum \gamma[k]\, e^{-j\omega k} \tag{6.42'''}$$

It is clear from the last four equations that normal frequencies introduce more symmetric relations, and that is the reason for their systematic use in what follows.

Let us now return to the *signal itself.* Using (6.37) in (6.15), we see that if the signal $x(t)$ is stationary, the increments $d\overline{X}(\nu)$ in its harmonic decomposition satisfy

$$E[d\overline{X}(\nu_1)\, d\overline{X}^*(\nu_2)] = 0 \tag{6.43}$$

if $\nu_1 \neq \nu_2$. As their expectation is zero for $\nu = 0$, because of (6.36), we deduce that increments corresponding to different frequencies are *uncorrelated*, which is a fundamental consequence of stationarity.

An interpretation of this result follows. We saw from (3.159) that (6.43) can be interpreted as a scalar product with the result that if the signal $x(t)$ is stationary the increments in (6.5) are *orthogonal.* As complex exponential signals can also be considered as orthogonal, the harmonic decompositon (6.5) is a *doubly orthogonal* expansion of stationary signals. This extends the result already discussed after (3.98).

At this point it is interesting to consider the consequences of stationarity on the other harmonic decompositions presented in Section 6.2.5. We shall start with the case of the Fourier series of a stationary signal $x(t)$ truncated in a finite interval as $[-T/2, +T/2]$. The resulting signal $x_T(t)$ is then equal to $x(t)$ for $-T/2 \leq t \leq T/2$ and to zero if $|t| > T/2$. The Fourier coefficients are given by (6.33) where the integral goes from $-T/2$ to $T/2$. Replacing $x(t)$ in that equation by its expansion (6.5), and after integration with respect to the variable t, we obtain

$$X_k = \int d\overline{X}(\nu)\ \mathrm{sinc}\{T(\nu - k/T)\} \tag{6.44}$$

where the sinc function is equal to $\sin(\pi x)/\pi x$ and represented in Figure 5.4. Let us now calculate the correlation between these Fourier coefficients. Using (6.15), assuming the stationarity which gives (6.37) and introducing the spectral density appearing in (6.39), we find

$$E(X_k X_l^*) = \int \Gamma(\nu)\ \mathrm{sinc}\left[\left(\nu - \frac{k}{T}\right)T\right] \mathrm{sinc}\left[\left(\nu - \frac{l}{T}\right)T\right] d\nu \tag{6.45}$$

which is an integral extended over the whole range of frequencies, or from minus to plus infinity.

The first result that appears in (6.45) is that the Fourier coefficients *are no longer orthogonal* for a stationary signal. In reality this can already be observed from (6.44); in fact the function sinc never vanishes, except for a countable set of frequencies. Thus X_k is obtained by a mixture of the increments $d\overline{X}(\nu)$ and another mixture is used for X_l, but the same increments $d\overline{X}(\nu)$ can appear in the two mixtures, which generates the non-orthogonality. But (6.45) is interesting because it gives the exact value of the correlation between Fourier coefficients. In particular it can help us to discover under what condition on $\Gamma(\nu)$ it is possible to relocate orthogonal Fourier coefficients. Noting that the function sinc x is null for x integer, we find that orthogonality is ensured if $\Gamma(\nu)$ can be written as

$$\Gamma(\nu) = \sum \Gamma_p \, \delta(\nu - p/T) \tag{6.46}$$

This means that $\Gamma(\nu)$ has frequency lines only at periodic frequencies, and by taking the inverse Fourier transform (6.40) we deduce that $\gamma(\tau)$ is periodic with period T. In conclusion, the orthogonality of the Fourier coefficients can be achieved if the signal $x(t)$ has a periodic correlation function and if the observation time T is equal to this period. We have already met this type of signal in our discussion on stationarity on p. 180.

But (6.45) can also provide asymptotic results which are particularly interesting in practical applications. It is well known that the function sinc$x$ is decreasing. Suppose then that the variations of $\Gamma(\nu)$ are very slow compared to those of the sinc function. This situation appears if the correlation time t_c introduced on p. 172 is much smaller than T, and then (6.45) becomes approximately

$$E(X_k \, X_l^*) \cong \Gamma(k/T) \int \text{sinc}\left[\left(\nu - \frac{k}{T} \right) T \right] \text{sinc}\left[\left(\nu - \frac{l}{T} \right) T \right] d\nu \tag{6.47}$$

Using the Parseval theorem, we can easily find that the last integral is null, which again yields the orthogonality. Thus the orthogonality of the Fourier coefficients is an asymptotic property, but it is clear that if $T \to \infty$, we again find the integral (6.5) instead of a Fourier series.

Using the same approximation we can discuss the case where $k = l$. As the integral of $\text{sinc}^2(\nu T)$ is equal to $1/T$, which can also be deduced from the Parseval theorem, we see that for large values of T the variance $E\{|X_k|^2\}$ is equal to $(1/T)\,\Gamma(k/T)$, which specifies the properties of the RVs X_k completely.

The same reasoning stands for the DFT introduced by (6.35), and this is given as a problem at the end of the chapter.

We shall conclude this section on the consequences of stationarity with some comments on *circularity*. This concept was introduced in the context of complex normal RVs and characterized by the fact that the matrix defined by (4.125) is null. We will now verify that this condition arises for the increments of the integral (6.5). Suppose first that $x(t)$ is real, in which case we have (6.24) and (6.43) simultaneously due to second order stationarity. Taking $\nu_1 = \nu$ and $\nu_2 = -\nu$ in this equation, we immediately obtain

$$E\{[d\bar{X}(\nu)]^2\} = 0 \tag{6.48}$$

which is the condition of circularity for scalar normal RVs. For complex signals we cannot arrive at this condition uniquely from second order stationarity; we need higher-order properties that will be discussed later.

6.4 FILTERING AND POWER SPECTRUM

As indicated in the introduction, this section can be read *independently* of the two previous sections. For those who are approaching it directly from the introduction, we will summarize the notation already introduced in order to achieve coherent reasoning. Others may wish to turn immediately to Section 6.4.2.

6.4.1 Summary of Previous Notations

Let $x(t;\ s)$, or simply $x(t)$, be a stationary second order and zeromean CT random signal. We assume that it can be complex, and its correlation function is defined by (5.17), or

$$\gamma(\tau) = E[x(t)\ x^*(t-\tau)] = E[x(t+\tau)\ x^*(t)] \tag{5.17}$$

Suppose that this correlation has a Fourier transform $\Gamma(\nu)$, and can be written as

$$\gamma(\tau) = \int \Gamma(\nu)\ e^{2\pi j \nu \tau}\ d\nu \tag{6.40}$$

which can be inverted into

$$\Gamma(\nu) = \int \gamma(\tau)\ e^{-2\pi j \nu \tau}\ d\tau \tag{6.42}$$

In the DT case these expressions can be written as

$$\gamma[k] = \int_{-1/2}^{+1/2} \Gamma(\nu)\ e^{2\pi j \nu k}\ d\nu \tag{6.40'}$$

$$\Gamma(\nu) = \sum \gamma[k]\, e^{-2\pi j \nu k} \tag{6.42'}$$

For those wishing to work with angular frequencies ω instead of normal frequencies, it is possible to use the relation $\omega = 2\pi\nu$ to arrive at the expressions indicated on p. 208.

6.4.2 The Interference Formula

The purpose of this section is to investigate the consequences in the *frequency domain* of the relations established in Section 5.4 for the *time domain.* In order to introduce spectral matrices discussed at the end of this chapter, we will consider a situation which is a little more general than that discussed in the previous chapter.

Consider two random signals $x_k(t)$, $1 \leq k \leq 2$, which are complex, zeromean, second order, and jointly stationary. By an extension of (5.17), we introduce the function

$$\gamma_{x;\,kl}(\tau) \triangleq E[x_k(t)\ x_l^*(t-\tau)] \tag{6.49}$$

If $k = l$, we obtain the correlation functions of $x_1(t)$ or $x_2(t)$, as in (5.17). These functions are called *autocorrelation functions*. On the other hand, if $k \neq l$, we obtain the *crosscorrelation function* of $x_1(t)$ and $x_2(t)$, by an extension of the terminology used on p. 59. Suppose now, that $x_k(t)$ and $x_l(t)$ are used as inputs of two linear filters F_k and F_l, as pictured in Figure 6.1. The corresponding outputs are $y_k(t)$ and $y_l(t)$.

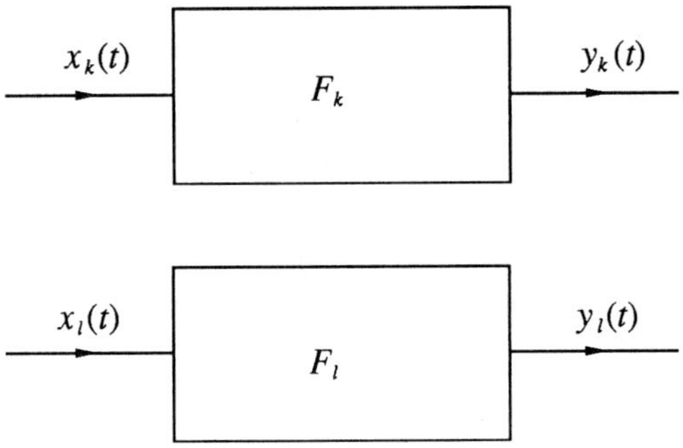

Figure 6.1 Linear filtering of two signals.

As the signals $x(t)$ are jointly stationary, the same property is valid for the signals $y(t)$ and we want to calculate their correlation functions. In the *time domain*, the calculation is the same as in Section 5.4, and (5.62) can now be written as

$$\gamma_{y;\,k\,l}(\tau) = \int h_k(\theta_1)\, h_l^*(\theta_2)\, \gamma_{x;\,kl}(\tau + \theta_2 - \theta_1)\, d\theta_1\, d\theta_2 \tag{6.50}$$

Let us now transform this equation in the *frequency domain*. To do this, suppose that $\gamma_{x;\,k\,l}(\tau)$ has a Fourier transform $\Gamma_{x;\,k\,l}(\nu)$ defined as in (6.42). Similarly, the Fourier transform of $\gamma_y(\tau)$ is called $\Gamma_y(\nu)$, and our purpose is to calculate $\Gamma_y(\nu)$ in terms of $\Gamma_x(\nu)$. To achieve this, we shall use the frequency responses of the filters defined by the classical expression

$$G(\nu) = \int h(\theta) \exp(-2\pi j \nu \theta)\, d\theta \tag{6.51}$$

and already used in (6.9). Starting from (6.50), the FT of $\gamma_{y;\,k\,l}(\tau)$ can be written as in (6.42) and takes the form

$$\Gamma_{y;\,k\,l}(\nu) = \iiint h_k(\theta_1)\, h_l^*(\theta_2)\, \gamma_{x;\,kl}(\tau + \theta_2 - \theta_1)\, e^{-2\pi j \nu \tau}\, d\tau\, d\theta_1\, d\theta_2 \tag{6.52}$$

Noting that

$$e^{-2\pi j \nu \tau} = e^{-2\pi j \nu(\tau + \theta_2 - \theta_1)} e^{-2\pi j \nu \theta_1} e^{2\pi j \nu \theta_2} \tag{6.53}$$

we can calculate (6.52) with the variables $\theta = \tau + \theta_2 - \theta_1$, θ_1, θ_2, which gives a product of three integrals. Using (6.51), we finally obtain

$$\boxed{\Gamma_{y,kl}(\nu) = G_k(\nu)\, G_l^*(\nu)\, \Gamma_{x,kl}(\nu)} \tag{6.54}$$

The first point to note is that this expression is strictly equivalent to (6.50) but much easier. The situation is similar to that observed in the input-output relationship of a filter: it is a convolution in the time domain and a product in the frequency domain.

The same procedure can be applied to discrete-time signals. The only difference is that all the integrals are replaced by sums and the final equation (6.54) remains the same. The consequence of the discrete-time property is that all the functions of the frequency are now periodical.

The result (6.54) has consequences of capital importance in what follows. This expression is sometimes called the *interference formula*, as it is used in the interpretation of interference experiments in optics. In Section 5.2.5(**a**) we discussed the principle of interference experiments, and we saw that the problem was to measure the power of a signal in the form $x(t) + x(t-d)$. It is the simplest possible situation as there is only one source and a pure delay d. In more sophisticated experiments it is possible to use two different sources, and the channels between the sources and the observation are no longer a simple delay but can be modelled by a linear filtering. The principle of the experiment is as in Figure 6.1, with the difference that the final result is the power of the sum $y(t)$ of the two signals, or

$$y(t) = y_k(t) + y_l(t)$$

The mean power of $y(t)$ is defined by (5.95), which gives

$$P_y = P_k + P_l + I_{kl}$$

where I_{kl} is the *interference term*. This term is given by

$$I_{kl} = E[\,y_k(t)\; y_l^*(t)] + E[\,y_l(t)\; y_k^*(t)] = \gamma_{y;\,kl}(0) + \gamma_{y;\,kl}^*(0)$$

The power spectrum of $y(t)$ corresponds to the frequency distribution of P_y, and its mathematical expression uses (6.54), which justifies the expression of the interference formula. It is clear that if $I_{kl} = 0$, there is no interference, which gives an incoherent addition of the powers of the two signals $y_k(t)$ and $y_l(t)$.

6.4.3 Consequences of the Interference Formula

(a) Power spectrum transformation by filtering. As seen in (6.40) and (6.42), the power spectrum is the FT of the autocorrelation function. We can then suppress the indices k and l or suppose that there is only one signal and one filter. In this case (6.54) becomes

$$\boxed{\Gamma_y(\nu) = |G(\nu)|^2\, \Gamma_x(\nu)} \tag{6.55}$$

which gives the power spectrum of the output in terms of that of the input and of the frequency response of the filter. We will find this expression again in the next section using a different argument.

(b) On the phase problem. We see in (6.55) that the phase of the filter plays no role in the calculation of $\Gamma_y(\nu)$. In other words, passing the signal $x(t)$ through various filters with the same value as the modulus of $G(\nu)$ but different phases $\phi(\nu)$, we obtain different signals but with the same power spectrum or correlation function. This is not surprising, since second order properties do not completely characterize a random signal. Note the similarity of this to the problem of the factorization of covariance matrices specified by (3.99). Just as there are several different matrices $\mathbb{A}$ giving $\mathbb{\Gamma}$ by (3.99), there are various vectors $\mathbf{Y}$ with the same covariance matrix. See the next section for more details on this point.

(c) Positivity of the power spectrum. We have already seen that the power spectrum is non-negative, a property which can also be deduced from (6.55) without using the earlier arguments. To do this we reason by contradiction. Suppose that $\Gamma_x(\nu)$ is negative for ν belonging to a frequency band $\Delta\nu$. Take a filter with a frequency response $G(\nu)$ equal to zero if ν does not belong to $\Delta\nu$. It follows that $\Gamma_y(\nu) < 0$, and by using (6.41) we obtain $P_y < 0$, which is impossible because it is a mean power. Thus $\Gamma_x(\nu)$ cannot have negative values.

(d) Orthogonality with non-overlapping filters. Let us now return to (6.54). It is clear that if the product $G_k G_l^* = 0$, then $\Gamma_{y,kl} = 0$. This means that the outputs of filters having no common frequencies are uncorrelated, or orthogonal. This result is true even if the two inputs of the filter are equal, and it is a very common way to obtain uncorrelated signals. In reality this result is a direct consequence of the orthogonality of the coefficients specified by (6.43).

It is worth repeating here that all the expressions in this section are valid without change for the discrete-time case, the only difference appearing when using integrals which are limited to the range $-1/2$, $+1/2$.

Example 6.2 Input-output correlation. Let us calculate the crosscorrelation between the input $x(t)$ and the output $y(t)$ of a filter. This situation can be represented by a scheme such as that in Figure 6.1 and given in Figure 6.2. The FT of this crosscorrelation is given by (6.54) in the specific case where $x_1 = x_2 = x$ and $y_1 = y$ and $y_2 = x$. Furthermore $G_k = G$ and $G_l = 1$. This gives

$$\Gamma_{yx}(\nu) = G(\nu)\,\Gamma_x(\nu) \tag{6.56}$$

which by Fourier transformation gives

$$\gamma_{yx}(\tau) = [h * \gamma_x](\tau) \tag{6.57}$$

meaning the convolution between γ_x and h. This is especially interesting in the case of *white noise*, an expression that will be explained more precisely later and characterized by $\Gamma_x(\nu) = 1$. If this is realized, it follows that $\gamma_{yx}(\tau)$

is equal to $h(\tau)$. It is an interesting way to deduce the impulse response of a filter by measuring the input-output crosscorrelation when the input is white.

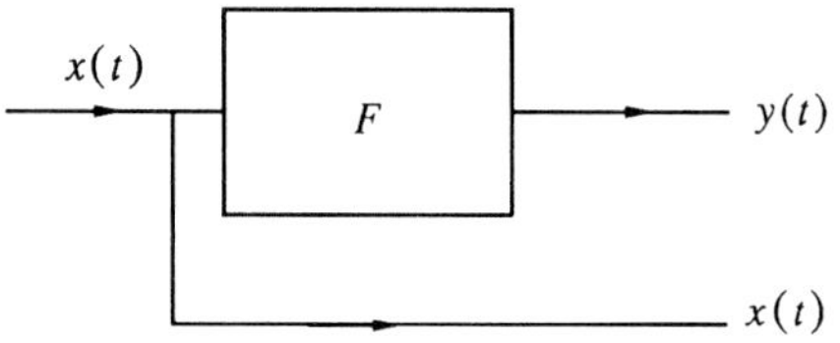

Figure 6.2 Input-output correlation.

Example 6.3 Derivative and integrals of stationary signals. It is well known that the derivative of a signal is obtained by passing it through a linear filter with the frequency response $G(\nu) = 2\pi j\nu$. Applying (6.54) and (6.55) and calling $d(t)$ the derivative of $x(t)$, we obtain the crosscorrelation between $x(t)$ and its derivative by

$$\gamma_{dx}(\tau) = \int 2\pi j\nu\, \Gamma(\nu)\, e^{2\pi j\nu\tau}\, d\nu \tag{6.58}$$

and

$$\gamma_{d}(\tau) = \int 4\pi^2\nu^2\, \Gamma(\nu)\, e^{2\pi j\nu\tau}\, d\nu \tag{6.59}$$

where $\Gamma(\nu)$ is the power spectrum of $x(t)$. As $\gamma(\tau)$ is given by (6.40), we obtain

$$\gamma_{dx}(\tau) = \dot{\gamma}(\tau) \;\; ; \;\; \gamma_{d}(\tau) = -\ddot{\gamma}(\tau) \tag{6.60}$$

which can also be deduced from (5.86) and (5.87) applied to $\gamma(t_1, t_2) = \gamma(t_1 - t_2)$. These equations are valid provided that the integrals (6.58) or (6.59) are finite. This is certainly the case if the signal $x(t)$ is band-limited, which means that the power spectrum is null outside a finite interval of frequencies. In fact, as $x(t)$ is second order, the integral of $\Gamma(\nu)$ over this frequency band is finite, and a multiplication by ν or ν^2 does not change the situation. By repeating the same arguments, or applying (6.54) with $\Gamma_{x;kl}(\nu) = \Gamma(\nu)$ and $G_1(\nu) = (2\pi j\nu)^m$ and $G_2(\nu) = (2\pi j\nu)^n$ we immediately find

$$\gamma_{m,n}(\tau) \triangleq E[x^{(m)}(t)\, x^{(n)*}(t-\tau)] = (-1)^n \int (2\pi j\nu)^{m+n}\, \Gamma(\nu)\, e^{2\pi j\nu\tau}\, d\nu$$

and using (6.40), we obtain

$$\gamma_{m,n}(\tau) = (-1)^n\, \gamma^{(m+n)}(\tau) \tag{6.60'}$$

where $a^{(k)}(t)$ means the derivative of order k of the function $a(t)$. These equations are, of course, valid insofar as these derivatives are finite.

These expressions are of particular interest for real signals and when $\tau = 0$. In this case it must be noticed that $\gamma^{(2k+1)}(\tau) = 0$ for $\tau = 0$ because $\gamma(\tau)$ is an even function. Thus the only non-null terms appear when $m + n$ is even, which means that m and n have the same parity. Using the same notations as in (2.65), (6.60') then gives

$$\gamma_{m,\,n}(0) = (-1)^m\ \gamma^{(m+n)}[0] = (-1)^n\ \gamma^{(m+n)}[0]$$

Let us discuss the validity of the previous calculations. Even if $x(t)$ is a second order random signal, its derivative in the mean square sense does not necessarily exist (see Section 5.8.2). We can verify this point either by using the proposition of this subsection or by a direct calculation on (6.59), the two methods being equivalent. In fact we observe in (6.59) that $\gamma_d(0)$ is not necessarily finite, meaning that the derivative is no longer second order. This is due to the presence of the factor ν^2 in the integral which can lead to a divergent integral. Such a situation especially occurs for signals whose correlation functions are in the form $\exp(-a|\tau|)$. The power spectrum is given by (5.39) and has a Lorentzian shape presented in Figure 5.2. It is clear that after multiplication by ν^2, the integral (6.59) is divergent. Some examples of these signals, which do not have mean square derivatives, will be discussed below.

The same discussion can be given for integration, where the filter $2\pi j\nu$ becomes $(2\pi j\nu)^{-1}$. In this case the possible divergence of the integrals is due to the low frequencies of the spectrum.

Example 6.4 Normalized power spectrum. A signal $x(t)$ is said to be normalized if it has a unit power. Its power spectrum is also said to be normalized and (6.41), where $P = 1$, means that $\Gamma(\nu)$ has the form of a PDF function as studied in Chapter 2. In this case (6.40) shows that $\gamma(\tau)$ is a characteristic function, and the characteristic functions presented in Chapter 2 can be considered as correlation functions of a unit power signal.

Example 6.5 Real random signals. The case of real random signals is probably the most important in practice. All previous results can be applied, and we shall summarize them here. The first consequence, characteristic of a real signal, is of course (6.24), which shows that the negative frequencies do not get additional information with respect to the positive ones. As for deterministic signals, it is possible to write (6.5) with only positive frequencies as

$$x(t) = \int_0^\infty d\bar{C}(\nu)\cos(2\pi\nu t) + d\bar{S}(\nu)\sin(2\pi\nu t) \tag{6.61}$$

where the increments $d\bar{C}$ and $d\bar{S}$ are directly deduced from $d\bar{X}$. We leave the final calculation to the reader. The correlation between the components $d\bar{C}$ and $d\bar{S}$ are analyzed in a problem.

Furthermore, if $x(t)$ is real, its correlation function is real and even, from (5.28). It follows that the power spectrum $\Gamma(\nu)$ is also even, and here also the negative frequencies can be suppressed. In this case (6.40) must be written as

$$\gamma(\tau) = \int_0^\infty 2\,\Gamma(\nu)\cos(2\pi\nu\tau)\,d\nu \tag{6.62}$$

6.5 SPECTRAL FACTORIZATIONS

As Fourier transforms are a particular case of more general transforms —namely Laplace transforms for continuous time and z-transforms for discrete time— it is useful to return to the previous expressions valid only in the Fourier domain. Although it is not of great interest to discuss the representation of the random signal in terms of general exponential signals, the questions treated in the previous section take on a new aspect and demonstrate more clearly the problem of factorization.

6.5.1 Filtering in the z-domain

We look here at the case of a zeromean discrete-time (DT) signal $x[k]$, k integer. Assuming second order stationarity, its correlation function is defined, as in (5.17), by

$$\gamma[k] = E\{x[p]x^*[p-k]\} \tag{6.63}$$

where p is also integer. The input-output relationship is given by (5.55) and the output correlation function by an extension to the DT case of (5.62), or

$$\gamma_y[k] = \sum_p \sum_q h[p]\, h^*[q]\, \gamma_x[k+q-p] \tag{6.64}$$

This relation between the input and output correlation functions is not very attractive, so below is the same expression after a z-transformation. Let us call $\Gamma(z)$ the z-transform of a correlation function defined by

$$\Gamma(z) = \sum_k \gamma[k]\, z^{-k} \tag{6.65}$$

This function is sometimes called the z-spectrum of the signal. If the concept of spectrum is associated with that of positiviy, this terminology is not at all appropriate because it is clear that $\Gamma(z)$ is a complex-valued function of z. Note also that the series (6.65) is calculated for k going from minus to plus infinity, since $\gamma[k]$ has no reason to be causal and we know, for example, that for a real signal $\gamma[k] = \gamma[-k]$. It follows that $\Gamma(z)$ is a two-sided z-transform, and we assume that its region of convergence contains the unit circle, which ensures the existence of a Fourier transform (see [Picinbono], p. 106). Using (6.64) and (6.65), we deduce

$$\Gamma_y[z] = \sum_k \sum_p \sum_q h[p]\, h^*[q]\, \gamma_x[k+q-p]\, z^{-k} \tag{6.66}$$

Let us suppose that the signals and the filter are real. Introducing the transfer function defined by

$$H(z) = \sum_k h[k]\, z^{-k} \tag{6.67}$$

and writing z^{-k} in the form $z^{-(k+q-p)}\, z^{-p}\, z^{q}$, we obtain

$$\Gamma_y(z) = H(z)\, H(z^{-1})\, \Gamma_x(z) \tag{6.68}$$

If we use the same procedure for complex signals and filters, we easily obtain

$$\Gamma_y(z) = H(z)H^*(1/z^*)\ \Gamma_x(z) \tag{6.69}$$

This expression corresponds to (6.55) valid for FT. To verify this point it is sufficient to replace z by $\exp(2\pi j\nu)$, and as

$$G(\nu) = H(e^{2\pi j\nu}) \tag{6.70}$$

we arrive at (6.55). To avoid confusion here, it would be preferable to write (6.65) as $\tilde{\Gamma}(z)$ rather than $\Gamma(z)$ because the power spectrum appearing in (6.55) is related to $\tilde{\Gamma}$ by

$$\Gamma(\nu) = \tilde{\Gamma}(e^{2\pi j\nu}) \tag{6.71}$$

But as the simultaneous use of Fourier and z-transforms does not occur very frequently, we prefer to use the simpler notation.

It is also clear that the same expressions are valid for crosscorrelation, which gives the formula corresponding to (6.54) as

$$\Gamma_{y,\,kl}(z) = H_k(z)\ H_l^*(1/z^*)\ \Gamma_{x,\,kl}(z) \tag{6.72}$$

Having arrived this far, we can reconsider the comments made after (6.54), and show, for example, the positivity of the spectrum which has already been demonstrated.

Let us conclude this section by repeating that the function $\Gamma(z)$ defined by (6.65) and used thereafter is sometimes called the z-spectrum, a usage which, although not fully satisfactory, will be followed here. In fact the concept of spectrum is usually associated with a positive function, while $\Gamma(z)$ is in general complex. In particular, the power P equal to $\gamma[0]$ can be deduced from (6.65) by the classical relation

$$P = \frac{1}{2\pi j}\int_C \Gamma(z)\ z^{-1}\ dz \tag{6.73}$$

which is an integral in the complex plane. If C is the unit circle, we can replace z by $\exp(2\pi j\nu)$ and obtain (6.41) where the integral is taken between $-1/2$ and $+1/2$.

6.5.2 Filtering in the Laplace Domain

We present here the same type of results for CT signals using the Laplace transformation. As the connecting links of the calculations are exactly the same, we will use a faster method. The relation between input and output correlation is given by (5.62) and the LT of a correlation function is given by

$$\Gamma(s) = \int_{-\infty}^{+\infty} \gamma(\tau)\ e^{-s\tau}\ d\tau \tag{6.74}$$

We assume that the imaginary axis is in the region of convergence, which ensures the existence of a Fourier transform obtained for $s = 2\pi j\nu$. The transfer function of a filter with impulse response $h(t)$ is

$$H(s) = \int_{-\infty}^{+\infty} h(t)\ e^{-st}\ dt \tag{6.75}$$

and we obtain

$$\Gamma_y(s) = \iiint h(\theta_1)\ h^*(\theta_2)\ \gamma_x(\tau + \theta_2 - \theta_1)\ e^{-s\tau}\ d\theta_1\ d\theta_2\ d\tau \qquad (6.76)$$

Writing $\exp(-s\tau)$ in the form $\exp[-s(\tau + \theta_2 - \theta_1)]\ \exp(-s\theta_1)\ \exp(s\theta_2)$, we obtain

$$\Gamma_y(s) = H(s)H^*(-s^*)\ \Gamma_x(s) \qquad (6.77)$$

By replacing s with $2\pi j\nu$, and as $G(\nu) = H(2\pi j\nu)$, we again obtain (6.55) with all its conclusions. For real filters we have $H^*(s) = H(s^*)$, and (6.77) becomes

$$\Gamma_y(s) = H(s)\ H(-s)\ \Gamma_x(s) \qquad (6.78)$$

We can also call $\Gamma(s)$ the Laplace spectrum of $x(t)$, although this name is not completely appropriate as it is a complex number. The power of the signal can be expressed by using the inversion formula of LT, which gives

$$P = \gamma(0) = \frac{1}{2\pi j} \int_B \Gamma(s)\ ds \qquad (6.79)$$

where B is a Bromwich path (see [Picinbono], p. 60). By taking $s = 2\pi j\nu$, we again find (6.41).

6.5.3 Spectral Factorizations for Discrete-Time Signals

Suppose that the input signal is a white noise. This expression, and its various meanings, will be discussed later, but at present it suffices it to say that a DT white noise is simply a sequence of uncorrelated RVs with the same mean and variance. Assuming, for further simplicity, that this variance is equal to one, which leads to the name *unit white noise*, we deduce that its correlation function is simply $\gamma[k] = \delta[k]$, where $\delta[k]$ is the Kronecker-delta signal. When we introduce this expression into (6.65), we obtain $\Gamma(z) = 1$. The power spectrum is also equal to one, which justifies the term *white noise* by analogy with the spectrum of white light. If we assume that the filter is real and introduce $\Gamma(z) = 1$ into (6.68), we obtain

$$\Gamma(z) = H(z)\ H(z^{-1}) \qquad (6.80)$$

where $\Gamma(z)$ now means the z-power spectrum of the output signal. This expression represents a factorization of $\Gamma(z)$, just as in (3.99).

The factorization problem is a direct consequence of (6.80) and can be stated as follows: starting from a given z-power spectrum $\Gamma(z)$, is it possible to find a transfer function $H(z)$ such that (6.80) holds? Physically this problem means: starting from a random signal $x[k]$, is it possible to find a filter such that if its input is a unit white noise its output has the same correlation function as $x[k]$? Note that we have not said that its output should be equal to $x[k]$. The following discussion takes into account the fact, indicated on p. 214, that completely different signals can have the same correlation function. As in the case of the correlation matrix, the factorization problem of $\Gamma(z)$ has no unique solution. The example below explains this point.

Example 6.6. Spectrum at the output of a first order FIR filter. Suppose that its transfer function is $H(z) = 1 - a\,z^{-1}$, which means that the only non-null coefficients of its impulse response are $h[0] = 1$ and $h[1] = -a$. It is clear that this filter has a zero for $z = a$. The corresponding z-spectrum given by (6.80) is

$$\Gamma(z) = -az^{-1} + (1 + a^2) - az \tag{6.81}$$

and the only non null values of the correlation function are $\gamma[0] = 1 + a^2$ and $\gamma[1] = \gamma[-1] = -a$. Consider now the filter with the transfer function $H(z^{-1})$. It is obvious that, inserted into (6.80), this gives the same z-spectrum. In reality any filter in the form $z^k\,H(z^{-1})$ will give the same result, because the term z^k disappears in (6.80). Thus taking $z^{-1}\,H(z^{-1})$ we define the filter $G(z) = -a + z^{-1}$. Its impulse response is $g[0] = -a$, $g[1] = 1$, and we see that the filters H and G are quite different but give the same z-spectrum. These results are summarized in the following figures. The power spectrum which corresponds to (6.81) is defined by (6.71), which gives

$$\Gamma(\nu) = 1 + a^2 - 2a\cos(2\pi\nu)$$

and is represented in Figure 6.3 for $0 \le \nu \le 1/2$, because $\Gamma(\nu)$ is an even function.

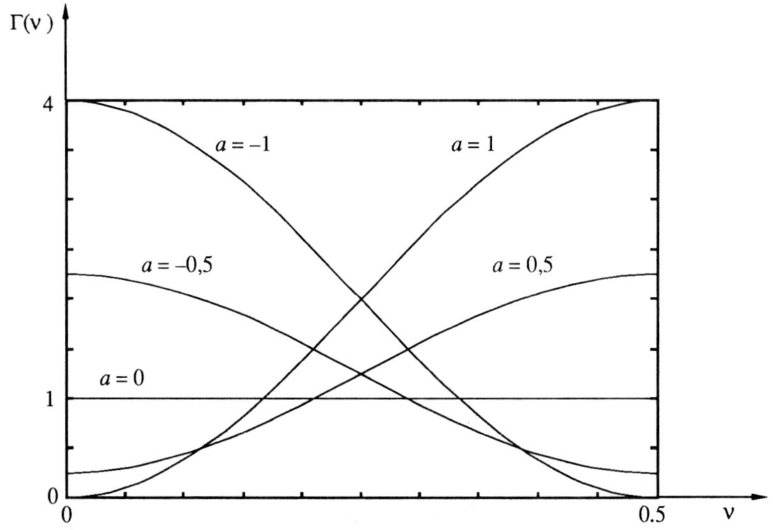

Figure 6.3 Power spectrum corresponding to (6.81).

It has a maximum value of $(1 - a)^2$ for $\nu = 0$ and a minimum value of $(1 + a)^2$ if $\nu = 1/2$ if $a < 0$, and the opposite for $a > 0$. In both cases the difference between the maximum and minimum is $4|a|$. The locations of the zeros and the corresponding impulse responses of the filters H and G are presented in

Figure 6.4. These two filters are causal but give the same spectrum $\Gamma(\nu)$ when the input is white.

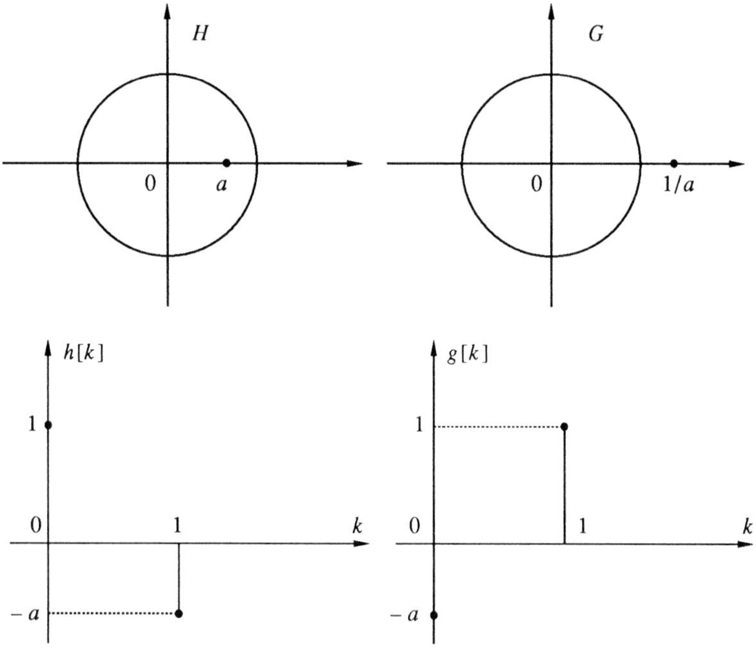

Figure 6.4 Zeros and impulse responses of H and G.

From this example we shall consider the case of rational filters (see [Picinbono], p. 139) and the degree of freedom that we have in changing $H(z)$ without changing $\Gamma(z)$ in (6.80). Firstly, there is a *sign ambiguity*, because replacing $H(z)$ by $-H(z)$ does not change $\Gamma(z)$. Secondly, there is a delay ambiguity. In fact, as noted above, replacing $H(z)$ by $z^{-k}H(z)$ does not change $\Gamma(z)$ in (6.80). In other words, we can translate the impulse response in the past or the future without generating any modification on the correlation function.

Consider now the structure of the transfer function which is a ratio between two polynomials $N(z)$ and $D(z)$. The numerator $N(z)$ can be written as a product of terms $(z - z_i)$ where the z_is are the zeros of the transfer function. The contribution of this term to $\Gamma(z)$ in (6.80) is then

$$\Gamma_i(z) = (z - z_i)(z^{-1} - z_i) = -z^{-1}z_i(z - z_i)(z - z_i^{-1}) \qquad (6.82)$$

These two zeros are represented in Figure 6.5.

This means that if z_i is a zero of $\Gamma(z)$, z_i^{-1} is also a zero. As a consequence there is the same number of zeros of $\Gamma(z)$ inside and outside the unit circle. As z_i is a zero of $H(z)$, if we replace the factor $(z - z_i)$ by $(z^{-1} - z_i)$ in $N(z)$, we change the filter but not the z-spectrum. Mathematically, the same procedure can be realized on the denominator, but some limitations may appear as a result of stability problems not discussed here. In conclusion, we can modify $H(z)$ by changing the sign, by multiplication by z^{-k} or by replacing any pole or zero by its inverse without changing the z-spectrum. This illustrates the non unicity of the spectral factorization (6.80).

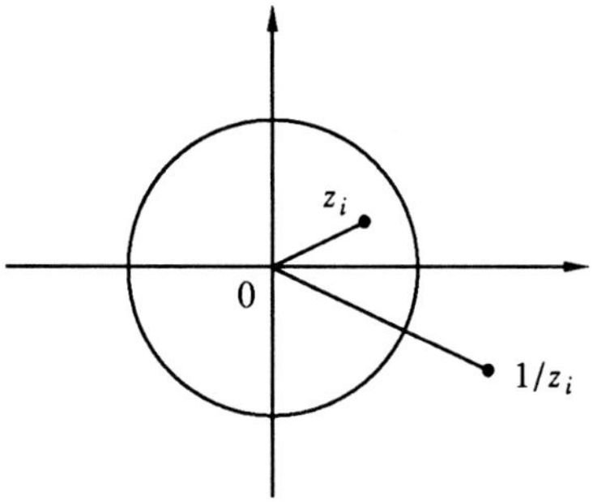

Figure 6.5 Location of the zeros z_i and $1/z_i$.

It is clear that the foregoing can be translated to the CT case with only a few differences. The concept of white noise is less obvious, because if $\Gamma(s) = 1$, the power P is infinite. This will be discussed later. Starting from (6.78), we see that the zeros or poles can be replaced not by their inverse but by their opposites.

6.5.4 Strong Factorization

In order to limit the ambiguity of the factorization problem, we can impose some conditions on the filter $H(z)$ in (6.80). We have seen, as in Example 6.6, that the causality condition is not sufficient to suppress the ambiguity. To explain the procedure of strong factorization we will begin by considering the case of rational filters. As seen above, this means that the z-power spectrum is a rational function with the same numbers of poles or zeros inside and outside the unit circle. We also assume that there is no pole or zero on this circle.

The *strong* (or canonical) *factorization* corresponds to (6.80) where $H(z)$ is a rational function with all poles and zeros inside the unit circle and such that the numerator and the denominator are polynomials of the same degree. We shall see that strong factorization is unique, except for a sign ambiguity. Physically, it means that the filter $H(z)$ used in (6.80) is causal and stable and that its inverse has the same property. It is sometimes said that $H(z)$ is a *minimum-phase* filter. In fact, a DT rational filter whose transfer function $H(z)$ is a ratio of two polynomials $N(z)$ and $D(z)$ is a *dynamical filter* if it is stable and causal (see [Picinbono], p. 157). This implies that all the poles of $H(z)$, which are the zeros of $D(z)$, are located inside the unit circle, and that the degree of the denominator is equal to or greater than that of the numerator. If we impose the condition that the filter $1/H(z)$ should also be dynamical, we arrive at the conclusion that $N(z)$ and $D(z)$ must have the same degree and that all the poles and zeros of $H(z)$ must be located inside the unit circle, which is the condition of strong factorization. The sign ambiguity comes from the fact that if $H(z)$ is a minimum-phase filter, $-H(z)$ has the same property. However, $H(z)$ and $-H(z)$ give the same result in (6.80).

The method of realizing strong factorization is relatively simple. The numerator and denominator of $\Gamma(z)$ are polynomials which can be decomposed as the products of terms $T_i(z)$ in the form $(z - a_i)(z - a_i^{-1})$. The root a_i is necessarily non-null, and we can assume that $|a_i| < 1$, because if $|a_i| > 1$ it suffices in our reasoning to take its inverse. Using (6.82), we can write

$$T_i(z) = -z\, a_i^{-1}(z - a_i)(z^{-1} - a_i) \tag{6.83}$$

If we perform this operation for all the terms $T_i(z)$ of the numerator and denominator of $\Gamma(z)$, we construct $H(z)$ in (6.80) by selecting in $\Gamma(z)$ only the terms in the form $(z - d_i)$ where $|d_i| < 1$. Finally, if the degrees of numerator and denominator of the filter are not the same, we introduce the appropriate numbers of zeros or poles at the origin which disappear in (6.80). This procedure, while a little laborious, presents no significant difficulties and is presented in more detail in Appendix 6.11.

Example 6.7. Strong factorization of a second order spectrum. Consider a real signal $x[k]$ with a correlation function $\gamma[k] = a^{|k|}$, $-1 < a < +1$. Its z-spectrum $\Gamma(z)$ can easily be obtained by using (6.65), which gives

$$\Gamma(z) = \frac{(1 - a^2)z}{(z - a)(1 - az)} = \frac{1 - a^2}{(z - a)(z^{-1} - a)} \tag{6.84}$$

Strong factorization is realized with the filter

$$H(z) = \sqrt{1 - a^2}\,\frac{z}{(z - a)} \tag{6.85}$$

which is obviously a minimum phase filter.

Let us now discuss briefly the problem of strong factorization without the rational function assumption. The question remains the same, i.e., find the factorization (6.80) in which the filter $H(z)$ and its inverse are causal. This problem is not necessarily solvable, for the power spectrum in the frequency domain $\Gamma(\nu)$ must satisfy a specific condition. In order to realize strong factorization, the spectral distribution function $\overline{\Gamma}(\nu)$ must be continuous in such a way that (6.39) holds. This means that there are *no spectral lines* in the signal. Furthermore, the power spectrum $\Gamma(\nu)$ must satisfy

$$\int_{-1/2}^{+1/2} |\ln\{\Gamma(\nu)\}|\, d\nu < +\infty \tag{6.86}$$

We will not give the proof for this condition, sometimes called, by analogy with continuous time, a Paley-Wiener condition. But it plays a very important role in prediction problems which is the true framework for its interpretation and is discussed later. We note, however, that if $\Gamma(\nu) = 0$ for some bandwidth, the condition is not satisfied, and strong factorization is impossible. This is especially the case for all band-limited signals.

Factorization can be obtained as follows. By introducing the function $A(\nu)$ defined by

$$A(\nu) = \ln\{\Gamma(\nu)\} \tag{6.87}$$

we see that the condition (6.86) ensures that $A(\nu)$ has a Fourier expansion equal to $A(\nu)$ if this function is of bounded variation. This can be written as

$$A(\nu) = \sum_{n=-\infty}^{+\infty} a_n \exp(-2\pi j \nu n) \tag{6.88}$$

and the a_ns are given by

$$a_n = \int_{-1/2}^{+1/2} A(\nu) \exp(2\pi j \nu n)\, d\nu \tag{6.89}$$

As the signal $x[k]$ under consideration is real, its correlation function is real and even. Then $\Gamma(\nu)$ and $A(\nu)$ are also even. This gives the condition $a_n = a_n^*$ and $a_n = a_{-n}$, deduced from (6.89). Let us introduce the function

$$A_+(\nu) = (1/2)\, a_0 + \sum_{n=1}^{\infty} a_n \exp(-2\pi j \nu n) \tag{6.90}$$

Its complex conjugate can be called $A_-(\nu)$, and we have $A(\nu) = A_+(\nu) + A_-(\nu)$. We now introduce the filter with the frequency response

$$G(\nu) = \exp[A_+(\nu)] \tag{6.91}$$

It is clear that $G^*(\nu) = \exp[A_-(\nu)]$, giving $\Gamma(\nu) = |G(\nu)|^2$, which is a factorization in the frequency domain. For the last step, to demonstrate that it is a strong factorization, we must show that the transfer function $H(z) = \exp[A_+(z)]$ where

$$A_+(z) = (1/2)\, a_0 + \sum_{n=1}^{\infty} a_n z^{-n} \tag{6.92}$$

has neither zero nor pole outside the unit circle. This is a consequence of the fact that $A_+(z)$ is analytic outside the unit circle because it is convergent on this circle and equal to $A(\nu)$, from (6.88). As $A_+(z)$ has no infinite value outside the unit circle, $H(z)$ cannot be equal to zero nor become infinite in this region, which means that $H(z)$ is a minimum-phase filter. For the practical calculation of $H(z)$ this method is not very convenient, as we have first to calculate the coefficients a_n with (6.89) and then $A_+(z)$ with (6.92).

6.6 ERGODICITY OF STATIONARY SIGNALS

Mean square ergodicity was introduced in Section 5.5, and a condition ensuring ergodicity appears if the last term of (5.65) tends to zero. We shall reconsider the problem to show that when the signal is stationary the condition ensuring ergodicity is much easier to interpret in the frequency domain than in the time domain. For this purpose let us write (5.63) in a slightly different form as

$$M_T(t) = \frac{1}{T} \int_{t-T}^{t} x(\theta)\, d\theta \tag{6.93}$$

We can easily verify that this expression can be written as (5.56), which means that $M_T(t)$ is the output at time t of a linear filter of impulse response $h_T(t)$ when the input is $x(t)$. This filter is called a causal time averager, and its impulse response is shown in Figure 6.6.

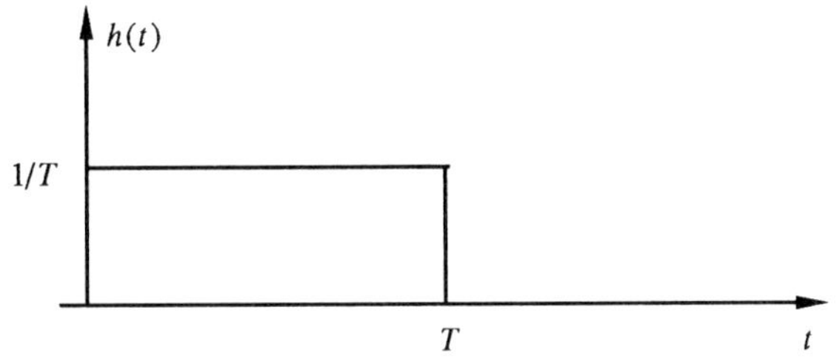

Figure 6.6 Impulse response of a time averager.

Its frequency response can be written as

$$G_T(\nu) = e^{-\pi j \nu T} \operatorname{sinc}(\nu T) \tag{6.94}$$

where sinc x is the sinus cardinal function, already used in (6.44). For our physical discussion it is important to note that $G_T(0) = 1$, and we deduce from (6.9) and (6.36) that the mean value of a stationary signal is unchanged by this filter. This can also be deduced directely from (6.93) by using (6.61). From the Parseval theorem we find that

$$\int |G_T(\nu)|^2\, d\nu = \int h_T^2(t)\, dt = \frac{1}{T} \tag{6.95}$$

which is the basic property for the discussion which follows. In fact, when T increases, the filter is such that $G_T(0)$ remains equal to one, but its equivalent bandwidth tends to zero as $1/T$.

We shall now calculate the second order properties of $M_T(t)$. If $x(t)$ has a constant mean value m, we have $E\{M_T(t)\} = m$, whatever the value of T. As the variance v_T of

$M_T(t)$ is the value for $\tau = 0$ of the correlation function, it is equal to the integral of the power spectrum of $M_T(t)$, and using (6.55) we obtain

$$v_T = \int |G_T(\nu)|^2 \, \Gamma_x(\nu) \, d\nu \tag{6.96}$$

where $\Gamma_x(\nu)$ is the power spectrum of $x(t)$. This expression is the spectral counterpoint of (5.65) and is of course much easier to discuss. But first, to avoid any confusion in the understanding of (6.96), note that throughout this chapter the correlation function is nothing other than the covariance function in the stationary case. In other words, the mean value has been subtracted to obtain $\gamma(\tau)$, and its Fourier transform $\Gamma(\nu)$. This is important, since the whole discussion revolves around mean value measurement, and, as indicated in Section 5.2.2 (**c**), the correlation function is sometimes defined without subtracting the mean value. So $\Gamma_x(\nu)$ in (6.96) is the Fourier transform of

$$\gamma_x(\tau) = E\{x(t)\, x^*(t-\tau)\} - |m|^2 \tag{6.97}$$

Returning to (6.96), we observe that a *sufficient condition* of mean square ergodicity is that $\Gamma(0) < +\infty$. In fact, as $\Gamma_x(\nu)$ is continuous, we can write v_T as

$$v_T \cong (1/T)\; \Gamma(0) \tag{6.98}$$

which decreases to zero when T tends to infinity. A stronger condition is to assume that $\Gamma(\nu)$ is a bounded function, which is realized especially if

$$\int |\gamma(\tau)| \, d\tau < +\infty \tag{6.99}$$

From a physical point of view mean square ergodicity is completely dependent on the behavior of the power spectrum in the *zero frequency* domain. This is of course related to the behavior of the correlation function for $|\tau|$ becoming infinite. In particular, the signal (5.32) cannot be ergodic; in fact, its power spectrum contains the term σ_Z^2 $\delta(\nu)$, which in (6.96) gives the contribution σ_Z^2 independent of T, and cannot then be suppressed by integration. In conclusion, we can broadly say that mean square ergodicity means *no frequency line* at the frequency zero. This is understandable because such a frequency line signifies a constant term in the signal, and we have seen in Example 5.6 that random variables do not exhibit ergodicity. Finally, note that all these points can be applied to the discrete time case, and that the filter in Figure 6.2 has an exact discrete time version.

Example 6.8. On the ergodicity of some signals. The signal (5.47) with a uniform phase has a zero mean value, and its correlation function is given by (5.48). The corresponding spectrum has only two spectral lines at the angular frequencies $\pm\omega$. We then have $\Gamma(0) = 0$ and the signal is ergodic. In reality it is obvious, since it is well known that the time average of a sinusoid is zero. Note that the condition (6.99) is not satisfied because of the existence of spectral lines in the spectrum. The same can be said for the signal (5.49). Consider now the case of a signal with a random frequency, such as in (5.50).

If ϕ is uniformly distributed the mean value is null and the correlation function is given by (5.52). It follows that the power spectrum is proportional to the PDF of the random frequency Ω. A spectral line to the frequency $\nu = 0$ means that there is a finite probability for having $\Omega = 0$. In this case a part of $Z(t)$ defined by (5.50) no longer depends on time and the time average cannot be null like the mean value. This explains the lack of ergodicity.

Although the above relates to mean variance measurement by time averaging, the same method can be transposed for any expected value. For example, the measurement of the correlation function can be realized by using the function

$$C_T(\tau;\, t) = \frac{1}{T} \int_{t-T}^{t} x(\theta)\, x(\theta - \tau)\, d\theta \tag{6.100}$$

as is done by any practical correlator working in real time. When T tends to infinity $C_T(\tau;\, t)$ becomes independent of t and tends to the correlation function if some conditions of ergodicity are met. If we want to transpose the previous discussion in this case, it turns out that we need a fourth-order moment of $x(t)$, which is outside the scope of this section, but will be examined later.

6.7 FROM CONTINUOUS TO DISCRETE TIME BY SAMPLING

Sampling of continuous-time random signals has already been dealt with, and (6.27) explains how a DT signal $x[k]$ can be obtained by sampling of a CT signal $x(t)$. We have also indicated how the harmonic decomposition of the two signals are related.

If the CT signal $x(t)$ is stationary with the correlation function $\gamma(\tau)$, the DT signal is also stationary and its correlation function $\gamma[k]$ is stationary and deduced from (6.27) by

$$\gamma[k] \triangleq E\{x[m]\, x^*[m-k]\} = E\{\, x(mT)\, x^*(\,[m-k]T\,)\,\} = \gamma(kT) \tag{6.101}$$

This means that the correlation function of $x[k]$ is simply obtained by sampling that of $x(t)$. Let us now calculate the power spectrum of $x[k]$. Starting from (6.40) we have only to replace τ by kT. By doing this the exponential term becomes a periodic function of ν with the period $1/T$. We can then translate all the spectral components in the interval $[-1/2T, +1/2T]$, which finally gives

$$\gamma[k] = \int_{-1/2T}^{1/2T} \sum_{n} \Gamma\left(\nu - \frac{n}{T}\right) \exp(2\pi j \nu k T)\, d\nu \tag{6.102}$$

The series inside the integral is then the power spectrum $\Gamma_d(\nu)$ of the DT signal $x[k]$, and we have the relations

$$\gamma[k] = \int_{-1/2T}^{1/2T} \Gamma_d(\nu)\, e^{2\pi j \nu k T}\, d\nu \tag{6.103}$$

$$\Gamma_d(\nu) = T \sum \gamma[k]\, e^{-2\pi j \nu k T} \tag{6.104}$$

It is sometimes more interesting to use the sampling operation defined by (6.28) that associates the function $\gamma_s(\tau) = \gamma(\tau)\ \delta_T(\tau)$ to the correlation function $\gamma(\tau)$. This introduces the function

$$\Gamma_s(\nu) = \frac{1}{T} \sum_n \Gamma\left(\nu - \frac{n}{T}\right) \tag{6.105}$$

and we clearly have $\Gamma_d(\nu) = T\,\Gamma_s(\nu)$, which can be inserted into (6.103) and (6.104) to obtain the relations between $\gamma[k]$ and $\Gamma_s(\nu)$.

One of the most important results within the framework of sampling is the *sampling theorem*. For deterministic signals it states that if a CT signal is band-limited in the frequency domain $[-B,\ +B]$, it can be written as

$$x(t) = \sum x(\theta_n)\ \mathrm{sinc}[2B(t - \theta_n)] \tag{6.106}$$

where the time instants θ_n are given by

$$\theta_n = \frac{n}{2B} + \theta \tag{6.107}$$

and θ is arbitrary (see [Picinbono], p. 81). Thus the question of whether (6.106) is still valid for random signals arises.

The first point to clarify is the transposition to the random case of the concept of band-limited signals. In the *deterministic case* we use the Fourier transform, which leads to spectral decomposition (6.5) for the random case. More precisely, a random signal $x(t;\, s)$ is said to be band-limited in the frequency range $[-B,\ +B]$ if the spectral increments $d\overline{X}(\nu)$ are null outside this range, or for $|\nu| > B$. In this case any trajectory appearing in (6.5) is also band-limited, and we can apply (6.106) whatever the value of s, which shows that it is valid with probability one.

In the case of *stationary signals* it is sometimes said that $x(t)$ is band-limited if its correlation function is band-limited in the sense of deterministic signals. This means that the power spectrum is null for $|\nu| > B$, and that the increments $d\overline{X}(\nu)$ are null with probability one, which returns us to the previous case.

An interesting situation appears when the signal is white and band-limited. Its power spectrum is constant for $|\nu| \le B$ and null for $|\nu| > B$. The correlation function is then $\sigma^2\ \mathrm{sinc}(2B\ \tau)$ where σ^2 is its variance. But it follows that the samples $x(\theta_n)$ appearing in (6.106) are uncorrelated, or orthogonal. As the sinc functions are also orthogonal (see [Picinbono], p. 91), we can conclude that (6.106) is *doubly orthogonal*, a situation discussed for random vectors on p. 67.

The sampling of CT signals introduces various questions well known for deterministic signals (see [Picinbono], Ch. 3). They can be transposed without difficulty to the random case signals as for example the study of signals, common in communication problems, which are band-limited around a carrier frequency. Some of their properties will be investigated in the next section. Note also that because of the frequency limitation, a band-limited signal can be differentiated at any order. This is a direct result of

(6.59), which by extension to higher-order derivatives shows that all these derivatives are second order. This does not occur, for example, if the power spectrum is a rational function. The originality of band-limited signals also appears in the DT case and we see that (6.86) is not satisfied. Strong factorization is thus not possible, a point which will be examined later within the framework of prediction problems.

6.8 NARROWBAND SIGNALS

Most signals used in practice, particularly in the context of communication, are narrowband. This expression means that the spectrum is practically limited to a frequency band $\Delta\nu$ around a carrier frequency f_0 and that $\Delta\nu << f_0$. Band-limited signals studied in the previous section do not belong to this class of signals because $f_0 = 0$. In this section we will restrict our attention to real signals and we wish to transpose to the random case the results valid for deterministic signals (see [Picinbono], p. 47). An example of the trajectory of a band-limited signal is represented in Figure 6.7. For a short observation period this trajectory is similar to that represented in Figure 1.1 and corresponds to a pure sinusoid signal. However, it appears as a slow modulation of the amplitude and of the phase, and a precise definition of these quantities must be introduced. A very powerful tool for this is the *analytic signal* related to the *Hilbert transform*.

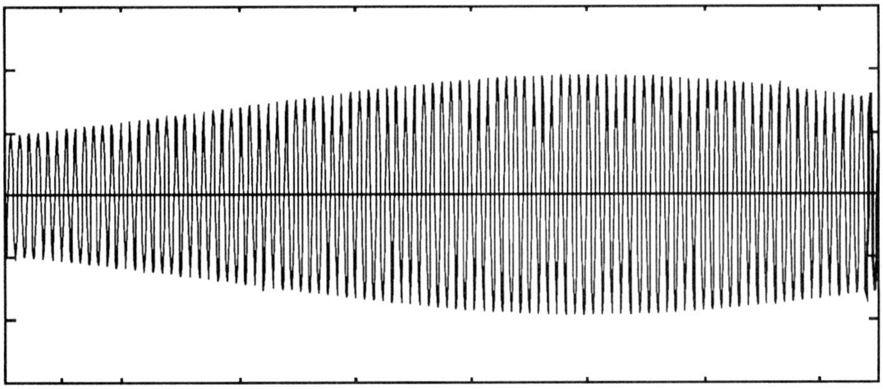

Figure 6.7 Trajectory of a narrowband signal.

We begin the discussion by considering the properties of the random analytic signal, and, for simplification, we shall assume that all signals considered are stationary, zeromean and second order.

6.8.1 Properties of the Random Analytic Signal

To any real signal $x(t)$ it is possible to associate a complex signal $z(t)$ called its *analytic signal* (AS) in the following manner. We use a filter with the frequency response

$$G_A(\nu) = 2u(\nu) \tag{6.108}$$

where $u(\nu)$ is the unit step function. If the input of this filter is $x(t)$, its output is the AS $z(t)$. Writing $G_A(\nu) = G_I(\nu) + j\,G_Q(\nu)$ with

$$G_I(\nu) = 1 \;;\; G_Q(\nu) = -j\,\nu/|\nu| = -j\mathrm{Sg}(\nu) \tag{6.109}$$

where $\mathrm{Sg}(\nu)$ means the sign of ν, we can express the impulse response of this filter by

$$h_A(t) = \delta(t) + j\,\frac{1}{\pi}\,\frac{pv}{t} \tag{6.110}$$

This last term represents the Cauchy *principal value*. As $z(t)$ is the convolution of $h_A(t)$ by $x(t)$, we can write

$$z(t) = x(t) + jy(t) \tag{6.111}$$

which signifies that the real part of $z(t)$ is the original signal $x(t)$ while the imaginary part $y(t)$, called a quadrature signal, is the *Hilbert transform* of $x(t)$.

Let us now investigate the second order properties of the signals $z(t)$ and $y(t)$. These are studied in the frequency domain by a systematic use of (6.54) and (6.55). Investigating firstly the second order properties of the two quadrature components $x(t)$ and $y(t)$, as $|G_Q(\nu)|^2 = 1$, it turns out that

$$\Gamma_y(\nu) = \Gamma_x(\nu) \tag{6.112}$$

which means that $x(t)$ and $y(t)$ have the same power spectrum. By Fourier transformation we deduce that their correlation functions are equal, or

$$\gamma_y(\tau) = \gamma_x(\tau) \tag{6.113}$$

Using (6.54) and noting that $G_Q^*(\nu) = -G_Q(\nu)$, we immediately obtain

$$\Gamma_{xy}(\nu) = -G_Q(\nu)\,\Gamma_x(\nu) \;;\; \Gamma_{yx}(\nu) = G_Q(\nu)\,\Gamma_x(\nu) \tag{6.114}$$

This shows that $\gamma_{yx}(\tau)$ is the Hilbert transform of $\gamma_x(\tau)$, and also that

$$\gamma_{xy}(\tau) = -\gamma_{yx}(\tau) \tag{6.115}$$

Consider now the AS $z(t)$. As $z(t) = x(t) + jy(t)$, it follows from the previous relations that

$$E\{z(t)\,z(t-\tau)\} = 0 \tag{6.116}$$

and finally that

$$\gamma_z(\tau) = 2\,\{\gamma_x(\tau) + j\,\gamma_{yx}(\tau)\} \tag{6.117}$$

The power spectrum of $z(t)$ is of course equal to

$$\Gamma_z(\nu) = 4u(\nu)\,\Gamma_x(\nu) \tag{6.118}$$

which shows, with (6.117), that $\gamma_z(\tau)$ is the AS of $2\,\gamma_x(\tau)$.

It is interesting to note what happens for $\tau = 0$. In this case we have $\gamma_{yx}(0) = E[y(t)\; x(t)] = \gamma_{xy}(0)$ and, as a result of (6.115), we obtain

$$\gamma_{xy}(0) = 0 \tag{6.119}$$

Thus, whatever the value of t, the random variables $x(t)$ and $y(t)$ have the same variance and are uncorrelated. This situation was already encountered in the discussion on circular normal RVs (see p. 120).

6.8.2 Properties of Amplitude and Phase

The definition of the amplitude and phase of a real signal is all but obvious, as noted in [Picinbono], p. 50, even though commonsense associates to the signal $a\cos(\omega_0 t + \phi)$ the amplitude a, the angular frequency ω_0, and the local phase ϕ. On the other hand, starting from a given real signal $x(t)$, there is no unique way to write this signal in the form $a(t)\cos\{\phi(t)\}$. This means that, without further assumptions, it is impossible to properly associate definite instantaneous amplitudes and phases to any real signal. In order to arrive at a definition without contradiction, we start from the AS $z(t)$ associated with a given real signal $x(t)$. It is always possible to introduce the modulus and the phase of $z(t)$, by writing

$$z(t) = a(t)\exp\{j\phi(t)\} \tag{6.120}$$

By definition the *instantaneous amplitude* of the real signal $x(t)$ is the modulus $a(t)$ of its analytic signal $z(t)$. This definition presents no ambiguity, because there is a one to one correspondence between a real signal $x(t)$ and its analytic signal $z(t)$. So, the instantaneous amplitude $a(t)$ is a non-negative number such that

$$a^2(t) = |z(t)|^2 \tag{6.121}$$

On the other hand, the *instantaneous phase* of a real signal $x(t)$ is the phase $\phi(t)$ of its analytic signal. This phase is defined mod(2π) and is usually taken as $0 \le \phi(t) < 2\pi$. But this convention is not always the best, particularly when considering phases such as $\omega_0 t$. In the case of narrowband signals the phase has a particular structure. As the bandwidth of the signal satisfies $\Delta\nu << f_0$, it is possible to use this frequency, or the angular frequency $\omega_0 = 2\pi f_0$, to write the instantaneous phase in the form

$$\phi(t) = \omega_0 t + \theta(t) \tag{6.122}$$

introducing the relative instantaneous phase $\theta(t)$. It should be noted that, starting from $\phi(t)$, the phase $\theta(t)$ is a function of the carrier frequency ω_0. This frequency is somewhat arbitrary, but must always be located in the frequency band $\Delta\nu$ where the spectrum of the signal is located. As soon as ω_0 is chosen, both $\theta(t)$ and the *complex envelope* of the signal $x(t)$, given by

$$\alpha(t) = z(t)\, e^{-j\,\omega_0 t} = a(t)\, e^{j\phi(t)}\, e^{-j\,\omega_0 t} \tag{6.123}$$

are defined. By reason of the above, it would be better to write the complex envelope as $\alpha(t;\ \omega_0)$ to indicate the dependence on the arbitrary frequency ω_0.

Beyond the mathematical equations, the physical meaning of these operations must be clearly understood. If $x(t)$ is narrowband, the Fourier transform of $z(t)$ is completely concentrated near the frequency ω_0 with a bandwidth $\Delta\,\nu$, and because of (6.108) there is no negative frequency in this spectrum. The last term of (6.123) signifies a frequency translation and $\alpha(t)$ becomes an approximately band-limited signal in the sense of the previous section. This appears in Figure 6.8.

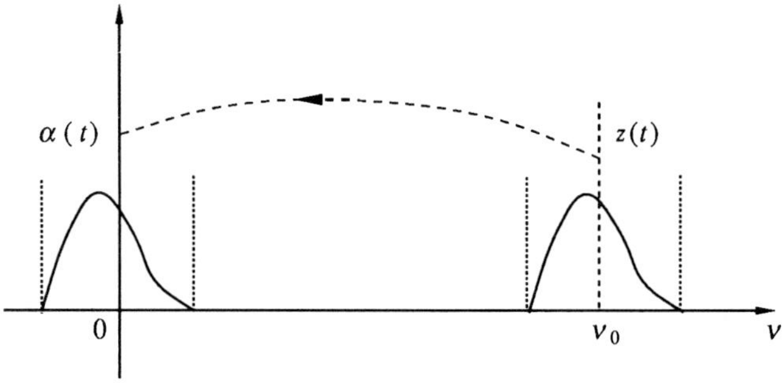

Figure 6.8 Fourier transforms of $z(t)$ and $\alpha(t)$.

Thus the variations of $\alpha(t)$ are very small during the period associated with ω_0. In other words, during some periods the signal $x(t)$ seems to be sinusoidal with amplitude $a(t)$ and phase $\phi(t)$. But after many such periods the amplitude and phase need not remain the same, as is seen in Figure 6.7. It is for this reason that narrowband signals are sometimes called quasisinusoidal or quasimonochromatic signals.

Instead of representing $z(t)$ as in (6.120) by using the modulus and phase, we can introduce the real and imaginary parts. For $z(t)$ these parts are of course $x(t)$ and $y(t)$. Let us now look at those of the complex envelope written as

$$\alpha(t) = m(t) + jn(t) \tag{6.124}$$

In consequence, it follows that $x(t)$, the real part of $z(t) = \alpha(t)\exp(j\,\omega_0 t)$, can be written

$$x(t) = m(t)\cos(\omega_0 t) - n(t)\sin(\omega_0 t) \tag{6.125}$$

which introduces the quadrature components $m(t)$ and $n(t)$.

Before calculating the statistical properties of the quantities introduced for representing narrowband signals, let us give some elementary ideas on the experimental devices allowing the practical calculation of these quantities.

The starting point is a real narrowband signal $x(t)$. There are two ways to obtain its analytical signal $z(t)$. The first makes use of the analytical filter (6.108) and the second uses the Hilbert transform $y(t)$. Of course, it is impossible to realize perfectly the filter which calculates the Hilbert transform of an arbitrary signal $x(t)$. However, as $x(t)$ is a narrowband signal around the mean frequency ν_0, we need to realize the quadrature filter only for these frequencies. For example, if the bandwidth is null, the filter must only transform $\cos(\omega_0 t)$ into $\sin(\omega_0 t)$. For discrete-time signals there are many approximate procedures for calculating the Hilbert transform.

The complex envelope $\alpha(t)$ can be deduced from the analytic signal $z(t)$ by a modulation procedure realizing (6.123). However, there is a more common way to obtain (6.124) directly from $x(t)$, called a *synchronous demodulation*. Using the representation of $x(t)$ given by (6.125), let us calculate $2x(t)\cos(\omega_0 t)$, which is a modulation by a real pure tone signal. The result is

$$2x(t)\cos(\omega_0 t) = m(t) + [m(t)\cos(2\omega_0 t) - n(t)\sin(2\omega_0 t)]$$

The last term of this equation is a narrowband signal around the frequency $2\omega_0$, and by using a low pass filter, it is possible to obtain $m(t)$. This procedure is pictured in Figure 6.9.

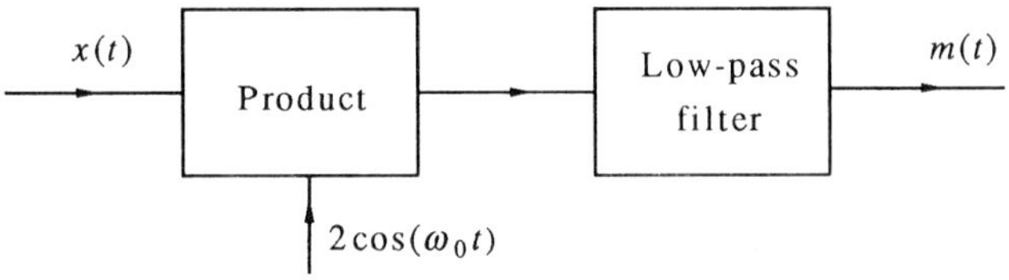

Figure 6.9 Principle of synchronous demodulation.

The same procedure can be used to calculate the component $n(t)$. It suffices to realize a modulation by the signal $2\sin(\omega_0 t)$.

Suppose now that the signal $x(t)$ is *random*. The complex envelope has almost the same second order properties as the analytical signal itself. In fact, we deduce from (6.123), (6.120) and (6.116) that

$$E\{\alpha(t)\ \alpha(t-\tau)\} = 0 \tag{6.126}$$

Similarly, as a result of (6.123), the correlation function of $\alpha(t)$ is

$$\gamma_\alpha(\tau) = \gamma_z(\tau)\ e^{-j\omega_0\tau} \tag{6.127}$$

Note especially that for $\tau = 0$ we obtain the same second order moments for $\alpha(t)$ and $z(t)$. Transposed into the frequency domain, (6.127) gives

$$\Gamma_\alpha(\nu) = \Gamma_z(\nu + \nu_0) \tag{6.128}$$

which, as mentioned above, is a frequency translation.

It is unfortunately impossible to give here the second order properties of $a(t)$ and $\phi(t)$ in (6.123), as these functions are deduced from $x(t)$ by non-linear operations and the study of their second order properties demands a greater knowledge of $x(t)$ than that given by the second order moments. This is particularly clear in (6.121) giving $a(t)$ by $|z(t)|$. In order to obtain properties on $a(t)$ and $\phi(t)$, other assumptions on the signal $x(t)$ must be introduced, and we will see later that if we introduce *normal* distribution, numerous results can be achieved.

Let us now consider the quadrature components. The relations between $\alpha(t)$, $\alpha^*(t)$ and $m(t)$ or $n(t)$ are of course

$$m(t) = (1/2)\ \{\alpha(t) + \alpha^*(t)\}\ ;\ n(t) = (1/2j)\ \{\alpha(t) - \alpha^*(t)\} \tag{6.129}$$

and using (6.126), we obtain

$$\gamma_m(\tau) = \gamma_n(\tau) = (1/4)\ \{\gamma_\alpha(\tau) + \gamma_\alpha^*(\tau)\} \tag{6.130}$$

$$\gamma_{mn}(\tau) = -\gamma_{nm}(\tau) = (-1/4j)\ \{\gamma_\alpha(\tau) - \gamma_\alpha^*(\tau)\} \tag{6.131}$$

By Fourier transformation we obtain the same relations in the frequency domain

$$\Gamma_m(\nu) = \Gamma_n(\nu) = (1/4)\ \{\Gamma_z(\nu_0 + \nu) + \Gamma_z(\nu_0 - \nu)\} \tag{6.132}$$

$$\Gamma_{mn}(\nu) = -\Gamma_{nm}(\nu) = (-1/4j)\ \{\Gamma_z(\nu_0 + \nu) - \Gamma_z(\nu_0 - \nu)\} \tag{6.133}$$

Let us now extract the most significant results from these equations. Firstly, recall that, starting from $z(t)$, AS of $x(t)$, we obtain a complex envelope depending on ω_0. This is demonstrated in the two last equations where the arbitrary carrier frequency ω_0 clearly appears because $\omega_0 = 2\pi\nu_0$.

Suppose now that $\tau = 0$. As for this value $\gamma_\alpha(\tau)$ is real, it results from (6.131) that

$$\gamma_{mn}(0) = \gamma_{nm}(0) = 0 \tag{6.134}$$

This means that, *considered at the same time instant t*, the RVs $m(t)$ and $n(t)$ are uncorrelated.

Furthermore, if the power spectrum of $z(t)$ is *symmetric* with respect to the carrier frequency ν_0, or if $\Gamma_z(\nu_0 - \nu) = \Gamma_z(\nu_0 + \nu)$, we deduce from (6.133) that $\Gamma_{mn}(\nu) = 0$. This implies that $\gamma_{mn}(\tau) = 0$ whatever τ, which means that the random signals $m(t)$ and $n(t)$ are *uncorrelated*, a very interesting property for various applications.

We also see that the power spectrum of $m(t)$ is deduced from that of $z(t)$, i.e., of $x(t)$, by a very simple operation adding the contributions at the same distance of the carrier frequency ν_0.

6.9 SPECTRAL MATRIX

All results obtained in this chapter for scalar complex random signals can be extended without difficulty to vector signals, the principal results of which will be only briefly outlined here. The covariance matrix of a random vector was defined by (5.14), and in the stationary case the correlation matrix is simply

$$\mathbb{\gamma}(\tau) = E\{\mathbf{X}(t)\,\mathbf{X}^H(t-\tau)\} - \mathbf{m}\,\mathbf{m}^H \tag{6.135}$$

For an arbitrary value of τ this matrix has no specific property. On the other hand, if $\tau = 0$ it is an NND matrix, as studied on p. 61.

The Fourier transform $\mathbb{\Gamma}(\nu)$ of $\mathbb{\gamma}(\tau)$ is called the *spectral matrix* of the random vector $\mathbf{x}(t)$. This is an NND matrix and thus is also Hermitian (see p. 61). In order to show this, let us consider the scalar complex signal $y(t)$ defined by

$$y(t) \triangleq \mathbf{u}^H \mathbf{x}(t) \tag{6.136}$$

where $\mathbf{u}$ is an arbitrary complex vector. The correlation function of $y(t)$ is

$$\gamma_y(\tau) = E\{y(t)\, y^*(t-\tau)\} - |E\{y(t)\}|^2 \tag{6.137}$$

and, using (6.136) and (6.135), we obtain

$$\gamma_y(\tau) = \mathbf{u}^H \mathbb{\gamma}(\tau)\mathbf{u} \tag{6.138}$$

The power spectrum of $y(t)$ is obtained by Fourier transformation, which gives

$$\Gamma_y(\nu) = \mathbf{u}^H \mathbb{\Gamma}(\nu)\mathbf{u} \tag{6.139}$$

As with any power spectrum, this quantity is non-negative and as $\mathbf{u}$ is an arbitrary complex vector, $\mathbb{\Gamma}(\nu)$ is an NND matrix.

The diagonal elements of $\mathbb{\Gamma}(\nu)$ are the spectral densities of the components of the vector $\mathbf{x}(t)$. The off-diagonal elements are the crossspectra, with the same reservation on the name as after (6.65), because they are complex-valued functions of the frequency.

Let us now consider the filtering of a vector signal. A vector filter is defined by a frequency response matrix $\mathbb{G}(\nu)$. If the input is a signal with n components while the output has m components, $\mathbb{G}(\nu)$ is an $m \times n$ matrix. In the frequency domain the filtering relation is a product, which gives

$$d\overline{\mathbf{Y}}(\nu) = \mathbb{G}(\nu)\, d\overline{\mathbf{X}}(\nu) \tag{6.140}$$

where the increments are vector increments as in (6.5) when the signals are vectors instead of scalars. Using exactly the same arguments as in the scalar case, we find that the random increments are uncorrelated when the signals are stationary and the spectral matrix is given by an expression similar to (6.37) while the l.h.s. term of (6.15) must be written $E\{d\overline{\mathbf{X}}(\nu_1)\; d\overline{\mathbf{X}}^H(\nu_2)\}$. This finally gives

$$\mathbb{\Gamma}_y(\nu) = \mathbb{G}(\nu)\; \mathbb{\Gamma}_x(\nu)\; \mathbb{G}^H(\nu) \tag{6.141}$$

an expression that generalizes (6.55) for random vectors. It is interesting to note that starting from this expression it is possible to show again that $\mathbb{\Gamma}(\nu)$ is an NND matrix by using the same reasoning for matrices as for scalars (see Section 6.4.3 (**c**)).

Spectral matrices play a very important role in many aspects of signal processing, especially in problems of spatial processing and antenna design.

6.10 HIGHER-ORDER SPECTRA

In many physical situations it is necessary to make use of higher-order moments or cumulants introduced in Section 5.2.6. Let us present examples of such situations. In Section 5.9, dealing with energy and power, we saw that the time-average power is given by (5.98). This expression is similar to (5.63) with the difference that, for real signals, the result is an integral of $x^2(t)$ instead of $x(t)$. If we want to transfer the condition of ergodicity for the power, valid for the mean and appearing in (5.65), we see that we have to calculate the covariance function of $x^2(t)$ defined by

$$\gamma(t, \theta) = E[x^2(t)\, x^2(\theta)] - E[x^2(t)]\, E[x^2(\theta)] \tag{6.142}$$

It is clear that this expression makes use of a fourth-order moment of $x(t)$. Now, if we want to discuss this question in the frequency domain, using the same approach as in Section 6.6, we see that we have to introduce Fourier transforms of fourth-order moments, or fourth-order spectra. The same problem arises for power measurement of a complex field. If this field is described by a complex signal $x(t)$, its instantaneous power is given by (5.93) and the fourth-order moment in (6.142) must be replaced by $E[x(t)x^*(t)x(\theta)x^*(\theta)]$, which is still a fourth-order moment of a complex signal.

The reasons for the interest of higher-order spectra are the same as for the second-order cases and result from the importance of linear filtering in signal processing problems. The transformation of higher-order moments or cumulants through linear filters are given by (5.59) and (5.60). These expressions make use of multiple convolution and are much simpler in the frequency domain. Consider, for example, (5.59) and suppose that $m_x(\mathbf{t})$ has a multidimensional Fourier transform written as

$$M_x(\mathbf{f}) = \int m_x(\mathbf{t}) \exp(-2\pi j \mathbf{f}^T \mathbf{t})\, d\mathbf{t} \tag{6.143}$$

which is a generalization of (6.1) with the same notations as in (3.56'). Introducing the frequency response $G(f)$ of the filter, (5.59) can be written as

$$M_y(\mathbf{f}) = G(f_1)G(f_2)\ldots G(f_n)\, M_x(\mathbf{f}) \tag{6.144}$$

which is similar to (6.9) and much simpler than (5.59). The same procedure is valid for cumulants and for the discrete-time case.

After this general introduction, let us now investigate more precisely some properties of higher-order spectra.

First, consider the case of a real continuous-time random signal $x(t)$. Its general kth-order moment is defined by (5.42) and is, for simplicity, written as $m_k(\mathbf{t})$, where $\mathbf{t}$ is the vector, the components of which are $t_1, t_2, \ldots, t_k$. If $x(t; s)$ is harmonizable, it can be expressed as in (6.5) and under very general assumptions $m_k(\mathbf{t})$ can be written as

$$m_k(\mathbf{t}) = \int \dots \int E[d\bar{X}(f_1)\, d\bar{X}(f_2) \dots d\bar{X}(f_k)] \exp(2\pi j \mathbf{f}^T \mathbf{t}) \tag{6.145}$$

where $\mathbf{f}$ is the vector the components of which are $f_1, f_2, \dots, f_k$. We are using the letter f instead of ν for the frequency only to differentiate between the one and the multi-dimensional cases. To simplify the discussion, we will assume that the expected value in (6.145) can be written in the form

$$E[d\bar{X}(f_1) \dots d\bar{X}(f_k)] = M_k(\mathbf{f})\, df_1\, df_2 \dots df_k \tag{6.146}$$

which is similar to (6.36) valid for $k = 1$. The function $M(\mathbf{f})$ defined in the frequency space can be considered in the distribution sense, exactly as in (6.36). This significantly simplifies (6.145) which becomes

$$m_k(\mathbf{t}) = \int M_k(\mathbf{f}) \exp(2\pi j \mathbf{f}^T \mathbf{t})\, d\mathbf{f} \tag{6.147}$$

By Fourier inversion, we find (6.143).

The same procedure can be applied to the kth-order cumulant, defined by (5.43) and noted $c_k(\mathbf{t})$. Its Fourier transform is written as $C_k(\mathbf{f})$ and there are the same relations between $c_k(\mathbf{t})$ and $C_k(\mathbf{f})$ as between $m_k(\mathbf{t})$ and $M_k(\mathbf{f})$. Finally, these notations can also be used in the discrete-time case. The only difference is that the vector $\mathbf{t}$, noted $\mathbf{k}$, has discrete components and the result in the frequency domain is that the functions $M_k(\mathbf{f})$ or $C_k(\mathbf{f})$ can be limited to the hypercube defined by $|f_i| \le 1/2$.

As in Section 6.3, it is now of great importance to investigate the consequences of the *stationarity* on the functions $m_k(\mathbf{t})$ or $c_k(\mathbf{t})$ and their Fourier transforms $M_k(\mathbf{f})$ and $C_k(\mathbf{f})$. This operation will introduce the concept of a higher-order spectrum, which is a generalization of that of the spectrum introduced as the Fourier transform of the correlation function.

Let us first discuss the continuous-time case. If $x(t)$ is strict-sense stationary, it satisfies (5.45). This means that $m_k(\mathbf{t})$ is invariant if we replace $\mathbf{t}$ by $\mathbf{t} + \tau\mathbf{u}$, where τ is arbitrary and $\mathbf{u}$ is the vector $[1, 1, \dots, 1]^T$. This also means that $m_k(\mathbf{t})$ is not a function of k independent variables $t_1, t_2, \dots, t_k$, but only of $k - 1$, as, for example, $t_2 - t_1$, $t_3 - t_1, \dots, t_k - t_1$. This has a consequence on its Fourier transform and, for example, if $k = 1$, this transform must satisfy (6.36), or $M_1(\nu) = m\,\delta(\nu)$. In fact, it is well-known that the Dirac distribution is the Fourier transform of the signal constant and equal to one. Writing $m_k(\mathbf{t}) - m_k(\mathbf{t} + \tau\mathbf{u}) = 0$, whatever the value of τ, and using (6.147), we obtain

$$\int M_k(\mathbf{f})\, [1 - \exp(2\pi j \tau \mathbf{f}^T \mathbf{u})] \exp(2\pi j \mathbf{f}^T \mathbf{t})\, d\mathbf{f} = 0 \tag{6.148}$$

which implies that

$$M_k(\mathbf{f})\, [1 - \exp(2\pi j \tau \mathbf{f}^T \mathbf{u})] = 0 \tag{6.149}$$

whatever the value of τ. As $M_k(\mathbf{f})$ is not a function of τ, it results from (6.149) that $M_k(\mathbf{f})$ can have non-null values only if

$$\mathbf{f}^T\mathbf{u} = \sum_{i=1}^{k} f_i = 0 \tag{6.150}$$

This equation characterizes a linear manifold of dimension $k - 1$ in the frequency space, which will be called the *stationary manifold.*

Conversely, if $M_k(\mathbf{f})$ can be written as

$$M_k(\mathbf{f}) = \Gamma_{k-1}(\mathbf{f})\ \delta(f_1 + f_2 + \ldots + f_k) \tag{6.151}$$

where $\Gamma_{k-1}(\mathbf{f})$ is only a function of f_i for $1 \le i \le k-1$, then $m_k(\mathbf{t}) = m_k(\mathbf{t} + \tau\mathbf{u})$, whatever the value of τ. In order to show this point, let us calculate $m_k(\mathbf{t})$ by using (6.147) and (6.151). Because of the Dirac distribution, one of the integrals of (6.147) disappears and f_k must be replaced by $-(f_1 + f_2 + \ldots + f_{k-1})$, which gives

$$m_k(\mathbf{t}) = \int \overset{k-1}{\cdots} \int \Gamma_{k-1}(f_1, f_2, \ldots, f_{k-1})\ \exp\{2\pi j[f_1(t_1 - t_k)$$

$$+ f_2(t_2 - t_k) + \ldots + f_{k-1}(t_{k-1} - t_k)]\}\ df_1\, df_2 \ldots df_{k-1} \tag{6.152}$$

It is clear that this function is invariant if each t_i is replaced by $t_i + \tau$, $1 \le i \le k$, which is the stationarity property. As a result, $m_k(\mathbf{t})$ is only a function of $k - 1$ independent variables $\theta_i = t_i - t_k$ for $1 \le i \le k-1$, and can be written as

$$m_k(t_1, t_2, \ldots, t_k) = \gamma_{k-1}(\theta_1, \theta_2, \ldots, \theta_{k-1}) =$$

$$\int \overset{k-1}{\cdots} \int \Gamma_{k-1}(f_1, f_2, \ldots, f_{k-1})\ \exp\left[2\pi j \sum_{i=1}^{k-1} f_i\, \theta_i\right] df_1\, df_2 \ldots df_{k-1} \tag{6.153}$$

The function $\Gamma_{k-1}(\mathbf{f})$ is called the *moment spectrum of order k* $- 1$. In order to illustrate these general calculations, let us first consider the case where $k = 2$. If the signal is zeromean-valued, $m_2(t_1, t_2)$ is the correlation function $\gamma(t_1 - t_2) = \gamma(\theta_1)$ and (6.153) becomes

$$\gamma(\theta_1) = \int \Gamma_1(f_1)\ \exp(2\pi j f_1 \theta_1)\ df_1 \tag{6.154}$$

which is exactly (6.40), and $\Gamma_1(f)$ is then the power spectrum or the spectral density used in the previous sections. Now, suppose that $k = 2$. In this case we have

$$m_3(t_1, t_2, t_3) = \gamma_2(\theta_1, \theta_2) =$$

$$\iint \Gamma_2(f_1, f_2)\exp[2\pi j(f_1\theta_1 + f_2\theta_2)]\ df_1\, df_2 \tag{6.155}$$

and $\Gamma_2(f_1, f_2)$ is called the moment bispectrum of $x(t)$. The same procedure can be used to introduce the moment trispectrum.

We can apply exactly the same method if we use the cumulants $c_k(\mathbf{t})$ instead of the moments, which introduces the concept of cumulant bi- or trispectrum, and more generally the cumulant spectrum of order k.

Let us now consider the problem of the transformation of higher-order spectra by linear filtering, when the stationarity assumption is introduced. The general expression is, of course, (6.144) which corresponds to (6.22). Now, suppose that the quantity $M_x(\mathbf{f})$ satisfies (6.151). It is obvious that the same property is valid for $M_y(f)$ and, because of the term $\delta(.)$, we obtain

$$\Gamma_{y,k-1}(\mathbf{f}) = \Gamma_{x,k-1}(\mathbf{f})\, G(f_1)\, G(f_2) \ldots G(f_{k-1})\, G(-f_1 - f_2 - \ldots - f_{k-1}) \quad (6.156)$$

As we are now working with real signals, it is appropriate to assume that the filter is also real and then satisfies $G(f) = G^*(-f)$, which gives

$$\Gamma_{y,k-1}(\mathbf{f}) = \Gamma_{x,k-1}(\mathbf{f})\, G(f_1)\, G(f_2) \ldots G(f_{k-1})\, G^*(f_1 + f_2 + \ldots + f_{k-1}) \quad (6.157)$$

For $k = 2$, this expression is exactly (6.55), which then becomes a particular case of a more general relation.

There is a very important application of this expression in the case of narrowband filters. A real narrowband filter satisfies $G(f) = G^*(-f)$ and is such that $G(f)$ is null outside a frequency band in the neighbourhood of the central frequency f_0. It can thus be characterized by $G(f) = 0$, if

$$f_0 - \Delta\nu < |f| < f_0 + \Delta\nu\,,\ \ 0 < \Delta\nu < f_0 \quad (6.158)$$

Let us first investigate the consequences of this expression on the bispectrum. In this case, (6.157) can be written

$$\Gamma_{y,2}(f_1, f_2) = \Gamma_{x,2}(f_1, f_2)\ G(f_1)\, G(f_2)\, G^*(f_1 + f_2) \quad (6.159)$$

We will show that this expression is null if $\Delta\nu / f_0$ is sufficiently small. In fact, if $(\Delta\nu / f_0) \ll 1$, the frequencies f_1 and f_2 must be approximately equal to $\pm f_0$ in order to obtain a non-null value of the product $G(f_1)G(f_2)$. However, in this case the sum $f_1 + f_2$ takes the values 0 and $\pm 2f_0$ and, as a result, the term $G(f_1 + f_2)$ is null because $f_1 + f_2$ does not satisfy (6.158). An exact calculation of the upperbound of $(\Delta\nu / f_0)$ ensuring that (6.159) is null is discussed in a problem.

The conclusion of this discussion is that the bispectrum of a narrowband signal is null. In fact, a narrowband signal can always be considered as the output of a narrowband filter, and the result of our discussion is only a consequence of the structure of $G(f)$.

It is clear that the same discussion can be repeated for all the spectra of the kind $\Gamma_{x,\,2n}$ corresponding to moments of cumulants of order $2n + 1$. The only point to note is that for any value of n there is a bandwidth $\Delta\nu_n$ such that, if it is inserted in (6.158), the resulting spectrum is null. In other words, for a given value $\Delta\nu$, all the odd moments or cumulants of $x(t)$ are null, up to an order depending on $\Delta\nu$.

On the other hand, the case of even moments will be studied in the context of complex signals.

Let us now consider the discrete-time case. The only difference in our discussion is that the frequency range for all the integrals in the frequency domain is the hypercube defined by $|f_i| \le 1/2$. To illustrate this point, the stationary manifold defined by (6.150) becomes, for $k = 2$, the straight line $f_1 + f_2 = 0$. In the discrete-time case, this manifold is limited to the square $|f_i| \le 1/2$ as represented in Figure 6.10. Conversely, expressions like (6.143) must be replaced by multiple series instead of integrals as in (6.42') or in (6.66).

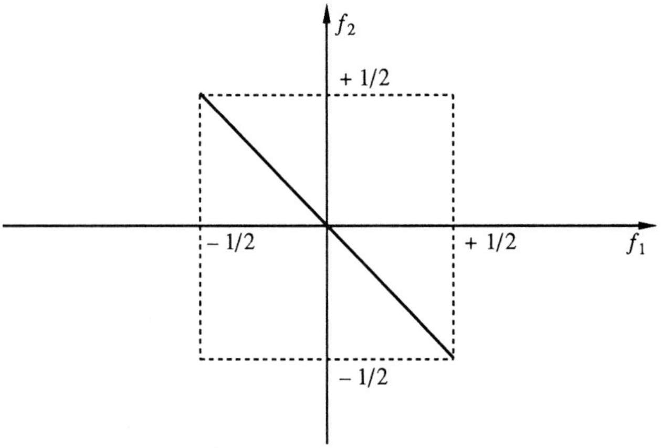

Figure 6.10 Stationary manifold for $k = 2$ and for discrete-time signals.

At the end of this section it will be of interest to discuss the case of *complex signals*. The situation is much more difficult because, as indicated on p. 122 in the discussion of complex RVs, there is not one moment for each order, but a great number, as a consequence of the possibility of complex conjugation. Therefore, there is not one manifold but various, and the discussion of this section must be adapted to each situation.

Nevertheless, there is one situation in which the discussion can exhibit interesting results: it corresponds to the case of a *random analytic signal*, introduced in Section 6.8.1. Suppose that $z(t)$ is the analytic signal of a real signal $x(t)$. We saw in Section 6.8.1 that $z(t)$ is deduced from $x(t)$ by a linear filter with the frequency response $G_A(\nu)$ defined by (6.108). As (6.144) is quite general and can be applied for real as well as for complex signals, we can write

$$M_z(\mathbf{f}) = G_A(f_1)\, G_A(f_2)\, \ldots\, G_A(f_k)\, M_x(\mathbf{f}) \tag{6.160}$$

Now, suppose that $x(t)$ is stationary. This implies that the frequencies f_i must satisfy (6.150). However, this condition cannot be satisfied if all the frequencies f_i are positive, and as $G_A(f) = 0$ if $f < 0$, we deduce that $M_z(\mathbf{f})$ is null. So any higher-order moment of a stationary analytic signal satisfies

$$E[z(t_1)\, z(t_2) \ldots\, z(t_k)] = 0 \tag{6.161}$$

It results from this expression that, in order to obtain some non-null moments, it is necessary to take some complex conjugates in any higher-order moments. Note at this step that for $k = 2$ the condition (6.161) is equivalent to that used to introduce circular complex normal random variables from (4.125) and was directly obtained in (6.116).

As the analytic signal is a tool mainly introduced to discuss properties of narrowband signals, we will now introduce this assumption. First, let us calculate the spectral representation of the analytic signal $z(t)$ of an harmonizable signal $x(t)$. As $z(t)$ is deduced from $x(t)$ by a linear filtering in a filter with frequency response $G_A(\nu)$ given by (6.108), we can apply (6.9), and the spectral representation of $z(t)$ becomes

$$z(t) = \int G_A(\nu)\, d\bar{X}(\nu)\, e^{2\pi j \nu t} \tag{6.162}$$

Taking the complex conjugate, using (6.24) and noting that $G_A(\nu)$ is a real function, we deduce that

$$z^*(t) = \int G_A(-\nu)\, d\bar{X}(\nu)\, e^{2\pi j \nu t} \tag{6.163}$$

In order to calculate the higher-order moments of $z(t)$, we will take the notation introduced on p. 122 and used in (4.153). As a result, we will define all the higher-order moments by

$$m_z(t_1, t_2, \ldots, t_k; \varepsilon_1, \varepsilon_2, \ldots, \varepsilon_k) \triangleq E[z^{\varepsilon_1}(t_1)\, z^{\varepsilon_2}(t_2) \ldots z^{\varepsilon_k}(t_k)] \tag{6.164}$$

Using (6.162) and (6.163) and the properties (6.144) and (6.146), we deduce that the Fourier transform of this moment is

$$M_z(f_1, f_2, \ldots, f_k; \varepsilon_1, \varepsilon_2, \ldots, \varepsilon_k) = G_A(\varepsilon_1 f_1)\, G_A(\varepsilon_2 f_2) \ldots G_A(\varepsilon_k f_k)\, M_x(\mathbf{f}) \tag{6.165}$$

In this expression $M_x(\mathbf{f})$ is the function defined by (6.146) or (6.143) and the terms ε_i must be equal to +1 or −1 for $z(t)$ or $z^*(t)$ respectively.

Let us now introduce the assumption of stationarity for the signal $x(t)$. This implies that, in order to have non-null values of $M_x(\mathbf{f})$, the condition (6.150) must be satisfied. To simplify the discussion, let us first consider the case where $k = 4$. The stationary manifold is defined by

$$f_1 + f_2 + f_3 + f_4 = 0 \tag{6.166}$$

and the structure of $G_A(\nu)$ requires that

$$\varepsilon_i f_i > 0\,, \qquad 1 \le i \le 4 \tag{6.167}$$

Furthermore, as $x(t)$ is a narrowband signal, all the frequencies f_i must satisfy (6.158), which means that they are approximatively equal to $\pm f_0$. The consequence of these constraints is clear: in (6.166) two frequencies f_i must be positive and two negative, and, as a result of (6.167), we must have two ε_is equal to 1 and the two others equal to −1. In conclusion, if $(\Delta\nu / f_0)$ is sufficiently small compared to one, the only non-vanishing fourth-order moment of $z(t)$ is

$$m_z(t_1, t_2, t_3, t_4) = E[z(t_1)\, z(t_2)\, z^*(t_3)\, z^*(t_4)] \tag{6.168}$$

which means that there is the same number of signals with and without complex conjugation. It is important to note that this property is only a consequence of the two assumptions of narrow-bandwidth and stationarity of the signal.

This result can be extended to higher-order even moments with the constraint that the bandwidth $\Delta\nu$ ensuring the property can depend on the order of the moment. Finally, note that if $x(t)$ is given by (5.47), its analytic signal is

$$z(t) = \exp\{j[\omega_0 t + \phi(s)]\} \tag{6.169}$$

It is a signal with zero bandwidth and it is obvious that all the odd moments are null, while the only non-null even moments take the form $E[z(t_1)\ldots z(t_k)z^*(t_{k+1})\ldots z^*(t_{2k})]$.

At the end of this section, it is worth noting the analogy with the results obtained on p. 123 on complex normal circular vectors. However, the result (6.168) is obtained without any normal assumption and, in consequence, there is no reason to obtain an expression such as (4.157).

6.11 APPENDIX ON STRONG FACTORIZATION

6.11.1 Additional Properties of the z-Spectrum

Let $H(z)$ be the transfer function of a rational filter. As a result $H(z)$ is a ratio of two polynomials $N(z)$ and $D(z)$ and the corresponding z-spectrum given by (6.80) is

$$\Gamma(z) = \frac{N(z)\,N(z^{-1})}{D(z)\,D(z^{-1})} \tag{6.170}$$

In order to realize the strong factorization of $\Gamma(z)$ we can operate with the numerator and the denominator separately, or simply with a polynomial instead of with the rational function itself. A polynomial z-spectrum is obtained only if the correlation function in (6.65) has a finite number of non null values. This gives

$$\Gamma(z) = \sum_{k=-n}^{+n} \gamma[k]\, z^{-k} \tag{6.171}$$

and as we are working with real signals we have $\gamma[k] = \gamma[-k]$, as seen in (5.28). This implies that $\Gamma(z)$ can be written in the form

$$\Gamma(z) = \gamma[0] + \sum_{k=1}^{n} \gamma[k]\, Z_k \tag{6.172}$$

where

$$Z_k = z^k + z^{-k} \tag{6.173}$$

Let us note $S(\omega)$ as the power spectrum written with the angular frequency $\omega = 2\pi\nu$. We deduce from (6.71) that

$$S(\omega) = \Gamma(e^{j\omega}) \tag{6.174}$$

and using (6.172) and (6.173), we find that $S(\omega)$ can be written as

$$S(\omega) = \gamma[0] + 2\sum_{k=1}^{n} \gamma[k]\cos(k\,\omega) \tag{6.175}$$

This shows that the power spectrum is a mixture of terms such as $\cos(k\,\omega)$. We will now show that it can also be written as a polynomial of degree n in $\cos\omega$. For this we return to the expression (6.172). Consider the term Z_2 defined by (6.173). It can be expressed as

$$Z_2 = z^2 + z^{-2} = (z + z^{-1})^2 - 2 = Z^2 - 2 \tag{6.176}$$

where $Z = Z_1$. Similarly

$$Z_3 = z^3 + z^{-3} = Z^3 - 3Z \tag{6.177}$$

and using the expansion of Z^k we see that Z_k is a polynomial of degree k in Z. Inserting this polynomial into (6.172), we deduce that $\Gamma(z)$ is a polynomial of degree n in Z defined by

$$Z = z + z^{-1} \tag{6.178}$$

If we take $z = e^{j\omega}$, we obtain $Z = 2\cos\omega$, which shows that $S(\omega)$ is a polynomial of degree n in $\cos\omega$.

Furthermore, as a result of (6.171),

$$\Gamma(z) = \Gamma(z^{-1}) \tag{6.179}$$

which makes the point indicated on p. 221, that if z_i is a root of $\Gamma(z)$, z_i^{-1} is also a root. It is thus easier for the factorization problem to work with the variable Z rather than z.

Before giving the general procedure for strong factorization, we will consider the cases of first and second order polynomials, working from specific examples which will give an idea of the general method.

6.11.2 Strong Factorization of a First Order Spectrum

Consider the spectrum

$$S(\omega) = 5 - 4\cos\omega = 5 - 2(e^{j\omega} + e^{-j\omega}) \tag{6.180}$$

It is clear that the corresponding z-spectrum is

$$\Gamma(z) = 5 - 2Z \tag{6.181}$$

We obtain $\Gamma(z) = 0$ for $Z = 5/2$ and the roots in z are deduced from (6.178), which gives

$$2z^2 - 5z + 2 = 0 \tag{6.182}$$

By an elementary calculation we obtain the two roots $z_1 = 1/2$ and $z_2 = 2$. For strong factorization, we take the root inside the unit circle, which gives the transfer function

$$H(z) = kz^{-1}(z - 1/2) \tag{6.183}$$

The term z^{-1} is added in order to obtain the same degree at the numerator and the denominator. Note that without this factor the filter $H(z)$ would not be causal. It now remains to calculate the coefficient k, which is obtained by noting that $\Gamma(z) = H(z)$ $H(z^{-1})$, and, as $\Gamma(z) = 5 - 2Z$, we deduce that $k^2 = 4$, or $k = \pm 2$. As indicated previously, a sign ambiguity always remains. Finally, the filters giving strong factorization are

$$H(z) = \pm z^{-1}(2z - 1) \tag{6.184}$$

This procedure can be extended to any first order polynomial, with an additional comment. One can ask whether a first order polynomial such as (6.181) is arbitrary. The answer is no, because the spectrum $S(\omega)$ must be non-negative. If we write $\Gamma(z)$ in the form

$$\Gamma(z) = \alpha(1 + aZ) \tag{6.185}$$

the corresponding spectrum is

$$S(\omega) = \alpha(1 + 2a\cos\omega) \tag{6.186}$$

and the condition $S(\omega) > 0$ implies $\alpha > 0$ and $4a^2 < 1$. This condition also implies that the roots r, solutions of (6.178), are real. In fact, the root of $\Gamma(z)$ is $Z = -1/a$, and we must now find the roots r_1 and r_2 such that

$$r + 1/r = -1/a \tag{6.187}$$

This gives the equation

$$ar^2 + r + a = 0 \tag{6.188}$$

and the condition $4a^2 < 1$ ensures that the two inverse roots r_1 and r_2 are real.

6.11.3 Strong Factorization of a Second Order Spectrum

Consider the spectrum

$$S(\omega) = 4(8\cos^2\omega - 12\cos\omega + 5) \tag{6.189}$$

The corresponding z-spectrum is

$$\Gamma(z) = 4(Z^2 - 6Z + 5) \tag{6.190}$$

where Z is still defined by (6.178). The roots of this polynomial are

$$Z = (1/2)(3 \pm j) \tag{6.191}$$

and are complex. These two roots in Z will generate four roots in z, but we know in advance that only one among these four must be calculated, the others being deduced by a direct procedure. This is due to the property that $\Gamma(z)$ is a polynomial of degree 4 in z with real coefficients and such that if z_i is a root, z_i^{-1} is another root. Thus from any complex root r of this polynomial we can deduce the three others which are r^*, $1/r^*$ and $1/r$. Let us arbitrarily choose the root $Z = (1/2)(3-j)$ and calculate one of the roots r solution of

$$r + r^{-1} = (1/2)(3 - j) \tag{6.192}$$

This gives

$$2r^2 - (3 - j)r + 2 = 0 \tag{6.193}$$

We note that the product of the two roots is equal to one, and the root located inside the unit circle is

$$r = (1/2)(1 + j) \tag{6.194}$$

Taking the other root located inside the unit circle which is r^*, we deduce that strong factorization introduces the filter

$$H(z) = kz^{-2}[(z-(1/2)(1+j)][(z-(1/2)(1-j)] = kz^{-2}[z^2 - z + 1/2] \tag{6.195}$$

Writing $\Gamma(z) = H(z)\,H(z^{-1})$, we deduce that $k^2 = 16$, and strong factorization of the spectrum (6.189) is realized with the minimum phase filters

$$H(z) = \pm 2z^{-2}(2z^2 - 2z + 1) \tag{6.196}$$

6.11.4 General Procedure

This is an obvious extension of the results obtained in the two previous examples. Starting from a spectrum $S(\omega)$, which is a polynomial in $\cos\omega$, we deduce $\Gamma(z)$ which is a polynomial of degree n in Z. The case of real roots is treated as in 6.11.2. To any real root Z_k we associate two real roots r_k and r_k^{-1}, and we select that which is located inside the unit circle. The case of complex roots is treated as in 6.11.3. We arbitrarily choose one complex root Z_k and deduce the three other complex roots r_k associated to Z_k. Selecting the two roots located inside the unit circle, we construct a filter such as (6.196), which has two complex conjugate roots, and introduce a minimum phase filter.

PROBLEMS

6.1 Let $x[k]$ be a stationary zeromean DT random signal. For various reasons it is interesting to work with the discrete Fourier transform (DFT) defined by (6.35). The purpose of this problem is to study the statistical properties of the coefficients $X[n]$ appearing in the DFT.

(a) Calculate the expected value of $X[n]$.

(b) Express the covariance between $X[n_1]$ and $X[n_2]$ in terms of the correlation function $\gamma[p]$ of $x[k]$ defined by (5.17").

(c) Express the same covariance as in (b) in terms of the power spectrum $\Gamma(\nu)$ of $x[k]$ and find a result similar to (6.45) but valid for DFT and not for the continuous Fourier Transform.

(d) Extend to the DFT the results given after (6.45) for the continuous FT.

6.2 Let $x(t)$ be a real, stationary, and zeromean CT random signal. We assume that the power spectrum of this signal is equal to s if $-B < \nu < +B$ and to 0 otherwise. Let F_1 be the linear filter whose impulse response is $1/T$ if $-T < t < +T$ and 0 otherwise. We introduce the filter F_2 which has the same structure as F_1 but in which T is replaced by $2T$. Let $y_1(t)$ and $y_2(t)$ be the outputs of the filters F_1 and F_2 respectively when the inputs are the same signal $x(t)$.

(a) Calculate the power spectra of $y_1(t)$ and $y_2(t)$.

(b) Calculate the power spectrum of $y_1(t) + y_2(t)$.

(c) Represent all these spectra in a figure in the particular case where $BT = 1$.

(d) Calculate the value of s ensuring that the power of $y_1(t) + y_2(t)$ is equal to one.

6.3 Let us consider the system represented in Figure 6.11. In this system the inputs are a signal s and a noise n, assumed to be uncorrelated. The box F represents a linear filter. This system can work as well in CT as in DT.

(a) Calculate the power spectrum $\Gamma_z(\nu)$ of z in terms of the power spectra $\Gamma_n(\nu)$ and $\Gamma_s(\nu)$ of n and s and of the frequency response $G(\nu)$ of the filter F.

(b) Assuming that $\Gamma_s(\nu) = \sigma_s^2\,\delta(\nu)$ and that the spectrum $\Gamma_n(\nu)$ is bounded, we use for F the filter represented in Figure 6.6. What becomes the output z when T tends to infinity? Give an interpretation of this result.

(c) Extend the previous result to the case where the spectrum $\Gamma_s(\nu)$ contains only a finite number of spectral lines at arbitrary frequencies f_i.

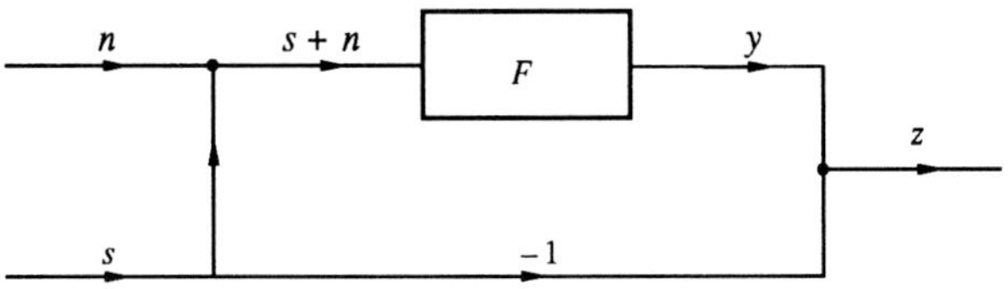

Figure 6.11 Filtering system.

6.4 Let $x_1(t)$ and $x_2(t)$ be two zeromean, real, and uncorrelated signals. Their spectral densities are $\Gamma_1(\nu)$ and $\Gamma_2(\nu)$ respectively. Calculate the power spectrum of the complex signal $z(t) = x_1(t) + jx_2(t)$ in terms of $\Gamma_1(\nu)$ and $\Gamma_2(\nu)$.

6.5 Let $\Gamma_1(\nu)$, $\Gamma_2(\nu)$, and $\Gamma_{12}(\nu)$ be the Fourier transforms of the auto and crosscorrelation functions of two zeromean real signals $x_1(t)$ and $x_2(t)$. Let $y_1(t)$ be the signal obtained by filtering $x_1(t)$ in a linear filter of frequency response $G_1(\nu)$

and $y_2(t)$ be the signal obtained by filtering $x_2(t)$ in a linear filter of frequency response $G_2(\nu)$. Calculate the power spectrum of the signal $s(t) = y_1(t) + y_2(t)$.

6.6 Consider the interference experiment introduced in Section 5.2.5 (**a**). This experiment is described by the signal $y(t) = x(t) + x(t - d)$, where $x(t)$ is a zeromean and stationary random signal with a correlation function $\gamma(\tau)$ and a power spectrum $\Gamma(\nu)$.

(**a**) Calculate the correlation function of $y(t)$ in terms of $\gamma(\tau)$.

(**b**) Deduce the power spectrum of $y(t)$ in terms of $\Gamma(\nu)$.

(**c**) Complete the calculations in the cases where $\gamma(\tau)$ is given by (5.38) or by (5.40).

6.7 Let $u(t)\exp(-at)$ be the impulse response of a causal linear filter F and let us call $x(t)$ and $y(t)$ the input and the output respectively.

(**a**) Assuming that the correlation function of $x(t)$ is $c\delta(\tau)$, which characterizes a weak white noise, calculate the correlation function of $y(t)$.

(**b**) Calculate the power spectrum of $y(t)$.

(**c**) Suppose that F_1 and F_2 are two filters similar to F, defined by the parameters a_1 and a_2 respectively. Assuming that these filters have the same input $x(t)$, which is still a white noise, calculate the crosscorrelation function $\gamma_{12}(\tau)$ between the outputs $y_1(t)$ and $y_2(t)$ and its Fourier transform $\Gamma_{12}(\nu)$.

6.8 Let $h_1(t)$ be the impulse response of a real linear filter and $G(\nu)$ its frequency response. We introduce two other filters defined by the impulse responses $h_2(t) = h_1(-t)$, $h_3(t) = h_1(t) + h_2(t)$. Suppose that these three filters are driven by the same input which is a white noise of power spectrum equal to 1. Let $x_1(t)$, $x_2(t)$ and $x_3(t)$ be the corresponding outputs.

(**a**) Calculate the power spectra $\Gamma_i(\nu)$, $1 \leq i \leq 3$, of the signals $x_i(t)$.

(**b**) Calculate the functions $\Gamma_{ij}(\nu)$, Fourier transforms of the crosscorrelation functions between $x_i(t)$ and $x_j(t)$.

(**c**) Let $\mathbb{\Gamma}(\nu)$ be the spectral matrix of elements $\Gamma_{ij}(\nu)$. Show that it can be written as $\mathbf{G}(\nu)\mathbf{G}^H(\nu)$ and determine the vector $\mathbf{G}(\nu)$. Calculate the determinant of this matrix and give an interpretation of the result obtained.

(**d**) Complete the calculations when $h_1(t) = u(t)\exp(-at)$, where $u(t)$ is the unit step function.

(**e**) Under the same assumption as the previous question, calculate all the auto and crosscorrelation functions of the signals $x_i(t)$.

6.9 Let $x(t)$ be a real stationary signal with a power equal to 1 and a spectrum $\Gamma(\nu)$ constant for $-B < \nu < B$ and zero otherwise. This is called a band-limited white signal.

(**a**) Show that all the derivatives of this signal are second order.

(**b**) Let $y(t) = x'(t)$ be the derivative of $x(t)$. Calculate the auto and crosscorrelation functions $\gamma_{xx}(\tau)$, $\gamma_{yy}(\tau)$ and $\gamma_{xy}(\tau)$.

(**c**) Calculate the spectral matrix $\mathbb{\Gamma}(\nu)$ which is the matrix whose elements are the Fourier transforms of the auto and crosscorrelation functions $\gamma_{xx}(\tau)$, $\gamma_{yy}(\tau)$ and $\gamma_{xy}(\tau)$. Calculate the determinant of this matrix and explain the origin of the result obtained.

6.10 Consider the signal $x(t) = \cos(2\pi Ft + \Theta)$ where F and Θ are two independent RVs. The RV F is characterized by its PDF $p(f)$ or by its characteristic function $\phi(u)$. The RV Θ is uniformly distributed over the interval $[0, 2\pi)$.

(a) Calculate the correlation function $\gamma(\tau)$ of $x(t)$ and express it in terms of $\phi(u)$.

(b) Deduce the expression of the power spectrum of $x(t)$ in terms of the PDF $p(f)$.

(c) Deduce from this expression that if the RV F has no finite moment of order $2n$, then $x(t)$ has no derivative of order n in the quadratic mean sense.

(d) Explain the apparent contradiction with the fact that any trajectory of $x(t)$ has a derivative of order n, whatever n. Give the expression of this derivative.

6.11 Consider the complex signal $x(t) = \exp[j(2\pi\nu_1 t + \phi_1)] + \exp[j(2\pi\nu_2 t + \phi_2)]$, where ϕ_1 and ϕ_2 are two random phases.

(a) Show that, whatever ϕ_1 and ϕ_2, each term of the sum giving $x(t)$ has a stationary covariance function.

(b) Show that if ϕ_1 and ϕ_2 are independent and uniformly distributed over $[0, 2\pi)$, $x(t)$ has a stationary covariance function.

(c) Show that the same property holds if $\phi_2 = \phi_1 + \Delta\phi$, where the phase difference $\Delta\phi$ is uniformly distributed over $[0, 2\pi)$.

(d) Show that, on the other hand, if $\phi_2 = \phi_1$ (phase coupling), $x(t)$ cannot have a stationary covariance function whatever the distribution of ϕ.

6.12 Suppose that the function $\bar{X}(\nu)$ appearing in the spectral representation (6.5) of a random signal $x(t)$ is zeromean valued and real. Suppose also that the increments $d\bar{X}(\nu)$ are uncorrelated, which is specified by (6.43).

(a) Show that the signal $x(t)$ cannot be a real function. Find the relation between $x(t)$ and $x(-t)$.

(b) We assume that $E\{|\bar{X}(\nu_2) - \bar{X}(\nu_1)|^2\}$ is an integral over the frequency range ν_1, ν_2 of a non-negative function $F(\nu)$ the integral of which is equal to 1 over the whole frequency range. Calculate in terms of $F(\nu)$ the second order moments $\gamma(t_1, t_2) = E\{x(t_1)x^*(t_2)\}$ and $c(t_1, t_2) = E\{x(t_1)x(t_2)\}$.

(c) By introducing the correlation function $\gamma(\tau)$ of $x(t)$, find the relation between $\gamma(t_1, t_2)$ and $c(t_1, t_2)$. Specify in which sense $x(t)$ is a stationary signal.

6.13 Let $x(t)$ be a real random signal. Its spectral representation can be written either as in (6.5) or as in (6.61) where only positive frequencies are used.

(a) Calculate the increments $d\bar{C}(\nu)$ and $d\bar{S}(\nu)$ in terms of the increments $d\bar{X}(\nu)$.

(b) Assuming that $x(t)$ is stationary, calculate the expected values $E\{d\bar{C}(\nu_1)\, d\bar{C}(\nu_2)\}$, $E\{d\bar{S}(\nu_1)\, d\bar{S}(\nu_2)\}$ and $E\{d\bar{C}(\nu_1)\, d\bar{S}(\nu_2)\}$.

6.14 The purpose of this problem is to show that the sum of some stationary signals can be non-stationary. Let $x(t)$ be a real, zeromean and second order stationary random signal. It is assumed that its power spectrum is constant for ν_0

$-B < |\nu| < \nu_0 + B$ with $\nu_0 > 2B$ and zero otherwise. Furthermore its power is equal to 1.

(a) Calculate its correlation function and represent its power spectrum in a figure.

(b) The spectral representation of this signal is given by (6.5) and the general properties of the increments $d\bar{X}(\nu)$ are specified by (6.15), (6.37) and (6.39). Deduce the frequency range of integration in (6.5) from these expressions and from the properties of the power spectrum of $x(t)$.

(c) Let $y(t)$ be the signal defined by $y(t) = \cos(2\pi\nu_0 t)x(t)$. Show that $y(t)$ is not second order stationary.

(d) Write the spectral representation of $y(t)$ and show that it is an integral like (6.5) calculated over 3 non-overlapping frequency bands B_1, B_2 and B_3. The integrals (6.5) associated with these intervals define 3 signals $y_1(t)$, $y_2(t)$ and $y_3(t)$. Show that these signals are stationary and calculate their correlation functions and their power spectra.

(e) Calculate the crosscovariance functions between these signals and show that they are not stationary.

(f) Using all these results, comment on the first sentence of this problem.

6.15 Let $x[k]$ be a complex signal and $H(z)$ the transfer function of a complex filter. Prove the relation (6.69) and deduce that (6.55) is valid for both real and complex signals or filters.

6.16 Use the same reasoning as in the previous problem to prove (6.77).

6.17 Let $x_1(t)$ and $x_2(t)$ be two independent zeromean stationary random signals with the same power spectrum. This spectrum is constant in the frequency band $-B$, $+B$ and zero otherwise.

(a) Let $y_1(t)$ be the signal $x_1(t)\cos(\omega_0 t)$, $\omega_0 > 2\pi B$. Calculate the covariance function of $y_1(t)$. What conclusion can be deduced from the result concerning the second order stationarity of $y_1(t)$?

(b) Make the same calculation and the same discussion concerning the signal $y_2(t) = x_2(t)\sin(\omega_0 t)$.

(c) Show that the signal $y(t) = y_1(t) + y_2(t)$ is second order stationary and explain the origin of this property. Calculate its correlation function and its power spectrum.

6.18 Suppose that $x(t)$ is a real narrowband white signal. This means that for $\nu > 0$ its spectrum is constant if $|\nu - \nu_0| \le B$ and zero otherwise, with $\nu_0 >> B$.

(a) Assuming that $x(t)$ is zeromean valued and that its variance is σ^2, calculate its power spectrum and its correlation function.

(b) Introducing the complex envelope defined by (6.123) and (6.124), calculate the correlation functions $\gamma_m(\tau)$, $\gamma_n(\tau)$, and the crosscorrelation function $\gamma_{mn}(\tau)$.

(c) Assuming that $x(t)$ is normal, calculate the correlation function of the square of its instantaneous amplitude $a(t)$.

6.19 **Hilbert transforms of the product of some signals.** The purpose of this problem is to study the relation $H[x_1(t)x_2(t)] = x_1(t)H[x_2(t)]$ where $x_1(t)$ and $x_2(t)$ are two real signals and $H[.]$ is the Hilbert transform introduced in

Section 6.8.1. This property is sometimes referred to as the Bedrossian theorem.

(a) Suppose that $x_1(t)$ is a deterministic (or non-random) low frequency signal, which means that its Fourier transform is zero if $|\nu| > B$, where B is some arbitrary frequency. Suppose also that $x_2(t)$ is a deterministic high frequency signal, which means that its Fourier transform is zero if $|\nu| < B$, where B is the same frequency as for $x_1(t)$. Let $z_2(t)$ be the analytic signal of $x_2(t)$ defined by (6.111). Show that $x_1(t)z_2(t)$ is an analytic signal, which means that its Fourier transform is zero for $\nu < 0$.

(b) Calculate the real and imaginary parts of $x_1(t)z_2(t)$ in terms of Hilbert transforms.

(c) Using the point that $x_1(t)z_2(t)$ is an analytic signal, deduce the relation $H[x_1(t)x_2(t)] = x_1(t)H[x_2(t)]$.

(d) Using the spectral representation of *random* signals, show that this property is still valid for these signals.

(e) In order to prove that the converse is not true, it suffices to prove that the relation $H[x_1(t)x_2(t)] = x_1(t)H[x_2(t)]$ can be valid for signals which do not satisfy the conditions given in (**a**). Make this verification by using the signals $x_1(t) = (t^2 + a^2)^{-1}$, $a > 0$, and $z_2(t) = (t - ja)[(t + jb)(t + jc)]^{-1}$, $b > 0$ and $c > 0$.

6.20 By using the results of the previous problem, calculate the analytic signal of the correlation function given by (5.41').

6.21 Let $G(\nu)$ be the frequency response of a real narrowband filter satisfying (6.158). Find the upper bound of $\Delta\nu$ ensuring that the last term of (6.159) is zero whatever the frequencies f_1 and f_2.

6.22 Calculate the z-spectrum associated with the correlation function $\gamma[k] = a^{|k|}$, $|a| < 1$. Find the filter $H(z)$ realizing the strong factorization of this spectrum.

6.23 Let $H(z) = z^2[(z - a)(z - b)]^{-1}$ be the transfer function of a dynamical second order filter and let us call $h[k]$ its impulse response.

(a) Find the conditions on the real numbers a and b ensuring that the filter is stable.

(b) Suppose that the input of this filter is a unit DT white noise. Calculate the z-spectrum and the power spectrum of the output signal.

(c) Using the results of the previous problem, calculate the correlation function $\gamma[k]$ of this output.

(d) Using the relation (6.64), calculate again the correlation function $\gamma[k]$.

(e) Calculate $\gamma[0]$ and deduce from the previous question that $\gamma[0] > 1$ whatever a and b.

(f) Suppose now that the input is only white. Calculate its variance in order to find an output with a unit power.

6.24 Consider the function $\gamma[k] = \alpha a^{|k|} + \beta b^{|k|}$ where α and β are non-negative and a and b are real numbers satisfying $|a| < 1$, $|b| < 1$.

(a) Show that this function is a correlation function.

(b) Calculate the z-spectrum $\Gamma(z)$ associated with this correlation function. Write this spectrum as a ratio of polynomials $N(z)/D(z)$, where $D(z)$ is a polynomial of degree 2 with respect to the variable Z defined by (6.178) and beginning with the term abZ^2.

(c) Find the filter $H(z)$ realizing the strong factorization of this spectrum.

6.25 As in the previous problem we consider the function $\gamma[k] = \alpha a^{|k|} + \beta b^{|k|}$ where α and β are no longer non-negative while a and b still satisfy $|a| < 1$, $|b| < 1$.

(a) Find α and β in such a way that $\Gamma(z)$ can be written as $1/D(z)$, where $D(z)$ is a polynomial of degree 2 with respect to the variable Z defined by (6.178) and beginning with the term abZ^2. Calculate the function $\gamma[k]$ and show that it is still a correlation function.

(b) Determine the function $H(z)$ realizing the strong factorization of $\Gamma(z)$.

(c) Compare the results of this problem with those obtained in Problem 6.23.

6.26 We consider the same transfer function as in Problem 6.23, but we assume that the roots a and b are complex. We introduce the notation $a = m\exp(j\phi)$, $m \geq 0$, $0 < \phi < \pi$, and we assume that the denominator of $H(z)$ is a real polynomial.

(a) Calculate the root b and find the condition on m ensuring that the filter is stable.

(b) Calculate the impulse response $h[k]$ of the filter.

(c) Assuming that the input of the filter is a unit white noise, calculate the power spectrum $\Gamma(\nu)$ of the output in terms of m and ϕ.

(d) Calculate the correlation function $\gamma[k]$ of the output signal and show that it satisfies the difference equation $\gamma[k] = 2m\cos\phi\,\gamma[k-1] - m^2\gamma[k-2]$.

6.27 Let $x(t)$ be a zeromean stationary signal and $\Gamma(\nu)$ its power spectrum. Its second order moment $m_2(t_1, t_2)$ is defined by (5.42) and the Fourier transform of this moment is noted as $M_2(\nu_1, \nu_2)$.

(a) Using the ideas presented in Section 6.10, calculate $M_2(\nu_1, \nu_2)$ in terms of $\Gamma(\nu)$. Represent in a figure the stationary manifold for a CT signal similar to that presented in the DT case in Figure 6.10.

(b) Consider the signal $y(t) = \cos(2\pi\nu_0 t)x(t)$. Calculate the second order moment $m_2(t_1, t_2)$ of $y(t)$ and its Fourier transform $M_2(\nu_1, \nu_2)$. Show that $M_2(\nu_1, \nu_2)$ is not only distributed on the stationary manifold and give the reason for this.

(c) Assuming that $\Gamma(\nu)$ is constant for $-B < \nu < B$ and zero otherwise, represent in a figure the function $M_2(\nu_1, \nu_2)$ of the signal $y(t)$.

6.28 Consider an antenna with n sensors and suppose that the outputs of all these sensors are the same signal $x(t)$ assumed to be zeromean and with a power spectrum $\Gamma(\nu)$. This signal is processed by n filters characterized by their frequency responses $G_i(\nu)$ and the n outputs of these filters are the components of a vector $\mathbf{y}(t)$. Let $\mathbf{G}(\nu)$ be the vector whose components are $G_i(\nu)$, $1 < i < n$.

(a) Calculate the spectral matrix of $\mathbf{y}(t)$ and give its expression in terms of $\mathbf{G}(\nu)$ and $\Gamma(\nu)$.

(b) Show that the rank of this matrix is equal to 1.

(c) In various problems it is important to use the eigenvectors and eigenvalues of a spectral matrix. Find these elements for the matrix calculated in (**a**).

(d) Complete the calculation when the frequency responses $G_i(\nu)$ correspond to pure delays, which means that $G_i(\nu) = \exp[2\pi j(i-1)\nu\tau]$.

6.29 Apply the method presented in Section 6.11 to find the strong factorization of the spectrum

$$S(\omega) = 4m^2\cos^2\omega - 4m(1+m^2)\cos\phi\cos\omega + m^4 + 2m^2\cos(2\phi) + 1.$$

6.30 To become accustomed to calculating strong factorization problems of polynomial spectra by using the procedure described in Section 6.11, choose arbitrarily a polynomial in z, say $H(z)$. From this polynomial calculate the z-spectrum $\Gamma(z) = H(z)H(z^{-1})$ and the power spectrum $S(\omega) = \Gamma(e^{j\omega})$. Make the reverse operation and find $H(z)$ again if it is a minimum phase filter or the minimum phase filter giving the same spectrum as $H(z)$.

Chapter 7

Statistical Models for Random Signals

7.1 INTRODUCTION

In this chapter we shall use our knowledge of random variables, vectors, and signals in the discussion of some statistical models. Modeling is one of the most important activities in the physical sciences. Natural phenomena generally appear very complex and their representation needs models. It is clear that there is almost always a difference between the reality and the model used for its description, but if this difference is too marked, another model must be used. Thus many physical signals can be described with a good approximation as Gaussian (or normal) signals, but in some cases deviation from the normal distribution is so evident that the Gaussian model must be rejected and another chosen. It is sometimes important to differentiate between models for representation and those for knowledge. A large number of models used in physics are introduced to understand the laws of physical phenomena. On the other hand, many models introduced in the engineering sciences do not claim to understand the reality, but only to describe the behavior of what is observed. For example, saying that the emission of photoelectrons by a metal can be described by a Poisson process does not explain the microscopic phenomenon of the photoelectric effect.

The topics studied below are the result of a selection, since it would not make sense to present all the statistical models of random signals. The selection is of course somewhat arbitrary, and those chosen were felt to be the most important in signal processing problems. Finally, note that the continuous and discrete time cases will be presented simultaneously and, where the transition from one to the other is straightforward, the discussion will not be repeated.

7.2 WHITE NOISES

The expression "white noise" is so widely used that we shall not replace it by that of "white signal." However, saying that a signal is white means that it is a white noise.

When describing white noises we meet a situation where the DT and CT cases are completely different, so we shall begin with the simplest, the DT case.

7.2.1 Discrete-Time White Noises

A zeromean stationary DT signal $x[k]$ is *weakly white* if its correlation function can be written as

$$\gamma[k] = \sigma^2 \delta[k] \tag{7.1}$$

where σ^2 is the mean power of the signal and $\delta[\ ]$ the Kronecker-delta symbol. Using (6.42') we deduce that $\Gamma(\nu) = \sigma^2$ and it results from (6.41) and (6.42') that $P = \sigma^2$. It is clear that (7.1) shows that $x[k]$ and $x[l]$ are uncorrelated for $k \neq l$, and a white noise signal is in reality a sequence of uncorrelated RVs with the same variance. Such sequences have often been used previously, particularly in Chapter 3 and in the corresponding problems. We thus already have some knowledge of weakly white noises. Note also that if in (7.1) $\sigma^2 = 1$, we say that we are dealing with a *unit white noise*.

The term *white* is due to an analogy with optics. We tend to talk about white light as opposed to colored light. The white light is a mixture of the fundamental colors whose spectrum is broad while in general colored light has a narrow-band spectrum. But in optics the spectrum for natural light is not constant, and natural white light does not correspond exactly to the concept of white noise defined by (7.1).

We have seen that there is a difference between non-correlation and independence, which leads to the concept of *strict white noise*.

A signal $x[k]$ is a strict white noise if it is a sequence of IID RVs. Such sequences have been considered previously (see Section 4.8.9). It is clear that a DT strict white noise is the simplest stochastic signal to be found. Its family of finite-dimensional distribution is completely specified by the one-dimensional distribution common to all the RVs $x[k]$ and it is also strict sense stationary.

It may be asked here whether other kinds of whiteness exist between weak and strict whiteness. This is indeed the case, and among the many other forms of whiteness we shall select only one which will be of particular interest in prediction problems to be studied later. The principle of prediction problems was broadly introduced on p. 7. As one of the basic properties of white noise is that there is no correlation between the past and the future, it seems natural that a white noise be unpredictable, because prediction uses such correlation to obtain information about the future from observation of the past.

A weakly white noise is said to be linearly unpredictable. This means that $x[k]$ is uncorrelated with any linear combination of the values of the signal in the past. This property results directly from (7.1). Suppose that we drop the term linear, and consider a signal such that $x[k]$ is uncorrelated with any combination of the values of the signal in the past. This situation was encountered on p. 97 in the discussion on the properties of the regression. In particular, we find from (4.39) that if $\mathbf{r}(\mathbf{y}) = \mathbf{0}$, the zeromean random vector $\mathbf{X}$ is uncorrelated with $\mathbf{g}(\mathbf{Y})$, whatever the vector-valued function $\mathbf{g}(.)$. This leads to the definition of a *strong white noise*.

A signal $x[k]$ is said to be strongly white if

$$E\{x[k] \mid P_k\} = 0 \tag{7.2}$$

where P_k means all the RVs $x[k - l]$, $l > 0$. From the three definitions we verify that a strict white noise is also strong white, because the independence implies the nullity of the conditional expectation value (7.2). Similarly, a strong second order white noise is a weak white noise because (7.2) and (4.39) indicate that $x[k]$ is uncorrelated with $x[k - l]$, which gives (7.1). But the opposite is not true, as seen in the examples below.

Example 7.1 Hierarchy between some white noises. Consider first the signal

$$x[k] = \exp[j(3k + 1)\,\Phi] \tag{7.3}$$

where Φ is an RV uniformly distributed in $[0, 2\pi)$ and k is an integer. It results from this assumption that $E\{x[k]\} = 0$ and that

$$E\{x[k]\, x^*[l]\} = \delta[k - l] \tag{7.4}$$

which shows that $x[k]$ is weakly white. Furthermore, we have

$$E\{x[k]\, x[l]\} = 0 \tag{7.5}$$

because it is impossible to find two integers positive or negative such that $3(k + l) + 2 = 0$ which is the condition required to obtain a non-null expectation value. On the other hand, it is clear that $E\{x[k]|P_k\} = x[k]$, because if $x[l]$ is known, Φ is also known and $x[k]$ is no longer random. Thus $x[k]$ is not strongly white nor, of course, strictly white. In fact, we have an example of a random signal which is weakly white, that is to say unpredictable by linear methods, but also completely predictable by non linear procedures. This signal will later be called a deterministic signal.

Suppose now that $x[k] = A(s)\, w[k]$ where $w[k]$ is a normal strict white noise, which means a sequence of IID normal RVs, and that $A(s)$ is an RV independent of the $w[k]$s. Because of the amplitude term $A(s)$, $x[k]$ is not strictly white. In fact, the RVs $x[k]$ and $x[l]$ are not independent. To verify this point, let us calculate the characteristic function of a random vector $\mathbf{X}$ the components of which are $x[k_1]$, $x[k_2]$, ..., $x[k_N]$, the k_is being arbitrary distinct integers. Assuming that the RVs $w[k]$ are $N(0, 1)$, the conditional characteristic function of $\mathbf{X}$ is

$$\Phi(\mathbf{u}|a) = \exp\left(-\frac{1}{2}\, a^2 \sum_{i=1}^{N} u_i^2\right) \tag{7.6}$$

which gives the marginal characteristic function by

$$\Phi(\mathbf{u}) = E\left[\exp\left(-\frac{1}{2}\, A^2 \sum_{i=1}^{N} u_i^2\right)\right] \tag{7.7}$$

It is clear that this function cannot be factorized as a product such as (4.70). But as the $w[k]$ are independent we can write

$$E\{x[k]|P_k,\, a\} = 0 \tag{7.8}$$

which gives (7.2) after taking the expectation value on a. The same reasoning gives (7.1), and $x[k]$ is not strictly white, but only strongly and weakly. We can relate this discussion to the properties of spherically invariant RVs. In fact, by taking the Fourier transform of (7.7), we find a function similar to (4.48) which, as seen in Section 4.8.2, gives a null regression.

Let us now consider the *experimental* aspect of discrete-time white noises. A DT signal is in reality a sequence of numbers and is difficult to represent by a curve such as the random signal in Figure 1.2. Furthermore, when observing a sequence of numbers, the question arises whether it is a particular realization of a white noise, and which of the three possible kinds of whiteness is represented. Although these questions are rather complex, we shall outline some of the simpler features.

There is one case where it is easy to generate a realization of a strict white noise. For this it suffices to take a coin and to associate 0 and 1 to heads and tails respectively. By repeating the experiment we construct a sequence of 0 and 1 which is the realization of a strict white noise. In fact, the independence between the successive trials can be considered as an *a priori* assumption. Let us now consider the reverse problem: starting from a sequence of 0 and 1, is it possible to conclude its whiteness? If there is good reason to state that this sequence is stationary, it is possible to measure its correlation function by using time averaging instead of ensemble averaging. The method is described by the transposition of (6.100) valid for CT signals to DT. In this way it is possible to verify (7.1) with an approximation depending on the length of the sequence of numbers observed. On the other hand, there is no simple way to verify experimentally that a weakly white sequence is in reality a sequence of *independent* RVs. There are statistical tests that allow us to admit the assumption of independence with some confidence, but their description is outside the scope of this book. The independence is more a hypothesis justified by physical arguments, as in the heads and tails experiment.

In many signal processing problems it is necessary to verify the behavior of some systems, which is often done by simulation. It is then also necessary to simulate a white noise. There are many computer programs that give a sequence of random numbers. By using the ideas given at the end of Chapter 2, these numbers can have an arbitrary probability distribution. The whiteness of the sequences of points presented in the Figures 7.1 to 7.3 has been carefully verified and the correlation function, measured with a much longer sequence, is given by (7.1). So the figures present three trajectories of different white noises. These trajectories are calculated for 200 successive time instants.

Figure 7.1 corresponds to a white noise taking only the values ± 1 with the same probability. The trajectory represented in Figure 7.2 corresponds to a sequence of uncorrelated RVs with a uniform distribution, as defined by (2.27). The trajectory of Figure 7.3 corresponds to a discrete-time Gaussian white noise, which in many applications is the most important case. It is a sequence of RVs characterized by the PDF given by (2.35) and represented in Figure 2.4. These figures are quite different but correspond to the same correlation function (7.1). In order to distinguish these noises, other moments are necessary.

Figure 7.1 White noise with discrete amplitude.

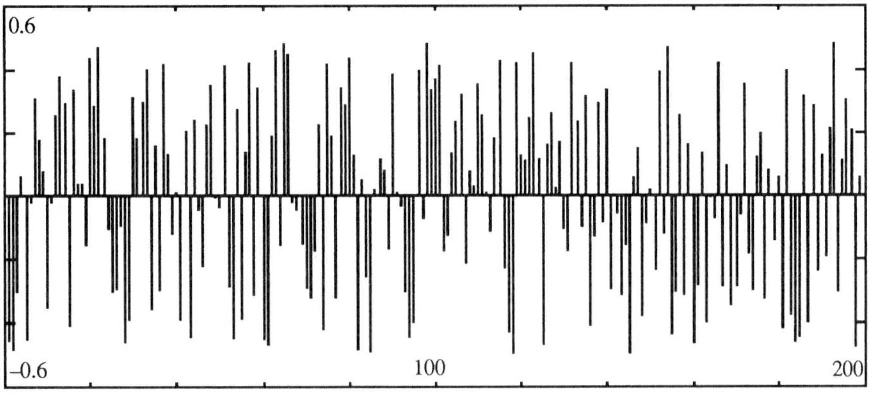

Figure 7.2 White noise with uniform distribution.

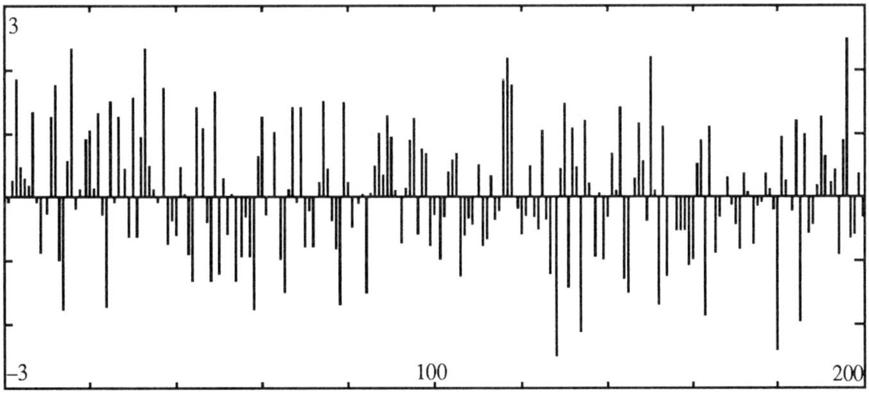

Figure 7.3 Normal white noise.

Let us then investigate *higher-order properties* of discrete-time white noises. Nothing can be said about weak white signals, because the definition (7.1) does not imply a higher-order property. On the other hand, as indicated previously, the assumption of strong white noise means that $x[k]$ is uncorrelated with any non-linear function of the past values of the signal. As $x[k]$ has a zeromean value, this gives

$$E\{x[k] \prod_i x^{p_i}[k-i]\} = 0 \tag{7.9}$$

where i and p_i are positive integers. This expression is valid provided that the p_is are sufficiently small to ensure that the expectation value of the product is finite.

However, the most important situation, and the simplest for our discussion, appears when $x[k]$ is a *strict white noise*. As the RVs $x[k]$ are now IID, we can apply the results of Chapter 4, especially those introducing cumulants. We deduce, for example, from (4.76) that the third and fourth order cumulants of a strict white noise satisfy

$$c[k_1, k_2, k_3] = c_3 \, \delta_3[\mathbf{k}] \tag{7.10}$$

$$c[k_1, k_2, k_3, k_4] = c_4 \, \delta_4[\mathbf{k}] \tag{7.11}$$

where $c[.]$ has the same meaning as in (5.44) applied to the DT case, and c_3 and c_4 are the third and fourth order cumulants of the distribution common to all the RVs $x[k]$. Furthermore, the terms $\delta[.]$ have the same meaning as in (4.81).

Let us now demonstrate some spectral consequences of (7.10). The tridimensional Fourier transform of this cumulant is given by (6.143) where the integrals are replaced by series. This gives

$$M_3(\mathbf{f}) = \sum_{k_1} \sum_{k_2} \sum_{k_3} c_3 \exp[-2\pi j \mathbf{f}^T \mathbf{k}] \, \delta_3[\mathbf{k}] \tag{7.12}$$

Because of the term $\delta_3[\mathbf{k}]$, we must have $k_1 = k_2 = k_3 = k$, which gives

$$M_3(\mathbf{f}) = c_3 \sum_k \exp[-2\pi j (f_1 + f_2 + f_3) \, k] \tag{7.13}$$

As the series of $\exp(-2\pi j \nu k)$ is equal to the Dirac distribution $\delta(\nu)$, we obtain

$$M_3(\mathbf{f}) = c_3 \, \delta(f_1 + f_2 + f_3) \tag{7.14}$$

and, comparing this with (6.151), we deduce that the cumulant bispectrum is

$$\Gamma_2(f_1, f_2) = c_3 \tag{7.15}$$

Exactly the same procedure can be applied to all the other cumulants, giving

$$\Gamma_{k-1}(f_1, f_2, \ldots, f_{k-1}) = c_k \tag{7.16}$$

The conclusion we draw from this calculation is very clear. A weak white noise is characterized by (7.1), which is equivalent to saying that the power spectrum is constant. This corresponds to (7.16) applied to $k = 2$, and the cumulant c_2 is effectively the variance σ^2, as seen in (4.77). Similarly, a strict white noise is characterized by the fact that all the higher-order cumulant spectra are constant provided that they are finite. In fact we have seen, for example in Problem 2.12, that some distributions do not have moments or cumulants. If the cumulants c_k are finite only for $k \leq N$, it is possible to say that the signal is strictly white until the order N. On p. 153 we already discussed the concept of random variables which are normal up to an order q, and the concept of whiteness up to a specific order is the same.

To conclude, let us note that the above discussion is irrelevant as soon as the normal assumption is introduced. We have seen that uncorrelated normal RVs are independent, which means that a normal signal that is weakly white is automatically strictly white.

7.2.2 Discrete-Time White Noises and Linear Systems

The assumption of whiteness greatly simplifies the calculation of statistical properties at the output of linear systems. Suppose, for example, that $u[k]$ is a unit zeromean weak white noise, and let $x[k]$ be the signal deduced from $u[k]$ by a linear filtering, as in (5.55). The correlation function of $x[k]$ can be written by using (5.60) with $n = 2$, and becomes

$$\gamma[k] = \sum_{l} h[l]\, h[l-k] \tag{7.17}$$

As a result we see that the correlation function of the random signal $x[k]$ is equal to the correlation function of the impulse response $h[k]$ as seen with deterministic signals. Similarly, as the power spectrum of $u[k]$ is equal to one, we find by using (6.55) that the power spectrum of $x[k]$ is simply

$$\Gamma(\nu) = |G(\nu)|^2 \tag{7.18}$$

This expression was in fact already introduced in Section 6.5.3 when dealing with spectral factorization.

In order to visualize the effect of linear filtering on a white noise, the following figures present some trajectories. All these figures are realized with the *same filter*, which is causal exponential. As a consequence of (7.18), the trajectories correspond to signals with the same correlation function. In order to compare them with the trajectories of a white noise, we use the same kinds of white noises as in the previous figures. As the trajectories are different, we verify once more that the second order properties, and especially the correlation function, are quite insufficient to characterize a random signal.

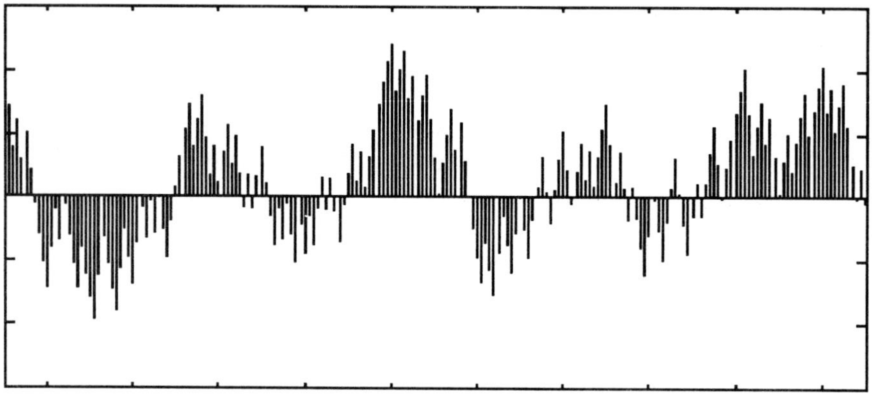

Figure 7.4 Filtering of a discrete amplitude white noise.

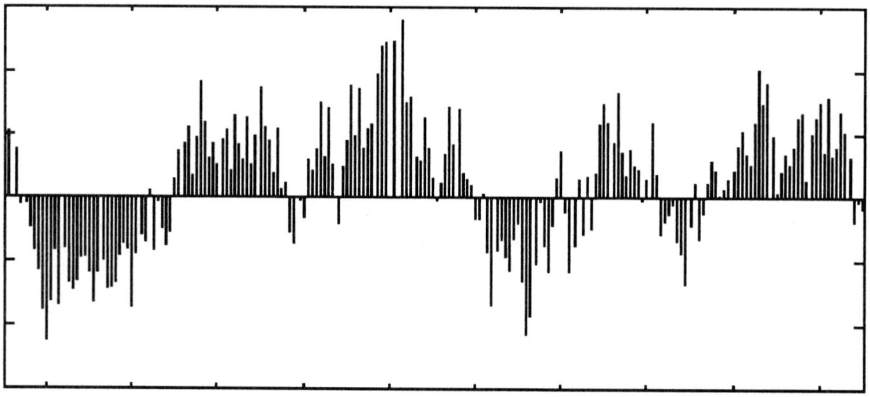

Figure 7.5 Filtering of a uniform white noise.

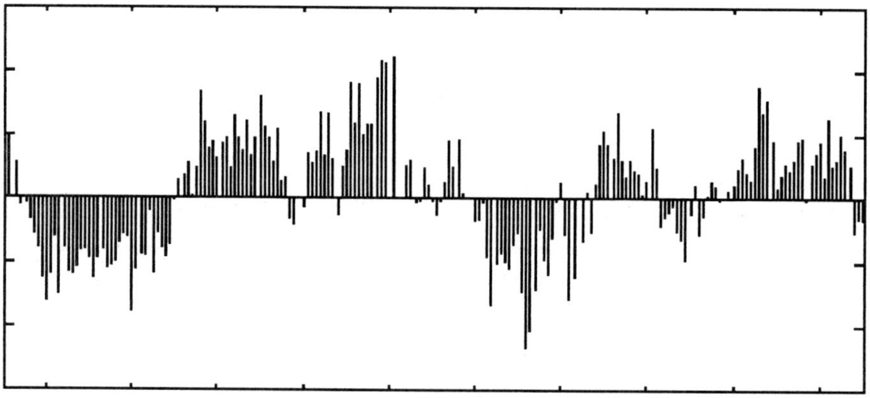

Figure 7.6 Filtering of a Gaussian white noise.

Let us present some comments on these figures. The first point is the filtering effect. The impulse response of the filter used in the three cases is the same and its value is $h[k] = a^k$ for $k \geq 0$ and $h[k] = 0$ for $k < 0$. The value of a corresponding to the figures is $a = 0.9$, which means that a great number of values of $u[k - l]$ contribute significantly to give $y[k]$ from the convolution (5.55). In other words, the filter has a long memory effect, and the result is that the signals $x[k]$ appearing in the three figures no longer have the chaotic behavior of the white noises presented in Figures 7.1, 7.2, and 7.3. On the other hand, the three trajectories are much more similar than those in the figures presenting trajectories of white noises. This shows that, because of the strong filtering effect, the influence of the statistical distributions of the input white noises does still exist, but is less important than for the trajectories of these white noises.

It should be noted that the covariance function at the output of a linear but nonstationary system can also be obtained without difficulty. Many examples were given in Chapter 3, notably in Problems 3.11, 3.12, and 3.13.

We shall now turn to higher-order statistics. Here the assumption of weak white noise is not sufficient, since it describes only second order properties. So we assume that $u[k]$ is a unit strict white noise. As it corresponds to a sequence of IID random variables, it seems more interesting to work with cumulants than with moments.

Our purpose is to transpose the relation (7.18), valid for the power spectrum, to higher-order spectra. Let us begin with the bispectrum in order to explain the principle of the calculation. The bispectrum of a strict white noise is given by (7.15) and the transformation of the bispectrum through linear filtering is given by (6.157) in the case of real signals. As a result of these expressions, the output bispectrum can be written as

$$\Gamma_2(f_1, f_2) = c_3\, G(f_1)\, G(f_2)\, G^*(f_1 + f_2) \tag{7.19}$$

Exactly the same procedure can be applied to higher-order spectra, giving

$$\Gamma_{k-1}(\mathbf{f}) = c_k\, G(f_1)\, G(f_2)\, \dots\, G(f_{k-1})\, G^*(f_1 + f_2 + \dots + f_{k-1}) \tag{7.20}$$

For $k = 2$, and assuming that $c_2 = 1$, we again find (7.18).

7.2.3 Transformation of Whiteness by Linear Filtering

It is clear in (7.18) that if the input of a filter is a discrete-time unit white noise, the output is no longer white. However, it remains white if the filter satisfies $|G(\nu)| = 1$, which characterizes a pure phase filter. Its frequency response takes the form $\exp[j\phi(\nu)]$, where $\phi(\nu)$ is a phase depending on the frequency. This is, for example, the case of a delay filter. As we have introduced the concept of whiteness up to a certain order, it is interesting to investigate its mode of transformation in linear filtering.

Consider a strict white noise, i.e., a sequence of IID RVs, and call c_k the cumulants of order k of these RVs. At the output of a phase filter the higher-order spectra are given by (7.20), which takes the form

$$\Gamma_{k-1}(\mathbf{f}) = c_k \exp\{j[\phi(f_1) + \ldots + \phi(f_{k-1}) - \phi(f_1 + \ldots + f_{k-1})]\} \quad (7.21)$$

In order to obtain a white noise at the output, $\Gamma_{k-1}(\mathbf{f})$ must be constant, as in (7.16). This can be realized either by a structure of the phase $\phi(f)$ or by a property of the cumulants c_k. The first situation demands that

$$\sum_{i=1}^{k-1} \phi(f_i) = \phi\left[\sum_{i=1}^{k-1} f_i\right] \mathrm{mod}(2\pi) \quad (7.22)$$

whatever the frequencies f_i. This can be realized only with a linear phase in the form $2\pi f\tau$, where τ is a time-constant. This corresponds to a filter realizing a delay of the signal, and it is obvious that if a sequence of RVs is IID, the delayed sequence has the same property, conserving the strict whiteness. If this phase condition is not realized, (7.21) can be constant only if $c_k = 0$ for $k > 2$. As seen after (4.87), this characterizes a *normal* distribution. Thus, apart from the obvious case of a time delay filter, a strict white noise can conserve this property after linear filtering only if it is normal. This exhibits a very important property of the Gaussian white noise, and the mode of reasoning and the result obtained are similar to the property discussed in Section 4.8.9.

7.2.4 Continuous-Time White Noise

By a direct extension of (7.1), we can say that a continuous-time signal $x(t)$ is weakly white if

$$\gamma(\tau) = k\ \delta(\tau) \quad (7.23)$$

which means that the power spectrum is constant and equal to k. But this definition immediately generates a problem: as $\gamma(0)$ is infinite, the variance of $x(t)$ is also infinite, which means that it is not a second-order signal. In consequence, a number of the results previously discussed which took for granted that the signal was second order cannot be applied, or need special care. Let us, for example, examine the properties studied in Section 5.8. From (7.23) we see that a continuous-time white noise is neither second order nor mean square continuous. Furthermore, as the derivative of the Dirac distribution is only defined as a distribution, we cannot apply the results of Section 5.8.2, and a continuous-time white noise has thus no mean square derivative.

On the other hand, the mean square integral is in general well defined. To verify this point it suffices to note that if we insert (7.23) in the integral (5.91), the result is in general well defined under very broad conditions on a, b and $h(t)$. Here are two important examples for what follows. Consider first an integral such as (5.90) where $h(t) = 1$. Using (7.23), we see that the integral (5.91) is equal to $k(b - a)$ and is finite as soon as a and b remain finite.

Let us now consider the output of a linear filter driven by a continuous-time white noise. The input-output relationship is given by (5.56), and using (5.90) and (5.91) we deduce that the condition ensuring that the output is second order takes the form

$$\int |h(t)|^2 \, dt < +\infty \tag{7.24}$$

which means that the filter has an impulse response of finite energy. The correlation function of the output signal is given by (5.62), and using (7.23), we obtain

$$\gamma_Y(\tau) = k \int h(\theta)\, h^*(\theta - \tau)\, d\theta \tag{7.25}$$

This last integral is the correlation function of the deterministic function $h(t)$. By Fourier transformation we obtain the spectrum, which is

$$\Gamma_Y(\nu) = k \, |G(\nu)|^2 \tag{7.26}$$

This expression has exactly the same structure as (7.18). In conclusion, note that even if a continuous-time white noise is not a second order signal, it generates a second order signal by passing through a linear filter of finite energy. This is a property which is very often used. We shall end with a comment on notation. The constant k appearing in (7.23) is often written as $(N/2)$, and N is called the unilateral spectral density of the white noise. If this noise is passed in a real filter of bandwidth W, the corresponding power is NW.

The parallel presentation of the properties of white noise in the case of continuous time would now require the discussion of the concept of strict white noise. Unfortunately this is much more difficult than in the discrete-time case, where we have a sequence of IID RVs. As a complete discussion would demand more abstract arguments, we will limit our study to normal or Poisson distributions presented later.

7.3 RANDOM WALK AND BROWNIAN MOTION

7.3.1 Random Walk

Let $u[k]$ be a discrete-time weak white noise with mean value m and variance σ^2. From this signal we construct the random walk $x[k]$ defined by

$$x[k] = \sum_{i=1}^{k} u[i] \tag{7.27}$$

Some trajectories of the random walk are presented in the following figures. These figures correspond to the three kinds of white noises in Figures 7.1 to 7.3. On each figure three distinct trajectories of 200 points corresponding to distinct samples of the $u[i]$s are represented. In order to make the figures more readable, the points are connected to obtain a curve. The main features which appear are the non-stationary aspect and the increase of the variance with time, specified below by calculation. It is also clear that the trajectories depend on the statistical nature of the white noise $u[i]$.

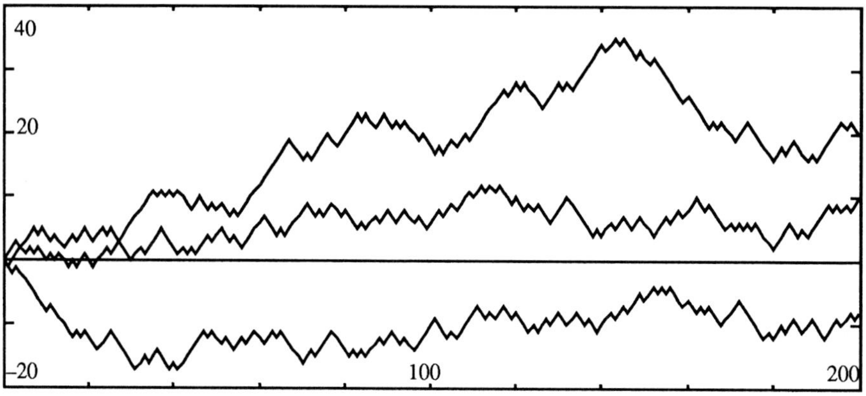

Figure 7.7 Three trajectories of random walk with discrete white noise.

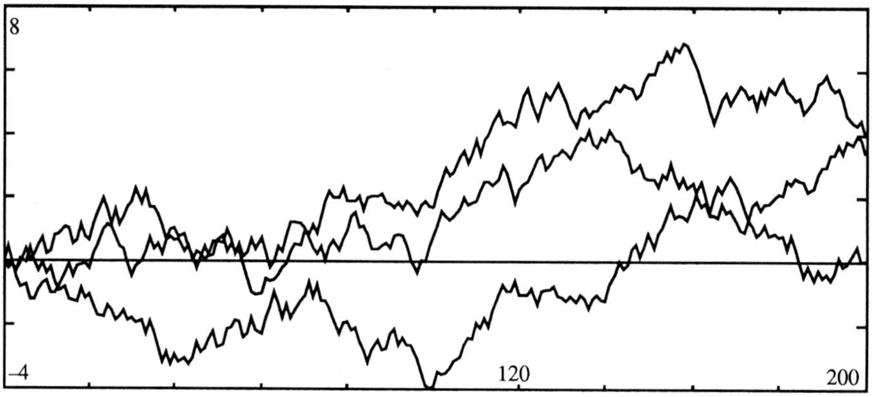

Figure 7.8 Three trajectories of random walk with uniform white noise.

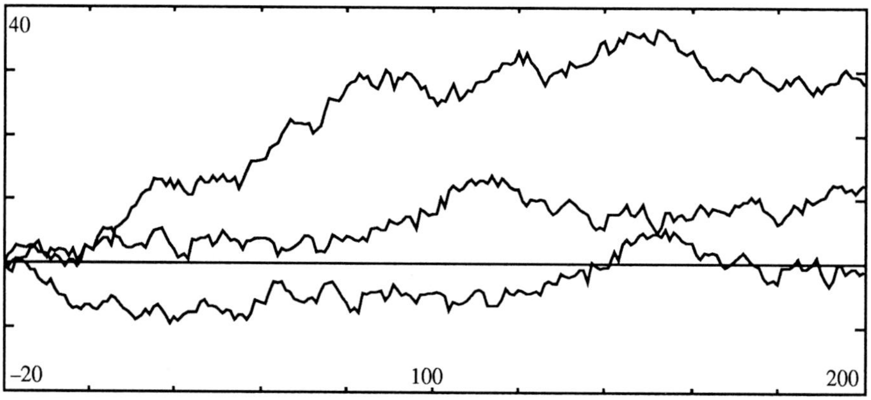

Figure 7.9 Three trajectories of normal random walk.

The random walk signal is sometimes called a discrete-time Brownian motion, for reasons explained below. It is clear that $x[k]$ is not stationary, since its mean value and covariance function are

$$m[k] = km \tag{7.28}$$

$$\gamma[k, l] = \sigma^2 k \wedge l \tag{7.29}$$

where $\wedge$ means the minimum. This notation was used in (3.121), and some properties of $x[k]$ are studied in Problem 3.11. The signal $x[k]$ satisfies the recursion

$$x[k] = x[k-1] + u[k] \tag{7.30}$$

which especially shows that even though $x[k]$ is not stationary, its increments $x[k] - x[k-1]$, equal to $u[k]$, are stationary. It is sometimes said that $x[k]$ is a signal with *stationary increments*. The covariance matrix, the elements of which are $\gamma[k, l]$ given by (7.29), is called a Brownian matrix and has interesting properties which are studied in Problem 3.11.

An important property of the random walk signal is that its variance is equal to $\sigma^2 k$ and increases linearly with time. The mean value and the variance thus have the same behavior in time.

Suppose now that $m = 0$. This condition does not change the fact that the variance of $x[k]$ is equal to $\sigma^2 k$. Therefore, if $x[k]$ represents the coordinate of a particle moving on an axis and observed at discrete time instants k, we deduce that the average position is zero while the variance is always increasing. This is a specific property of random walk.

Suppose now that the signal $u[k]$ is a strict white noise or a sequence of IID random variables. In this case $x[k]$ becomes a signal with *stationary and independent increments*. This expression can be taken in a more general sense; in fact, it results directly from (7.27) and from the assumption on the $u[k]$s that the increments $x[k] - x[l]$ and $x[p] - x[q]$ are independent as soon as we have $q < p < l < k$.

The family of finite-dimensional distribution functions of $x[k]$ is completely specified by the probability distribution of the $u[k]$s. As we are working with sums of IID random variables, it is convenient to use the characteristic function. For example, the characteristic function of $x[k]$ is

$$\Phi(u\ ;\ k) = [\phi(u)]^k \tag{7.31}$$

where $\phi(u)$ corresponds to $u[k]$. An interesting situation arises in the case of the probability distributions which are stable by addition, as discussed in Problem 4.26. In those cases the distribution of $x[k]$ and $u[k]$ are of the same nature and only a parameter is changed. Let us consider the three classical examples of this situation. In the normal case $\phi(u)$ is given by (2.84), and it follows that $x[k]$ is an RV with the distribution $N(0, k\sigma^2)$. If $u[k]$ is a binomial RV defined by p and n, it results from (2.89) and (7.31) that $x[k]$ is still binomial defined by p and nk. Finally, if $u[k]$ is a Poisson RV with mean value m, (2.91) shows that $x[k]$ is still Poisson, but with mean value mk. At this stage we note that the expression random walk is sometimes reduced to the case where $u[k]$ is a binomial RV.

It is interesting to discuss here some properties of the regression used above. For this purpose let us call X the RV $x[k]$ and $\mathbf{Y}$ the vector with p components Y_i defined by $Y_i = x[k - l_i]$ where $0 < l_1 < l_2 < \ldots\ l_p$. From (7.27) and from the independence of the RVs $u[i]$ we find that $X - Y_1$ is independent of the random vector $\mathbf{Y}$ or $E[(X - y_1)|\mathbf{y}] = 0$. This immediately gives

$$r(\mathbf{y}) = E(X|\mathbf{y}) = y_1 \tag{7.32}$$

So the regression is linear, as in the case of normal RVs [see (4.109)] or of other distributions [see (4.190)]. Apart from the property of linearity, it appears that the regression is equal to y_1 which is the value of the signal at the time instant $k - l_1$ and does not depend on the other values of $x[k]$ corresponding to time instants anterior to $k - l_1$. This property is characteristic of a discrete-time *martingale* and will appear later in relation to other signals.

In reality we can calculate not only the regression but the complete probability distribution. In fact, we deduce from the structure (7.27) that X can be written as

$$X = Y_1 + (X - Y_1) = Y_1 + V \tag{7.33}$$

where V is independent of the vector $\mathbf{Y}$. The conditional PDF of X for a given value $\mathbf{y}$ of $\mathbf{Y}$ can thus be written

$$p_X(x|\mathbf{y}) = p_V(v - y_1) \tag{7.34}$$

where $p_V(v)$ is the PDF of the RV V. As V is zeromean-valued, this gives (7.32). Finally, as V is a sum of l_1 RVs $u[i]$ which are IID, we deduce that

$$p_V(v) = [p_U(v)]^{*l_1} \tag{7.35}$$

which means a multiple convolution of order l_1. The same reasoning can be applied to discrete RVs by using the DF instead of the density.

Note finally that the results in this section can be extended to complex RVs, which introduces a two-dimensional random walk, or a random walk in a plane. In particular, if the increments $u[k]$ are normal circular, $x[k]$ is also normal circular. In consequence the modulus of $x[k]$ is an RV with a Rayleigh distribution defined by (4.217) where σ^2 must be replaced by $k\sigma^2$.

7.3.2 Brownian Motion

The Brownian motion signal, or the Wiener-Lévy process, is the continuous-time counterpart of the discrete-time random walk. Instead of starting from a constructive definition such as (7.27) that would require stochastic integrals, we will start from moments, as in (7.28) and (7.29). The Brownian motion signal is the signal $x(t)$ defined for $t \geq 0$ by the following properties:

(*a*) It is a Gaussian signal.

(*b*) Its mean value is $m(t) = \alpha t$.

(*c*) Its covariance function is

$$\gamma(t, \theta) = c.(t \wedge \theta), \; c > 0 \tag{7.36}$$

As a result we have $x(0) = 0$ with probability one. Let us comment on this definition. First, note that the precise definition of a Gaussian signal will be introduced later. At present the point to be noted is that the family of finite-dimensional distributions introduced in Section 5.1 is a family of normal distributions, as defined in Section 4.6. A vector normal distribution $N(\mathbf{m}, \mathbb{\Gamma})$ is completely specified by the moments of first and second order, as seen in (4.82). Similarly, a normal, or Gaussian, signal is completely specified as soon as its mean value $m(t)$ and its covariance function $\gamma(t, \theta)$ are known, which is the case from (*b*) and (*c*) for the Brownian motion.

Second, if $\alpha = 0$, we have $m(t) = 0$, which introduces a zeromean Brownian signal.

Third, we can ask whether (7.36) is really a covariance function. For this we must show that $\gamma(t, \theta)$ is a non-negative definite function as seen in Section 5.2.3 (**g**). This is realized if and only if for any set of arbitrary time instants t_i, $1 \le i \le N$, and any integer N, the quantities $\gamma(t_i, t_j)$ are the matrix elements M_{ij} of a non-negative definite matrix. This property can be obtained by an elementary extension of (7.27). Suppose that the t_is are ordered in such a way that $t_1 < t_2 < \ldots < t_N$ and consider a unit white noise $u[k]$. Let us introduce the N RVs X_k defined as follows: $X_1 = (t_1)^{1/2}\, u[1]$ and

$$X_k = X_1 + \sum_{i=2}^{k} (t_i - t_{i-1})^{1/2} u[i], \; k > 1 \tag{7.37}$$

It is clear that the covariance matrix of the vector $\mathbf{X}$ with components X_k is positive definite and its matrix elements are simply $t_i \wedge t_j$, which shows that (7.36) is a positive definite function.

Note that it follows from (7.36) that the variance of the Brownian motion signal is

$$\sigma^2(t) = ct \tag{7.38}$$

In physical problems the constant c is called the *diffusion constant* for reasons explained below.

Let us now investigate some properties of the Brownian motion signal. In the following we assume $\alpha = 0$, which implies a zeromean value.

(a) Continuity. Using the mean square continuity criterion given in Section 5.8.1, we see that $x(t)$ is mean square continuous at any time instant t, $t > 0$. In fact, it is clear that the function $t \wedge \theta$ is continuous at any point $t = \theta = t_0$.

Concerning continuity, there is another important result, the proof of which can be found in more specialized works (see for example [Blanc-Lapierre], [Cramér], [Wong] or

[Loève]). This result states that every trajectory of $x(t)$ with probability one is a continuous function. But even with this nice property we will see that the trajectories of a Brownian motion signal are extremely singular.

(b) Increments. We now introduce the increments of $x(t)$ defined by

$$\Delta x(t;\ \tau) = x(t + \tau) - x(t),\ \ \tau > 0 \tag{7.39}$$

Using (7.36) we immediately obtain

$$E[\Delta x^2(t;\ \tau)] = c\tau \tag{7.40}$$

Consider two nonoverlapping intervals $[t, t + \tau]$ and $[\theta, \theta + \tau']$, or such that $t < t + \tau < \theta < \theta + \tau'$, and the two increments $\Delta x(t;\ \tau)$ and $\Delta x(\theta;\ \tau')$. From (7.39) we deduce

$$E[\Delta x(\theta;\ \tau')\ \Delta x(t;\ \tau)] = E\{[x(\theta + \tau') - x(\theta)]\ [x(t + \tau) - x(t)]\} \tag{7.41}$$

Using (7.36) we obtain

$$E[\Delta x(\theta;\ \tau')\ \Delta x(t;\ \tau)] = c\,[\,t + \tau - (t + \tau) + t - t\,] = 0 \tag{7.42}$$

This means that the increments are uncorrelated. As they are normal RVs, these increments are *independent*. From this we see that the Brownian motion signal has *independent increments*. This was also a property of the strict random walk signal.

(c) Differentiation. The application of the criterion of Section 5.8.2 to the covariance function (7.36) shows that $x(t)$ has no derivative in the quadratic mean sense. This is in connection with (7.40), since a function $x(t)$ with a second order derivative satisfies $\Delta x(t;\ \Delta t) \cong x(t)\ \Delta t$ and the variance of Δx is of the order of Δt^2, while for a Brownian motion signal it is of the order of Δt.

As for continuity questions it is possible to show more abstract properties concerning differentiability. More precisely, any trajectory of $x(t)$ is nowhere differentiable and has an unbounded variation in any interval with probability one. The trajectories of a Brownian motion signal are thus very singular functions with an extremely chaotic behavior.

(d) Integral representation and relation with white noise. Consider any partition of the interval $[0, t)$ with increasing time instants t_i. The signal $x(t)$ can be written as sum of increments, or

$$x(t) = \sum_{i=1}^{n} [x(t_i) - x(t_{i-1})] \tag{7.43}$$

assuming that $t_0 = 0$ and $t_n = t$. For any value of n, the increments $x(t_i) - x(t_{i-1})$ are independent. By passing to the limit we can write

$$x(t) = \int_0^t dx(\theta) \tag{7.44}$$

where the increments $dx(\theta)$ are normal, independent and with the variance $E[dx^2(\theta)] = c\, d\theta$. This can be summarized as

$$E[dx(\theta)\, dx(\theta')] = c\,\delta(\theta - \theta')\, d\theta\; d\theta' \tag{7.45}$$

This type of expression has already been used for the discussion of point processes in Section 5.7.4, and (7.45) is also fairly similar to (6.15), (6.37), and (6.39). Furthermore, (7.44) is very similar to (7.27) valid for discrete-time signals. From all this we conclude that the Gaussian white noise must be represented by the increments $dx(\theta)$ of the Brownian motion signal. In particular, the output of a continuous-time linear filter driven by a Gaussian white noise must be written as

$$y(t) = \int h(t - \theta)\, dx(\theta) \tag{7.46}$$

where $dx(\theta)$ is the increment appearing in (7.44). The covariance function is given by (5.79) where M is replaced by x, and using (7.45) we again find (7.25) where k is replaced by c.

The fact that $dx(\theta)$ cannot be written as $u(\theta)\, d\theta$ is due to the lack of derivative and illustrates the point that a continuous-time white noise has an infinite variance, discussed in Section 7.2.4. Manipulation of the continuous-time white noise thus calls for special caution.

(e) Martingale property. Let τ and t be two positive time instants such that $\tau < t$. Using (7.39) we can write

$$x(t) = x(\tau) + \Delta x(\tau;\, t - \tau) \tag{7.47}$$

and, from (7.44), $\Delta x(\tau; t - \tau)$ is a random variable independent of the random signal $x(\theta)$ for $\theta \leq \tau$. Let us call $A(x;\, \tau)$ any event depending only on $x(\theta)$ for $\theta \leq \tau$. As Δx and A are independent, and as $x(t)$ is a zeromean signal, we deduce from (7.47) that

$$E[x(t) | A(x;\, \tau)] = x(\tau) \tag{7.48}$$

which is the martingale property already obtained for discrete time in (7.32). We will consider this property again in the context of estimation problems.

(f) Time-varying Brownian motion. The coefficient c appearing in (7.36) and (7.45) can be a function of time. In that case the latter equation can be written

$$E[dx(\theta)\; dx(\theta')] = c(\theta)\delta(\theta - \theta')\, d\theta\; d\theta' \tag{7.49}$$

Most of the previous results remain valid; for example, the covariance function (7.36) becomes

$$\gamma(t,\, \theta) = \int_0^{t \wedge \theta} c(\theta)\, d\theta \tag{7.50}$$

Furthermore the increments remain independent but, instead of (7.40), their variance is given by

$$E[\Delta x^2(t;\tau)] = \int_t^{t+\tau} c(\theta)\, d\theta \qquad (7.50')$$

Of course we again find (7.40) when $c(\theta)$ is constant. Similarly, the output of a linear filter given by (7.46) is no longer stationary, and by inserting (7.49) in (5.79) we find the covariance function of $y(t)$ given by

$$\gamma_y(t_1, t_2) = \int h(t_1 - \theta)\, h(t_2 - \theta)\, c(\theta)\, d\theta \qquad (7.51)$$

It is clear that this function does not depend only on $t_1 - t_2$ as in the case where $c(\theta)$ is constant. This latter case again gives (7.25). This is the reason why time-varying Brownian motion is sometimes called non-stationary Brownian motion. If $c(t) = c$, the Brownian motion is not stationary, but its increments are stationary and $y(t)$ given by (7.46) is stationary. This is no longer true when c is time-varying.

(g) Quadratic variation. Consider again two positive time instants τ and t satisfying $\tau < t$ and suppose that the interval $[\tau, t]$ is divided into N equal intervals of length $(t - \tau)/N$ introducing the time instants $t_0 = \tau$ and $t_i = t_0 + i(t - \tau)/N$. This of course gives $t_N = t$. Let us call Δx_n the increments $x(t_n) - x(t_{n-1})$ and consider the RV defined by

$$Q_N \triangleq \sum_{i=1}^{N} [\Delta x_i]^2 \qquad (7.52)$$

The RVs Δx_i are $N(0, v_i)$ and the variances v_i are given by (7.50') which becomes

$$v_i = \int_{t_{i-1}}^{t_i} c(\theta)\, d\theta \qquad (7.53)$$

Using the results of Section 4.8.6, we will calculate the mean value and the variance of Q_N. The mean value is simply given by

$$E[Q_N] = \sum_{i=1}^{N} v_i = \sum_{i=1}^{N} \int_{t_{i-1}}^{t_i} c(\theta)\, d\theta \qquad (7.54)$$

However, as the Δx_i are independent, the variance of Q_N is the sum of the variances of the increments Δx_i. Using the fact that the variance of the square of an RV which is $N(0, \sigma^2)$ is $2\sigma^4$, we obtain

$$V[Q_N] = 2 \sum_{i=1}^{N} (v_i)^2 \qquad (7.55)$$

Let us now consider the situation which appears when $N \to \infty$. From (7.54) we obtain

$$\lim_{N\to\infty} E[Q_N] = \int_\tau^t c(\theta)\, d\theta \qquad (7.56)$$

Furthermore, the variances v_i can be written $c(\theta_i)\,\Delta t_i$ with the result that if we insert this in (7.55), we obtain

$$\lim_{N\to\infty} V[Q_N] = 0 \tag{7.57}$$

From this discussion we see that the RV Q_N tends in the quadratic mean sense to a non-random number given by (7.56) which is called the quadratic variation of $x(t)$ in the interval $[\tau, t]$. Without too much difficulty it is also possible to show that this convergence appears with probability one.

This result can be summarized by saying that the quantity

$$Q = \int_\tau^t [dx(\theta)]^2 \tag{7.58}$$

is non-random and equal to the integral appearing in (7.56). By a similar reasoning it can be seen that

$$\int_\tau^t dx(\theta)\, dx(\theta + \tau) = 0 \tag{7.59}$$

Let us now interpret these results, which again exhibit the very chaotic behavior of the continuous-time Brownian motion. First, it is clear that if a function $x(t)$ has a derivative $f(t)$, we have $dx(\theta) = f(\theta)\, d\theta$. In this case the quantity Q is equal to zero because of the term $d\theta^2$. This is not the case for the Brownian motion because $E\{[dx(\theta)]^2\} = c(\theta)\, d\theta$ and is of the order of $d\theta$ and not of $d\theta^2$.

Second, suppose that we are in the stationary case, which is characterized by $c(\theta) = c$. We then have $Q = c(t - \tau)$. By sampling $x(t)$ at the time instants t_i defined previously, we can estimate without error the diffusion constant c by taking an arbitrary high sampling rate for given values of τ and t. This means that c, which is the only parameter appearing in the definition of $x(t)$, can be determined from an observation of finite duration, whatever the value of this duration. This is due to the extremely chaotic behavior of any trajectory of the Brownian motion signal. It is clear that this property cannot be verified in any real experiment, which means that Brownian motion is only a mathematical model. In fact the property of the quadratic variation is valuable in a continuous-time context but disappears in discrete time. Any experiment in the real world introduces a limitation in the sampling rate or a lower bound ΔT of the intervals between two time instants t_i and t_j, which makes it impossible to take a limit for $N \to \infty$ giving the property of quadratic variation.

(h) Extension to the complex case. Complex Brownian motion signals can easily be introduced, as can multidimensional Brownian motion. An interesting situation arises in the circular Gaussian case. This appears if $E[dx(\theta)dx(\theta')] = 0$, whatever the values of θ and θ', and if (7.45) is written with $dx^*(\theta')$ instead of $dx(\theta')$. At this point it is interesting to note that the increments that appear in the spectral representation of a stationary complex signal satisfy these relations, and the function $\overline{X}(\nu)$ introduced in

(6.5) can be a complex circular Brownian motion. This will be studied later within the framework of Gaussian signals.

7.3.3 Physical Approach to Brownian Motion

The term Brownian motion is related to the studies of the English botanist Robert Brown (1773-1858) on the motion of particles suspended in a liquid. The main features of this motion are a very erratic behavior and a diffusion property. The erratic behavior can be described as a random signal, and the diffusion property shows that the variance of this signal is proportional to time, introducing a diffusion constant c. This is the property seen in (7.40). The Brownian motion studied in the previous section is a mathematical model of this physical phenomenon, but in reality there are many other models which can approximately describe the diffusion process.

We shall discuss this class of models first in order to choose the most appropriate for the description of Brownian motion. Consider a stationary, zeromean and second order random signal $u(t)$ with the power spectrum $\Gamma(\nu)$. By integration we deduce the signal

$$x_T(t) = \int_{t-T}^{t} u(\theta)\, d\theta \tag{7.60}$$

which is very similar to $M_T(t)$ defined by (6.93). Using exactly the same method as that following this equation, and assuming that $\Gamma(\nu)$ is continuous and bounded for $\nu = 0$, we can write the variance of $x_T(t)$ for large values of T as

$$v_T = \Gamma(0)\, T \tag{7.61}$$

As $u(t)$ is stationary, the variance of $x_T(t)$ is the same as that of the signal

$$x(T) = \int_0^T u(\theta)\, d\theta \tag{7.62}$$

This means that whatever the signal $u(t)$, provided that its power spectrum is continuous, we have for $T >> t_c$, where t_c is the correlation time of $u(t)$, a variance proportional to T. The great difference is that this property is only asymptotic, while it was valid whatever T for the Brownian motion signal. But if we remember that the correlation time of a white noise is null, we can reconcile the two situations.

In conclusion, the Brownian motion signal introduces a diffusion linear in time at any time scale, while this property appears only after a transient period for the function defined by (7.62).

Among the variety of possible models which introduce a Brownian diffusion, the simplest is a consequence of the *Langevin equation.* This equation describes the one-dimensional motion of a particle of mass m in a medium introducing a friction force proportional to the velocity and submitted to a random force $F(t)$. Newton's law gives

$$m\dot{v} + fv = F(t) \tag{7.63}$$

which is the Langevin equation.

Assuming that the force $F(t)$ is due to collisions, it can be represented as a white noise with a flat spectrum and a spectral density γ. As (7.63) introduces a relation of linear filtering with a frequency response

$$G(\omega) = (f + jm\,\omega)^{-1} \tag{7.64}$$

we deduce that the power spectrum of the velocity is

$$\Gamma(\nu) = \frac{\gamma}{f^2 + 4\pi^2\,m^2\,\nu^2} \tag{7.65}$$

which gives $\Gamma(0) = \gamma/f^2$. As the position $x(t)$ of the particle is the integral of the velocity, it can be written as in (7.62). It follows from (7.61) that, for large values of t, which means in this problem $t >> m/f$, the variance of the position of the particle increases linearly with time and the diffusion constant is $c = \gamma/f^2$. By using the same results of statistical physics, and especially the principle of equipartition of energy, it can be shown that the power spectrum γ is equal to $2kTf$, where k and T are the Boltzmann constant and T the temperature respectively. Using this value, the constant of diffusion becomes

$$c = 2kT/f \tag{7.66}$$

7.4 GAUSSIAN SIGNALS

Gaussian signals, or Gauss-Laplace, or normal, signals are certainly the most important random signals both theoretically and from the point of view of applications. The reason for this was given when studying random variables, and is also valid for random signals: the normal distribution is the simplest possible, the second order properties are sufficient to completely specify the distribution, and the central limit theorem and all its extensions justify the practical importance of this distribution in physical phenomena.

As random signals are here defined by their family of finite-dimensional distributions, the study of normal signals can be reduced to that of normal random vectors presented in Chapter 4. We shall therefore concentrate our attention in this section on points concerning the time dependence of signals not discussed when dealing with random variables. As a separate book could be devoted to the study of all signals that can be constructed from normal signals, we shall restrict ourselves to only a few examples in the text and given as problems.

7.4.1 Definition of Gaussian Signals

A signal $x(t)$ is said to be Gaussian, or normal, if its family of finite-dimensional distributions introduced in Section 5.1 is normal. More precisely, whatever n and whatever the t_is, the vector $\mathbf{X}$ introduced in (5.1) is $N(\mathbf{m}, \mathbb{\Gamma})$. Thus a normal signal is defined by its mean-value $m(t)$ and its covariance function $\gamma(t, \theta)$, which can be expressed by the notation $N[m(t), \gamma(t, \theta)]$. The same definition can be used for complex random and vector signals. Particular mention should be made of the case of normal complex circular random signals written with the notation $N_c[m(t), \gamma(t, \theta)]$. In this expression $m(t)$ and $\gamma(t, \theta)$ are complex functions. It results from this definition and the discussion in Section 4.6.3 that any linear combination of the form

$$X = \sum_{i=1}^{n} \alpha_i \, x(t_i) \tag{7.67}$$

is a normal random variable. This property is valid when n tends to infinity if the series is convergent in the quadratic mean sense, and, using the same arguments, for integrals. The fundamental consequence of this property is that the normal character is preserved when a signal is transformed into another one by any linear operation. A normal signal thus remains normal if it is processed by a linear filter.

Note that all these definitions and results are the same in the case of a discrete-time signal and the normality is specified by the notation $N(m[k], \gamma[k, l])$.

The fact that a normal signal remains normal after linear filtering is one of its most important properties in applications. In fact it is in general very difficult to calculate the probability distribution of a signal after linear transformation, as already indicated in Section 5.4. On the other hand, as a normal signal is completely specified by its mean value and its covariance function, the mode of transformation of these quantities completely specifies the mode of transformation of the probability distributions between input and output. This is an easy task, and it suffices to use the formulae given in Section 5.4. Note finally that all these results are valid in the complex case.

The normal property also introduces simplifications in the discussion of some properties introduced above. For example, in Section 5.3 we made a distinction between strict-sense and second-order stationarity. This distinction disappears for normal signals. If the mean value and the covariance function are invariant by changing the origin of time, which characterizes second-order stationarity, the same is valid for all probability distributions that are only a function of these first two moments. Similarly, as noncorrelation implies independence for normal distribution, the distinction between weak white noise and strict white noise becomes irrelevant for Gaussian signals.

7.4.2 Spectral Representation of Normal Signals

Let $x(t)$ be a complex zeromean normal random signal. For the discussion which follows we assume that t is continuous, but the transposition of the results to the discrete-time case is in general straightforward. The spectral representation of $x(t)$ is given by (6.5) which uses the increments $d\overline{X}(\nu)$. It results from (6.11) that if $\overline{X}(-\infty) = 0$, the function $\overline{X}(\nu)$ is deduced from $x(t)$ by a linear filter the frequency response of which is given by (6.10). Consequently $\overline{X}(\nu)$ and its increments $d\overline{X}(\nu)$ are normal and their statistical properties are completely specified by their second order moments. The increments $d\overline{X}(\nu)$ have interesting properties only when the signal $x(t)$ is stationary, which we will assume for the following discussion.

Suppose first that $x(t)$ is complex but satisfies the condition $E[x(t)\ x(t-\tau)] = 0$. This is the case for the analytic signal of a real signal, as seen in (6.116), and also for the complex envelope $\alpha(t)$, as seen in (6.126). The result of this assumption is that

$$E[d\overline{X}(\nu)\ d\overline{X}(\nu')] = 0 \tag{7.68}$$

Furthermore, the result of (6.15) and (6.37) is that

$$E[d\overline{X}(\nu)\ d\overline{X}^*(\nu')] = d\overline{\Gamma}(\nu)\ \delta(\nu-\nu')\,d\nu' \tag{7.69}$$

The consequences of these equations are that the RVs $d\overline{X}(\nu)$ and $d\overline{X}(\nu')$ are normal circular and independent for $\nu \neq \nu'$. The independence of the increments means that the function $\overline{X}(\nu)$ appearing in the spectral representation of a harmonizable normal signal is a complex Brownian motion, as defined in Section 7.3.2 (**h**).

Suppose now that $x(t)$ is *real*. In this case (7.68) can no longer be valid because, as a result of the Hermitian property (6.24) and from (7.69), the moment (7.68) has a non-negative value for $\nu = -\nu'$. In this case it is more interesting to use the spectral representation (6.61) and to discuss the properties of the increments of the functions $\overline{C}(\nu)$ and $\overline{S}(\nu)$ that appear in this equation. Comparing (6.5) and (6.61), and using the property (6.24), we easily obtain

$$d\overline{C}(\nu) = d\overline{X}(\nu) + d\overline{X}^*(\nu) \tag{7.70}$$

$$d\overline{S}(\nu) = j\,[d\overline{X}(\nu) - d\overline{X}^*(\nu)] \tag{7.71}$$

Note that the integral (6.61) uses only positive frequencies, and only these frequencies must be considered in the calculations that follow. The starting point is to note that as a result of (7.69) and (6.24), the relation (7.68) is again valid for positive frequencies ν and ν'. Using this property, we deduce from (7.70) and (7.71) that if $\nu \neq \nu'$, we have

$$E[d\overline{C}(\nu)\,d\overline{C}(\nu')] = E[d\overline{S}(\nu)\,d\overline{S}(\nu')] = E[d\overline{C}(\nu)\,d\overline{S}(\nu')] = 0 \tag{7.72}$$

On the other hand, for $\nu = \nu'$ we find

$$E[d\overline{C}(\nu)\,d\overline{S}(\nu)] = 0 \tag{7.73}$$

$$E[d\overline{C}^2(\nu)] = E[d\overline{S}^2(\nu)] = 2\,E[|dX(\nu)|^2] \tag{7.74}$$

The last point to note is that the functions $\overline{C}(\nu)$ and $\overline{S}(\nu)$ are normal, because they are obtained from $\overline{X}(\nu)$ by a linear operation. Using all these arguments, we can conclude that $\overline{C}(\nu)$ and $\overline{S}(\nu)$ are two independent Brownian signals with the same distribution because of (7.74). This conclusion is valid only if we use positive frequencies. If we take into account the zero frequency, the two Brownian motions are not strictly independent because $d\overline{X}(0)$ is real and $d\overline{S}(\nu)$ becomes null. The simplest way to suppress this point is to assume that $d\overline{X}(0) = 0$. This is the case of narrowband signals, which will now be considered.

7.4.3 Gaussian Narrowband Signals

The general properties of narrowband signals were presented in Section 6.8. The assumption of normality introduced now will allow us to make many calculations which are almost impossible without this assumption. This is especially the case for statistical properties of the amplitude and the phase which are obtained from the signal by a nonlinear operation.

The first point is to study the consequences of normality on the analytic signal examined in Section 6.8.1. One property which appears is that the analytic signal is complex normal, as introduced by (5.6). In fact, if $x(t)$ is normal, the quadrature signal $y(t)$ is also normal because it is deduced from $x(t)$ by a linear transformation which is the Hilbert transform. In reality this transform is a linear filtering using the filter with the frequency response $G_Q(\nu)$ defined by (6.109). But as a result of (6.116), if $x(t)$ is second order stationary, we then have

$$E[z(t_1)\ z(t_2)] = 0 \tag{7.75}$$

whatever the values of t_1 and t_2. So if we consider the complex vector $\mathbf{Z}$ with components $z(t_i)$ where the t_is are n arbitrary time instants, this vector satisfies the condition $\mathbb{C} = \mathbb{O}$ introduced in Section 4.6.6 and which characterizes the circularity. Thus the vector $\mathbf{Z}$ is normal circular and its PDF is given by (4.152). This density is completely defined by the covariance matrix $\mathbb{\Gamma}$ with matrix elements defined by $\gamma_z(t_i - t_j)$ where $\gamma_z(\tau)$ is the correlation function of the analytic signal defined by (6.117). In conclusion, the analytic signal of a normal zeromean real stationarity signal is normal circular, which can be specified by the notation $N_c[0, \gamma_z(\tau)]$. It would seem that the whole discussion on normal circular random vectors was introduced to be applied to the analytic signal and from there to the study of narrowband signals.

Consider now the case of the complex envelope defined by (6.123). As it is deduced from $z(t)$ by multiplication by the phase factor $\exp(-j\omega_0 t)$, this envelope is still normal. Using the same arguments as for the analytic signal, we deduce from (6.126) that $\alpha(t)$ is also normal circular, which can be noted $N_c[0, \gamma_\alpha(\tau)]$.

It is now of interest to discuss the statistical properties of some signals used for the representation of narrowband signals. These signals are related to their complex representation which introduces real and imaginary parts or amplitude and phase. More precisely, we will investigate the properties of the pairs $[x(t), y(t)]$ and $[m(t), n(t)]$, real and imaginary parts of $z(t)$ and $\alpha(t)$, and those of the pairs $[a(t), \phi(t)]$, amplitude and phase of $z(t)$ and $\alpha(t)$. For this we shall use extensively the results in Section 4.8.4.

(a) Statistical properties of the analytic signal. At a given time instant, $z(t)$ is an RV Z which is $N_c(0, \sigma_z^2)$ or $N_c[0, \gamma_z(0)]$. Its real and imaginary parts $x(t)$

and $y(t)$ are two RVs X and Y which are independent and $N(0, \sigma^2)$ where $\sigma^2 = (1/2)\sigma_z^2$. This last relation, already used when passing from (4.137) to (4.138), also appears as a consequence of (6.117) and (6.119). The amplitude and phase given by (4.214) are two RVs with the PDF given by (4.216), which shows that these RVs are independent. The marginal PDF of the amplitude is given by (4.217) which introduces a Rayleigh distribution.

Consider now two time instants t_1 and t_2 and let us call Z_1 and Z_2 the RVs $z(t_1)$ and $Z(t_2)$. The statistical properties of this pair of RVs are discussed on p. 137. If we are interested in real and imaginary parts we must use the PDF defined by (4.223) to (4.226). In those equations the parameters c and θ are defined by

$$\gamma_z(t_1 - t_2) = \gamma_z(0)\, c e^{j\theta} \tag{7.76}$$

which means that c and θ are functions of $t_1 - t_2$. If, however, we are interested in amplitude and phase representation, we must use the PDFs defined by (4.227) and (4.228). Note that as $\gamma_z(\tau)$ is the analytic signal of $2\gamma_x(\tau)$, a result indicated after (6.118), it cannot be a real function. In fact the Fourier transform of a real function must satisfy a Hermitian symmetry similar to (6.24), which is in contradiction to (6.118). Therefore the comment made after (4.228) cannot be applied, with the result that the real and imaginary parts of Z_1 and Z_2 are generally not independent RVs.

In some problems it is interesting to have the marginal PDF of the amplitudes. This is obtained by integration of (4.228) over the variables ϕ_1 and ϕ_2. Using (4.219) we easily find

$$p(a_1, a_2) = \frac{a_1 a_2}{\sigma^4(1 - c^2)} I_0\left[\frac{c a_1 a_2}{\sigma^2(1 - c^2)}\right] \exp\left[\frac{-(a_1^2 + a_2^2)}{2\sigma^2(1 - c^2)}\right] \tag{7.77}$$

For the values of $t_1 - t_2$ such that $\gamma_z(t_1 - t_2)$ is null we have $c = 0$, and thus $p(a_1, a_2)$ is the PDF of two independent RVs with two identical Rayleigh PDFs.

(b) Statistical properties of the complex envelope. The calculations are exactly the same, as they are a direct result of the property of circularity common to $z(t)$ and $\alpha(t)$. The first difference is that (7.76) must now be replaced by

$$\gamma_\alpha(t_1 - t_2) = \gamma_\alpha(0)\, c'\, e^{j\theta'} \tag{7.78}$$

But, as a result of (6.127), we have

$$c' = c \quad ; \quad \theta' = \theta - \omega(t_1 - t_2) \tag{7.79}$$

The second difference is that even if $\alpha(t)$ is circular, because of (6.126), there is no reason for it to remain an analytic signal because of the frequency translation defined by (6.128). In other words, $\gamma_\alpha(\tau)$ can be a real function which implies $\sin\theta' = 0$. Consequently, the last term of (4.226) disappears and the real and imaginary parts of the complex envelope at two different time instants are independent. Let us now investigate the cases when this situation arises.

The correlation function $\gamma_\alpha(\tau)$ is real if and only if its Fourier transform has Hermitian symmetry. But as it is positive, this is equivalent to saying that

$$\Gamma_\alpha(\nu) = \Gamma_\alpha(-\nu) \tag{7.80}$$

We therefore deduce that

$$\gamma_\alpha(\tau) = \gamma_\alpha(-\tau) \tag{7.81}$$

and using (6.128) we can also deduce that

$$\Gamma_z(\nu_0 + \nu) = \Gamma_z(\nu_0 - \nu) \tag{7.82}$$

Physically, this means that the spectrum of $z(t)$ is symmetric with respect to the central frequency ν_0. However, by using (6.118), we deduce that this symmetry property is also valid for the power spectrum of $x(t)$. A classic case when this situation appears is in the consideration of a *white band-limited* signal, which means that $\Gamma(\nu)$ is constant in some frequency band and null otherwise. As indicated at the end of Section 6.8, this symmetry condition implies that $m(t)$ and $n(t)$ become two *independent* normal signals. This is the property of independence noted after (4.228).

(c) Sum of sinusoidal and narrowband normal signals. This situation appears very often in communication problems where the narrowband normal signal can be considered as a disturbing noise. We can again take the previous discussion and note that as a sinusoidal signal is not random, the only change is that the mean value of the sum is no longer null but is equal to that signal. In particular, if we consider only one time instant, the amplitude and phase have a PDF given by (4.218) and the marginal PDF of the amplitude is a Rayleigh-Rice distribution defined by (4.220) and represented in Figure 4.9. The consideration of two time instants does not introduce conceptual difficulties, but the mathematical expressions become a little more complicated.

7.4.4 Higher-Order Spectra

Let $x(t)$ be a *real* zeromean normal signal. The higher-order moments of $x(t)$ are defined by (5.42). But as $x(t)$ is normal, the RVs $x(t_i)$ in this expression are jointly normal, and we can then apply the general expressions (4.90) and (4.95). The first equation means that all the odd moments of this signal are null. The second expression allows us to express all the even moments in terms of the covariance function defined by (5.12) where the last term disappears because of the assumption on the mean values. For $n = 4$, this can be written as

$$m(t_1, t_2, t_3, t_4) = \gamma(t_1, t_2)\, \gamma(t_3, t_4) + \gamma(t_1, t_3)\, \gamma(t_2, t_4) + \gamma(t_1, t_4)\, \gamma(t_2, t_3) \tag{7.83}$$

and more generally we have

$$m(t_1, t_2, \ldots, t_{2k}) = \sum_G \gamma(t_{i_1}, t_{i_2}) \ldots \gamma(t_{i_{2k-1}}, t_{i_{2k}}) \tag{7.84}$$

where the sum is extended to all the Gaussian permutations introduced after (4.95).

Suppose now that $x(t)$ is *stationary*. In this case (7.83) and (7.84) must be changed by replacing $\gamma(t_i, t_j)$ by $\gamma(t_i - t_j)$. All the higher-order moments of $x(t)$ can thus be expressed explicitly in terms of the correlation function of the signal.

Let us now study the properties of the higher-order spectra of a zeromean stationary normal signal. In order to explain the structure of the calculation we shall begin with the fourth-order moments and use the notation introduced in Section 6.10. The moment introduced by (7.83) will be called $m_4(\mathbf{t})$ and its Fourier transform noted $M_4(\mathbf{f})$. Because of the stationarity assumption, (6.151) can be written

$$M_4(\mathbf{f}) = \Gamma_3(\mathbf{f})\ \delta(f_1 + f_2 + f_3 + f_4) \tag{7.85}$$

and $M_4(\mathbf{f})$ is then related to the trispectrum. Similarly we can write a relation in the time domain between $\gamma_3(\theta_1, \theta_2, \theta_3)$ and $m_4(\mathbf{t})$ in the form

$$m_4(\mathbf{t}) = \gamma_3(t_1 - t_4,\ t_2 - t_4,\ t_3 - t_4) \tag{7.86}$$

As will now appear, it is easier in the normal case to work with the pair $[m, M]$ than with the pair $[\gamma, \Gamma]$.

Returning to (7.83), we must now calculate the Fourier transform of $m_4(\mathbf{t})$. It is the sum of three terms, the first being the Fourier transform of $\gamma(t_1 - t_2)\ \gamma(t_3 - t_4)$. It is easy to verify that

$$\iiiint \gamma(t_1 - t_2)\ \gamma(t_3 - t_4) \exp[-2\pi j(f_1 t_1 + f_2 t_2 + f_3 t_3 + f_4 t_4)\ dt_1\, dt_2\, dt_3\, dt_4 =$$

$$\Gamma(f_1)\,\Gamma(f_3)\ \delta(f_1 + f_2)\ \delta(f_3 + f_4) \tag{7.87}$$

In fact the fourth-order integral can be factorized in a product of two second-order integrals which directly introduce the power spectra, as in (6.37). There are two differences with this last equation. The first is that we assume that (6.39) is valid, which means that a spectral density exists. If necessary, it is not difficult to use (6.38) instead of (6.40). The second difference is due to the fact that we are working with real signals so there is no complex conjugate, and $-\nu_2$ is replaced by ν_2. Making similar calculations for the other two terms of (7.83), we finally find

$$M_4(\mathbf{f}) = \Gamma(f_1)\,\Gamma(f_3)\ \delta(f_1 + f_2)\ \delta(f_3 + f_4) + \Gamma(f_1)\,\Gamma(f_2)\ \delta(f_1 + f_3)\ \delta(f_2 + f_4)$$

$$+ \Gamma(f_1)\,\Gamma(f_2)\ \delta(f_1 + f_4)\ \delta(f_2 + f_3) \tag{7.88}$$

In spite of its complex form, this expression has a very simple interpretation. In the fourth-order frequency space we have already introduced the stationary manifold defined by (6.150). Let us now introduce the three *normal manifolds*. They are defined by the three equations

$$f_1 + f_2 = 0 \quad ; \quad f_3 + f_4 = 0 \tag{7.89}$$

$$f_1 + f_3 = 0 \quad ; \quad f_2 + f_4 = 0 \tag{7.90}$$

$$f_1 + f_4 = 0 \quad ; \quad f_2 + f_3 = 0 \tag{7.91}$$

and it is obvious that they belong to the stationary manifold. If we compare this with (7.88) we deduce that $M_4(\mathbf{f})$ is null outside the normal manifolds and on these manifolds is simply a product of values of the spectral density of the signal. To summarize all these points we can say that $M_4(\mathbf{f})$ is distributed only on the normal manifolds and has a *normal density* on these manifolds. Note that the reverse calculation can be made, and if this spectral property is realized, the fourth order moment takes the form (7.83).

This discussion can be transposed without difficulty to the discrete-time case. In (7.87) the integrals must be replaced by series, but the final result remains the same. The only point to note is that in the Fourier domain all the functions are periodic, and it is then possible to reduce the study to the frequencies f_i such that $|f_i| \le 1/2$, exactly as in Figure 6.10.

Example 7.2 Some examples of moment trispectra. In order to illustrate these concepts, let us study some examples of structures of fourth-order Fourier transforms of moments. In particular we wish to investigate with more precision the structure of the fourth-order function $M_4(\mathbf{f})$ for several kinds of random signals.

The first example that arises is of course that of white signals. The fourth-order cumulant of a strict white noise is given by (7.11) while its correlation function is given by (7.1). In (7.11) the cumulant c_4 can be expressed in terms of moments by using (4.80). This expression can be written as

$$c_4 = m_4 - 3\sigma^4 \tag{7.92}$$

where $m_4 = E\{x^4[k]\}$. Using these expressions, we can calculate the fourth-order moment $m_4[\mathbf{k}]$ that is the discrete-time version of $m_4(\mathbf{t})$ defined by (7.86). Combining all these relations, we obtain

$$m_4[\mathbf{k}] = \sigma^4\{\delta[k_1 - k_2]\,\delta[k_3 - k_4] + \delta[k_1 - k_3]\,\delta[k_2 - k_4] + \delta[k_1 - k_4]\,\delta[k_2 - k_3]\} + c_4\,\delta_4[\mathbf{k}] \tag{7.93}$$

The Fourier transform $M_4(\mathbf{f})$ is defined as in (6.42') or (7.12) by

$$M_4(\mathbf{f}) = \sum_{\mathbf{k}} m_4[\mathbf{k}] \exp[-2\pi j \mathbf{f}^T \mathbf{k}] \tag{7.94}$$

and using (7.93), we obtain

$$M_4(\mathbf{f}) = \sigma^4\{\delta(f_1 + f_2)\,\delta(f_3 + f_4) + \delta(f_1 + f_3)\,\delta(f_2 + f_4) + \delta(f_1 + f_4)\,\delta(f_2 + f_3)\} + c_4\,\delta(f_1 + f_2 + f_3 + f_4) \tag{7.95}$$

Let us now comment on this expression. First, if $x[k]$ is not only white but also normal, we have $c_4 = 0$, which is a specific property of normal random variables, as noted after (4.87). In this case (7.95) is a particular form of (7.88) corresponding to a power spectrum equal to σ^2, characterizing the spectral density of a white signal. If $c_4 \neq 0$, we see that $\mathrm{M}_4(\mathbf{f})$ is a sum of two kinds of terms: the first introduces a normal density on the normal manifolds and the second is a constant density equal to c_4 on the stationary manifold defined by (6.150).

Let us now give an example of a signal with a non-normal density on the normal manifolds. Consider the signal $y(t)$ defined by

$$y(t) = A x(t) \tag{7.96}$$

where $x(t)$ is a normal signal and A a positive RV, independent of $x(t)$ and describing a random modulation of the normal signal. This situation was introduced, for example, in some questions in Problem 4.22. The correlation function of $y(t)$ is of course

$$\gamma_y(\tau) = E(A^2)\ \gamma_x(\tau) \tag{7.97}$$

where $\gamma_x(\tau)$ is the correlation function of $x(t)$. Using the property of independence between A and $x(t)$, we can write

$$m_{4,y}(\mathbf{t}) = E(A^4)\ m_{4,x}(\mathbf{t}) \tag{7.98}$$

or by Fourier transformation

$$M_{4,y}(\mathbf{f}) = E(A^4)\ M_{4,x}(\mathbf{f}) \tag{7.99}$$

where $M_{4,x}(\mathbf{f})$ is given by (7.88) because $x(t)$ is normal. As a result $M_{4,y}(\mathbf{f})$ is null outside the normal manifolds and the density on these manifolds is $E(A^4)\Gamma_x(f_i)\Gamma_x(f_j)$. The normal density is $\Gamma_y(f_i)\Gamma_y(f_j)$ or $[E(A^2)]^2\ \Gamma_x(f_i)\Gamma_x(f_j)$. Therefore, in order to obtain a normal density on the normal manifolds, we must have

$$E(A^4) = [E(A^2)]^2 \tag{7.100}$$

which means that the variance of A^2 is null. Using the result discussed after (2.22), this means that A^2 is non-random with probability one. But as A is positive, it is non-random and consequently $y(t)$ is normal. Thus, if A is effectively random, the Fourier transform $M_4(\mathbf{f})$ is null outside the normal manifolds, but has a non-normal density on these manifolds.

It is clear that this discussion can be extended to any even order moment. If $x(t)$ is zeromean-valued stationary and normal, the function $m_{2k}(\mathbf{t})$ defined by

$$m_{2k}(\mathbf{t}) = E[x(t_1)\ x(t_2)\ \ldots\ x(t_{2k})] \tag{7.101}$$

can be expressed as in (7.84), or in the form

$$m_{2k}(\mathbf{t}) = \sum_{G} \gamma(t_{i_1} - t_{i_2}) \ldots \gamma(t_{i_{2k-1}} - t_{i_{2k}}) \tag{7.102}$$

In this sum there are $(2k-1)!!$ distinct terms $t_i(\mathbf{t})$ and the first can be written as

$$t_1(\mathbf{t}) = \gamma(t_1 - t_2)\ \gamma(t_3 - t_4)\ \ldots\ \gamma(t_{2k-1} - t_{2k}) \tag{7.103}$$

The Fourier transform $M_{2k}(\mathbf{f})$ of $m_{2k}(\mathbf{t})$ is also a sum of $(2k-1)!!$ terms and the Fourier transform of $t_1(\mathbf{t})$ has the form

$$T_1(\mathbf{f}) = \Gamma(f_1)\ \Gamma(f_3)\ \ldots\ \Gamma(f_{2k-1})\ \delta(f_1 + f_2)\ \delta(f_3 + f_4)\ \ldots\ \delta(f_{2k-1} + f_{2k}) \tag{7.104}$$

This term corresponds to a distribution on the normal manifold

$$f_1 + f_2 = 0 \; ; \; f_3 + f_4 = 0 \; ; \; \dots ; \; f_{2k-1} + f_{2k} = 0 \tag{7.105}$$

and there are $(2k-1)!!$ distinct normal manifolds. On these manifolds the density is a product of spectral densities such as $\Gamma(f_1)\,\Gamma(f_3)\,\dots\,\Gamma(f_{2k-1})$. The general conclusion is the same: a normal signal is characterized by the fact that all the even order spectra are null outside the normal manifolds and have a normal density on these manifolds. The normal density is simply a product of spectral densities defined by (6.42) or (6.42'). Furthermore, do not forget that if $x(t)$ is zeromean, all the odd order moments are null.

To complete this section it is interesting to examine the case of normal *complex* signals. In the general case there is no simple result because of the complexity of the relation (4.157). On the other hand the situation is simplified if we introduce the assumption of *circularity*, valid for example for the analytic signal or for the complex envelope. In this case the only non-null higher-order moment analog to (7.101) takes the form

$$m_{2k}(t_1, \dots, t_k, t_{k+1}, \dots, t_{2k}) = E[x(t_1) \dots x(t_k)\, x^*(t_{k+1}) \dots x^*(t_{2k})] \tag{7.106}$$

It can be expressed in terms of correlation function by an extension of (4.158), which gives (7.102), but the number of Gaussian circular permutations is now equal to $k!$ instead of $(2k-1)!!$. This is discussed in Problem 4.39. We obtain the same result for the Fourier transform and the first circular manifold analog to (7.105) is

$$f_1 + f_{k+1} = 0 \; ; \; f_2 + f_{k+2} = 0 \; ; \; \dots ; \; f_k + f_{2k} = 0 \tag{7.107}$$

Clearly there are $k!$ distinct circular manifolds.

7.4.5 Instantaneous Nonlinear Transformations

A system is said to be instantaneous or without memory if the output $y(t)$ at time t is only a function of the input $x(t)$ at the same time instant. It is clear that a linear filter defined by an input-output relationship such as (5.55) or (5.56) is not instantaneous. On the other hand, the quadratic system $y(t) = x^2(t)$ is obviously instantaneous.

It is in general very difficult to calculate the statistical properties of the output of an instantaneous nonlinear system in terms of those of the input. Formal expressions can be obtained by using the general results given in Section 4.4 concerning the functions of random vectors. However, when the assumption of normality is introduced numerous calculations become possible, as discussed in Section 4.8.5. The purpose of this section is to give some examples among many possibilities.

(a) Quadratic transformation. As mentioned above, the output of the corresponding system is $y(t) = x^2(t)$. This appears in many physical devices and particularly in all systems that depend only on instantaneous power, given by (5.93). For example, the signal delivered by a photoelectric cell is proportional to the power of the optical field.

Suppose first that $x(t)$ is real, zeromean, stationary, and normal. The expected value of $y(t)$ is

$$E[y(t)] = E[x^2(t)] = \gamma(0) = \sigma^2 \tag{7.108}$$

where $\gamma(\tau)$ is the correlation function of $x(t)$. The correlation function of $y(t)$ is defined by (5.17') which gives

$$\gamma_y(\tau) = E[x^2(t)\, x^2(t-\tau)] - \gamma^2(0) \tag{7.109}$$

As $x(t)$ is normal, we can use (4.94), which gives

$$\gamma_y(\tau) = 2\,[\gamma(\tau)]^2 \tag{7.110}$$

It is now interesting to transpose this expression into the frequency domain. As the Fourier transform of a product of functions is equal to the convolution of their Fourier transforms, we obtain the power spectrum of $y(t)$ by

$$\Gamma_y(\nu) = 2 \int \Gamma(f)\, \Gamma(\nu - f)\, df \tag{7.111}$$

where $\Gamma(\nu)$ is the power spectrum of $x(t)$. Let us to illustrate these simple relations with some examples.

Suppose first that $x(t)$ is a white band-limited signal, as introduced in Section 5.2.4 (**c**). Its power spectrum is represented in Figure 7.10. The value of the spectral density is $\sigma^2/2B$ in order to satisfy (6.41). It is well known, and easily verified, that the convolution of a rectangular signal by itself gives a triangular signal. So the power spectrum of $y(t)$ takes the form of Figure 7.11. The value for $\nu = 0$ is deduced from (7.111) where $\Gamma(\nu)$ is replaced by $\sigma^2/2B$.

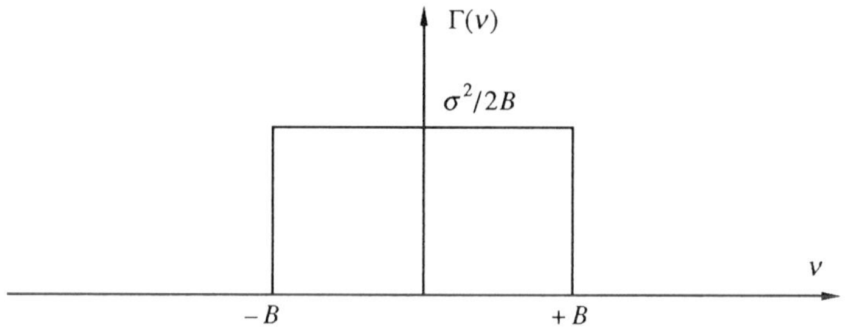

Figure 7.10 Power spectrum of $x(t)$.

Suppose now that $x(t)$ is still white and band-limited but in the frequency band $-\nu_0 \pm B$ and $\nu_0 \pm B$. As $\Gamma(\nu)$ is symmetric, the power spectrum of $x(t)$ is presented for positive frequencies only in Figure 7.12. When calculating the convolution (7.111), a contribution similar to that represented in Figure 7.11 appears but there is another one in the neighborhood of frequencies $\pm 2\,\nu_0$. This is due to an interference between positive and negative frequencies of $x(t)$. If we again use only positive frequencies because of the symmetry of $\Gamma_y(\nu)$, the result is as shown in Figure 7.13.

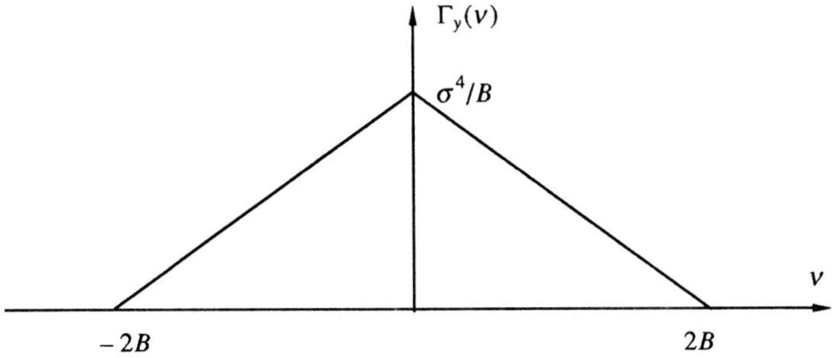

Figure 7.11 Power spectrum of $y(t)$.

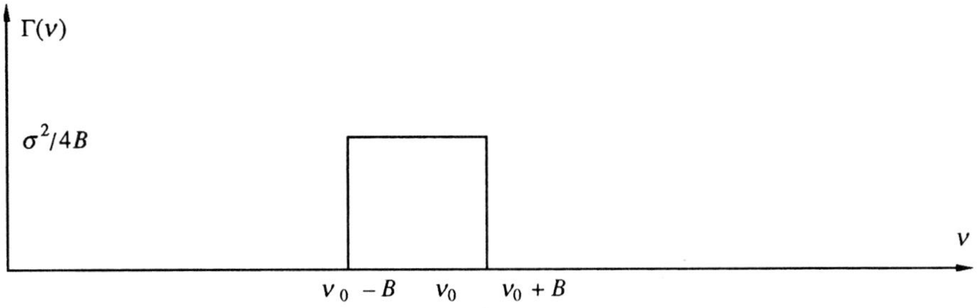

Figure 7.12 Power spectrum of $x(t)$.

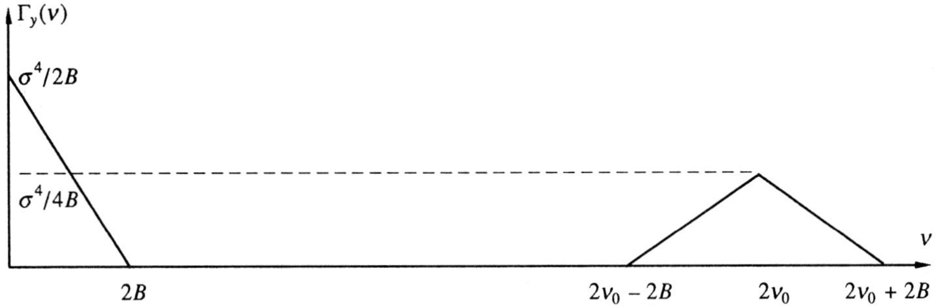

Figure 7.13 Power spectrum of $y(t)$.

The interesting point which appears in this figure is that there are both a low frequency contribution and a high frequency contribution in this spectrum, the latter of which can be removed by a low pass filter. This is not really surprising and also occurs for sinusoidal signals. Thus the Fourier transform of the signal $\cos(2\pi\nu_0 t)$ is

$$\mathrm{FT}[\cos(2\pi\nu_0 t)] = \frac{1}{2}[\delta(\nu - \nu_0) + \delta(\nu + \nu_0)] \tag{7.112}$$

which exhibits two spectral lines at the frequencies $\pm \nu_0$. However, as $2\cos^2\alpha = 1 + \cos(2\alpha)$, we have

$$\mathrm{FT}[\cos^2(2\pi\nu_0 t)] = \frac{1}{2}\,\delta(\nu) + \frac{1}{4}\,[\delta(\nu - 2\nu_0) + \delta(\nu + 2\nu_0)] \tag{7.113}$$

This shows that the spectrum of $\cos^2(2\pi\nu_0 t)$ has three spectral lines at the frequencies 0 and $\pm 2\nu_0$.

Using the results in Figure 7.13, it is possible to solve the problem presented at the end of Section 6.6 concerning *ergodicity* for power measurements. Power (or variance) measurement is a very important operation in various applications as it is the first parameter that can give information in a zeromean random signal. It is sometimes the only operation that can be made. It corresponds to the logical process presented in Section 5.2.1 concerning the description of a random signal by some moments. As the mean value (5.7) is null, the moment which must be known first is the variance (5.9) equal to the power, from (5.95). The principle of the device used for power measurement is presented in Figure 7.14.

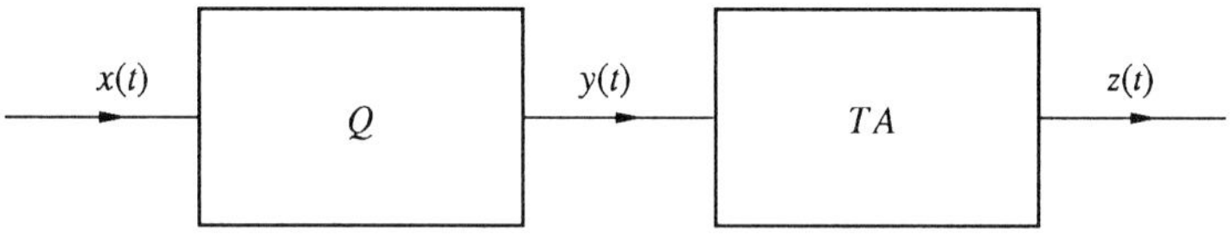

Figure 7.14 Principle of power measurement using a quadratic system and a time averager.

The output of the quadratic system in Q is obviously $y(t) = x^2(t)$, which is filtered by a time averager. The perfect time averager was introduced in Section 6.6. Its impulse response is presented in Figure 6.6 and its frequency response is given by (6.94). In reality it is not necessary to use a *perfect* time averager to realize power measurements. Any filter with a frequency response $G_T(\nu)$ such that $G_T(0) = 1$ and that satisfies (6.95) can be used. This is especially so in the case of the RC integrator, so called because it uses only a resistor R and a capacitor C. Its frequency response is

$$G_T(\nu) = [1 + j2\pi RC\nu)]^{-1} \tag{7.114}$$

and in order to obtain (6.95) we take $T = 2\,RC$.

At this point we can use the same reasoning as in Section 6.6, replacing $x(t)$ by $y(t)$. The mean value of $y(t)$ is σ^2, from (7.108). Thus the mean value of $z(t)$, output of the time averager, is also σ^2 and its variance is given by (6.98) where $\Gamma(0)$ is now $\Gamma_y(0)$. Using (7.111) and noting that $\Gamma(\nu)$ is even, we obtain

$$v_T \cong \frac{2}{T} \int [\Gamma(\nu)]^2\, d\nu \tag{7.115}$$

where $\Gamma(\nu)$ is obviously the power spectrum of $x(t)$. In conclusion, the ergodicity for the variance measurement of a normal signal is *ensured* as soon as the power density of the signal is square integrable, as particularly in the spectra in Figures 7.10 and 7.12. It is clear that the condition specified in the frequency domain can be translated into the time domain. Using the Parseval theorem, it becomes

$$\int [\gamma(\tau)]^2 \, d\tau < +\infty \tag{7.116}$$

which must be compared with (6.99) valid for mean value estimation. If there is at least one spectral line in the power spectrum of $x(t)$, which means a term in the form $\alpha\delta(\nu - \nu_0)$, then $\Gamma(\nu)$ cannot be square integrable. We then have the important result that a normal signal which is ergodic for power measurement cannot have spectral lines, meaning that the spectral distribution function $\bar{\Gamma}(\nu)$ used in (6.39) is continuous. In reality, using a more abstract definition of ergodicity, it is possible to show that this condition characterizes ergodicity for a real normal signal (see [Cramér], p. 157).

Example 7.3 Some comments on ergodic problems. It is important to keep clearly in mind that all the previous results are valid provided the signal $x(t)$ is *normal*. These results can be untrue if the assumption of normality is removed. To make this point clear, let us return to the signal defined by (7.96) to examine its ergodic properties. As this signal is zeromean-valued and its correlation function given by (7.97), we see that the condition (6.99) is the same for $\gamma_x(\tau)$ and $\gamma_y(\tau)$. So if $x(t)$ is ergodic for mean value measurements, $y(t)$ has the same property. To study ergodicity for power measurements, let us calculate the correlation function of $s(t) = y^2(t)$. The mean value of $s(t)$ is given by

$$E[s(t)] = E(A^2)\,\gamma_x(0) = E(A^2)\sigma_x^2 \tag{7.117}$$

where σ_x^2 is the variance of $x(t)$. Using the property (4.94), we deduce that

$$E[s(t)\,s(t-\tau)] = E(A^4)\,[2\gamma_x^2(\tau) + \gamma_x^2(0)] \tag{7.118}$$

in such a way that the correlation function of $s(t)$ defined by (5.17') becomes

$$\gamma_s(\tau) = 2E(A^4)\,\gamma_x^2(\tau) + \sigma_x^4\,[E(A^4) - E^2(A^2)] \tag{7.119}$$

In conclusion, even if $x(t)$ is ergodic for the power, that is, even if (7.116) is satisfied, the condition (6.99) cannot be valid for $\gamma_s(\tau)$ because of the last term of (7.119). It would only be possible if this term disappeared, giving the condition (7.100), which means that A is no longer random. The fact that $y(t)$ defined by (7.96) is not ergodic for the power is not surprising, as it complements the results discussed in Example 5.6 showing that the presence of random variables in a signal introduces an element independent of time and in general destroys the ergodic properties.

The extension of the above results to the complex case introduces no difficulties. It is particularly simple in the case of complex normal *circular* signals, such as the analytic signal. Let us call $z(t)$ an analytic signal and $y(t)$ the output of a quadratic device given by $y(t) = |z(t)|^2$. The relation (7.108) is still valid while (7.110) must be replaced by

$$\gamma_y(\tau) = |\gamma_z(\tau)|^2 \tag{7.120}$$

because of (4.158). It is interesting to note that the phase of the correlation function is dropped and, as a result of (6.127), the analytic signal and its complex envelope will give the same correlation function after a quadratic device.

(b) Perfect clipping. This transformation is characterized by the relation

$$y(t) = \mathrm{Sg}[x(t)] = \frac{x(t)}{|x(t)|} \tag{7.121}$$

which makes use of the sign function already met in Section 4.8.5. The interest of this transformation is that it introduces a perfect regulation of the amplitude of the signal. In fact, a multiplication of $x(t)$ by any amplitude factor a, $a > 0$, does not change the output. The second point is that $y(t)$ takes only two possible values, namely ± 1, which greatly simplifies many numerical calculations. For example, the product $y(t)\,y(t-\tau)$ which appears in correlation measurements is very simple to realize and also has only two possible values. The calculation of the correlation function of $y(t)$ is in general very difficult, but is much simplified if the normal assumption is introduced. In this case it is obvious that $y(t)$ remains zeromean-valued. Furthermore, using (4.247), we can write the correlation function of $y(t)$ in the form

$$\gamma_y(\tau) = \frac{2}{\pi}\,\mathrm{Arc\ sin}\left[\frac{\gamma_x(\tau)}{\gamma_x(0)}\right] \tag{7.122}$$

On the other hand, the power spectrum of $y(t)$ cannot be expressed in closed form because of the complexity of the structure of the Arc sin function. To give an idea of the result of this transformation, let us first analyze the case of a band-limited white Gaussian noise. Its spectrum is represented in Figure 7.10 and the corresponding correlation function is $\sigma^2\,\mathrm{sinc}(2B\,\tau)$. As this function is even, we can consider only the case $\tau > 0$. The curve 1 appearing in Figure 7.15 is the function $\mathrm{sinc}(x)$, already represented in Figure 5.4. The curve 2 is the function $(2/\pi)$ Arc $\sin[\mathrm{sinc}(x)]$, associated with the correlation function (7.122). It appears that the general shape of the two curves is similar. However, there are some differences, and especially for $x = 0$. It is clear that the derivative of the function $\mathrm{sinc}(x)$ is null for $x = 0$. By a simple limited expansion it can be shown that the derivative of the function $(2/\pi)$ Arc $\sin[\mathrm{sinc}(x)]$ is non-null. As this function is even, this means that it has no well defined derivative at the origin. Using the result of Section 5.8.2, this implies that the signal $y(t)$ defined by (7.121) has no mean square derivative. This is quite natural and is a consequence of the structure of the transformation (7.121). The same kind of results appear in Figure 7.16 which corresponds to an exponential correlation function similar to that presented in Figure 5.1. In this case $x(t)$ and $y(t)$ have no mean square derivative.

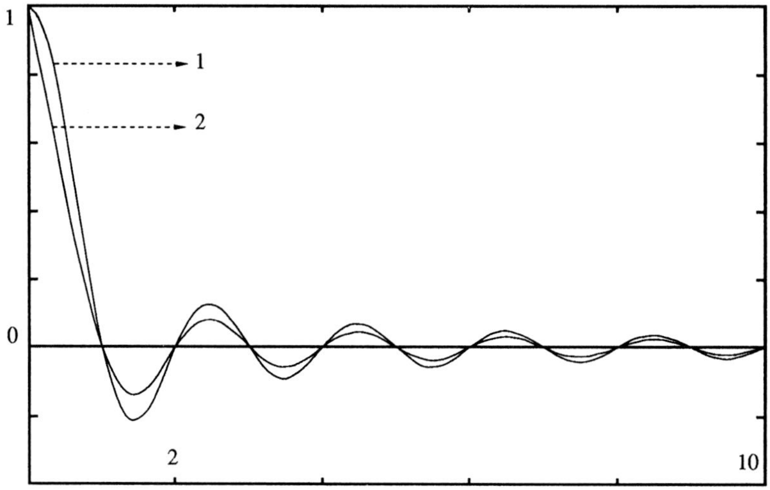

Figure 7.15 Correlation functions of a band-limited white normal noise before (1) and after (2) perfect clipping.

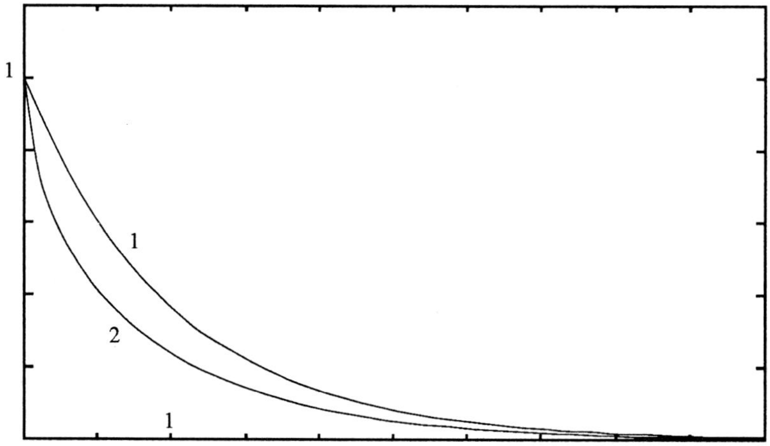

Figure 7.16 Correlation functions before (1) and after (2) clipping of a normal signal with exponential correlation function.

(c) Other transformations. There is no general method valid for any nonlinear transformation, but there are many specific problems to which the concepts given in Section 4.8.5 can be applied. All examples in that section show the importance of the assumption of normality.

7.5 SIGNALS WITH STATIONARY INCREMENTS

We have already met signals which are not stationary, but the *increments* of which are stationary. This was the case of the random walk studied in Section 7.3.1, and the Brownian motion in Section 7.3.2. These signals have increments which are not only stationary but also independent. In this section we shall relax this last assumption to study a broader class of random signals. But first, why are such signals interesting?

Many communication systems use frequency or phase modulation. Under very general conditions, instantaneous angular frequency is defined as the derivative of the phase. This means that the instantaneous phase of an amplitude stabilized oscillator can be written as

$$\Phi(t) = \Phi_0 + \int_{t_0}^{t} F(\theta)\, d\theta \tag{7.123}$$

where Φ_0 is the phase at t_0. If the frequency is constant the phase takes the form of (1.1). Because of many instabilities the instantaneous frequency often appears as the sum of a deterministic and a random term considered as frequency noise. The phase noise generated by such a frequency noise is given by (7.123). It is clear that if the frequency term $F(t)$ is stationary, this is no longer true for the phase. For example, if the mean value of $F(t)$ is F, that of $\Phi(t)$ is a linear function of time and is thus not constant. On the other hand, the *increments* of $\Phi(t)$ defined by

$$\Phi(t) - \Phi(t-T) = \int_{t-T}^{t} F(\theta)\, d\theta \tag{7.124}$$

are stationary, because they are obtained by linear filtering of $F(t)$. This filtering is similar to that introduced in (6.93) and shown in Figure 6.6.

Another way to introduce signals with stationary increments is to note that (7.123) is the solution of the differential equation

$$d\Phi/dt = F(t) \tag{7.125}$$

and that random signals with stationary increments appear naturally when studying such differential equations driven by a random signal. Similarly, the basic equation in mechanics relating motion and force $m\ddot{x} = F$ will introduce the same kind of situation. If the driving force is random, the velocity is a process with stationary first increments and the position is a process with stationary second increments. As these equations are common in many areas of physical science, signals with stationary increments also appear very frequently.

7.5.1 Definition and General Properties

The increments of a signal were introduced on p. 164, and defined by

$$\Delta x(t;\ \tau) = x(t) - x(t - \tau) \tag{7.126}$$

This increment is called a causal increment because at time t it only uses past values of the signal when $\tau > 0$. However, it is possible to use anti-causal increments defined by

$$\Delta_a x(t; \tau) = x(t + \tau) - x(t) \tag{7.127}$$

which were used, for example, in (5.83) for the definition of the derivative or also in (7.39). It is obvious that

$$\Delta_a x(t; \tau) = \Delta x(t + \tau; \tau) \tag{7.128}$$

and we will restrict our study here to causal increments.

A random signal is said to be a *signal with stationary increments* if the increments $\Delta x(t; \tau)$ defined by (7.126) are stationary functions of time, whatever the value of τ. In the same way as there are many possible interpretations of the concept of stationarity, as seen in Section 5.3, there are also a variety of interpretations of the concept of signals with stationary increments. We shall now discuss some elementary properties of these signals, for which in general second order stationarity is sufficient.

It is evident that any stationary signal is also a signal with stationary increments. This results from the fact that the difference of two stationary signals remains stationary.

Second, it is clear that if a signal with stationary increments $x(t)$ has a first order derivative in the mean square sense, it also has an integral representation as in (7.123). But there are many cases where this derivative does not exist, the best example being signals with independent increments. In this case the integral representation (7.123) must be replaced by equations like (7.44).

7.5.2 Increments and Linear Filtering

Let us call $x(t)$ a signal with stationary increments. The increment defined by (7.126) can be written

$$\Delta x(t; \tau) = \int_{t-\tau}^{t} dx(\theta) \tag{7.129}$$

and if $x(t)$ has a derivative called $v(t)$, by analogy to velocity, this becomes

$$\Delta x(t; \tau) = \int_{t-\tau}^{t} v(\theta)\, d\theta \tag{7.130}$$

which corresponds to (7.124). Note that (7.129) was already used in the context of point processes. In fact, if $x(t)$ is the number of points in the interval $[0, t)$, $\Delta x(t; \tau)$ represents the number of points in $[t - \tau, t)$, as seen in Section 5.7.1. As indicated previously, the relations (7.129) and (7.130) are input-output relationships in a linear filter dependent on τ. The impulse response of this filter is a rectangular function similar to that represented in Figure 6.6, and the corresponding frequency response is

$$G(\nu; \tau) = (2\pi j\nu)^{-1}\ (1 - e^{-2\pi j\nu\tau}) \tag{7.131}$$

Using the sinc function defined by $\operatorname{sinc} x = \sin(\pi x)/\pi x$ we easily obtain

$$G(\nu;\ \tau) = \tau\ \mathrm{sinc}(\nu\tau)\ \exp(-\pi j \nu\tau) \tag{7.132}$$

This filter has the particular property

$$G(0;\ \tau) = \tau \tag{7.133}$$

which will be used below.

7.5.3 First and Second Order Properties of the Increments

As the increments $\Delta x(t;\ \tau)$ are stationary, their mean value does not depend on t and if $E[dx(\theta)] = m\,d\theta$, or if m is the mean value of $v(t)$ in (7.130), we obtain

$$E[\Delta x(t;\ \tau)] = m\,\tau \tag{7.134}$$

which can also be deduced from (7.133) by using (6.9) and (6.36). Without loss of the generality, we will now assume that the increments $\Delta x(t;\ \tau)$ are zeromean valued.

Let us now consider the *variance* $\sigma^2(\tau)$ of $\Delta x(t;\ \tau)$. If the power spectrum of $v(t)$ in (7.130) is noted $\Gamma_v(\nu)$, the application of (6.41) and of (6.55) with (7.132) gives

$$\sigma^2(\tau) = \tau^2 \int \Gamma_v(\nu)\ \mathrm{sinc}^2(\nu\tau)\ d\nu \tag{7.135}$$

This expression shows that $\Delta x(t;\ \tau)$ can have a finite variance even if $v(t)$ is not second order, in which case (7.130) must be replaced by (7.129). This is especially the case if $\Gamma_v(\nu) = \alpha$, which corresponds to a white signal $v(t)$. Using the Parseval theorem, as in (6.95), we find

$$\sigma^2(\tau) = \alpha\tau \tag{7.136}$$

which is the variance of a Brownian motion, as seen in (7.38).

It is also interesting to find the conditions on $\Gamma_v(\nu)$ ensuring that $\sigma^2(\tau)$ is bounded. Suppose that there exists a spectral density $\Gamma_s(\nu)$ with a finite integral, or which corresponds to a second order stationary signal $s(t)$ such that we have

$$\Gamma_v(\nu) = (1 + 4\pi^2\nu^2)\,\Gamma_s(\nu) \tag{7.137}$$

It is clear that $\Gamma_v(\nu)$ is not necessarily integrable because of the contribution of the infinite frequencies in (7.137). By inserting $\Gamma_v(\nu)$ in (7.135) we again find a finite integral, because of the term $\mathrm{sinc}^2(\nu\tau)$. As a result the variance $\sigma^2(\tau)$ is finite and the increment $\Delta x(t;\ \tau)$ is second order. Conversely, if the integral (7.135) is finite and if we define $\Gamma_s(\nu)$ by (7.137), it appears that $\Gamma_s(\nu)$ is integrable. But (7.137) is a particular form of the general relation (6.55) and corresponds to a filter with the frequency response

$$G(\nu) = 1 + 2\pi j \nu \tag{7.138}$$

This filter can be decomposed into a sum of two filters. The first is the identity filter already used in (6.109), while the second is the derivative filter studied in Example 6.3.

As a result, the increment $\Delta x(t;\ \tau)$ can be considered as the output of a filter with frequency response

$$G'(\nu;\ \tau) = G(\nu;\ \tau)\, G(\nu) \tag{7.139}$$

where $G(\nu;\ \tau)$ and $G(\nu)$ are defined by (7.131) and (7.138) respectively and the corresponding input is now a second order signal $s(t)$. This is the advantage over (7.130) using a function $v(t)$ which is not necessarily second order.

Let us illustrate this point by taking the example of Brownian motion. In this case $v(t)$ is a Gaussian white noise, which is not a second order signal. The signal $s(t)$, however, is normal and its power spectrum deduced from (7.137) is

$$\Gamma_s(\nu) = c[1 + 4\pi^2\nu^2]^{-1} \tag{7.140}$$

where c is the diffusion constant which appears in (7.38). The spectral density (7.140) has a Lorentzian shape, as discussed on p. 216, where it was pointed out that this type of signal does not have a second order derivative. This explains why $v(t)$ in (7.130) is not second order. The use of the second order signal $s(t)$ will be of interest below when discussing the spectral representation of signals with stationary increments.

7.5.4 Asymptotic Properties of the Increments

Let us suppose that the conditions ensuring that the increments $\Delta x(t;\ \tau)$ are second order whatever the value of τ are fulfilled. It is now interesting to discover the behavior of the variance $\sigma^2(\tau)$ when $\tau \to \infty$. For example, the variance $\sigma^2(\tau)$ given by (7.136) and corresponding to Brownian motion is finite but increases to infinity with τ. There are other situations where $\sigma^2(\tau)$ is bounded by a constant, and we shall look at those problems that depend on the form of $\Gamma_v(\nu)$ or $\Gamma_s(\nu)$ given in (7.137). Inserting this expression in (7.135), we obtain

$$\sigma^2(\tau) = \tau^2 \int \Gamma_s(\nu)\, \text{sinc}^2(\nu\tau)\, d\nu + 4 \int \Gamma_s(\nu) \sin^2(\pi\nu\tau)\, d\nu \tag{7.141}$$

where $\Gamma_s(\nu)$ has a finite integral. This expression is the sum of two terms noted $A(\tau)$ and $B(\tau)$. As $\sin^2(\pi\nu\tau) \le 1$, the second term satisfies

$$B(\tau) \le 4 \int \Gamma_s(\nu)\, d\nu = 4\,\sigma_s^2 \tag{7.142}$$

and remains bounded by a quantity independent of τ. In reality $B(\tau)$ can be written in the form

$$B(\tau) = 2 \int \Gamma_s(\nu)\, d\nu - 2 \int \Gamma_s(\nu) \cos(2\pi\nu\tau)\, d\nu \tag{7.143}$$

and by using twice (6.40), we obtain

$$B(\tau) = 2[\gamma_s(0) - \gamma_s(\tau)] \tag{7.144}$$

where $\gamma_s(\tau)$ is the correlation function of $s(t)$.

As seen in (5.31), the last term tends in general to zero and in this case

$$\lim_{\tau \to \infty} B(\tau) = 2\,\sigma_s^2 \tag{7.145}$$

where σ_s^2 is the variance of $s(t)$.

Let us now examine the behavior of the term $A(\tau)$ in (7.141) defined by

$$A(\tau) = \tau^2 \int \Gamma_s(\nu)\,\mathrm{sinc}^2(\nu\tau)\,d\nu \tag{7.146}$$

For this we shall introduce the filter with frequency response

$$G_P(\nu) \triangleq [2\pi j\nu]^{-1} \tag{7.147}$$

If we apply the signal $s(t)$ to the input of this filter, the variance of the output is

$$\sigma_P^2 = \int \frac{\Gamma_s(\nu)}{4\pi^2\nu^2}\,d\nu \tag{7.148}$$

If this quantity is finite, the output $P(t)$ is a second order signal and its derivative is precisely $s(t)$, because the inverse of $G_P(\nu)$ is the derivative filter. In this case we can say that $s(t)$ has a second order integral. On the other hand, if (7.148) is not bounded, it is said that $s(t)$ has no second order integral. It is clear that the existence of such an integral is only due to the properties of $\Gamma_s(\nu)$ in the neighborhood of the frequency $\nu = 0$, which will be the fundamental frequency in the following discussion. This is quite natural, since the behavior of $\sigma^2(\tau)$ for large values of τ depends on that of $\Gamma_s(\nu)$ for small values of ν. In our discussion of the properties of $A(\tau)$ we can now distinguish two different cases.

First case: s(t) has a second order integral. This means that the spectral density

$$\Gamma_P(\nu) = (4\pi^2\nu^2)^{-1}\,\Gamma_s(\nu) \tag{7.149}$$

is integrable and its integral can be denoted σ_P^2. In this case $A(\tau)$ given by (7.146) can be written as

$$A(\tau) = 4 \int \Gamma_P(\nu)\sin^2(\pi\nu\tau)\,d\nu \tag{7.150}$$

and the discussion concerning $B(\tau)$ can be repeated. In other words, we have

$$A(\tau) = 2[\gamma_P(0) - \gamma_P(\tau)] \le 4\sigma_P^2 \tag{7.151}$$

If the function $P(t)$ has regular behavior, its correlation function tends to zero when τ tends to infinity, and

$$\lim_{\tau \to \infty} A(\tau) = 2\,\sigma_P^2 \tag{7.152}$$

In this case, using (7.142), we obtain

$$\lim_{\tau \to \infty} \sigma^2(\tau) = 2(\sigma_s^2 + \sigma_P^2) \tag{7.153}$$

but, more generally, the variance $\sigma^2(\tau)$ is bounded by $4(\sigma_s^2 + \sigma_P^2)$. The case where (7.153) does not hold corresponds to the situation where the power spectrum $\Gamma_s(\nu)$ has some spectral lines, so that the corresponding correlation function has no limit when τ tends to infinity.

In conclusion, it appears that the fact that the signal $s(t)$ has a second order integral ensures that the variance of the increments $\Delta x(t; \tau)$ remain bounded when τ tends to infinity.

But this assumption also introduces a particular form of these increments. As a result of (7.137), if the integral (7.148) is bounded, the same integral where $\Gamma_s(\nu)$ is replaced by $\Gamma_v(\nu)$ is also bounded. This introduces the function $Q(t)$, integral of $v(t)$; its spectral density is of course

$$\Gamma_Q(\nu) = (4\pi^2 \nu^2)^{-1} \Gamma_v(\nu) \tag{7.154}$$

and its variance, noted σ_Q^2, is finite. Inserting this expression in (7.135) and using the same reasoning as before, we obtain

$$\sigma^2(\tau) = 2[\gamma_Q(0) - \gamma_Q(\tau)] \leq 4\sigma_Q^2 \tag{7.155}$$

and under general conditions on the spectrum

$$\lim_{\tau \to \infty} A(\tau) = 2\sigma_Q^2 \tag{7.156}$$

We shall now demonstrate that the function $Q(t)$ appears directly in the expression of the increments. This is suggested by (7.155), which allows us to write $\Delta x(t; \tau)$ as

$$\Delta x(t; \tau) = Q(t) - Q(t - \tau) \tag{7.157}$$

where $Q(t)$ is second order and stationary. It is clear that the variance of $\Delta x(t; \tau)$ is given by (7.155). In order to prove (7.157) from (7.130), we recall that the increment is deduced from the function $v(t)$ by passing it through the filter defined by (7.131). But this filter can be decomposed as a product of the filter defined by (7.147), and the filter with frequency response

$$G_\tau(\nu) = 1 - \exp(-2\pi j \nu \tau) \tag{7.158}$$

By passing $v(t)$ through the filter (7.147) we obtain a second order signal $Q(t)$ called the integral of $v(t)$ and with the power spectrum (7.154). If we pass this signal $Q(t)$ through the filter defined by (7.158), we obtain (7.157). The same reasoning can be used if we start from (7.129) instead of (7.130), which completes the proof.

We shall now summarize and comment on these results. In order to model the increments $\Delta x(t; \tau)$ we start from a second order stationary signal $s(t)$ with power spectrum $\Gamma_s(\nu)$. The increment $\Delta x(t; \tau)$ is deduced from $s(t)$ by passing it through a linear filter the frequency response of which is given by (7.139). This filter can be decomposed into many other filters, as shown in Figure 7.17, which also exhibits the signals $v(t)$ and $Q(t)$ mentioned above.

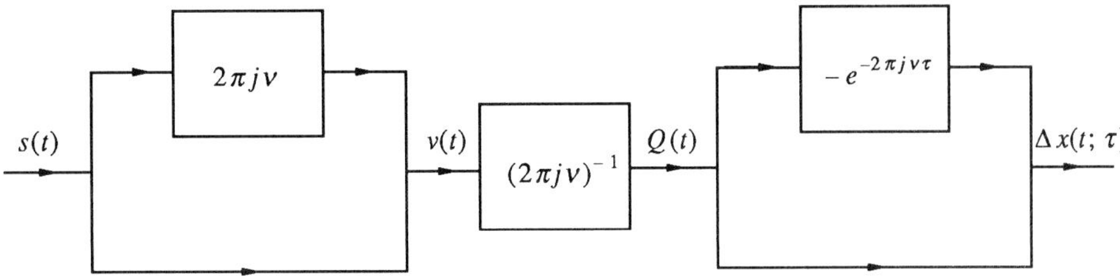

Figure 7.17 Decomposition of the filter $G'(\nu;\tau)$.

The starting assumption to obtain increments $\Delta x(t;\tau)$ with bounded variance when τ tends to infinity is that the signal $s(t)$ has a second order integral, meaning that (7.148) is finite. As a result of this assumption, the function $Q(t)$ in the figure at the output of the integrator is also stationary and second order. The function $v(t)$, however, is not necessarily second order, which introduces the two forms (7.129) and (7.130).

Example 7.4 Generation of some increments. Suppose first that the signal $s(t)$ is band-limited in the frequency range $[B_1, B_2]$. This means that for frequencies outside this interval the power spectrum is null. In consequence, σ_P^2 given by (7.148) is finite, and $\Gamma_v(\nu)$ given by (7.137) is integrable, so that $v(t)$ is second order. Furthermore, $v(t)$ is also band-limited in the same frequency band $[B_1, B_2]$ and $Q(t)$ is second order, which gives the formula (7.157) for $\Delta x(t;\tau)$.

Suppose now that the second order signal $s(t)$ has a power spectrum given by (7.140) for $|\nu| \geq B$ and null for $|\nu| < B$. As a result of (7.137), $\Gamma_v(\nu)$ is equal to c for $|\nu| \geq B$ and null for $|\nu| < B$, and $v(t)$ is a high frequency white noise. Therefore it does not have a finite variance, and (7.130) cannot be used. However, $Q(t)$ is second order and its variance is given by

$$\sigma_Q^2 = \frac{2c}{4\pi^2}\int_B^{\infty}\frac{d\nu}{\nu^2} = \frac{2c}{4\pi^2 B} \tag{7.159}$$

Thus the increments $\Delta x(t;\tau)$ are still given by (7.155) and their variance satisfies (7.156). It is clear in these examples that the basic property is that the power spectrum of $s(t)$ is null in the neighborhood of the null frequency, which ensures that (7.148) is finite.

Second case: s(t) has no second order integral. This means that whereas $s(t)$ is second order, the spectral density (7.149) is not integrable. As $\Gamma_s(\nu)$ has a finite integral equal to the variance of $s(t)$ which is finite because $s(t)$ is second order, the divergence of $\Gamma_P(\nu)$ is due to the low frequencies of $\Gamma_s(\nu)$. More precisely, $\Gamma_s(\nu)$ does not tend to zero for $\nu \to 0$ in order to ensure that $\Gamma_P(\nu)$ is integrable.

Suppose that for ν sufficiently small we can write

$$\Gamma_s(\nu) = \nu^m f(\nu), \ f(0) > 0 \tag{7.160}$$

In order to obtain a signal $s(t)$ with a finite variance, we must have $m > -1$. Furthermore, $s(t)$ has a second order integral as soon as $m > 1$. As a result, the case considered here concerns a power spectrum satisfying (7.160) with

$$-1 < m < +1 \tag{7.161}$$

In particular, the spectrum given by (7.140) and giving the increments of Brownian motion takes the form (7.160) with $m = 0$.

Let us now return to the variance $\sigma^2(\tau)$ given by (7.141) when τ tends to infinity. As the term $B(\tau)$ remains bounded because (7.142) is still valid, we will concentrate our attention on the term $A(\tau)$ defined by (7.146). Introducing the variables $x = \pi\nu\tau$ and $\alpha = 1 - m/2$, we obtain

$$A(\tau) = \pi^{2\alpha-3}\tau^{2\alpha-1}\int\left[\frac{\sin x}{x^{\alpha}}\right]^2 f\left(\frac{x}{\pi\tau}\right)dx \tag{7.162}$$

The condition (7.161) states that $1 < 2\alpha < 3$. Let us define the constant $k(m)$ by

$$k(m) \triangleq \int \frac{\sin^2 x}{x^{2\alpha}}\,dx \tag{7.163}$$

The condition on α ensures that this integral is convergent because the function under integration is bounded by $|x|^{-2\alpha}$ for $x \to \infty$ and by $|x|^{2-2\alpha}$ for $x \to 0$. So for $\tau \to \infty$ we can write

$$A(\tau) \cong \pi^{-(m+1)}\,k(m) f(0)\,\tau^{1-m} \tag{7.164}$$

As $A(\tau)$ is the first term appearing in (7.141), the variance $\sigma^2(\tau)$ increases to infinity with τ and its behavior takes the form

$$\sigma^2(\tau) \cong c\,\tau^{1-m} \tag{7.165}$$

where m satisfies (7.161). This term is proportional to τ^k where $0 < k < 2$. A case of particular interest appears when $m = 0$, which simply means that $\Gamma_s(0)$ is positive. In this case the *asymptotic* behavior of the variance $\sigma^2(\tau)$ is the same as that of Brownian motion. Do not forget that for Brownian motion (7.136) is valid whatever the value of τ, while (7.165) is only valid when τ tends to infinity.

Let us conclude by returning to the representation in Figure 7.17. The fact that $s(t)$ is not integrable means that $v(t)$ also is not integrable, and the function $Q(t)$ does not exist, which makes it impossible to write $\Delta x(t;\,\tau)$ as in (7.157).

7.5.5 Properties of the Signal

Up to this point we have mainly discussed properties of the increments $\Delta x(t;\,\tau)$. We shall now deduce some properties of the signal $x(t)$ itself, resulting from those of its increments. For this let t_0 be an arbitrary time instant. It is possible to write

$$x(t) = x(t_0) + \Delta x(t;\, t - t_0) \tag{7.166}$$

Suppose now that the increments correspond to the first case studied in the previous section. Then (7.157) can be applied, giving

$$x(t) = Q(t) - [Q(t_0) - x(t_0)] \tag{7.167}$$

Introducing the RV X_0 equal to $x(t_0) - Q(t_0)$, we deduce that

$$x(t) = Q(t) + X_0 \tag{7.168}$$

It then appears that $x(t)$ is the sum of a second order stationary function $Q(t)$ and an RV X_0. It is important to note that, as $Q(t)$ and X_0 are correlated, $x(t)$ is in general not stationary. But $x(t)$ is almost stationary because the difference $x(t) - X_0$ is stationary. Furthermore, for $|t - t_0|$ sufficiently large, the influence of the RV disappears and $x(t)$ becomes stationary. As $Q(t)$ and X_0 are second order it is also obvious that the variance of $x(t)$ is bounded even when $|t|$ tends to infinity.

In conclusion, the first class of signals $x(t)$ with stationary increments are those deduced from a stationary signal by addition of an RV correlated with this signal. This RV disappears when the increments are introduced.

Suppose now that the increments correspond to the second class studied in the previous section. In this case (7.157) is no longer valid and $x(t)$ cannot be written as (7.167), while (7.166) is always valid. It is then clear that $x(t)$ is essentially nonstationary, which means that it cannot be transformed into a stationary signal only by the addition of an RV. Furthermore, the variance of $x(t)$ tends to infinity with behavior such as $(t - t_0)^k$, $0 < k < 2$. We then say that $x(t)$ is a *diffusion process* with stationary increments, by an extension of the term "diffusion" often used in Brownian motion (see Section 7.3.3). Note here that if we are only interested in the behavior of the variance of the signal for large values of t, a behavior in $(t - t_0)$ corresponding to $k = 1$ is not specific to Brownian motion but can appear in other cases.

7.5.6 Spectral Representations

Again we are obliged to distinguish the representation of the increments $\Delta x(t;\, \tau)$ and that of the signal itself.

Let us begin with the case of the increments. We have seen that they can be deduced from a second order stationary signal by using the filter represented in Figure 7.17. Suppose that $s(t)$ has a spectral representation given by (6.5) or

$$s(t) = \int d\bar{S}(\nu)\, e^{2\pi j \nu t} \tag{7.169}$$

where the increments $d\bar{S}(\nu)$ are *uncorrelated* because of the stationarity, or satisfy (6.15) and (6.37), which especially implies (6.43). Using (6.9), we can write the spectral representation of $\Delta x(t;\, \tau)$ in the form

$$\Delta x(t;\, \tau) = \int d\bar{S}(\nu)\, [1 + 2\pi j \nu][\, 2\pi j \nu]^{-1} [1 - e^{-2\pi j \nu \tau}]\; e^{2\pi j \nu t} \tag{7.170}$$

This expression is quite general and implies the relation (7.141) concerning the variance of the increments.

Consider now the case of the signal itself. From (7.166) and (7.170) we deduce that

$$x(t) = x(t_0) + \int d\bar{S}(\nu) \frac{1 + 2\pi j\nu}{2\pi j\nu} \left[1 - e^{-2\pi j\nu(t - t_0)}\right] e^{2\pi j\nu t} \tag{7.171}$$

Suppose that $s(t)$ has a finite integral, or that (7.148) is finite. In this case it is possible to introduce a second order stationary signal $Q(t)$ defined by

$$Q(t) = \int d\bar{S}(\nu) \frac{1 + 2\pi j\nu}{2\pi j\nu} e^{2\pi j\nu t} \tag{7.172}$$

and $x(t)$ can be written in terms of $Q(t)$ with (7.168) introducing a random variable X_0 correlated with $Q(t)$ and then with $d\bar{S}(\nu)$.

If, however, (7.148) is not finite, $Q(t)$ is not a second order signal and is not harmonizable, which makes a spectral representation *with uncorrelated increments* impossible. This does not mean that there is no spectral representation, but that it is of lesser interest. The main problem is that the arbitrary time instant t_0 cannot be remote from infinity in the past, because in that case $x(t)$ becomes infinite. To overcome this difficulty, suppose that the signal $x(t)$ is null for $t \leq t_0$ and given by

$$x(t) = \int_{t_0}^{t} v(\theta)\, d\theta \tag{7.173}$$

for $t > t_0$, where $v(t)$ can be considered either as stationary or semistationary as defined by 5.3.3 (**a**). Thus $x(t)$ is a signal with stationary increments only if t and $t - \tau$ in (7.126) are greater than t_0. Introducing the unit step function $u(t - t_0)$ equal to zero for $t \leq t_0$, and to one otherwise, we can write

$$x(t) = \int_{-\infty}^{t} u(t - t_0)\, v(\theta)\, d\theta \tag{7.174}$$

valid whatever t. Suppose that $v(t)$ is harmonizable, meaning that it has a representation with uncorrelated increments as in (6.5) . It is clear that $u(t - t_0)\, v(t)$ is then also harmonizable, but that the increments are no longer uncorrelated. Inserting this representation in (7.174) gives us a spectral representation, but with correlated increments, which is not of great interest. In conclusion, the spectral representation of signals with stationary increments is interesting only in the first case considered in the previous section.

7.5.7 Extension to Higher-Order Increments

The increment $\Delta x(t;\, \tau)$ defined by (7.126) can be called a first order increment of $x(t)$. By taking the increment of $\Delta x(t;\, \tau)$ we can introduce the second order increment of $x(t)$, defined as

$$\Delta_2 x(t; \tau_1, \tau_2) = \Delta x(t; \tau_1) - \Delta x(t - \tau_2; \tau_1) \tag{7.175}$$

It can be expressed in terms of $x(t)$ by using (7.126), which gives

$$\Delta_2 x(t; \tau_1, \tau_2) = x(t) - x(t - \tau_1) - x(t - \tau_2) + x(t - \tau_1 + \tau_2) \tag{7.176}$$

This procedure can easily be generalized up to the order n. In particular, if all the delays τ_i are equal to τ, we obtain

$$\Delta_n x(t; \tau) = \sum_{k=0}^{n} \binom{n}{k} (-1)^k \, x(t - k\tau) \tag{7.177}$$

Similarly, the expression of the increments corresponding to (7.130) becomes

$$\Delta_n x(t; \{\tau_i\}) = \int_{t-\tau_1}^{t} d\theta_1 \int_{\theta_1 - \tau_2}^{\theta_1} d\theta_2 \ldots \int_{\theta_{n-1} - \tau_n}^{\theta_{n-1}} d\theta_n \, v(\theta_n) \tag{7.178}$$

which is simpler to write in the frequency domain, because (7.131) is replaced by

$$G_n(\nu; \{\tau_i\}) = (2\pi j \nu)^{-n} \prod_{i=1}^{n} \left[1 - e^{-2\pi j \nu \tau_i} \right] \tag{7.179}$$

We deduce from this expression that (7.133) must be replaced by

$$G_n(0; \{\tau_i\}) = \tau_1 \, \tau_2 \ldots \tau_n \tag{7.180}$$

Finally, the expression (7.166), where $\Delta x(t; t - t_0)$ is written with (7.130), becomes

$$x(t) = \sum_{k=0}^{n-1} v_k(t_0) \frac{(t - t_0)^k}{k!} + \int_{t_0}^{t} d\theta_1 \int_{t_0}^{\theta_1} d\theta_2 \ldots \int_{t_0}^{\theta_{n-1}} d\theta_n \, v_n(\theta) \tag{7.181}$$

where $v_i(t)$ is the derivative of the order i of $x(t)$. In particular, if $v_n(\theta) = 0$, we have

$$x(t) = \sum_{k=0}^{n-1} A_k \frac{(t - t_0)^k}{k!} \tag{7.182}$$

where the A_ks are second order RVs.

The principles of the discussion concerning the asymptotic properties of the variance and of the spectral representation of the signal are similar to the case of first order increments. Part of this discussion appears in problems proposed at the end of this chapter.

7.6 SPHERICALLY INVARIANT AND CIRCULAR SIGNALS

The concept of spherically invariant RVs was mentioned in previous chapters, especially in Chapter 4 where their PDF was defined by (4.48) and their regression was analyzed

(Section 4.8.2). The concept of *normal* circular complex RVs was defined by (4.152), and the concept of circularity was extended to non normal RVs in a broader context when studying higher-order spectra in Section 6.10. The purpose of this section is to apply these ideas more systematically to random signals, rather than only to random variables.

7.6.1 Real Spherically Invariant Signals

These signals are a natural extension of normal signals extensively studied in Section 7.4. Their definition exactly follows the pattern used in 7.4.1 and starts from the definition of spherically invariant (SI) *random vectors* which will be precisely stated here. The definition introduced by (4.48) is too narrow for the discussion which follows. To understand this clearly, suppose that the function $f(.)$ in (4.48) has an exponential form. In this case $p(\mathbf{x})$ becomes a normal distribution in the form $N(\mathbf{0}, \mathbb{I})$. In other words, the PDF (4.82) where $\mathbf{m} = \mathbf{0}$ characterizing an $N(\mathbf{0}, \mathbb{\Gamma})$ RV cannot be considered as belonging to the class of SI random vectors. It is for this reason that we must broaden the definition (4.48), and it will be more interesting to work with the characteristic function than with the PDF.

A zeromean random vector of $\mathbb{R}^n$ is said to be *spherically invariant of type* 1, or SI 1, if its characteristic function defined by (4.59) is given by

$$\phi_n(\mathbf{u}) = F\left(\frac{1}{2}\,\mathbf{u}^T\mathbb{\Gamma}\mathbf{u}\right) \tag{7.183}$$

where $\mathbb{\Gamma}$ is an $n \times n$ symmetric and positive definite matrix. The function $F(s)$ in this definition is not at all arbitrary. Firstly, it must satisfy $F(0) = 1$, which results from the definition (4.59). But the most important constraint on $F(s)$ is that $\phi(\mathbf{u})$ must be a non-negative definite function, in order to ensure that its Fourier transform given by (4.61) is non-negative. This is a very restrictive constraint. It is clear that (4.86) with $\mathbf{m} = \mathbf{0}$ is a particular case of (7.183).

If the function $F(s)$ has an integral representation in the form

$$F(s) = \int_0^\infty f(a)\,e^{-as}\,da \tag{7.184}$$

where $f(a)$ is a real function, the random vector $\mathbf{X}$ is said to be SI 2. It is clear that $F(s)$ appears as the one-sided Laplace transform of $f(a)$. The condition $F(0) = 1$ implies that $f(a)$ is a normalized function.

If the function $f(a)$ in (7.184) is non-negative, the random vector $\mathbf{X}$ is said to be SI 3. In this case it becomes obvious that $\phi_n(\mathbf{u})$ defined by (7.183) is non-negative definite. In fact it can be written as

$$\phi_n(\mathbf{u}) = \int_0^\infty f(a)\exp\left[-\frac{1}{2}\,a\,\mathbf{u}^T\mathbb{\Gamma}\mathbf{u}\right]da \tag{7.185}$$

The exponential term in this integral is non-negative definite, because it is the characteristic function of a vector $N(\mathbf{0}, a\,\mathbb{\Gamma})$. As $f(a) \geq 0$, the integral is also non-negative definite. Another way to verify this point is to calculate the PDF associated with (7.185). As the Fourier transform of the exponential term is given by (4.82), we immediately deduce that

$$p(\mathbf{x}) = (2\pi)^{-n/2}\,\{\det(\mathbb{\Gamma})\}^{-1/2} \int_0^\infty a^{-n/2} \exp\left\{-\frac{1}{2a}\,\mathbf{x}^T\mathbb{\Gamma}^{-1}\mathbf{x}\right\} f(a)\,da \quad (7.186)$$

It is clear that if $f(a) \geq 0$ we also have $p(\mathbf{x}) \geq 0$. Finally, there is a very simple representation of the SI 3 random vectors. Consider a random vector $\mathbf{Y}(\omega)$ which is $N(\mathbf{0}, \mathbb{\Gamma})$ and a positive random variable $A(\omega)$ independent of $\mathbf{Y}(\omega)$ and defined by the PDF $f(a)$ appearing in (7.185) and (7.186). Let $\mathbf{X}(\omega)$ be the vector defined by

$$\mathbf{X}(\omega) = A^{1/2}(\omega)\,\mathbf{Y}(\omega) \quad (7.187)$$

which is similar to (7.96). It is clear that, conditionally to $A(\omega) = a$, the vector $\mathbf{X}(\omega)$ is $N(\mathbf{0}, a\,\mathbb{\Gamma})$ and (7.186) is similar to (4.24) giving the PDF of the random vector $\mathbf{X}(\omega)$. In conclusion, an SI 3 random vector can be represented as the product of a normal vector by a positive scalar RV.

However, it is perfectly possible to construct examples of PDF $p(\mathbf{x})$ written as in (7.186) and using a function $f(a)$ taking negative values. The method is similar to that introduced in Problem 2.8 concerning mixtures of Poisson distributions.

It should be noted that (7.186) can be written as

$$p(\mathbf{x}) = g(\mathbf{x}^T\,\mathbb{\Gamma}^{-1}\mathbf{x}) \quad (7.188)$$

and if $\mathbb{\Gamma} = \mathbb{I}$ we have a form similar to (4.48) justifying the term "spherically invariant." In fact, the surfaces defined by $p(\mathbf{x}) = c$ are hyperspheres of $\mathbb{R}^n$, which is no longer true in (7.188). But it is always possible to make a linear transformation similar to (4.50), transforming (7.188) into (4.48). This transformation corresponds to a whitening procedure as described in Section 3.5.3.

Finally, it is important to see how SI random vectors are transformed by linear operations. Consider first an SI 1 random vector $\mathbf{X}$ defined by a characteristic function such as (7.183). Suppose that $\mathbf{X}$ is transformed into $\mathbf{Y}$ by a linear transformation $\mathbf{y} = \mathbb{A}\mathbf{x}$. The characteristic function of $\mathbf{Y}$ is given by (4.68), and by inserting this result into (7.183), we obtain the characteristic function of $\mathbf{Y}$ in the form

$$\phi_Y(\mathbf{v}) = F\left(\frac{1}{2}\,\mathbf{v}^T\mathbb{A}\mathbb{\Gamma}\mathbb{A}^T\mathbf{v}\right) = F\left(\frac{1}{2}\,\mathbf{v}^T\mathbb{\Gamma}_y\mathbf{v}\right) \quad (7.189)$$

This shows that, as for normal distribution, an SI 1 random vector remains SI 1 after linear transformation.

This is, however, quite different in the normal case. It is obvious that two independent normal random vectors $\mathbf{X}$ and $\mathbf{Y}$ are jointly normal. This is no longer true for

SI random vectors. In fact, the joint characteristic function defined as in (4.63) takes the form

$$\phi(\mathbf{u}, \mathbf{v}) = F_X\left(\frac{1}{2}\,\mathbf{u}^T\Gamma_X\mathbf{u}\right)F_Y\left(\frac{1}{2}\,\mathbf{u}^T\Gamma_Y\mathbf{v}\right) \tag{7.190}$$

which in general cannot be written as in (7.183). This is, of course, possible if F_X and F_Y take an exponential form, which means that **X** and **Y** are normal. In consequence the sum of two independent SI random vectors is in general no longer SI.

Let us now define an SI *random signal.* The procedure is the same as in 7.4.1. A signal $x(t)$ is said to be SI if all the random vectors appearing in the family of finite-dimensional distributions are SI. That is, whatever the value of n, the characteristic functions are given by (7.183) where F does not depend on n. This definition immediately introduces a constraint on the function $F(s)$ in (7.183). In fact, as Γ is assumed to be positive-definite, it can be factorized as in (3.99) and its inverse becomes (3.104). Introducing the vector $\mathbf{v} = (\mathbb{A}^T)^{-1}\mathbf{u}$, and noting that $\mathbb{A}^T = \mathbb{A}^H$ because Γ is real, we can write (7.183) in the form

$$\phi_n(\mathbf{v}) = F\left(\frac{1}{2}\,\mathbf{v}^T\mathbf{v}\right) \tag{7.191}$$

So the function $F(s)$ is such that, whatever the value of n, $F(\mathbf{v}^T\mathbf{v})$ is a non-negative definite function. As a result of a theorem due to Schoenberg(1), this property can appear whatever n only if $F(s)$ can be written as in (7.184) with a non-negative function $f(a)$. In conclusion, an SI random signal necessarily introduces SI 3 random vectors, and, by using the interpretation of these vectors which appear in (7.187), we can say that an SI random signal $x(t)$ can be written as

$$x(t) = A^{1/2}y(t) \tag{7.192}$$

where $y(t)$ is a normal signal and A a positive RV independent of $y(t)$. It is interesting to note that this type of signal was introduced for (7.96) and in the discussion following that expression.

From a physical point of view, SI random signals can be considered to be obtained by a random amplitude modulation of normal signals.

7.6.2 Complex Circular Signals

Spherically invariant signals can be considered as extensions of real normal signals. Using exactly the same procedure, complex circular signals can be introduced as extensions of normal circular signals. As the reasoning is similar, we shall present only the principal ideas.

---

(1) I. J. Schoenberg, "Metric spaces and completely monotone functions," *Annals of Mathematics*, 39, pp. 811-848, 1938.

Consider first the case of random vectors. The characteristic function of a normal circular vector is given by (4.134). Consequently the characteristic function of a circular random vector can be written as

$$\phi(\mathbf{u}, \mathbf{v}) = \phi(\mathbf{w}) = F\left(\frac{1}{4}\,\mathbf{w}^H \Gamma \mathbf{w}\right) \tag{7.193}$$

As previously, the function $F(s)$ can have an integral representation, and the corresponding PDF analog to (7.186) takes the form deduced from (4.152)

$$p(\mathbf{z}) = (\pi)^{-n}[\det(\Gamma)^{-1}] \int_0^\infty a^{-n} \exp\left\{-\frac{1}{a}\,\mathbf{z}^H \Gamma^{-1} \mathbf{z}\right\} f(a)\, da \tag{7.194}$$

If $f(a) \geq 0$, the vector $\mathbf{Z}$ can be written as

$$\mathbf{Z}(\omega) = A^{1/2}(\omega)\,\mathbf{U}(\omega) \tag{7.195}$$

where $\mathbf{U}$ is $N_c(\mathbf{0}, \Gamma)$ and $A(\omega)$ is a positive RV, independent of $\mathbf{U}(\omega)$.

The same procedure can be applied to circular signals, and if we again use Schoenberg's result, we deduce that a complex circular signal can be written as

$$z(t) = A^{1/2} u(t) \tag{7.196}$$

where $u(t)$ is complex normal circular. Finally, the physical interpretation is exactly the same as for real SI signals.

Note that the *higher-order moments* of a circular signal have the same properties as those of normal signals. In fact, as $u(t)$ is normal circular, the only non null moments of $z(t)$ take the form of (7.106). This is a direct consequence of (7.196). But this property is not specific to characteristic functions as in (7.193). Using (4.177), we can write

$$\mathbf{w}^H \Gamma \mathbf{w} = \mathrm{Tr}(\Gamma \mathbf{w} \mathbf{w}^H) \tag{7.197}$$

and $\phi(\mathbf{w})$ defined by (7.193) is then only a function of $\mathbf{w}\mathbf{w}^H$. It is possible to show that all the characteristic functions $\phi(\mathbf{w})$ which depend only on $\mathbf{w}\mathbf{w}^H$ have the property that the only non-null moments have the form which appears in (7.106). This is especially the case of the higher-order moments of a signal $z(t) = z_1(t) + z_2(t)$, the sum of two independent circular signals. However $z(t)$ is in general no longer complex circular, but its only non-null moments take the form (7.106).

PROBLEMS

7.1 Complex white noises. Let $x(t)$ be a complex signal written as $x_1(t) + jx_2(t)$, where $x_1(t)$ and $x_2(t)$ are two real, zeromean and jointly stationary signals. These two signals can also be considered as the two components of a real vector $\mathbf{x}(t)$. The second order properties of $x(t)$ are specified by the correlation function $\gamma(\tau) = E[x(t)x^*(t - \tau)]$ and by the second order moment $m(\tau) =$

$E[x(t)x(t - \tau)]$. Also let $\mathbb{\Gamma}(\tau)$ be the correlation matrix of the vector $\mathbf{x}(t)$ defined by $\mathbb{\Gamma}(\tau) = E[\mathbf{x}(t)\mathbf{x}^T(t - \tau)]$ and whose matrix elements are written as $\gamma_{ij}(\tau)$, $1 \leq i$ or $j \leq 2$.

(a) Calculate $\gamma(\tau)$ and $m(\tau)$ in terms of the $\gamma_{ij}(\tau)$s.

(b) Conversely calculate $\mathbb{\Gamma}(\tau)$ in terms of $\gamma(\tau)$ and $m(\tau)$.

(c) Calculate the spectral matrix $\mathbb{\Gamma}(\nu)$ in terms of the power spectrum $\Gamma(\nu)$ of $x(t)$ and of the Fourier transforms $M_1(\nu)$ and $M_2(\nu)$ of $m_1(\tau)$ and $m_2(\tau)$, real and imaginary parts of $m(\tau)$.

(d) What is the form of $\mathbb{\Gamma}(\nu)$ when $x(t)$ is real?

(e) Show that $M_1(\nu)$, $M_2(\nu)$, and $\Gamma(\nu)$ must satisfy a condition, and explain it.

(f) A complex signal is said to be white if $\Gamma(\nu)$ is constant. Give the most general form of the matrix $\mathbb{\Gamma}(\nu)$ associated with a complex white signal.

(g) A vector signal is said to be white if $\mathbb{\Gamma}(\nu)$ is a constant matrix. What are the relationships between a complex white signal and a white vector?

(h) What is the structure of a complex white signal for which $m(\tau) = 0$?

7.2 Let $z(t)$ be the complex signal defined for $t > 0$ and equal to $\exp[j\omega_0 t + \phi(t)]$, where $\phi(t)$ is a random phase. Suppose that $\phi(t)$ is a zeromean Brownian motion characterized by a diffusion constant c.

(a) Calculate $E[z(t)]$ and its limit for large values of t.

(b) Calculate the covariance function $\gamma(t_1, t_2)$ and show that it tends to become a correlation function for large values of t_1 and t_2.

(c) Extend the previous results to the case of the signal $x(t) = \cos[\omega_0 t + \phi(t)]$, $t > 0$.

7.3 Let $x(t)$ be a zeromean Brownian motion satisfying $x(0) = 0$ and with a diffusion coefficient $c(t)$. Let $y(t)$ be the signal defined by $y(t) = tx(t)$ for $t > 0$.

(a) Calculate the covariance function of $y(t)$.

(b) Calculate the conditional expected values $E[y(t_2)|y(t_1)]$ and $E[y^2(t_2)|y(t_1)]$ for $t_1 < t_2$.

(c) Consider the two increments $\Delta y(t_1; \tau_1)$ and $\Delta y(t_2; \tau_2)$, as defined by (7.39), and suppose that $t_2 > t_1 + \tau_1$. Calculate the covariance between these two increments.

(d) Is it possible to find a function $c(t)$ ensuring that $x(t)$ is second order and that $y(t)$ is still a signal with independent increments?

7.4 Suppose that $x(t)$ and $y(t)$ are defined as in the previous problem and that the diffusion coefficient $c(t)$ is time-independent.

(a) Calculate the characteristic function of the pair of RVs defined by $X = y(t_1)$ and $Y = y(t_2)$, $t_1 < t_2$.

(b) Calculate the regression $r(x) = E(Y|x)$.

(c) Compare this result with (7.48) valid for a Brownian motion.

7.5 Let $x(t)$ be a zeromean Brownian motion satisfying $x(0) = 0$ and with a diffusion constant equal to 1. Suppose that $x(T) = \lambda$, where T is a given positive instant and λ an arbitrary number. For any positive t, calculate the distribution of $x(t)$ conditional to the fact that $x(T) = \lambda$.

7.6 Let $x(t)$ be the signal studied in Problem 6.12 and suppose that the function $\bar{X}(\nu)$ appearing in this problem is a non-stationary Brownian motion with a diffusion coefficient $f(\nu)$. Let $y(t) = |x(t)|^2 - E\{|x(t)|^2\}$ and let $z_T(t)$ be the time average of $y(t)$ defined by (6.93).

(a) Calculate the covariance function $\gamma_y(t_1, t_2)$ of $y(t)$. Express this covariance in terms of the correlation function $\gamma(\tau)$ of $x(t)$.

(b) Calculate $E\{z_T(t)\}$ and $E\{z_T^2(t)\}$.

(c) As in Section 6.6, discuss the behavior of $z_T(t)$ when T tends to infinity. It may be appropriate to express the variance of $z_T(t)$ in terms of $f(\nu)$.

7.7 **Some properties of Gaussian narrowband signals.** Let $x(t)$ be a zeromean, real, stationary, narrowband, and normal signal, characterized by its correlation function $\gamma(\tau)$ or its power spectrum $\Gamma(\nu)$. We assume that this spectrum is zero outside a frequency band $[\nu_0 - \Delta\nu, \nu_0 + \Delta\nu]$, with $\Delta\nu/\nu_0 << 1$, and symmetric with respect to the mean frequency ν_0.

(a) Calculate the PDF of its instantaneous amplitude $a(t)$ and phase $\phi(t)$. Calculate the PDF of the square of its amplitude, sometimes called instantaneous intensity.

(b) Let $\alpha(t)$ be the complex envelope of $x(t)$ defined by (6.124). Calculate the PDF of the 4 dimensional RV $m(t_1)$, $m(t_2)$, $n(t_1)$, $n(t_2)$, where $m(t)$ and $n(t)$ are the quadrature components of $x(t)$ appearing in (6.125).

(c) Deduce from the previous calculation the PDF of $a(t_1)$, $a(t_2)$ and of the RV $\phi(t_2) - \phi(t_1)$.

7.8 Let $x(t)$ and $y(t)$ be two zeromean, real, stationary, and normal signals. We assume that they are independent and have the same correlation function $\gamma(\tau)$.

(a) What is the condition on $\gamma(\tau)$ ensuring the existence of the derivatives $x'(t)$ and $y'(t)$?

(b) Calculate the probability distributions of the four RVs $[x(t), y(t), x'(t), y'(t)]$.

(c) Deduce from this result the probability distribution of the pair $[a(t), a'(t)]$, where $a(t)$ is the amplitude $[x^2(t) + y^2(t)]^{1/2}$ and $a'(t)$ its derivative.

7.9 Let $x(t)$ be a zeromean, real, stationary, and normal signal, characterized by its power spectrum $\Gamma(\nu)$. Calculate the power spectrum $S(\nu)$ of the signal $y(t) = [x(t) + a\cos(\omega_0 t + \phi)]^2$, where ϕ is an RV independent of $x(t)$ and uniformly distributed in 0, 2π. Represent this spectrum when, for $\nu > 0$, $\Gamma(\nu)$ is constant in the frequency band $\nu_0 - B$, $\nu_0 + B$ and zero otherwise, $\omega_0 = 2\pi\nu_0$.

7.10 Let $x(t)$ be a zeromean, real, stationary, and normal signal, characterized by its power spectrum $\Gamma(\nu)$. From this signal we deduce $y(t) = x^3(t)$.

(a) Calculate the mean value $E[y(t)]$ of $y(t)$.

(b) Calculate the correlation function of $y(t)$ and express it in terms of the correlation function of $x(t)$. For this calculation use the expression of the higher-order moments of a normal zeromean RV and show first that there are only 2 distinct terms in the correlation function of $y(t)$.

(c) Calculate the power spectrum of $y(t)$ when $\Gamma(\nu)$ is equal to c for $|\nu| < B$ and zero otherwise. Represent the result in a figure.

7.11 Consider the same signal as in the previous problem. Let $z(t)$ be its analytic signal and $s(t) = z^3(t)$.

(a) Calculate the correlation function and the power spectrum of $s(t)$. Is $s(t)$ an analytic signal?

(b) Calculate the second order moment $E[s(t)s(t-\tau)]$ and compare the result with that obtained for an analytic signal.

7.12 Let $x(t)$ be a zeromean, real, stationary, and normal signal, characterized by its correlation function $\gamma(\tau)$. Calculate the mean value and the correlation function of the signal $y(t)$ deduced from $x(t)$ by a rectifier and defined by $y(t) = |x(t)|$. For this, use the results of Problem 4.45.

7.13 Make the same calculation as in the previous problem when $y(t) = u[x(t)]x(t)$, where $u(x)$ is the unit step function. Represent in a figure the general behavior of the signals $y(t)$ studied in this and in the previous problem.

7.14 Let $x(t)$ be a zeromean, real, stationary, and normal signal and $y(t) = x^2(t) - E[x^2(t)]$. Calculate the third and fourth order moments of $y(t)$, as defined by (5.42). Calculate the Fourier transforms $M_3(\mathbf{f})$ and $M_4(\mathbf{f})$ defined by (6.143) and show that they are distributed only on the stationary manifold (6.150). Show that the density associated with $M_4(\mathbf{f})$ on the normal manifolds is a normal density, as defined in Section 7.4.4.

7.15 Ergodicity of normal signals. It is shown in Section 6.6 that mean square ergodicity for a mean value experiment means that there is no spectral line at the frequency zero.

(a) Using the results of the previous problem, show that the power spectrum of $y(t)$ has no spectral line at zero frequency if $x(t)$ has no spectral line at any frequency.

(b) Deduce that this is the condition of ergodicity for the measurement of the variance of $x(t)$.

(c) Generalizing this point, show that a normal signal is ergodic for the measurement of any moment such $x^k(t)$ if its power spectrum contains no spectral line, which means that its spectral distribution in (6.38) has no stepwise discontinuity.

7.16 Ergodicity and normal manifolds. Let $x(t)$ be a stationary signal with a bounded power spectrum, which ensures ergodicity for the expectation. We want to study conditions ensuring ergodicity for the variance. For this, we assume that the function $M(\mathbf{f})$ defined by (6.143) can be decomposed as a sum of two terms. The first is a distribution on the stationary manifold defined by (6.150) with a continuous density $A(f_1, f_2, f_3)$ such that $|A(f_1, f_2, f_3)|$ has a finite integral. The second is given by (7.88) where $\Gamma(f)$ is replaced by $A(f)$, which is an even, continuous, and square integrable function. Let $S(f)$ be the power spectrum of $y(t) = x^2(t)$.

(a) Show that ergodicity for the variance is ensured if $S(f)$ has no spectral line at the zero frequency.

(b) Calculate the contribution of the term $A(f_1, f_2, f_3)$ to $S(f)$ and show that it does not introduce a spectral line at the zero frequency.

(c) Make the same calculation for the contributions of the manifolds (7.90) and (7.91) and deduce the same conclusion.

(d) Calculate the contribution to $S(f)$ of the manifold (7.89) and show that it can introduce a spectral line at the zero frequency.

(e) Find a condition on $A(f)$ ensuring that this spectral line disappears and show that this condition is verified if $x(t)$ is a normal signal.

(f) Compare this result with those obtained in the previous two problems.

7.17 Continuous time level crossing. Let $x(t)$ be a random signal similar to that represented in Figure 1.2. Consider an arbitrary number x_0. The difference $x(t) - x_0$ can be positive or negative and a crossing of the level x_0 appears each time this difference changes its sign. Our purpose is to find the mean number of level crossings in an arbitrary time interval.

(a) Let $a(t; x_0)$ be the signal defined by $u[x(t) - x_0]$, where $u(x)$ is the unit step function. Show that the level crossings can be deduced from the derivative $a'(t; x_0)$ of $a(t; x_0)$. Calculate this derivative.

(b) Show that the number $N(T)$ of level crossings in the interval $[0, T]$ can be written as an integral from 0 to T of the function $b(t) = |a'(t; x_0)|$.

(c) Assuming that $x(t)$ is stationary, show that the expected value of $N(T)$ is $TE[b(t)]$.

(d) To complete the calculation we assume that $x(t)$ is zeromean, stationary, normal and with a second order first derivative $x'(t)$. Taking $X = x(t)$ and $Y = x'(t)$, show that X and Y are two zeromean, independent, and normal RVs characterized by their variances σ_x^2 and σ_y^2. Deduce from this property the expected value of $N(T)$.

(e) Express this expectation in terms of the power spectrum of $x(t)$.

7.18 Discrete time zero crossing. Let $x[k]$ be a random signal similar to those represented in Figures 7.1 to 7.6. By extension of the terminology used in the CT case, it is said that there is a zero crossing of $x[k]$ at time k if $x[k]$ and $x[k-1]$ do not have the same sign. In order to analyze the problem we introduce the function $a[k] = x[k]\,x[k-1]$.

(a) Show that there is a crossing at k if $a[k] < 0$.

(b) Let $b[k]$ be the signal equal to 1 if $a[k] < 0$ and to 0 otherwise. Express the number $N(T)$, T integer, of zerocrossings in the interval $[0, T]$ in terms of $b[k]$.

(c) Assuming that $x[k]$ is stationary, calculate the expected value $E[N(T)]$. Taking $X = x[k]$ and $Y = x[k-1]$, express $E[N(T)]$ in terms of the PDF of the pair of RVs X and Y.

(d) To complete the calculation we assume that $x[k]$ is zeromean, stationary, and normal. Using the result of Problem 4.44(a), calculate $E[N(T)]$. Give an interpretation of the result in terms of the value $\gamma[1]$ of the correlation function of $x[k]$.

7.19 Consider the complex signal $z(t) = \exp j\left[\omega_0 t + \int_0^t x(\theta)\, d\theta + \phi\right]$, where $x(t)$ is a zeromean, real, stationary, and normal signal, characterized by its correlation function $\gamma_x(\tau)$ or its power spectrum $\Gamma_x(\nu)$ and where ϕ is a random phase independent of $x(t)$ and uniformly distributed over 0, 2π.

(a) Show that $z(t)$ is a zeromean signal.

(b) Calculate the correlation function $\gamma_z(\tau)$ of $z(t)$ in terms of the power spectrum $\Gamma_x(\nu)$.

(c) Calculate the power spectrum $\Gamma_z(\nu)$ of $z(t)$ in the case where $\Gamma_x(\nu) = c$.

(d) Make the same calculation in the opposite situation, or when $\Gamma_x(\nu) = c\delta(\nu)$.

(e) Calculate the fourth order moments of $z(t)$ defined by extending (5.42) to the complex case. Express these moments in terms of $\gamma_x(\tau)$.

7.20 Statistical optics experiments. Suppose that the optical field in a given point of the space is represented by a real function $x(t)$. In statistical optics this function must be considered as random, the randomness coming from various origins. It also appears very frequently that this function is narrowband and we assume here that its power spectrum $\Gamma(\nu)$ is constant for $\nu_0 - B < |\nu| < \nu_0 + B$. Most optical receivers measure the light intensity $I(t)$. This intensity is defined as $|z(t)|^2$ where $z(t)$ is the analytic signal of $x(t)$.

(a) Calculate the correlation functions $\gamma_x(\tau)$ of $x(t)$ and $\gamma_z(\tau)$ of $z(t)$.

(b) *Interference experiments.* By some physical device we can realize a superposition defined by $y_d(t) = x(t) + x(t-d)$, where d is a delay depending on the interference system. Calculate the analytic signal of $y_d(t)$ and express the expectation of the light intensity in terms of the correlation function of the field. Represent in a figure the evolution of the mean intensity as a function of the delay d.

(c) *Intensity correlation experiments.* In these experiments we measure the expectation of the product $I(t)I(t-d)$ or the intensity correlation function $\gamma_I(d)$. Assuming that the field $x(t)$ is a normal signal, calculate $\gamma_I(d)$ in terms of the correlation function of the field. Represent in a figure the evolution of $\gamma_I(d)$ as a function of the delay d.

(d) By comparing the figures obtained in the two cases, explain the main differences between the two kinds of experiments.

(e) In order to make this comparison clearer, apply the above results when the analytic signal of the optical field can be expressed as the function $z(t)$ studied in the previous problem.

7.21 Some properties of higher-order increments. The nth order increment of a signal $x(t)$ is defined by (7.178). In all that follows we assume that $\tau_i = \tau$, $1 \le i \le n$.

(a) Calculate the mean value of $\Delta x_n(t;\tau)$ and express the result as in (7.134).

(b) Calculate the variance $\sigma_n^2(\tau)$ and find an expression similar to (7.135).

(c) Express this variance in terms of the power spectrum $\Gamma_s(\nu)$ defined by (7.137) where $(1 + 4\pi\nu^2)$ is replaced by $(1 + 4\pi\nu^2)^n$.

(d) Extend the discussion presented in Section 7.5.4 and find for higher-order increments the two kinds of behavior of the variance when τ tends to infinity.

Chapter 8

Poisson Processes and Affiliated Signals

General concepts concerning point processes were introduced in Section 5.7, but no specific example was analyzed in detail. Among all possible point processes used in many areas of the information sciences and physics, the Poisson processes are certainly the most important. To some extent they play the same role as normal processes and for similar reasons. The first is that they are certainly the simplest to analyze, and also they appear as good statistical models for many physical phenomena. Furthermore, they can be the origin of many other signals of great practical interest, as will be seen in this chapter.

8.1 DEFINITIONS AND BASIC PROPERTIES OF POISSON PROCESSES

As indicated on p. 4, a point process is a random distribution of points on an axis which, in most cases studied below, is the time axis. Extensions to other spaces will be analyzed in following sections. As a result of this assumption, a point process can also be called a random distribution of time instants, or a stream of random events. These random instants or events will be simply noted by t_i, which are the random time instants when the events appear.

A *Poisson process* is defined by the two following conditions:

(*a*) The number of points (or of time instants) t_i appearing in non-overlapping time intervals are *independent random variables*;

(*b*) The number of points appearing in the interval $[t, t + \Delta T)$ is a Poisson random variable defined, as in (2.42), by its mean value

$$m(t, \Delta T) = \int_t^{t+\Delta T} \lambda(\theta)\, d\theta \tag{8.1}$$

The function $\lambda(t)$ is called the *density*, or the *rate*, of the process and is of course a nonnegative function. If $\lambda(t)$ is constant, the mean value does not depend on t and the point process becomes *stationary*.

Let us now investigate some consequences of this definition.

8.1.1 Distribution of Points in Arbitrary Small Intervals

Suppose that ΔT, now written Δt, tends to zero. As a result of (8.1), the mean value of the Poisson distribution takes the form

$$m(t, \Delta t) \cong \lambda(t) \Delta t \tag{8.2}$$

Using (2.42), we deduce that

$$p_0 \cong 1 - \lambda(t) \Delta t + O(\Delta t^2) \tag{8.3}$$

$$p_1 \cong \lambda(t) \Delta t + O(\Delta t^2) \tag{8.4}$$

$$p_k \cong O(\Delta t^2) \tag{8.5}$$

In these expressions the symbol $O(\Delta t^2)$ means a quantity which tends to zero when $\Delta t \to 0$ with a behavior like $(\Delta t)^\alpha$, where $\alpha \geq 2$. Consequently, when $\Delta t \to 0$, the time interval $[t, t + \Delta t)$ can contain only zero or one point t_i. It is sometimes said that the process has no accumulation of points. This also means that for $\Delta t \to 0$ the number of points in the interval $[t, t + \Delta t)$ is a heads and tails RV, with of course a probability p_1 tending to zero. Note that, as a result of (8.4), the probability of finding a point of the process at a given time instant t is null. As an obvious consequence, the number of points in the intervals $[t, t + \Delta T)$ or $(t, t + \Delta T]$ are Poisson RVs with the same mean value (8.1).

These properties can be taken as an *alternative definition* of a Poisson process. More precisely, if we assume that the numbers of points in non overlapping intervals *are independent* and if (8.3), (8.4) and (8.5) are valid, then the number of points in any finite interval is a Poisson RV with the mean value (8.1). The principle of the proof follows the arguments presented in Section 4.7.3. The interval $[t, t + \Delta T)$ is divided into an arbitrary number p of non-overlapping intervals Δt_i, and the number of random points in Δt_i is called N_i. From this assumption, the number of random points $N(t, \Delta T)$ which appear in $[t, t + \Delta T)$ can be written as

$$N(t, \Delta T) = \sum_{i=1}^{p} N_i \tag{8.6}$$

When $p \to \infty$, we obtain a sum similar to (4.165), and its distribution becomes a Poisson distribution.

This result has a very important physical meaning, which explains the importance of Poisson processes in many phenomena. The assumptions (8.3), (8.4), and (8.5) are quite common and are realized for most of the physical processes. Relation (8.4) means that no time instant t_i can appear at a deterministic position. Here is an example where it is not satisfied. Consider an infinite sequence of time instants $t_i = iT$, where T is given. This type of sequence appears, for example, when sampling signals, as in Section 6.7. Suppose that each point t_i can be dropped from this sequence with a probability p and independently of the situation of all the other points. With this operation we construct a point process, and it is evident that (8.4) is not satisfied. In fact, if the interval $[t, t + \Delta t)$, $\Delta t < T$, contains an instant kT, $p_1 = 1 - p$ and $p_1 = 0$ otherwise.

On the other hand, relation (8.5) means that several points cannot arrive in a small interval, which is of course quite obvious for a distribution of time instants.

But the true origin of the Poisson distribution is the assumption of *independence* at any scale of time. It is sometimes said that a Poisson process is characterized by a complete absence of memory. This assumption is often introduced at the microscopic scale, even though we have good reason to think that there is a limit, when working with very short time intervals, where it must no longer be valid.

8.1.2 Time Intervals between Points

In Section 5.7.2 we saw that there is another way of describing point processes, using the concepts of *lifetime* or *survival time*. Let us first consider the latter, and introduce the random variable $L(t)$ which is the time interval between an arbitrary time instant t and the first instant of the point process that comes after t. We wish to calculate the PDF of $L(t)$. For this we note that the event $l < L(t) \leq l + dl$ means that there is no point of the process in the interval $(t, t + l]$, and one point in $(t + l, t + l + dl]$. The probability of this event can thus be written as

$$p(l;\ t)\ dl = e^{-m(t,\ l)}\ \lambda(t + l)\ dl \tag{8.7}$$

where $m(t, l)$ is defined by (8.1). As a result the PDF of the RV $L(t)$ can be expressed uniquely in terms of the density of the process as

$$p(l;\ t) = \exp\left[-\int_t^{t+l} \lambda(\theta)\ d\theta\right] \lambda(t + l) \tag{8.8}$$

This PDF takes a very interesting form when the Poisson process is stationary. In fact, as $\lambda(t) = \lambda$ which does not depend on time, $p(l;\ t)$ is also independent of time and can be written as

$$p(l) = \lambda e^{-\lambda l} \tag{8.9}$$

which is a one-sided exponential distribution, as in (2.30).

Let us now consider the *lifetime* which is the time interval between two points of the Poisson process. In order to calculate its PDF we have to take into account the fact that there is a point of the process at the time instant t. But as a Poisson process has no memory, the existence of this point does not change the previous calculation. This means that the lifetime and the survival time have the same PDF. We saw in Section 5.7.2 that this characterizes a point process which does not age.

In conclusion, we can say that a Poisson process does not age and, in the stationary case, the time interval between two successive events has a one-sided exponential distribution. The statistical properties of the one-sided distribution were investigated in Chapter 2. For example, the corresponding mean value is given by (2.31) and is equal to

$1/\lambda$. This has a simple physical meaning: if the Poisson process has a density λ, the number of points per unit of time is λ. Conversely, the mean value of the time interval between successive points is $1/\lambda$.

8.1.3 Poisson Processes as Renewal Processes

A renewal point process is one in which the time intervals between two successive points are independent RVs. It is clear from the above discussion that a Poisson process is one of renewal. As a renewal process is completely specified by the probability distribution of the successive intervals, we deduce that a stationary Poisson process can be defined as a renewal process with an exponential distribution for successive time intervals. This can also be taken as an alternative definition of a Poisson process.

8.1.4 Time Intervals between Non-Successive Points

In order to simplify the following discussion, let us assume that the Poisson process is stationary with a density λ. Let us call L_k the RV describing the interval between one point of the process and the next kth that appears. The RV L_k is represented in Figure 8.1.

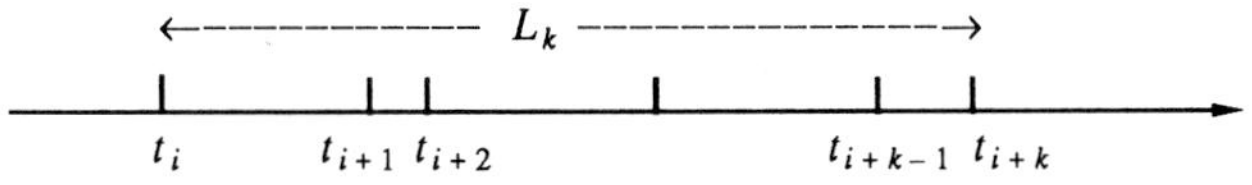

Figure 8.1 Representation of the RV L_k.

It is clear that the interval between t_i and t_{i+k} contains $k-1$ other points of the Poisson process. Thus L_1 corresponds to the lifetime studied above. In order to calculate the PDF of L_k we can use exactly the same method with the only difference that the interval $(t, t+l]$ must now contain $k-1$ points instead of zero, so as to arrive at (8.7). As the number of points in this interval is a Poisson RV with the mean value λl, deduced from (8.1), we obtain from (2.42)

$$p_k(l) = \lambda e^{-\lambda l} \left[\frac{(\lambda l)^{k-1}}{(k-1)!} \right] \tag{8.10}$$

which again gives (8.9) for $k = 1$. It is easy to verify, either by direct calculation or by using the characteristic function, that (8.10) is the PDF of the sum of k IID RVs with the PDF (8.9). This was the subject of Problem 4.25.

This has an interesting application in some methods for the reduction of the density of a Poisson process. Suppose that, starting from a stationary Poisson process of density λ, we regularly remove $k-1$ points after each point of the initial process. Let us start at an arbitrary time instant t_0 which is not a point of the process. The first point t_1 of the process after t_0 is preserved. The following points t_2, t_3, ..., t_k are dropped and t_{k+1} is preserved, and so on. We thus obtain a new process of points θ_i with a density λ/k.

The construction of the new process of points θ_i is explained in Figure 8.2. Note there that the initial point t_0 is an arbitrary origin and is not a point of the Poisson process.

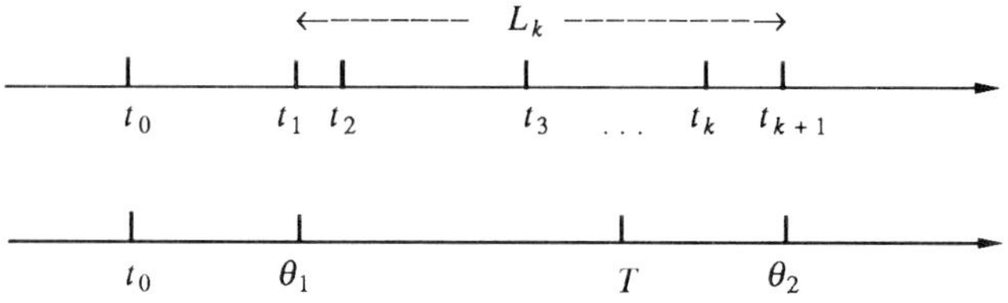

Figure 8.2 Construction of a point process θ_i by regularly erasing points of a Poisson process.

It is clear that this point process is still a renewal process, because there is no memory effect in the initial Poisson process. However, as seen in Figure 8.2, the distance between successive points in the new process is the RV L_k studied above and characterized by the PDF given by (8.10). As this PDF is not exponential, we deduce that the new process is no longer a Poisson process, and we may wonder where this property comes from. The reason is that the deletion procedure introduces a memory effect which destroys the basic property of a Poisson process. This memory effect is disclosed by certain properties. At first it is clear that for the point process θ_i the lifetime and survival times are different. In fact, the lifetime is represented by L_k while the survival time cannot be so simply calculated. This appears in Figure 8.2 where an arbitrary time instant T is indicated. It is clear here that the distance between T and θ_2 is not the RV L_k but another RV. The calculation of its PDF cannot be made as simply as for the lifetime. Another aspect of the memory effect can be illustrated by considering some conditional probabilities. For this purpose, let us call $dN(t)$ and $dM(t)$ the number of points of the processes t_i and θ_i respectively in the time interval $[t, t + dt)$. For the Poisson process we deduce that, because of the independence between non-overlapping intervals, we have

$$P\{[dN(t) = 1][dN(\tau) = 1]\} = \lambda^2 \, dt \, d\tau \tag{8.11}$$

which is a direct result of (8.4). This can be written as

$$P[dN(\tau) = 1 | dN(t) = 1] = \lambda \, d\tau \tag{8.12}$$

which is the result of the absence of memory in the Poisson process. This expression is no longer valid for the process with reduction. In fact, suppose that $dM(t) = 1$, which also means that $dN(t) = 1$. The event that $dM(\tau) = 1$ under the condition that $dM(t) = 1$ means that for the original Poisson $dN(\tau) = 1$ and that the interval $\tau - t$ contains a number of points ensuring that the point in the interval $(\tau, d\tau)$ is preserved. This means that $dM(t)$ and $dM(\tau)$ are not independent and the process described by $M(t)$ cannot be a Poisson process. In other words, the memory effect is created by the procedure of point erasure. This point is analyzed in a problem. However we will see below that other erasure procedures can preserve the Poisson character.

A final aspect of the cancellation of the Poisson character for the process of point θ_i can be deduced from the form of the PDF of the distance between successive time instants. This PDF is given by (8.10) and is represented in Figure 8.3 for several values of k.

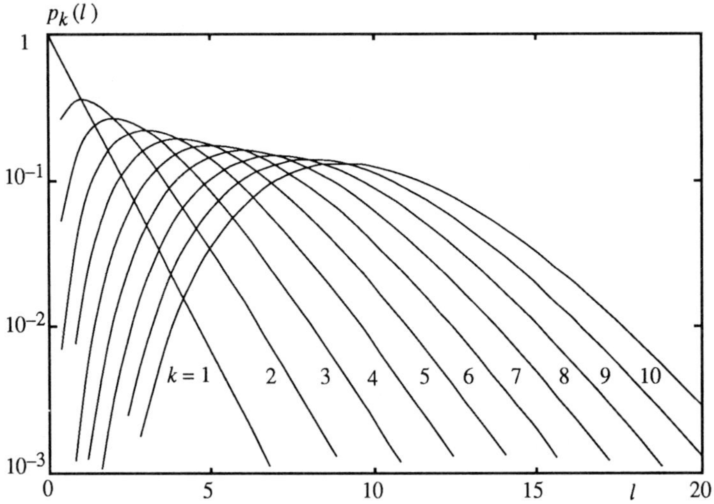

Figure 8.3 Probability density function of time intervals.

Let us now consider the structure of the point process θ_i for large values of k. Here it is necessary to introduce a normalization procedure, as in (4.170). In fact, the RV L_k in Figure 8.2 is the sum of k IID random variables L_1. Thus the mean value of L_k is k/λ, because $1/\lambda$ is the mean value of L_1, as indicated at the end of Section 8.1.2. Thus in order to compare point processes with the same density we must introduce the renewal process such that its lifetime, or the distance between successive points, is $(1/k)L_k$. As L_k is the sum of IID random variables, we deduce from (4.171) that the variance of the distance of points tends to zero. This means that starting from a Poisson process which is quite irregular, we arrive at a process of points which are almost periodically distributed. This process is extremely regular, and is no longer a Poisson process.

8.1.5 Conditional Distributions of Points

Suppose, as above, that the Poisson process is stationary with density λ. Consider the interval $[0, T)$ where the time origin 0 is arbitrary. Our purpose is to calculate the conditional distribution of the points of the process, with the condition that the number of points recorded in $[0, T)$ is known. To introduce the method, let us first suppose that this number is equal to one. Our purpose is to calculate $P[dN(t) = 1 | N(T) = 1]$, where $dN(t)$ and $N(T)$ are the numbers of random points in the time intervals $[t, t + dt)$ and $[0, T)$ respectively. In this calculation we assume that $t < T$.

As the two previous intervals overlap, we cannot directy apply the condition (*a*) appearing in the definition of Poisson point processes. To do so we divide the interval $[0, T)$ into three non overlapping intervals I_j defined by

$$I_1 = [0, t) \,;\, I_2 = [t, t + dt) \,;\, I_3 = [t + dt, T) \tag{8.13}$$

Let us call N_1, N_2, and N_3 the random numbers of points which appear in these intervals and E the event defined by

$$E = [dN(t) = 1]\ [N(T) = 1] \tag{8.14}$$

It can be written in terms of N_i in the form

$$E = [N_1 = 0]\ [N_2 = 1]\ [N_3 = 0] \tag{8.15}$$

As dt is arbitrarily small, the length of the time interval which corresponds to N_1 and N_3 is T, which gives

$$P[(N_1 = 0)\ (N_3 = 0)] = e^{-\lambda T} \tag{8.16}$$

and combining this with (8.15) and (8.4), we obtain

$$P(E) = P\{[dN(t) = 1]\ [N(T) = 1]\} = \lambda\, dt\, e^{-\lambda T} \tag{8.17}$$

But as $P[N(T) = 1]$ deduced from (2.42) is $(\lambda T)\exp(-\lambda T)$, we deduce from (4.1) that

$$P[dN(t) = 1 \,|\, N(t) = 1] = dt/T \tag{8.18}$$

This has a very simple interpretation: if we know that there is one point in the interval $[0, T)$, this point is uniformly distributed in this interval.

Let us now see the changes in this result when the Poisson process is no longer stationary, but has a density $\lambda(t)$. The principle of the calculation remains the same, but we replace λT by m given by (8.1), or

$$m = \int_0^T \lambda(\theta)\, d\theta \tag{8.19}$$

With this procedure (8.18) must be replaced by

$$P[dN(t) = 1 \,|\, N(t) = 1] = [\lambda(t)\, dt/m] = \lambda'(t)\, dt \tag{8.20}$$

The term m introduces a normalization effect that ensures that the integral of $\lambda'(t)$ over the interval $(0T)$ is equal to one.

Let us now consider p arbitrary time instants $t_1, t_2, \ldots, t_p$ and the corresponding intervals $[t_i + dt_i)$. Using exactly the same procedure, we obtain

$$P\{[dN(t_1) = 1]\ [dN(t_2) = 1] \ldots [dN(t_p) = 1] \,|\, N(t) = p\} =$$

$$p!\ \lambda'(t_1)\ \lambda'(t_2) \ldots \lambda'(t_p)\, dt_1\ dt_2 \ldots dt_p \tag{8.21}$$

where $\lambda'(t)$ is introduced in (8.20). The coefficient $p!$ is due to the fact that

$$P[N(T) = p] = e^{-m} m^p/p!$$

In particular, if $\lambda(t) = \lambda$, $\lambda' = 1/T$ and (8.20) introduces a very simple interpretation. Suppose that we take p independent points uniformly distributed over the interval $[0T)$. The probability of obtaining one of these points in $[t_1, t_1 + dt_1)$, another in $[t_2, t_2 + dt_2)$,..., another in $[t_p, t_p + dt_p)$ is given by (8.21) where $\lambda'(t) = 1/T$. The coefficient $p!$ is due to all the possible permutations before arriving at the final result. This result can be used to simulate a stationary Poisson process by using q uniformly distributed random numbers.

The above calculation can also be realized when the intervals are not infinitesimal. In this case we arrive at the same calculation as presented in Example 4.9 introducing the multinominal distribution.

8.2 POISSON PROCESSES IN OTHER SPACES

Although the discussion above was devoted to Poisson processes on an axis, and the terminology related to time instants, the concept of the Poisson process is not restricted to the description of random events in time. For example, point processes in a plane can be used to describe the random structure of photographic plates due to the random position of silver grains. There are several other examples of modelizations by point processes in a plane.

To remain general, we shall define the concept of Poisson processes in the space $\mathbb{R}^n$. It is easy to transpose to this space the definition given in Section 8.1 and valid in $\mathbb{R}^1$, the only difference being that time intervals are replaced by volumes in $\mathbb{R}^n$.

Thus a point process in $\mathbb{R}^n$ is a Poisson process if

(*a*) The number of random points P_i appearing in non overlapping volumes are independent RVs;

(*b*) The number of points P_i appearing in the volume V is a Poisson RV defined by its mean value

$$m(V) = \int_V \lambda(\mathbf{x})\, d\mathbf{x} \tag{8.22}$$

where $\lambda(\mathbf{x})$ is the density of the process.

The fundamental difference between Poisson processes on an axis, taken at the time axis, and Poisson processes on other spaces is that the concepts of order in time, and especially of past and future, disappear when working in $\mathbb{R}^n$ instead of $\mathbb{R}^1$. Thus the concepts of time intervals between points or survival time disappear for spaces of dimension greater than one. On the other hand, all the other consequences of the definition given in the previous section remain valid. Let us briefly mention these points.

The discussion concerning distribution in arbitrary small intervals can be transposed unchanged except for a modification of the vocabulary: the concept of memory effect is valid only in time, and not in space. The same applies to the conditional distribution of points. The concept of renewal process, however, cannot be transposed directly in space.

However we shall now see that there is a kind of transposition of the concept of distance. For simplification, we will study the case of point processes in a plane.

Example 8.1 Distance between points for a Poisson process in a plane. Consider an arbitrary point M and call R the distance between M and the point P of the process which is at the minimum distance of M. Let $p(r)$ be the PDF of R. The event $r < R \leq r + dr$ is realized if there is no point of the process in the circle of center M and of radius r and if there is one point in the annulus defined by the two circles of center M and of radii r and $r + dr$. As the two surfaces are non-overlapping and have the values πr^2 and $2\pi r \, dr$, we deduce that

$$p(r) \, dr = \lambda 2\pi r \exp(-\lambda \pi r^2) \, dr$$

which gives

$$p(r) = \lambda 2\pi r \exp(-\lambda \pi r^2) \tag{8.23}$$

This PDF is a Rayleigh distribution defined by (4.217) where σ^2 is replaced by $(2\pi\lambda)^{-1}$. The mean value of this distribution, calculated in Problem 4.43, is

$$E(R) = (1/2) \, \lambda^{-1/2} \tag{8.24}$$

Projection of a Poisson process. Consider a point process in $\mathbb{R}^2$. It is a random distribution of points, an example of which is represented in Figure 8.4. Each point M_i of the point process has two coordinates x_i and y_i. Consequently the random points located in the x-axis and defined by the x_is introduce a point process in $\mathbb{R}^1$. This process can be called the *projection* of the initial point process onto $\mathbb{R}^1$. The same procedure can be used for other dimensions, and starting from a point process in $\mathbb{R}^n$ it is possible to generate by projection a point process in $\mathbb{R}^p$, $p < n$.

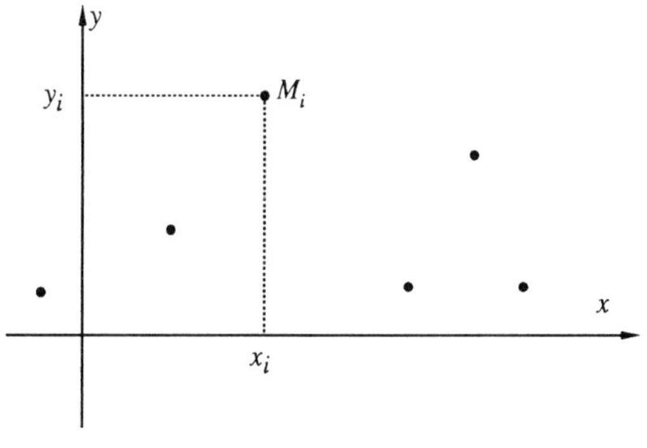

Figure 8.4 Example of a Poisson process in a plane.

Suppose now that the initial point process in $\mathbb{R}^2$ is a Poisson process defined by a density $\lambda(x, y)$. We will show that the process of points x_i in $\mathbb{R}^1$ is also generally a

Poisson process. For this purpose let us use the definition introduced at the beginning of Section 8.1. The number of random points x_i in the interval $[x, x + a)$ is the number of points M_i located between the straight lines parallel to the y-axis and defined by x and $x + a$. But, as the process of points M_i is a Poisson process, this number is a Poisson RV whose mean value is given by (8.22), or

$$m(x, a) = \int_{x}^{x+a} d\xi \int_{-\infty}^{+\infty} \lambda(\xi, y)\, dy \tag{8.25}$$

This equation introduces a condition on the point process M_i. In fact it is necessary for $m(x, a)$ to be finite, which implies that the density $\lambda(x, y)$ of the process M_i satisfies

$$\lambda'(x) \triangleq \int_{-\infty}^{+\infty} \lambda(x, y)\, dy < +\infty \tag{8.26}$$

whatever x. For example, this is not the case for a Poisson process in the plane which is stationary, because then $\lambda(x, y) = \lambda$. Furthermore, if we take several non overlapping intervals on the x-axis, it corresponds to non-overlapping domains in the plane $\mathbb{R}^2$ and as the process of points M_i is a Poisson process, the number of points in these domains are independent. In conclusion, provided that (8.26) is fulfilled, the process of points x_i satisfies the definition (*a*) and (*b*) of Section 8.1, and is thus a Poisson process. Its density $\lambda'(x)$ is of course defined by (8.26).

This procedure of projection can be extended in various directions, as for example in the case of non orthogonal projections.

8.3 POISSON PROCESSES AND SIGNALS WITH INDEPENDENT INCREMENTS

The concept of signals with independent increments was introduced in Chapter 7. In particular, Brownian motion, studied in Section 7.3.2, is a continuous-time signal with independent increments. A Poisson process in time is also a signal with independent increments, and it is interesting to compare their properties and to explore in more detail this class of continuous-time signals with independent increments. We shall first briefly recall the main properties of Brownian motion.

Let $x(t)$ be a *Brownian motion* defined for $t > 0$ and suppose that $x(0) = 0$ with probability one. As a result of the definition in Section 7.3.2, at any time instant t, $x(t)$ is an RV which has the distribution $N(\alpha t, ct)$. Its characteristic function can thus be written as

$$\phi_B(u) = \exp\left[ju\alpha t - \frac{1}{2}u^2 ct\right] \tag{8.27}$$

Furthermore, it was indicated, without proof, that every trajectory of $x(t)$ is a continuous function with probability one.

Consider now the case of a *Poisson process*. Let us call $N(t)$ the number of random points in the interval $[0, t)$, and suppose that $N(0) = 0$ with probability one. We assume at the moment that the process is stationary, which means that the density λ appearing in (8.1) is constant. In this case $N(t)$ is a Poisson-distributed RV with the mean value λt, and its characteristic function, given by (2.91), is

$$\phi_P(u) = \exp[\lambda t(e^{ju} - 1)] \tag{8.28}$$

Furthermore, like $x(t)$, $N(t)$ is a signal with independent increments. However, the main difference with $x(t)$ is that any trajectory of $N(t)$ is discontinuous with probability one. This is due to the fact that the only variations of $N(t)$ are steps of amplitude one appearing at the random points of the Poisson process. Thus $N(t)$ is a signal with independent increments with constant steps at random points.

Let us now introduce a *generalization* of this situation. Consider the signal $M(t)$ which also has only stepwise variations such that the steps appear at t_i, time instants of a Poisson process, but also so that the amplitudes of these steps are random. More precisely, suppose that A_i is the RV that describes the step of $M(t)$ at time t_i. We assume that all these RVs are independent and independent of the Poisson process. As a result of these assumptions it appears that $M(t)$ is still a signal with independent increments and that any trajectory is discontinuous with probability one, the signal $N(t)$ being a particular case corresponding to $A_i = 1$.

Let us now calculate the characteristic function of $M(t)$. We divide the time interval $[0, t)$ into p small intervals of amplitude Δt_i, as in (8.6). Let us call ΔM_i the variation of $M(t)$ in the interval corresponding to Δt_i and $\phi_i(u)$ the characteristic function of A_i. As ΔN_i takes only the values 0 and 1 with probabilities given by (8.3) and (8.4), we deduce that

$$\begin{aligned} E[\exp(ju\Delta M_i)] &= 1 - \lambda\Delta t_i + E[\exp(juA_i)]\,\lambda\Delta t_i \\ &= 1 + [\phi_i(u) - 1]\,\lambda\Delta t_i \end{aligned} \tag{8.29}$$

If we assume that the RVs A_i have the same distribution, the characteristic function of A_i does not depend on i, and $\phi_i(u)$ must be replaced by $\phi_A(u)$ in (8.29). But as the RVs ΔM_i are independent, we can apply the same reasoning as in Section 4.7.3. Consider the second characteristic function of ΔM_i which is

$$\psi_i(u) \cong [\phi_A(u) - 1]\;\lambda\Delta t_i \tag{8.30}$$

Because of the independence of the ΔM_is, the second characteristic function of $M(t)$ becomes

$$\psi_M(u) = [\phi_A(u) - 1]\;\lambda t \tag{8.31}$$

This corresponds to the characteristic function $\phi_M(u)$ given by

$$\phi_M(u) = \exp\{\lambda t\;[\phi_A(u) - 1]\} \tag{8.32}$$

If the steps A_i are no longer random but equal to one, $\phi_M(u) = \exp(j\,u)$, and we return to (8.28). By using methods that are outside the scope of this book, it is possible to show that (8.32) is the most general expression of the characteristic function of a signal with stationary and independent increments, which is a discontinuous function with probability one, the steps of which do not appear at non-random time instants.

By an expansion of $\phi_A(u)$ to the order two, it is possible to calculate the mean value and the variance of $M(t)$. This simple calculation gives

$$E(M) = E(A)\,\lambda t \quad ; \quad \sigma_M^2 = E(A^2)\,\lambda t \tag{8.33}$$

The fundamental property here, which also appears for Brownian motion, is that the mean value and the variance increase linearly with t.

All previous results can be extended without difficulty to the case of signals with increments that are independent but *no longer stationary*. Time-varying Brownian motion was introduced in Section 7.3.2.(**f**), and if α and c are time-dependent, αt and ct in (8.27) must be replaced by expressions such as (7.50').

Let us suppose that the Poisson process used for $N(t)$ and $M(t)$ has a density $\lambda(t)$ depending on time, and that the distribution of the RV A is also a function of time in such a way that $\phi_A(u)$ must be replaced by $\phi_t(u)$. In this case (8.30) must be replaced by

$$\psi_t(u) \cong [\phi_t(u) - 1]\,\lambda(t)\,\Delta t \tag{8.34}$$

and (8.32) becomes

$$\phi_M(u) = \exp\left\{ \int_0^t [\phi_\theta(u) - 1]\,\lambda(\theta)\,d\theta \right\} \tag{8.35}$$

which is the most general expression possible.

There is one last category of signals with independent increments and with discontinuous trajectories. These signals are constructed as follows. Consider a given sequence of time instants t_i which are *not random*. As a consequence the function $N(t)$ defined as above is a deterministic step-wise function, which is of course not continuous. However, the function $M(t)$ obtained, as previously, by using random amplitudes A_i at the points t_i is random, and if the RVs A_i are independent, the increments of $M(t)$ are also independent. But the main feature of this function is that its discontinuities appear at *deterministic* time instants. It is this property which limits its interest in the description of many physical phenomena, because the randomness is too artificial.

We shall now discuss the comparisons between the trajectories of Brownian motion and of a Poisson-type process with independent increments. It has been mentioned before that the trajectories of Brownian motion are extremely irregular, from which we derived the properties of quadratic variation presented in Section 7.3.2 (**g**). The origin of this fact is that the increments $dx(\theta)$ used in (7.44) satisfy $E[dx(\theta)] = 0$, $E[dx^2(\theta)] = c\,d\theta$ and $E[dx^4(\theta)] = 3(c\,d\theta)^2$, because of the normal property. Conversely, the trajectories of $N(t)$ and $M(t)$ are discontinuous but extremely regular, because these functions are constant in any interval where no point of the point process appears. We have seen that for a Poisson process described by $N(t)$, the increments $dN(\theta)$ can only take the values 0 or 1. We then have $dN(\theta) = dN^k(\theta)$ and, for example, the quadratic variation Q defined by (7.58) where $dx(\theta)$ is replaced by $dN(\theta)$ is simply equal to $N(t)$

$-N(\tau)$. This function is thus random, although it is no longer random for Brownian motion. The same reasoning can be followed for the function $M(t)$, and the basic result is that $E[dM^k(\theta)]$ remains in the order of $d\theta$, which is not the case for Brownian motion.

8.4 SHOT NOISES

8.4.1. Poisson Shot Noise, Second Order Properties

In Section 5.7.3 we saw that a point process can also be described as a distribution process, or as a sequence of pulses mathematically described by Dirac distributions. All these ideas can be applied to Poisson processes, and the signal $x(t)$ defined by (5.71) can be considered as the derivative of the counting signal $N(t)$ used in the previous section, since it is well known that the Dirac distribution is the derivative of the unit step function. Similarly, (5.72) is the derivative of $M(t)$ considered just above.

These distribution processes can be observed through linear filters, as discussed in Section 5.7.4. More precisely, (5.73) represents the effect of random pulses of random amplitudes on a linear filter of impulse response $h(t)$. If the amplitudes A_i are equal to one, and if the time instants t_i correspond to a Poisson process, we obtain the *shot noise* of a Poisson process.

Let us first briefly comment on this signal, written as

$$x(t) = \sum_i h(t - t_i) \tag{8.36}$$

where the time instants t_i are the points of a stationary Poisson process of density λ. For this, let T be the time constant of the filter. If $\lambda T << 1$, the mean value of the distance between successive points is much greater than T and there is almost no overlapping between the functions $h(t - t_i)$ and $h(t - t_j)$. An example of this type of situation is given in Figure 8.5. It is clear here that the presence of the impulse function $h(t)$ does not prevent the location of the time instants t_i. On the other hand, when $\lambda T >> 1$, there is such a strong mixture between the overlapping functions $h(t - t_i)$ that it is apparently impossible to locate the time instants t_i and to reconstruct the shape of $h(t)$ from the observation of the signal. An example of this is given in Figure 8.6. Remember that the signal $x(t)$ is uniquely random because the points t_i are random.

Let us now explain the term *shot noise*. Consider again the experiment on the photoelectric effect described on p. 4. The random time instants t_i represent the instants when photoelectrons are emitted from the photocathode. The electronic device used to study the phenomenon can be approximated by a linear filter, and its impulse response is the function $h(t)$. In other words, the system is driven by the "shots" due to the arrival of the photoelectrons on the anode. The impression of individual shots is clear for $\lambda T << 1$, while it disappears for greater values of λT, and the phenomenon appears as a continuous noise.

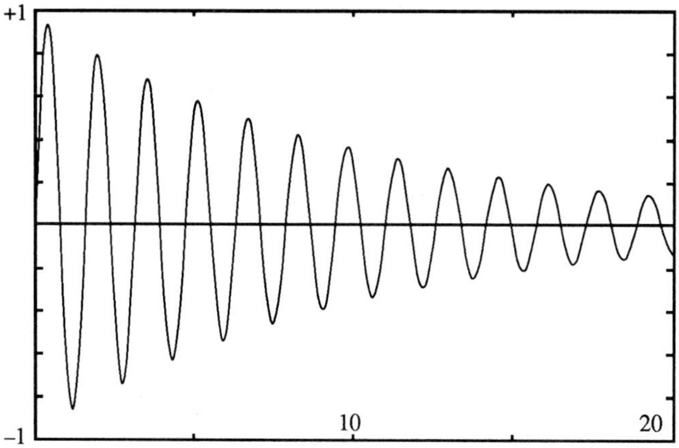

Figure 8.5 Example of trajectory of shot noise for $\lambda T << 1$.

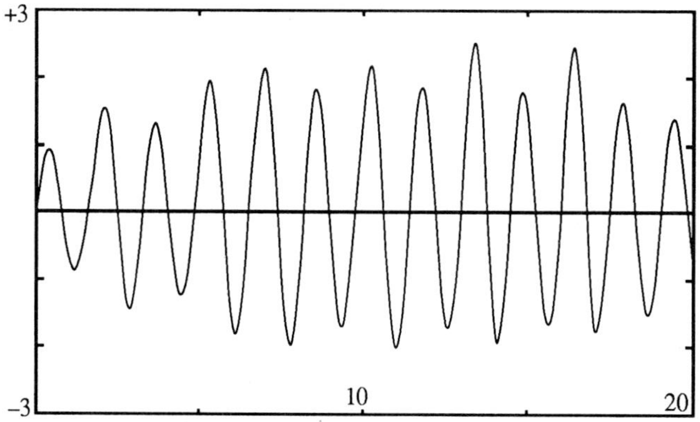

Figure 8.6 Example of trajectory of shot noise for $\lambda T >> 1$.

In order to calculate the statistical properties of $x(t)$, it is much more convenient to use (5.74) instead of (5.73). Thus $x(t)$ can be written as

$$x(t) = \int h(t - \theta)\, dN(\theta) \tag{8.37}$$

where $N(t)$ is the function considered in the previous section. Thus $dN(\theta)$ is the number of points in the interval $[\theta, \theta + d\theta)$. As in (5.77), we have

$$m_x = \lambda G(0) \tag{8.38}$$

where λ is the density of the Poisson process defined after (8.1).

If we wish to calculate the covariance function of $x(t)$, we can apply (5.79). We must calculate the expected value $E[dN(\theta_1)\, dN(\theta_2)]$ which appears in this equation because we have $dM(\theta) = dN(\theta)$ for the Poisson shot noise. If $\theta_1 \neq \theta_2$, the RVs $dN(\theta_1)$ and $dN(\theta_2)$ are independent, as a result of the definition of a Poisson process. So we deduce from (5.78) that

$$E[dN(\theta_1)\, dN(\theta_2)] = \lambda^2\, d\theta_1\, d\theta_2 \tag{8.39}$$

If, however, $\theta_1 = \theta_2$, we must calculate $dN^2(\theta)$. Using the results of Section 8.1.1, and especially (8.3) and (8.4), we obtain

$$E[dN^2(\theta)] = \lambda\, d\theta \tag{8.40}$$

By combining (8.39) and (8.40), we can write

$$E[dN(\theta_1)\, dN(\theta_2)] = [\lambda^2 + \lambda\,\delta(\theta_1 - \theta_2)]\, d\theta_1\, d\theta_2 \tag{8.41}$$

and when we insert this expression into (5.79), we obtain

$$\gamma_x(t_1,\, t_2) = \lambda \int h(t_1 - \theta)\, h(t_2 - \theta)\, d\theta \tag{8.42}$$

It is obvious that this function remains unchanged if t_1 and t_2 are replaced by $t_1 + t$ and $t_2 + t$ respectively. This infers that $\gamma_x(t_1, t_2)$ is only a function of $t_1 - t_2$, and after simple algebraic manipulations we obtain the correlation function of $x(t)$ defined by (5.16) or (5.17), in the form

$$\gamma(\tau) = \lambda \int h(\theta)\, h(\theta - \tau)\, d\theta \tag{8.43}$$

By Fourier transformation we obtain the power spectrum in the form

$$\Gamma(\nu) = \lambda |G(\nu)|^2 \tag{8.44}$$

The correlation function (8.43) has the same shape as that obtained by passing a CT white noise through a linear filter, given by (7.25). In order to obtain exactly the same function, it suffices to take $k = \lambda$. This situation requires some further comments.

As a result of the above discussion, we see that the signal $y(t)$ defined by (7.46) and the signal $x(t)$ defined by (8.36) or (8.37) have the same correlation function if $k = \lambda$. This does not mean that these two signals have the same statistical properties. We have often indicated (see Section 6.4.3(**b**)) that two completely different signals can have the same second order statistical properties. In particular, it was indicated that $y(t)$ is a Gaussian signal, while $x(t)$ has no reason to have this property. We will verify this fact below by calculating the characteristic function of $x(t)$.

However, (8.44) means that $x(t)$ can be considered as the output of the filter with the impulse function $h(t)$ driven by a white noise. But this input is precisely the distribution process defined by (5.71) where the t_i are the random points of a Poisson point process with density λ. So this signal is a CT white noise and its correlation function is given by (7.23) where $c = \lambda$. Physically this signal is a sequence of pulses arriving at the random time instants of a Poisson process. This allows us to interpret the points discussed after (7.23). The signal (5.71) is a CT signal but has no derivative because the derivative of the Dirac distribution is not a regular function. Furthermore, like any other sequence of pulses this is, clearly, not a continuous signal.

Example 8.2 Counting signal. Suppose that the filter used in (8.37) has an impulse response $h(t)$ equal to one for $0 \leq t < \tau$ and zero otherwise. This filter is proportional to that represented in Figure 6.6 and was used in (7.129). Using this last expression, we can write

$$x(t) = \int_{t-\tau}^{t} dN(\theta). \tag{8.45}$$

which means that $x(t)$ is the number of points of the Poisson process arriving in the interval $[t - \tau, t)$. In other words, $x(t)$ represents the output of a causal counter of time constant τ. Using (7.133) and (8.38), we deduce that

$$m_x = \lambda \tau \tag{8.46}$$

which is effectively the mean number of points in an interval of duration τ. The corresponding power spectrum can be deduced from (8.44) and (7.132), which gives

$$\Gamma(\nu) = \lambda \tau^2 \operatorname{sinc}^2(\nu\tau) \tag{8.47}$$

If we are interested in the variance, we can use (8.43) directly, and, remembering that $\sigma^2 = \gamma(0)$, we obtain

$$\sigma^2 = \lambda \tau \tag{8.48}$$

This result is not surprising, since it is clear that $x(t)$ is a Poisson RV for which the mean value and variance are equal. The fact that $x(t)$ is a Poisson RV results from the definition 8.1(b).

8.4.2 Family of Distributions of a Poisson Shot Noise

In this section we shall show that, because of the Poisson distribution of the time instants t_i, it is possible to calculate, at least in principle, the family of finite distributions of the signal $x(t)$ defined by (8.37) describing the shot noise due to a Poisson process. In order to do so, we shall first establish a result which will be used several times hereafter.

Let X be an RV defined by

$$X = \int f(\theta)\, dN(\theta) \tag{8.49}$$

where $f(.)$ is an arbitrary function, provided that X is second-order, which means that $|f(t)|$ and $f^2(t)$ are integrable. It is clear that $x(t)$ defined by (8.37) has the same form as X. Let us now calculate the characteristic function $\phi_X(u)$ of X defined by (2.53). We use exactly the same procedure as we did to arrive at (8.32). The RV X can be considered as a limit for $n \to \infty$ of sums of the type

$$X_n = \sum_{i=1}^{n} f(\theta_i)\, \Delta N(\theta_i) \tag{8.50}$$

Using the same ideas as in (8.29), we can write

$$E\{\exp[juf(\theta_i)\,\Delta N(\theta_i)]\} = 1 + \{\exp[juf(\theta_i)] - 1\}\ \lambda \Delta\,\theta_i \tag{8.51}$$

As the RVs $\Delta N(\theta_i)$ are independent whatever n, we can use the same procedure as in (8.30) and (8.31), and passing to the limit when n tends to infinity, we obtain, as in (8.31) and (8.32), the second and first characteristic functions of X by

$$\psi_X(u) = \lambda \int\ [e^{juf(\theta)} - 1]\, d\theta \tag{8.52}$$

$$\phi_X(u) = \exp\left\{ \lambda \int\ [e^{juf(\theta)} - 1]\ d\,\theta \right\} \tag{8.53}$$

By expanding $\psi_X(u)$ up to u^2 we obtain

$$\psi_X(u) \cong \lambda ju \int f(\theta)\, d\theta - \frac{1}{2}\lambda u^2 \int f^2(\theta)\, d\theta \tag{8.54}$$

By comparison with (2.93), we deduce

$$E(X) = \lambda \int f(\theta)\, d\theta \tag{8.55}$$

$$\sigma_X^2 = \lambda \int f^2(\theta)\, d\theta \tag{8.56}$$

which can be deduced directly from (8.49) and (8.41). In particular, if $f(t)$ is a function taking only the values 0 and 1, X is a Poisson RV, and we verify from (8.55) and (8.56) that $E(X) = \sigma_X^2$, as indicated in (2.47).

Let us now apply these results to the RV $x(t)$ defined by (8.37). Replacing $f(\theta)$ by $h(t - \theta)$, we deduce that

$$\phi(u) = \exp\left\{ \lambda \int\ [e^{juh(t-\theta)} - 1]\ d\theta \right\} \tag{8.57}$$

By changing $t - \theta$ into τ, we find that $\phi(u)$ does not depend on t, a situation similar to that described by (5.4). This shows that $x(t)$ is stationary in the same sense as (5.4).

Unfortunately, the calculation of the PDF $p(x)$ of $x(t)$ defined by (2.60) is in general very complex, except for some specific forms of the impulse function $h(t)$. If explicit forms of $p(x)$ are difficult to obtain, numerical evaluations are always possible.

Let us now consider a distribution function such as (5.1). To calculate the corresponding characteristic function, we introduce the vector $\mathbf{X}$, the components of which are $x(t_1)$, $x(t_2)$, ..., $x(t_n)$, where $x(t)$ is always defined by (8.37). The characteristic function of $\mathbf{X}$ is defined by (4.59). In this expression the scalar product $\mathbf{u}^T\mathbf{X}$ can be written

$$\mathbf{u}^T\mathbf{X} = u_1\ x(t_1) + u_2\ x(t_2) + \dots + u_n\ x(t_n) \tag{8.58}$$

and with (8.37) this becomes

$$\mathbf{u}^T\mathbf{X} = \int \sum_{i=1}^{n} u_i\, h(t_i - \theta)\; dN(\theta) \tag{8.59}$$

Introducing the function $f_n(\theta;\, \mathbf{u})$ defined by

$$f_n(\theta;\, \mathbf{u}) = \sum_{i=1}^{n} u_i\, h(t_i - \theta) \tag{8.60}$$

we see that $\mathbf{u}^T\mathbf{X}$ has the form (8.49), and using the same procedure, we obtain

$$\phi_n(\mathbf{u}) = \exp\left\{ \lambda \int [e^{jf_n(\theta;\, \mathbf{u})} - 1]\; d\theta \right\} \tag{8.61}$$

which again gives (8.57) when $n = 1$. As n and the t_is are arbitrary, (8.61) completely specifies all the statistical properties of $x(t)$. In principle, by Fourier transformation (which is not an easy task), we can obtain the family of finite PDFs of the shot noise of a Poisson process. Furthermore, because of the integration over the variable θ, $\phi_n(\mathbf{u})$ satisfies the same relation as (5.45), which means that $x(t)$ is strict-sense stationary.

8.4.3 Shot Noise with Random Amplitudes

As explained on p. 187, it is sometimes necessary for physical reasons to study the signal defined by (5.72) rather than that described by (5.71). In other words, the amplitudes associated with each time instant of the Poisson process are no longer constant as they are for the Poisson shot noise. Our purpose is to study the consequences of this fact on the statistical properties of a signal such as (5.73). The signal under consideration is no longer described by (8.37) but by (5.75), or

$$x(t) = \int h(t - \theta)\; dM(\theta) \tag{8.62}$$

Its mean value is given by (5.76) and its correlation function by (5.79). Using the results of Section 8.3, we note that the increments $dM(\theta)$ are independent RVs and we have

$$E[dM(\theta)] = E(A)\; \lambda\, d\theta \tag{8.63}$$

The mean value (8.38) must now be replaced by

$$m_x = \lambda E(A)\; G(0) \tag{8.64}$$

and (8.40) is replaced by

$$E[dM^2(\theta)] = E(A^2)\; \lambda\, d\theta \tag{8.65}$$

As a result (8.41) is replaced by

$$E[dM(\theta_1)\, dM(\theta_2)] = [E^2(A)\lambda^2 + E(A^2)\; \lambda\, \delta(\theta_1 - \theta_2)]\; d\theta_1\, d\theta_2 \tag{8.66}$$

Inserting these results in (5.79), we deduce that the correlation function $\gamma(\tau)$ is given by (8.43) where λ is replaced by $E(A^2)\lambda$. The discussion following (8.44) remains valid here provided that the density λ is replaced by $\lambda' = E(A^2)\lambda$. In other words, the first and second order properties remain the same and the change of density shows that the random amplitudes have been taken into account. Thus λ must be replaced by $E(A)\lambda$ for the mean value and $E(A^2)\lambda$ for the correlation function.

The higher-order properties, however, are changed in a most significant way. To understand this, let us return to (8.49), where $dN(\theta)$ is replaced by $dM(\theta)$. Comparing (8.51) with (8.29), we can write

$$E\{\exp[juf(\theta_i)\,\Delta M(\theta_i)]\} = 1 + \{\phi_A[uf(\theta_i)] - 1\}\,\lambda\Delta t_i \tag{8.67}$$

where $\phi_A(u)$ is, as in (8.32), the characteristic function of the random amplitude A. Consequently (8.52) must be replaced by

$$\psi_X(u) = \lambda \int \{\phi_A[uf(\theta)] - 1\}\,d\theta \tag{8.68}$$

and $\phi_X(u)$ is simply $\exp[\psi_X(u)]$. Similarly, (8.61) becomes

$$\phi_n(\mathbf{u}) = \exp\left\{\lambda \int \{\phi_A[f_n(\theta;\mathbf{u})] - 1\}\,d\theta\right\} \tag{8.69}$$

which completely specifies the statistical properties of $x(t)$. It is clear that if the amplitudes A_i are equal to one, $\phi_A(u) = \exp(ju)$, and (8.69) becomes (8.61). It is easy to extend this expression to the case where the initial Poisson process is no longer stationary, but has a non-constant density. For this it is sufficient to replace λ by $\lambda(\theta)$ in (8.69). Another generalization can also be introduced. The random amplitude A can have a probability distribution depending on time and thus a characteristic function $\phi_\theta(u)$, as in (8.34). In this case the characteristic function (8.69) takes the form

$$\phi_n(\mathbf{u}) = \exp\left\{\int \lambda(\theta)\,\{\phi_\theta[f_n(\theta;\mathbf{u})] - 1\}\,d\theta\right\} \tag{8.70}$$

which is similar to (8.35).

There is a last degree of generalization that can be introduced. There are some situations in which not only the amplitudes A_i are random but also the filters $h(t - t_i)$. This leads to the replacement of (5.73) by

$$x(t) = \sum_i A_i\,h_i(t - t_i)$$

which means that the filters driven by the pulses can depend on the arrival times t_i of these pulses. If the filters $h_i(t)$ are independent and also independent of the Poisson process, the same procedure can be used. There are several reasons which physically justify the use of random filters. However, the assumption of independence, which is necessary in order to arrive at a possible calculation, is the most difficult to justify.

Finally, note that the above calculations can be extended without difficulty to the non stationary Poisson process. In this case it suffices to replace the density λ by $\lambda(\theta)$, as seen for example in (8.70). The introduction of a non constant density does not change the structure of the equations but only introduces more complex expressions.

8.4.4 Asymptotic Properties of Shot Noises

We saw in Section 8.4.1 and in Figures 8.5 and 8.6 that the behavior of the Poisson shot noise is different for low and high densities of the Poisson process. It is thus interesting to find out whether this can be explained more exactly. For this purpose we shall return to (8.49). When λ tends to infinity the mean value (8.55) and variance (8.56) also tend to infinity, and nothing more can be said. But consider the RV Y defined by

$$Y = \frac{1}{\sqrt{\lambda}} [X - E(X)] \tag{8.71}$$

It is clear that Y is zeromean-valued. Furthermore, its variance is

$$\sigma_Y^2 = E(Y^2) = \int f^2(\theta)\, d\theta \tag{8.72}$$

which is a direct consequence of (8.56). Let us now calculate the characteristic function of Y in terms of $\phi(u)$, the characteristic function of X defined by (8.57). Starting from

$$\phi_Y(u) \triangleq E[e^{juY}] = E\{\exp[ju\lambda^{-1/2}(X - m)]\} \tag{8.73}$$

where $m = E(X)$, we deduce that

$$\phi_Y(u) = \phi_X[u\lambda^{-1/2}] \,.\, \exp[-ju\lambda^{-1/2}m] \tag{8.74}$$

So the second characteristic function of Y takes the form

$$\psi_Y(u) = \psi_X(u\lambda^{-1/2}) - ju\lambda^{-1/2}m \tag{8.75}$$

Using (8.52) and (8.55) giving m, we finally obtain

$$\psi_Y(u) = \lambda \int \{\exp[ju\lambda^{-1/2}f(\theta)] - 1 - ju\lambda^{-1/2}f(\theta)\}\, d\theta \tag{8.76}$$

This expression can be expanded in power series of u, which gives

$$\psi_Y(u) = -\frac{u^2}{2} \int f^2(\theta)\, d\theta + O(\lambda^{-1/2}) \tag{8.77}$$

This means that when λ tends to infinity, $\psi_Y(u)$ tends to $-(1/2)u^2\sigma^2$, where

$$\sigma^2 = \int f^2(\theta)\, d\theta \tag{8.78}$$

Comparing this with (2.93), we can say that $\psi_Y(u)$ tends to the second characteristic function of an $N(0, \sigma^2)$ RV. In conclusion, when $\lambda \to \infty$, the RV X, normalized in the appropriate way, converges in distribution to the normal law. This is similar to the central limit theorem presented in Section 4.7.2.

Let us now apply the same method to the signal $x(t)$ defined by (8.37). From this signal we deduce $y(t)$ defined by

$$y(t) = \lambda^{-1/2}[x(t) - m] \tag{8.79}$$

where m is given by (8.38). The correlation function of $y(t)$ can be deduced from (8.43) and becomes

$$\gamma_y(\tau) = \int h(\theta)\, h(\theta - \tau)\, d\theta \tag{8.80}$$

and the variance is of course equal to $\gamma_y(0)$. We shall now calculate the nth order characteristic function of $y(t)$. To do this we introduce the scalar product $\mathbf{u}^T\mathbf{Y}$ defined as in (8.58) where $x(t)$ is replaced by $y(t)$. Following the same procedure as above, we find that the second characteristic function of $\mathbf{Y}$ tends to

$$\psi(\mathbf{u}) = \frac{-1}{2}\mathbf{u}^T \mathbb{\Gamma} \mathbf{u} \tag{8.81}$$

where $\mathbb{\Gamma}$ is the matrix the elements of which are

$$\Gamma_{ij} = \int h(t_i - \theta)\, h(t_j - \theta)\, d\theta = \gamma_y(t_i - t_j) \tag{8.82}$$

As this result is valid whatever the value of n in (8.58), we deduce that the signal $y(t)$ has a distribution which tends to the normal distribution $N[0, \gamma(\tau)]$.

And now some comments on these results. It is clear that either (8.37) or (8.62) introduces signals which appear as a sum of independent RVs. The result is thus similar to the central limit theorem in Section 4.7.2. It is one of the extensions of the central limit theorem discussed on p. 126, because the sum is now an integral and the RVs are no longer identically distributed.

From a physical point of view, the concept of large values of λ should be mentioned. This concept is clear for integers in (4.160) and can be written as $n >> 1$, but not for λ, which represents the mean number of points of the Poisson process per unit of time. To understand this question, we shall return to the RV X defined by (8.49) and Y defined by (8.71).

If we wish to arrive at a distribution $N(0, 1)$, we must replace λ in (8.71) by μ defined by

$$\mu = \lambda \int f^2(\theta)\, d\theta \tag{8.83}$$

as a result of which Y becomes a zeromean RV with a variance equal to one. The parameter μ is without dimension and the normal distribution appears if $\mu >> 1$. But the integral (8.83) is one possible definition of the time constant of a filter, and as a result the normality appears for $\lambda T >> 1$, as in the interpretation of Figure 8.6.

Let us now extend these results to the case of the shot noise with *random amplitude* discussed in Section 8.4.3. As the discussion needs a zeromean signal, instead of subtracting the mean value as in (8.71) we will simply assume that $E(A) = 0$. This gives $m_x = 0$ because of (8.64). So the random amplitudes do not have the same sign, which is an extension of the concept of shot effect. In our study of asymptotic properties there is another parameter which can be used, and which is due to a scaling effect of the random amplitudes A_i.

To explain this point further, we shall start from an RV defined, as in (8.49), by

$$X = \int f(\theta)\, dM(\theta) \tag{8.84}$$

In this expression $dM(\theta)$ is defined as before, but the random amplitudes A_i are equal to

$$A_i = c\, B_i \tag{8.85}$$

where the B_is are IID *zeromean* and with *unit variance* RVs. Their characteristic function is denoted $\phi_B(u)$ in such a way that $\phi_A(u)$ in (8.67) can be written as

$$\phi_A(u) = \phi_B(cu) \tag{8.86}$$

As the RVs A_i are zeromean, $E[dM(\theta)]$ given by (8.63) is null, and X is also a zeromean RV. Furthermore, $E(A^2) = c^2$, because the B_is have a unit variance and we deduce from (8.66) that

$$E(X^2) = \lambda c^2 \int h^2(\theta)\, d\theta = \lambda c^2 T \tag{8.87}$$

Finally, the second characteristic function of X is still given by (8.68) which takes the form

$$\psi(u) = \lambda \int \left\{ \phi_B[cuf(\theta)] - 1 \right\} d\theta \tag{8.88}$$

Suppose that all the moments of B are finite. As B is zeromean-valued and has a unit variance, the expansion of $\phi_B(v)$ has the form

$$\phi_B(v) = 1 - \frac{1}{2} v^2 + O(v^3) \tag{8.89}$$

As a result of all these assumptions, the RV

$$Y \triangleq (\lambda c^2)^{-1/2} X \tag{8.90}$$

is zeromean-valued and has a variance equal to T. Its second characteristic function is

$$\psi_Y(u) = \lambda \int \left\{ \phi_B[\lambda^{-1/2}\, uf(\theta)] - 1 \right\} d\theta \tag{8.91}$$

The main point of this equation is that this function does not depend on c. If the $\lambda \to \infty$, we deduce that $\psi_Y(u)$ tends to the limit $-1/2\ T\ u^2$, which characterizes a normal distribution. In this procedure it is perfectly possible to maintain $\lambda\, c^2 = 1$, which means that $Y = X$. Thus the RV X defined by (8.84) may tend to have a normal distribution when $\lambda \to \infty$ and $c \to 0$ in such a way that $\lambda\, c^2 = 1$. Physically this means that X is constructed by a large number of shots with small amplitudes. These results can be applied without difficulty to the signal (8.62).

8.5 HIGHER-ORDER PROPERTIES

We have seen above, and especially in Section 8.3, that a Brownian motion or a generalized Poisson process are continuous-time signals *with independent increments*. Similarly a discrete-time signal such as (7.27) representing a random walk is also a signal with independent increments if $u[k]$ is a strict white noise. In other words, when we compare the statistical properties of DT and CT white noises we must use the increments in the CT case. As this point is often misunderstood, we shall carefully compare the properties of these increments, beginning by a review of some results concerning only their second order properties, in introduction to the study of higher-order properties.

Let $u[k]$ be a DT strict sense stationary white noise, as defined in Section 7.2, and suppose that m_1 and m_2 are the first two moments of $u[k]$. As a result of the definition, the second order moment $m[k,\ k']$ defined by (5.43) can be written as

$$m[k,\ k'] \triangleq E\{u[k]\ u[k']\} = m_1^2 + (m_2 - m_1^2)\ \delta[k - k'] \tag{8.92}$$

where $\delta[k - k']$ is the Kronecker-delta function equal to one if $k = k'$ and to zero otherwise. In fact, if $k \neq k'$, $m[k,\ k'] = m_1^2$ because of the independence between $u[k]$ and $u[k']$, and if $k = k'$, we have $m[k, k] = m_2$. Note that $m_2 - m_1^2$ is the variance of $u[k]$, and also the cumulant c_2, and if $m_1 = 0$, we find an expression similar to (7.10).

Consider now a CT signal $u(t)$ with independent increments. The mean value of these increments is

$$E[du(t)] = \mu_1\, dt \tag{8.93}$$

In the case of Brownian motion, the result from point (*b*) of the definition in Section 7.3.2 is that $\mu_1 = \alpha$. However, for a generalized Poisson process, the result from (8.33) or (8.63) is that $\mu_1 = \lambda\, E(A)$.

The main difficulty arises when considering second order moments. In fact, we can write

$$E\{\,[du]^2\} = \mu_2\, dt + O(dt^2) \tag{8.94}$$

This can be deduced from (7.40) for Brownian motion which gives $\mu_2 = c$, or from (8.65) which gives $\mu_2 = \lambda\, E(A^2)$. If we wish to combine (8.93) and (8.94) in one expression, we must write, as in (8.66),

$$E[du(t)\, du(t')] = \mu(t, t')\, dt\, dt' \tag{8.95}$$

where

$$\mu(t, t') = \mu_1^2 + \mu_2\, \delta(t-t') \tag{8.96}$$

In this expression $\delta(.)$ is the Dirac distribution or the unit impulse distribution. The main difference between (8.92) and (8.96) is that the coefficients μ_i no longer have the meaning of moments or cumulants.

It does not make sense to verify (8.96) "directly," because for $t = t'$, $\mu(t, t')$ becomes infinite, and this expression has a meaning only after integration. For example, consider the signal $u(t)$ defined, as in (7.44), by

$$u(t) = \int_0^t du(\theta) \tag{8.97}$$

Its second order moment can be written

$$m(t_1, t_2) \triangleq E[u(t_1)\, u(t_2)] = \int_0^{t_1} \int_0^{t_2} \mu(t, t')\, dt\, dt' \tag{8.98}$$

and with (8.96) we obtain

$$m(t_1, t_2) = \mu_1^2\, t_1\, t_2 + \mu_2\, t_1 \wedge t_2 \tag{8.99}$$

As $m(t) = \mu_1 t$, we find that the covariance function of $u(t)$ is given by (7.36), also valid for a Poisson process.

In conclusion, $\mu(t, t')$ defined by (8.96) must be considered as a distribution in the plane $t \times t'$. This distribution is the sum of a uniform distribution with the density μ_1^2 and a distribution on the manifold $t = t'$ with a constant density μ_2.

The discussion which follows considers the extension to higher-order dimensions of distributions such as $\mu(t, t')$ valid for second order moments. Note that we no longer have to consider Brownian motion, for, as it is a *normal* signal, all results in Section 7.4 can be applied.

8.5.1 Higher-Order Moments in the Time Domain

Consider the function $M(t)$ defined in Section 8.3, and corresponding to a Poisson process with random amplitudes. To simplify the discussion we will assume that $E[dM(t)] = 0$. This can be achieved either by taking random steps in such a way that $E(A) = 0$, or by considering the increments $dM'(t) = dM(t) - E[dM(t)]$. It is obvious that if the increments $dM(t)$ are independent, this is also true for the increments $dM'(t)$.

Considering now the third order properties of the increments, and using the same procedure as in (8.40) or (8.65), we can deduce that

$$\mu_3 = E(A^3)\, \lambda \tag{8.100}$$

which is obtained as (8.65). This allows us to write

$$\mu(t_1, t_2, t_3) = \mu_3\, \delta(t_1 - t_2)\, \delta(t_1 - t_3) \tag{8.101}$$

This means that the moment $E[dM(t_1)\ dM(t_2)\ dM(t_3)]$ is null if t_1, t_2 and t_3 are not equal. In other words, $\mu(t_1, t_2, t_3)$ is a distribution which is reduced to the manifold $t_1 = t_2 = t_3$, where the density is constant and equal to μ_3.

Let us now consider the fourth order moment. It is clear that $E[dM(t_1)\ dM(t_2)\ dM(t_3)\ dM(t_4)]$ can be different from zero only if the four time instants are coupled by pair or are equal, which gives

$$\mu(t_1, t_2, t_3, t_4) = \mu_2^2\ [\ \delta(t_1 - t_2)\ \delta(t_3 - t_4) + \delta(t_1 - t_3)\ \delta(t_2 - t_4) + \delta(t_1 - t_4)\ \delta(t_2 - t_3)] + \mu_4\ \delta(t_1 - t_2)\ \delta(t_2 - t_3)\ \delta(t_3 - t_4) \qquad (8.102)$$

This corresponds to a distribution on the three manifolds appearing in the normal case with a density of μ_2^2 and to a distribution on the manifold $t_1 = t_2 = t_3 = t_4$. It is interesting here to compare this expression to that of the DT case given by (7.93). There are two main differences. The first is that the coefficient of the last term of (8.102) is neither a moment nor a cumulant. This was already seen in the case of second order moments. The second appears in the *normal case*. If $x[k]$ is normal, $c_4 = 0$ which means that the last term of (7.93) disappears. The same phenomenon occurs in (8.102), but for another reason. In fact as the cumulants are not used it is no longer possible to introduce their properties. In reality, when $u(t)$ in (8.93) and (8.94) is a Brownian motion, $\mu_4 = 0$ for another reason. We saw that for Brownian motion, because of the normal distribution, we obtain $E[dx^2(t)] = c\,dt$ and $E[dx^4(t)] = 3c^2(dt)^2$. As μ_4 is defined by $E\{[du]^4\} = \mu_4\ dt$, we see that $\mu_4 = 0$ for Brownian motion.

It is easy to extend these results to higher-order moments. The only difficulty is in writing expressions such as (8.102), because the number of terms increases rapidly.

8.5.2 Higher-Order Moments in the Frequency Domain

Higher-order moment spectra are defined from functions such as $M(\mathbf{f})$ which appeared in (6.143). In this equation the moments $m(\mathbf{t})$ are defined as in (5.42). In this section we will follow the same ideas, but instead of taking the Fourier transforms of moments, we shall use those of the distribution $\mu(\mathbf{t})$ introduced in the previous section. Furthermore, as stationarity is assumed, we will systematically use (6.151). The first order spectrum deduced from (8.96) is thus

$$\Gamma_1(f) = \mu_1^2\ \delta(f) + \mu_2 \qquad (8.103)$$

The first term disappears if $\mu_1 = 0$ and we obtain a constant spectrum, introducing the expression *white noise*. Similarly, the Fourier transform of μ given by (8.101) is

$$M_3(\mathbf{f}) = \mu_3\ \delta(f_1 + f_2 + f_3) \qquad (8.104)$$

When we compare this with (6.151), we deduce that

$$\Gamma_2(f_1, f_2) = \mu_3 \qquad (8.105)$$

which also introduces a constant bispectrum. However, this property of constant value of the spectrum, valid for the bispectrum, is no longer true for the trispectrum, as it can easily be obtained from (8.102). But it is clear in this expression that the trispectrum generates a normal density on the normal manifolds, as discussed in Section 7.4.4.

8.5.3 Application to the Shot Noise

Starting from (8.62), it is not difficult to calculate the higher-order properties of $x(t)$ from those of the increments studied in 8.5.1. Similarly, if we are interested in these properties in the frequency domain, we can apply the general expressions (6.156) to the higher-order spectra obtained in 8.5.2. As the results are not hard to obtain, we shall only discuss a simple example concerning fourth order moments.

In the time domain we deduce from (8.62) that the moment (5.42) can be written as

$$m_x(t_1, t_2, t_3, t_4) = \mu_2^2 \left[\gamma_x(t_1 - t_2)\, \gamma_x(t_3 - t_4) + \gamma_x(t_1 - t_3)\, \gamma_x(t_2 - t_4) + \right.$$

$$\left. \gamma_x(t_1 - t_4)\, \gamma_x(t_2 - t_3) \right] + \mu_4 \int h(t_1 - \theta)\, h(t_2 - \theta)\, h(t_3 - \theta)\, h(t_4 - \theta)\, d\theta \quad (8.106)$$

where $\gamma_x(\tau)$ is the correlation function defined by (8.42) in which λ is replaced by μ_2. It is clear that the first three terms of (8.106) correspond to the moment of a normal distribution. The last term appears because $x(t)$ is not a normal signal.

In the frequency domain we can calculate the Fourier transform of (8.106). The Fourier transforms of the three normal terms are the same as in (7.88). The Fourier transform of the last term of (8.106) is

$$T(\mathbf{f}) = \mu_4\, G(f_1)\, G(f_2)\, G(f_3)\, G(f_4)\, \delta(f_1 + f_2 + f_3 + f_4) \qquad (8.107)$$

8.6 SOME AFFILIATED SIGNALS

We now turn to some examples of signals which can be constructed from a Poisson process. It is not our purpose to be exhaustive, and the choice of examples is rather arbitrary. Note that the shot noise which is perhaps the most important example was discussed in previous sections.

8.6.1 Random Telegraph Signal

Consider a stationary Poisson process with density λ and an arbitrary origin of time 0. Let $N_+(t)$, $t \geq 0$, and $N_-(t')$, $t' < 0$, be the number of points of the process in the intervals $[0, t)$ and $[t', 0)$ respectively. Finally, let $x(0)$ be an RV independent of the point process and taking only the values ± 1, thus verifying $x^2(0) = 1$. The random telegraph signal is defined by

$$x(t) = x(0)\, (-1)^{N_+(t)}, \quad t \geq 0 \qquad (8.108)$$

$$x(t) = x(0)\, (-1)^{N_-(t)}, \quad t < 0 \qquad (8.108')$$

It is clear from this definition that any trajectory of this signal is a function which takes only the values ± 1, and the switches between these two values appear at the points of the Poisson process. An example of the trajectory of such a signal appears in Figure 8.7.

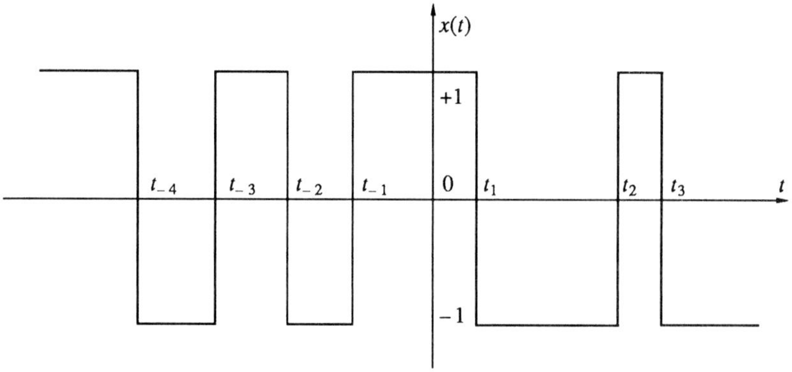

Figure 8.7 Example of the trajectory of the random telegraph signal.

Let us first calculate the *mean value* of $x(t)$. We need only note that $N_+(t)$ and $N_-(t)$ are Poisson distributed RVs with mean value $m = \lambda |t|$. Furthermore, as $\exp(j\pi) = -1$, we deduce from the characteristic function (2.91) that

$$E\{(-1)^{N(t)}\} = E\{\exp[j\pi N(t)] = \exp[\lambda |t| (e^{j\pi} - 1)] = \exp(-2\lambda |t|) \quad (8.109)$$

Using the fact that $N(t)$ and $x(0)$ are independent, we obtain

$$E[x(t)] = E[x(0)] \exp(-2\lambda |t|) \quad (8.110)$$

It appears that $E[x(t)] \longrightarrow 0$ if $|t| \longrightarrow \infty$, but the only way to obtain a zeromean signal, whatever the value of t, is to assume that $E[x(0)] = 0$. This means that for any value of t, $x(t)$ takes the value +1 and −1 with the probability 1/2.

On the other hand, if $x(0) = 1$, and if we introduce the probabilities $p_+(t)$ and $p_-(t)$ to obtain $x(t) = +1$ and $x(t) = -1$ respectively, we deduce from (8.110) that

$$p_+(t) - p_-(t) = e^{-2\lambda |t|} \quad (8.111)$$

and as $p_+(t) + p_-(t) = 1$ we obtain

$$p_+(t) = \text{ch}(\lambda t)\, e^{-\lambda |t|}\,;\; p_-(t) = \text{sh}(\lambda |t|)\, e^{-\lambda |t|} \quad (8.112)$$

We shall now calculate the *second order* properties of $x(t)$. In order to find the second order moment defined by (5.43), suppose first that $t_1 < t_2$. As a result of this assumption and of (8.108),

$$x(t_2) = x(t_1)\,(-1)^{\Delta N_2} \quad (8.113)$$

where ΔN_2 is defined by

$$\Delta N_2 = \Delta N(t_1;\ t_2 - t_1) \tag{8.114}$$

In this expression the increment ΔN has the same meaning as in (7.39). Noting that $[x(t_1)]^2 = 1$, as a consequence of the definition (8.108), we obtain

$$m(t_1,\ t_2) = (-1)^{\Delta N_2} = \exp(j\pi\Delta N_2) \tag{8.115}$$

As ΔN_2 is a Poisson RV of mean value $\lambda(t_2 - t_1)$, we deduce from (8.109) that

$$m(t_1, t_2) = \exp[-2\lambda(t_2 - t_1)] \tag{8.116}$$

The same calculation can be made for $t_2 < t_1$, and we can write the general result as

$$m(t_1,\ t_2) = \exp[-\ 2\lambda|t_1 - t_2|] \tag{8.117}$$

This expression shows that, whatever the probability distribution of $x(0)$, the second order moment $m(t_1,\ t_2)$ is stationary. This is due to the fact that when calculating the product $x(t_1)x(t_2)$ the value $x(0)$ disappears. Using the definition given in Section 5.3.2, this does not mean that $x(t)$ is second order stationary, because we must also assume that $m(t) = m$. As seen before, $E[x(t)]$ must equal 0, and in this case the correlation function of $x(t)$ can be written

$$\gamma(\tau) = \exp[-2\lambda|\tau|] \tag{8.118}$$

Exactly the same procedure can be applied to higher-order moments. Consider *odd moments*, where, for example, three time instants $t_1,\ t_2,\ t_3$ satisfy $0 < t_1 < t_2 < t_3$. We can always write (8.113), and

$$x(t_3) = x(t_2)\ (-1)^{\Delta N_3} = x(t_1)\ (-1)^{\Delta N_2 + \Delta N_3} \tag{8.119}$$

We then obtain

$$x(t_1)\ x(t_2)\ x(t_3) = [x(t_1)]^3\ (-\ 1)^{2\Delta N_2}\ (-\ 1)^{\Delta N_3} \tag{8.120}$$

and as $[x(t_1)]^2 = 1$, this gives

$$x(t_1)\ x(t_2)\ x(t_3) = x(t_1)\ (-\ 1)^{\Delta N_3} \tag{8.121}$$

As the Poisson process has independent increments, $x(t_1)$ and ΔN_3 are independent RVs, and by using (8.110) and (8.109), we obtain

$$m(t_1,\ t_2,\ t_3) = E[x(0)]\ \exp[-\ 2\lambda(t_1 + t_3 - t_2)] \tag{8.122}$$

It is clear that, as for $m(t)$, the only way to obtain a stationary moment is to assume that $E[x(0)] = 0$. If we introduce this assumption it can be easily verified that all the odd moments are null, or

$$m(t_1,\ t_2, \ldots,\ t_{2k-1}) = 0 \tag{8.123}$$

Let us now consider the even moments, starting with the fourth order one. Using expressions like (8.113) or (8.119), and simplifying the calculation by noting that $[x(t)]^4 = 1$ and that $(-1)^{2k\Delta N} = 1$, we obtain for $t_1 < t_2 < t_3 < t_4$

$$m(t_1,\ t_2,\ t_3,\ t_4) = \exp\{-2\lambda[(t_2 - t_1) + (t_4 - t_3)]\} \tag{8.124}$$

It is clear that this expression is invariant when replacing t_i by $t_i + \tau$, and then this moment is stationary. By an obvious generalization, we obtain the moment of order $2k$. Let $\mathbf{t}$ be a vector with components t_i, $1 \le i \le 2k$. Let θ_i be the time instants deduced by the permutation of the t_is such that $\theta_1 \le \theta_2 \le \ldots \le \theta_{2k}$. We then have

$$m(\mathbf{t}) = \exp\left\{-2\lambda \sum_{i=1}^{k} (\theta_{2i} - \theta_{2i-1})\right\} \tag{8.125}$$

Applying this expression to $k = 2$, and using (8.118), we obtain

$$m(t_1,\ t_2,\ t_3,\ t_4) = \gamma(\theta_2 - \theta_1)\ \gamma(\theta_4 - \theta_3) \tag{8.126}$$

It is interesting to compare this with (7.83) obtained for normal signals. There is an analogy in that even higher-order moments can be expressed only in terms of the correlation function. But the main differences are that there is only one term in (8.126) and, above all, that we have first to arrange the time instants t_i so as to obtain the permutation which gives the θ_is. It is this operation which complicates the calculation of the higher-order spectra of the random telegraph signal.

Example 8.3 Extensions of the random telegraph signal. Instead of using the number of points of the process described by a function such as $N(t)$ in (8.108), we can take a function $M(t)$ which introduces the random amplitude A, provided that A has only integer values. Thus a new parameter, which is the characteristic function $\phi_A(u)$ introduced in (8.30), appears in the description of $x(t)$. Using this expression, (8.109) must be changed to

$$E\{(-1)^{M(t)}\} = \exp\{\lambda|t|\ [\phi_A(\pi) - 1]\} \tag{8.127}$$

which again gives (8.109) where $\phi_A(u) = \exp(ju)$. The calculation of the first moment is similar to the previous one, and yields

$$E[x(t)] = E[x(0)]\ \exp\{\lambda|t|\ [\phi_A(\pi) - 1]\} \tag{8.128}$$

This function is independent of t if either $E[x(0)] = 0$ or if $\phi_A(\pi) = 1$. The first condition is the same as that discussed above, and means that $x(t)$ takes the two possible values with probability (1/2). The second condition can appear only if the integer random number A takes only even values. This means that $(-1)^{\Delta M} = 1$ or that there is no longer a change of sign for $x(t)$. Thus $x(t)$ becomes a constant signal equal to $x(0)$. This degenerate case is of no interest.

Consider now the second order moments. Using the same procedure, we easily find a result similar to (8.118), and written as

$$\gamma(\tau) = \exp[-\lambda'|\tau|] \tag{8.129}$$

where λ' takes the form

$$\lambda' = \lambda[1 - \phi_A(\pi)] \tag{8.130}$$

As $|\phi_A(u)| \le 1$, we have

$$0 \leq \lambda' \leq 2\lambda \tag{8.131}$$

The bound $\lambda' = 0$ is reached if $\phi_A(\pi) = 1$, which means that the signal $x(t)$ must be constant. The bound $\lambda' = 2\lambda$, obtained in (8.118), also appears if $\phi_A(\pi) = -1$. This means that A takes only odd values, and in fact it is clear that the shifts of the random telegraph signal are the same if $A = 1$ or if A takes only odd values.

The above discussion refers to the stationary Poisson process. Another possible extension is to introduce the random telegraph signal generated by a non stationary Poisson process. This will be considered in some problems.

8.6.2 Switchboard Signals

This is a class of signals the simplest of which can be used to describe aspects of the behavior of a telephone switchboard. In explanation we shall begin with a simple model of a switchboard. This switchboard receives a large number of telephone calls, which of course arrive randomly. In the absence of any proof to the contrary, we can assume that these calls are not inter-related, so that there is no memory effect, and their times of arrival can be considered as a stream of events described as a Poisson process. Within a finite period, it is also possible to state that this process is stationary. Suppose also that each call has the same duration, which is an approximation of reality, and let us call the duration T. The switchboard is considered *in operation* if at least one call is processed, and not in operation otherwise. This can be described by a signal taking only two values, for example $x(t) = 0$ if there is no point of the process in the interval $[t - T, t)$, and $x(t) = 1$ otherwise. A more realistic description would need to take into account the random duration of each call, but the principle remains the same.

Note that this signal can be deduced by a clipping operation in a particular example of shot noise. To explain this, consider the counting signal studied in Example 8.2, which takes only positive integer values. Passing this signal into a device giving the output 1 if $x(t) \neq 0$ and 0 if $x(t) = 0$, we arrive at the signal introduced above. In other words, (8.45) indicates the number of calls processed at time t while $x(t)$ describes the actual state of the switchboard between its two possible states.

After these preliminaries, let us calculate some statistical properties of $x(t)$. Let $p_1(t)$ and $p_0(t)$ be the probabilities that $x(t) = 1$ or $x(t) = 0$ respectively. It is clear that

$$p_0(t) = e^{-\lambda T} \tag{8.132}$$

which is the probability that there is no point of the process in the interval $[t-T, t)$. As $x(t)$ is a heads or tails RV, as studied in Example 2.5, we deduce that

$$E[x(t)] = 1 - e^{-\lambda T} \tag{8.133}$$

It is obvious that for large values of T, $E[x(t)] = 1$, and its variance is null, which means that the switchboard is always in operation.

To calculate the second order moment $m(t_1, t_2)$ we note that the product $x(t_1)$ $x(t_2)$ also only takes the values 0 or 1. Furthermore, if $|t_1 - t_2| > T$, the RVs $x(t_1)$ and

$x(t_2)$ are independent, as a consequence of the Poisson process assumption. So let us now suppose that $|t_1 - t_2| \leq T$ and also that $t_1 < t_2$. Let N_1, N_2 and N_3 be the RVs corresponding to the number of points in the intervals $[t_1 - T,\ t_2 - T)$, $[t_2 - T,\ t_1)$ and $[t_1,\ t_2)$ respectively, as seen in Figure 8.8.

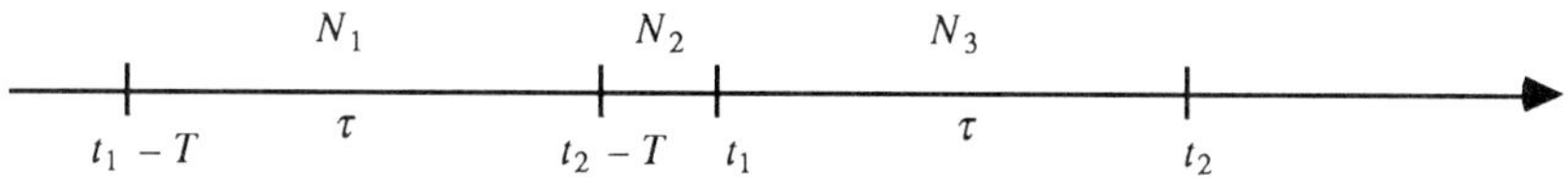

Figure 8.8 Intervals introducing the RVs N_i.

To simplify the notation, we also state that $t_2 - t_1 = \tau$.

If $N_2 = 0$, the event $x(t_1)\ x(t_2) = 1$ appears if and only if $N_1 \neq 0$ and $N_3 \neq 0$. If $N_2 \neq 0$ we have $x(t_1)\ x(t_2) = 1$ whatever the values of N_1 and N_3. As a result,

$$P[x(t_1)\ x(t_2) = 1] = e^{-\lambda(T - \tau)}(1 - e^{-\lambda\tau})^2 + 1 - e^{-\lambda(T - \tau)} \tag{8.134}$$

As $\gamma(\tau) = m(t_1,\ t_2) - m^2$, we deduce from (8.133) that for $\tau < T$

$$\gamma(\tau) = e^{-\lambda T}(e^{-\lambda|\tau|} - e^{-\lambda T}) \tag{8.135}$$

and $\gamma(\tau) = 0$ for $\tau \geq T$. In particular, the variance of $x(t)$ is

$$\sigma^2 = \gamma(0) = e^{-\lambda T}(1 - e^{-\lambda T}) \tag{8.136}$$

and, for large values of λ, $x(t)$ tends to one in the quadratic mean sense.

The above calculations were made with the proviso that the duration of the calls was constant and equal to T. The results can also be extended to the case where the calls have a random duration. This introduces a more complex calculation that will not be presented here.

Let us now examine an example of a signal with the same kind of structure but corresponding to a completely different physical problem. The purpose of this signal is to describe the random noise due to the *granularity* of photographic plates. Consider a black and white photographic plate. At a given point $\mathbf{r}$ the transparency $T(\mathbf{r})$ is a function satisfying $0 \leq T(\mathbf{r}) \leq 1$. If $T(\mathbf{r}) = 0$, no light can be transmitted and the result is a black print. If, however, $T(\mathbf{r}) = 1$, the plate is completely transparent and we have a white print. As the plate is realized with microscopic silver grains randomly distributed, the transparency is random. For example, saying that the transparency of the whole plate is equal to $1/2$ means that $E[T(\mathbf{r})] = 1/2$. But there are some fluctuations which can be described by a random function.

A simple model providing a good representation of reality can be constructed as follows. Consider a two-dimensional stationary Poisson process and let us call $\mathbf{r}_i$ the random points. To each point $\mathbf{r}_i$ describing the position of a silver grain we associate a circle of center $\mathbf{r}_i$ and of radius l, which is constant. This circle corresponds to a black area, which means that the transparency $T(\mathbf{r})$ associated to any point $\mathbf{r}$ inside the circle is null. The function $T(\mathbf{r})$ thus takes only two values: $T(\mathbf{r}) = 1$ if there is no random point $\mathbf{r}_i$ inside the circle of center $\mathbf{r}$ and radius l, and $T(\mathbf{r}) = 0$ otherwise.

Assuming that the density of the Poisson process is λ, we deduce that

$$P_1 = P[T(\mathbf{r}) = 1] = \exp(-\lambda\pi l^2) \tag{8.137}$$

and, as for $x(t)$, $E[T(\mathbf{r})] = P_1$. To calculate the second order moment we note that, because of the stationarity, $m(\mathbf{r}_1, \mathbf{r}_2)$ is only a function of $\mathbf{x} = \mathbf{r}_1 - \mathbf{r}_2$. Furthermore, as all the statistical properties are invariant in rotation, $m(\mathbf{x})$ is only a function of the modulus x of $\mathbf{x}$. If $x > 2l$, the RVs $T(\mathbf{r}_1)$ and $T(\mathbf{r}_2)$ are independent. If $x < 2l$ we have $T(\mathbf{r}_1)\,T(\mathbf{r}_2) = 1$ if there is no point on the surface represented in Figure 8.9.

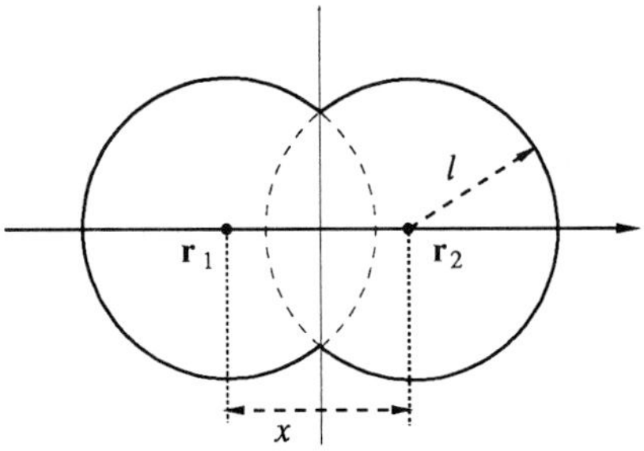

Figure 8.9 Surface without silver grains ensuring $T(\mathbf{r}_1)\,T(\mathbf{r}_2) = 1$.

After a simple geometric calculation of this surface we find

$$\gamma(x) = \left[e^{-\pi\lambda l^2} e^{-l^2(\theta + \sin\theta)} - e^{-\pi\lambda l^2}\right], \quad x \le 2l \tag{8.138}$$

and $\gamma(x) = 0$ for $x \ge 2l$, with

$$\theta = 2 \text{ Arc } \sin(x/2l) \tag{8.139}$$

To obtain a more realistic model, it would be possible to consider grains of random dimensions, which would introduce more advanced calculations.

8.6.3 Point Processes Deduced from a Poisson Process

Starting from a Poisson process, it is possible to construct several other point processes which can be useful in the representation of some physical phenomena. We shall give only a few examples, leaving it up to the imagination of the reader to extend the list.

(a) Erasing effect. Any user of the telephone system has experienced his call being suddenly cut off. As this phenomenon is generally unpredictable, the simplest approach is to assume that it is random. Similarly, when studying the behavior of a point process, it appears that, due to some imperfection in the electronic device, some points are dropped. If these phenomena do not introduce a memory effect, the process obtained is still a Poisson process. To show this, let us call $N(t)$ and $N'(t)$ the functions which describe the initial Poisson process and the process obtained after erasure. We have already seen that if $N(t)$ is a process with independent increments and if (8.4) and (8.5) hold, then $N(t)$ describes a Poisson process. It is clear that $N'(t)$ is still a process with independent increments, because of the assumption on the erasing effect. Furthermore, (8.5) is unchanged, and (8.4) must be changed by replacing $\lambda(t)$ by $[1 - \mu(t)]\,\lambda(t)$, where $\mu(t)$ is the probability that a point appearing in $[t, t + dt)$ be erased. The conclusion is that the new point process is still a Poisson process with a smaller density.

Note that this procedure of random erasing is quite different from that described in Section 8.1.4, which destroys the Poisson structure to introduce a renewal process.

(b) Dead time. Many electronic systems introduce dead-time effects which can be explained as follows. When a point t_i of the initial Poisson process appears, the system is blocked for the duration T in such a way that no other point of the process can be registered during the interval $[t_i, t_i + T)$. At the output of this device we obtain a new process which obviously cannot be a Poisson process because of the memory effect due to dead time. However, because of the lack of memory of the input process, the output is a renewal process and its statistical properties are completely specified by the probability distribution of the interval between successive points. The principle of the calculation of this distribution is similar to that presented in Section 8.1.2, and we will use the same notation here. To simplify the calculation, let us assume that the input Poisson process is stationary. With this assumption the event $l < L \leq l + dl$ means that there is an arbitrary number of points of the Poisson process in $(t, t + T]$ (as all these points are erased because there is a point at t), no point in $(t + T, t + l]$ and one point in $(t + l, t + l + dl]$. This immediately gives

$$p(l) = u(l - T)\, e^{-\lambda(l - T)} \tag{8.140}$$

Comparing this with (8.9), we observe that it is still an exponential distribution but with a translation to the right. The mean value and variance of this distribution is very easy to obtain.

(c) Random transformation. Let us start from a Poisson process, the points of the process being time instants t_i. Suppose that to each point of this process we associate another point process in another space. This can for example be the space $\mathbb{R}^2$. By this transformation, pictured in Figure 8.10, we generate a new point process of points M_i in $\mathbb{R}^2$ and, by introducing some appropriate assumptions on the transformation, it is possible to calculate its statistical properties.

Consider an arbitrary domain D in $\mathbb{R}^2$ and let us call $N_T(D;\ t)$ the number of points M_i belonging to D and generated through the transformation T by a point of the initial process arriving at t. The basic assumption of the following calculation is that $N_T(D;\ t)$ is an RV independent of the initial Poisson process and that the RVs $N_T(D;\ t_i)$ and $N_T(D;\ t_j)$ are independent whatever t_i and t_j. It is clear that this assumption is exactly the same as that introduced for the random amplitudes A_i used in Section 8.3.

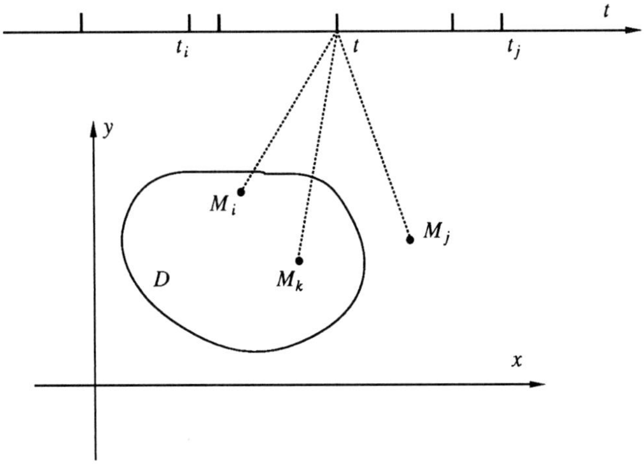

Figure 8.10 Generation of a point process in $\mathbb{R}^2$ from a time Poisson process.

The total number of points M_i belonging to D can be written as

$$N(D) = \sum_i N_T(D;\ t_i) \tag{8.141}$$

and this expression is similar to those used for describing the shot noise with random amplitude, for example (8.62). For this reason we can use exactly the same procedure in order to calculate the characteristic function of $N(D)$. For this we write (8.141), as (8.62), in the form

$$N(D) = \int N_T(D;\ \theta)\ dN(\theta) \tag{8.142}$$

Decomposing the time axis into small time intervals Δt_i we deduce that $N(D)$ is the sum of independent RVs and its characteristic function can be written, like (8.70), in the form

$$\phi_N(u; D) = \exp\left\{ \int \lambda(\theta) \left\{ \phi_\theta(u; D) - 1 \right\} d\theta \right\} \tag{8.143}$$

where $\phi_\theta(u; D)$ is the characteristic function of $N_T(D;\ \theta)$, or

$$\phi_\theta(u; D) = E\left\{ \exp\left[juN_T(D;\ \theta) \right] \right\} \tag{8.144}$$

Let us give a simple application of these expressions. Suppose that, instead of what appears in Figure 8.10, each point t_i *generates only one point* M_i, and let us call $p(D;\ t)$ the probability that this point belongs to D when it is generated at t. In this case $N_T(D;\ t)$ is a heads and tails RV and its characteristic function is given by (2.87) where p is replaced by $p(D;\ t)$. Inserting this expression into (8.143), we obtain

$$\phi_N(u; D) = \exp[m(D)(e^{ju} - 1)] \tag{8.145}$$

where

$$m(D) = \int \lambda(\theta) p(D; \theta) \, d\theta \tag{8.146}$$

Assuming that $\lambda(t)$ and $p(D; t)$ are such that this last expression is finite, we deduce that (8.145) has the form (2.91), which shows that the number of points belonging to D is a Poisson RV with a mean value defined by (8.146).

In order to verify if the process of points M_i is a Poisson process, it suffices now to verify that the number of points belonging to two non overlapping domains D_1 and D_2 are independent. Calling N_1 and N_2 the corresponding RVs, and using exactly the same procedure, we easily find

$$\phi_N(u_1, u_2) = \exp\left\{ \int \lambda(\theta) \left\{ \phi_\theta(u_1, u_2) - 1 \right\} d\theta \right\} \tag{8.147}$$

where $\phi_\theta(u_1, u_2)$ is the characteristic function of the RVs $N_T(D_1, \theta)$ and $N_T(D_2, \theta)$. Let us introduce the probabilities $p(D_1, \theta)$ and $p(D_2, \theta)$ defined as above. Using the assumption that each t_i *generates only one point* M_i, $p(D_1, \theta)$ is the probability of the event $N_T(D_1, \theta) = 1$ and $N_T(D_2, \theta) = 0$, $p(D_2, \theta)$ is the probability of the event $N_T(D_1, \theta) = 0$ and $N_T(D_2, \theta) = 1$ and finally the probability of the event $N_T(D_1, \theta) = 0$ and $N_T(D_2, \theta) = 0$ is $1 - p(D_1, \theta) - p(D_2, \theta)$. Thus the characteristic function $\phi_\theta(u_1, u_2)$ is

$$\phi_\theta(u_1, u_2) = 1 + p(D_1, \theta)(e^{ju_1} - 1) + p(D_2, \theta)(e^{ju_2} - 1) \tag{8.148}$$

Inserting this expression in (8.147) we deduce that

$$\phi_N(u_1, u_2) = \phi_N(u_1, D_1)\, \phi_N(u_2, D_2) \tag{8.149}$$

where $\phi_N(u_1, D)$ is given by (8.145) and $m(D_1)$ and $m(D_2)$ are given by (8.146). This shows that the RVs $N(D_1)$ and $N(D_2)$ defined by (8.142) are independent.

In conclusion, the point process deduced from the t_is by the transformation T satisfies the two conditions introduced in Section 8.1 to define a Poisson process. It is then a Poisson process in $\mathbb{R}^2$ and its density can be deduced from (8.146). More precisely, let us introduce the function $f(x, y; \theta)$ such that the probability introduced above can be written

$$p(D; \theta) = \iint_D f(x, y; \theta) \, dx \, dy \tag{8.150}$$

It results from (8.146) that the density of the Poisson process of points M_i is

$$\lambda(x, y) = \int \lambda(\theta) f(x, y; \theta) \, d\theta \tag{8.151}$$

Let us present some comments on this calculation. It is realized in the case where the initial point process is a distribution of time instants, or a point process in $\mathbb{R}^1$. It is clear that the same calculation can be realized if the initial process is a point process in

$\mathbb{R}^m$. Similarly, the process of points M_i is presented in $\mathbb{R}^2$, as seen in Figure 8.10. But the same calculation can be realized if the final process is a distribution of points in $\mathbb{R}^n$.

Furthermore, all the previous calculations are valid provided that the mean value $m(D)$ defined by (8.146) is finite. This introduces the condition

$$\int d\theta \iint_D \lambda(\theta) f(x, y; \theta)\, dx\, dy \tag{8.152}$$

Finally, the erasing effect considered in (**a**) is a particular case of this general method. It corresponds to the case where the space of the points M_i is $\mathbb{R}^1$, or the same as that of the points t_i. The probability $p(D;\ \theta)$ is extremely simple. If θ belongs to D this probability is $1 - \mu(\theta)$, and if θ does not belong to D this probability is 0.

(d) Random delays. Starting from a Poisson process of points t_i, suppose that we generate a point process of points τ_i in such a way that each t_i is transformed into one τ_i by a random delay L. Assuming that the PDF of this delay is $p(l;\ \theta)$ if the initial point arrives at the time θ, we can apply the previous reasoning and deduce that the point process of points τ_i is a Poisson process with the density

$$\lambda'(\tau) = \int \lambda(\theta) p(\tau - \theta;\ \theta) d\theta \tag{8.153}$$

(e) Random amplitude. Starting from a Poisson process in time, we construct a point process in the plane by an operation similar to that pictured in Figure 8.10. To each point t_i of the initial process we associate one point M_i such that $x_i = t_i$ and $y_i = A_i$, where A_i is a random amplitude satisfying all the requirements introduced previously. It then follows from the previous calculations that the point process of points M_i is a Poisson process. Introducing the probability density $p(a;\ t)$ of the random variables A_i, we deduce that the function $f(x,\ y;\ \theta)$ in (8.150) is

$$f(x,\ y;\ \theta) = \delta(x - \theta)\ p(y;\ \theta) \tag{8.154}$$

and as a result of (8.150), the density of the process of points M_i is

$$\lambda(x,\ y) = \lambda(x)\ p(y;\ x) \tag{8.155}$$

If the amplitudes A_i are IID random variables, we obtain

$$\lambda(x,\ y) = \lambda(x)\ p(y) \tag{8.156}$$

which has an obvious interpretation.

To conclude this section, it is worth pointing out the diversity of signals which can be constructed from a Poisson process. The assumption of absence of memory is so strong that calculations which would be unmanageable otherwise are made possible. Several other examples can be discussed, including the generation of signals by random systems driven by Poisson processes.

8.7 COMPOUND POISSON PROCESSES

A Poisson process is completely defined by its density λ, as indicated in Section 8.1. The only requirements for this density are, first, to be non-negative and, second, to be such that the mean number of points in any finite interval be finite. The compound Poisson processes introduce the idea that the density is *random*, or can be written $\lambda(t, s)$, where s refers to the randomness, as indicated in Section 5.1. In a sense, such a process is doubly random, because of the randomness resulting from the Poisson process and that resulting from its density. This is why such a process is sometimes called a doubly Poisson process. Whatever the terminology, a compound Poisson process must be seen as a Poisson process whose density is random.

Compound Poisson processes are not only introduced for the purpose of sheer mathematical generalization. They are important for the description of various physical phenomena. The most common is certainly the photoelectric effect discussed in Section 2.1 and in Figure 1.3. It is possible to show that the pulses observed at the output of a photomultiplier constitute a Poisson process whose density is proportional to the light intensity of the optical field impinging on the photocathode. In many situations this optical field is random, as described in Statistical Optics, and therefore the point process observed is a compound Poisson process. Randomness in density can clearly be introduced in many other situations, for example in traffic problems.

Let us briefly indicate what changes when the density of a Poisson process becomes random. For a given experiment, or for $s = s_0$, the density is equal to $\lambda(t, s_0)$ and all the results of the previous sections can be applied. However, they must be considered as results conditionally to the value $s = s_0$. Consequently, all the probabilities or all the expected values calculated previously must be considered as conditional probabilities or conditional expected values. The *a priori* probabilities or expected values are calculated as in (1.38), (4. 24) or (4.33). The main consequence of the randomness of the density is that the fundamental property of microscopic independence in Poisson processes disappears and some memory effect may appear.

8.7.1 Definition of Compound Poisson Processes

Let us use the approach and notation introduced in Section 5.7.1. A point process is defined by the probabilities

$$p[\{n_i\}] = P\left\{ \prod_{i=1}^{k} (N = n_i) \right\} \tag{8.157}$$

In the case of a Poisson process these probabilities are

$$p[\{n_i\}] = \prod_{i=1}^{k} e^{-m_i} \frac{(m_i)^{n_i}}{n_i!} \tag{8.158}$$

where

$$m_i = \int_{\Delta T_i} \lambda(\theta)\, d\theta \tag{8.159}$$

If the density is a random non-negative function $\lambda(t, s)$, we can introduce the RVs

$$M_i = \int_{\Delta T_i} \lambda(\theta; s)\, d\theta \tag{8.160}$$

and the probabilities defining a compound Poisson process take the form

$$p[\{n_i\}] = E\left\{ \prod_{i=1}^{k} e^{-M_i} \frac{(M_i)^{n_i}}{n_i!} \right\} \tag{8.161}$$

where the expectation is taken with respect to the RVs M_i.

Following the ideas presented in Section 8.1.1, it is possible to write (8.161) in the case where the intervals ΔT_i are arbitrary small. To do this, let us call $dN(\theta_i)$ the number of points in the interval $[\theta_i, \theta_i + d\theta_i)$. This number can take only the values 0 or 1, and we deduce from (8.161) that

$$P\left\{ \prod_{i=1}^{k} [dN(\theta_i) = 1] \right\} = E[\lambda(\theta_1)\lambda(\theta_2)\ldots\lambda(\theta_k)]\, d\theta_1 \ldots d\theta_k \tag{8.162}$$

For $k = 2$ and $\lambda(\theta)$ non-random and constant we return to (8.11). The probability in (8.162) is sometimes called the *coincidence probability*, because it is experimentally recorded by using a coincidence system. It is a system which analyzes the events appearing in the intervals $[\theta_i, \theta_i + d\theta_i)$ and the coincidence is realized if one point is recorded in each interval. This expression is especially interesting for $k = 2$, and can be written as

$$P\left\{ [dN(\theta_1) = 1]\ [dN(\theta_2) = 1] \right\} = E[\lambda(\theta_1; s)\lambda(\theta_2; s)]\, d\theta_1\, d\theta_2 \tag{8.163}$$

Because of the expectation appearing either in (8.162) or in (8.163), the result is that the RVs $dN(\theta_i)$ are no longer independent as in the case of a pure Poisson process. There is thus a memory effect which can be analyzed by using the conditional probabilities such as (8.12). In fact, by using (8.162) for $k = 1$ and the rule (1.24) we deduce that

$$P\left\{ [dN(\theta_2) = 1] \mid [dN(\theta_1) = 1] \right\} = \frac{E[\lambda(\theta_1)\lambda(\theta_2)]}{E[\lambda(\theta_1)]}\, d\theta_2 \tag{8.164}$$

For further simplification, suppose now that $\lambda(t, s)$ is stationary. In this case the denominator of (8.164) is constant and the numerator is a function of $\theta_2 - \theta_1$. This function is maximum for $\theta_1 = \theta_2$, as seen in Section 5.2.3, and this property used in (8.164) introduces a *bunching effect* which is characteristic of a compound Poisson process. This effect of course disappears if $\lambda(t, s)$ is no longer random.

8.7.2 Number of Points in a Given Interval

Let ΔT be an arbitrary time interval and $\Delta N(T)$ be the RV equal to the number of points of the process recorded in this interval. In this case (8.161) takes the form

$$p[n] \triangleq P\{N(\Delta T) = n\} = E\left\{e^{-M} \frac{M^n}{n!}\right\} \tag{8.165}$$

where M is the RV defined by (8.160). This probability is sometimes called the counting probability because it is used in counting experiments and recorded by a counter. Suppose that the RV M is continuous and defined by a PDF $p_M(m)$. Since $M \geq 0$, we can write (8.165) as

$$p[n] = \int_0^\infty p_M(m)\, e^{-m} \frac{m^n}{n!}\, dm \tag{8.166}$$

This relation is sometimes called the Poisson transform. Some of its properties were discussed in Problem 2.8 and others appear as problems at the end of this chapter. Similarly the characteristic function of $N(\Delta T)$ is deduced from (2.91) by taking the expected value, which gives

$$\phi_N(u) = E\left\{\exp[M(e^{ju} - 1)]\right\} = \int_0^\infty p_M(m) \exp[m(e^{ju} - 1)]dm \tag{8.167}$$

By a limited expansion we easily deduce the first moments of N, which are

$$E[N] = E[M] \tag{8.168}$$

$$E[N^2] = E[M] + E[M^2] \tag{8.169}$$

Thus the variance of N takes the form

$$\sigma_N^2 = E[M] + \sigma_M^2 \tag{8.170}$$

This expression is especially interesting. It shows that the variance of a compound Poisson process cannot be smaller than that of a pure Poisson process with the same mean value. This latter case appears if M is no longer random, which gives $\sigma_M^2 = 0$, and (8.170) becomes equivalent to (2.47).

For some problems it is interesting to work with the generating function and the factorial moments studied in Problem 2.4. From the calculation presented in that problem we find that

$$E[N(N-1)(N-2)\ldots(N-k+1)] = \int_0^\infty p_M(m) m^k dm \tag{8.171}$$

Some examples of compound RVs are discussed as problems at the end of this chapter. Note also that Problem 2.8 is a good introduction to other properties of these RVs.

8.7.3 Time Intervals between Points

Because of the memory effect of compound Poisson processes it is necessary to distinguish between the lifetime and the survival time introduced in Section 5.7.2.

(a) Lifetime. The principle of the calculation is the same as in Section 8.1.2. Consider the three non-overlapping intervals $[t,\ t+dt)$, $[t+dt,\ t+l)$, $[t+l,\ t+l+dl)$. The probability of finding one point in the first interval, zero point in the second and one point in the third is deduced from (8.161) and (8.162), which gives

$$P(t,\ dt\ ;\ l,\ dl) = E\left\{\lambda(t)\lambda(t+l)\ \exp\left[-\int_t^l \lambda(\theta)d\theta\right]\right\} dt\ dl \tag{8.172}$$

The PDF of the lifetime is related to the conditional probability of finding zero point in $[t,\ t+l)$, one point in $[t+l,\ t+l+dl)$ conditionally to the event of finding one point in $[t,\ t+dt)$. The probability of this last event is $E[\lambda(t)]\ dt$. This gives the PDF $p_L(l;\ t)$ by

$$p_L(l;\ t) = \frac{1}{E[\lambda(t)]} \cdot E\left\{\lambda(t)\lambda(t+l)\ \exp\left[-\int_t^{t+l} \lambda(\theta)d\theta\right]\right\} \tag{8.173}$$

If $\lambda(t)$ is stationary this gives

$$p_L(l) = (1/\lambda_m)\, E\left\{\lambda(0)\lambda(l)\ \exp\left[-\int_0^l \lambda(\theta)d\theta\right]\right\} \tag{8.174}$$

where $\lambda_m = E[\lambda(t)]$. If $\lambda(t)$ is not random and constant we return to (8.9).

(b) Survival time. The calculation is the same but without the condition of a point at the beginning of the interval. By the same procedure we easily obtain

$$p_S(l;\ t) = E\left\{\lambda(t+l)\ \exp\left[-\int_t^{t+l} \lambda(\theta)d\theta\right]\right\} \tag{8.175}$$

which in the stationary case gives

$$p_S(l) = E\left\{\lambda(l)\ \exp\left[-\int_0^l \lambda(\theta)d\theta\right]\right\} \tag{8.176}$$

It is obvious that $p_L(l)$ and $p_S(l)$ are different, which was not the case with a pure Poisson process. Some examples of these PDs are discussed as problems.

It is possible to extend to compound Poisson processes the whole discussion concerning the shot noise effects. The basic idea remains the same. For example the characteristic function (8.70) related to a pure Poisson process and a shot noise with random amplitudes can be considered as a conditional expectation when $\lambda(t)$ is random.

The *a priori* characteristic function is simply deduced from (8.70) by taking the expectation with respect to the random density $\lambda(t)$.

On the other hand, the asymptotic properties are not the same as those appearing in Section 8.4.4. The asymptotic shot noise is not necessarily normal and the situation is similar to that seen in Section 4.8.8 for the central limit theorem of random sums. For further information on these problems see [Picinbono, 1970].

PROBLEMS

8.1 By using the same procedure as that used to obtain (8.8), extend the result of Section 8.1.4 to non-stationary Poisson processes, or to Poisson processes with non-constant density $\lambda(t)$.

8.2 Calculate the conditional probability (8.12) for the point process obtained from a Poisson process by the method described in Figure 8.2. Begin the calculation by taking $k = 2$ and generalize the procedure for arbitrary k. Explain the situation which appears when k becomes very large.

8.3 Extend the results of Section 8.1.5 as indicated at the end of this section and show the relation between Poisson and multinomial distribution.

8.4 Starting from a stationary Poisson process of density λ we construct the random signal $x(t)$ defined as follows. Let t_k and t_{k+1} be two successive points of the Poisson process. In the interval $t_k \leq t \leq t_{k+1}$ the signal $x(t)$ is equal to $t - t_k$ and this procedure of definition is extended over the complete time axis.
(**a**) Calculate $E[x(t)]$.
(**b**) Calculate the correlation function of the signal $x(t)$.
(**c**) Deduce its power spectrum.

8.5 Consider a stationary Poisson process of density λ and let $N(t)$ be the number of points belonging to the time interval $[0, t)$. Calculate the mean value of $N(t)$ and its covariance function. Calculate the characteristic function of the two-dimensional random variables $N(t_1), N(t_2)$, $0 < t_1 < t_2$.

8.6 Consider the same questions as in the previous problem if $N(t)$ is replaced by the function $M(t)$ introduced in Section 8.3. Consider the specific case where the RVs A_i have an one-sided exponential PDF.

8.7 Consider the signal $x(t) = \int_0^t \theta \, dM(\theta)$ where $M(t)$ is the function introduced in Section 8.3 and used in (8.62). Calculate the mean value and the covariance function of $x(t)$.

8.8 Consider a Poisson process in time with a constant density λ and let $M(t)$ be the signal introduced in Section 8.3. More precisely, $M(t)$ is defined for $t \geq 0$, $M(0) = 0$ and $M(t)$ varies by steps of amplitudes A_i at the points t_i of the Poisson process. The assumptions of independence are the same as in Section 8.3 and we also assume that $E[M(t)] = 0$. Furthermore, all the RVs A_i have the same distribution function $F(a)$.

(a) Calculate $E(A)$.

(b) Deduce $E(A^2)$ from the variance of $M(t)$ and from the density λ.

(c) From $M(t)$ we construct another point process of points θ_i by retaining only the points t_i when A_i satisfies $a_1 < a \leq a_2$. Calculate the PDF of the distance between successive points θ_i.

8.9 At each point t_i of a stationary Poisson process of density λ we associate an RV ϕ_i. These RVs are IID with a uniform distribution over 0, 2π and are independent of the Poisson process. Let $x(t)$ be the signal defined by $x(t) = \exp j\phi_i$ for $t_i \leq t < t_{i+1}$. This means that $x(t)$ is constant between the points of the Poisson process and its value changes randomly at each such points.

(a) Calculate the covariance function of the signal $x(t)$.

(b) Calculate the second order moment $m(t_1, t_2) = E[x(t_1)x(t_2)]$.

(c) Calculate the fourth order moment $E[x(t_1)x(t_2)x^*(t_3)x^*(t_4)]$.

8.10 Starting from a stationary Poisson process of density λ we construct the signals $x_i(t)$ defined by (8.37) where $h(t)$ is replaced by $h_i(t)$.

(a) Calculate the crosscorrelation function $\gamma_{ij}(\tau)$ between $x_i(t)$ and $x_j(t)$.

(b) Assuming that $h_1(t) = 0$ for $0 \leq t < T$ and 0 otherwise, and that $h_2(t) = h_1(-t)$, calculate the autocorrelation functions $\gamma_1(\tau), \gamma_2(\tau)$ and the crosscorrelation functions $\gamma_{12}(\tau)$ and $\gamma_{21}(\tau)$.

(c) Make the same calculations for $h_3(t) = u(t)\exp(-at)$, where $u(t)$ is the unit step function, and for $h_4(t) = h_3(-t)$.

(d) Introducing $y_1(t) = x_1(t) + x_2(t)$ and $y_2(t) = x_3(t) + x_4(t)$, calculate the autocorrelation functions of $y_1(t)$ and $y_2(t)$.

8.11 Let $x(t)$ be the signal defined by (8.37) in which $h(t) = \exp(-at)$ and the integral is only calculated over the finite interval T_1, T_2. It is also assumed that the Poisson process defining $N(t)$ is stationary with a density λ.

(a) Calculate $E[x(t)]$ and indicate if $x(t)$ is stationary.

(b) Calculate the covariance function of $x(t)$ noted as $\gamma(t_1, t_2)$.

(c) In mean square estimation problems it is necessary to calculate a conditional expected value or a regression. Assuming that $t_1 < t_2$ and that $x(t_1) = x_1$, calculate the regression $E[x(t_2)|x_1]$ and show that this regression is linear.

8.12 **Counting signal.** This signal is defined by (8.45) and its power spectrum is given by (8.47).

(a) Calculate the characteristic function and the probability distribution of the pair of RVs $x(t_1)$ and $x(t_2)$.

(b) Calculate the third and fourth order moments of $x(t)$, as defined by (5.42).

8.13 By using the properties indicated in 8.5.3, calculate the power spectrum of the signal $y(t) = x^2(t) - E[x^2(t)]$, where $x(t)$ is the shot noise signal deduced from a Poisson process with random amplitudes.

8.14 Let $x(t)$ be the random signal studied in Problem 6.12 and suppose that the function $\bar{X}(\nu)$ appearing in this problem is the function $N(\nu)$ describing a

Poisson process of density $F(\nu)$, where $F(\nu)$ is a non-negative function with a Fourier transform $f(t)$ such that $f(0)$ is finite.

(a) Give the physical interpretation of this kind of situation in terms of the signal $x(t)$.

(b) Show the relation between $f(t)$ and the expectation $E[x(t)]$. Indicate whether the signal $x(t)$ is first order stationary or not.

(c) Calculate the covariance function of $x(t)$ and show that it is stationary. Explain its relation to $f(t)$.

(d) Show that the centred signal $x_c(t) = x(t) - E[x(t)]$ is zeromean valued and second order stationary.

(e) Let $y(t) = |x_c(t)|^2$ and $y_T(t)$ be its time average defined by (6.93). By using the property $dN^2(\nu) = dN(\nu)$, show that $y_T(t)$ tends to a random limit L when T tends to infinity. Specify the probability distribution of this limit.

(f) Give a physical interpretation of all these results.

8.15 Let $x(t)$ be the random telegraph signal studied in Section 8.6.1 and such that $E[x(t)] = 0$. Let $y(t)$ be the signal defined by $y(t) = \exp[jax(t)]$, where a is a real parameter. Calculate the expected value and the correlation function of $x(t)$.

8.16 Let $X(t)$ be the random telegraph signal studied in Section 8.6.1. Suppose that $t_1 < t_2$ and let $X(t_1) = X_1$ and $X(t_2) = X_2$. Calculate the regression $r(x_1) = E[X_2|x_1]$.

8.17 Let $x(t)$ be the random telegraph signal studied in Section 8.6.1 and such that $E[x(t)] = 0$. The derivative $x'(t)$ of this signal is a sequence of alternated Dirac pulses appearing at the time instants of the Poisson process when $x(t)$ changes its value. Let $y(t)$ be the signal equal to $2\lambda x(t) + x'(t)$. Calculate the correlation function and the power spectrum of $y(t)$. Indicate if $y(t)$ is a signal with independent increments.

8.18 The purpose of this problem is to study a generalization of the random telegraph signal. We start from a stationary Poisson process of density λ. Let $x(t)$ be the signal defined as follows. For $t = 0$, $x(0)$ is an RV taking the values $\exp[2\pi j(k/K)]$, k and K integers satisfying $0 \leq k < K$, with the same probability $1/K$. At each point of the Poisson process, $x(t)$ changes in such a way that $x(t) = x(t_i)\exp[2\pi j/K]$ for $t_i < t \leq t_{i+1}$.

(a) Show that the random telegraph signal is obtained for $K = 2$.

(b) Show that $x(t)$ is second order stationary and calculate its correlation function $\gamma(\tau) = E[x(t)x^*(t - \tau)]$. Make this calculation for $\tau > 0$ and extend the result for $\tau < 0$ by using the symmetry relations (5.27).

(c) Calculate the power spectrum $\Gamma(\nu)$ of $x(t)$. Show that this spectrum has a Lorentzian shape, as presented in Figure 5.2. Give the central frequency and the bandwidth of this spectrum in terms of λ and K.

(d) Explain the situation which appears when K tends to infinity.

8.19 As a result of the discussion after (8.44), the distribution process (5.71) associated with a stationary Poisson process can be considered as a white noise. Consider now the distribution process $x(t) = \sum(-1)^i \delta(t - t_i)$ which corresponds to a succession of random pulses of alternated signs.

(a) Starting from the properties of the random telegraph signal, calculate the power spectrum of $x(t)$ and show that it is no longer a white noise. What is the origin of this property?

(b) Calculate the power spectrum of the shot noise deduced from $x(t)$ which is the signal $y(t) = \sum(-1)^i h(t - t_i)$ in terms of the frequency response of the filter $h(t)$.

8.20 Let $x(t)$ be the random telegraph signal studied in Section 8.6.1 but defined from a non-stationary Poisson process with density $\lambda(t)$.

(a) Calculate the mean value of $x(t)$ and compare the result obtained with (8.110).

(b) Find the condition ensuring that $E[x(t)] = 0$ whatever t. This condition is assumed to be valid in what follows.

(c) Calculate the covariance function of $x(t)$ and compare the result obtained with (8.117) or (8.118).

(d) Give an expression of the higher-order moments of $x(t)$.

8.21 We want to extend the statistical model of photographic plate granularity. However, in order to simplify the calculations we will restrict the analysis to the one-dimensional case. Let $\{x_i\}$ be the random points of a Poisson process of density λ on an axis. To each point x_i of the process we associate the grain g_i which is the domain $[x_i - a, x_i + a]$.

(a) Suppose first that the transparency associated with a grain is 0 or 1. This implies that the transparency $T(x)$ at any arbitrary point x is 1 if no grain reaches the point x and 0 otherwise. Calculate the expectation, the correlation function and the power spectrum of $T(x)$.

(b) Make the same calculations in the more sophisticated situation where each grain introduces a transparency equal to r, $0 < r < 1$. In this case the transparency $T(x)$ at any arbitrary point x is equal to r^n if n grains reach the point x.

8.22 Make the calculation giving the correlation function (8.138).

8.23 **Oscillator with random phase shifts.** In order to represent the phase fluctuations of certain oscillators, it is of interest to study a random signal $x(t)$ in which the phase changes randomly at the time instants of a stationary Poisson process. More precisely, for $t > 0$ we can write $x(t) = \cos[\omega_0 t + \Phi + \sum_{i=1}^{n} \Delta\phi_i]$ with the following assumptions. The random phase Φ is uniformly distributed over $0, 2\pi$. The phase shifts $\Delta\phi_i$ appear at the time instants of a stationary Poisson process of density λ. These shifts are IID random variables, characterized by their characteristic function $f(u)$, and are independent of the Poisson process. Finally the number n is the number of random points of the process appearing in the time interval $0, t$.

(a) Calculate the expected value of $x(t)$.

(b) Calculate the correlation function and the power spectrum of $x(t)$.

(c) Discuss the results in terms of the function $f(u)$. Study especially the modification of the central frequency of the oscillator and its bandwidth.

8.24 Let $\{t_j\}$ be the points of a Poisson process of density $\lambda(t)$. Suppose that $\lambda(t) = \lambda$ for $t \leq 0$ and $\lambda(t) = 0$ for $t > 0$. To each point t_i we associate a point $\tau_j = t_i + L_i$, where L_i is an RV whose PDF is $p(l) = u(l)\exp(-l)$. In this expression $u(l)$ is the unit step function. It is assumed that the RVs L_i are IID and independent of the Poisson process.

(a) Calculate the density $\lambda'(t)$ of the process of points τ_j.

(b) Calculate the probability that no point τ_j satisfies $\tau_j > 0$.

8.25 Consider the renewal process introduced by dead time effect in a stationary Poisson process and described in Section 8.6.3(**b**). Calculate the mean value and the variance of the RV representing the random distance between successive points of this renewal process. Explain the results obtained when the density λ of the Poisson process tends to infinity.

8.26 Let $p[n]$ be the probability of the RV N representing the number of points of a compound Poisson process counted in a given time interval, and $p_M(m)$ the PDF of the RV M appearing in (8.165). We assume that M has an exponential distribution or that $p_M(m) = (1/a)\exp(-m/a)$. Calculate $p[n]$, its characteristic function, the mean value, the variance, and the factorial moments of the RV N.

8.27 Let $p[n]$ be the probability of the RV N representing the number of points of a compound Poisson process counted in a given time interval, and $p_M(m)$ the PDF of the RV M appearing in (8.165). We assume that $p_M(m)$ is equal to $1/a$ for $0 \leq m \leq a$ and zero otherwise. Calculate the factorial moments of N, its mean value, its variance, and its characteristic function.

8.28 Consider a compound Poisson process and suppose that its random density is a constant RV M defined by the PDF $p_M(m) = (1/a)\exp(-m/a)$.

(a) Calculate the coincidence probability (8.163) and show that there is no longer microscopic independence as in the case of a pure Poisson process.

(b) Calculate the conditional probability (8.164) and find the same conclusion as in the previous question.

(c) Calculate the PDF of the lifetime and compare the result with that obtained for a pure Poisson process.

(d) Calculate the PDF of the survival time.

(e) Calculate the probability of counting n points in an arbitrary time interval.

(f) Make the same calculation as in the previous question with the additional assumption that there is one point of the process at the beginning of the counting interval.

8.29 **Poisson transform**. It is defined by (8.165) and transforms the PDF $p_M(m)$ of an RV M into the probabilities $p[n]$ of a discrete RV appearing in counting experiments. In many problems it is interesting to realize the inverse transformation or to calculate the PDF of M from the results of counting experiments.

This situation appears in various problems of statistical optics but introduces some difficulties analyzed hereafter.

(a) Show that if $p_M(m)$ is a PDF, which means it is a non-negative normalized function, the $p[n]$s constitute a set of probabilities, which means a set of non-negative numbers with a sum equal to 1.

(b) It was seen in Problem 2.8 that the inverse is not true. Let us present here another example. Consider the function $p_M(m) = c(2cm - 1)\exp(-cm)$, $c > 0$. Show that it is normalized and that it takes negative values whatever c. Calculate its Poisson transform and find the values of c ensuring that $p[n] > 0$.

(c) In order to calculate the inverse Poisson transform, let us introduce the function $f(m) = \exp(-m)p_M(m)$. Let $\phi(u)$ be the characteristic function of $f(m)$. Show that $\phi(u)$ can be expressed in terms of the $p[n]$s.

(d) Present the complete procedure allowing the inversion of the Poisson transform and explain the step of this procedure which cannot ensure that $p[n] \geq 0$.

Chapter 9

Random Signals and Dynamical Systems

9.1 INTRODUCTION

In many signal processing problems it is more convenient to describe signals by their process of generation than by the family of finite-dimensional distribution introduced in Section 5.1. In reality this is not a new perspective, and has already been used earlier, for example in the study of shot noise in the previous chapter. The shot noise is the output of a linear filter driven by impulses arriving at the time instants of a Poisson process. Looking at (8.36) or (8.37), we see that this signal is completely defined by the impulse function $h(t)$ and by the statistical properties of the input impulses. If they correspond to a stationary Poisson process with constant amplitude, these statistical properties depend only on the density λ of the process. As a result, the family of finite-dimensional distributions is completely specified when λ and $h(t)$ are known, as given in the calculation in Section 8.4.2.

In this chapter we introduce a new concept, that of the *dynamical system.* Roughly speaking, it corresponds to systems satisfying some physical laws. For example, in the class of all *linear filters*, the dynamical filters are stable, causal and their transfer function is a rational function as indicated in [Picinbono], p. 141. They are closely connected to linear differential equations with constant coefficients which describe most mechanical and electrical phenomena. In the discrete-time case, rational functions are connected to difference equations which have perhaps a less evident physical origin, but are very common in the computer world. In this case the concept of causality is related to that of real time calculation, which implies that the result of a calculation cannot be obtained before the arrival of the data used in that calculation.

There is another point which should be noted: many dynamical systems can be described by a finite number of parameters. This is especially the case for linear dynamical filters, since it is obvious that a rational function, the ratio of two polynomials, is defined by the coefficients of those polynomials. Consequently it is sometimes said that the approach discussed in this chapter introduces a *parametric description* (or modeling) of signals.

Finally, note that this chapter is almost entirely devoted to discrete-time signals, and continuous-time signals are only briefly mentioned at the end. In fact, as we will promptly verify, dynamical systems introduce a large number of numerical algorithms which require the computer processing of discrete data. However, if the physical systems used for signal generation are continuous-time systems, as in the case of the shot noise signal, it is always possible to use the results concerning the sampling of stochastic signals introduced in Chapter 6, especially in Section 6.7.

9.2 AUTOREGRESSIVE SIGNALS

9.2.1 The AR(1) Signal

Consider the first order purely recursive filter with the transfer function $H(z)$ defined by

$$H(z) = \frac{z}{z - a} = [1 - az^{-1}]^{-1} \tag{9.1}$$

and suppose that a is real, satisfies $|a| < 1$ and that the unit circle is in the region of convergence. As a result the frequency response of this filter is defined by (6.70). Similarly, by expansion of $H(z)$ in the region of convergence in terms of z^{-1} we find

$$H(z) = \sum_{k=0}^{\infty} a^k z^{-k} \tag{9.2}$$

which means that the impulse function of this filter is null for $k < 0$ and for $k \geq 0$ defined by

$$h_k = h[k] = a^k \tag{9.3}$$

The input-output relationship (5.55) thus takes the form

$$x[k] = \sum_{l=0}^{\infty} a^l u[k-l] \tag{9.4}$$

where u and x are the input and output respectively.

Suppose now that the input $u[k]$ is a *zeromean weakly white noise* of variance σ^2, as defined in Section 7.2.1. As a result of (5.57), $x[k]$ is also zeromean-valued and its correlation function, given as in (7.17), takes the form

$$\gamma[k] = \sigma^2(1-a^2)^{-1} a^{|k|} \tag{9.5}$$

The z-spectrum of $x[k]$ can be calculated either by introducing (9.5) into (6.65) or by using (6.68), which gives

$$\Gamma(z) = \sigma^2[(z-a)(z^{-1}-a)]^{-1} = \sigma^2[-az + (1+a^2) - az^{-1}]^{-1} \tag{9.6}$$

which is the inverse of (6.81) if $\sigma^2 = 1$. The power spectrum is simply obtained by replacing z by $\exp(j2\pi\nu)$. This gives

$$\Gamma(\nu) = \frac{\sigma^2}{(1+a^2) - 2a\cos 2\pi\nu} \tag{9.6'}$$

It appears that the denominator of this spectrum is the function studied in Example 6.6 and represented in Figure 6.3. The spectrum (9.6') appears in Figure 9.1 where $\sigma^2 = 1$ and several values of a are used. It is impossible to represent the power spectrum corresponding to $a = \pm 1$ because it tends to infinity for the frequencies 0 or 1/2. Assuming that $\sigma^2 = 1$, the function $\Gamma(\nu)$ is represented in Figure 9.1 with logarithmic scale for several values of a. Because of the symmetry with respect to the frequency 0.25 appearing in Figure 6.3, only positive values of a are used.

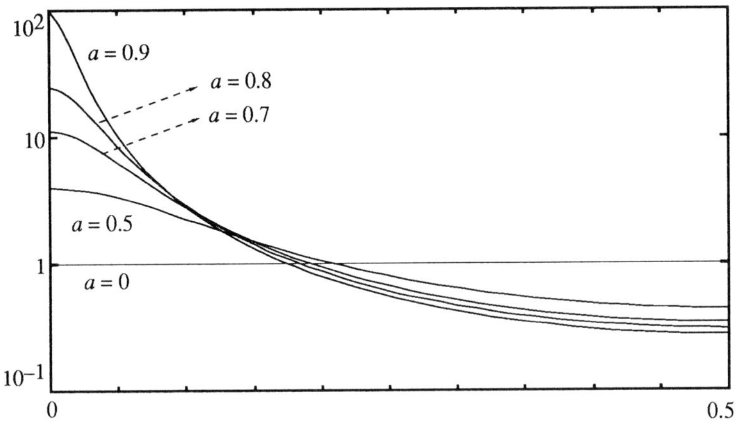

Figure 9.1 Spectra of AR(1) signals for several values of a.

The signal $x[k]$ obtained by (9.4) where $u[k]$ is a zeromean weakly white noise is called an *autoregressive signal of order* 1 or an AR(1) signal. Note immediately that its second order properties are completely specified by two parameters only, namely a and σ^2. The first is called the *regression coefficient* and the second is the *variance* of the driving white noise. Let us now present some features concerning the trajectories of an AR(1) signal. If a is small, for example equal to 10^{-1}, a^l in (9.4) decreases rapidly and $x[k]$ is similar to a white noise. Trajectories of various white noises are presented in Figures 7.1 to 7.3 and show very chaotic behavior. On the other hand, if a is large, for example equal to 0.9, there is a strong smoothing effect and the signal appears more regular. This can be seen in Figures 7.4 to 7.6 representing AR(1) signals with $a = 0.9$ and various kinds of PDF of driving white noises. These figures show the importance of the smoothing effect. Furthermore, even though the PDFs of white noises play no role in the form of the correlation function they have an influence on the trajectories.

Let us now explain the term of *autoregression*. Starting from (9.4), we deduce that, for any k, we have

$$x[k] = ax[k-1] + u[k] \tag{9.7}$$

which means that $x[k]$ is obtained from its past value $x[k-1]$ and $u[k]$ by a linear equation, or a linear autoregression. The physical meaning of this equation is that in order to evaluate the signal at time instant k in terms of its output, it suffices to keep in a memory its past value $x[k-1]$. On the other hand it is clear that the direct calculation of the output in terms of the input by using (9.4) would require storing in a memory all the past values of the input, as in any infinite impulse response (IIR) filter. The flow graph of the system realizing (9.7) is represented in Figure 9.3. We use the classical convention presented in Figure 9.2 showing that a black box system is simply represented by a line indicating its transfer function. The nodes at the extremities of the lines indicate the value of the signal. If several lines arrive at a node, the signal assigned at this node is the sum of all the signals arriving at this node. Finally, note that a given linear filter, characterized by its transfer function, can have very different but equivalent, flow graphs. With these conventions one of the possible realizations of the recursion (9.7) appears in Figure 9.3. Note that the transfer function z^{-1} corresponds to a pure delay system transforming $x[k]$ into $x[k-1]$.

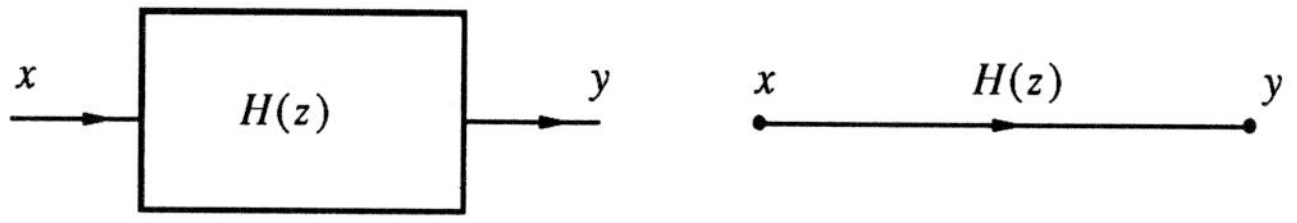

Figure 9.2 Equivalence between black box and flow graph.

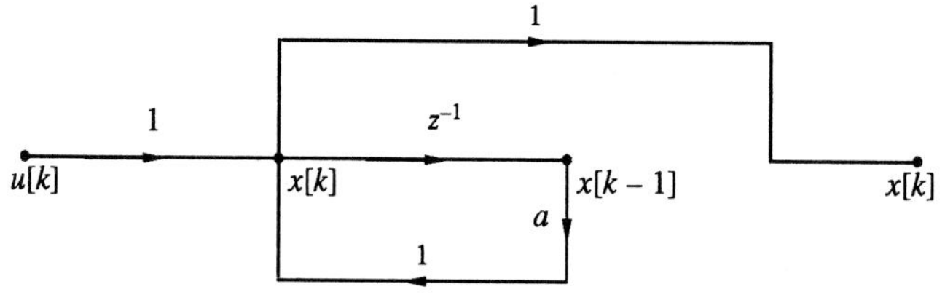

Figure 9.3 Filter generating an AR(1) signal.

We shall introduce here the concept of *normal equations*, sometimes called the Yule-Walker equations. As a result of the whiteness of $u[k]$ and from (9.4), $E\{x[k-1]\,u[k]\} = 0$. Taking the expected value of the square of (9.7) and noting that because of the stationarity, $x[k]$ and $x[k-1]$ have the same variance $\gamma[0]$, we deduce

$$\gamma[0] = a^2\gamma[0] + \sigma^2 \tag{9.8}$$

Furthermore, multiplying (9.7) by $x[k-1]$ and taking the expected value, we deduce

$$\gamma[1] = a\gamma[0] \tag{9.9}$$

It is obvious that these equations can be deduced directly from (9.5) calculated for $k = 0$ and $k = 1$. But the method using (9.7) is more interesting for the generalization which will follow. Equations (9.8) and (9.9) are the two *normal equations* and give the pair of parameters $[a,\ \sigma^2]$ in terms of $\{\gamma[0],\ \gamma[1]\}$. This means that the first two values of the correlation function $\gamma[k]$ corresponding to $k = 0$ and $k = 1$ completely specify the second order properties of the signal characterized by the two parameters a and σ^2. Furthermore note that (9.9) is a linear equation for a.

Going one step further, let us assume that $u[k]$ is a *strict white noise*, as defined in Section 7.2.1. By using the comment made after (2.3), let us write (9.7) in capital letters in order to avoid any confusion with the possible values of $X[k]$ and $U[k]$, also written X_k and U_k. A strict white noise is completely specified by the distribution function

$$F_U(u) = P(U_k \leq u) \tag{9.10}$$

which does not depend on k because of the stationarity of $U[k]$. Let us now calculate the conditional distribution function

$$F_X(x_k | x_{k-1}, x_{k-2}, \ldots) = P[X_k \leq x_k | (X_{k-1} = x_{k-1})\ (X_{k-2} = x_{k-2}) \ldots] \tag{9.11}$$

If $X_{k-1} = x_{k-1}$, we have from (9.7)

$$X_k = a x_{k-1} + U_k \tag{9.12}$$

which is the sum of an RV U_k and a non-random term. As a consequence, (9.11) can be written with (9.10) in the form

$$F(x_k | x_{k-1}, x_{k-2}, \ldots) = F_U(x_k - a x_{k-1}) \tag{9.13}$$

This means that $F_X(.)$ is a function of x_k and x_{k-1} only and not of $x_{k-2}, x_{k-3}, \ldots$ This property characterizes a *Markov process* of order 1, and will also be called a Markovian property. The physical meaning of this property is that the statistical properties of X_k at time k when all the past is known depend only on the "past of order 1," that is x_{k-1}. If we note as P_k the past at time k, this can be written as

$$F_X(x_k | P_k) = F_X(x_k | x_{k-1}) \tag{9.14}$$

We deduce from (9.13) and (9.14) that the conditional expected value satisfies

$$E[X_k | x_{k-1}, x_{k-2}, \ldots] = E[X_k | x_{k-1}] = a x_{k-1} \tag{9.15}$$

because $F_U(u)$ is the distribution function of a zeromean RV. Comparing this with (4.34), we see that the term $a x_{k-1}$ is the regression, which sheds light on the expression "autoregression" already introduced. But there is a more interesting fact: (9.15) shows that the *regression is linear*, a property discussed many times before, and especially in

Sections 4.6.4 and 4.8.2. Finally, note that if $a = 1$, (9.15) exhibits the *martingale property* which appears in (7.32). This is easy to understand. The case $a = 1$ corresponds to the limit of the stability, and is sometimes referred to as the singular case in stability problems. If $a = 1$, the recursive equation (9.7) becomes (7.30), characterizing a DT random walk. So this signal is the limit aspect of an AR(1) signal.

Before moving on, let us make some additional comments. The first is related to some notation problems. In many books or papers (9.7) is written in the form

$$x[k] + a'x[k-1] = u[k] \tag{9.16}$$

which introduces $a' = -a$. The advantage of our notation is that the pole of the transfer function (9.1) is a instead of $-a'$ and that the regression equation (9.15), which is the basis of many problems discussed later, is more convenient than with the term $-a'x_{k-1}$. There is of course no difficulty in changing a into a' in any calculation.

Second, until now we have been discussing the case of real signals. This can be extended to $u[k]$ and a complex. Many results remain the same, but some caution is nevertheless necessary, as when passing from real to complex normal RVs. In order to avoid these problems, we will continue to work with real signals and systems and treat the complex cases in a specific section.

We have not hesitated to discuss at great length the presentation of the properties of AR(1) signals. This is because this very simple signal contains all the ingredients necessary to the understanding of AR(p) signals. We shall continue at a faster pace for the general case.

9.2.2 The AR(p) Signal

Let $\mathbf{a}$ be a real vector with components a_i, $1 \le i \le p$, and consider the filter whose transfer function is

$$H(z) = [1 - (a_1 z^{-1} + a_2 z^{-2} + \dots + a_p z^{-p})]^{-1} \tag{9.17}$$

This is a rational function with the numerator z^p and the denominator

$$D(z) = z^p - \sum_{i=1}^{p} a_i z^{p-i} \tag{9.18}$$

We assume that the vector $\mathbf{a}$ is such that all the roots of $D(z)$ are located inside the unit circle, so this circle is inside the region of convergence. In this case $H(z)$ is the transfer function of a causal and stable filter and the input-output relationship can be written as

$$x[k] = \sum_{l=0}^{\infty} h[l]\, u[k-l] \tag{9.19}$$

where $h[k]$ is the impulse response, inverse z-transform of $H(z)$.

Suppose now that $u[k]$ is a zeromean weakly white noise of variance σ^2. As the filter is stable, $x[k]$ is a zeromean second order stationary signal called AR(p) signal. We deduce immediately from (9.17) that it satisfies the difference equation

$$x[k] = a_1 x[k-1] + a_2 x[k-2] + \dots + a_p x[k-p] + u[k] \qquad (9.20)$$

Looking at this equation, it is obvious that vector notations would be convenient; we thus introduce the vector

$$\mathbf{X}[k] \triangleq [x[k-1], x[k-2], \dots, x[k-p]]^T \qquad (9.21)$$

For obvious reasons this vector can be called the vector of past values of x, or simply the *past vector* of $x[k]$. This allows us to write (9.20) in the form

$$x[k] = \mathbf{a}^T \mathbf{X}[k] + u[k] = \mathbf{X}^T[k]\mathbf{a} + u[k] \qquad (9.22)$$

where $\mathbf{a}$ is the vector whose components are the a_i s. This vector is called the *regression vector.*

The physical realization of the recursion (9.20) or (9.22) is shown in Figure 9.4 where the same notation is used as in Figure 9.3. For simplification we assume that p = 3, the generalization for arbitrary values of p being obvious. Note that this filter uses three delay systems noted by their transfer function z^{-1}. It also appears that the three components of the vector $\mathbf{X}[k]$ are precisely the outputs of three delay systems.

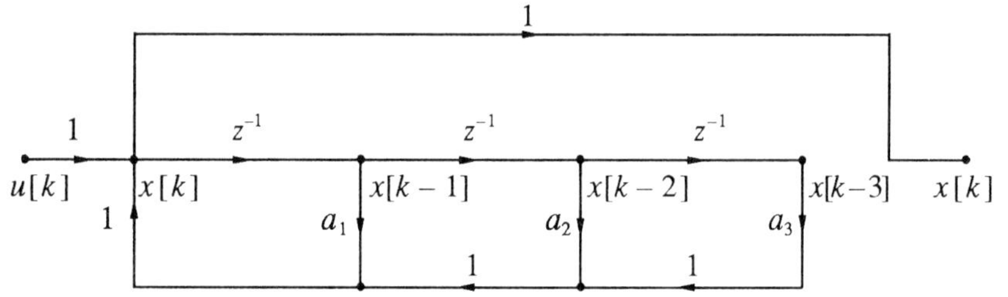

Figure 9.4 Filter generating an AR(3) signal.

From (9.22) we deduce that the second order properties of an AR(p) signal are completely specified by the variance σ^2 and the vector $\mathbf{a}$. For the following discussion we must introduce two second order quantities. The first is called the *correlation vector*, and is defined by

$$\mathbf{c} \triangleq E\{x[k]\,\mathbf{X}[k]\} \qquad (9.23)$$

The second is simply the *covariance matrix* of $\mathbf{X}[k]$, defined by (3.63) and written as

$$\mathbb{R} \triangleq E\{\mathbf{X}[k]\,\mathbf{X}^T[k]\} \qquad (9.24)$$

Note that because of the stationarity assumption $\mathbb{R}$ and $\mathbf{c}$ do not depend on k.

In order to arrive at the *normal equations* the basic property which will be used is

$$E\{u[k]\,\mathbf{X}[k]\} = 0 \tag{9.25}$$

This is a direct consequence of (9.19). In fact, any component of $\mathbf{X}[k]$ is a linear combination of past values of $u[k]$, and no component of $\mathbf{X}[k]$ contains the term $u[k]$, the first term in $u[.]$ present being $u[k-1]$ appearing in $x[k-1]$. As the $u[k]$s are uncorrelated RVs, we immediately deduce (9.25).

The *first normal equation* is obtained by taking the expected value of the square of (9.22). Using (9.25) and (3.91) read backwards with H replaced by T, we obtain

$$\gamma[0] = \mathbf{a}^T\mathbb{R}\,\mathbf{a} + \sigma^2 \tag{9.26}$$

which corresponds to (9.8).

The *second normal equation* is obtained by multiplying (9.22) by $\mathbf{X}[k]$ and taking the expected value. Using (9.25), this gives

$$\mathbb{R}\,\mathbf{a} = \mathbf{c} \tag{9.27}$$

Let us now comment on these equations. We assume that $\mathbb{R}$ has no null eigenvalues, and is then PD, as defined in 3.5.2. We can thus write

$$\mathbf{a} = \mathbb{R}^{-1}\mathbf{c} \tag{9.28}$$

and inserting this relation in (9.26), we obtain

$$\sigma^2 = \gamma[0] - \mathbf{c}^T\mathbb{R}^{-1}\mathbf{c} = \gamma[0] - \mathbf{a}^T\mathbf{c} \tag{9.29}$$

To understand these equations, we must calculate the components of $\mathbf{c}$ and the matrix elements of $\mathbb{R}$. Noting, as before, $\gamma[k]$ as the correlation function of $x[k]$, we obtain, from the definitions (9.21), (9.23) and (9.24),

$$c_i = \gamma[i], \qquad 1 \le i \le p \tag{9.30}$$

$$R_{ij} = \gamma[|i-j|], \qquad 1 \le i \le p,\ 1 \le j \le p \tag{9.31}$$

In other words, $\mathbb{R}$ and $\mathbf{c}$ are known once $\gamma[k]$ is known for $0 \le k \le p$. The conclusion is thus very similar to that encountered for the AR(1) signal: the regression vector $\mathbf{a}$ and the variance σ^2 of $u[k]$ are deduced from the first $(p+1)$ values of the correlation function by linear equations (9.28) and (9.29).

It is important to realize the full implication of this result. If the parameters $[\sigma^2, \mathbf{a}]$ are known, all the second order properties of $x[k]$ are also known. In fact, $\mathbf{a}$ defines the filter (9.17) and σ^2 defines the second order properties of the input signal. As the impulse response of the filter is known, the correlation function of $x[k]$ can be completely calculated by using (7.17) where σ^2 is inserted. Similarly, the z-spectrum is given by (6.68), or $\sigma^2 H(z)\,H(z^{-1})$. The same can be said if the parameters $[\gamma[0], \mathbf{c}]$ are known, because they can give σ^2 and $\mathbf{a}$ by (9.28) and (9.29). An interesting consequence of this is that the *power spectrum*, or the complete correlation function of an AR(p) signal, can be deduced only from the first $(p+1)$ values of the correlation function, namely $\gamma[0]$, $\gamma[1]$, ..., $\gamma[p]$. The exact method of calculation will be studied later in the context of algorithms used for autoregressive modeling.

In order to calculate the *correlation function* we start from (9.20). As $\gamma[q]$ is known for $0 \le q \le p$, let us calculate this function for $q > p$. Multiplying (9.20) by $x[k - q]$ and taking the expectation, we note that $E\{u[k]\ x[k-q]\} = 0$, which gives

$$\gamma[q] = a_1\,\gamma[q-1] + a_2\,\gamma[q-2] + \ldots + a_p\,\gamma[q-p] \tag{9.32}$$

This is a recursion similar to that valid for the signal itself. By using this recursion, we can construct the correlation function of $x[k]$, as soon as $\gamma[k]$ is known for $0 \le k \le p$.

Using this complete correlation function, it is possible to calculate its Fourier transform by using (6.42'), which gives the power spectrum. This is the principle of the so-called autoregressive spectral analysis, or parametric spectral analysis. In fact, the spectrum of the signal is entirely determined by $p + 1$ parameters used in the normal equations.

The covariance matrix $\mathbb{R}$ defined by (9.24) and used in the normal equations is symmetric and positive-definite, as is any regular covariance matrix. The singular case where $\mathbb{R}$ cannot be inverted to calculate **a** by (9.28) needs specific treatment which will be presented later. But $\mathbb{R}$ also has a very specific structure which is the origin of many consequences discussed later. We deduce from (9.31) that all the elements R_{ij} such that $|i-j|$ is constant are equal. This structure is known as that of a *Toeplitz matrix*. Thus a symmetric Toeplitz matrix is completely defined by its first row or its first column, and is noted in what follows as

$$\mathbb{R} = T\{\gamma[0],\ \gamma[1],\ ...,\ \gamma[p-1]\} \tag{9.33}$$

This matrix has the following form

$$\mathbb{R} = \begin{bmatrix} \gamma_0 & \gamma_1 & \gamma_2 & \cdots\cdots & \gamma_{p-1} \\ \gamma_1 & \gamma_0 & \gamma_1 & \cdots\cdots & \gamma_{p-2} \\ \gamma_2 & \gamma_1 & \gamma_0 & \cdots\cdots & \gamma_{p-3} \\ & \cdot & \cdot & \cdot\ \cdot\ \cdot & \\ \gamma_{p-1} & \gamma_{p-2} & \gamma_{p-3} & & \gamma_0 \end{bmatrix} \tag{9.33'}$$

where $\gamma_i = \gamma[i]$.

Note also that if **a** is the solution of the normal equations, (9.22) is satisfied. So we have

$$x[k] - \mathbf{a}^T\,\mathbf{X}[k] = u[k] \tag{9.34}$$

and using (9.25), we can say that $x[k] - \mathbf{a}^T\,\mathbf{X}[k]$ is uncorrelated to the vector $\mathbf{X}[k]$. This has an important interpretation in prediction problems discussed later.

Finally, we will show that the two normal equations can be expressed in a condensed form using only one equation. For this we introduce the two vectors

$$\mathbf{A}^T \triangleq [1, \ -\mathbf{a}^T] \quad ; \quad \mathbf{Y}^T[k] \triangleq [x[k], \ \mathbf{X}^T[k]] \tag{9.35}$$

As a result (9.34) takes the form

$$\mathbf{A}^T\mathbf{Y}[k] = u[k] \tag{9.36}$$

As a result of (9.22) and (9.25), we have

$$E\{u[k]\ x[k]\} = \sigma^2 \tag{9.37}$$

which allows us to write

$$E\{u[k]\ \mathbf{Y}^T[k]\} = [\sigma^2, \ \mathbf{0}^T] \triangleq \mathbf{C}^T \tag{9.38}$$

Furthermore, let us call $\mathbb{R}_y$ the covariance matrix of $\mathbf{Y}[k]$. Because of the definition of this vector, it is clear that the matrix elements of $\mathbb{R}_y$ are still given by (9.31) where p is replaced by $p+1$. So $\mathbb{R}_y$ has the same form as (9.33) with one additional row and column. Using (9.38), we immediately obtain

$$\mathbb{R}_y\mathbf{A} = \mathbf{C} \tag{9.39}$$

which is the normal equation giving (9.26) and (9.27). In order to visualize this point, we can deduce from (9.35) a block decomposition of the matrix $\mathbb{R}_y$, and (9.39) can be written as

$$\left[\begin{array}{c|c} \gamma_0 & \mathbf{c}^T \\ \hline \mathbf{c} & \mathbb{R} \end{array}\right] \left[\begin{array}{c} 1 \\ \hline -\mathbf{a} \end{array}\right] = \left[\begin{array}{c} \sigma^2 \\ \hline \mathbf{0} \end{array}\right] \tag{9.40}$$

By developing this equation we immediately obtain (9.26) and (9.27).

Let us now discuss the *Markovian property* already mentioned in connection with AR(1) signals. We shall assume that the driving noise is now strict sense white. Using exactly the same procedure as previously, we can deduce from the regression equation that

$$F_X(x_k|P_k) = F_U(x_k - \mathbf{a}^T\mathbf{X}_k) = F_X(x_k|\mathbf{X}_k) \tag{9.41}$$

which introduces the Markovian property of order p. This means that the probability distribution of $x[k]$ conditionally to all its past is in reality only a function of a finite past. Similarly, as $u[k]$ is zeromean-valued, we deduce from (9.41) that

$$E\{x[k]|\mathbf{X}[k]\} = \mathbf{a}^T\mathbf{X}[k] \tag{9.42}$$

which again introduces a *linear regression*, as in the normal case and in other distributions. But it is important to note that this linearity appears whatever the distribution function of $u[k]$, and is indeed a general property of AR signals.

Example 9.1 AR(1) signals as a particular case of AR(2) signals. Suppose that the correlation function $\gamma[k]$ is exponential, or that $\gamma[k] = a^k$, $k \geq 0$. Let us construct the AR(2) signal which corresponds to this correlation function. In order to find the regression vector we must solve the normal equations, in which the matrices $\mathbb{R}$ and $\mathbb{R}^{-1}$ and the vector $\mathbf{c}$ are

$$\mathbb{R} = \begin{bmatrix} 1 & a \\ a & 1 \end{bmatrix} \; ; \; \mathbb{R}^{-1} = \frac{1}{1 - a^2} \begin{bmatrix} 1 & -a \\ -a & 1 \end{bmatrix} \; ; \; \mathbf{c} = \begin{bmatrix} a \\ a^2 \end{bmatrix}$$

Applying (9.28) and (9.29), we obtain

$$\mathbf{a}^T = [a, 0] \quad ; \quad \sigma^2 = 1 - a^2$$

which is (9.5) where $\sigma^2(1 - a^2)^{-1} = 1$. We thus arrive at the conclusion that an AR(1) signal can also be considered as an AR(2) signal for which the second component of the regression vector is null. This result is quite natural, and if in (9.17) all the coefficients a_i are null for $i > 1$, we return to (9.1) characterizing an AR(1) signal.

Example 9.2 Other characterizations of an AR(2) signal. The denominator of the transfer function (9.17) is given by (9.18). If $p = 2$, it depends only on two coefficients a_1 and a_2 that are also the components of the regression vector $\mathbf{a}$. But it is well known that a polynomial is also defined by its roots, and if a_1 and a_2 are real, the two roots are either real or complex conjugate. In this latter case these roots can be written as

$$z_1 = m\exp(j\phi) \quad ; \quad z_2 = m\exp(-j\phi) \quad ; \quad 0 < \phi < \pi$$

and the denominator $D(z)$ is also characterized by m and ϕ instead of a_1 and a_2. We can then write $D(z)$ as

$$D(z) = (z - z_1)\,(z - z_1^*) = z^2(1 - z^{-1} z_1)(1 - z^{-1} z_1^*)$$

Replacing z by $\exp(2\pi j\nu)$, this gives

$$D(e^{2\pi j\nu}) = e^{4\pi j\nu}\left[1 - me^{-j(2\pi\nu - \phi)}\right]\left[1 - me^{-j(2\pi\nu + \phi)}\right]$$

As $\Gamma(\nu) = [|D(e^{2\pi j\nu})|^2]^{-1}$, we can write the power spectrum of the AR(2) signal as $\Gamma(\nu) = 1/\Delta(\nu)$, where

$$\Delta(\nu) = \left\{[(1 + m^2) - 2m\cos(2\pi\nu - \phi)]\,[(1 + m^2) - 2m\cos(2\pi\nu + \phi)]\right\}^2$$

This power spectrum is equal to the square of the magnitude of a second order system which is given in textbooks on linear systems (see [Oppenheim], p. 354). The main feature of this spectrum is the possibility of a resonance frequency and an example of a spectrum is given in Figure 9.5. The curves corresponding to $\phi = \pi/4$ are calculated for several values of m and represented in logarithmic scale. The resonance frequency is approximately equal to 0.125 and for larger values of m the peak at this frequency is even greater.

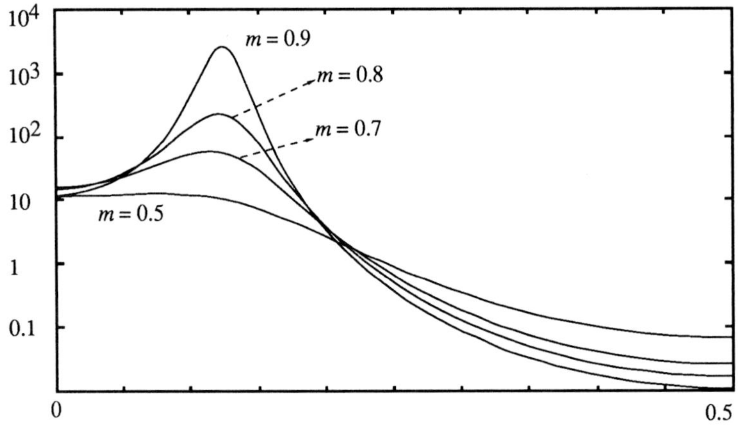

Figure 9.5 Example of second order spectra.

9.2.3 Additional Comments on AR Signals

Here follow some points of interest which are, however, not fundamental to the understanding of the basic properties of AR signals.

(a) Transfer function and linear difference equations. In the definition of AR(p) signals our starting point was the transfer function (9.17) and its region of convergence instead of the linear difference equation (9.20). There are two reasons for this. The first is related to the definition of the transfer function. Starting from the difference equation (9.20) we arrive at the rational function (9.17), but this rational function does not define a specific filter if its region of convergence is not specified. The transfer function, z-transform of the impulse response, is thus a pair made up of one rational function and one region of convergence. As the region of convergence is not specified in the difference equation, this equation does not define a specific filter.

The second reason is that this discussion is presented within the framework of stationary signals. As seen in Chapters 5 and 6, the use of the transfer function is perfectly adapted to this assumption. On the other hand, a difference equation such as (9.20) completely defines $x[k]$ only if some initial conditions are known. The usual way is to consider that the initial conditions are given in the infinite remote past to avoid their influence to the present. It is not necessary to introduce this precision when working with the transfer function.

(b) Complex AR signals. Suppose now that the coefficients a_i in (9.17) are complex and that the driving white noise is also complex. Its correlation function is still given by (7.1) where $\sigma^2 = E[|u[k]|^2]$. Starting from the definition of the correlation function given by (5.17), the vector $\mathbf{c}$ whose components are given by (9.30) is

$$\mathbf{c} = E\{x[k]\ \mathbf{X}^*[k]\} \tag{9.43}$$

Similarly the matrix $\mathbb{R}$ defined by

$$\mathbb{R} \triangleq E\{\mathbf{X}[k]\ \mathbf{X}^H[k]\} \tag{9.44}$$

is Toeplitz Hermitian and the first row is given by (9.33). Because of the Hermitian character, the elements of the first column are $\gamma^*[i]$, $0 \le i \le p-1$.

Finally, the basic equation introducing the normal equations is (9.25) written for complex elements, or

$$E\{u[k]\ \mathbf{X}^*[k]\} = 0 \tag{9.45}$$

Following the same steps as in the real case, we can first write

$$\gamma[0] = E\left\{ (\mathbf{a}^T\mathbf{X}[k])(\mathbf{X}^T[k]\mathbf{a})^* \right\} + \sigma^2 = \mathbf{a}^T\mathbb{R}\,\mathbf{a}^* + \sigma^2 \tag{9.46}$$

which corresponds to (9.26). Furthermore, writing (9.43) with (9.22) and (9.45) gives

$$\mathbb{R}^*\mathbf{a} = \mathbf{c} \tag{9.47}$$

It is clear that we have almost the same Yule-Walker equation when working with complex AR signals.

However, these equations do not completely characterize the second order properties of the model as when working with real signals. The situation is similar to that met when passing from real to complex random variables, as seen in Section 4.6.6. In fact, a complex weak white noise is characterized by (7.1) and by

$$c[k] \triangleq E\{u[n]\ u[n-k]\} = c\,\delta[k] \tag{9.48}$$

where c can be a complex number. If the assumption that $c = 0$, which is analogous to a circularity property, is introduced, then σ^2 completely characterizes the white noise. If that is not the case, we can proceed as we do for real AR signals. This gives

$$E\{x^2[k]\} = \mathbf{a}^T\mathbb{C}\,\mathbf{a} + c$$

where $\mathbb{C}$ is the matrix $E(\mathbf{X}\,\mathbf{X}^T)$. Note that for complex signals this matrix, already introduced in (4.125), is no longer a covariance matrix.

(c) Sign of the regression vector. As in the first order case, the regression equation (9.22) can be written as (9.16), which gives

$$x[k] + \mathbf{X}^T[k]\ \mathbf{a}' = u[k]$$

It is clear that $\mathbf{a}' = -\,\mathbf{a}$, but this change of sign does not change the previous discussion.

(d) Limiting behavior. In the above discussion it was assumed that the poles of the transfer functions introducing AR signals were located inside the unit circle. This was given to ensure the stability of the filters. However, it is possible, with care, to extend the results when some poles are located on the unit circle, and we shall now discuss the consequences of this.

Consider first the AR(1) case and suppose that $a = 1$ in (9.7). In this case this equation is equivalent to (7.30), and $x[k]$ becomes a DT *random walk signal*. It is clear, for example from (7.29), that such a signal is not stationary and its properties were analyzed in Section 7.3. However, it can become stationary if we assume that the driving noise is null, since in this case $x[k]$ is either null or constant. Leaving the first case as without interest, we can consider that a random *constant* signal, analyzed in Example 5.1, is a limit aspect of an AR(1) signal. We saw in Example 5.6 that such a signal is not ergodic and this can be understood within the framework of the results in Section 6.6. In fact its correlation function is constant and it then has a spectral line at the frequency zero which makes ergodicity for the mean value impossible.

Suppose now that $a = -1$. In this case if we assume $x[0] = 0$, we deduce from (9.7) that $x[k]$ is given for $k > 0$ by

$$x[k] = (-1)^{k+1}\{u_1 - u_2 + u_3 - \dots + (-1)^k u_k\}$$

and its covariance function is

$$\gamma[k, k+p] = (-1)^p k\,\sigma^2, \qquad p > 0$$

which is also not stationary. On the other hand if the driving noise is null and if $x[0]$ is not null we obtain

$$x[k] = (-1)^k x[0] = x[0]\cos(\pi k)$$

which is a constant signal with alternating signs. Its correlation function $\gamma[p]$ takes the form

$$\gamma[p] = E\{x[k]\,x[k+p]\} = \sigma_x^2(-1)^p = \sigma_x^2\cos(pk)$$

where σ_x^2 is its variance. In reality this signal can also be considered as the real part of a DT *complex exponential* signal with the frequency $\nu = 1/2$. Its spectrum is reduced to a spectral line at the frequency $1/2$.

This idea can be generalized by using *complex* AR signals discussed in (**b**). For example, the DT complex exponential signal $x[k] = A(\omega)\exp(2\pi jfk)$, where $A(\omega)$ is a random amplitude, satisfies the recursion

$$x[k] = e^{2\pi jf}\,x[k-1]$$

which characterizes an AR(1) signal with null driving noise and a complex regression coefficient equal to $\exp(2\pi jf)$.

This principle can be extended to second order signals. For example, it is easy to verify that the real sinusoidal signal $x[k] = A(\omega)\cos(2\pi fk)$ satisfies the recursion

$$x[k] - 2\cos(2\pi f)\,x[k-1] + x[k-2] = 0 \tag{9.49}$$

This is similar to (9.20) with $u[k] = 0$ and the roots associated to $D(z)$ in (9.18) are $\exp(2\pi jf)$ and $\exp(-2\pi jf)$. These two roots are located on the unit circle, or at the limit of the stability.

(e) Semistationary AR signals. Suppose that the input signal $u[k]$ applied in (9.4) is no longer strictly stationary, but is a sequence of uncorrelated RVs $u[k]$ beginning at $k = 1$. Using the concept of semistationarity introduced in Section 5.3.3 (a), we can say that $u[k]$ is a semistationary white noise. This means that if $k > 0$, the covariance function $\gamma[k,\ k + p] = \sigma^2 \delta[p]$. In order to evaluate the consequences of this assumption on the signal $x[k]$, let us start from the recursion (9.7). To evaluate $x[1]$ we must know the initial condition $x[0]$, and we deduce easily that

$$x[k] = a^k x[0] + \sum_{l=0}^{k-1} a^l u[k - l] \ ,\ k > 0$$

As $|a| < 1$, for large values of k the memory of the initial condition disappears and $x[k]$ tends to be described by (9.4).

Let us now study the consequences on the *variance* and the *correlation function.* We assume that the initial condition $x[0]$ is an RV uncorrelated with the $u[k]$s. Noting as $\sigma_x^2[k]$ the variance of $x[k]$, we deduce from (9.7) that

$$\sigma_x^2[k] = a^2 \sigma_x^2[k - 1] + \sigma^2$$

where σ^2 is the variance of $u[k]$. Using this recursion we easily obtain

$$\sigma_x^2[k] = a^{2k} \sigma_x^2[0] + \frac{1 - a^{2k}}{1 - a^2} \sigma^2$$

It is clear that $\sigma_x^2[k]$ is in general a function of k, which means that $x[k]$ is not stationary. But if $k \to \infty$, $\sigma_x^2[k]$ tends to the limit given by (9.5) for $k = 0$. However, if $\sigma_x^2[0] = \sigma^2(1 - a^2)^{-1}$, we find that $\sigma_x^2[k] = \sigma_x^2[0]$ and the variance is constant. In conclusion, the variance of the signal is stationary as soon as the initial condition $x[0]$ has the variance of a stationary AR(1) signal.

Let us now consider the covariance function $\gamma[k,\ k - p] = E\{x[k]\ x[k - p]\}$, where $k > 0$ and $k - p > 0$. We also deduce from (9.7) that

$$\gamma[k,\ k - p] = a\gamma[k - 1, k - p]$$

Repeating this operation, we obtain

$$\gamma[k,\ k - p] = a^p \gamma[k - p, k - p] = a^p \sigma_x^2[k - p]$$

This function depends in general on k and p, which means that $x[k]$ is not second order stationary. However, if we take the same assumption as above concerning $x[0]$, we deduce that $\gamma[k,\ k - p]$ becomes $\sigma^2(1 - a^2)^{-1} a^p$, and is no longer a function of k. This means that the AR signal is *semistationary*, as the driving signal $u[k]$. It is clear that this reasoning can be generalized without great difficulty for AR(p) signals with the initial conditions $x[0]$, $x[1]$, ..., $x[p]$.

9.3 MOVING AVERAGE SIGNALS

Consider a finite impulse response (FIR) filter inducing the input-output relationship

$$x[k] = \sum_{l=0}^{q} b[l]\, u[k-l] \tag{9.50}$$

Note that, as stated previously, $b[l]$ and b_l have the same meaning. It is clear that this filter is stable and causal and its transfer function $H(z)$ takes the form

$$H(z) = \sum_{l=0}^{q} b[l]\, z^{-l} \tag{9.51}$$

The physical realization of this filter is presented in Figure 9.6 in the case where $q = 3$. The extension to other values of q is straightforward. Note that, as for the filters used in the previous section, there are only 3 delay systems in this realization. It is sometimes called a transversal realization of an FIR filter.

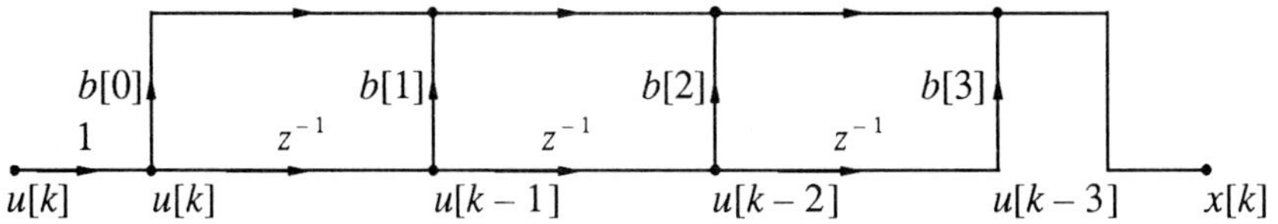

Figure 9.6 Transversal realization of an FIR filter.

If $u[k]$ is a zeromean weak white noise, then $x[k]$ is said to be an MA(q) signal. In many cases it is also assumed that $b[0] = 1$. If all the $b[k]$s are equal to $1/q + 1$, the filter is the discrete version of the time averager presented in Section 6.6. This justifies the terminology "moving average."

The relation (9.50) seems much simpler than (9.20) which defines an AR signal. In fact there is no restriction on the $b[k]$s because the transfer function has only a pole at the origin and introduces a filter which is always stable. However, we will see that equations solving the same problems as normal equations are much more complex.

Assuming that the coefficients $b[k]$ and the signal $u[k]$ are real, the correlation function of $x[k]$ is immediately obtained by an equation such as (5.62) transposed in the DT case, which gives

$$\gamma[k] = \sigma^2 \sum_{l=0}^{q-|k|} b[l]\, b[l+|k|] \tag{9.52}$$

Note that this correlation function is invariant if we multiply σ^2 by α^2 and all the $b[k]$s by α^{-1}. It is then impossible to completely separate the action of $u[k]$ and of the $b[k]$s.

This becomes possible if we impose $b[0] = 1$, which is the case for AR signals. In fact (9.18) shows that $a[0] = 1$.

The first obvious fact which appears in (9.52) is that $\gamma[k] = 0$ if $|k| > q$. This property characterizes a signal with a *finite memory*. It is important to avoid any confusion with the Markovian property indicated for AR signals. The Markovian property means that the conditional DF, given all the past of the signal, is only a function of the present, or of a finite past. The finite memory property means that if the delay k is sufficiently great, the signals $x[n]$ and $x[n-k]$ are uncorrelated, or independent when the driving noise is a strict white noise. One can also ask whether it is possible for a signal to be simultaneously AR and MA. The answer is no. This point is obvious for AR(1) signals; in fact they can be expressed in terms of $u[k]$ by (9.4), which does not introduce an FIR filter. The property can be generalized without difficulty, since no stable and causal filter with a transfer function such as (9.17) can have a finite impulse response, except in the trivial case where the a_is are null.

We now come to the problem solved by normal equations in the autoregressive case. Suppose that we are given the first q values of the correlation function $\gamma[k]$. Is it possible to determine the variance σ^2 and the coefficients $b[k]$? Writing $q+1$ equations deduced from (9.52) when k varies from 0 to q, we arrive at a system which is *non-linear* in b. Here is the difference with the normal equation (9.27) which gives **a** for an AR(p) signal. The consequence of the non-linearity of the system is that the solution is not unique. This is not surprising, as it relates to the discussion in Section 6.5.3. We saw in Example 6.6 that two different FIR filters can provide the same correlation function, due to the fact that there is no constraint on the zeros of an FIR filter while the poles of a purely recursive filter must be located inside the unit circle. However, if we impose the condition that this be a minimum phase filter, meaning that all the zeros are located inside the unit circle, the solution becomes unique with only an indetermination on the sign.

Example 9.3 Minimum phase FIR filters deduced from the correlation function. Let us start from a correlation function such that $\gamma[0] = 1 + a^2$, $\gamma[1] = -a$, and $\gamma[k] = 0$ for $|k| > 1$. This correlation function was introduced in Example 6.6. Using the remark presented just after (9.52), let us suppose that $\sigma^2 = 1$. In this case (9.52) gives two equations which are

$$b^2[0] + b^2[1] = 1 + a^2$$

$$b[0]\,b[1] = -a$$

After elimination of $b[1]$ by the second equation and taking $x = b^2[0]$, we obtain the equation

$$x^2 - (1 + a^2)\,x + a^2 = 0$$

which has two solutions $x_1 = 1$ and $x_2 = a^2$. As a result there are four solutions for the coefficients $b[k]$, which are

$b[0]$:	a	$-a$	1	-1
$b[1]$:	-1	1	$-a$	a

We observe the sign ambiguity noted after Example 6.6 and if we eliminate this, there is, whatever a, only one minimum phase filter. The method is general, although the explicit calculation rapidly becomes complex when the order of the filter increases.

It is important to note that if the *crosscorrelation* function between the input $u[k]$ and the output $y[k]$ is known, then the coefficients $b[k]$ of the MA signal are easily obtained. In fact, starting from (9.50) and noting that $u[k]$ is white, we immediately obtain

$$\gamma_{xu}[p] \triangleq E\{x[k]\, u[k-p]\} = \sigma^2\, b[p] \tag{9.53}$$

This equation is the discrete-time version of the result indicated in Example 6.2, and is very interesting in system identification, where the input and output can often be observed. Unfortunately that is rarely the case in signal problems, because the signal $u[k]$ is unknown and we only have the output $x[k]$ at our disposal. This is why it is always necessary to solve non-linear equations, in order to identify an MA signal from its second order properties.

9.4 ARMA SIGNALS

As the name implies, ARMA signals are a combination of AR and MA signals. The general procedure for introducing these signals is similar. We shall start from the rational function $H(z) = N(z)/D(z)$ with

$$N(z) = 1 + \sum_{i=1}^{q} b_i\, z^{-i} \tag{9.54}$$

$$D(z) = 1 - \sum_{j=1}^{p} a_j\, z^{-j} \tag{9.55}$$

and we suppose that the poles of $H(z)$ are located inside the unit circle, which implies some constraints on the coefficients a_i. Assuming that the region of convergence of $H(z)$ contains the unit circle, we introduce a dynamical filter, or a rational filter which is causal and stable. Note that the use in (9.55) of the sign + instead of – gives a more symmetric form. If the input and output are noted $u[k]$ and $x[k]$ respectively, they satisfy

$$x[k] = \mathbf{a}^T\mathbf{X}[k] + \mathbf{b}^T\mathbf{U}[k] + u[k] \tag{9.56}$$

where $\mathbf{a}$ and $\mathbf{X}$ have the same meaning as in (9.22), while $\mathbf{b}$ and $\mathbf{U}$ are defined by

$$\mathbf{b}^T = [b_1, b_2, \ldots, b_q] \tag{9.57}$$

$$\mathbf{U}^T[k] = [u_{k-1}, u_{k-2}, \ldots, u_{k-q}] \tag{9.58}$$

If $u[k] = u_k$ is a zeromean weak white noise, the output signal is said to be an ARMA(p, q) signal. It is also sometimes said that $x[k]$ is described by an ARMA (p, q) model.

Let us now examine the same problem as that discussed for AR and MA signals, namely the relation between vectors **a** and **b** and the *correlation function* of the signal $x[k]$. The basic relation similar to (9.25) is

$$E\{\mathbf{U}[k]\,\mathbf{X}^T[k-l]\} = 0 \quad \text{if} \quad l \geq q \tag{9.59}$$

The principle of the proof is the same as for (9.25). Because of the causality, the vector $\mathbf{X}[k-l]$ is a linear function of the $u[i]$s up to $i = k - l - 1$. On the other hand we see in (9.58) that $\mathbf{U}[k]$ is a linear function of $u[i]$ till $i = k - q$. If $l \geq q$, there is no $u[i]$ common to $\mathbf{U}[k]$ and $\mathbf{X}[k-l]$ and as the $u[i]$s are zeromean and uncorrelated, we deduce (9.59). As, for the same reason, $E\{u[k]\,\mathbf{X}[k-l]\} = 0$, by multiplying (9.56) by $\mathbf{X}[k-l]$ and taking the expected value, we obtain, if $l \geq q$,

$$E[x_k\,\mathbf{X}_{k-l}] = E[\mathbf{X}_{k-l}\,\mathbf{X}_k^T]\mathbf{a} \tag{9.60}$$

which can be written as

$$\mathbb{R}_l\,\mathbf{a} = \mathbf{c}_l \tag{9.61}$$

with

$$\mathbb{R}_l \triangleq E[\mathbf{X}_{k-l}\,\mathbf{X}_k^T] \tag{9.62}$$

$$\mathbf{c}_l \triangleq E[x_k\,\mathbf{X}_{k-l}] \tag{9.63}$$

In conclusion, the vector **a** corresponding to the AR part of the signal is still deduced from the correlation function of $x[k]$ by a linear equation. This equation (9.61) seems to be similar to (9.27), but is rather different in reality.

The main difference is that the matrix $\mathbb{R}_l$ is *no longer a covariance matrix* because it takes the form (3.65) rather than (3.63). It is thus a crosscovariance matrix which has no reason to be an NND matrix. However, $\mathbb{R}_l$ remains a Toeplitz matrix, but is no longer symmetric. In order to verify this point it is sufficient to calculate its matrix elements. We deduce immediately from (9.62) and (9.21) that the matrix elements of $\mathbb{R}_l$ are

$$\mathbb{R}_l[i, j] = \gamma[j - i - l] = \gamma[i - j + l] \tag{9.64}$$

These elements are only functions of $i - j$, and are equal along any diagonal. But as $l \neq 0$, $\mathbb{R}_l[i, j]$ is not necessarily equal to $\mathbb{R}_l[j, i]$, and the matrix $\mathbb{R}_l$ has no reason to be symmetric. In consequence the conditions ensuring that $\mathbb{R}_l$ is invertible are not the same as for Toeplitz covariance matrices. Finally, note that if $q = 0$, that is, if $x[k]$ is an AR(p) signal, it is possible to take $l = 0$ in (9.61), and we return to (9.27).

Let us now move onto the calculation of the MA part. To do this we introduce the signal $y[k]$ defined by

$$y[k] = x[k] - \mathbf{a}^T\mathbf{X}[k] \tag{9.65}$$

It is clear that in order to calculate $y[k]$ from $x[k]$ we must know the vector **a**. This vector is the solution of (9.61), but in practice this solution is not strictly equal to **a** and can contain some errors due either to the measurements of the correlation function or to the numerical solution of (9.61). Consequently, $y[k]$ is rarely equal to its true value. From (9.56) we deduce that the signal $y[k]$ satisfies

$$y[k] = \mathbf{b}^T \mathbf{U}[k] + u[k] \tag{9.66}$$

which is similar to (9.50). If we want to apply the method described in Section 9.3 for MA signals we must investigate if the correlation function of $y[k]$ can be deduced from that of $x[k]$. Using (9.65), this correlation function can be written as

$$\gamma_y[k] \triangleq E[y_n y_{n-k}] = E[(x_n - \mathbf{a}^T \mathbf{X}_n)(x_{n-k} - \mathbf{a}^T \mathbf{X}_{n-k})] =$$

$$\gamma[k] - \mathbf{a}^T \mathbf{c}_k - \mathbf{a}^T \mathbf{c}_{-k} + \mathbf{a}^T \mathbb{R}_k \mathbf{a} \tag{9.67}$$

If **a** is considered as known, this relation shows that $\gamma_y[k]$ can be expressed in terms of $\gamma[i]$, and $\gamma_y[k]$ can then be considered as known.

It remains now to note that (9.66) can be written, as (9.50), in the form

$$y[k] = \sum_{i=0}^{q} b[i]\, u[k-i] \tag{9.68}$$

where $b[0] = 1$. From this expression we deduce, as in (9.52), that

$$\gamma_y[k] = \sigma^2 \sum_{l=0}^{q-|k|} b[i]\, b[i + |k|] \tag{9.69}$$

Writing this equation for $k = 0, 1, ..., q$, we obtain $q + 1$ non-linear equations for the determination of σ^2 and the q coefficients $b[i]$. But the same situation arises as for MA(q) signals: as these equations are non-linear, the solution has no reason to be unique. In fact, the solution is not unique because there is no constraint on the localization of the zeros of $H(z)$, and we saw in Section 6.5 that several different transfer functions can be factored and provide the same correlation function.

9.5 RANDOM SIGNALS AND STATE REPRESENTATION OF SYSTEMS

It is well known that a linear filter can be described either by its external representation using the impulse response or the frequency response, or by its internal representation using the state and observation equations. The latter representation is also called a *state representation*. We shall discuss the main features of this representation (for further information see [Picinbono], Chap. 7).

In the discrete time case and for scalar inputs and outputs the internal representation of a linear time-invariant filter is defined by

$$\mathbf{x}[k+1] = \mathbb{A}\mathbf{x}[k] + \mathbb{B}u[k] \tag{9.70}$$

$$y[k] = \mathbb{C}\mathbf{x}[k] + Du[k] \tag{9.71}$$

The first equation is called the state equation and $\mathbf{x}[k]$ is the *state vector* at time k. The square matrix $\mathbb{A}$ is called the state or dynamical matrix because its eigenvalues are the poles of the system. The second is the observation equation which relates the state vector to the output with a possible direct connection between u and y through the last term of (9.71).

The internal representation introduces three main comments. First, the state equation (9.70) is the origin of the deterministic Markovian property of the state. As its name implies, this property has nothing to do with the stochastic aspect of systems. It simply tells us that if the state is known at time k, it is possible to calculate the state at any posterior time by using only the values of the input signal at times posterior to k. In other words, all the past actions on the system due to the input signal are summarized in the state at time k.

Second, if the eigenvalues of $\mathbb{A}$ are inside the unit circle, the two equations (9.70) and (9.71) introduce a stable and causal linear filter and its impulse response is given by

$$h[0] = D \quad ; \quad h[k] = \mathbb{C}\mathbb{A}^{k-1}\mathbb{B} \quad \text{for } k > 0 \tag{9.72}$$

Third, the transfer function of this filter is simply given by

$$H(z) = D + z^{-1}\mathbb{C}(\mathbb{I} - \mathbb{A}z^{-1})^{-1}\,\mathbb{B} \tag{9.73}$$

which is convergent on the unit circle.

Suppose now that $u[k]$ is a zeromean *weak white noise*. Let us calculate the second order properties of the state and of the output. By comparing (9.70) to (9.7), we can say that the state vector can be considered as an AR(1) vector signal. There is a slight difference concerning the origin of time, but this does not change our discussion. It is then tempting to apply to the state equation the methods and results applied to AR(1) signals. As $\mathbf{x}[k]$ is a linear combination of the $u[i]$s up to $i = k - 1$, we deduce that $E\{u[k]\ \mathbf{x}[k]\} = 0$. Further, as $\mathbf{x}[k]$ is stationary, its covariance matrix $\mathbb{\Gamma}[0]$ is

$$\mathbb{\Gamma}[0] = E\{\mathbf{x}[k]\ \mathbf{x}^T[k]\} = E\{\mathbf{x}[k+1]\ \mathbf{x}^T[k+1]\} \tag{9.74}$$

and using (9.70) we obtain

$$\mathbb{\Gamma}[0] = \mathbb{A}\mathbb{\Gamma}[0]\mathbb{A}^T + \sigma^2\mathbb{B}\mathbb{B}^T \tag{9.75}$$

which corresponds to (9.8) valid in the scalar case and with $\mathbb{B} = B = 1$. Similarly, multiplying (9.70) by $\mathbf{x}^T[k]$ and taking the expected value, we obtain

$$\mathbb{\Gamma}[1] \triangleq E\{\mathbf{x}[k]\,\mathbf{x}^T[k-1]\} = E\{\mathbf{x}[k+1]\,\mathbf{x}^T[k]\} = \mathbb{A}\mathbb{\Gamma}[0] \qquad (9.76)$$

which corresponds to (9.9). Repeating the same procedure, we obtain

$$\mathbb{\Gamma}[p] = \mathbb{A}^p\mathbb{\Gamma}[0] \qquad (9.77)$$

In order to obtain $\mathbb{\Gamma}[p]$ for $p < 0$ note that, as $\mathbb{\Gamma}[p] = E\{\mathbf{x}[k]\,\mathbf{x}^T[k-p]\}$ and as $\mathbf{x}[k]$ is stationary, we obtain $\mathbb{\Gamma}[p] = \mathbb{\Gamma}^T[-p]$. Furthermore note that this matrix is not a covariance matrix but a cross-covariance matrix, as defined by (3.65).

Let us now calculate the correlation function of the scalar signal $y[k]$. Starting from the definition

$$\gamma[p] = E\{y[k]\,y[k-p]\} \qquad (9.78)$$

and using (9.71) twice, we obtain

$$\gamma[p] = \mathbb{C}\mathbb{\Gamma}[p]\,\mathbb{C}^T + \mathbb{C}E\{\mathbf{x}[k]\,u[k-p]\}D \qquad (9.79)$$

In order to calculate the last term we note by iterating (9.70) that

$$E\{\mathbf{x}[k]\,u[k-p]\} = \sigma^2\mathbb{A}^{p-1}\mathbb{B} \qquad (9.80)$$

This finally gives, for $p > 0$

$$\gamma[p] = \mathbb{C}\mathbb{\Gamma}[p]\,\mathbb{C}^T + \sigma^2\mathbb{C}\mathbb{A}^{p-1}\mathbb{B}D \qquad (9.81)$$

and for $p = 0$

$$\gamma[0] = \mathbb{C}\mathbb{\Gamma}[0]\,\mathbb{C}^T \qquad (9.82)$$

As $y[k]$ is real and scalar we have of course $\gamma[p] = \gamma[-p]$. The main result of these calculations is that the correlation function of the output depends entirely on the correlation matrix of the state vector $\mathbf{x}[k]$.

Suppose now that the input is a strict white noise. As a result, the state vector is a vector Markov process, just as an AR(1) signal is a Markov signal. The proof is exactly the same and is a result of the fact that (9.70) is, like (9.7), a first order recursive equation. Although this Markovian property does not have the same meaning as seen previously for deterministic signals, it has the same origin, which is a first order recursion. However note that $y[k]$ has no reason to be a Markov signal and this point is discussed at the end of the next section. This leads to a more detailed analysis of Markov processes.

9.6 MARKOV PROCESSES

The concept of discrete-time Markov signals was introduced in Section 9.2 when discussing autoregressive signals. In this section we want to describe more precisely the class of Markov processes or Markov signals by giving an exact definition and by exhibiting their most important properties. The discussion is limited to DT signals, leaving the study of the continuous-time case until the end.

9.6.1 First Order Markovian Signals

Let $x[k;\, \omega]$ be a DT random signal. In order to distinguish the signal itself and its possible values, we will return to the notation of the first chapters (see especially (2.3) and the discussion which follows), using capital and lowercase letters. The random signal at k is thus written either as $X[k]$ or X_k and its possible values are written x_k.

In order to define a first order Markovian signal, noted M(1), we must first introduce some notations. Taking an arbitrary integer N and N arbitrary distinct integers $n_1, n_2, \ldots, n_N$ which are all greater than one, we introduce the random vector $\mathbf{Y}$ defined as follows. It has N components Y_i defined by

$$Y_i = X[k - n_i], \qquad 1 \le i \le N \tag{9.83}$$

which are the values of the signal at the time instants $k - n_i$. As we have to work with conditional distributions, we will present the discrete and continuous cases separately following the scheme of Section 4.2.3.

(a) Discrete case. This means that the possible values x_k of the signal X_k are taken in a discrete set. Similarly, the values of the components of the vector $\mathbf{Y}$ are also so taken. Under these conditions a signal X_k is said to be M(1) if

$$P(x_k | x_{k-1},\, \mathbf{y}) = P(x_k | x_{k-1}) \tag{9.84}$$

In this expression the first term is the *conditional probability* that $X_k = x_k$, given that $X_{k-1} = x_{k-1}$ and $\mathbf{Y} = \mathbf{y}$. The meaning of this expression is very simple: the values taken by the vector $\mathbf{Y}$ play no role in the conditional probability. In our discussion of AR signals k is the present, $k - 1$ is the past of order one and $\mathbf{Y}$ is related to values of the signal in a more distant past. In the state representation discussed in the previous section, $k - 1$ is the present and x_k and $\mathbf{y}$ are related to the future and to the past respectively. From this perspective the Markovian property means that the past plays no role in the conditional probability.

If the possible values taken by the signal are the same irrespective of k, the last term of (9.84) can be considered as an element of a square matrix called the matrix of *transition probabilities*. Note that all the elements of this matrix are non-negative.

(b) Continuous case. In this situation an M(1) signal is still characterized by (9.84), but the latter has a different meaning because it relates to the *probability density functions* as defined by (4.21) rather than to probabilities. Most of the following discussion refers to the continuous case, but the transposition to the discrete case is straightforward.

Finally, note that the previous definitions do not require any stationarity, but can of course be applied to stationary signals.

9.6.2 Some Consequences of the Definition

It is possible to give a more symmetric definition of M(1) signals. To introduce this point let us consider three continuous RVs X, Y, and Z defined by their PDF $p(x, y, z)$. By a direct extension of (4.22), we can write

$$p(x, y, z) = p(z)\, p(x|z)\, p(y|x, z) \tag{9.85}$$

The Markovian structure introduces the relation

$$p(y|x, z) = p(y|z) \tag{9.86}$$

and as we also have

$$p(x, y, z) = p(z)\, p(x, y|z) \tag{9.87}$$

we deduce

$$p(x, y|z) = p(x|z)\, p(y|z) \tag{9.88}$$

Comparing this with (4.29), it can be said that the RVs X and Y are *conditionally independent.* Conversely, (9.88) implies the Markovian structure (9.86).

Applying this result to (9.84), it can be said that conditionally to the present, the past and the future are independent. This explains the difference between AR(1) and MA(1) signals. It is clear that AR(1) signals driven by strict white noise are also M(1), which means that X_{k-1} and X_{k+1} *for a given value* x_k of X_k are conditionally independent. On the other hand, for MA(1) signals generated by strict white noise, X_{k-1} and X_{k+1} are independent, which was expressed by the terminology of *finite memory.*

There are some obvious consequences of (9.84) and (9.85). For example, choosing an arbitrary origin, we can write

$$p(x_1, x_2, x_3) = p(x_1)\, p(x_2|x_1)\, p(x_3|x_2) \tag{9.89}$$

and more generally

$$p(\mathbf{x}) = p(x_1) \prod_{i=2}^{n} p(x_i|x_{i-1}) \tag{9.90}$$

which means that, apart from the situation at $n = 1$, the probability density of the signal is determined by the transition probability densities.

Similarly, starting from

$$p(x_{k+1}, x_k|x_{k-1}) = p(x_k|x_{k-1})\, p(x_{k+1}|x_k, x_{k-1}) = p(x_k|x_{k-1})\, p(x_{k+1}|x_k) \tag{9.91}$$

and by integration with respect to the variable x_k, we obtain

$$p(x_{k+1}|x_{k-1}) = \int p(x_{k+1}|x_k)\, p(x_k|x_{k-1})\, dx_k \tag{9.92}$$

which is called the Chapman-Kolmogorov equation.

9.6.3 Regression, Covariance and Normal Distribution

The Markovian property induces a consequence on the regression defined by (4.38). In particular, from the definition of the regression and from (9.84) defining the Markovian structure, we have

$$E[X_k | x_{k-1}, \mathbf{y}] = E[X_k | x_{k-1}] = r(x_{k-1}) \tag{9.93}$$

Generally this regression has no particular structure. Suppose now that it is *linear*, or that

$$r(x_{k-1}) = a\, x_{k-1} \tag{9.94}$$

In this case it is said that the signal $X[k]$ is *linear* M(1), or linear Markovian of order one.

This affects the *covariance function* or the *correlation function*, if we assume that the signal X_k is stationary. In fact, if (9.94) holds, we see from (4.39) and (9.93) that $X_k - a\, X_{k-1}$ is uncorrelated with any function of X_{k-1}, $\mathbf{Y}$. This implies that

$$E[(X_k - a\, X_{k-1})\, X_{k-p}] = 0, \quad p \geq 1 \tag{9.95}$$

which gives

$$\gamma[k, k-p] = a\gamma[k-1, k-p], \quad p \geq 1 \tag{9.96}$$

If the signal X_k is stationary this can be written as

$$\gamma[p] = a\, \gamma[p-1], \quad p \geq 1 \tag{9.97}$$

which gives

$$\gamma[p] = \gamma[0]\, a^{|p|} \tag{9.98}$$

The modulus is due to the fact that $\gamma[p]$ is an even function. In conclusion we deduce that if $X[k]$ is stationary and linear-Markovian of order one, it has an *exponential correlation function* defined by (9.98). This is especially the case if $X[k]$ is normal and M(1), because the normal distribution implies a linear regression, as seen in (4.109).

Conversely a stationary *normal* signal with a correlation function given by (9.98) is M(1). To show this let us introduce the signal

$$U_k = X_k - aX_{k-1}$$

Its correlation function can be written in terms of that of X_k by

$$\gamma_U[p] = \gamma_X[p] - a\gamma_X[p+1] - a\gamma_X[p-1] + a^2\gamma_X[p] \tag{9.99}$$

Consequently, if $\gamma_X[p]$ satisfies (9.98), we deduce that $\gamma_U[p] = 0$ for $p \neq 0$. This means that U_k is a weak white noise. But as X_k is normal, U_k is also normal and becomes a strict white noise. As the relation between X_k and U_k is (9.7), X_k is an AR(1) signal driven by a strict white noise. As seen on p. 359, it is then an M(1) signal.

It must be clear that an M(1) signal with an exponential correlation function is not necessarily normal. To verify this, it suffices to call to mind that any AR(1) signal driven by a strict white noise is linear M(1). It then has an exponential correlation function. But as the distribution of $U[k]$ is not specified, $X[k]$ has no reason to be normal.

Similarly it must be clear that a stationary M(1) signal does not necessarily have an exponential correlation function. In fact, if $r(x_{k-1})$ in (9.93) is not linear, $X[k]$ is not an AR(1) signal, and its correlation function has no reason to be exponential. In order to construct an example of this situation, let us start with a *normal* M(1) signal $X[k]$. It is obvious that the signal $Y[k] = X^3[k]$ is still M(1) and an elementary calculation shows that its correlation function is not exponential.

Example 9.4 Let us discuss in more detail the properties of the signal $Y[k] = X^3[k]$ where $X[k]$ is an AR(1) signal defined by (9.7) where $U[k]$ is strictly white. As the transformation $y = x^3$ is monotone it is clear that the conditional PDF of $Y[k]$ given all its past is only a function of $Y[k-1]$, which characterizes the M(1) property. In fact, as there is a one to one correspondence between x and y, knowing the past of Y is similar to knowing the past of X. Let us now calculate the regression of Y_k defined by

$$r_Y(y_{k-1}) = E[Y_k | y_{k-1}]$$

Starting from (9.7) we can write

$$Y_k = X_k^3 = a^3 X_{k-1}^3 + 3a^2 X_{k-1}^2 U_k + 3a X_{k-1} U_k^2 + U_k^3$$

$$= a^3 Y_{k-1} + 3a^2 (Y_{k-1})^{2/3} U_k + 3a (Y_{k-1})^{1/3} U_k^2 + U_k^3$$

If we suppose that U_k is zeromean-valued and that its second and third order moments are m_2 and m_3 respectively, we deduce that the expectation of Y_k for a given value y_{k-1} of Y_{k-1} is

$$r(y_{k-1}) = a^3 y_{k-1} + 3a y_{k-1}^{1/3} m_2 + m_3$$

We note that, in contradiction to (9.93), this regression is not linear. Suppose now that $U[k]$ is $N(0, \sigma^2)$. In this case $m_2 = \sigma^2$ and $m_3 = 0$, which gives

$$r(y_{k-1}) = a^3 y_{k-1} + 3a y_{k-1}^{1/3} \sigma^2$$

It is clear that (4.33) is satisfied because $E(Y_{k-1}) = 0$ and as $Y_{k-1}^{1/3} = X_{k-1}$, we also have $E(Y_{k-1}^{1/3}) = 0$. It is also possible to calculate the correlation function of $Y[k]$ which is given by

$$\gamma_Y[p] = E\{Y[k]\,Y[k-p]\} = E\{X^3[k]\,X^3[k-p]\}$$

Using the general expression of higher order moments of a normal distribution given by (4.95), we deduce that

$$\gamma_Y[p] = 6\gamma_X^3[p] + 9\gamma_X^2[0]\,\gamma_X[p]$$

and it appears that this correlation function is not exponential. In particular, if the input signal $U[k]$ is $N(0, 1 - a^2)$, we deduce from (9.5) that $\gamma_X[p] = a^{|p|}$, which gives

$$\gamma_Y[p] = 6a^{3|p|} + 9a^{|p|}$$

Using the regression given above where $\sigma^2 = 1 - a^2$, it is easy to verify that

$$\gamma_Y[1] = E\left[r(Y_{k-1})\, y_{k-1}\right]$$

Let us conclude this section with a few additional comments. Suppose that $X[k]$ is an AR(1) signal driven by a weak white noise $U[k]$, i.e. a sequence of uncorrelated RVs. The correlation function has an exponential form, but $X[k]$ has no reason to be M(1). However, it is sometimes said that $X[k]$ is a weak linear M(1) signal.

Suppose now that $U[k]$ is a strong white noise, i.e., satisfies (7.2). In this equation P_k denotes all the RVs $U[k-l]$, $l > 0$, i.e., all the past of U_k. We can then replace P_k by $P_U[k]$ to be more explicit. Consider the quantity $P_X[k]$ which has the same meaning provided that U is replaced by X. Using the recursion (9.7) we obtain

$$E\{X_k|P_X[k]\} = E\{ax_{k-1} + U_k|P_X[k]\} = ax_{k-1} + E\{U_k|P_X[k]\}$$

Let us now show that the last term of this equation is null. In fact the recursion (9.7) can be written

$$U_{k-p} = X_{k-p} - aX_{k-p-1}$$

which implies that if all the RVs $X[k-p]$, $p > 0$ are known, all the RVs $U[k-p]$, $p > 0$, are also known. As a result we have $P_U[k] = P_X[k]$, and applying (7.2) we deduce that $E\{U_k|P_X[k]\}$ is null. We thus have a Markovian property *only on the expected value.* In this case we can say that $X[k]$ is a strong linear M(1) signal, which of course has an exponential correlation function.

Finally, if $U[k]$ is a strict white noise we obtain a linear M(1) signal, which by extension can be called a strict linear M(1) signal. These concepts are all equivalent in the normal case.

9.6.4 Markov Signals of Order p

As seen in the previous section, M(1) signals are closely connected to AR(1) signals. It is thus quite natural to associate the concept of Markov signals of order p, or M(p) signals, to that of AR(p) signals This was already touched on at the end of Section 9.2.2. As the general reasoning is similar to that for M(1) signals, we will deal only with the main points. The basic equation is similar to (9.84) in which we must introduce the past of order p.

For this we again use the vector $\mathbf{X}[k]$ or $\mathbf{X}_k$ defined by (9.21). The vector $\mathbf{Y}$ used in Section 9.6.1 has its components defined by (9.83), but now all the N integers n_i must be greater than p, which means that the vector $\mathbf{Y}$ belongs to the past with respect to $\mathbf{X}[k]$. With this notation the M(p) property is characterized by

$$p(x_k|\mathbf{x}_k,\ \mathbf{y}) = p(x_k|\mathbf{x}_k) \tag{9.100}$$

which has the same meaning as (9.84). In particular, it can be applied either in the continuous case with PDF or in the discrete case with probabilities. In the latter case it is clear that the transition probability $p(x_k|\mathbf{x}_k)$ does not introduce a square matrix.

Following the comments in Section 9.6.2, the Markovian property can take a symmetric form so that, conditionally to $\mathbf{X}_k = \mathbf{x}_k$, the random variable X_k and the random vector $\mathbf{Y}$ are independent. This is the main difference between M(p) and MA(p) signals introducing a finite memory.

The definition equation (9.100) introduces a consequence on the conditional expectation value which is, as in (9.93),

$$E(X_k|\mathbf{x}_{k-1},\ \mathbf{y}) = E[X_k|\mathbf{x}_k] = r(\mathbf{x}_k) \tag{9.101}$$

If this regression takes the form

$$r(\mathbf{x}_k) = \mathbf{a}^T\mathbf{x}_k \tag{9.102}$$

the signal becomes *linear* M(p). This is the case when $x[k]$ is normal, where $\mathbf{a}$ is given by (4.109). This gives a specific form of the correlation function which is more complicated than a simple exponential form. In reality it is the correlation function of an AR(p) signal which is completely defined by the regression vector $\mathbf{a}$. The structure of this correlation function will be studied in a following chapter dealing with the solution of normal equations in the framework of prediction problems.

Finally, the discussion concerning weak, strong, and strict linear Markovian signals can easily be transposed from order one to an arbitrary order.

9.6.5 Vector Markovian Signals

This concept was introduced at the end of Section 9.5 dealing with the state vector. If the input vector of the state equation is strictly white, the state vector is M(1). All the properties of scalar M(1) signals can be transposed without difficulty to vector signals.

But there is one point that must be made. If a vector $\mathbf{X}[k]$ is Markovian, this does not mean that its components are separately Markovian.

To explain this, consider a two-dimensional vector with components x and y. The transposition of (9.90) to this case gives

$$p(\mathbf{x},\ \mathbf{y}) = p(x_1,\ y_1)\prod_{i=2}^{n} p(x_i,\ y_i|x_{i-1},\ y_{i-1}) \tag{9.103}$$

If we are interested only in the component $\mathbf{x}$ we may calculate its marginal distribution with (3.11), or

$$p(\mathbf{x}) = \int p(\mathbf{x}, \mathbf{y})\, d\mathbf{y} \tag{9.104}$$

It is clear that without specific conditions on the transition probabilities, this PDF will not satisfy (9.90) characterizing a sequence $x[k]$ which is Markovian.

This has a specific application to the solution of the state equation discussed in Section 9.5. Looking at (9.70) and (9.71) we see that $\mathbf{x}[k]$ is M(1) when $u[k]$ is strict white. But this does not mean that $y[k]$ is still Markovian, and in general this is not so. It is sometimes said that the output $y[k]$ has a Markovian representation.

9.7 SIGNALS GENERATED BY VOLTERRA FILTERS

In our discussion of ARMA signals we saw the interest of signals generated by *linear systems* driven by white noise. It is now tempting to transpose the same procedure to *non-linear systems*. Unfortunately, the question immediately becomes much more difficult. Although the concept of linear systems is clearly defined, there are many ways for a system to be non-linear, thus making a general theory impossible. In this section we will consider only filters defined by a limited Volterra expansion. To understand this, let us first consider the case of quadratic filters. A *quadratic filter* is defined by the following input-output relationship

$$x[k] = \sum_{p,\,q} h[p,\, q]\; u[k-p]\; u[k-q] \tag{9.105}$$

$$= \sum_{p,\,q} h[k-p,\, k-q]\; u[p]\; u[q] \tag{9.105'}$$

It is called quadratic because it resembles a quadratic form as, for example, (3.69). Furthermore, as the structure is similar to (3.119), the function $h[p, q]$ will be called the impulse response or the kernel of the filter. In reality, the output associated to the input $\delta[k]$ is $h[k, k]$. The relations (9.105) are well adapted to the introduction of the concepts of causality or of FIR.

A quadratic filter is causal if $h[p, q] = 0$ for negative values of p and q. It has a finite impulse response (FIR) if there is only a finite number of terms in (9.105) or (9.105').

Comparing this with (9.50), it is possible to introduce the concept of quadratic MA(N) signals. This is a signal defined by Eqs. (9.105) when the input is a *white noise* and when p and q are non-negative and smaller or equal to N. Following the same procedure as for MA signals, we will now calculate the first and second order properties of a quadratic MA signal. Note that if the input is stationary, so is the output, because of the structure of (9.105).

Taking the mean value of (9.105) and using (7.1), we obtain

$$E\{x[k]\} = \sigma^2 \sum_{p=0}^{N} h[p, p] \tag{9.106}$$

If we consider the quantities $h[p, q]$ as the elements of an $N \times N$ matrix $\mathbb{H}$, (9.106) shows that

$$E\{x[k]\} = \sigma^2 \mathrm{Tr}(\mathbb{H}) \tag{9.107}$$

which is exactly (4.254), because the matrix $\mathbb{\Gamma}$ present in this expression is now $\sigma^2\, \mathbb{I}$.

In order to calculate the correlation function of $x[k]$ the assumption that $u[k]$ is a weak white noise is not sufficient, for we need the fourth-order moments of this input. We are then led to introduce the assumption that $u[k]$ is strictly white, at least up to the fourth order, as explained on p. 259. In this case, using (7.11) and (4.79), we find

$$m(k_1, k_2, k_3, k_4) = \sigma^4 \{ \delta[k_1 - k_2]\, \delta[k_3 - k_4] + \delta[k_1 - k_3]\, \delta[k_2 - k_4] + \delta[k_1 - k_4]\, \delta[k_2 - k_3] \} + c_4\, \delta[\mathbf{k}] \tag{9.108}$$

If, furthermore, $u[k]$ is normal, $c_4 = 0$ and the last term of this equation disappears. As $x[k]$ is not zeromean-valued, its correlation function is

$$\gamma[k] = E\{x[n]\, x[n-k]\} - E^2\{x[n]\} \tag{9.109}$$

Using (9.105), (9.106) and (9.108), and after simple algebra, we obtain

$$\gamma[k] = \sigma^4 \sum_{p,\, q} \{ h[p, q]\, h[p-k, q-k] + h[p, q]\, h[q-k, p-k] \} + c_4 \sum_{p} h[p, p]\, h[p-k, p-k] \tag{9.110}$$

If the input signal is normal, the last term disappears. Further, if the kernel $h[p, q]$ is symmetric, which means that $h[p, q] = h[q, p]$, we find for $k = 0$ the result (4.260) obtained in another way. There is a strong analogy between (9.110) and (9.52) obtained for MA signals, but the expressions are more complex for quadratic filters. It is possible to give a more compact form to these expressions by using the appropriate vector notation, but this procedure is beyond the scope of this section.

A pure Volterra filter of order p is given by a direct extension of (9.105) which becomes

$$x[k] = \sum_{\mathbf{k}} h[k_1, k_2, \ldots, k_p]\; u[k - k_1]\, u[k - k_2] \ldots u[k - k_p] \tag{9.111}$$

where the sum is extended to all the integers k_i. And finally a Volterra filter of order p is a sum of pure Volterra filters of order 1, 2, ... , p. It is clear that in order to obtain the correlation function of the output it is necessary to use the moments of the input up to the order $2p$.

We saw in the previous section that linear Markovian signals are related to the structure of the correlation function and that an AR(1) signal generated by a strict white signal is linear M(1). We can ask the same question for Volterra filters. There are some extensions of the same property, but which use more abstract arguments (see [Blanc-Lapierre], p. 319).

9.8 EXTENSIONS TO THE CONTINUOUS-TIME CASE

The difficulty in extending the results of the previous sections to the CT case is related to the concept of CT white noise studied in Chapter 7. For example, the Langevin equation (7.63) can be considered as the CT version of (9.7) introducing AR(1) signals. But there is a great difference. In (9.7) the driving signal $u[k]$ is a DT white noise which is a second order signal with a very simple definition, whatever the nature of its whiteness. On the other hand, the signal $F(t)$, assumed to be white in (7.63), is not a second order signal, and only its increments are well defined. In reality it would be better to write this equation in the form

$$m\,dv + f v\,dt = dF(t) \tag{9.112}$$

where $F(t)$ is a process either with uncorrelated or with independent increments.

This equation introduces a filter with the frequency response (7.64) and with an impulse response $h(t)$, the inverse Fourier transform of $G(\omega)$. As a result the input-output relationship must be written as

$$v(t) = \int h(t-\theta)\,dF(\theta) \tag{9.113}$$

which is, for example, the form of (5.75) or (7.46). As a dynamical filter is causal, we will now discuss the properties of signals written in the form

$$x(t) = \int_{-\infty}^{t} h(t-\theta)\,du(\theta) \tag{9.114}$$

$$= \int_{0}^{\infty} h(\theta)\,du(t-\theta) \tag{9.114'}$$

where the increments $du(\theta)$ are uncorrelated (weak whiteness) or independent (strict whiteness). Our main interest is in the properties of finite memory or of Markovian behavior.

Suppose that the increments $du(\theta)$ are only *uncorrelated.* If they are zeromean-valued, they satisfy a relation such as (7.45) and as a result the correlation function of $x(t)$ is

$$\gamma(\tau) = c \int h(t)\, h(t - \tau)\, dt \tag{9.115}$$

In the case of an FIR and causal filter, $h(t)$ is null if t does not belong to the interval $[0, T]$, and (9.115) becomes

$$\gamma(\tau) = c \int_{\tau}^{T} h(t)\, h(t - \tau)\, dt \tag{9.116}$$

It is thus clear that $\gamma(\tau) = 0$ if $|\tau| > T$, which characterizes an MA signal in the CT. As in the DT case, we will say that the signal has a *finite memory.*

Suppose now that the filter in (9.114) is first order dynamical with the transfer function

$$H(s) = (s + a)^{-1}, \qquad a > 0 \tag{9.117}$$

which corresponds to (9.1), the starting point of the introduction of AR(1) signals. Its impulse function is

$$h(t) = u(t) \exp(-at) \tag{9.118}$$

where $u(t)$ is now the unit step function. Inserting this expression in (9.115), we obtain

$$\gamma(\tau) = c(2a)^{-1} \exp[-a|\tau|] \tag{9.119}$$

which corresponds to (9.5). We then deduce that

$$\gamma(0) = c/2a \tag{9.120}$$

which corresponds to (9.8) and for $\tau > 0$

$$\frac{d\gamma}{d\tau} = -a\gamma \tag{9.121}$$

As a result of these equations, and as in AR(1) signals, the value of $\gamma(0)$ and of its first derivative at $\tau = 0$ can provide the two parameters of the problem, i.e. a and c. This can be extended to transfer functions with several poles instead of just one, as in (9.117). There is thus an analogy with normal equations, although their structure is changed.

We now come to the concept of CT *Markov processes.* In the CT case, only first order Markovian signals can be easily introduced, simply called CT Markovian (M) signals. The basic equation is similar to (9.84), and we consider here only a continuous-valued signal $X(t)$, leaving the transposition to signals taking discrete values to the reader. Calling $x(t)$ the possible values of the RV $X(t)$, the PDF of the set of RVs $X(t_1)$, $X(t_2)$, ..., $X(t_n)$ is $p[x(t_1), x(t_2), \ldots, x(t_n)]$.

A random signal $X(t)$ is said to be a *Markovian signal* if, for all n and for all the ordered time instants $t_1 < t_2 < \ldots < t_n$, the conditional PDF satisfies

$$p[x(t_n)|x(t_{n-1}), x(t_{n-2}), \ldots, x(t_1)] = p[x(t_n)|x(t_{n-1})] \tag{9.122}$$

As in the DT case, it is possible to interpret this result by saying that conditionally to the present, the past and the future are independent.

The discussion of the regression can also be transposed, and if a signal $X(t)$ is M, we have

$$E[X(t_n)|x(t_{n-1}),\ x(t_{n-2}),\ \ldots,\ x(t_1)] = E[X(t_n)|x(t_{n-1})] \tag{9.123}$$

If this regression is *linear*, the signal $X(t)$ is said to be linear M and satisfies

$$E[X(t_n)|x(t_{n-1})] = a(t_n,\ t_{n-1})\ x(t_{n-1}) \tag{9.124}$$

This is especially the case for normal M signals, because the normal distribution implies a linear regression. Note that in (9.124) the coefficient a is a function of t_n and t_{n-1}, which was not the case in (9.94). In fact, in the DT case the difference between t_k and t_{k-1} is constant and equal to one if $t_k = k$, and furthermore the discussion in Section 9.6.3 concerns the stationary case.

Using the properties of the regression we will now calculate the coefficient $a(t_{n-1}, t_n)$ in terms of the covariance function. For this we use (4.39) in the form

$$E\{\,[X(t_n) - a(t_n,\ t_{n-1})X(t_{n-1})]\ X(t_{n-1})\,\} = 0 \tag{9.125}$$

which gives

$$a(t_n,\ t_{n-1}) = \gamma(t_n,\ t_{n-1})\ [\gamma(t_{n-1},\ t_{n-1})]^{-1} \tag{9.126}$$

If we write (4.39) in the form

$$E\{\,[X(t_n) - a(t_n,\ t_{n-1})X(t_{n-1})]\ X(t_{n-2})\,\} = 0 \tag{9.127}$$

and use (9.126), we obtain

$$\gamma(t_n,\ t_{n-1})\ \gamma(t_{n-1},\ t_{n-2}) = \gamma(t_n,\ t_{n-2})\ \gamma(t_{n-1},\ t_{n-1}) \tag{9.128}$$

Let us now assume that $X(t)$ is *stationary* and that $t_n = t + dt$, $t_{n-1} = t$, $t_{n-2} = t - \tau$, where dt and τ are positive to ensure the order of the t_is. With these values (9.128) becomes

$$\gamma(dt)\ \gamma(\tau) = \gamma(\tau + dt)\ \gamma(0) \tag{9.129}$$

which introduces the differential equation

$$\dot{\gamma}(\tau) = \dot{\gamma}(0)\,[\,\gamma(0)]^{-1}\ \gamma(\tau) \tag{9.130}$$

and the solution for $\tau > 0$ is

$$\gamma(\tau) = \sigma^2 \exp(-\alpha\tau) \tag{9.131}$$

where α is deduced from (9.124). The final result is then that a stationary linear M signal has an *exponential correlation function.*

This appears especially in the *normal case*, but not only in this case. All the signals, such as (9.114) where $h(t)$ is given by (9.118) and where the increments are independent, are linear M. If $u(t)$ is a Brownian signal, we find the normal case again, but we saw in Chapter 8 that there are several other processes with independent increments which are not normal.

Finally, note that, as in the DT case, an exponential correlation function does not imply a Markovian signal, and there are stationary Markovian signals which do not have an exponential correlation function. The simplest example is still $y(t) = x^3(t)$ where $x(t)$ is a Gauss-Markov signal, or a signal that is M and normal.

PROBLEMS

9.1 Let $H(z)$ be the transfer function defined by (9.17) in the case where $p = 2$. Let z_1 and z_2 be the roots of the polynomial $D(z)$ defined by (9.18). Calculate in terms of z_1 and z_2 the impulse response $h[k]$ of the filter whose transfer function is $H(z)$. When the roots are complex, use the notation introduced in Example 9.2 and express $h[k]$ in terms of m and ϕ. Calculate the output (9.19) when the input is the DT unit step function.

9.2 **Solution of a second order difference equation.** The filter studied in the previous problem is directly connected to the difference equation (9.20) written for $p = 2$. The purpose of this problem is to find the general solution to this equation.

(a) Show that if the input of the filter is null for $k \leq 0$, this property also holds for the output.

(b) Show that the convolution between the input and the impulse function $h[k]$ calculated in the previous problem satisfies the difference equation.

(c) Suppose now that the input is null and that the outputs at times $k = 0$ and $k = -1$ are given. Calculate the output for $k > 0$. This output is called the zero input solution, or the solution of the homogeneous equation.

(d) Combining the results of the two previous questions, calculate the output for $k > 0$ corresponding to the input $u[k]$ null for $k \leq 0$, provided that the output at times 0 and -1 are given. Compare the result to that for a first order recursion obtained in Section 9.2.3(e).

(e) Complete the previous calculations in the case where the roots of the polynomial $D(z)$ defined by (9.18) are complex.

9.3 Let $x[k]$ be a *complex* AR(1) signal defined by a complex regression coefficient a and a white driving noise characterized by its power σ_u^2 and its second order moment $E[u^2] = c$, where c is a complex number.

(a) Calculate the correlation function $\gamma[k]$ of $x[k]$ and the moment $E\{x[n]x[n-k]\}$.

(b) Conversely, calculate the quantities a, c and σ_u^2 from $\gamma[k]$ and $E\{x[n]x[n-k]\}$.

9.4 The purpose of this problem is clearly to demonstrate the difference between real and complex AR(1) signals. Complex AR signals are frequently used in the framework of spectral analysis where it is often more convenient to manipulate a signal such that $\exp[j\omega k]$ instead of $\cos[\omega k]$. Consider the transfer function (9.1) where the pole a is assumed to be complex or written as $a = m\exp(j\Omega)$ with $0 < m < 1$.

(a) Calculate the impulse response of this filter and compare the result with (9.3).

(b) Suppose that the input of this filter is a complex zeromean white noise with variance σ^2. By using the relation (5.58) adapted to the DT case, calculate for $k > 0$ the correlation function $\gamma[k]$ of the output. Express this correlation function in terms of m and Ω. By using the symmetry relation of correlation functions specified by (5.27), give the expression of $\gamma[k]$ valid for any k and deduce that (9.5), valid in the real case, does not hold in the complex case.

(c) Without any calculation and referring only to the properties of Fourier transforms, deduce the power spectrum from (9.6').

(d) In order to verify this result analytically, use the relation (6.69) for calculating the z-spectrum, and, using (6.71), calculate the power spectrum.

(e) Let $x[k]$ be the signal studied in the preceding questions. Show that the signal $y[k] = x[k]\exp[-j\Omega k]$ satisfies a difference equation with real coefficients and can then be considered as a real AR(1) signal.

(f) Give a physical interpretation of the results of this whole problem.

9.5 Suppose that $x[k] = s[k] + n[k]$, where $s[k]$ is a real AR(1) signal with a regression coefficient m and driven by a real white noise of variance $1 - m^2$, and $n[k]$ is a real white noise of variance σ^2.

(a) Calculate the correlation function $\gamma[k]$ of $x[k]$ and deduce its z-spectrum. Show that this spectrum is that of an ARMA signal.

(b) By using the strong factorization of the spectrum, deduce the parameters of the minimum phase ARMA model.

(c) By using the method presented in Section 9.4, solve the same question starting from the correlation function $\gamma[k]$.

9.6 Let $x[k]$ be a real AR(2) signal defined by a regression vector $\mathbf{a}^T = [0, a]$ and a driving white noise $u[k]$ of unit variance. We want to study the properties of its correlation function and of its power spectrum.

(a) Calculate the transfer function $H(z)$ of the filter generating $x[k]$ from $u[k]$. Present the results separately according to the sign of a.

(b) Deduce from the transfer function the z-spectrum and decompose this spectrum as a sum of two terms corresponding to AR(1) spectra.

(c) Deduce from the previous question the correlation function and write this function according to the sign of a.

(d) Calculate the power spectrum either from the correlation function or from the z-spectrum calculated in **(b)**.

9.7 **Whiteness and quadratic filters.** Consider the signal $x[k]$ defined by (9.105) where the input $u[k]$ is assumed to be a strictly white signal. The problem is to study the whiteness of the output.

(a) Suppose first that the input is normal. Find the condition on the kernel $h[p, q]$ ensuring that the output remains weakly white.

(b) Solve the same question when the cumulant c_4 of the input is non-zero.

9.8 Let $x(t)$ be the random telegraph signal discussed in Section 8.6.1. Show that it is a Markovian signal and calculate the transition probability appearing in (9.122). Calculate the regression (9.123) and show that it is linear. Verify the property indicated after (9.131).

9.9 Consider the signal $x(t) = \exp[j(2\pi Ft + \Phi)]$, where F is a random frequency and Φ a random phase uniformly distributed over $0,\ 2\pi$.

(a) Show that it is a strictly stationary Markovian signal.

(b) Show that the conditional expectation (9.123) is linear.

(c) Show that any positive definite function can be its correlation function and explain why the property indicated after (9.131) is not necessarily satisfied.

9.10 Consider the signal $x(t)$ defined by (9.114) in which the increments $du(\theta)$ are independent and $h(t)$ is given by (9.118).

(a) Express $x(t + \tau)$, $\tau > 0$, in terms of $x(t)$ and of an RV independent of $x(t)$.

(b) Deduce from the preceding result that $x(t)$ is a linear Markovian signal.

(c) Verify the property indicated after (9.131).

Chapter 10

Mean Square Estimation

10.1 INTRODUCTION

Mean square estimation is one of the most important methods of statistical signal processing. Although it is not the main purpose of this work to deal with signal processing methods, mean square estimation does nevertheless require careful study. This theory is so closely connected to the statistical properties of random signals that it can illustrate various points discussed in the previous chapters and also provide new results on the structure of random signals. This is especially the case in prediction theory, a specific aspect of mean square estimation presented in another chapter. For example, we saw in Example 7.1 that some random signals can be perfectly predictable by linear or non-linear methods. This is a very important property of signals which cannot be understood without an extensive presentation of both prediction theory and estimation theory.

Before examining the problem in detail, we shall give an overview with some examples, which introduce the relevant notation and terminology.

Let us return to the signal presented in Figure 1.2. This can be considered as a particular trajectory of a random signal $x(t, \omega)$. In reality, this trajectory is limited in time and the figure represents $x(t, \omega)$ for $0 \leq t \leq T$, where 0 and T mean the times of the beginning and end of the recording or of the experiment. In the terminology used hereafter we will say that the figure represents an *observation* of $x(t, \omega)$ for $0 \leq t \leq T$. The next question is: is it possible to deduce from this observation some values of this signal for $t < 0$ or $t > T$? This introduces the problem of the *extrapolation* of the signal. If, on the other hand, we are only interested in values of $x(t, \omega)$ for $t > T$, we encounter a *prediction* problem, which means that we hope to predict some future values of the signal from observations concerning the present or the past. The vocabulary of prediction was briefly mentioned in Section 1.3, and we will now examine it more closely.

Let $x[k, \omega]$ be a discrete-time (DT) random signal and suppose, for the purpose of simplification, that its statistical properties are known. This means, in our approach, that the family of finite-dimensional distributions introduced in Section 5.1 is known. Suppose that this signal was recorded, or observed, at the time instants $k-1, k-2, \ldots, k-n$. From this observation we want to *estimate* the signal at time k, or $x[k, \omega]$. This estimate can be called the one step prediction of $x[k, \omega]$ in terms of a finite past. The term "one step" comes from the fact that the last observation is $x[k-1, \omega]$ and the term "finite past" from the fact that we will use a finite number of observations in the past for the estimation.

In order to simplify the notation let us introduce the past vector, or the vector $\mathbf{X}[k, \omega]$ defined by

$$\mathbf{X}^T[k, \omega] = [x[k-1, \omega], x[k-2, \omega], \ldots, x[k-n, \omega]] \tag{10.1}$$

Note that this vector was introduced in (9.21) when discussing properties of AR signals. This vector will now be called the *observation vector.* It is clear that for a specific experiment characterized by $\omega = \omega_i$ this vector is not random. The situation is the same as when we say that a specific trajectory of a random signal is no longer random. However, if we repeat the experiment under the same conditions, we will obtain another value $\mathbf{X}[k, \omega_j]$ of the vector, and, before its recording, the observation vector $\mathbf{X}[k, \omega]$ can be considered as a random vector with a known probability distribution function.

From the observation vector $\mathbf{X}[k, \omega]$ we want to make an *estimation* of $x[k, \omega]$. So it is logical to call this value $x[k, \omega]$ the *estimandum,* which means the parameter which has to be estimated. As this parameter is a random variable, the estimandum is random. In order to realize the estimation, we introduce an *estimator,* which is a function of the observation written as

$$\hat{x}[k, \omega] = S\{\mathbf{X}[k, \omega]\} \tag{10.2}$$

It is the output of a system $S\{.\}$ associated with the input $\mathbf{X}[k, \omega]$. In statistical literature, $S\{.\}$ is called a statistic, but in engineering literature it is simply a system, and more generally a filter. It is important to note two points concerning this filter. First, it has no reason to be linear, even though linear systems play an important role in what follows. Secondly, the filter is deterministic, and for our purposes it contains no random element.

We can now state the estimation problem: find the optimum system $S_o\{\ \}$ in such a way that $\hat{x}[k, \omega]$ will be the *best approximation* of the estimandum $x[k, \omega]$.

Unfortunately, our task is not completed by using the word "best," which is too indefinite. To be more precise we need an *optimality criterion.* Many criteria are possible, and much time could be spent in the discussion of their respective merits and defects. But among this family of criteria one stands out as being particularly important, the *minimum mean square error* criterion, which explains the title of this chapter. This criterion means that we want to find the optimal filter $S_o\{.\}$ such that the mean square error

$$\varepsilon^2 = E\left\{\left(x[k, \omega] - \hat{x}[k, \omega]\right)^2\right\} \tag{10.3}$$

is minimum. Let us present some arguments in favor of this criterion. As it is a quadratic quantity, it has a physical interpretation as the minimum of the energy error, discussed in Section 5.9. From a mathematical point of view, it can be interpreted as a distance criterion, as discussed in Section 3.7, and we will use this interpretation extensively. Finally, in the specific but very important case where $S\{.\}$ is linear, the calculation of ε^2 requires only *second order* properties of signals, dealt extensively with in Chapters 3, 5, and 6.

Using this criterion, suppose now that the problem is solved. The solution

$$\hat{x}[k, \omega] = S_o\{\mathbf{X}[k, \omega]\} \tag{10.4}$$

is called the mean square estimate of $x[k, \omega]$ in terms of the observation $\mathbf{X}[k, \omega]$.

It is important to understand the meaning of this expression fully . It is clear that, as the optimum system $S_o\{.\}$ is not random, any realization ω_i of $\mathbf{X}[k, \omega]$ will give a non-random output $\hat{x}[k, \omega_i]$. In practical experiments it is what appears at the output of the estimation filter. This particular output $\hat{x}[k, \omega_i]$ has no specific relation of optimality with $x[k, \omega_i]$. It is quite possible that for this specific value ω_i another filter, say $S'\{.\}$, will give a null error, or that the output $S'\{x[k, \omega_i]\}$ will be equal to $x[k, \omega_i]$. This is not in contradiction to the fact that $S_o\{.\}$ is better than $S'\{.\}$ in the mean. This is a direct consequence of the criterion (10.3), which needs an averaging over all possible realizations ω. Thus for some specific realizations ω_i, $S'\{.\}$ can provide better results, but not in the mean sense, where all the possible values of ω are taken into account. The same situation was encountered on p. 71 when discussing convergence in the quadratic mean sense.

As a result of the above, even though in practical applications we start from a specific observation $\mathbf{X}[k, \omega_i]$, the calculation of the optimum filter for mean square estimation assumes that the estimandum $x[k, \omega]$ and the observation $\mathbf{X}[k, \omega]$ are random. For this reason we will omit the symbol ω in all calculations below.

10.2 MEAN SQUARE ESTIMATION AND REGRESSION

We shall now present the estimation problem in its most general form. Let $\mathbf{x}$ be a second order real random vector called the *observation vector*. Consider now a real second order scalar RV y, called the *estimandum*. The problem is to find the estimate noted

$$\hat{y}(\mathbf{x}) = \mathrm{MSE}[y|\mathbf{x}] = S(\mathbf{x}) \tag{10.5}$$

in such a way that the mean square error

$$\varepsilon^2 \triangleq E\{[y - \hat{y}(\mathbf{x})]^2\} \tag{10.6}$$

is minimum. To solve this we suppose that the statistical properties of the pair $[\mathbf{x}, y]$ are known.

There are many ways of finding the result, and we shall first give an analytical and then a geometric method. The first is perhaps the more direct, but the second is much more significant and is central to various extensions discussed in other sections of this chapter. In reality it is the true framework of all mean square estimation problems.

10.2.1 Analytical Solution of the Problem

The starting point of this method is the mean square error (10.6) expressed in terms of the PDF $p(\mathbf{x}, y)$ of the pair $[\mathbf{x}, y]$, which gives

$$\varepsilon^2 = \iint [y - S(\mathbf{x})]^2 \, p(\mathbf{x}, y) \, d\mathbf{x} \, dy \tag{10.7}$$

Using (4.28), this can be written as

$$\varepsilon^2 = \iint [y - S(\mathbf{x})]^2 \, p_X(\mathbf{x}) \, p_Y(y|\mathbf{x}) \, d\mathbf{x} \, dy \tag{10.8}$$

The problem is to find the function $S(.)$ such that this expression is minimized. As $p_X(\mathbf{x})$ is non-negative, we can search for the minimum at each point $\mathbf{x}$ and thus minimize the quantity $q(\mathbf{x})$ defined by

$$q(\mathbf{x}) = \int [y - S(\mathbf{x})]^2 p_Y(y|\mathbf{x}) \, dy \tag{10.9}$$

By calculating the integral, this expression becomes

$$q(\mathbf{x}) = S^2(\mathbf{x}) - 2S(\mathbf{x}) \, E(y|\mathbf{x}) + E(y^2|\mathbf{x}) \tag{10.10}$$

where the conditional expectations are defined as in (4.31) by permuting $\mathbf{x}$ and y only. As a result the minimization problem becomes extremely simple: for each value of $\mathbf{x}$, find the function $S(\mathbf{x})$ giving the minimum value of $q(\mathbf{x})$ defined by (10.10). Because of the structure of the polynomial (10.10), there is one minimum that is obtained for

$$\boxed{\hat{y}(\mathbf{x}) = S_o(\mathbf{x}) = E(y|\mathbf{x}) = r(\mathbf{x})} \tag{10.11}$$

where $r(\mathbf{x})$ is the regression defined by (4.34) or (4.38) and encountered many times in previous chapters.

There is one case where the above reasoning does not hold. This is the situation where there exists a domain D of points $\mathbf{x}$ such that if $\mathbf{x} \in D$, then $p_X(\mathbf{x}) = 0$. It is clear that this means that the probability measure of D is null. But if $\mathbf{x} \in D$ we can take any arbitrary value for $S(\mathbf{x})$, because the contribution of D to the integration with respect to $\mathbf{x}$ in (10.8) is null. In other words, this means that the optimum system is given by (10.11) for $\mathbf{x} \notin D$ and is arbitrary for $\mathbf{x} \in D$. This does not change the general result, because the probability that the observation vector $\mathbf{x}$ belongs to D is null, and we can also omit the domain D of the observation space. We assume that this takes place in the following.

In conclusion, the solution to the mean square estimation problem is the *conditional expectation* or the *regression* given by (10.11).

Let us now study more carefully the value of the *minimum error* which can be obtained in the mean square estimation. Here we introduce the *innovation* $\tilde{y}(\mathbf{x})$ defined by

$$\boxed{\tilde{y}(\mathbf{x}) \triangleq y - \hat{y}(\mathbf{x})} \tag{10.12}$$

The expression "innovation" is used because $\tilde{y}(\mathbf{x})$ represents what is new in y when $\mathbf{x}$ is observed. In other words, $\mathbf{x}$ provides information about y which is completely contained in the regression. But this information is in general not sufficient to determine y, and the difference between y and $\hat{y}(\mathbf{x})$ is the innovation.

The first property of the innovation to be noted is that it is a *zeromean* RV. In fact the mean value of y is m_Y and it results from (4.33) that $E[\hat{y}(\mathbf{x})] = m_Y$.

Consequently, the minimum mean square error (10.6) is nothing but the *variance* of the innovation. It is interesting to write this error in another form. Expanding (10.6), we find

$$\varepsilon^2 = E(y^2) - 2E[y\,\hat{y}(\mathbf{x})] + E[\hat{y}^2(\mathbf{x})] \tag{10.13}$$

But we deduce from (4.39) that

$$E\{[y - r(\mathbf{x})]\ \hat{y}(\mathbf{x})\} = 0 \tag{10.14}$$

and as $\hat{y}(\mathbf{x}) = r(\mathbf{x})$, it follows that

$$\varepsilon^2 = E(y^2) - E[\hat{y}^2(\mathbf{x})] = E(y^2) - E[r^2(\mathbf{x})] \tag{10.15}$$

This expression can also be written in terms of variances of y and $r(\mathbf{x})$. As $E[r(\mathbf{x})] = m_Y$, the variance of $r(\mathbf{x})$ is

$$v_R = E[r^2(\mathbf{x})] - m_Y^2 \tag{10.16}$$

and the minimum mean square error takes the form

$$\varepsilon^2 = v_Y - v_R \tag{10.17}$$

From these expressions we can deduce the two extreme situations of mean square estimation. If $r(\mathbf{x}) = y$, the innovation (10.12) is null and $\varepsilon^2 = 0$. We then obtain a case of *perfect* mean square estimation, sometimes called *singular estimation.*

On the other hand, if $r(\mathbf{x}) = m_Y$, its variance is null and $\varepsilon^2 = v_Y$. This case corresponds to the *null estimation*, an expression that is particularly clear when y is zeromean, which gives $r(\mathbf{x}) = m_Y = 0$. Let us discuss this case in more detail. It is clear that null estimation appears if the RVs $\mathbf{x}$ and y are independent, for in this case we have

$$r(\mathbf{x}) = \int y\,p_Y(y|\mathbf{x})\,dy = \int y\,p_Y(y)\,dy = m_Y \tag{10.18}$$

This result is not surprising because if the observation $\mathbf{x}$ and the estimandum y are independent, the observation cannot provide any information on the estimandum and the estimation problem is no longer of interest. In fact the basic idea of the estimation problem is to use the statistical relations between $\mathbf{x}$ and y to estimate y in terms of $\mathbf{x}$. But null estimation can appear even without independence between $\mathbf{x}$ and y. Examples of this are given in Section 4.8.2. Thus, if the PDF of the pair $\mathbf{x}$ and y has the form (4.184), or if

$$p(\mathbf{x}, y) = f(\mathbf{x}^T\mathbf{x} + y^2) \tag{10.19}$$

we have $m_Y = 0$ and $r(\mathbf{x}) = 0$, which gives a null estimation situation even when $\mathbf{x}$ and y are not independent. This problem is very similar to that met in Example 7.1, in the discussion concerning the hierarchy between white noises, and a similar discussion is given, in Problem 4.17. Using the same ideas, we can say that null estimation appears in all such situations where the RVs y and $h(\mathbf{x})$ are uncorrelated whatever the function $h(.)$. In fact this can be written as

$$E[y\,h(\mathbf{x})] = m_Y\ E[h(\mathbf{x})] \tag{10.20}$$

But we deduce from (4.39) that, for any $h(.)$,

$$E[y\,h(\mathbf{x})] = E[r(\mathbf{x})\,h(\mathbf{x})] \tag{10.21}$$

Combining these two equations, we obtain

$$E\{[m_Y - r(\mathbf{x})]\,h(\mathbf{x})\} = 0 \tag{10.22}$$

whatever $h(.)$. This gives $r(\mathbf{x}) = m_Y$, which defines the null mean square estimation.

Before leaving this section, it is important to extend the previous results to the case where the estimandum is no longer scalar but is a *vector* RV. The problem is then to estimate a vector $\mathbf{y}$ belonging to $\mathbb{R}^m$ in terms of an observation vector $\mathbf{x}$ belonging to $\mathbb{R}^n$. The simplest way is to apply the previous results to each component y_i of $\mathbf{y}$. The best mean square estimation of y_i in terms of $\mathbf{x}$ is given by (10.11), and finally, the best estimate of $\mathbf{y}$ can be written as

$$\hat{\mathbf{y}}(\mathbf{x}) = E[\mathbf{y}|\mathbf{x}] = \mathbf{r}(\mathbf{x}) \tag{10.23}$$

where $\mathbf{r}(\mathbf{x})$ is still the regression defined by (4.38). The innovation is also a vector of $\mathbb{R}^m$ defined by

$$\tilde{\mathbf{y}}(\mathbf{x}) = \mathbf{y} - \mathbf{r}(\mathbf{x}) \tag{10.24}$$

It is still a zeromean vector and we can introduce its covariance matrix defined by (3.36), or

$$\Gamma_\varepsilon = E(\tilde{\mathbf{y}}\ \tilde{\mathbf{y}}^T) \tag{10.25}$$

This matrix is sometimes called the *error matrix*, because its diagonal elements are simply the variances of the innovations, or the errors in the estimation of each component. Discussion on the extreme cases of estimation is similar to the scalar case: if the error matrix is null the estimation is singular, and if this matrix is equal to the covariance matrix of $\mathbf{y}$ the estimation is null. Note finally that the off diagonal elements of the error matrix have no direct bearing in our development.

10.2.2 Geometric Solution

Using the geometric concepts introduced in Section 3.7, we note that the mean square error (10.6) is the square of the distance between y and $\hat{y}$ ($\mathbf{x}$), which leads to the idea that the mean square estimation problem can be converted into a problem of *minimum dis-*

tance in an appropriate space. We saw in Section 3.7.1 that this problem can be solved using the *projection theorem*, and we will now apply this approach to estimation problems.

As in the previous section, we assume that the estimandum is a real second order scalar RV, and we want to calculate its mean square estimation in terms of a random vector **x** belonging to $\mathbb{R}^n$.

Let us call L^2 the Hilbert space of real second order RVs in which the scalar product between two RVs $u(\omega)$ and $v(\omega)$ is defined by

$$(u,\ v) = E[u(\omega)\ v(\omega)] \tag{10.26}$$

This space is complete, as shown in Section 3.7.2. The scalar product (10.26) allows us to define orthogonal RVs and it is important to avoid any confusion with uncorrelated RVs. Two RVs are *orthogonal*, which is written $u \perp\!\!\!\perp v$, if

$$E(uv) = 0 \tag{10.27}$$

and they are *uncorrelated* if $c = 0$, as seen on p. 52, which means from (3.26) that

$$E(uv) = E(u)\ E(v) \tag{10.28}$$

It is clear that the two concepts are equivalent if at least one of these RVs is zeromean. This is precisely the case in (4.35) because $X - r(Y)$ is zeromean-valued. Thus $X - r(Y)$ is both orthogonal and uncorrelated with any function of Y such as $g(Y)$.

On the other hand, for non zeromean-valued RVs the two concepts of orthogonality and uncorrelation are quite different. As a result of (10.28) two uncorrelated RVs are not orthogonal if their mean values are non-zero. Conversely, it is perfectly possible to have orthogonal RVs which are correlated. As an example of this, let U and V be two zeromean RVs such that $E(U^2) = E(V^2) = 1$ and $E(UV) = c < 0$. It is clear that c is the correlation coefficient of U and V defined by (3.33). Let X and Y be defined by

$$X = U - \sqrt{|c|} \quad ; \quad Y = V - \sqrt{|c|} \tag{10.29}$$

As a result we obtain $E(XY) = c + |c| = 0$, because $c < 0$. Thus X and Y are orthogonal. However, their correlation coefficient is still equal to c, which is non-zero. In fact, we saw on p. 52 that the correlation coefficient is unchanged by linear operations such as (10.29). So X and Y are correlated and orthogonal RVs.

Let us now introduce the concept of *observation subspace*. In order to realize an estimation of y in terms of the observation **x** we use a filter $h(\mathbf{x})$, similar to $S(\mathbf{x})$ in (10.5). It is clear that $h(\mathbf{x})$ is a scalar RV, defined as in (4.40) where $n = 1$. However $h(\mathbf{x})$ does not necessarily belong to L^2 because it can have an infinite second order moment. In the following we shall restrict our attention to filters $h(.)$ such that $h(\mathbf{x})$ is second order, and then belongs to L^2. It is clear that the set of all the RVs such as $h(\mathbf{x})$ is a subspace of L^2 that is called the observation subspace H_X defined by

$$H_X = \{h(\mathbf{x}) | h(\mathbf{x}) \in L^2\} \tag{10.30}$$

In reality it can only be said that H_X is a pre-Hilbert subspace of L^2, because H_X may not be complete, as discussed on p. 77. This point has no importance, as will be seen later. Note also that the regression $r(\mathbf{x})$ defined by (10.11) belongs to H_X, because $r(\mathbf{x})$ is second order. In fact, as a result of (10.17), $v_R \le v_Y$ and v_Y is finite.

Let us now state the estimation problem in this geometric framework. Let y be an RV belonging to L^2. The problem is to find an RV belonging to the observation subspace H_X such that its distance from y is minimum. Using (3.140), the solution is obviously the projection of y onto the observation subspace

$$\hat{y}(\mathbf{x}) = \text{Proj}[y | H_X] \tag{10.31}$$

Let us now show that this projection is precisely the regression $r(\mathbf{x}) = E(y | \mathbf{x})$. In fact we can write (4.39) in the form

$$E\{[y - r(\mathbf{x})]\, h(\mathbf{x})\} = 0 \tag{10.32}$$

whatever $h(\mathbf{x})$. This shows that $y - r(\mathbf{x})$ is orthogonal to H_X, and as $r(\mathbf{x})$ belongs to H_X it is the projection of y onto this subspace. We then have

$$\hat{y}(\mathbf{x}) = r(\mathbf{x}) \tag{10.33}$$

and (10.32) can be written either as

$$y - \hat{y}(\mathbf{x}) \perp\!\!\perp H_X \tag{10.34}$$

or as

$$E\{[y - \hat{y}(\mathbf{x})]\, h(\mathbf{x})\} = 0, \quad \forall h(.) \tag{10.35}$$

which exhibits the *orthogonality principle*, already introduced in (3.142). Note finally that as $r(\mathbf{x})$ belongs to H_X, we can apply the projection theorem without proving that the subspace H_X is complete. We are now in the case discussed on p. 79.

The geometric situation can be pictured in Figure 10.1, which is similar to Figure 3.4.

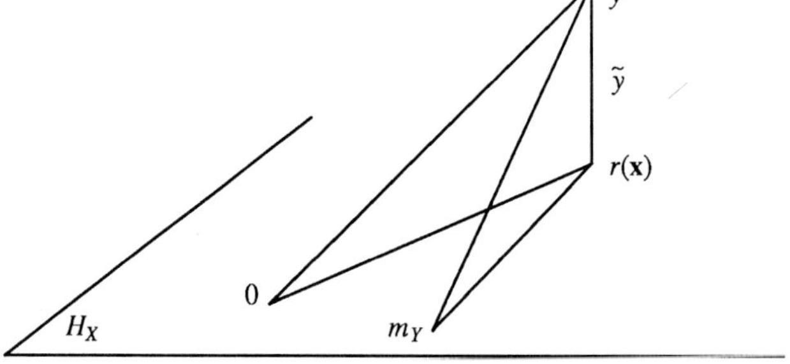

Figure 10.1 Mean square estimation and projection.

Let us make some comments on this figure. The orthogonality principle can be stated in terms of the innovation defined by (10.12) and appearing in the figure. In fact (10.34) means that the *innovation is orthogonal to the observation subspace*, which is the basis of all estimation problems. Note also that, as the innovation is zeromean-valued, the orthogonality principle means that the innovation is *uncorrelated* with any element of the observation subspace.

Suppose now that $\mathbf{x}$ and y are *independent*, which leads to the null estimation. In this case we have

$$E\{yh(\mathbf{x})\} = m_Y \; E[h(\mathbf{x})] \tag{10.36}$$

for any function $h(.)$. This means that $y - m_Y$ is orthogonal to H_X, or that the projection of y onto the subspace H_X is m_Y. So in the case of independence between $\mathbf{x}$ and y we have

$$\hat{y}(\mathbf{x}) = m_Y \tag{10.37}$$

The same situation of null estimation appears if $r(\mathbf{x}) = m_Y$, which means that the regression has a null variance, even with dependent RVs $\mathbf{x}$ and y, as discussed earlier. The situation of null estimation is characterized by the fact that $y - m_Y$ is orthogonal to H_X, or that y is equal to the sum of its expectation and its innovation.

If, however, y belongs to H_X, the innovation is null and we are in the case of *singular estimation*, which means estimation with null error.

Furthermore, as the innovation is zeromean-valued, its variance is the estimation error ε^2 and the relation (10.17) defining the estimation error is a direct application of the Pythagorean theorem. The same theorem gives the relation (10.15). Note finally that the whole discussion is simplified when y is zeromean-valued, because in this case $m_Y = 0$, which also simplifies Figure 10.1.

All these results can be extended, as in Section 10.2.1, to vector RVs $\mathbf{y}$, and the orthogonality principle (10.35) must be written for each component of $\mathbf{y}$, which gives

$$E\{\,[y_i - \hat{y}_i(\mathbf{x})]\; h(\mathbf{x})\,\} = 0, \qquad 1 \le i \le m \tag{10.38}$$

This can be written as a relation between matrices in the form

$$E\{\,[\mathbf{y} - \hat{\mathbf{y}}(\mathbf{x})]\; \mathbf{h}^T(\mathbf{x})\,\} = \mathbb{0} \tag{10.39}$$

whatever the function $\mathbf{h}(\mathbf{x})$. It is clear that (10.39) introduces an m x m matrix. Comparing this orthogonality equation with (4.39) we deduce that $\hat{\mathbf{y}}\,(\mathbf{x})$ is given by the regression in (10.23).

We can conclude by saying that the analytic and geometric solutions of the estimation problem can be considered as equivalent and there is no reason to select one method rather than the other. However, we shall now show that the geometric method is much better adapted to other estimation problems which can also be solved with the projection theorem.

10.3 CONSTRAINED MEAN SQUARE ESTIMATION

In many physical problems we have to solve a mean square estimation of y in terms of the observation vector $\mathbf{x}$ by using filters $h(\mathbf{x})$ which must satisfy some constraints. We will see below many examples of such constraints, and will give only one example here. For several well known reasons *linear systems* play a particularly important role in filter theory. The principal reason is probably because they are the simplest to realize. We may want to estimate y by using only filters in the form

$$h(\mathbf{x}) = c + \mathbf{h}^T\mathbf{x} \tag{10.40}$$

which are defined by a constant c and a vector $\mathbf{h}$ of $\mathbb{R}^n$. This constraint introduces the problem of *linear mean square estimation* of y in terms of $\mathbf{x}$. In reality (10.40) is not exactly a linear operation but an affine operation. It is also sometimes called a linear non-homogeneous operation. However, if $c = 0$, we have a homogeneous operation or a true linear system.

Let us now consider the *general problem* of mean square estimation with a constraint C on the filter. The main point to be noted is that any constraint on the filter used for the estimation introduces a *reduction* of the observation space defined without constraint by (10.30). Let us now introduce the *set* of constrained filters defined by

$$H_C = \{ h(\mathbf{x}) | h(\mathbf{x}) \in L^2 \text{ and } h(\mathbf{x}) \text{ satisfies the constraint } C \} \tag{10.41}$$

At this stage this set has no reason to be a Hilbert subspace of H_X because H_C is not necessarily a linear space. But this set satisfies the inclusion relation

$$H_C \subset H_X \subset L^2 \tag{10.42}$$

In spite of this constraint the estimation problem remains the same - i.e., to find in H_C an element of minimum distance from y. Depending on the structure of H_C there may be no solution, one solution, or many solutions to the problem. There are some constraints for which H_C takes a structure allowing a solution. This is especially the case when the set H_C has the structure of a vector space, and thus becomes a Hilbert subspace of H_X. In these cases we say that we have a *linear constraint*. It is important to note that the term "linear constraint" does not mean that the filter $h(\mathbf{x})$ is linear, as in (10.40), but means that if $h_1(\mathbf{x})$ and $h_2(\mathbf{x})$ satisfy the constraint, $\lambda_1 h_1(\mathbf{x}) + \lambda_2 h_2(\mathbf{x})$ also satisfies the constraint whatever the real numbers λ_1 and λ_2. For example, the purely quadratic filters in the form

$$h(\mathbf{x}) = \mathbf{x}^T \mathbb{M} \mathbf{x} \tag{10.43}$$

where $\mathbb{M}$ is a symmetric matrix, introduce a linear constraint because

$$\lambda_1 h_1(\mathbf{x}) + \lambda_2 h_2(\mathbf{x}) = \mathbf{x}^T [\lambda_1 \mathbb{M}_1 + \lambda_2 \mathbb{M}_2]\ \mathbf{x} \tag{10.44}$$

At this point of the discussion, note that a linear constraint means only that H_C is a pre-Hilbert subspace, because H_C can be non-complete. We will discuss this point later, and assume for the moment that H_C is complete. In this case it is said that the constraint is *linear and complete.*

The mean square estimation problem with a linear and complete constraint is exactly the same as that studied in the previous section, and consists in finding the element of H_C which is at the minimum distance from y. As a result the solution is the same, given by the projection theorem

$$\hat{y}_C(\mathbf{x}) = \text{Proj}[\, y \,|\, H_C\,] \tag{10.45}$$

which must be compared to (10.31). This solution can be found once again by applying the orthogonality principle, or

$$y - \hat{y}_C(\mathbf{x}) \perp\!\!\!\perp H_C \tag{10.46}$$

which means that for any filter $h(\mathbf{x})$ *satisfying the constraint* we have

$$E\{[\, y - \hat{y}_C(\mathbf{x})]\, h(\mathbf{x})\} = 0 \tag{10.47}$$

One can wonder whether there is a relation between the unconstrained solution $\hat{y} = r(\mathbf{x})$ given by (10.31) and (10.33) and the constrained solution $\hat{y}_C(\mathbf{x})$ defined by (10.45). This is effectively the case, the relation being

$$\hat{y}_C(\mathbf{x}) = r_C(\mathbf{x}) \triangleq \text{Proj}[\, r(\mathbf{x}) | H_C] \tag{10.48}$$

In order to prove this result note that (10.33) and (10.34) show that $y - r(\mathbf{x})$ is orthogonal ($\perp\!\!\!\perp$) to H_X. Let us remember that two vectors u and v are orthogonal ($\perp\!\!\!\perp$) if they satisfy (10.27). As H_C is a subspace of H_X, $y - r(\mathbf{x})$ is then orthogonal ($\perp\!\!\!\perp$) to H_C. Now the function $r_C(\mathbf{x})$ defined by (10.48) is also defined by the orthogonality principle

$$r(\mathbf{x}) - r_C(\mathbf{x}) \perp\!\!\!\perp H_C \tag{10.49}$$

As a result $y - r(\mathbf{x}) + r(\mathbf{x}) - r_C(\mathbf{x}) = y - r_C(\mathbf{x})$ is orthogonal ($\perp\!\!\!\perp$) to H_C, and comparing this with (10.46), we obtain (10.48). This shows the virtue of geometric reasoning.

Let us now discuss the effect of the constraint on the *estimation error*. In principle the constraint must imply an increase in the error, which means that $\varepsilon_C^2 \geq \varepsilon^2$. If $\varepsilon_C^2 = \varepsilon^2$, the optimum filter in the general problem, which is the regression $r(\mathbf{x})$, satisfies the constraint, or belongs to H_C. We will encounter such situations later.

It is now of interest to find an expression for the *additional error* due to the constraint. For this we again use the projection theorem. This states that any vector u of H_X can be decomposed in a unique way as a sum of two vectors of H_X such as

$$u = u_C + u_C^{\perp\!\!\!\perp} \tag{10.50}$$

where u_C belongs to H_C and $u_C^{\perp\!\!\!\perp}$ is orthogonal ($\perp\!\!\!\perp$) to H_C. Applying (10.50) to the solution $\hat{y}(\mathbf{x}) = r(\mathbf{x})$ of the unconstrained problem, and using (10.48), we obtain

$$\hat{y}(\mathbf{x}) = \hat{y}_C(\mathbf{x}) + \hat{y}_C^{\perp\!\!\!\perp}(\mathbf{x}) \tag{10.51}$$

The constrained error ε_C^2 is given by

$$\varepsilon_C^2 = E\{[y - \hat{y}_C]^2\} \tag{10.52}$$

which with (10.51) can be expressed in the form

$$\varepsilon_C^2 = E\{[(y-\hat{y}) + (\hat{y} - \hat{y}_C)]^2\} = E\{[(y-\hat{y}) + \hat{y}_C^{\perp\!\!\perp}]^2\} \tag{10.53}$$

But $y - \hat{y}$ is orthogonal ($\perp\!\!\perp$) to H_X, from (10.34), while $\hat{y}_C^{\perp\!\!\perp}$ belongs to H_X. So $(y - \hat{y})$ and $\hat{y}_C^{\perp\!\!\perp}$ are orthogonal ($\perp\!\!\perp$) and (10.53) takes the form

$$\varepsilon_C^2 = \varepsilon^2 + E\{[\hat{y}_C^{\perp\!\!\perp}(\mathbf{x})]^2\} \tag{10.54}$$

where ε^2 is the error of the unconstrained problem given by (10.6). The last term of (10.54) is then the *increment* of the mean square error due to the constraint C. It should be noted that this increment is *bounded* because $\varepsilon_C^2 \le E(y^2)$. This appears in (10.52), but can also be seen in Figure 10.1. In fact, if we are working with a constraint we can replace H_X by H_C in this figure. There may be a difference depending on whether H_C contains the constant function or not. For example, the constraint (10.40) introduces filters which can be constant, while this is not the case for the filters satisfying (10.43). If $m_Y \in H_C$, then we have $\varepsilon_C^2 \le v_Y$, where v_Y is the variance of y, while if m_Y does not belong to H_C we have only $\varepsilon_C^2 \le E(y^2)$.

However, there are also some constraints for which the increment of the error vanishes. Geometrically this means that the optimum solution, which is the regression (10.11), belongs to the subspace H_C. The best example of this situation appears in the *normal case*. In fact we have seen that if $\mathbf{x}$ and $\mathbf{y}$ are zeromean-valued and jointly normal, the regression given by (4.109) is linear. As a result the regression belongs to the space H_C defined by the constraint (10.40), and the introduction of this constraint does not affect the performance of the estimation.

Finally, it is interesting to note that the constrained mean square estimation can be solved by an *indirect method* using the following ideas. Suppose that the solution to the unconstrained problem is known. Then (10.51) shows that in order to calculate $y_C(\mathbf{x})$ we can write

$$\hat{y}_C(\mathbf{x}) = \hat{y}(\mathbf{x}) - \hat{y}_C^{\perp\!\!\perp}(\mathbf{x}) \tag{10.55}$$

and calculate $\hat{y}_C^{\perp\!\!\perp}(\mathbf{x})$ in order to obtain $\hat{y}_C(\mathbf{x})$. To do this, let us introduce the subspace $H_C^{\perp\!\!\perp}$ which is the complementary space of H_C with respect to H_X. This means that

$$H_X = H_C \oplus H_C^{\perp\!\!\perp} \tag{10.56}$$

and that H_C and $H_C^{\perp\!\!\perp}$ are orthogonal. We saw in (3.149) that the projection of any vector onto the sum of two orthogonal subspaces is the sum of the projections. This yields

$$\text{Proj}[y|H_X] = \text{Proj}[y|H_C] + \text{Proj}[y|H_C^{\perp\!\!\perp}] \tag{10.57}$$

Using (10.31), (10.45) and (10.51) we deduce that

$$\hat{y}_C^{\perp\!\!\perp} = \text{Proj}[y|H_C^{\perp\!\!\perp}] \tag{10.58}$$

and $\hat{y}_C^{\perp\!\!\perp}$ is then also determined by the orthogonality principle

$$y - \hat{y}_C^{\perp\!\!\perp} \perp\!\!\perp H_C^{\perp\!\!\perp} \tag{10.59}$$

If we compare (10.46) and (10.59), we see that $\hat{y}_C(\mathbf{x})$ and $\hat{y}_C^{\perp\!\!\perp}(\mathbf{x})$ are determined by exactly the same procedure. In some situations the projection onto $H_C^{\perp\!\!\perp}$ is much easier to obtain than that onto H_C. In this case it is recommended to use the indirect method by calculating $\hat{y}(\mathbf{x})$ and $\hat{y}_C^{\perp\!\!\perp}(\mathbf{x})$ to arrive at $\hat{y}_C(\mathbf{x})$ using (10.51). Some examples of this will be given later.

Before concluding we should discuss the case where the constraint is linear but not *complete*. This means that H_C is only a pre-Hilbert subspace of H_X. In this case the procedure is as follows. As the projection theorem used in (10.45) requires that H_C be complete, it is not guaranteed that a solution to the problem exists. If one does, it will satisfy the orthogonality principle (10.46) or (10.47). If, therefore, there is a function $\hat{y}_C(\mathbf{x})$ satisfying those orthogonality equations, it is the solution of the minimum distance problem, or of the mean square estimation with constraint. On the other hand, if there is no function satisfying the constraint and the orthogonality principle, there is no solution to the constrained problem. An example of this very exceptional situation is discussed in Problem 3.18. However, as the question is of no great interest in signal processing problems, we shall not pursue it further.

10.4 LINEAR MEAN SQUARE ESTIMATION

We shall now consider the mean square estimation problem with the constraint that the filters have the form (10.40). It is obvious that this constraint is linear because if $h_1(\mathbf{x})$ and $h_2(\mathbf{x})$ have the form (10.40), the same is true for $\lambda_1 h_1(\mathbf{x}) + \lambda_2 h_2(\mathbf{x})$. So we will replace H_C of the previous section by H_L defined by

$$H_L = \{ h(\mathbf{x}) | h(\mathbf{x}) = c + \mathbf{h}^T\mathbf{x}, \mathbf{h} \in \mathbb{R}^n \} \tag{10.60}$$

It is obvious that if $\mathbf{h}$ belongs to $\mathbb{R}^n$ and if the vector $\mathbf{x}$ is second order, as defined at the beginning of Section 3.5, then $h(\mathbf{x})$ belongs to L^2. For the moment we do not know if H_L is complete, and in order to find the linear mean square estimate $\hat{y}_L$ of y, we will apply the orthogonality principle (10.46) or (10.47). Let us use this latter equation. Our aim is to find the linear estimate

$$\hat{y}_L(\mathbf{x}) = \alpha + \mathbf{a}^T\mathbf{x} \tag{10.61}$$

such that for any function $h(\mathbf{x}) = c + \mathbf{h}^T\mathbf{x}$ we have

$$E\{ [y - \hat{y}_L(\mathbf{x})]\, h(\mathbf{x}) \} = 0 \tag{10.62}$$

This equation can be developed by introducing the auto- and crosscovariance matrices given by (3.63) and (3.65), which gives

$$c(m_Y - \alpha - \mathbf{a}^T\mathbf{m}_X) + \mathbf{h}^T(\Gamma_{xy} + m_Y\mathbf{m}_X - \alpha\mathbf{m}_X - \Gamma_x\,\mathbf{a} - \mathbf{m}_X\mathbf{m}_X^T\,\mathbf{a}) = 0 \quad (10.63)$$

As this equation must be valid whatever c and $\mathbf{h}$, it can be decomposed in two independent equations giving α and $\mathbf{a}$ by

$$\alpha = m_Y - \mathbf{a}^T\mathbf{m}_X \quad (10.64)$$

$$\Gamma_x\,\mathbf{a} = \Gamma_{xy} \quad (10.65)$$

If Γ_x is positive definite, or has non-zero eigenvalues, (10.65) has a unique solution, which is

$$\mathbf{a} = \Gamma_x^{-1}\,\Gamma_{xy} \quad (10.66)$$

By inserting this value in (10.64) and the result in (10.61), we obtain the best linear mean square estimation of y in terms of $\mathbf{x}$ with

$$\hat{y}_L = m_Y + \Gamma_{xy}^T\,\Gamma_x^{-1}\,(\mathbf{x} - \mathbf{m}_X) \quad (10.67)$$

We can apply exactly the same reasoning when y is no longer scalar but vector, and by applying (10.67) to each component of $\mathbf{y}$ we finally obtain

$$\boxed{\hat{\mathbf{y}}_L = \mathbf{m}_Y + \Gamma_{yx}\,\Gamma_x^{-1}\,(\mathbf{x} - \mathbf{m}_X)} \quad (10.68)$$

This very general expression motivates several comments.

(a) If the mean values of $\mathbf{x}$ and $\mathbf{y}$ are null, we obtain

$$\hat{\mathbf{y}}_L = \Gamma_{yx}\,\Gamma_x^{-1}\,\mathbf{x} \quad (10.69)$$

and comparing this with (4.109), we see that $\mathbf{y}_L$ is the regression $\mathbf{r}(\mathbf{x})$ obtained when the random vectors $\mathbf{x}$ and $\mathbf{y}$ are zeromean and *jointly normal.* In reality it is easy to show that if $\mathbf{x}$ and $\mathbf{y}$ are jointly normal but with non-zero mean values, the regression $\mathbf{r}(\mathbf{x})$ is given by (10.68). Thus we arrive at a very important conclusion. In the case where the vectors $\mathbf{x}$ and $\mathbf{y}$ are jointly normal the constraint of linearity does not increase the estimation error. This means, as seen above, that the regression, being the solution of the mean square estimation problem, belongs to the subspace H_L. Apart from its simplicity, this is the reason why linear estimation is so often used instead of regression. We saw in Section 4.8.2 and in (9.102) that this linearity of the regression is not a property characteristic only of the normal law but also appears for some other probability distributions or signals.

(b) It is clear from (10.68) that

$$E\{\hat{y}_L\} = m_Y \quad (10.70)$$

a property already encountred for the regression $\mathbf{r}(\mathbf{x})$.

(**c**) Let us now show that it is not necessary to prove that H_L is *complete* to solve the mean square estimation problem. In fact, if $\mathbb{\Gamma}_x$ has no null eigenvalue, the linear equation (10.65) has a unique solution, which means that there is a unique element of H_L, which is (10.68) and that it is at the minimum distance from y, which gives thus the solution of the estimation problem. If $\mathbb{\Gamma}_x$ has some null eigenvalues, we return to the discussion in Section 3.5.2 (**e**). This means that the observation vector is not a vector of $\mathbb{R}^n$ but takes its value in a subspace of $\mathbb{R}^n$. In this case the first task is to reduce the observation vector $\mathbf{x}$ to $\mathbf{x}'$ with a full rank covariance matrix. In this operation no information is lost, and we return to a problem giving a unique solution.

(**d**) Let us comment further on (10.65). If we take $\mathbb{\Gamma}_x = \mathbb{R}$ and $\Gamma_{xy} = \mathbf{c}$, we obtain the *normal equation* (9.27) introduced in the context of autoregressive signals. This relation between AR signals and estimation problems will be analyzed in more depth in a later chapter.

Let us now calculate and discuss the *estimation error* in the linear mean square estimation problem. The error matrix is the covariance matrix of the linear innovation

$$\tilde{\mathbf{y}}_L = \mathbf{y} - \hat{\mathbf{y}}_L \tag{10.71}$$

where $\hat{\mathbf{y}}_L$ is given by (10.68). In order to calculate this matrix we can refer again to Figure 10.1 in which $\tilde{y}$ and $r(\mathbf{x})$ are simply replaced by $\tilde{\mathbf{y}}_L$ and $\hat{\mathbf{y}}_L$ respectively. As $\tilde{\mathbf{y}}_L$ is zeromean, this gives

$$\mathbb{\Gamma}_{\varepsilon,L} \triangleq E[\tilde{\mathbf{y}}_L\ \tilde{\mathbf{y}}_L^T] = \mathbb{\Gamma}_y - E[(\mathbf{y}_L - \mathbf{m}_Y)\,(\mathbf{y}_L - \mathbf{m}_Y)^T] \tag{10.72}$$

Using (10.68) and the definition of the covariance matrix of $\mathbf{x}$ in (3.63), we obtain

$$\mathbb{\Gamma}_{\varepsilon,L} = \mathbb{\Gamma}_y - \mathbb{\Gamma}_{yx}\,\mathbb{\Gamma}_x^{-1}\,\mathbb{\Gamma}_{xy} \tag{10.73}$$

which has exactly the same structure as (4.111), giving the conditional covariance matrix of $\mathbf{y}$ given $\mathbf{x}$ for normal vectors $\mathbf{x}$ and $\mathbf{y}$. This is not at all surprising, because in the normal case the regression is linear.

Suppose now that y is scalar. In this case $\mathbb{\Gamma}_y$ becomes the variance v_Y and $\mathbb{\Gamma}_{xy}$ is a vector called $\mathbf{c}$. We then obtain (10.73) in the form

$$\varepsilon_L^2 = v_Y - \mathbf{c}^T\,\mathbb{\Gamma}_x^{-1}\,\mathbf{c} \tag{10.74}$$

which has the same structure as (9.29) obtained for AR signals. This will also be discussed in more detail within the context of prediction problems.

We shall now investigate some extreme cases of estimation problems. First, if $\mathbb{\Gamma}_{xy} = \mathbb{O}$ we deduce from (10.68) that $\hat{y}_L = m_Y$ and from (10.74) that $\varepsilon^2 = v_Y$. This corresponds to the null linear mean square estimation. It should be noted that null linear mean square estimation does not mean null mean square estimation. In fact it is perfectly possible to find situations where $\varepsilon_L^2 = v_Y$ but $\varepsilon^2 < v_Y$. Some examples are discussed as problems.

On the other hand, let us consider the case of singular linear mean square estimation. Looking at Figure 10.1, this means that the estimandum belongs to H_L or can be written

$$\mathbf{y} = \mathbf{k} + \mathbb{G}\mathbf{x} \tag{10.75}$$

For a given value of $\mathbf{m}_X$ and $\mathbb{\Gamma}_x$, this equation completely specifies $\mathbf{m}_Y$, $\mathbb{\Gamma}_y$ and $\mathbb{\Gamma}_{yx}$, and it is easy to verify that using these values in (10.73) we obtain $\mathbb{\Gamma}_{\varepsilon,L} = \mathbb{0}$.

There is one last point concerning the difference between *linear* and *affine* systems. If $\mathbf{x}$ and $\mathbf{y}$ are zeromean-valued, we saw that the best linear mean square estimation of $\mathbf{y}$ in terms of $\mathbf{x}$ is given by (10.69). This is because (10.64) gives $\alpha = 0$, which means that the constant c in (10.60) may be ignored. Thus, when working with zeromean elements, we can reduce the space H_L by considering only elements in the form $\mathbf{h}^T\mathbf{x}$. On the other hand, if we use this space when $\mathbf{x}$ and $\mathbf{y}$ are not zeromean-valued, we commit an error and find an estimate which is not equal to (10.69) but takes another form, analyzed in a problem. It is obvious that the corresponding error is greater than that obtained with an affine filter. It is thus important, before solving a linear mean square estimation problem, to know whether or not we are working with zeromean-valued elements.

10.5 CONSTRAINED LINEAR MEAN SQUARE ESTIMATION

Within the framework of linear mean square estimation it is often necessary to add some constraints on the filter. This means that the vector $\mathbf{a}$ in (10.61) cannot be arbitrary but must satisfy some constraints. These cases occur frequently. For example, if the components of $\mathbf{x}$ refer to the time evolution of a signal, the constraint of *causality* means that some components of the vector $\mathbf{a}$ must be null. In some problems of signal processing the vector $\mathbf{a}$ of estimation must be orthogonal to some vectors of $\mathbb{R}^n$, examples of which will be given later. But it is also possible to impose the condition that the components of $\mathbf{a}$ be bounded or satisfy $|a_i| < M$. This can be due to physical limitations on the coefficients of the filters. We shall now see how constraints can greatly modify the linear mean square estimation problem. It is clear that our method will be to adapt to the linear context the results presented in Section 10.3.

Any constraint introduces a reduction of the subspace H_L in which we have to find the best estimation. As in Section 10.3, we will say that the *constraint is linear* if the set of linear systems satisfying the constraint remains a Hilbert subspace of H_L defined by (10.60). This is the case of the causality constraint introduced above. However, the constraint on the bounded components is obviously not linear, since the set

$$S = \{h(\mathbf{x}) | h(\mathbf{x}) = c + \mathbf{h}^T\mathbf{x},\ |h_i| < M\} \tag{10.76}$$

is not a linear subspace of H_L. In fact if $h_1(\mathbf{x})$ and $h_2(\mathbf{x})$ belong to S, the sum $h_1(\mathbf{x}) + h_2(\mathbf{x})$ may not belong to S. As in Section 10.3, we will restrict our study to linear constraints, leaving the study of non-linear constraints for later.

In order to simplify calculations, we assume in this section that the RVs y and $\mathbf{x}$ are *zeromean*. The extension to non-zeromean valued RVs is left to problems, as it is not difficult. Consequently we shall use only linear systems and not affine systems as previously. The problem can then be stated as follows: to find the best linear mean square estimate of y in terms of $\mathbf{x}$, or in the form $\mathbf{h}^T\mathbf{x}$, where $\mathbf{h}$ is constrained to belong to a vector subspace S of $\mathbb{R}^n$. The observation subspace is then

$$H_S = \{\mathbf{h}^T\mathbf{x} \mid \mathbf{h} \in S\} \tag{10.77}$$

The solution is still obtained by projection written as

$$\hat{y}_S = \mathbf{a}_S^T\mathbf{x} = \text{Proj}[\,y \mid H_S] \tag{10.78}$$

which is equivalent to (10.45) obtained in a more general case. We again apply the orthogonality principle which, as in (10.46), gives

$$y - \hat{y}_S \perp\!\!\!\perp H_S \tag{10.79}$$

This means that for every vector $\mathbf{h}$ belonging to S we have

$$\mathbf{h}^T\mathbb{R}(\mathbf{a}_S - \mathbf{a}) = 0 \tag{10.80}$$

where $\mathbb{R} = \Gamma_x$ and $\mathbf{a}$ is the vector used in the unconstrained solution and defined by (10.66). It is possible to set (10.80) in a more elegant form by using the scalar product between two vectors introduced by (3.131) and used in Section 4.8.7. In fact (10.80) can be written as

$$(\mathbf{h}, \mathbf{a}_S - \mathbf{a})_{\mathbb{R}} = 0 \tag{10.81}$$

whatever $\mathbf{h}$ belonging to S. Since any scalar product introduces a specific orthogonality and then a *specific projection*, (10.81) means that the vector $\mathbf{a} - \mathbf{a}_S$ is orthogonal ($\mathbb{R}$), which means orthogonal with the scalar product (3.131) where $\mathbb{M} = \mathbb{R}$, to the subspace S. As $\mathbf{a}_S$ belongs to this subspace, we can write

$$\mathbf{a}_S = \text{Proj}(\mathbb{R})[\mathbf{a} \mid S] \tag{10.82}$$

It is clear that equations (10.78) and (10.82) are equivalent.

To solve this problem it is sometimes more convenient to use the *indirect method* introduced at the end of Section 10.3. Note that, as a result of the projection theorem, the vector $\mathbf{a}$ defined by (10.66) and defining the solution of the unconstrained linear mean square estimation can be decomposed in a unique manner in the form

$$\mathbf{a} = \mathbf{a}_S + \mathbf{a}_S^\perp \tag{10.83}$$

where $\mathbf{a}_S$ is defined by (10.82) and $\mathbf{a}_S^\perp$ is orthogonal ($\mathbb{R}$) to $\mathbf{a}_S$. It is obvious that $\mathbf{a}_S^\perp$ is defined by

$$\mathbf{a}_S^\perp = \text{Proj}(\mathbb{R})[\mathbf{a} \mid S^\perp] \tag{10.84}$$

where $S^\perp$ is the subspace of $\mathbb{R}^n$ orthogonal ($\mathbb{R}$) to S. It may sometimes be more convenient to calculate (10.84) than (10.82). This is especially the case when the dimension of

$S^{\perp}$ is much smaller than that of S. Having calculated $\mathbf{a}_S^{\perp}$ from (10.84) we deduce $\mathbf{a}_S$ by using (10.83).

It is important to note that (10.83) can be transposed in terms of a linear estimate, and we have

$$\hat{y}_L(\mathbf{x}) \triangleq \mathbf{a}^T\mathbf{x} = \hat{y}_S(\mathbf{x}) + \hat{y}_S^{\perp\!\!\perp}(\mathbf{x}) \tag{10.85}$$

The orthogonality ($\mathbb{R}$) between the last two terms of (10.83) implies the orthogonality of the RVs in the r.h.s. of (10.85). To verify this point, we note that

$$E[\,\hat{y}_S(\mathbf{x})\,\hat{y}_S^{\perp\!\!\perp}(\mathbf{x})] = E[\mathbf{a}_S^T\,\mathbf{x}\mathbf{x}^T\mathbf{a}_S^{\perp}\,] = \mathbf{a}_S^T\,\mathbb{R}\,\mathbf{a}_S^{\perp} = 0 \tag{10.86}$$

With the indirect method we calculate the constrained estimator $\hat{y}_S(\mathbf{x})$ by first calculating $\mathbf{a}_S^{\perp}$, which gives $\hat{y}_S^{\perp\!\!\perp}(\mathbf{x})$ and as $\hat{y}_L(\mathbf{x})$ is known, we deduce $\hat{y}_S(\mathbf{x})$ from (10.85).

As in the general case, the indirect method can provide an expression of the increments of the error due to the constraint. For this purpose we note that the constrained error is

$$\varepsilon_S^2 = E[(y - \hat{y}_S)^2] \tag{10.87}$$

which, with (10.85), becomes

$$\varepsilon_S^2 = E\{\,[(y - \hat{y}_L) + \hat{y}_S^{\perp\!\!\perp}]^2\,\} \tag{10.88}$$

But as a result of (10.62), $y - \hat{y}_L$ is orthogonal to H_L defined by (10.60) with $c = 0$. Furthermore, $\hat{y}_S^{\perp\!\!\perp}$ defined by $\mathbf{x}^T\,\mathbf{a}_S^{\perp}$ belongs by definition to H_L. Thus (10.88) takes the form

$$\varepsilon_S^2 = \varepsilon_L^2 + E[(\hat{y}_S^{\perp\!\!\perp})^2] \tag{10.89}$$

and the last term of this equation is the increment of the error due to the constraint. Note that the term ε_L^2 is still given by (10.74). The last term of (10.89) can be expressed with the vector $\mathbf{a}_S^{\perp}$ obtained by (10.84) in the form

$$E[(\hat{y}_S^{\perp\!\!\perp})^2] = (\mathbf{a}_S^{\perp})^T\mathbb{R}\,\mathbf{a}_S^{\perp} = (\mathbf{a}_S^{\perp}, \mathbf{a}_S^{\perp})_{\mathbb{R}} \tag{10.90}$$

Let us now discuss a very simple example of the application of these results. Suppose that we want to estimate an RV y in terms of a vector $\mathbf{x}$ which has the form (3.106). It is the sum of a noise term $\mathbf{N}$ which is zeromean-valued and has a known covariance matrix and a signal $A\mathbf{s}$ with unknown amplitude A. As the mean value of A is unknown, the simplest way to avoid an error due to the mean value is to impose a condition on the vector $\mathbf{a}$ of the estimation so that it satisfies

$$\mathbf{a}^T\mathbf{s} = 0 \tag{10.91}$$

As a consequence, the quantity $\mathbf{a}^T\mathbf{x}$ used in the estimation is independent of the amplitude of the signal. Note that the function of the condition (10.91) is exactly opposite to that of the condition (4.271) used in a similar context. It can better be compared to the second equation (4.296) which must also eliminate the effect of the signal.

The condition (10.91) means that the space S is the space of vectors of $\mathbb{R}^n$ orthogonal ($\mathbb{I}$), which means orthogonal with the scalar product (10.91), to $\mathbf{s}$. As a consequence the orthogonal subspace is of dimension one, and we are immediately led to use the indirect method.

In order to use (10.84) we have to define the subspace $S^{\perp}$. As all orthogonality equations are expressed with the scalar product (3.131) where $\mathbb{M} = \mathbb{R}$, we must first write (10.91) with this scalar product. Introducing the vector $\mathbf{s}_R = \mathbb{R}^{-1}\mathbf{s}$, we can write this equation as

$$\mathbf{a}^T \mathbb{R}\,\mathbb{R}^{-1}\mathbf{s} = (\mathbf{a}, \mathbf{s}_R)_{\mathbb{R}} = 0 \tag{10.92}$$

which means that the space S is the space of vectors of $\mathbb{R}^n$ which are orthogonal ($\mathbb{R}$) to $\mathbf{s}_R$. As a result, the space $S^{\perp}$ is the space of vectors proportional to $\mathbf{s}_R$. Using the same procedure as in (4.299) and (4.300), we can decompose any vector $\mathbf{v}$ of $\mathbb{R}^n$ in the form

$$\mathbf{v} = \mathbf{v}_S + k_v\,\mathbf{s}_R = \mathbf{v}_S + k_v\,\mathbb{R}^{-1}\mathbf{s} \tag{10.93}$$

where $\mathbf{v}_S$ belongs to S and $\mathbf{s}_R$ to $S^{\perp}$. From this orthogonality property we deduce that

$$k_v = (\mathbf{v}, \mathbf{s}_R)_{\mathbb{R}} \,/\, (\mathbf{s}_R, \mathbf{s}_R)_{\mathbb{R}} = \mathbf{v}^T\mathbf{s} \,/\, \mathbf{s}^T\mathbb{R}^{-1}\mathbf{s} \tag{10.94}$$

where $\mathbb{R} = \Gamma_x$.

Applying this to the vector $\mathbf{a}$ defined by (10.66), we deduce

$$\mathbf{a}_S^{\perp} = \frac{\mathbf{a}^T\mathbf{s}}{\mathbf{s}^T\mathbb{R}^{-1}\mathbf{s}}\,\mathbb{R}^{-1}\mathbf{s} = \frac{\mathbf{c}^T\mathbb{R}^{-1}\mathbf{s}}{\mathbf{s}^T\mathbb{R}^{-1}\mathbf{s}}\,\mathbb{R}^{-1}\mathbf{s} \tag{10.95}$$

where $\mathbf{c} = \Gamma_{xy}$. Using (10.85), we arrive at the constrained estimate which can be expressed as

$$\hat{y}_S(\mathbf{x}) = \left[\mathbf{c} - \frac{\mathbf{c}^T\mathbb{R}^{-1}\mathbf{s}}{\mathbf{s}^T\mathbb{R}^{-1}\mathbf{s}}\,\mathbf{s}\right]^T \mathbb{R}^{-1}\mathbf{x} \tag{10.96}$$

where

$$\mathbb{R} = E(\mathbf{x}\mathbf{x}^T) \quad ; \quad \mathbf{c} = E(\mathbf{x}y) \tag{10.97}$$

The additional error due to the constraint is given by (10.90), and using (10.95), we obtain

$$E[(\hat{y}_S^{\perp})^2] = (\mathbf{c}^T\mathbb{R}^{-1}\mathbf{s})^2 / (\mathbf{s}^T\mathbb{R}^{-1}\mathbf{s}) \tag{10.98}$$

We see that this term vanishes if $\mathbf{c}^T\mathbb{R}^{-1}\mathbf{s} = 0$, which is equivalent to $\mathbf{a}^T\mathbf{s} = 0$. This result is natural because the equation shows that the vector $\mathbf{a}$ defining the optimal filter without constraint satisfies (10.91). In this case it is clear that $\hat{y}_S(\mathbf{x}) = \hat{y}_L(\mathbf{x})$, and the constraint (10.91) does not introduce an increment of the estimation error.

This example is one of the most significant to show the relevance of the indirect method. Note the analogy of these results with those obtained on p. 147 in a different problem but also illustrating the interest of geometric methods.

10.6 LINEAR-QUADRATIC MEAN SQUARE ESTIMATION

10.6.1 General Equations

We have seen that in normal cases the regression is linear, which implies that the constraint of linearity does not introduce a loss of performance. However, there are some probability distributions for which the regression is not linear but is too complex to be used. In order to approximate this regression, one can use linear estimation, and if that is not sufficient, add a quadratic term which introduces linear-quadratic estimation. A linear-quadratic filter is characterized by

$$h(\mathbf{x}) = c + \mathbf{h}^T\mathbf{x} + \mathbf{x}^T\mathbb{M}\mathbf{x} \tag{10.99}$$

and then defined by the triplet $[c, \mathbf{h}, \mathbb{M}]$ containing a scalar, a vector of $\mathbb{R}^n$, and a square matrix $\mathbb{M}$. It is obvious in (10.99) that the set of all the linear-quadratic filters is a vector space, and so the constraint (10.99) is a linear constraint. It introduces a subspace of the observation space which is

$$H_{LQ} = \{h(\mathbf{x}) | h(\mathbf{x}) = c + \mathbf{h}^T\mathbf{x} + \mathbf{x}^T\mathbb{M}\mathbf{x}\} \tag{10.100}$$

and should be compared to (10.60) valid for linear systems.

The linear-quadratic estimate of y in terms of $\mathbf{x}$ can be written

$$\hat{y}_{LQ}(\mathbf{x}) = \alpha + \mathbf{a}^T\mathbf{x} + \mathbf{x}^T\mathbb{A}\mathbf{x} \tag{10.101}$$

and the triplet $[\alpha, \mathbf{a}, \mathbb{A}]$ must be obtained by writing the orthogonality principle

$$E\{[y - \hat{y}_{LQ}(\mathbf{x})]\, h(\mathbf{x})\} = 0 \tag{10.102}$$

whatever the filter $h(\mathbf{x})$ in the form (10.99). By applying this equation three times to filters in the forms $h(\mathbf{x}) = c$, $h(\mathbf{x}) = \mathbf{h}^T\mathbf{x}$ and $h(\mathbf{x}) = \mathbf{x}^T\mathbb{M}\,\mathbf{x}$, we obtain the following three equations

$$\alpha + \mathbf{a}^T\mathbf{m}_X + \mathrm{Tr}[\mathbb{A}(\Gamma_x + \mathbf{m}_X\mathbf{m}_X^T)] = m_Y \tag{10.103}$$

$$\alpha\mathbf{m}_X + (\Gamma_x + \mathbf{m}_X\mathbf{m}_X^T)\,\mathbf{a} + E(\mathbf{x}\mathbf{x}^T\mathbb{A}\mathbf{x}) = \mathbf{c} \tag{10.104}$$

$$\alpha(\Gamma_x + \mathbf{m}_X\mathbf{m}_X^T) + E(\mathbf{x}\mathbf{x}^T\mathbf{x}^T)\,\mathbf{a} + E(\mathbf{x}\mathbf{x}^T\mathbf{x}^T\mathbb{A}\mathbf{x}) = E(y\mathbf{x}\mathbf{x}^T) \tag{10.105}$$

In these equations $\mathbf{c}$ is, as before, the vector $E(\mathbf{x}y)$ and Γ_x is the covariance matrix of $\mathbf{x}$. Note that (10.103) implies that $E(\hat{y}_{LQ}) = m_Y$.

The first point to note in these equations is that they make use of *higher-order* moments of the observation vector. In (10.104) we have a third order moment in $\mathbf{x}$ and in (10.105) we have a fourth order moment in $\mathbf{x}$ and a third order moment in y, $\mathbf{x}$. The philosophy of the procedure is simple to understand: the results obtained in linear-quadratic estimation are better than in linear estimation, but at the cost of greater knowledge of the statistical properties of $\mathbf{x}$ and y.

The second point to note is that the three equations are coupled, and as a result the vector **a** obtained from them has no reason to be equal to the vector **a** obtained by (10.66) and valid for linear estimation. To explain this further, the best mean square estimate of y in terms of $\mathbf{x}$ is the regression $r(\mathbf{x})$ given by (10.23). When this function is too complex to be used one can consider using a Taylor expansion limited to terms either linear or quadratic in $\mathbf{x}$. When this expansion is used, the term linear in $\mathbf{x}$ is not changed when adding a quadratic term. This is however not the case when passing from $y_L(\mathbf{x})$ to $y_{LQ}(\mathbf{x})$. For this reason there should be no confusion between (10.101) and a Taylor expansion of the regression. In fact, (10.101) is a *Volterra filter* of order 2, and by extending this sum to a higher-order term we may obtain a Volterra filter of arbitrary order.

10.6.2 The Scalar Case

In order to exhibit some consequences of equations (10.103) – (10.105), and because of their complexity, we shall first discuss the case where x and y are zeromean-valued and where x is scalar. This considerably simplifies these equations, in this case causing (10.104) and (10.105) to become scalar equations. Note also that in this case (10.101) takes the form introduced in Problem 3.20. Introducing the quantities

$$m_k = E(x^k) \quad ; \quad c = E(xy) \quad ; \quad c_2 = E(x^2 y) \tag{10.106}$$

the system of equations defining (10.101) becomes

$$\alpha + m_2 A = 0 \tag{10.107}$$

$$m_2 a + m_3 A = c \tag{10.108}$$

$$m_2 \alpha + m_3 a + m_4 A = c_2 \tag{10.109}$$

The matrix associated with this system is as in (3.109') where $m_1 = 0$ and its determinant is given by (3.111), or

$$\Delta = m_2(m_4 - m_2^2) - m_3^2 \tag{10.110}$$

We assume that $\Delta > 0$, and the specific situation which appears when $\Delta = 0$ is analyzed in Problem 3.19. After simple calculations we obtain

$$a = (1/\Delta)\left[(m_4 - m_2^2)c - m_3 c_2\right] \tag{10.111}$$

$$A = (1/\Delta)\left[-m_3 c + m_2 c_2\right] \tag{10.112}$$

For our purpose note here that the linear mean square estimation corresponding to the same problem is simply

$$\hat{x}_L = (c/m_2)\, x \tag{10.113}$$

and the linear-quadratic estimation (10.101) is

$$\hat{x}_{LQ} = ax + A(x^2 - m_2) \tag{10.114}$$

where a and A are given by (10.111) and (10.112). We verify in these relations that, in general, the linear term of (10.114) is not (10.113). However, this situation occurs if $m_3 = 0$ because in this case (10.108) and (10.109) are uncoupled and the solution to (10.108) takes the form (10.113). If, furthermore, we assume that x is $N(0, m_2)$, and that x and y are jointly normal, then $m_3 = 0$ and $c_2 = 0$, which implies $A = 0$. We find once again the result, already known to us, that in the normal case the linear estimation is the best possible estimation. However, this is not a characteristic property of normal distribution, and as soon as $m_3 = c_2 = 0$, without the normal assumption, the quadratic term in (10.114) is zero, and $\hat{x}_{LQ}$ is equivalent to $\hat{x}_L$.

Let us now look at the *errors* obtained in these estimation problems. The error brought about by using the linear estimator (10.113) is

$$\varepsilon_L^2 = v - c^2/m_2 \tag{10.115}$$

where $v = E(y^2)$ is the variance of the estimandum y. This error is null if $c^2 = m_2 v$. This means that the Schwarz inequality (3.29) where A and B are replaced by x and y becomes an equality. We have seen that this implies (3.30), or $y = a\,x$. In this case it is clear that $\hat{y}_L = y$, which gives a null linear estimation error. Using the same reasoning it is clear that if

$$y = ax + A(x^2 - m_2) \tag{10.116}$$

where a and A are given, then $\hat{y}_{LQ} = y$, and the error in the linear-quadratic estimation is null, while by using only linear estimation the error would be given by (10.115) where v and c^2 are deduced from (10.116). This finally gives

$$\varepsilon_L^2 = (1/m_2)\,\Delta A^2 \tag{10.117}$$

where Δ is given by (10.110). This error obviously becomes null if $A = 0$.

In general cases the advantage of linear-quadratic over linear systems can be characterized by the quantity

$$\delta = \varepsilon_L^2 - \varepsilon_{LQ}^2 \tag{10.118}$$

When $m_3 = 0$, it is easy to find that

$$\delta = c_2^2 = [E(x^2 y)]^2 \tag{10.119}$$

10.6.3 White Observation Case

We shall now return to the equations (10.103) – (10.105). Another case where these equations can be easily solved is when the observation vector $\mathbf{x}$ is zeromean-valued and *white*. Of course, as we have to use moments up to order 4, we must assume that $\mathbf{x}$ is strictly white at least up to the fourth order. This means that the components of $\mathbf{x}$ satisfy (4.79) and (4.81) or (7.10) and (7.11). Using these equations, it can be shown that the system (10.101) can be written as

$$\hat{y}_{LQ}(\mathbf{x}) = \mathbf{a}^T\mathbf{x} + \mathbf{x}^T \mathbb{A}\mathbf{x} - m_2\,\mathrm{Tr}(\mathbb{A}) \tag{10.120}$$

where $m_2 = E(x_i^2)$. Assuming that $E(x_i y) = c$ and that $E(x_i^2 y) = c_2$, while $E(x_i x_j y) = c_2[i, j]$, we find that the off diagonal elements of $\mathbb{A}$ satisfy

$$A_{ij} + A_{ji} = c_2[i, j]/m_2^2 \tag{10.121}$$

This shows that the solution of the problem is unique if $\mathbb{A}$ is a symmetric matrix, which can be expected since it was noted in (4.251) and (4.252) that any quadratic form using a non symmetric matrix can be written by using a symmetric matrix.

Furthermore, the diagonal elements A_{ii} and the components a_i are determined by the system (10.107) – (10.109) discussed in the case of scalar observations. The global conclusion is similar to that discussed just above.

Linear-quadratic systems are clearly the simplest Volterra filters. The extension of the above results to higher-order Volterra filters can be realized but is outside the scope of this discussion. For further information see [Picinbono, 1990].

10.7 EXTENSIONS TO THE COMPLEX CASE

Most of the results above were obtained by using geometric reasoning, especially the projection theorem in appropriate subspaces. In these spaces orthogonality was deduced from the scalar product (10.26). However, we saw in Chapter 3, especially in Section 3.7, that when considering *complex random* variables it is also possible to introduce Hilbert subspaces with the scalar product (3.159) instead of (10.26). From these ideas it may seem that the extension of the above results working with complex RVs is straightforward, and can in some sense be considered as an exercise for the reader.

We shall show, however, that this is not so, for the introduction of complex random elements in estimation problems needs very careful treatment. Let us begin with the case of *linear* mean square estimation which is the simplest and most important case. Some ideas on non-linear estimation problems will be given at the end of this section.

Before entering into the details of the problem, it is important to note that this is not a purely mathematical question. Estimation with complex observations arises very often in signal processing problems, especially when working in the *frequency* domain. We saw in Chapter 6 that the spectral representation of random signals requires complex random elements, and as spectral analysis is important, a careful analysis of estimation with complex elements is necessary.

10.7.1 Classical Presentation of Complex Linear Estimation

Our purpose is to show that the results given in Section 10.4 can be easily extended to the complex case by using geometric methods applied to Hilbert spaces and introduced in Section 3.7.2. To simplify the discussion, we shall assume that all the random elements in this section are zeromean-valued. This allows us to consider strictly linear operations, and not affine ones as in (10.67). Starting from a complex observation $\mathbf{x}$ we can introduce, as in (10.60), the linear observation subspace defined by

$$H_L = \{ h(\mathbf{x}) | h(\mathbf{x}) = \mathbf{h}^H \mathbf{x}, \mathbf{h} \in \mathbb{C}^n \} \tag{10.122}$$

In this space the scalar product of elements which are complex RVs is given by (3.159). Suppose now that we want to estimate linearly the complex RV y in terms of the observation in such a way that the estimate

$$\hat{y}_L(\mathbf{x}) = \mathbf{a}^H \mathbf{x} \tag{10.123}$$

gives the minimum mean square error. As this error can be written $E[\,|y - \hat{y}(\mathbf{x})|^2]$ we deduce from (3.161) that

$$\hat{y}_L(\mathbf{x}) = \text{Proj}[y | H_L] \tag{10.124}$$

This projection is determined by applying the orthogonality principle, as in (10.62), but with the scalar product (3.159). This gives

$$E\{[y - \hat{y}_L(\mathbf{x})]^* \, h(\mathbf{x})\} = 0 \tag{10.125}$$

whatever the function $h(\mathbf{x})$ belonging to H_L. This can be written as

$$\mathbf{h}^H \{ E(\mathbf{x}\mathbf{x}^H)\mathbf{a} - E(\mathbf{x}y^*) \} = 0 \tag{10.126}$$

for all vectors $\mathbf{h}$ of $\mathbb{C}^n$. This immediately gives (10.65) where

$$\Gamma_x = E(\mathbf{x}\mathbf{x}^H) \quad ; \quad \Gamma_{xy} = E(\mathbf{x}y^*) \tag{10.127}$$

are the auto- and crosscovariance matrices defined by (3.90) and (3.65) adapted for the complex case. The error in this complex linear mean square estimation is still the variance of the innovation given by (10.74) in the real case and which in the complex case becomes

$$\varepsilon_L^2 = v_Y - \mathbf{c}^H \Gamma_x^{-1} \mathbf{c} \tag{10.128}$$

where $v_Y = E(|y|^2)$ and $\mathbf{c}$ is the complex vector equal to $E(\mathbf{x}y^*)$. The same reasoning can be applied to the vector random element $\mathbf{y}$, and in the complex case (10.69) remains valid with Γ_x given by (10.127) and Γ_{yx} by

$$\Gamma_{yx} = E(\mathbf{y}\mathbf{x}^H) \tag{10.129}$$

The error matrix is still the covariance matrix of the innovation $\mathbf{y} - \hat{\mathbf{y}}_L$, and it is still given by (10.73) where Γ_y is $E(\mathbf{y}\mathbf{y}^H)$.

Summing up, we see that the extension of linear mean square estimation to the complex case is straightforward and the only difference is that we must remember to take the complex conjugate in the definition of the auto- and crosscovariance elements, following the ideas given in Section 3.7.2. However, we shall see that the situation is not as simple as it may seem.

10.7.2 Insufficiency of the Previous Presentation

To explain the problem in simple terms we shall consider the case where x and y are complex *scalar* random variables. In this case (10.123) can be written

$$\hat{y}_{LC} = A_c x \tag{10.130}$$

where $A_c = a^*$ of (10.123) and the subscripts C and c indicate that we are using the classical complex theory presented above. The optimum scalar filter A_c is given by the orthogonality equation (10.126) which can be written

$$E(|x|^2)A_c = E(yx^*) \tag{10.131}$$

or

$$\gamma_c A_c = r_c \tag{10.132}$$

where γ_c and r_c are extracted from (10.131). The solution is of course

$$A_c = r_c / \gamma_c = A_1 + jA_2 \tag{10.133}$$

where A_1 and A_2 are the real and imaginary parts of A_c. Noting that γ_c is real, we obtain

$$A_1 = r_1 / \gamma_c \quad ; \quad A_2 = r_2 / \gamma_c \tag{10.134}$$

where r_1 and r_2 are the real and imaginary parts of r_c defined by

$$r_c \triangleq E(yx^*) = r_1 + jr_2 \tag{10.135}$$

The error obtained in this estimation procedure is given by (10.128) which takes the form

$$\varepsilon_{LC}^2 = E(|y|^2) - (r_1^2 + r_2^2)/\gamma_c \tag{10.136}$$

However, we have noted repeatedly, and especially in (3.38), that a complex RV is none other than a pair of real RVs, or a random vector of $\mathbb{R}^2$. So we can apply here the theory of linear mean square estimation developed in Section 10.4. To do this we introduce the two random vectors

$$\mathbf{X} = \begin{bmatrix} x_1 \\ x_2 \end{bmatrix} \quad ; \quad \mathbf{Y} = \begin{bmatrix} y_1 \\ y_2 \end{bmatrix} \tag{10.137}$$

and the linear mean square estimation of $\mathbf{Y}$ in terms of $\mathbf{X}$ can be written as

$$\hat{\mathbf{Y}} = \mathbb{A}\mathbf{X} \tag{10.138}$$

where $\mathbb{A}$ is a 2 x 2 matrix with matrix elements A_{ij}. The matrix $\mathbb{A}$ appears in (10.69), which gives

$$\mathbb{A}\mathbb{\Gamma} = \mathbb{R} \tag{10.139}$$

where

$$\mathbb{\Gamma} \triangleq E(\mathbf{X}\mathbf{X}^T) \quad ; \quad \mathbb{R} \triangleq E(\mathbf{Y}\mathbf{X}^T) \tag{10.140}$$

Before continuing, let us compare (10.130) and (10.138). Using (10.133), we can write (10.130) in the form

$$\hat{y}_1 + j\hat{y}_2 = (A_1 + jA_2)(x_1 + jx_2) \tag{10.141}$$

This expression can take the form of (10.138) with the matrix $\mathbb{A}_c$ given by

$$\mathbb{A}_c = \begin{bmatrix} A_1 & -A_2 \\ A_2 & A_1 \end{bmatrix} \tag{10.142}$$

This shows that the complex estimate (10.130) can be written as (10.138) provided that the matrix $\mathbb{A}$ has the structure (10.142). In general there is no need to realize this form, and thus the complex estimate presented in the previous section is not necessarily the best linear mean square estimate for complex random elements. It is interesting to note the analogy between (10.142) and (4.140), which appears in the discussion on the circularity of normal RVs.

At this point it is important to note that the purpose of (10.130) and of (10.138) is to *solve the same problem.* In fact, there is no difference in the error criterion used. In (10.130) the criterion is the minimum of $E(|\tilde{y}_c|^2)$ where $\tilde{y}_c$ is the complex innovation equal to $y - \hat{y}_{LC}$, and this error is equal to $E(\tilde{y}_1^2) + E(\tilde{y}_2^2)$. In (10.138) the criterion is the minimum of $E(\tilde{y}_1^2)$ and $E(\tilde{y}_2^2)$, which obviously minimizes the sum.

In order to compare the two approaches, let us now calculate the estimate (10.138) by solving (10.139). Introducing the matrix elements γ_{ij} and r_{ij} of the matrices defined by (10.140), we obtain the matrix elements a_{ij} of $\mathbb{A}$ by

$$a_{11} = (r_{11}\gamma_{22} - r_{12}\gamma_{12})(\gamma_{11}\gamma_{22} - \gamma_{12}^2)^{-1} \tag{10.143}$$

$$a_{12} = (r_{12}\gamma_{11} - r_{11}\gamma_{12})(\gamma_{11}\gamma_{22} - \gamma_{12}^2)^{-1} \tag{10.144}$$

and similar expressions for a_{21} and a_{22}, by permuting 1 and 2.

It is now of interest to write the complex solution by using the elements γ_{ij} and r_{ij} also. Note that γ_c in (10.132) is

$$\gamma_c \triangleq E(|x|^2) = \gamma_{11} + \gamma_{22} \tag{10.145}$$

Furthermore, starting from (10.135), we find

$$r_c = r_{11} + r_{12} + j(r_{21} - r_{12}) = r_1 + jr_2 \tag{10.146}$$

Inserting these relations in (10.134) we obtain

$$A_1 = (r_{11} + r_{22})(\gamma_{11} + \gamma_{22})^{-1} \tag{10.147}$$

$$A_2 = (r_{21} - r_{12})(\gamma_{11} + \gamma_{22})^{-1} \tag{10.148}$$

and it is clear that, without any further assumption, the matrices $\mathbb{A}$ and $\mathbb{A}_c$ are quite different. As the complex estimation (10.130) introduces a constraint on the matrix $\mathbb{A}$, the mean square error is in general greater than when (10.138) is used.

We shall compare the two approaches more closely by considering two examples.

Suppose first that x and y are *circular complex* RVs. As seen in Section 4.6.6 this means that

$$E(x^2) = E(xy) = 0 \tag{10.149}$$

Writing $x = x_1 + jx_2$ and $y = y_1 + jy_2$, this equation gives

$$\gamma_{11} = \gamma_{22} = \gamma \quad ; \quad \gamma_{12} = \gamma_{21} = 0 \tag{10.150}$$

and

$$r_{11} = r_{22} = r \quad ; \quad r_{12} = -r_{21} = r' \tag{10.151}$$

Inserting these values in (10.143) and (10.144) we obtain

$$a_{11} = a_{22} = r/\gamma \tag{10.152}$$

$$a_{12} = -a_{21} = r'/\gamma \tag{10.153}$$

and both (10.147) and (10.148) provide exactly the same result. Thus when the assumption of second order circularity characterized by (10.149) is introduced the complex approach characterized by (10.130) and the real approach (10.138) give the same estimator. We have often noted the importance and origin of circularity, confirmed here by this result in linear mean square estimation.

Suppose now that the estimandum y is *real*, while the data are complex. This situation often occurs when we are working with real signals but using the complex spectral components as observation.

If y is real, its imaginary part y_2 is null. In consequence the matrix $\mathbb{R}$ defined by (10.140) has a second row equal to zero, or

$$r_{21} = r_{22} = 0 \tag{10.154}$$

We then deduce from the equations similar to (10.143) and (10.144) giving a_{21} and a_{22} that

$$a_{12} = a_{22} = 0 \tag{10.155}$$

This is logical because the second component of $\mathbf{Y}$ defined by (10.137) is no longer random and its estimate $\hat{y}_2$ is null. As a result, the estimator (10.138) provides a real estimate for y which is equal to $\hat{y}_1$. On the other hand the second row of $\mathbb{A}_c$ cannot be null and the complex estimate (10.130) necessarily provides a complex estimate of a real estimandum, which is not a satisfactory result. Let us now add some further assumptions to see what the consequences are.

Suppose that x is also *real*. Consequently the only non-zero elements γ_{ij} and r_{ij} are γ_{11} and r_{11}. In this case we cannot use (10.143) and (10.144) because the matrix $\mathbb{\Gamma}$ which appears in (10.139) is singular. This is an obvious result, since in this situation the vectors $\mathbf{X}$ and $\mathbf{Y}$ do not belong to $\mathbb{R}^2$ but to $\mathbb{R}^1$. By applying the linear mean square estimation in $\mathbb{R}^1$ we easily find

$$\hat{y}_1 = (r_{11}/\gamma_{11})\, x_1 \tag{10.156}$$

The complex estimator (10.130) gives the same result because in (10.135) $r_2 = 0$, which implies $A_2 = 0$. The matrix (10.142) becomes diagonal but as $x_2 = 0$, we also obtain (10.156).

Suppose now that x is *complex circular,* with y still real. In this case (10.150) holds, and as y is real, we also have (10.154) and (10.155). Using (10.143) and (10.144), we deduce that

$$a_{11} = r_{11}/\gamma \quad ; \quad a_{12} = r_{12}/\gamma \tag{10.157}$$

and (10.147) and (10.148) give

$$A_1 = (1/2)\, r_{11}/\gamma \quad ; \quad A_2 = -(1/2)\, r_{12}/\gamma \tag{10.158}$$

The two estimators corresponding to (10.138) and (10.130) are

$$\hat{y}_{LR} = (1/\gamma)\,(r_{11}x_1 + r_{12}x_2) \tag{10.159}$$

$$\hat{y}_{LC} = (1/2\gamma)\,[(r_{11}x_1 + r_{12}x_2) + j(-r_{12}x_1 + r_{11}x_2)] \tag{10.160}$$

As indicated above, these two estimators are quite different, the second giving a complex estimate of a real estimandum. For a better comparison let us calculate the corresponding errors. For the complex estimator (10.160) the error is given by (10.136) which becomes

$$\varepsilon_{LC}^2 = v_Y - (1/2\gamma)\,(r_{11}^2 + r_{12}^2) \tag{10.161}$$

On the other hand the estimator (10.159) gives an error

$$\varepsilon_{LR}^2 = v_Y - (1/\gamma)\,(r_{11}^2 + r_{12}^2) \tag{10.162}$$

and we see that the reduction of the variance by the estimation is twice as great with the estimator (10.159) as with (10.160). This clearly shows why, in this particular case, the complex estimator (10.130) should not be used.

10.7.3 General Discussion of Estimation in the Complex Case

In the above discussion x and y are complex but scalar RVs. The extension of these results to complex vectors does not introduce any special difficulties. The notation is more complicated: for example, in (10.142) A_1 and A_2 become matrices instead of scalars. But the general result is the same. The estimate (10.123) calculated in Section 10.7.1 is not necessarily the best one. It is easy to prove that it is effectively the best one when the vectors introduced are second order circular, which is specified by extension of (10.149) to vector cases, such as

$$E(\mathbf{x}\mathbf{x}^T) = \mathbb{O} \quad ; \quad E(\mathbf{x}\mathbf{y}^T) = \mathbb{O} \tag{10.163}$$

But the importance of the circular assumption in many practical problems should not allow us to forget the fact that there are many situations where this assumption does not hold. In those cases it can be dangerous to use (10.123) instead of a more general expression obtained by stating the estimation problem in a real context, where real and imaginary parts are treated independently. When doing this we use (10.138) which can be written as

$$\hat{\mathbf{y}}_1 = \mathbb{A}_{11}\mathbf{x}_1 + \mathbb{A}_{12}\mathbf{x}_2 \tag{10.164}$$

$$\hat{\mathbf{y}}_2 = \mathbb{A}_{21}\mathbf{x}_1 + \mathbb{A}_{22}\mathbf{x}_2 \tag{10.165}$$

By combining these two equations we obtain

$$\hat{\mathbf{y}} = \mathbb{A}_{11}\mathbf{x}_1 + \mathbb{A}_{12}\mathbf{x}_2 + j(\mathbb{A}_{21}\mathbf{x}_1 + \mathbb{A}_{22}\mathbf{x}_2) \tag{10.166}$$

This can also be written as

$$\hat{\mathbf{y}} = \mathbb{M}\mathbf{x} + \mathbb{N}\mathbf{x}^* \tag{10.167}$$

where $\mathbb{M}$ and $\mathbb{N}$ are two complex rectangular matrices. The complex estimation, such as (10.123), uses only the term $\mathbb{M}\mathbf{x}$, but in general it is necessary to take into consideration the last term of (10.167) to obtain the best performance that can be reached by using only linear estimators.

The discussion can be extended without difficulty to non-linear estimators. For example, a purely quadratic filter in the real case takes the form $\mathbf{x}^T \mathbb{M} \mathbf{x}$ as in (10.43). Its obvious extension in the complex case would be $\mathbf{x}^H \mathbb{M} \mathbf{x}$, where $\mathbb{M}$ is a complex square matrix. However, that is not sufficient for all problems in the complex case. For example, in the scalar case a quadratic receiver must not only use the term $|x|^2$ but also x^2. The principle of the general discussion remains the same although the details are more complicated than in the linear case.

Finally we return to the opening remarks on mean square estimation problems. Suppose that we want to estimate a complex scalar random variable y in terms of a complex random vector observation $\mathbf{x}$. It is always possible to decompose the problem so that it is stated with real random elements. In that case the solution is the same as in Section 2, obtained either by an analytical or geometric procedure. This solution is of course the regression. On the other hand the regression (10.11) is much more difficult to define in the complex case because the concept of probability density $p(\mathbf{z})$ for complex vectors must be defined with great care, as seen in Section 4.6.6 for the normal case.

PROBLEMS

10.1 Consider N zeromean and uncorrelated RVs x_n and another zeromean RV y. The second order properties of this set of RVs are specified by the relations

$$E(y^2) = \sigma_y^2 \quad ; \quad E(yx_n) = c_n \quad ; \quad E(x_n^2) = \gamma_n$$

(a) Calculate the best LMSE $\hat{y}$ of y in terms of the x_ns, $1 \le n \le N$.

(b) Calculate the corresponding estimation error and express this error in terms of σ_y^2 and of the n correlation coefficients between y and x_n.

(c) Deduce the condition on the x_ns ensuring that the error is non-negative.

(d) Show directly that it is a necessary and sufficient condition for the existence of an RV y corresponding to the given second order properties.

10.2 Let Φ be a random phase uniformly distributed in the interval 0, π. Consider the RVs X and Y defined by $X = \cos\Phi$ and $Y = \sin\Phi$.

(a) Calculate the mean values and the variances of X and Y.

(b) Calculate the covariance of X and Y.

(c) Calculate the best LMSE of Y in terms of X.

(d) Calculate the MSE of Y in terms of X and compare the result with that obtained in the previous question. By calculating the estimation errors compare the two estimators and relate this result to the discussion in Section 10.4(**d**).

10.3 The previous problem is a particular case of a more general situation illustrating the difference between MSE and LMSE. This can be explained as follows. Let x be an RV which, for simplification, is assumed to be zeromean valued. Let y be the RV defined by $y = h(x)$, where $h(.)$ is a deterministic transformation satisfying only the constraint ensuring that y is second order. The problem is to compare various possible estimators of y in terms of x.

(a) MSE. Using the results of Section 4.8.3, show that the conditional expectation $E[y|x]$ is equal to $h(x)$ and deduce that the estimation problem is singular, or reaches a zero error.

(b) Calculate the best LMSE of y in terms of x. Show that it also gives a zero error when $h(x) = ax + b$. Calculate the estimation error and show that the estimation is zero if $E[xh(x)] = 0$. Show that if the LMSE is singular, then $h(x) = ax + b$.

(c) It is possible to extend the previous method to LQMSE, where LQ means "linear quadratic." In order to calculate the best LQMSE, we use the method discussed in Problem 3.20. Show that the coefficients α_0 and α_1 in this expression are not those obtained in the best LMSE. Explain the origin of this fact and, more generally, prove that the expansion appearing in Problem 3.20 is not a finite Taylor expansion of the function $h(x)$.

10.4 We want to extend the previous discussion to a more general situation. Let x be an RV which, for simplification, is assumed to be zeromean valued. Let y be the RV defined as follows. Let $h_i(x)$, $1 \le i \le n$, be a set of n functions satisfying only the constraint ensuring that the RVs $h_i(x)$ are second order. We assume that the RV y is equal to $h_i(x)$ with the probability p_i. The sum of these p_i is of course equal to 1.

(a) Extending the results of the previous problem and using a relation similar to (2.62), calculate the conditional expectation $E[y|x]$ which is the best MSE of y in terms of x.

(b) Calculate the estimation error and express this error in terms of the expectations of first and second order of $h_i(x)$.

(c) Make the same calculation for the LMSE.

(d) In order to compare the two estimation procedures let us consider the situation discussed in Problem 10.2. Suppose now that the phase Φ is uniformly distributed over the interval 0, 2π. Show that y can be described with two functions $h_i(x)$ and two equal probabilities. Complete the calculations and compare the mean square errors of the MSE and LMSE procedures.

10.5 Let X and Y be two independent normal RVs with the same distribution $N(0, 1)$. We introduce the RV $A = |X^2 + Y^2|^{1/2}$.

(a) Calculate the expected value and the variance of A. Calculate the expectations $E[AX]$ and $E[A|X|]$.

(b) Calculate the LMSE of A in terms of X, as defined by (10.68), and the corresponding error. What is the particularity of this estimation?

(c) Make the same calculation when the observation is $|X|$ instead of X.

(d) Compare the two estimation procedures and discuss the result.

10.6 **Orthogonalization of the observation.** Let $\mathbf{x}$ be a real zeromean random vector with components x_i, $1 \le i \le N$, and y a zeromean RV. Suppose that the covariance matrix $\mathbb{\Gamma}$ of $\mathbf{x}$ is factored as $\mathbb{\Gamma} = \mathbb{A}\mathbb{A}^T$, and let $\mathbb{B}$ be the matrix equal to $\mathbb{A}^{-1}$.

(a) By introducing the random vector $\mathbf{u} = \mathbb{B}\mathbf{x}$, calculate the covariance matrix of $\mathbf{u}$. What is the fundamental property of $\mathbf{u}$?

(b) By introducing the vector $\mathbf{c} = E(y\mathbf{x})$ and the vector $\mathbf{r}$ whose components r_i are the correlation coefficients between y and u_i, calculate $\mathbf{r}$ in terms of $\mathbf{c}$, in terms of the variance of y and of the matrix $\mathbb{B}$.

(c) Calculate the best LMSE of y in terms of $\mathbf{u}$ and express the optimum filter in terms of the variance of y and of the vector $\mathbf{r}$.

(d) Calculate the estimation error and deduce a condition which must be satisfied by the vector $\mathbf{c}$.

(e) By returning to the vector $\mathbf{x}$, calculate the best LMSE of y in terms of $\mathbf{x}$, $\mathbf{c}$ and $\mathbb{B}$. Calculate the corresponding error.

(f) By using the results of Section 3.5.3, explain the interest of this procedure with respect to the standard method.

10.7 We want to apply the method indicated in the previous problem to the following situation. The matrix elements of $\mathbb{\Gamma}$ are $\gamma[m, n] = m \wedge n$. The variance of y is equal to one and the components of the vector $\mathbf{r}$ are α^i, $1 \le i \le N$.

(a) Find the condition on α making possible the existence of y, whatever N.

(b) Express the optimum LMSE of y in terms of $\mathbf{x}$.

(c) Calculate the corresponding error.

10.8 **Optimum ordering for estimation.** Consider the situation described in Problem 10.1. Suppose that, for various reasons, it is not possible to take all the RVs x_i to make the estimation of y but only n $(n < N)$ of these RVs. The problem is then to choose among the N RVs x_i the ns which give the best estimation.

(a) Assuming, as in Problem 10.1, that the x_is are uncorrelated, and using the expression of the mean square error calculated in (**b**), give the solution of the problem.

(b) Relaxing the assumption of non-correlation between the x_is, solve first the problem for $n = 1$.

(c) In order to calculate the solution for $n > 1$ we introduce the following constraint: when the problem is solved for k, its solution for $k + 1$ is obtained by searching among the $N - k$ remaining RVs x_i the RV which makes the greatest reduction of the error estimation of y. This means that we solve the problem recursively and we must pass from k to $k + 1$. For this let us introduce the appropriate notation. Call $\mathbf{x}_n$ the vector whose components are the k x_is already calculated, and z one of the $N - k$ remaining x_is. Show first that the LMSE $\hat{y}$ of y in terms of $\mathbf{x}_n$ and z is a linear combination of these two elements. Express this LMSE in terms of $\mathbf{x}_n$ and $\tilde{z}$, which is the innovation in the LMSE of z in terms of $\mathbf{x}_n$. Using the fundamental property of the innovation, calculate the contribution of the term $\tilde{z}$ to the estimation error of y and indicate how to find z among the $N - k$ remaining RVs x_i in order to obtain the minimum estimation error of y.

10.9 Let x be a random phase uniformly distributed in $[-\pi, +\pi)$, and let y be the RV equal to $\sin x + n$, where n is a noise component assumed to be $N(0, 1/2)$ and independent of x.

(a) Calculate the best MSE of y in terms of x and the corresponding error.

(b) We want to realize a polynomial estimation of y similar to that studied in Problem 3.20. This estimation is written as

$$\hat{y}_n = h_0 + h_1 x + \ldots. + h_n x^n$$

Write the general equation allowing the calculation of the coefficients h_i by using the moments $m_k = E(x^k)$ and $c_k = E(x^k y)$.

(c) Calculate the estimates $\hat{y}_n$ for $n = 1, 2, 3$ and the corresponding errors.

(d) Compare these errors with those obtained by approximating y in (a) by a limited expansion in terms of x^k. Give the meaning of the results obtained.

10.10 Using the same notation as in the previous problem, suppose now that the RV x is uniformly distributed in the interval $[-a, +a]$. Make the same calculations and study the limit when a tends to infinity. Give an interpretation of the results obtained.

10.11 **Noise canceling.** A typical goal in signal processing is to cancel noise. In this problem we discuss a particular situation using a noise alone reference signal. Suppose that the observation is composed of two vectors $\mathbf{x}_1$ and $\mathbf{x}_2$ belonging to $\mathbb{R}^p$ and $\mathbb{R}^{N-p}$ respectively, where N and p are arbitrary integers satisfying $p < N$. These vectors can be written as

$$\mathbf{x}_1 = \mathbf{s} + \mathbf{b}_1 \quad ; \quad \mathbf{x}_2 = \mathbf{b}_2$$

where $\mathbf{s}$ is the estimandum signal, while $\mathbf{b}_1$ and $\mathbf{b}_2$ are noise terms. It is sometimes said that $\mathbf{x}_2$ is a "noise alone reference."

(a) Show that the linear mean square estimation (LMSE) of $\mathbf{s}$ in terms of $\mathbf{x}_1$ and $\mathbf{x}_2$ can be written as

$$\hat{\mathbf{s}} = \mathbb{M}_1 \mathbf{x}_1 + \mathbb{M}_2 \mathbf{x}_2$$

and determine the number of rows and columns of $\mathbb{M}_1$ and $\mathbb{M}_2$.

(b) It is assumed that $\mathbf{s}$ is uncorrelated to $\mathbf{b}_1$ and $\mathbf{b}_2$. By introducing the matrices

$$\mathbb{\Gamma}_s = E[\mathbf{s}\,\mathbf{s}^T] \ ; \ \mathbb{\Gamma}_i = E[\mathbf{b}_i \mathbf{b}_i^T] \ ; \ \mathbb{\Gamma}_{12} = E[\mathbf{b}_1 \mathbf{b}_2^T]$$

write the equations allowing the calculation of the matrices $\mathbb{M}_1$ and $\mathbb{M}_2$.

(c) Solve these equations and express $\hat{\mathbf{s}}$ in terms of the matrices $\mathbb{\Gamma}$ and of the observation vectors $\mathbf{x}_1$ and $\mathbf{x}_2$.

(d) The action of the matrix $\mathbb{M}_2$ can be interpreted as a noise cancellor. For this purpose let us introduce

$$\hat{\mathbf{b}}_1 = \mathbb{A}\,\mathbf{b}_2$$

where $\hat{\mathbf{b}}_1$ is the LMSE of $\mathbf{b}_1$ in terms of $\mathbf{b}_2$. Express $\hat{\mathbf{s}}$ in terms of $\mathbf{x}_1$ and $\hat{\mathbf{b}}_1$ by using only the matrices $\mathbb{\Gamma}_s$ and $\mathbb{W}$, where $\mathbb{W}$ is the error matrix of the estimation giving $\hat{\mathbf{b}}_1$.

(e) Give the structure of the complete system for the LMSE of $\hat{\mathbf{s}}$ in terms of the observation $\mathbf{x}_1$ and $\mathbf{x}_2$. Explain why this system is sometimes referred to as noise subtraction.

(f) Study the two cases where the LMSE appearing in **(d)** is either singular or null.

10.12 **Estimation in multiplicative noise.** Let $x(t)$ and $n(t)$ be two zeromean independent signals. The observation is the signal $y(t) = [1 + n(t)]x(t)$ and the problem is to study the LMSE of $x(t)$ in terms of the time average $y_T(t)$ of $y(t)$ given by (6.93).

(a) Give the expression of the estimator in terms of the correlation functions of $x(t)$ and $n(t)$. Calculate the minimum error.

(b) We assume that $n(t)$ is the random telegraph signal introduced in Section 8.6.1 with an amplitude equal to a instead of 1. To simplify the problem, we also assume that the estimandum $x(t)$ is time-independent and is then an RV of variance σ^2. Complete the calculation and explain the situation which appears when $\lambda T << 1$, where λ is the density of the Poisson process and T the time constant of the time averager calculating $y_T(t)$.

10.13 The purpose of this problem is to study the effect of the mean values in the performance of LMSE. Let x and y be two second order RVs characterized by their mean values m_x and m_y, their variances σ_x^2 and σ_y^2 and their correlation coefficient c. We want to estimate y in terms of x either by an affine estimator $\hat{y}_A = a_0 + a_1 x$ or by a linear estimator $\hat{y}_L = bx$. Let H_A and H_L be the Hilbert subspaces of the observation subspace spanned by the RVs $h_0 + h_1 x$ and hx respectively.

(a) By using the projection theorem, or the orthogonality principle, calculate the coefficients a_0, a_1, and b.

(b) Show that $\hat{y}_L$ is the projection of $\hat{y}_A$ onto H_L.

(c) Show that $\hat{y}_A - \hat{y}_L$ is orthogonal to $y - \hat{y}_A$. Make a geometrical figure representing the situation.

(d) Calculate the estimation error obtained when using the affine estimator.

(e) Calculate the increment of this error when using the linear estimator. For this it is recommended to use the result of question **(c)**.

(f) Study and give an interpretation of all the cases where the use of the linear estimator does not induce an increment of the estimation error.

(g) Extend all these results to the case where x is no longer a scalar but a second order random vector $\mathbf{x}$. For the calculation introduce the vector $\mathbf{c} = E(y\mathbf{x})$ and use the mean value $\mathbf{m}_x$ of $\mathbf{x}$ and its covariance matrix Γ.

10.14 The purpose of this problem is to study the effect of the mean values in the performance of constrained LMSE presented in Section 10.5. Let H_{AS} be the observation subspace defined by the extension of (10.77), or by $\{c + \mathbf{h}^T\mathbf{x} \mid \mathbf{h} \in S\}$. The optimum affine constrained estimator of y is now $\hat{y}_{AC} = \alpha + \mathbf{a}_S^T \mathbf{x}$. By using the orthogonality principle, calculate the coefficient α and show that the linear part of the estimator is still given by (10.82).

10.15 Interpolation of a component of a random vector. Let $\mathbf{X}$ be a zeromean random vector of $\mathbb{R}^n$ with components x_i, $1 \leq i \leq n$ and let Γ be its covariance matrix. We want to realize the LMSE of the component x_i in terms of the other components x_j, $j \neq i$, which is an interpolation problem when the components x_i represent the values of a DT signal. The LMSE of x_i is written $\hat{x}_i = \mathbf{h}_i^T\mathbf{X}$.

(a) Determine the constraint which must be satisfied by the vector $\mathbf{h}_i$. Show that this constraint can be written as in (10.91) by using the vector $\mathbf{e}_i$ whose all of components are zero except the ith component which is equal to 1.

(b) Express the component x_i in terms of the vectors $\mathbf{e}_i$ and $\mathbf{X}$. Deduce from this expression the LMSE without constraint of x_i in terms of $\mathbf{X}$ and show that this estimation is singular, or gives an error equal to zero.

(c) By using the procedure presented in Section 10.5, show that the vector $\mathbf{h}_i$ is given by $[\mathbb{I} - \alpha_i \Gamma^{-1}]\mathbf{e}_i$ where $\alpha_i = [\Gamma^{-1}]_{ii}$ or is the element in position i,i of the matrix Γ^{-1}.

(d) Calculate the constrained innovation $\tilde{x}_i$ and deduce that the estimation error is $1/[\Gamma^{-1}]_{ii}$.

(e) Indicate how to find the component x_i giving the smaller LMSE error.

Chapter 11

Estimation for Stationary Signals

11.1 INTRODUCTION TO STATISTICAL FILTERING

The purpose of this chapter is to apply the ideas developed in the previous chapter to stationary signals. In order to understand this thoroughly, we shall adapt the general presentation used in Section 10.2 to the present situation.

Instead of an observation vector $\mathbf{x}$ we start from an *observation signal* $x[k]$, $k \in K$. This means that this signal is observed for values of k belonging to a set of integers K. This set can be finite or infinite, and for the time being no particular property is introduced on K. From this observation signal we want to estimate another signal $y[k]$ called the *estimandum signal* in such a way that the mean square error is minimum. The estimated signal, or the estimate, $\hat{y}[k]$, can be written, as in (10.2) or (10.5),

$$\hat{y}[k] = S\{x[l]\ ;\ l \in K\} \tag{11.1}$$

and the problem is to determine the system $S[.]$ minimizing the error

$$\varepsilon^2[k] = E\{(y[k] - \hat{y}[k])^2\} \tag{11.2}$$

The general solution of this problem was calculated in Section 10.2 and is the regression or the conditional expected value

$$\hat{y}[k] = E\{y[k] | x[l],\ l \in K\} \tag{11.3}$$

The performance of this estimator is characterized by the variance of the innovation defined by

$$\tilde{y}[k] = y[k] - \hat{y}[k] \tag{11.4}$$

One result of the discussion in the previous chapter was that mean square estimation problems are best solved within a geometric framework. We wish to apply the same spirit in this chapter. So the first concept to be introduced is that of the *observation subspace*. This is obtained by a simple extension of (10.30). By *filtering* the observation $x[k]$ we obtain a signal $s[k]$ written as

$$s[k] = h\{x[l],\ \ l \in K\} \tag{11.5}$$

and the observation subspace H_X is the space containing all the RVs such as $s[k]$ defined by (11.5), provided that they are second order. Following the method in the previous chapter, we find that the best estimate $\hat{y}[k]$ is obtained by projection of $y[k]$ onto H_X, and the simplest way to calculate this projection effectively is to apply the *orthogonality principle* (10.35).

We shall apply all these ideas in the present chapter. But since in general the regression (11.3) is too complex to be calculated, we will consider filters (11.1) which have a simpler structure. As before, this corresponds to linear or Volterra filters. Furthermore, the introduction of stationary properties will greatly simplify the solution of the problem. Note also that, for simpler terminology, questions discussed below will be described as *statistical filtering*, which, unless otherwise indicated, means linear filtering, the most important procedure. Similarly, all signals and systems considered below are *real* unless otherwise specified.

11.2 LINEAR STATISTICAL FILTERING WITHOUT CONSTRAINT

The problem here is similar to that presented in Section 10.4. As we use linear filters, and because of the absence of any constraint, the general expression (11.1) becomes

$$\hat{y}[k] = \alpha + \sum a[l]x[k-l] \tag{11.6}$$

and in order to determine the constant α and the impulse response $a[k]$ of the filter we will simply apply the orthogonality principle. As in (10.62), this can be written

$$E\{(y[k] - \hat{y}[k])\, s[k]\} = 0 \tag{11.7}$$

where $s[k]$ is

$$s[k] = c + \sum h[l]x[k-l] \tag{11.8}$$

As (11.7) must be valid for any c and $h[l]$, we immediately deduce that $y[k] - \hat{y}[k]$ must be orthogonal to c and to $x[k-p]$, whatever the values of c and p. This gives two equations which are

$$E\{y[k] - \hat{y}[k]\} = 0 \tag{11.9}$$

$$E\{(y[k] - \hat{y}[k])\, x[k-p]\} = 0 \tag{11.10}$$

Introducing the mean values and the correlation functions defined in Section 5.2, we can write these equations as

$$m_y - \alpha - m_x \sum a[l] = 0 \tag{11.11}$$

$$\gamma_{yx}[p] + m_x m_y - \alpha m_x - \sum a[l]\, \{\gamma_x[p-l] + m_x^2\} = 0 \tag{11.12}$$

Inserting (11.11) in (11.12), we obtain

$$\boxed{\sum a[l]\ \gamma_x[p-l] = \gamma_{yx}[p]} \tag{11.13}$$

which defines the impulse response $a[k]$ of the filter. If we use this value in (11.11), we deduce the coefficient α.

It is clear that (11.13) is similar to (10.65), while (11.11) corresponds to (10.64). The only, but important, difference is that (11.13) introduces a linear system of infinite dimension making it impossible to write (10.66). Fortunately, the structure of (11.13) makes its solution with the use of a transform method very easy. In fact, the first term of this equation is the convolution between $a[k]$ and $\gamma_x[k]$. Consequently, by using z-transforms defined by (6.4) we deduce

$$\Gamma_x\{z\}A\{z\} = \Gamma_{yx}\{z\} \tag{11.14}$$

which gives

$$A\{z\} = [\Gamma_x\{z\}]^{-1}\Gamma_{yx}\{z\} \tag{11.15}$$

Taking the inverse z-transform of this equation we deduce $a[k]$ which can be inserted in (11.11) in order to obtain α. Using this value in (11.6) we obtain the expression of the estimator which is

$$\hat{y}[k] = m_y + \sum a[l]\ \{x[k-l] - m_x\} \tag{11.16}$$

In this equation the impulse response $a[k]$ of the estimation filter is deduced from its z-transform (11.15) and (11.16) corresponds to (10.68) obtained for vector signals.

Let us comment on this result. First we deduce from (11.16) that the mean value of $\hat{y}[k]$ is m_y, because the expectation of the last term of this equation is null.

Second it is important for various applications to use the *frequency response* instead of the impulse response. This frequency response is the Fourier transform of $a[k]$, given by an equation similar to (6.3). Comparing (6.3) and (6.4) we can write

$$X(\nu) = X\{e^{2\pi j\nu}\} \tag{11.17}$$

and we will now use this notation, instead of (6.71). So from now a z transform will be indicated as $X\{z\}$ and a Fourier transform by $X(\nu)$. Note that this implies that

$$X(0) = X\{1\}$$

Using this notation we can write (11.15) as

$$\boxed{A(\nu) = [\Gamma_x(\nu)]^{-1}\Gamma_{yx}(\nu)} \tag{11.18}$$

In this expression $A(\nu)$ is the *frequency response* of the filter giving the best estimation, $\Gamma_x(\nu)$ is the power spectrum of the signal $x[k]$ and $\Gamma_{yx}(\nu)$ is the Fourier transform of the crosscorrelation function $\gamma_{yx}[k]$. Note that this transform is in general complex. Returning to the validity of (11.18), we note that this expression requires the power spectrum of $x[k]$ to be positive. In reality we can still use (11.18) if for some frequencies $\Gamma_x(\nu) = \Gamma_{yx}(\nu) = 0$, because these frequencies do not contribute either to x or to y and can *a priori* be suppressed. This appears especially when x and y are band limited in the same frequency domain. This corresponds exactly to the situation discussed in (c) of Section 10.4. The general procedure to realize optimum filtering is presented in Figure 11.1, in which F is the linear filter characterized by the frequency response (11.18).

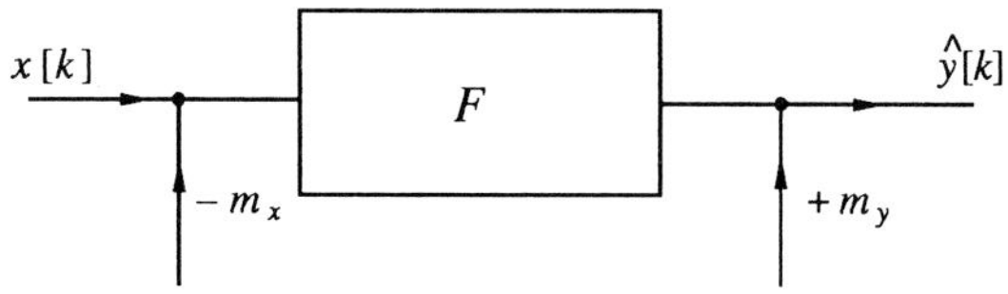

Figure 11.1 Optimum linear filtering for stationary signals.

Let us now calculate the minimum *mean square error* ε^2 obtained with the optimum filter. This error is the variance of the innovation $y[k] - \hat{y}[k]$ and as this innovation is zeromean valued we have

$$\varepsilon^2 = E\{(y[k] - \hat{y}[k])^2\} = E\{(\tilde{y}[k])^2\} \tag{11.19}$$

This error does not depend on k because of the assumption of stationarity. As a result of the orthogonality equation (11.10), the innovation $\tilde{y}[k]$ is orthogonal to the estimation $\hat{y}[k]$, which gives

$$\varepsilon^2 = E\{y^2[k]\} - E\{\hat{y}[k]\, y[k]\}$$

Writing $y[k] = \hat{y}[k] + \tilde{y}[k]$ and applying the orthogonality property, we deduce that

$$\varepsilon^2 = E\{y^2[k]\} - E\{\hat{y}^2[k]\} \tag{11.20}$$

As the optimum filter is calculated in the frequency domain, we will calculate this error in the same domain. Using (5.17') we note that

$$E\{y^2[k]\} = m_y^2 + \gamma_y[0] \tag{11.21}$$

and as $\gamma_y[k]$ is the Fourier transform of the power spectrum $\Gamma_y(\nu)$, we obtain

$$E\{y^2[k]\} = m_y^2 + \int_{-1/2}^{+1/2} \Gamma_y(\nu)\, d\nu \tag{11.22}$$

In order to calculate the last term of (11.20) we write (11.6) as

$$\hat{y}[k] = \alpha + d[k] \tag{11.23}$$

and (11.11) as

$$\alpha = m_y - m_x T \tag{11.24}$$

It is clear that $E\{d[k]\} = m_x T$, which gives

$$E\{\hat{y}^2[k]\} = \alpha^2 + 2\alpha m_x T + E\{d^2[k]\} \tag{11.25}$$

Using (11.24), we deduce that

$$E\{\hat{y}^2[k]\} = m_y^2 - (m_x T)^2 + E\{d^2[k]\} \tag{11.26}$$

As $d[k]$ is deduced from $x[k]$ by a linear filtering with the frequency response $A(\nu)$, we deduce from (6.55) that

$$E\{d^2[k]\} = m_d^2 + \int_{-1/2}^{+1/2} |A(\nu)|^2 \Gamma_x(\nu)\, d\nu \tag{11.27}$$

where m_d is the mean value of $d[k]$. Using (5.61) and (6.67), we find that

$$m_d = m_x A(0) = m_x T \tag{11.28}$$

which finally gives, with (11.17),

$$E\{\hat{y}^2[k]\} = m_y^2 + \int_{-1/2}^{+1/2} [\Gamma_x(\nu)]^{-1} |\Gamma_{yx}(\nu)|^2\, d\nu \tag{11.29}$$

If we combine (11.22) and (11.29), it follows that

$$\varepsilon^2 = \int_{-1/2}^{+1/2} \left\{\Gamma_y(\nu) - \Gamma_{yx}(\nu)\, [\Gamma_x(\nu)]^{-1}\, \Gamma_{xy}(\nu)\right\} d\nu \tag{11.30}$$

whose form is similar to (10.73). It is important to note that the mean values m_x and m_y do not appear in the expression of the error. Furthermore, when $m_x = m_y = 0$, the calculation of ε^2 from (11.20) is greatly simplified.

In fact this presentation becomes much simpler if we work with *zeromean signals*, which is the most common case. Let us briefly review the equations of the problem and their solutions. For zeromean signals the term α in (11.6) must be dropped. The only orthogonality equation of the problem is thus (11.10), which leads directly to (11.13). The corresponding solution is (11.15) or (11.17). The calculation of the error from (11.20) is straightforward because $E(y^2)$ and $E(\hat{y}^2)$ are the variances of $y[k]$ and $\hat{y}[k]$ respectively, which are easily expressed from the power spectra of $x[k]$ and $y[k]$.

Let us now investigate the meaning of the two extreme values of the error ε^2. The maximum value of this error is obviously the variance of the signal $y[k]$, which is obtained if $\Gamma_{xy}(\nu) = 0$. This means that the observation and the estimandum are uncorrelated, a perfectly natural situation. To investigate the singular case, it is convenient to write ε^2 in the form

$$\varepsilon^2 = \int_{-1/2}^{+1/2} [\Gamma_x(\nu)]^{-1}\, [\Gamma_x(\nu)\, \Gamma_y(\nu) - |\Gamma_{xy}(\nu)|^2]\, d\nu$$

The last term in brackets is the determinant of the spectral matrix of the vector signal of components $x(t)$ and $y(t)$. We saw in Section 6.9 that this spectral matrix is NND, which implies that the determinant is non-negative. It follows that ε^2 can be null if and only if

$$\Gamma_x(\nu)\ \Gamma_y(\nu) - |\Gamma_{xy}(\nu)|^2 = 0$$

for all ν. This is equivalent to (6.9), which means that the estimandum is deduced by a linear filtering from the observation. It is clear that this situation implies a null estimation error.

Finally, a last question can be asked. Because of the stationarity of the signals the last term of (11.6) corresponds to a time-invariant linear system. One may wonder whether the use of a function $a[k, l]$ instead of $a[k]$ could introduce another optimal filter. If we write the orthogonality equation again, we can easily show that the solution does not depend on k, which justifies (11.6). The same problem will be analyzed in more detail in the next section.

The above calculations can be applied to various situations but we shall give here only the example of the extraction of a signal from a corrupting noise.

Example 11.1 Filtering of a signal in noise. Suppose that the observation $x[k]$ is the sum of a signal $s[k]$ and a disturbance noise $n[k]$, as for example in (3.106). From this observation we want to estimate the signal. This means that the estimandum $y[k]$ is equal to $s[k]$. This problem of signal filtering is sometimes called a problem of noise cancelling. To simplify, we assume that the signal and noise are uncorrelated and we denote m_S, m_N, $\Gamma_S(\nu)$, and $\Gamma_N(\nu)$ the mean values and power spectra of s and n respectively. The filtering (11.6) must now be written as

$$\hat{s}[k] = \alpha + \sum a[l]\ x[k-l] \tag{11.31}$$

where

$$x[k] = s[k] + n[k] \tag{11.32}$$

The frequency response of the filter (11.18) now takes the form

$$A(\nu) = \left[\Gamma_S(\nu) + \Gamma_N(\nu)\right]^{-1} \Gamma_S(\nu) \tag{11.33}$$

and the constant α given by (11.11) can be written as

$$\alpha = m_S - (m_S + m_N)\ \left[\Gamma_S(0) + \Gamma_N(0)\right]^{-1} \Gamma_S(0) \tag{11.34}$$

The optimum system for filtering a signal in a background noise is presented in Figure 11.2, in which A means the linear filter defined by (11.33).

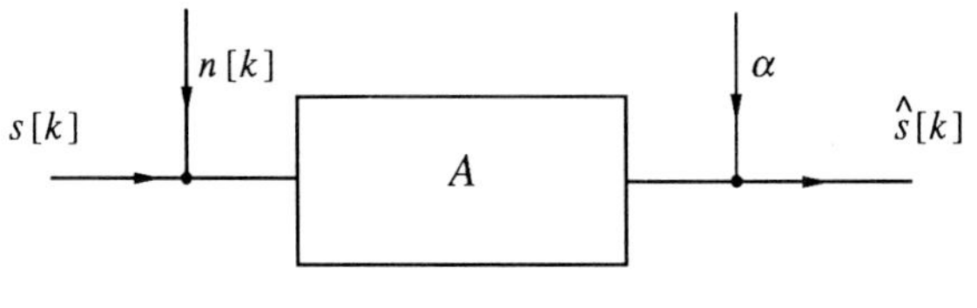

Figure 11.2 Filtering of a signal in noise.

The mean square error is given by (11.30) which, after simplification, takes the form

$$\varepsilon^2 = \int_{-1/2}^{+1/2} \frac{\Gamma_S(\nu)\,\Gamma_N(\nu)}{\Gamma_S(\nu) + \Gamma_N(\nu)}\, d\nu \qquad (11.35)$$

Let us consider these results. We have seen that the correlation function of real signals is symmetric. So the Fourier transform has the same property, and we deduce from (11.33) that $A(\nu) = A(-\nu)$. The result is that $a[k] = a[-k]$ and the filter (11.31) has the same effect on the past and the future of the observation. In particular, this means that it cannot be a causal filter. Another interesting point is that of a possible *singular* estimation. It is clear in (11.35) that $\varepsilon^2 = 0$ if, and only if, the product $\Gamma_S(\nu)\,\Gamma_N(\nu)$ is null. This means that the frequency bands of the signal and noise do not overlap, and in this case $A(\nu)$ takes only the values 0 or 1, this latter value appearing when $\Gamma_S(\nu) \neq 0$. In other words, $A(\nu)$ does not change the signal but completely cancels the noise. This is a characteristic of singular estimation.

11.3 SAMPLING AS AN ESTIMATION PROBLEM

11.3.1 General Theory

Consider a *zeromean* continuous-time signal $x(t)$ and suppose that it is observed at some discrete time instants t_n where n belongs to a set S. The sampling problem, sometimes called the interpolation problem, is to estimate $x(t)$ from its samples $x(t_n)$. If $t = t_n$ we find a singular estimation problem, or an estimation problem with a null estimation error. If we use only linear methods, the estimate can be written

$$\hat{x}(t) = \sum_{n \in S} h(t;\, t_n)\, x(t_n) \qquad (11.36)$$

and the problem is to find the function h such that the mean square error is minimum. Because of the previous remark, it is clear that the function h must satisfy $h(t_n;\, t_m) = \delta[m, n]$.

The orthogonality equation is

$$E[\hat{x}(t)\, x(t_k)] = E[x(t)\, x(t_k)],\ \forall\, k \in S \qquad (11.37)$$

and using the correlation function of $x(t)$, this gives

$$\sum_{n \in S} h(t;\, t_n)\, \gamma(t_n - t_k) = \gamma(t - t_k),\ \forall\, k \in S \qquad (11.38)$$

The solution of this equation obviously depends on the position of the time instants t_n and the structure of the set S. In particular, if S is finite, we have a system of linear equations which can be solved by using standard techniques.

The problem is more complex if S is not finite, and we will consider this case with the restriction that the time instants t_n are periodic, or satisfy $t_n = n\,T$, as in (6.27). In this case (11.38) becomes

$$\sum_n h(t;\, t_n)\, \gamma(t_n - t_k) = \gamma(t - t_k) \quad , \forall\, k \tag{11.39}$$

valid for any integer k. Replacing t and t_n by $t + t_p$ and $t_n + t_p$ respectively, we obtain

$$\sum_n h(t + t_p;\, t_n + t_p)\, \gamma(t_n + t_p - t_k) = \gamma(t + t_p - t_k) \tag{11.40}$$

If we introduce $t_q = t_k - t_p$, we deduce that the functions h in (11.39) and (11.40) satisfy the same equation, and are then equal, or

$$h(t;\, t_n) = h(t + t_p;\, t_n + t_p) \tag{11.41}$$

The interpretation of this equation from (11.36) is obvious. In fact, because of the stationarity of the problem and of the periodicity of the time instants t_n, the solution of the problem giving the estimate at t is the same as at $t + t_p$. If we use $p = -n$ in (11.41), we deduce that

$$h(t;\, t_n) = h(t - t_n;\, 0) \triangleq a(t - t_n) \tag{11.42}$$

in such a way that (11.36) becomes

$$\hat{x}(t) = \sum_n a(t - t_n)\, x(t_n) \tag{11.43}$$

where $a(t)$ is defined by

$$\sum_n a(t - t_n)\, \gamma(t_n - t_k) = \gamma(t - t_k) \tag{11.44}$$

As this equation has a structure similar to that of a convolution, we will solve it by using a Fourier transformation. Let us call $A(\nu)$ the Fourier transform of $a(t)$ and $\Gamma(\nu)$ the power spectrum of $x(t)$. Taking the Fourier transforms of the two terms of (11.44), we obtain

$$A(\nu) \sum_n \gamma(t_n - t_k)\, e^{-2\pi j \nu t_n} = \Gamma(\nu)\, e^{-2\pi j \nu t_k} \tag{11.45}$$

which can be expressed as

$$A(\nu) \sum_p \gamma(pT)\; e^{-2\pi j \nu p T} = \Gamma(\nu) \tag{11.46}$$

Using (6.101) and (6.104), this gives

$$A(\nu) = T\, \Gamma(\nu) / \Gamma_d(\nu) \tag{11.47}$$

where $\Gamma_d(\nu)$ is defined by (6.104). We can also use $\Gamma_s(\nu)$ defined by (6.105) to arrive at the following three equivalent relations

$$A(\nu) = \frac{T\,\Gamma(\nu)}{\Gamma_d(\nu)} = \frac{\Gamma(\nu)}{\Gamma_s(\nu)} = \frac{T\,\Gamma(\nu)}{\sum_n \Gamma(\nu - \frac{n}{T})} \tag{11.48}$$

The optimum sampling is thus realized with (11.43) using a function $a(t)$ whose Fourier transform is given by (11.48). This transform $A(\nu)$ is the ratio between the power spectrum of the signal $x(t)$ and that of the signal obtained by sampling $x(t)$ at the time instants nT. As those spectra are even functions of ν, $a(t)$ is also even, which means that the past and the future plays the same role in the calculation of the best mean square linear estimator. We have seen that the function $a(t)$ must satisfy the constraint which ensures that $\hat{x}(t_n) = x(t_n)$, whatever n. This constraint implies that $a(0) = 1$ and $a(t_n) = 0$ if $n \neq 0$. Using the comb distribution $\delta_T(t)$ introduced in (6.28), this property of $a(t)$ is characterized by

$$a(t)\,\delta_T(t) = \delta(t) \tag{11.49}$$

In the Fourier domain this can be expressed as

$$\frac{1}{T}\sum_n A\left(\nu - \frac{n}{T}\right) = 1 \tag{11.50}$$

Using the last term of (11.48), we see immediately that this property is verified.

Let us now calculate the sampling error $\varepsilon^2(t)$ defined by

$$\varepsilon^2(t) = E\left\{ [x(t) - \hat{x}(t)]^2 \right\} \tag{11.51}$$

The first point to note is that although the problem is stationary, the error remains a function of time. This is due to the fact that (11.43) is a linear operation which is not time invariant because of the position of the sampling times given *a priori*. Furthermore, we have noticed that the error is by construction null for $t = t_n$, but it has no reason to be null for other values of t.

Using the same procedure as before when passing from (11.18) to (11.20), we can write $\varepsilon^2(t)$ as

$$\varepsilon^2(t) = E[x^2(t)] - E[\hat{x}^2(t)] \tag{11.52}$$

In the Fourier domain the term $E(x^2)$ is given by (6.41) and we must now calculate the last term of (11.52). Starting from (11.43) we obtain

$$E[\hat{x}^2(t)] = \sum_m \sum_n a(t - t_m)a(t - t_n)\,\gamma(t_m - t_n) \tag{11.53}$$

Using (11.44), we can perform the sum on m, which gives

$$E[\hat{x}^2(t)] = \sum_n a(t - t_n)\,\gamma(t - t_n) \tag{11.54}$$

where $a(t)$ is defined by (11.48). This leads to the use of the Fourier transforms $A(\nu)$ and $\Gamma(\nu)$, giving

$$E[\hat{x}^2(t)] = \sum_n \iint A(\nu)\,\Gamma(\nu')\,e^{2\pi j(\nu+\nu')(t-t_n)}\,d\nu\,d\nu' \tag{11.55}$$

The sum over n can be obtained by using the relation

$$\sum_n e^{2\pi j\nu nT} = \frac{1}{T}\sum_n \delta\left(\nu - \frac{n}{T}\right) \tag{11.56}$$

which states that the Fourier transform of $\delta_T(t)$ is still a Dirac comb, as used in (6.30). Inserting (11.56) in (11.55), using the parity of $\Gamma(\nu)$ and the definition of t_n, we obtain

$$E[\hat{x}^2(t)] = \frac{1}{T}\sum_n \int A(\nu)\,\Gamma\left(\nu - \frac{n}{T}\right) e^{2\pi j\frac{n}{T}(t+\theta)}\,d\nu \tag{11.57}$$

Finally, if we use the first equation (11.48), we obtain

$$\boxed{\varepsilon^2(t) = \int \frac{\Gamma(\nu)}{\Gamma_d(\nu)}\left\{\sum_n \Gamma\left(\nu - \frac{n}{T}\right)\left[1 - e^{2\pi j\frac{n}{T}(t+\theta)}\right]\right\} d\nu} \tag{11.58}$$

Let us now investigate the possibility of obtaining a *singular* estimation characterized by $\varepsilon^2(t) = 0$ for all t. This is possible if $\Gamma(\nu)$ is null for $|\nu| > 1/2T$. In fact, it means that the integral (11.58) is limited to the interval $[-1/2T, +1/2T]$, a limitation which reduces the sum in n to the term $n = 0$. This gives a null contribution because of the bracket $\{1 - \exp[2\pi j(n/T)(t+\theta)]\}$.

If $\Gamma(\nu)$ is limited to the frequency band $[-1/2T, 1/2T]$, the function $A(\nu)$ given by (11.48) is rectangular, and $a(t)$ is a sine cardinal function, which gives the sampling formula (6.106). These remarks thus show that the sampling theorem can be considered as a particular case of a more general sampling procedure presented within an estimation framework.

11.3.2 Direct Proof of the Sampling Theorem

If we are *interested only in the proof* of (6.106), it is not necessary to go through the above discussion which gives (6.106) as a particular case. A *direct method* is quite possible, with important results. The starting point is that the sampling theorem for deterministic functions, a property not used in the earlier calculation, is taken for granted. This theorem states that if a function $\Gamma(\nu)$ is null for $|\nu| > B$, its Fourier transform $\gamma(t)$ can be written as

$$\gamma(t) = \sum_n \gamma(t_n)\ \text{sinc}[2B(t-t_n)] \tag{11.59}$$

where $t_n = nT$. Consider now a zeromean signal $x(t)$ such that its power spectrum is band-limited so that its correlation function satisfies (11.59). As the function $\gamma(t - t_k)$ also has a band-limited spectrum, we can write

$$\gamma(t-t_k) = \sum_n \gamma(t_n - t_k)\ \text{sinc}[2B(t-t_n)] \tag{11.60}$$

We see that this equation is the orthogonality equation (11.44), and we can then write

$$\hat{x}(t) = \sum_n x(t_n)\ \ \text{sinc}[2B(t-t_n)] \tag{11.61}$$

Now we need only to show that the estimation is singular, which means that $x(t) = \hat{x}(t)$ with probability one. For this purpose we use (11.52) and (11.54), and we can then write

$$E[\hat{x}^2(t)] = \sum_n \gamma(t-t_n)\ \text{sinc}[2B(t-t_n)] \tag{11.62}$$

If we take $t - t_n = \theta_n$, as in (6.107), we deduce from (11.59) that $E[\hat{x}^2(t)] = \gamma(0) = E[x^2]$, which implies that $\varepsilon^2 = 0$.

Let us further comment on the sampling theorem expressed by (11.61) where $\varepsilon^2 = 0$, or by (6.106). The latter expression is given for deterministic signals and is now also valid for random signals. Within the framework of the sampling theory of deterministic signals, one can show that the functions sinc are orthogonal (see [Picinbono], p. 91), expressed by

$$\int \text{sinc}[2B(t-t_m)]\ \text{sinc}[2B(t-t_n)]\, dt = 0 \tag{11.63}$$

if $m \neq n$. So (6.106) can be considered as an expansion of the signal $x(t)$ in terms of orthogonal functions, similar to (3.97) valid for vectors. It is then natural to extend the concept of doubly orthogonal expansion and to investigate the properties of a signal $x(t)$ such that the coefficients $x(t_n)$ are orthogonal, or uncorrelated since we work with zeromean signals.

The orthogonality of the coefficients $x(t_n)$ means that the correlation function $\gamma(\tau)$ of $x(t)$ satisfies $\gamma(kT) = 0$ for $k \neq 0$. Using (11.59), we immediately deduce that

$$\gamma(\tau) = \gamma(0)\ \text{sinc}(2B\tau) \tag{11.64}$$

which shows that $x(t)$ is a *band-limited white noise*, because the Fourier transform of $\text{sinc}(2B\tau)$ is a rectangular function in $[-B, +B]$.

Suppose now that the samples $x(t_n)$ are not only orthogonal but also independent. We will show that this additional assumption means that $x(t)$ is *normal*. The reasoning is similar to that used in Sections 4.8.9 or 7.2.3, and uses cumulants. Basically, because of the stationarity assumption, the cumulants of $x(t)$ are independent of time, and thus the cumulants of $x(t)$ and $x(\theta_n)$ in (6.106) are equal. Using the same rule as in (5.60), and applying the same equation as (7.11) for p instead of 4 we obtain

$$c_p = c_p \sum \text{sinc}^p[2B(t-t_n)] \tag{11.65}$$

Applying (6.106) to $x(t) = \text{sinc}(2B\,t)$, we obtain

$$x(0) = \sum \text{sinc}^2(2B\,\theta_n) = 1 \tag{11.66}$$

and when we apply this expression for $t - t_n = \theta_n$ we obtain $c_2 = c_2$ in (11.65). On the other hand, as

$$\left|\sum \text{sinc}^p(2B\,\theta_n)\right| \leq \sum \text{sinc}^2(2B\,\theta_n)\,|\text{sinc}^{p-2}(2B\,\theta_n)| \tag{11.67}$$

and as $|\text{sinc}(2B\,\theta_n)| < 1$ if $\theta_n \neq 0$, we deduce that (11.65) means $c_p = 0$ for $p > 2$, which gives the normal distribution.

11.4 LINEAR STATISTICAL FILTERING WITH CONSTRAINTS

11.4.1 General Theory

The problems discussed in this section are similar to those in Section 10.5. As before, to simplify matters we will assume that all signals are zeromean, and concentrate on problems arising from the constraint. The problem is to estimate a signal $y[k]$ in terms of an observation $x[k]$ by a system such as

$$\hat{y}[k] = \sum a[l]\,x[k-l] \tag{11.68}$$

where the impulse response $a[k]$ must satisfy some constraints. The use of a time invariant system, or a discrete-time convolution, is justified by the fact that $x[k]$ and $y[k]$ are jointly stationary.

Let us denote examples of such constraints. One of the most important, studied in detail below, is the *causality constraint.* This means that the filter used in (11.68) must be causal, characterized by the condition $a[k] = 0$ if $k < 0$. There are a number of constraints of the same kind characterized by the property that $a[k] = 0$ for some *a priori* given values of k.

As the filter (11.68) is often realized by a physical device with a finite dynamical range, a common constraint on $a[k]$ is that it must be bounded or satisfy a relation such as $|a[k]| < M$. There are also some norm constraints such as

$$\sum a^2[k] = 1 \tag{11.69}$$

It is clear that even though the general problem is to find a *linear filter* as in (11.68) minimizing the mean square error, the introduction of a constraint on the impulse response $a[k]$ completely changes the equation of the problem and the solution is in general different from that obtained in Section 11.2.

A constraint C on the impulse response of a linear filter is said to be a *linear constraint* if the set S of the impulse responses $h[k]$ satisfying this constraint is a vector space. We shall assume in what follows that all constraints introduced are linear. This is obviously the case of the causality constraint. However, (11.69) introduces a constraint which is not linear.

By a direct transposition of (10.77) we can introduce the observation subspace defined by

$$H_S = \left\{ \sum h[l]\; x[k-l] \,|\, h[k] \in S \right\} \tag{11.70}$$

This space contains all the RVs which are the outputs at time k of linear filters satisfying the constraint C when the input signal is the observation $x[k]$. The optimum filter is characterized by the orthogonality equation similar to (10.79) and written as

$$y[k] - \hat{y}[k] \perp\!\!\!\perp \sum_p h[p]\; x[k-p] \tag{11.71}$$

for all $h[k]$ belonging to S. Using (11.68), where $a[l]$ is replaced by $a_S[l]$ in order to specify that it must belong to S, this can be written as

$$\sum_p h[p] \left\{ \gamma_{yx}[p] - \sum_l a_S[l]\; \gamma_x[p-l] \right\} = 0 \tag{11.72}$$

The problem now is to find the function $a_S[k]$ belonging to S such that, for all $h[k]$ belonging to S, (11.72) holds.

Before embarking on the geometric method to solve this problem, we shall give some examples to illustrate the equation.

Suppose first that *there is no constraint* at all. Thus (11.72) must be valid for any impulse function $h[k]$. Taking all the function $\delta[k-k_0]$ successively, we deduce that the bracket of (11.71) must be null, or

$$\sum a[l]\, \gamma_x[p-l] = \gamma_{yx}[p], \quad \forall\, p \tag{11.73}$$

which again gives (11.13) defining the unconstrained optimum filter.

Consider now the *causality constraint.* In this case $h[k]$ and $a_S[k]$ must be null for $k < 0$. Using successively the functions $h[k] = \delta[k-k_0]$ with $k_0 \geq 0$, we deduce that (11.72) can be written

$$\sum_{l=0}^{\infty} a_S[l]\, \gamma_x[p-l] = \gamma_{yx}[p], \quad p \geq 0 \tag{11.74}$$

We have then to find the impulse response $a_S[k]$ null for $k < 0$ and satisfying (11.74) only for $p \geq 0$. This equation is sometimes called the *Wiener-Hopf equation.* Because of its importance this equation will be solved by a specific procedure presented in the next section.

Consider now a nullity constraint which appears in some interpolation problems below. This means that

$$h[k] = 0, \quad k \in K \tag{11.75}$$

where K is a finite set of integers. Using the same procedure, we find that (11.72) becomes

$$\sum_{l \notin K} a_S[l] \; \gamma_x[p-l] = \gamma_{yx}[p], \;\; p \notin K \tag{11.76}$$

where the series is extended over all the integers which do not belong to K.

In order to find a *geometric solution* to the problem, we want to write (11.72) in the form of (10.81) by introducing an appropriate scalar product.

Let us call F the set of impulse responses $h[k]$ with a Fourier transform $H(\nu)$ defined by (6.67) and (6.70), or

$$H(\nu) = \sum_k h[k]\, e^{-2\pi j \nu k} \tag{11.77}$$

This means that the z-transform of $h[k]$ is convergent on the unit circle. It is obvious that F is a vector space. Consider now a positive even function $R(\nu)$, $-1/2 \leq \nu \leq +1/2$ and the quantity

$$< h, g >_R \triangleq \int_{-1/2}^{+1/2} H^*(\nu) R(\nu) \; G(\nu)\, d\nu \tag{11.78}$$

It is clear that this quantity can be considered as a scalar product between the real functions $h[k]$ and $g[k]$. In particular, if $R(\nu) = 1$ we deduce from the Parseval theorem that

$$< h, g > = \sum_k h[k]\, g[k] \tag{11.79}$$

which is similar to (3.68) where $\mathbb{M} = \mathbb{I}$. Let us now transpose (11.78) in the time domain. We call $r[k]$ the Fourier transform of $R(\nu)$ given by

$$r[k] = \int_{-1/2}^{+1/2} R(\nu)\, e^{2\pi j \nu k} \; d\nu \tag{11.80}$$

As $R(\nu) > 0$, $r[k]$ is positive-definite, or is a correlation function. Furthermore, as $R(\nu)$ is even, we have $r[k] = r[-k]$. Introducing the Fourier transform of $H(\nu)$ and $G(\nu)$, (11.78) becomes

$$< h, g >_R \; = \int_{-1/2}^{+1/2} \sum_{k,\, l,\, p} h[k]\, r[p]\, g[l] \; e^{2\pi j \nu(-k+p+l)}\, d\nu \tag{11.81}$$

Integration over ν gives $\delta[-k+p+l]$, which implies

$$< h, g >_R = \sum_k \sum_l h[k] r[k-l] \; g[l] \tag{11.82}$$

where $r[k-l]$ can also be replaced by $r[l-k]$. Note the analogy between this expression and (3.131) and (3.68) valid for complex vectors.

Returning to (11.72), let us write this expression in the frequency domain. The first term can be written as

$$\sum_p h[p] \; \gamma_{yx}[p] = \int_{-1/2}^{+1/2} H^*(\nu) \Gamma_{yx}(\nu) \, d\nu =$$

$$\int_{-1/2}^{+1/2} H^*(\nu) \; \Gamma_x(\nu) A(\nu) \, d\nu \tag{11.83}$$

where $A(\nu)$ is given by (11.18) and is the frequency response of the *unconstrained optimum filter*. Using (11.82), we can finally write (11.72) as

$$\int_{-1/2}^{+1/2} H^*(\nu) \Gamma_x(\nu) [A(\nu) - A_S(\nu)] \, d\nu = 0 \tag{11.84}$$

or

$$< h, \; a - a_S >_\Gamma = 0 \tag{11.85}$$

where the scalar product is defined as in (11.78) by taking $R(\nu)$ as $\Gamma_x(\nu)$. This is exactly the same equation as (10.81) and therefore the solution is also the same.

The constrained problem can thus be stated as follows: to find a function $a_S[k]$ belonging to S in such a way that, for any function $h[k]$ of S, (11.85) holds. This means that $a[k] - a_S[k]$ is orthogonal to S, and then

$$a_S[k] = \text{Proj}\{ a[k] | S \} \tag{11.86}$$

To avoid any errors, it is important to note that $a[k]$ is the Fourier transform of $A(\nu)$, the solution of the unconstrained problem, and that the projection, or orthogonality, is taken with the scalar product (11.78) where $R(\nu) = \Gamma_x(\nu)$.

It is sometimes more interesting to use the space $S^\perp$ complementary of S, the orthogonality being always defined with the same scalar product. In this case we apply the projection theorem to $a[k]$, or

$$a[k] = a_S[k] + a_S^\perp[k] \tag{11.87}$$

where

$$a_S^\perp[k] = \text{Proj}\{ a[k] | S^\perp \} \tag{11.88}$$

Having calculated $a_S^\perp[k]$, we deduce $a[k]$ from (11.87).

It is clear that $a_S[k]$ and $a_S^\perp[k]$ define two signals $\hat{y}_S[k]$ and $\hat{y}_S^\perp[k]$ by using (11.68). As in (10.86) it is easy to show that the two random variables $\hat{y}_S[k]$ and $\hat{y}_S^\perp[k]$ are uncorrelated. Furthermore, the *error* is still given by (10.89) where the last term (10.90) now becomes

$$E[(\hat{y}_S^{\perp})^2] = \int_{-1/2}^{+1/2} [A_S^{\perp}(\nu)]^* \, \Gamma_x(\nu) \, A_S^{\perp}(\nu) \, d\nu \tag{11.89}$$

11.4.2 Application to Interpolation Problems

Interpolation problems are the simplest case of the application of the above results, because the space $S^{\perp}$ is of dimension one and thus the calculation of (11.88) is almost self-evident. Let us now state the most elementary interpolation problem for discrete-time signals. We shall call $x[k]$ such a signal. At the time instant k the values $x[l]$ for $l < k$ belong to the *past* and those for $l > k$ belong to the *future*. In interpolation problems we want to estimate $x[k]$ in terms of all its past and future values. Thus (11.68) becomes

$$\hat{x}(k) = \sum_{l \neq 0} a[l] \, x[k-l] \tag{11.90}$$

and we wish to calculate $a[l]$ giving the smallest mean square error. This problem is similar to that of prediction which can be written as

$$\hat{x}(k) = \sum_{l=1}^{\infty} a[l] \, x[k-l] \tag{11.91}$$

and introduces the constraint of causality. The constraint introduced by (11.90) is simply $h[0] = 0$ and the set S of filters satisfying the constraint is therefore

$$S = \{h[k] \,|\, h[0] = 0\} \tag{11.92}$$

Instead of using the impulse response $h[k]$ to define the filter, we can use its frequency response so that (11.92) then becomes

$$S = \left\{ H[\nu] \;\middle|\; \int_{-1/2}^{+1/2} H(\nu) d\nu = 0 \right\} \tag{11.93}$$

Let us now calculate $S^{\perp}$ in order to apply (11.88). The integral in (11.93) can be written as

$$\int_{-1/2}^{+1/2} H^*(\nu) \, \Gamma(\nu) \, [\Gamma(\nu)]^{-1} d\nu = 0 \tag{11.94}$$

where $\Gamma(\nu)$ is the power spectrum of $x[k]$.

Let us call $c[k]$ the signal whose Fourier transform is $C(\nu) = [\,\Gamma(\nu)]^{-1}$. Noting that (11.94) is the scalar product (11.78) between h and c, we deduce that (11.92) can be written

$$S = \{h[k] \,|\, \langle h, c \rangle_{\Gamma} = 0\} \tag{11.95}$$

This means that S is the space of filters orthogonal in the sense of (11.78) to $c[k]$. This implies that $S^{\perp}$ is the space of filters proportional to $c[k]$, or

$$S^{\perp} = \{h[k] \,|\, h[k] = \alpha c[k]\} \tag{11.96}$$

where $c[k]$ is defined by its Fourier transform

$$C(\nu) = [\Gamma(\nu)]^{-1} \tag{11.97}$$

As a result, these calculations are valid provided $\Gamma(\nu) > 0$ for $|\nu| \leq 1/2$. Note here that even though S is the space of filters such as $h[0] = 0$, the identical filter defined by $h[k] = \delta[k]$ is not orthogonal to S. In fact, for any element $h[k]$ of S we have

$$\sum h[k]\delta[k] = 0 \tag{11.98}$$

which is a scalar product (11.78) corresponding to $R(\nu) = 1$ and not to $R(\nu) = \Gamma(\nu)$ as in (11.94).

In order to apply (11.88) we must know $a[k]$ which is the optimum filter without constraint. If there is no constraint the estimation problem becomes singular and

$$a[k] = \delta[k] \tag{11.99}$$

In fact, using this function in (11.90) without the constraint, we obtain $\hat{x}[k] = x[k]$.

In conclusion, we can write (11.87) as

$$a_S[k] = \delta[k] - \alpha c[k] \tag{11.100}$$

where α is deduced from the orthogonality of the last terms of (11.87). This gives

$$\alpha = < \delta, c >_\Gamma / < c, c >_\Gamma \tag{11.101}$$

Using (11.78) with $R(\nu) = \Gamma(\nu)$, and noting that the Fourier transforms of $\delta[k]$ and $c[k]$ are 1 and $C(\nu)$ respectively, we deduce that

$$< \delta, c >_\Gamma = 1 \tag{11.102}$$

$$< c, c >_\Gamma = \int_{-1/2}^{+1/2} C(\nu)\, d\nu = c[0] \tag{11.103}$$

This gives the impulse response of the interpolation filter by

$$a_S[k] = \delta[k] - (c[0])^{-1} c[k] \tag{11.104}$$

and its frequency response is

$$A_S(\nu) = 1 - (c[0])^{-1} C[\nu] = 1 - (c[0])^{-1} \frac{1}{\Gamma(\nu)} \tag{11.105}$$

where $c[0]$ is given by (11.103), or

$$c[0] = \int_{-1/2}^{+1/2} \frac{1}{\Gamma(\nu)}\, d\nu \tag{11.106}$$

Note finally that $a_S[0] = 0$, which is precisely the constraint of the problem.

Let us now calculate the error in the interpolation problem. This is especially simple in view of the fact that, because the problem without constraint is singular, the corresponding error is null. The interpolation error is thus given by (11.89) where $A_S^{\perp}(\nu)$ is the last term of (11.105). This gives

$$\varepsilon^2 = \left[\int_{-1/2}^{+1/2} \frac{1}{\Gamma(\nu)}\, d\nu \right]^{-1} \tag{11.107}$$

It is clear that if the integral is infinite the interpolation error is null. This appears especially for all those signals whose power spectrum is null in some frequency band. We shall make additional comments on this question after having considered the prediction error.

11.5 CAUSALITY CONSTRAINT, WIENER FILTERING

As often indicated above, the causality constraint is one of the most important because all physical systems obey the causality principle. This states that the effect cannot precede its cause. Linear mean square estimation with causality constraint leads to the Wiener-Hopf equation (11.74) and its solution defines the so-called *Wiener filter*. Because of the importance of this problem, we will discuss here two ways of obtaining the solution of this equation. The first uses the geometric method previously introduced and starts from (11.86). The second is a more physical approach starting directly from the Wiener-Hopf equation, and does not use a projection of the unconstrained solution.

11.5.1 Geometric Method

Calling S_C the subspace of F containing all the causal filters and $a_C[k]$ the solution of (11.74), we can apply (11.86), which gives

$$a_C[k] = \text{Proj}\{\, a[k] \mid S_C \,\} \tag{11.108}$$

In this equation $a[k]$ is the solution of the problem without constraint and the projection is defined by using the scalar product (11.78) where $R(\nu) = \Gamma_x(\nu)$. This gives

$$< f,\ g >_\Gamma = \int_{-1/2}^{+1/2} F^*(\nu)\, \Gamma_x(\nu)\, G(\nu)\, d\nu \tag{11.109}$$

For the discussion which follows it is of interest to write this scalar product in terms of the z-transform of $f[k]$, $\gamma[k]$, and $g[k]$. Calling $F\{z\}$, $\Gamma_x\{z\}$, and $G\{z\}$ these transforms given by (6.65) or (6.67), we find

$$< f,\ g >_\Gamma = \frac{1}{2\pi j} \int_C F\{z^{-1}\}\Gamma_x\{z\}G\{z\}z^{-1}\, dz \tag{11.110}$$

where C is the unit circle. In order to verify that this is equivalent to (11.109) it suffices to take $z = \exp(2\pi j \nu)$.

The projection (11.108) can be obtained by stating that $a[k] - a_C[k]$ is orthogonal in the sense of the scalar product (11.110) to any signal of S_C. This space is spanned by all the signals of the form

$$\delta_p[k] = \delta[k-p], \qquad p \geq 0 \tag{11.111}$$

and their z-transforms are given by

$$\Delta_p\{z\} = z^{-p}, \qquad p \geq 0 \tag{11.112}$$

As a result the orthogonality equation

$$< a - a_C,\ \delta_p >_\Gamma = 0 \tag{11.113}$$

takes the form

$$\int_C z^p \,\Gamma_x\{z\}[A\{z\} - A_C\{z\}]\, z^{-1}\, dz = 0 \tag{11.114}$$

But the unconstrained solution is given by (11.15), and if we insert this relation in (11.114), we obtain

$$\int_C z^{p-1}[\Gamma_{yx}\{z\} - \Gamma_x\{z\}\ A_C\{z\}]dz = 0 \tag{11.115}$$

which must be valid for all $p \geq 0$. Let us now interpret this equation. Remember that if $H\{z\}$ is the z-transform of $h[k]$ and is convergent on the unit circle C, then

$$\int_C z^{p-1} H\{z\}dz = 2\pi j h[p] \tag{11.116}$$

Thus if this expression is null for $p \geq 0$, this means that the signal $h[p]$ is *anticausal* and by extension of the terminology, we can say that $H\{z\}$ is anticausal. As a result, the term

$$H\{z\} = \Gamma_{yx}\{z\} - \Gamma_x\{z\}\ A_C\{z\} \tag{11.117}$$

must be anticausal. Let us now introduce the *strong factorization* of $[\Gamma_x\{z\}]^{-1}$, as introduced in Section 6.5.4 and written as

$$[\Gamma_x\{z\}]^{-1} = F\{z\}\ F\{z^{-1}\} \tag{11.118}$$

where $F\{z\}$ and $[F\{z\}]^{-1}$ are causal. Note that there is a sign ambiguity and this point will be discussed later. Furthermore if the starting point is the power spectrum instead of the z-spectrum, it must satisfy (6.86) and does not contain spectral lines. It is clear that if $F\{z\}$ is causal, then $F\{z^{-1}\}$ is anticausal. Furthermore, the product of two anticausal transfer functions is anticausal, and this is the case of the function $H'\{z\}$ defined by

$$H'\{z\} \triangleq F\{z^{-1}\}H\{z\} \tag{11.119}$$

Let us now introduce the function

$$G\{z\} \triangleq F\{z^{-1}\}\Gamma_{yx}\{z\} \tag{11.120}$$

As a result of (11.117) and (11.118), we can write

$$H'\{z\} = G\{z\} - [F\{z\}]^{-1}A_C\{z\} \tag{11.121}$$

The function $G\{z\}$ is *a priori* neither causal nor anticausal and can be written as

$$G\{z\} = \sum_{k=-\infty}^{+\infty} g[k]\, z^{-k} \tag{11.122}$$

which is convergent on the unit circle. It is possible to decompose this series into its causal and anticausal parts by writing

$$G\{z\} = [G\{z\}]_- + [G\{z\}]_+ \tag{11.123}$$

where

$$[G\{z\}]_- = \sum_{k=-\infty}^{-1} g[k]\, z^{-k} \tag{11.124}$$

This function is anticausal and as $[G\{z\}]_+$ is a series from 0 to $+\infty$, it is causal. Note that the functions $[G\{z\}]_+$ and $[G\{z\}]_-$ have no reason to be orthogonal in the sense of the scalar product (11.110). Inserting (11.123) into (11.121), we obtain

$$H'\{z\} = [G\{z\}]_+ + [G\{z\}]_- - [F\{z\}]^{-1}A_C\{z\} \tag{11.125}$$

Note that the last term is causal because $A_C\{z\}$ is by definition causal and because the inverse of $F\{z\}$ is causal, as a consequence of the strong factorization (11.118). We thus deduce that $H'\{z\}$ is anticausal if and only if

$$[G\{z\}]_+ - [F\{z\}]^{-1}A_C\{z\} = 0 \tag{11.126}$$

which defines the solution $A_C\{z\}$ by

$$A_C\{z\} = F\{z\}\ [F\{z^{-1}\}\ \Gamma_{yx}\{z\}]_+ \tag{11.127}$$

At this step we see that the sign ambiguity noted after (11.118) disappears in $A_C\{z\}$, because if we replace $F\{z\}$ by $-\{F(z)\}$ there is no change in $A_C\{z\}$. It is clear that this solution of the estimation problem with causality constraint is quite different from the unconstrained solution given by (11.15).

Let us now present some comments on this solution.

(a) Let us summarize the operations necessary to calculate it. The functions $\Gamma_x\{z\}$ and $\Gamma_{yx}\{z\}$ are known or are obtained from a z-transform of $\gamma_x[k]$ and $\gamma_{yx}[k]$ appearing in the Wiener-Hopf equation (11.74). The first step is to calculate the strong factorization of $\Gamma\{z\}$ or $[\Gamma\{z\}]^{-1}$ giving $[F\{z\}]^{-1}$ or $F\{z\}$ by (11.118). The techniques used for this operation are given in Section 6.11. The second step is to expand $[F\{z\}]^{-1}$ $\Gamma_{yx}\{z\}$ in terms of z^{-k} and to extract the *causal* part. The product of this causal part by

$F\{z\}$ gives the causal solution $A_C\{z\}$. By a new expansion in terms of z^{-k} we deduce the impulse response $a_C[k]$ of the filter, the solution of (11.74). None of these operations presents any specific difficulties, but in practice the most difficult task is to find the strong factorization of the z-spectrum or of its inverse.

(b) The filter $F\{z\}$ has a particular physical meaning. If its input is the observation signal $x[k]$, the output is white. In fact, by applying (6.68) and (11.118) we obtain a z-spectrum equal to one. But we have seen that there are many ways to whiten a signal and the filter $F\{z\}$ gives the only procedure that is causal and whose inverse is also causal. This will be used in the other solution below.

(c) The mean square error in the estimation problem is still a sum of two terms. The first is the unconstrained error given by (11.30). The additional term is given by (11.89) which can also be written in terms of z-transforms instead of Fourier transforms. By taking the z-transforms of (11.87), it is clear that

$$A_S^{\perp}\{z\} = A\{z\} - A_C\{z\} \tag{11.128}$$

11.5.2 Direct Solution of the Wiener-Hopf Equation

The basic idea is that (11.74) has an obvious solution when the observation $x[k]$ is weakly white with a variance equal to one. This implies that its correlation function is given by (7.1) where $\sigma^2 = 1$. Inserting this equation into (11.74) we obtain

$$a_C[k] = \gamma_{yx}[k], \quad k \geq 0 \tag{11.129}$$

and of course $a_C[k] = 0$ for $k < 0$ because the filter is causal. The transfer function can be written as

$$A_C\{z\} = \sum_{k=0}^{\infty} a_C[k] z^{-k} = [\Gamma_{yx}\{z\}]_+ \tag{11.130}$$

where the notation $[\ ]_+$ was already introduced in (11.122), (11.123), and (11.124). It is clear that this result also appears with (11.127), since when $x[k]$ is white with a unit variance, $\Gamma_x\{z\} = 1$ and $F\{z\} = 1$.

Suppose now that $x[k]$ is no longer white and let $F\{z\}$ again be the transfer function of the filter realizing the strong factorization (11.118). If we pass the signal $x[k]$ through this filter we obtain the output $v[k]$ and the z-spectrum of $v[k]$ is given by (6.68) or

$$\Gamma_v\{z\} = F\{z\}F\{z^{-1}\}\Gamma_x\{z\} = 1 \tag{11.131}$$

because of (11.118). Furthermore, as $F\{z\}$ and its inverse are causal, we can write

$$v[k] = \sum_{l=0}^{\infty} f[l]x[k-l] \tag{11.132}$$

and also

$$x[k] = \sum_{l=0}^{\infty} \bar{f}[l]\, v[k-l] \tag{11.133}$$

where $\bar{f}[k]$ is the inverse z-transform of $[F\{z\}]^{-1}$. Thus at time k there is exactly the same information in the pasts of $x[k]$ and $v[k]$. This means that the Hilbert spaces spanned by the RVs $x[l]$ and $v[l]$ for $l \le k$ are the same. The advantage of using the $v[l]$s is that they are orthogonal while the $x[l]$s are not. We can therefore apply (11.130) to the signal $v[k]$, and the frequency response of the corresponding filter is

$$\bar{A}_c\{z\} = [\Gamma_{yv}\{z\}]_+ \tag{11.134}$$

But as $v[k]$ is deduced from $x[k]$ by a linear filtering, we deduce from (6.72) that

$$\Gamma_{yv}\{z\} = 1\ F\{z^{-1}\}\Gamma_{yx}\{z\} \tag{11.135}$$

in which $F^*\{1/z^*\} = F\{z^{-1}\}$because all the signals and systems are real. Finally, the filter giving the estimate from the observation is the product of two filters $F\{z\}$ and $\bar{A}_c\{z\}$ which are causal, and the result is (11.127). It appears in Figure 11.3.

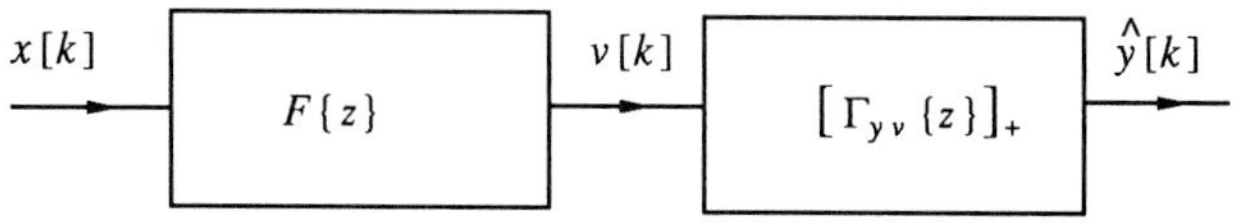

Figure 11.3 Estimation by using a whitening procedure.

This method of obtaining the optimum filter is of course much more direct and faster than the previous one. But it does not show clearly that the result is the projection of the unconstrained solution. On the other hand, the idea of transforming a sequence of correlated RVs into another where the RVs are uncorrelated is a very powerful method which will often be used below in another context.

11.5.3 Examples of Applications

There are many possible ways of applying the above results, but only those that are of more specific interest are selected here. Others will be presented as problems, while some of the most interesting applications, dealing with prediction problems, will be extensively studied in the next chapter.

Let us reconsider the problem of signal filtering in noise solved for the non-causal case in Example 11.1. In order to simplify the calculations, we assume that the noise is white, with a variance σ^2, which gives

$$\Gamma_N\{z\} = \Gamma_N(\nu) = \sigma^2 \tag{11.136}$$

From (11.32) and (11.118) we deduce that

$$\Gamma_x\{z\} = \Gamma_S\{z\} + \sigma^2 = [F\{z\}\,F\{z^{-1}\}]^{-1} \tag{11.137}$$

As the signal and the noise are uncorrelated, we have

$$\Gamma_{yx}\{z\} \triangleq \Gamma_{Sx}\{z\} = \Gamma_S\{z\} \tag{11.138}$$

and with (11.137) this becomes

$$\Gamma_{yx}\{z\} = [F\{z\}\ F\{z^{-1}\}]^{-1} - \sigma^2 \tag{11.139}$$

With these notations the noncausal solution takes the form

$$A\{z\} = 1 - \sigma^2\, F\{z\}\ F\{z^{-1}\} \tag{11.140}$$

Using (11.137) this solution can easily be expressed as in (11.33).

Let us now calculate the *causal solution* defined by (11.127). The bracket appearing in this expression is

$$\begin{aligned}[F\{z^{-1}\}\ \Gamma_{yx}\{z\}]_+ &= [F\{z^{-1}\}\{\,[F\{z\}\ F\{z^{-1}\}]^{-1} - \sigma^2\}\,]_+ \\ &= [\,\{F\{z\}\}^{-1} - \sigma^2 F\{z^{-1}\}\,]_+ \end{aligned} \tag{11.141}$$

As the inverse of $F\{z\}$ is causal, we have

$$[\,\{F\{z\}\}^{-1}\,]_+ = [\,F\{z\}\,]^{-1} \tag{11.142}$$

Furthermore, as $F\{z\}$ is causal, we can write

$$F\{z\} = \sum_{n=0}^{\infty} f_n\, z^{-n} \tag{11.143}$$

and this implies

$$[\,F\{z^{-1}\}\,]_+ = f_0 \tag{11.144}$$

Combining all these results, we obtain from (11.127)

$$A_C\{z\} = 1 - \sigma^2 f_0 F\{z\} \tag{11.145}$$

Let us now make some comments on the two solutions (11.140) and (11.145).

(a) It is clear that these solutions are quite different, and the introduction of the causality constraint completely changes the structure of the optimum filter. However, these solutions are identical when *the signal is white*. In fact, introducing its variance σ_S^2, we deduce from (11.137) that

$$F\{z\} = (\sigma_S^2 + \sigma^2)^{-1/2} \tag{11.146}$$

and it follows that

$$A\{z\} = A_C\{z\} = \sigma_S^2 (\sigma_S^2 + \sigma^2)^{-1} \tag{11.147}$$

This result is quite natural because when the signal and the noise are white, their sum is also white, and the optimum filter makes use only of the present. There is thus no memory effect.

(b) Let us now suppose that the signal is an AR(1) signal with a correlation function such as

$$\gamma[k] = \sigma_S^2 a^{|k|} \tag{11.148}$$

The corresponding z-spectrum can be deduced from (9.5) and (9.6), which give

$$\Gamma_S\{z\} = \sigma_S^2 (1 - a^2) [(z - a)(z^{-1} - a)]^{-1} \tag{11.149}$$

Inserting this expression into (11.137), we obtain

$$\Gamma_x\{z\} = \sigma^2 \frac{(z - a)(z^{-1} - a) + r(1 - a^2)}{(z - a)(z^{-1} - a)} \tag{11.150}$$

where r is the signal to noise ratio σ_S^2/σ^2. The non-causal solution is given by

$$A\{z\} = \frac{\Gamma_S\{z\}}{\Gamma_x\{z\}} = r(1 - a^2) \frac{1}{(z - a)(z^{-1} - a) + r(1 - a^2)} \tag{11.151}$$

and can also be written as in (11.140), provided that $F\{z\}$, from the strong factorization of $\Gamma_x\{z\}$, is calculated. Let us solve this problem, which is necessary to obtain the causal solution with (11.145). We apply the method presented in Section 6.11. Using the same notation as in (6.170), we deduce that $D\{z\} = z - a$, because $|a| < 1$. This means that the strong factorization of the denominator of $\Gamma\{z\}$ is obvious. Similarly, we note that the numerator of (11.150) takes the form (6.185), and we can then write the function $N\{z\}$ appearing in (6.170) in the form

$$N\{z\} = \sigma(z - b) \tag{11.152}$$

It now remains to calculate the root b. This number is the root of the polynomial similar to (6.188) and such that $|b| < 1$. Starting from the numerator of (11.150), we must now calculate the solutions of the second order equation

$$az^2 - [(1 + a^2) + r(1 - a^2)] z + az = 0 \tag{11.153}$$

The discriminant of this equation can be written as

$$\Delta = f(a^2) = (r - 1)^2 a^4 - 2(r^2 + 1) a^2 + (r + 1)^2 \tag{11.154}$$

and it is easy to verify that it is always positive for $a^2 < 1$. As a consequence b is the root of (11.153) satisfying $|b| < 1$. Returning to (11.137) we obtain

$$F\{z\} = \frac{1}{\sigma} \cdot \frac{z - a}{z - b} \tag{11.155}$$

and as a result we have $f_0 = 1/\sigma$. Inserting this expression in (11.145), we deduce the transfer function of the causal filter, which is

$$A_C\{z\} = \frac{a - b}{z - b} \tag{11.156}$$

Comparing this result with (11.151) we again find that the causality constraint completely changes the structure of the optimum filter. It is also interesting to note that, as a result of (11.156), $a_C[0] = 0$ and the causal optimum filter does not use the present but only the past.

(c) This last property is general for all AR signals. In fact, the z-spectrum of an AR(p) signal can be expressed as

$$\Gamma_S\{z\} = \sigma_u^2 \, [A\{z\}\, A\{z^{-1}\}]^{-1} \tag{11.157}$$

where σ_u^2 is the power of the driving noise appearing in (9.20) and $A\{z\}$the polynomial (9.18). Inserting this expression into (11.137), we obtain

$$\Gamma_x\{z\} = \sigma^2 \frac{A\{z\}\, A\{z^{-1}\} + \rho}{A\{z\}\, A\{z^{-1}\}} \tag{11.158}$$

where ρ is the ratio σ_u^2/σ^2. The strong factorization of the numerator of (11.158) can be written as $A'\{z\}\, A'\{z^{-1}\}$, where $A'\{z\}$ is a polynomial with the same degree as $A\{z\}$ and with roots inside the unit circle. As a result, we obtain

$$F\{z\} = \frac{1}{\sigma} \cdot \frac{A\{z\}}{A'\{z\}} \tag{11.159}$$

Noting that $A\{z\}$ and $A'\{z\}$ are polynomial beginning with z^p, we deduce that $f_0 = 1/\sigma$, which gives $a_C[0] = 0$.

(d) The previous calculation can be extended to the case when the noise is no longer white. The most difficult step in any calculation always remains the strong factorization of $\Gamma_x\{z\}$.

11.6 STATISTICAL FILTERING OF CONTINUOUS-TIME SIGNALS

All the previous calculations on statistical filtering were presented in the discrete-time case. However, continuous-time signals appear in Section 11.3 in the framework of sampling problems. The purpose of this section is to present the calculation of the optimum filtering when continuous-time signals are used. In reality the calculations are almost the same, and in many cases we will only outline what changes when passing from continuous- to discrete-time signals. In order to facilitate the reading we will use, as far as is possible, the same notations.

11.6.1 Linear Statistical Filtering without Constraint

Starting from an observation $x(t)$, we want to estimate a CT signal $y(t)$, assuming that $x(t)$ and $y(t)$ are jointly stationary. The mean square linear estimate can be written as in (11.6), or

$$\hat{y}(t) = \alpha + \int a(\theta)\, x(t-\theta)\, d\theta \tag{11.160}$$

Applying the orthogonality principle we find two equations to calculate α and $a(t)$. The first is (11.9). The second can be written as

$$E\left\{ [\, y(t) - \hat{y}(t)]\; x(t-\tau) \right\} = 0, \quad \forall \tau \tag{11.161}$$

Using the same procedure as in the DT case, we find, as in (11.16),

$$\hat{y}(t) = m_y + \int a(\theta)[x(t-\theta) - m_x]\, d\theta \tag{11.162}$$

where $a(t)$ is the solution of the integral equation

$$\int a(\theta)\, \gamma_x(\tau-\theta)\, d\theta = \gamma_{yx}(\tau) \tag{11.163}$$

As there is no constraint on the problem, this equation, which introduces a convolution between $a(t)$ and $\gamma_x(t)$, can be solved either by Laplace or by Fourier transformations. This latter transformation gives the solution (11.18), and the frequency response of the optimum filter takes the same form in both DT and CT signals. The Laplace transform is also given by (11.15) where z is replaced by s.

The calculation of the error in the filtering procedure is the same as in the DT case, and the final result is given by (11.30), where the limits of integration are now $-\infty$ and $+\infty$. The following discussion can be presented without change.

11.6.2 Linear Statistical Filtering with Constraint

The discussion at the beginning of Section 11.4 can be repeated, provided that the series are replaced by integrals. For example, the Wiener-Hopf equation (11.74), from the causality constraint, becomes in the CT case

$$\int_0^{\infty} a_S(\theta)\, \gamma_x(\tau-\theta)\, d\theta = \gamma_{yx}(\tau), \qquad \tau \geq 0 \tag{11.164}$$

and the solution $a_S(t)$ must be causal, or null for $t < 0$.

The principle of the geometric solution introduced to solve (11.76) is the same for continuous-time signals with the exception that (11.77) is now a Fourier integral and that there is no limit of integration in (11.78). In order to practice the method some examples are discussed as problems.

Let us now investigate more carefully the *causality* constraint, which is the most important in applications. The geometric method can be given as in the DT case, and for brevity we shall only present the direct solution introduced in Section 11.5.2.

Whatever the method used, the first step is to realize the strong factorization such as (11.118). This is also valid when considering continuous-time signals. It is then necessary to transpose to CT signals the ideas and methods given in Chapter 6 concerning DT signals.

11.6.3 Strong Factorization in the Continuous-Time Case

Let $\Gamma(s)$ be the Laplace spectrum of a CT signal $x(t)$, or the Laplace transform of its correlation function. We assume hereafter that the imaginary axis is in the region of convergence of $\Gamma(s)$, which ensures the existence of a power spectrum. Assuming that $x(t)$ is real, any factorization of $\Gamma(s)$ takes the form

$$\Gamma(s) = H(s)\, H(-s) \tag{11.165}$$

This equation corresponds to (6.80) valid for DT signals, and from (6.78) we see that this factorization means that $x(t)$ can be considered to be generated by a white noise passing through the filter $H(s)$.

The factorization (11.165) becomes strong if the filter $H(s)$ is *minimum phase*, which means that both it and its inverse are causal. For rational filters this is equivalent to saying that all the poles and zeros of $H(s)$ are located in the left complex plane, that is, to the left of the imaginary axis.

The strong factorization is not always possible, and there is a condition similar to (6.86). Let $S(\nu) = \Gamma(2\pi j \nu)$ be the power spectrum of $x(t)$. In order to attain a possible strong factorization, this spectrum must satisfy the Paley-Wiener condition

$$\int \frac{|\ln S(\nu)|}{1 + \nu^2}\, d\nu < +\infty \tag{11.166}$$

Note immediately that this condition cannot be satisfied by band-limited signals.

We shall now restrict our discussion to rational filters, as in Section 6.11. As a rational function is the ratio of two polynomials, we can limit our analysis to the strong factorization of polynomials. Suppose then that the function $H(s)$ in (11.165) takes the form

$$H(s) = s^n + c_1 s^{n-1} + \ldots + c_n \tag{11.167}$$

Introducing the roots a_i of this polynomial, we can write (11.167) in the form

$$H(s) = \prod_{i=1}^{n} (s - a_i) \tag{11.168}$$

in which some a_is can be equal if all the roots are not distinct. Let us first analyze the contribution in (11.167) of a term such as $(s - a)$ where a is *real*. This contribution can be written as

$$\Gamma_a(s) = (s - a)\,(-s - a) = a^2 - s^2 \tag{11.169}$$

As the power spectrum is expressed in the frequency domain, or by replacing s by $j\omega$, the contribution to the power spectrum of the real root a becomes

$$S_a(\omega) = \omega^2 + a^2 \tag{11.170}$$

Suppose now that a is complex. As we are working with polynomials with real coefficients c_i, the fact that a is a complex root implies that a^* is also a root of $H(s)$. Consequently, the contribution of the root a to $\Gamma(s)$ becomes

$$\begin{aligned}\Gamma_a(s) &= (s - a)\,(s - a^*)(-s - a)\,(-s - a^*)\\ &= (s^2 - a^2)\,(s^2 - a^{*2})\\ &= s^4 - 2\mathrm{Re}(a^2)s^2 + |a|^4\end{aligned} \tag{11.171}$$

Replacing s by $j\omega$, we obtain

$$S_a(\omega) = \omega^4 + 2\mathrm{Re}(a^2)\omega^2 + |a|^4 \tag{11.172}$$

Applying this procedure to all the roots of $H(s)$, we deduce that the polynomial corresponding to the power spectrum is a polynomial in ω^2, or can be expressed as

$$S(\omega) = \omega^{2n} + k_1\,\omega^{2(n-1)} + \ldots + k_n \tag{11.173}$$

This expression must be compared with (6.175) valid in the DT case. This is the starting point of the strong factorization problem, when the second order properties of the observation signal are specified by the power spectrum, which is assumed to be a rational function. From (11.173) we must deduce $\Gamma(s)$ in order to write (11.165). This step is very simple: we simply replace ω^{2n} by $(-1)^n s^{2n}$.

Let us illustrate this procedure with two examples. Consider the polynomial in ω

$$S(\omega) = \omega^2 + 9 \tag{11.174}$$

The corresponding polynomial in s is

$$\Gamma(s) = 9 - s^2 \tag{11.175}$$

and the corresponding strong factorization is realized with

$$H(s) = s + 3 \tag{11.176}$$

which has a single root in the left complex plane.

Consider the term

$$S(\omega) = \omega^4 + 6\omega^2 + 25 \tag{11.177}$$

The corresponding polynomial in s is

$$\Gamma(s) = s^4 - 6s^2 + 25 \tag{11.178}$$

and its roots are $s^2 = 3 \pm 4j$, or

$$s = \pm (2 \pm j) \tag{11.179}$$

The corresponding strong factorization is realized with

$$H(s) = (s + 2 + j)(s + 2 - j) \tag{11.180}$$

which has two complex conjugate roots located in the left complex plane.

From these two examples we can now deduce the general methodology. The starting point is the polynomial $S(\omega)$ given by (11.173). Replacing ω^{2n} by $(-1)^n s^{2n}$ we deduce the polynomial

$$\Gamma(s) = S^n + k'_1 S^{n-1} + k'_2 S^{n-2} + \dots + k'_n \tag{11.181}$$

where $S = s^2$. As a result of (11.169), the real roots in S are positive. Let a^2 be a real root. The corresponding term in the strong factorization is then $(s + a)$, $a > 0$. Suppose now that $\alpha + j\beta$ is a complex root in S. By calculating its square root, we obtain two complex numbers. If $a + jb$ is the square root such that $a > 0$, the corresponding term in the strong factorization is similar to (11.180), or

$$H(s) = (s + a + jb)(s + a - jb) = (s + a)^2 + b^2 \tag{11.182}$$

By repeating this procedure with all the roots of (11.182), we complete the strong factorization of $\Gamma(s)$ in the form of (11.165), where $H(s)$ is a product of terms corresponding to roots in the left complex plane.

Let us explain this procedure with an example. Suppose that the polynomial (11.173) is

$$S(\omega) = \omega^6 + \omega^4 + 4\omega^2 + 4 \tag{11.183}$$

The corresponding polynomial in s is

$$\Gamma(s) = -s^6 + s^4 - 4s^2 + 4 = -S^3 + S^2 - 4S + 4 \tag{11.184}$$

There is an obvious root $S = 1$, which, in (11.165), introduces the term $s + 1$. The other roots of (4.184) are $\pm 2j$. Taking the square root, we find $\pm (1 \pm j)$, and the roots in the left plane are $-1 \pm j$. The corresponding term in (11.165) is $(s + 1 + j)(s + 1 - j)$, and the strong factorization is realized with

$$H(s) = (s + 1)(s^2 + 2s + 2) = s^3 + 3s^2 + 4s + 2 \tag{11.185}$$

After the *calculation* of the strong factorization, it is useful to give its physical interpretation. For this, let us start from (11.165). Using (6.78), this equation means that the spectrum of $x(t)$ is that of a signal generated by a white noise filtered by the filter $H(s)$. But we have noted many times that in the continuous-time case a white noise is not a second-order signal, and the input-output relationship must be written as

$$x(t) = \int h(t - \theta)\, dw(\theta) \tag{11.186}$$

introducing the increments of a function $w(t)$ without derivative. This expression was used, for example, in (7.46), where $w(t)$ corresponds to a stationary Brownian motion, and in (8.62), where $w(t)$ is deduced from a Poisson process with random amplitude.

Let us now investigate the consequences of strong factorization. As the filter is causal, the integral (11.186) takes the form

$$x(t) = \int_{-\infty}^{t} h(t - \theta)\, dw(\theta) \tag{11.187}$$

But as the inverse filter of $h(t)$ is causal, there is the same information in $x(.)$ observed from $-\infty$ to t as in the increments $dw\,(.)$ observed in the same period of time. The only, but essential, difference is that the increments $dw\,(\theta)$ are uncorrelated, while $x(t)$ has a non zero correlation time.

In particular, if $x(t)$ is Gaussian, the increments $dw(t)$ are also Gaussian, and as they are uncorrelated, they are independent. So the increments $dw(t)$ are those of a Brownian motion studied in Section 7.3.2. Similarly, if $x(t)$ is a shot noise such as (8.37) or (8.62), the increments $dw(t)$ are those of a Poisson process, or of its generalization by random amplitude.

To examine the structure of the inverse of the filter in (11.187), we will consider a particular example which is sufficient to understand the general situation. Consider a signal $x(t)$ whose correlation function is exponential, or given by (5.38). The corresponding power spectrum is given by (5.39) and the Laplace spectrum is $2a(a^2 - s^2)^{-1}$. Applying the results given previously, the strong factorization of this spectrum is realized with the filter

$$H(s) = \sqrt{2a}\,(s + a)^{-1}, \qquad a > 0 \tag{11.188}$$

and the inverse filter is

$$F(s) = (2a)^{-1/2}\,(a + s) \tag{11.189}$$

The term in the bracket is the sum of a constant filter which multiplies by a and a filter calculating the derivative, as discussed in Example 6.3. Let us apply this procedure when the signal $x(t)$ is the shot noise of a Poisson process given by (8.37). If the impulse response of the filter in (8.37) is

$$h(t) = u(t)\exp(-at) \tag{11.190}$$

and if the density of the Poisson process is $\lambda = 2a$, the power spectrum (8.44) is equal to $\Gamma(\nu)$ given by (5.39). Consequently, the strong factorization is given by (11.188) and the inverse filter is (11.189). Let us examine the action of this filter on the shot noise

$x(t)$ given by (8.37). It is simpler to start from (8.36), showing that $x(t)$ is a sum of signals $s_i(t)$ such as

$$s_i(t) = u(t - t_i) \exp[-a(t - t_i)] \tag{11.191}$$

The derivative of this signal is

$$s'_i(t) = \delta(t - t_i) \exp[-a(t - t_i)] - a s_i(t) \tag{11.192}$$

and using the well known relation $\delta(t) f(t) = \delta(t) f(0)$, we obtain

$$s'_i(t) = \delta(t - t_i) - a s_i(t) \tag{11.193}$$

in such a way that the action of the filter $(a + s)$ gives the impulse $\delta(t - t_i)$. As a result the action of the inverse filter $F(s)$ given by (11.189) regenerates a signal like (5.71), i.e., a sequence of pulses arriving at the time instants of a Poisson process. The relation $\lambda = 2a$ is only there to ensure that the power spectrum of (5.71) is equal to 1. These results are in accordance with the comments made after (8.44) concerning the point that a sequence of Poissonian pulses such as those in Figure 1.3 can be considered as a white noise.

Let us now take a less obvious example. Suppose that $x(t)$ is the random telegraph signal discussed in Section 8.6.1 and represented in Figure 8.7. Its correlation function is given by (8.118), and if we take $a = 2\lambda$, the strong factorization and its inverse are given by (11.188) and (11.189). It is clear that the filter $F(s)$ generates a CT white noise which is not a sequence of Poissonian pulses. In fact, the derivative of $x(t)$ is a sequence of pulses of alternate signs, as seen in Figure 8.7. Consequently the sum of the signal $ax(t)$ and the sequence of alternating Poisson pulses is a white noise. Therefore, if the increments $dw(\theta)$ in (11.187) are uncorrelated, they are no longer independent.

11.6.4 Solution of the Wiener-Hopf Equation

Our purpose is to find the function $a(t)$ vanishing for $t < 0$ and solution of

$$\int_0^\infty a(\theta)\, \gamma_x(\tau - \theta)\, d\theta = \gamma_{yx}(\tau)\ , \qquad \tau \geq 0 \tag{11.194}$$

The method used is the same as that given for the DT in Section 11.5.2.

Suppose first that $x(t)$ has a power spectrum equal to 1, or that $\gamma_x(\tau) = \delta(\tau)$. Inserted in (11.194), this gives

$$a(t) = u(t)\ \gamma_{yx}(t) \tag{11.195}$$

The corresponding transfer function can be written as

$$A(s) = [\Gamma_{yx}(s)]_+ \tag{11.196}$$

which is the one-sided Laplace transform of $\gamma_{yx}(\tau)$ defined by

$$[\Gamma_{yx}(s)]_+ \triangleq \int_0^{\infty} e^{-st}\, \gamma_{yx}(t)\, dt \tag{11.197}$$

The corresponding estimation error is given by (11.20). The last term of this equation can be calculated by using (5.62), which gives

$$E[\hat{y}^2(t)] = \iint a(\theta_1)\, a(\theta_2)\, \delta(\theta_2 - \theta_1)\, d\theta_1\, d\theta_2 = \int_0^{\infty} a^2(\theta)\, d\theta \tag{11.198}$$

Using (11.195), the estimation error becomes

$$\varepsilon^2 = \gamma_y(0) - \int_0^{\infty} \gamma_{yx}^2(t)\, dt \tag{11.199}$$

Let us now suppose that the observation $x(t)$ has a non-constant power spectrum $\Gamma_x(\nu)$. Starting with the strong factorization of $\Gamma_x(s)$ given by (11.165), we can use the filter $F(s) = [H(s)]^{-1}$. This filter transforms the observation $x(t)$ into a white noise $v(t)$, as represented in Figure 11.3. The Laplace transform of the crosscorrelation between $x(t)$ and $v(t)$ is given by the extension of (6.78) similar to (6.72), or by

$$\Gamma_{yv}(s) = 1\,.\, F(-s)\,.\, \Gamma_{yx}(s) \tag{11.200}$$

Finally, the complete causal filter is given by

$$A(s) = F(s)\,[F(-s)\, \Gamma_{yx}(s)]_+ \tag{11.201}$$

which corresponds to (11.127), valid in the DT case. It is interesting to note that the bracket has the same meaning as in (11.197), which means that $[G(s)]_+$ is the one-sided Laplace transform of $g(t)$. In other words, to calculate (11.201) it is, in general, necessary to calculate the inverse two-sided transform of $F(-s)\,\Gamma_{yx}(s)$, or $\gamma_{yv}(t)$, and then the one-sided transform of this function, not always an easy task.

Finally, the error is always given by (11.199), which becomes

$$\varepsilon^2 = \gamma_y(0) - \int_0^{\infty} \gamma_{yv}^2(t)\, dt \tag{11.202}$$

where $\gamma_{yv}(t)$ is the inverse Laplace transform of $\Gamma_{yv}(s)$.

11.6.5 Signal Estimation in a White Noise

Let us reconsider the problem of estimating a signal in a white noise, discussed in Section 11.5.3 in the DT case. The CT observation signal is written as in (11.32), or

$$x(t) = s(t) + n(t) \tag{11.203}$$

where $s(t)$ and $n(t)$ represent the signal to be estimated and the background noise.

The non-causal solution of this problem is given by (11.33) and the estimation error by (11.35), where the integration goes from $-\infty$ to $+\infty$.

For the causal solution, let us assume that the noise is white with a correlation function given by (7.23) where k is replaced by N. Let us call $\Gamma_S(s)$ the Laplace spectrum of the signal $s(t)$. As a result, the Laplace spectrum of the observation becomes

$$\Gamma_x(s) = \Gamma_S(s) + N = [F(s)\, F(-s)]^{-1} \tag{11.204}$$

where $F(s) = 1/H(s)$ appearing in (11.165) for the strong factorization of $\Gamma_x(s)$. Finally, the Laplace transform of the crosscorrelation function $\gamma_{yx}(\tau)$ is, as in (11.138) and (11.139),

$$\Gamma_{yx}(s) = \Gamma_S(s) = [F(s)\, F(-s)]^{-1} - N \tag{11.205}$$

With these expressions the non-causal solution can be written as

$$A(s) = 1 - N F(s)\, F(-s) \tag{11.206}$$

which corresponds to (11.140) valid in the DT case.

The causal solution $A_C(s)$ is given by (11.201), and it takes the form

$$A_C(s) = F(s)\ \left[\, F(-s)\ \{\Gamma_x(s) - N\}\,\right]_+$$

Using (11.204) and noting that $[A(s) + B(s)]_+ = [A(s)]_+ + [B(s)]_+$, we obtain

$$A_C(s) = 1 - N F(s)\ [F(-s)]_+ \tag{11.207}$$

It remains to calculate the last term of this equation. We assume that $\Gamma_S(s)$ is a rational function of s and that the degree of the denominator is greater than that of the numerator. To simplify the discussion, let us first assume that the signal $s(t)$ has an exponential correlation function. Its Laplace spectrum can then be expressed as

$$\Gamma_S(s) = \frac{\alpha^2}{a^2 - s^2} \tag{11.208}$$

If the variance of the signal is σ_S^2, it results from (5.38) and (5.39) that $\alpha^2 = 2a\sigma_S^2$. Inserting (11.208) into (11.204), we obtain

$$\Gamma_x(s) = N\ \frac{(s+b)\,(s-b)}{(s+a)\,(s-a)} \tag{11.209}$$

where b is such that $b^2 = a^2 + \alpha^2/N$. This immediately gives the strong factorization of $\Gamma_x(s)$, and $F(s)$ in (11.204) becomes

$$F(s) = N^{-1/2}\ \frac{s+a}{s+b} \tag{11.210}$$

From this expression we deduce that $F(-s)$ in (11.207) can be written as

$$F(-s) = N^{-1/2}\left[\ \frac{s}{s-b} - \frac{a}{s-b}\ \right] \tag{11.211}$$

In order to calculate $[F(-s)]_+$ we must first calculate the function $f_-(t)$, the two-sided inverse Laplace transform of $F(-s)$. By elementary calculation, where it is assumed that the imaginary axis is in the region of convergence, we obtain

$$N^{1/2} f_-(t) = \delta(t) + (a - b)\, u(-t) \exp(bt) \tag{11.212}$$

where $\delta(t)$ and $u(t)$ are the Dirac distribution and the unit step function respectively. To obtain the value of $[F(-s)]_+$, we calculate the monolateral causal Laplace transform of $f_-(t)$. As the last term of (11.212) is anticausal, it remains $[F(-s)]_+ = N^{-1/2}$, which in (11.207) gives

$$A_c(s) = 1 - N^{1/2} F(s) \tag{11.213}$$

This is very similar to (11.145).

Looking at the proof, it is easy to verify that this result is general and comes from the two assumptions introduced: white noise and the property of the rational spectrum of the signal. In fact, if the degree of the denominator of $\Gamma_S(s)$ is greater than that of the numerator, as in (11.208), it appears that $\Gamma_x(s) \to N$ when $|s| \to \infty$. This introduces the term $N^{-1/2}\,\delta(t)$ in the inverse Laplace transform of $F(-s)$. As the other terms of $f_-(t)$ are anticausal, because $F(s)$ is causal, we deduce that only the term $N^{-1/2}$ remains in $[F(-s)]_+$.

Let us close this section by looking at the structure of $A_C(s)$ when (11.208) holds. Using (11.210) we obtain

$$A_c(s) = 1 - \frac{s + a}{s + b} = \frac{b - a}{s + b} \tag{11.214}$$

This is a first order system with a stable pole for $s = -b$. As a result of the definition of b, $b^2 - a^2 = \alpha^2/N$, and then $b > a$.

11.7 TAYLOR EXPANSIONS AND ESTIMATION

Taylor series are a very important tool in the analysis of deterministic signals. Suppose that $x(t)$ is a deterministic signal and that all its derivatives exist. In this case it is possible to write

$$x(t + \tau) = \sum_{k=0}^{\infty} x^{(k)}(t)\, \frac{\tau^k}{k!} \tag{11.215}$$

If $t = 0$, we obtain a McLaurin series often used previously, and especially in the calculation of the moments of the characteristic function. The series (11.215) is of great practical importance, and its meaning is that the function $x(t + \tau)$ can be considered as a polynomial in τ, but of infinite degree. In practice, we replacc the series by a sum of a finite number of terms, making an approximation which is better the greater the number.

Let us now consider the situation when $x(t)$ is random. The concept of the derivative of a random signal has been introduced before, and our purpose here is twofold: first, we want to know if an expansion such as (11.215) is possible, and second, how to approximate a signal when a finite number of derivatives is known. As it is a question of approximating a random object, a criterion must be chosen, and as we are dealing with mean square estimation problems, the criterion will be the mean square error.

11.7.1 Taylor Series

Suppose that $x(t)$ is a stationary, zeromean-valued, and real signal. The basic assumption is that all its derivatives are second order signals. The condition ensuring this property is given in the time domain by extensions of (5.87), but it is much easier to express these conditions in terms of the power spectrum $\Gamma(\nu)$. This is discussed in Example 6.3. The first order derivative of $x(t)$ exists if $\Gamma(\nu)$ is such that

$$\int_{-\infty}^{+\infty} \Gamma(\nu)\, 4\pi^2\nu^2\, d\nu < +\infty \tag{11.216}$$

and, more generally, the derivative of order k exists if

$$\int_{-\infty}^{+\infty} \Gamma(\nu)\, (4\pi^2\nu^2)^k\, d\nu < +\infty \tag{11.217}$$

Note that, if $x(t)$ is a second order signal, the condition is satisfied for $k = 0$ because the integral yields the variance of $x(t)$.

It is clear that the existence of derivatives of any order is entirely due to the behavior of $\Gamma(\nu)$ when $\nu \to \infty$. Several situations can be indicated. If $x(t)$ is a band-limited signal, which means that $\Gamma(\nu) = 0$ as soon as $|\nu| > B$, (11.217) holds for any k, and all the derivatives exist. The same situation appears if $\Gamma(\nu)$ has an exponential behavior for large values of ν, and especially if $\Gamma(\nu)$ has a Gaussian shape. However, if $\Gamma(\nu)$ is a rational function of ν, $x(t)$ can only have a finite number of second order derivatives. This number can even be null, since we saw that a signal with an exponential correlation function or a Lorentzian spectrum has no first order derivative.

Suppose now that all the derivatives are second order, or that (11.217) holds whatever k. We can then state the problem of estimating $x(t+\tau)$ in terms of the derivatives of $x(t)$. Writing this estimation as

$$\hat{x}(t+\tau) = \sum_{k=0}^{\infty} a_k(\tau)\, x^{(k)}(t) \tag{11.218}$$

we want to calculate the coefficients $a_k(\tau)$ giving the minimum mean square error.

The solution to this problem is deduced from the orthogonality principle

$$x(t+\tau) - \hat{x}(t+\tau) \perp\!\!\!\perp x^{(l)}(t), \quad \forall l \tag{11.219}$$

This is the same as writing

$$E[\hat{x}(t+\tau)\, x^{(l)}(t)] = E[x(t+\tau)\, x^{(l)}(t)] \tag{11.220}$$

and inserting (11.218) in this equation, we obtain

$$\sum_{k=0}^{\infty} a_k(\tau)\ E[x^{(k)}(t)x^{(l)}(t)] = E[x(t+\tau)\ x^{(l)}(t)] \tag{11.221}$$

Using (6.60'), applied for real signals, this gives

$$\sum_{k=0}^{\infty} a_k(\tau)\ \gamma^{(k+l)}(0) = \gamma^{(l)}(\tau) \tag{11.222}$$

But (11.217) implies that all the derivatives of $\gamma(\tau)$ exist, and (11.222) appears as a McLaurin series of $\gamma^{(l)}(\tau)$, which implies that

$$a_k(\tau) = \frac{\tau^k}{k!} \tag{11.223}$$

Consequently, the application of the orthogonality principle gives the same coefficients as those appearing in the Taylor series of a deterministic signal.

Let us now calculate the error in this estimation problem. The error is always given by the same expression,

$$\varepsilon^2 = E[x^2(t+\tau)] - E[\hat{x}^2(t+\tau)] = \gamma[0] - E[\hat{x}^2(t+\tau)] \tag{11.224}$$

Using (11.223), the last term of this equation can be written as

$$E[\hat{x}^2(t+\tau)] = \sum_k \sum_l \frac{\tau^{k+l}}{k!\ l!}\ E[x^{(k)}(t)x^{(l)}(t)] \tag{11.225}$$

Expressing the last term of this equation in terms of $\Gamma(\nu)$, we obtain

$$E[\hat{x}^2(t+\tau)] = \sum_k \sum_l \frac{\tau^{k+l}}{k!\ l!} \int (2\pi j\nu)^k (-2\pi j\nu)^l \Gamma(\nu)\ d\nu$$

Calculating the series yields $\exp(2\pi j\nu\tau)\exp(-2\pi j\nu\tau)$ and then

$$E[\hat{x}^2(t+\tau)] = \int \Gamma(\nu)\ d\nu = \gamma(0) \tag{11.226}$$

By inserting this value in (11.224) we obtain $\varepsilon^2 = 0$.

In conclusion, if $x(t)$ is a random signal with second order derivatives of any order, the Taylor series (11.215) is valid, as for a deterministic signal.

11.7.2 Limited Taylor Expansion

Suppose now that instead of (11.215), we want to make a limited expansion of $x(t+\tau)$, written in the form

$$\hat{x}_n(t+\tau) = \sum_{k=0}^{n} a_k^n(\tau)\, x^{(k)}(t) \tag{11.227}$$

The limitation of the Taylor series to the order n can be either to limit the complexity of the calculation of the sum, or because the derivatives of order greater than n do not exist. Note that the coefficients of the limited expansion can depend on n, which is not the case when the series (11.215) is simply truncated at n.

Let us call $\mathbf{a}_n(\tau)$ the vector with components $a_k^n(\tau)$, $0 \le k \le n$, and $\mathbf{X}(t)$ the random vector whose components are $x^{(k)}(t)$, $0 \le k \le n$. With these notations (11.227) can be expressed as

$$\hat{x}_n(t+\tau) = \mathbf{a}_n^T(\tau)\mathbf{X}(t) \tag{11.228}$$

This equation is similar to (10.61), and the orthogonality principle yields the equation

$$\Gamma\, \mathbf{a}_n(\tau) = \mathbf{c} \tag{11.229}$$

where Γ is the matrix defined by

$$\Gamma_{ij} = E[x^{(i)}(t)x^{(j)}(t)] \tag{11.230}$$

and $\mathbf{c}$ the vector with components

$$c_i = E[x(t+\tau)\, x^{(i)}(t)] \tag{11.231}$$

As a result of (6.60'), we find

$$\Gamma_{ij} = (-1)^j\, \gamma^{(i+j)}(0) \tag{11.232}$$

$$c_i = (-1)^i\, \gamma^{(i)}(\tau) \tag{11.233}$$

To simplify our discussion let us introduce the coefficients

$$\gamma_{2k} \triangleq \gamma^{(2k)}(0) \tag{11.234}$$

Note also that, as the correlation function $\gamma(\tau)$ is an even function, all the odd derivatives satisfy

$$\gamma^{(2k+1)}(0) = 0 \tag{11.235}$$

With these notations the matrix Γ corresponding to $n = 3$ takes the form

$$\Gamma = \begin{bmatrix} \gamma_0 & 0 & \gamma_2 & 0 \\ 0 & -\gamma_2 & 0 & -\gamma_4 \\ \gamma_2 & 0 & \gamma_4 & 0 \\ 0 & -\gamma_4 & 0 & -\gamma_6 \end{bmatrix} \tag{11.236}$$

It is clear that this matrix is completely defined by its first row and its last column. This property is valid whatever n.

The structure of this matrix explains why the components of the vector $\mathbf{a}_n(\tau)$, solution of (11.229), are functions of n. This is because $\mathbb{\Gamma}$ is not a diagonal matrix. Consequently, when the order of the approximation increases, there is no reason for the components of $\mathbf{a}$ to remain the same.

As an example, let us give the first three solutions of (11.227). By an easy matrix inversion, we find

$$\hat{x}_0(t+\tau) = [\gamma(\tau)/\gamma_0]\, x(t) \tag{11.237}$$

$$\hat{x}_1(t+\tau) = [\gamma(\tau)/\gamma_0]\, x(t) + \gamma^{(1)}(\tau)/\gamma_2]\, x^{(1)}(t) \tag{11.238}$$

$$\begin{aligned}\hat{x}_2(t+\tau) = {} & \Delta^{-1}\, [\gamma_4\, \gamma(\tau) - \gamma_2\, \gamma^{(2)}(\tau)]\, x(t) + [\, \gamma^{(1)}(\tau)/\gamma_2]\, x^{(1)}(t) \\ & + \Delta^{-1}\, [-\gamma_2\, \gamma(\tau) + \gamma_0\, \gamma^{(2)}(\tau)]\, x^{(2)}(t)\end{aligned} \tag{11.239}$$

with

$$\Delta = \gamma_0\, \gamma_4 - \gamma_2^2 \tag{11.240}$$

In conclusion, if we want to approximate $x(t+\tau)$ from knowledge of its first derivatives $x^{(k)}(t)$ and using the mean square error criterion, the solution is not the truncation of the Taylor series (11.215). The fundamental reason for this result is that the derivatives of $x(t)$ are correlated with each other, with the result that the matrix $\mathbb{\Gamma}$ defined by (11.130) is not diagonal.

Finally, note that if τ is small, we can expand $\gamma(\tau)$ and its derivative by a limited McLaurin expansion, and it is easy to find that (11.239) becomes

$$\hat{x}_2(t+\tau) = x(t) + \tau x'(t) + \frac{\tau^2}{2}\, x''(t) \tag{11.241}$$

With this approximation we return to (11.215) limited to $n = 3$, which is then only valid for $\tau \to 0$.

PROBLEMS

11.1 Let $x[k]$ be a zeromean and stationary DT random signal. Suppose that its power is equal to 1 and let $\gamma[k]$ be its correlation function. Let $\hat{x}[k+p]$ be the LMSE of $x[k+p]$ in terms of $x[k]$ written $\hat{x}[k+p] = h[p]x[k]$.

(a) Calculate $h[p]$ in terms of $\gamma[k]$.

(b) Find the condition on $\gamma[k]$ ensuring that the innovation $\hat{x}[k+p] - x[k+p]$ is uncorrelated (or orthogonal) with $x[k-q]$ whatever $q > 0$.

(c) Show that this condition implies a recursion between $\gamma[k+1]$ and $\gamma[k]$. Show that this recursion completely specifies $\gamma[k]$ as soon as $\gamma[1]$ is known. Give the expression of $\gamma[k]$ in terms of $\gamma[1]$.

11.2 Let $x[k]$ be a DT signal and $\mathbf{x}$ the vector whose components are $x[k-i]$, $1 \le i \le n$. Suppose that $x[k] = s[k] + n[k]$, where $n[k]$ is a zeromean and stationary weak white noise of variance v_N. We assume that $s[k] = A$, where A is a zeromean RV uncorrelated with $n[k]$ and with the variance v_S.

(a) Calculate the correlation function $\gamma_x[n]$ of $x[k]$.

(b) Write the covariance matrix $\mathbb{\Gamma}$ of the vector $\mathbf{x}$ in terms of the variances v_N and v_S and using the vector $\mathbf{r}$ all of whose components are equal to 1.

(c) The inverse matrix $\mathbb{\Gamma}^{-1}$ can be written in the form $\alpha\mathbb{I} + \beta\mathbf{r}\mathbf{r}^T$. By introducing the parameter $\mu = v_N/v_S$, express α and β in terms of v_S and μ.

(d) We want to estimate the signal $s[k]$ from the observation vector $\mathbf{x}$. Find the best LMSE $\hat{s}[k]$ and calculate the corresponding error. Study the behavior of this error when n is increasing and explain the limit obtained when n tends to infinity.

(e) Solve the same problem as in the previous question when the estimandum is $x[k]$. This problem is called the one-step prediction with finite past.

(f) Let $y[k]$ be the signal equal to $(1/n)\{x[k-1] + x[k-2] + \ldots + x[k-n]\}$. Show that this signal is deduced from $x[k]$ by a linear filtering and calculate the impulse response of this filter.

(g) Calculate the variance of the signal $z[k] = s[k] - y[k]$ and compare this variance with the error introduced in the LMSE of $x[k]$. Study especially the case where n tends to infinity and explain the result.

11.3 The purpose of this problem is to answer the question stated at the end of Section 11.2. Suppose that in (11.6) the coefficient α and the function $a[l]$ also depend on k. Write again the orthogonality equations and show that the result is that α and $a[l]$ are independent of k and given by (11.11) and (11.12).

11.4 Let $x(t)$ be a zeromean stationary random signal. By using this signal as the input of two linear filters F_1 and F_2 we obtain two different output signals $y_1(t)$ and $y_2(t)$.

(a) Calculate the frequency response of the non-causal filter giving the best LMSE of $y_2(t)$ in terms of $y_1(t)$. Calculate the corresponding error and explain the result.

(b) Assuming that F_1 and F_2 are minimum phase filters, solve the same problem when the causality constraint is introduced. Explain the result.

11.5 Consider the same question as in the previous problem in the case where $y_1(t)$ and $y_2(t)$ are replaced by $z_1(t)$ and $z_2(t)$ where $z_i(t) = m_i\, y_i(t)$. In this equation m_1 and m_2 are two zeromean RVs such that

$$E(m_1^2) = \sigma_1^2 \ ; \ E(m_2^2) = \sigma_2^2 \ ; \ E(m_1 m_2) = c\sigma_1\sigma_2$$

Answer the questions (**a**) and (**b**) of the previous problem and explain why the results are different. Explain the situation when $c = 0$.

11.6 **Filtering of a signal in noise.** Consider the same problem as in Example 11.1 but without the assumption that the signal and the noise are uncorrelated. This point is described by the introduction of the crosscorrelation function between the signal and the noise or of its Fourier transform $\Gamma_{sn}(\nu)$.

(a) Calculate the frequency response of the optimum non-causal filter giving the best LMSE of the signal.

(b) Calculate the corresponding estimation error.

(c) Discuss the properties of this error and explain the cases where the error is null.

11.7 **Noise canceling for stationary signals.** The purpose of this problem is to analyze the same situation as that presented in Problem 10.11. We want to estimate a signal $s(t)$ from the observation of two signals $x_1(t)$ and $x_2(t)$. These observations are described by

$$x_1(t) = s(t) + n_1(t) \quad ; \quad x_2(t) = n_2(t)$$

where $n_1(t)$ and $n_2(t)$ are two disturbing noises. It is assumed that the signal and the noises are zeromean and jointly stationary, that the signal is uncorrelated with the noises but that the crosscorrelation between the two noises is not necessarily zero. It is sometimes said that the signal $x_2(t)$ is a noise alone reference. The best LMSE of $s(t)$ can be written as

$$\hat{s}(t) = \int h_1(\theta)x_1(t - \theta)d\theta + \int h_2(\theta)x_2(t - \theta)d\theta$$

introducing two linear filters of impulse responses $h_1(t)$ and $h_2(t)$.

(a) Write the orthogonality equations allowing the calculation of $h_1(t)$ and $h_2(t)$.

(b) Show that these equations can easily be solved in the frequency domain. For this, introduce the power spectra of $s(t)$, $n_1(t)$ and $n_2(t)$ and the Fourier transform of the crosscorrelation function between $n_1(t)$ and $n_2(t)$. Give explicitly the expressions of the frequency responses $G_1(\nu)$ and $G_2(\nu)$ of the two filters used in the estimation procedure.

(c) Calculate the estimation error. Show that it can be null in two quite different situations. The first can appear even when the crosscorrelation function between $n_1(t)$ and $n_2(t)$ is null and the latter comes from a particular structure of this crosscorrelation. Give a physical interpretation of these two situations.

11.8 Let $x(t)$ be a real, zeromean and CT random signal and call $\gamma(\tau)$ its correlation function.

(a) Calculate the best LMSE $\hat{x}(t)$ of $x(t)$ in terms of $x(t - \theta)$ and $x(t + \theta)$.

(b) Complete the calculation when $\gamma(\tau) = \exp(-a|\tau|)$.

11.9 Let $x(t)$ be a real, zeromean and CT random signal with a power spectrum $\Gamma(\nu)$ equal to $1/2B$ when $-B < \nu < B$ and to zero otherwise.

(a) Calculate its correlation function $\gamma(\tau)$.

(b) Calculate the LMSE $\hat{x}(t)$ of $x(t)$ in terms of $x(t - 1/4B)$ and $x(t + 1/4B)$. Determine the corresponding error.

(c) Suppose that $y(t) = x^2(t)$ and that $x(t)$ is normal. Calculate the affine MSE $\hat{y}(t)$ of $y(t)$ in terms of $y(t - 1/4B)$ and $y(t + 1/4B)$. Determine the corresponding error.

11.10 Let $x(t)$ be a normal, zeromean and stationary random signal of variance σ^2 and with a power spectrum constant for $|\nu| \leq B$ and zero otherwise. Let $y(t)$ be the signal $x^2(t) - \sigma^2$.

(a) Calculate the correlation functions $\gamma_x(\tau)$ and $\gamma_y(\tau)$ of $x(t)$ and $y(t)$ respectively.
(b) Calculate the best LMSE of $y(t)$ in terms of the observations $x(t_n)$, $t_n = n/2B$ and $-N \le n \le N$, where n and N are integers.
(c) Calculate the estimation error.
(d) Show that if $t \neq t_n$, whatever n, the error cannot tend to zero when N tends to infinity.

11.11 The DT Wiener-Hopf equation has been solved by two methods presented in Sections 11.5.1 and 11.5.2 respectively. In the CT case the solution (11.201) is obtained by using a whitening approach. Find the same solution again by adapting to the CT case the geometric method presented for the DT case in Section 11.5.1.

11.12 **Continuous-time prediction**. Let $x(t)$ be a CT signal and suppose that we want to predict $x(t + T)$ from the observation of the signal from $-\infty$ to t. In order to apply the general theory we define the estimandum signal by $y(t) = x(t + T)$.
(a) Calculate the crosscorrelation function $\gamma_{yx}(\tau)$ used in the Wiener-Hopf equation (11.194).
(b) By using the general solution of this equation, calculate the optimum predictor.
(c) Give a general expression of the prediction error.
(d) Apply the results of the previous questions in the case where the correlation function of $x(t)$ is $\sigma^2 \exp(-a|\tau|)$. Give an explanation of the result. Represent the curve of the prediction error as a function of T.
(e) Make the same calculation as in the previous question when the power spectrum of $x(t)$ is $1/S(\omega)$ where $S(\omega)$ is given by (11.177).

11.13 **Signal estimation in a white noise.** Show that equation (11.213) which is proved in a particular case specified by (11.208) is in reality quite general for signals with rational z-spectrum.

11.14 **Statistical filtering with vector signals.** We want to extend the method of linear statistical filtering without constraints in the case of vectors signals. The observation signal is a CT vector signal $\mathbf{x}(t)$ with n components. These components can for example be the outputs of the n sensors of an antenna. The estimandum signal is a vector signal $\mathbf{y}(t)$ with m components. We assume, for simplification, that $\mathbf{x}(t)$ and $\mathbf{y}(t)$ are zeromean and real. The second order properties of the vectors $\mathbf{x}(t)$ and $\mathbf{y}(t)$ are known and defined by the matrices $E[\mathbf{x}(t)\mathbf{x}^T(t)]$, $E[\mathbf{y}(t)\mathbf{y}^T(t)]$, and $E[\mathbf{y}(t)\mathbf{x}^T(t)]$ or by their Fourier transforms. The optimum filter can be written as in (5.100) and is characterized by a matrix-valued impulse response $\mathbb{a}(t)$. Its Fourier transform $\mathbb{A}(\nu)$ is the frequency response to be determined.
(a) Write the integral equation similar to (11.163) which allows one to determine the impulse response $\mathbb{a}(t)$.
(b) By using a Fourier transform calculate the matrix frequency response $\mathbb{A}(\nu)$ of the filter which, for scalar signals, is given by (11.18).

(c) The error in the filtering procedure is defined by the covariance matrix of the innovation. Calculate this matrix and give an expression similar to (11.30), valid in the scalar case.

11.15 Filtering of a signal in a vector noise. The notations in this problem are the same as in the previous one. Suppose that the observation is the sum of a signal vector and a noise vector which are assumed to be uncorrelated. It can then be written as $\mathbf{x}(t) = \mathbf{s}(t) + \mathbf{n}(t)$ where the noise component is characterized by its spectral matrix $\mathbb{C}(\nu)$. The vector signal is given by

$$\mathbf{s}(t) = \int \mathbf{h}(\theta)s(t - \theta)d\theta$$

where $s(t)$ is a scalar signal with the power spectrum $\Gamma(\nu)$. As a consequence the signal vector is characterized by $\Gamma(\nu)$ and by the frequency response $\mathbf{G}(\nu)$ of the spatial filter defining $\mathbf{s}(t)$. The estimandum is equal to $s(t)$ and then scalar. As a result of this assumption the impulse response of the statistical filter is a row vector and its Fourier transform is noted $\mathbf{B}^H(\nu)$.

(a) Give a physical interpretation of the situation corresponding to these assumptions in the framework of spatial statistical filtering or communication.

(b) Using the results of the previous problem, write the equation which allows one to determine the vector $\mathbf{B}(\nu)$ characterizing the optimum statistical filter.

(c) Calculate the solution of this equation when the spectral matrix $\mathbb{C}(\nu)$ is PD.

(d) Using this last assumption again, give the expression of the estimation error obtained in the filtering procedure.

(e) Show that, as in the scalar case, the estimation error is zero if the frequency bands of the signal and of the noise do not overlap.

(f) There are other situations giving a zero error, which are due to the spatial structure of the signal and the noise. As an example show that the error is zero if the noise vector can be written as the signal vector with a frequency response $\mathbf{N}(\nu)$ satisfying $\mathbf{N}^H(\nu)\mathbf{G}(\nu) = 0$. Give an interpretation of this situation of singular estimation.

Chapter 12

Prediction for Stationary Signals

12.1 INTRODUCTION

Prediction theory is one of the most illuminating applications of the results given in the previous two chapters. There are various reasons for this. The first is that the calculations, which are rather complex in the general case, are greatly simplified in prediction, mainly due to the fact that the estimandum signal $y[k]$ used in the previous chapter is only a shifted version of the observation signal. The second reason is that prediction theory of stationary signals is a very powerful method for the analysis and classification of these signals. For example, the concept of deterministic signals, already introduced in the study of white noises (see Example 7.1), cannot be fully understood outside the framework of prediction theory. Finally, the calculation of filters introduced in prediction theory, called *predictors*, introduces a very powerful class of filters known as lattice filters. Although these filters can be studied separately, they have a natural explanation within the prediction framework.

Let us now introduce the basic *vocabulary* used in this chapter. Let $x[k]$ be a discrete-time (DT) stationary signal. The one-step predictor with finite past p is the filter which calculates the best mean-square estimation of $x[k]$ in terms of its finite past, or $x[k-1], x[k-2], \ldots, x[k-p]$. If the filter is linear, which is the most frequent case in the following, the predictor is a one-step linear predictor with a finite past p. If p tends to infinity, we talk of predictors with an infinite past. Finally, using an observation such as $x[k-1], x[k-2], \ldots, x[k-p]$, it can be interesting to estimate $x[k+s-1]$. In this case we encounter the problem of an s-step predictor with finite or infinite past. If $s = 1$ we obviously return to the one-step case.

It is clear that prediction is by definition a *causal* operation. It is useless to predict an event which has already happened, and we always want to predict the future in terms of past events. Consequently the results of Section 11.5 concerning causality constraint will be fully applied in prediction problems.

12.2 PREDICTION WITH INFINITE PAST

12.2.1 Calculation of the Linear Predictor

Let $x[k]$ be a real, zeromean, and stationary DT signal. We want to calculate the impulse response $a[k]$ of a causal filter such that

$$\hat{x}[k] = \sum_{l=1}^{\infty} a[l]\, x[k-l] \tag{12.1}$$

is the best linear mean square estimation of $x[k]$. It would be possible here to apply the results of the previous chapter. However, the prediction problem is so specific that a direct method allows us to arrive at the result more simply. We shall present the application of the general method at the end of this section.

In all prediction problems the *error filter* plays a role almost as important as the predictor filter. This filter calculates the innovation

$$\tilde{x}[k] = x[k] - \hat{x}[k] \tag{12.2}$$

which, when necessary, is called the innovation of one-step prediction with infinite past. It is clear that knowledge of the prediction or of the innovation is equivalent, as shown in Figure 12.1.

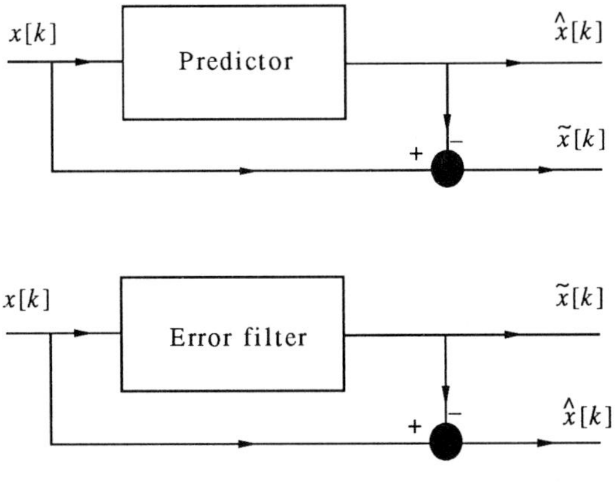

Figure 12.1 Equivalence between predictor and error filters.

The input-output relationship in the error filter is deduced from (12.1) and can be written as

$$\tilde{x}[k] = \sum_{l=0}^{\infty} b[l]\, x[k-l] \tag{12.3}$$

and, as a result of (12.2),

$$b[0] = 1 \quad ; \quad b[k] = -a[k] \tag{12.4}$$

The value $b[0] = 1$ is an essential property of the error filter and it is clear that either the $a[k]$s or the $b[k]$s can be calculated in order to solve the prediction problem.

Furthermore, the mean square error in the prediction problem is the variance of the output of the error filter. Using the concept of power introduced in Section 5.9, and especially (5.95), we can say that the mean square error is the mean power of the output of the error filter.

In order to determine the error filter we can simply apply the *orthogonality principle* which states that the innovation is orthogonal to the observation. As any element of the observation subspace is a linear combination of past values of $x[k]$, the orthogonality principle means that $\tilde{x}[k]$ is orthogonal to $x[k-l]$ for $l > 0$, or

$$\tilde{x}[k] \perp\!\!\!\perp x[k-l], \quad l > 0 \tag{12.5}$$

This relation implies that $\tilde{x}[k]$ is a *weak white noise*. In fact, as a result of (12.3), $\tilde{x}[k-p], p > 0$, is a linear combination of values of the signals at time instants $k-i$, $i \geq p$. But (12.5) indicates that these values of the signal are orthogonal to $\tilde{x}[k]$. Consequently, $\tilde{x}[k]$ is orthogonal to $\tilde{x}[k-p]$. As these RVs are zeromean-valued, we deduce that $\tilde{x}[k]$ and $\tilde{x}[k-p]$ are uncorrelated, which characterizes a weak white noise. If we now consider the values of the innovation $\tilde{x}[k]$ and $\tilde{x}[k+p], p > 0$, it is sufficient to reason in reverse, stating that $\tilde{x}[k+p]$ is orthogonal to $x[k+p-l], l > 0$, which implies that $\tilde{x}[k+p]$ is orthogonal to $\tilde{x}[k]$.

The result of this reasoning is that the error filter is a *whitening filter* of the signal. Introducing the transfer function $B\{z\}$ of this filter and the z-spectrum $\Gamma\{z\}$ of the signal $x[k]$, we can apply (6.68), which gives

$$\varepsilon_\infty^2 = \Gamma\{z\}\, B\{z\}\, B\{z^{-1}\} \tag{12.6}$$

In fact, as a result of (6.65) and (7.1), a white noise has a constant z-spectrum, and the value of the constant is the output variance. But we indicated above that this variance is the prediction error, noted ε_∞^2 to show that it corresponds to an infinite past.

Unfortunately, (12.6) does not characterize a specific filter since we saw, especially in Section 6.5, that many different filters can verify (12.6). We also noticed that the assumption of causality, valid for the error filter, as seen in (12.3), is not sufficient to suppress the ambiguity of the problem. This ambiguity will disappear when we show that the error filter is a *minimum phase filter*, which means, as seen in Section 6.5.4, that both it and its inverse are causal.

We must now show that $x[k]$ can be calculated in terms of its innovation by using a causal procedure. For this purpose we start from (12.2), which can be written as

$$x[k] = \tilde{x}[k] + \hat{x}[k] \tag{12.7}$$

As a result of (12.1), $\hat{x}[k]$ is only a function of past values of $x[k]$. If we repeat the same procedure indefinitely, or write

$$x[k-i] = \tilde{x}[k-i] + \hat{x}[k-i] \tag{12.8}$$

we can express $x[k]$ in terms of the past values of $\tilde{x}[k]$, which means that the inverse filter is also causal. In conclusion, the filter in (12.6) is a minimum phase filter. Writing this equation in the form

$$[\Gamma\{z\}]^{-1} = \frac{1}{\varepsilon_\infty^2} B\{z\} \, B\{z^{-1}\} \tag{12.9}$$

we deduce that it represents the strong factorization of the inverse of the z-spectrum of $x[k]$ almost exactly as in (11.118). The only difference is due to the additional constraint on the filter $B\{z\}$(1). In fact, the filter $F\{z\}$ which appears in the strong factorization of $[\Gamma\{z\}]^{-1}$ given by (11.118) is unique, except that $-F\{z\}$ gives the same result. The ambiguity of the sign is removed for $B\{z\}$, for we saw in (12.4) that $b[0] = 1$. As the similar coefficient $f[0]$ of $F\{z\}$ has no reason to satisfy the same condition, there is in the factorization a positive coefficient which is related to the prediction error in (12.9).

Let us now summarize the method which gives the *error filter*. We begin with a DT signal $x[k]$ with a z-spectrum equal to $\Gamma\{z\}$. If this z-spectrum is deduced from the power spectrum $\Gamma(\nu)$, this power spectrum must satisfy (6.86) and must not contain spectral lines. By using the techniques in Section 6.5, we calculate the strong factorization of $[\Gamma\{z\}]^{-1}$, as in (11.118). This gives the two filters $\pm F\{z\}$. The value of their impulse response for the null delay is $\pm f[0]$. The error filter is then defined by

$$B\{z\} = \frac{1}{f[0]} F\{z\} \tag{12.10}$$

and by construction it satisfies the condition $b[0] = 1$. Writing (11.118) again, we deduce from (12.10) that

$$[\Gamma\{z\}]^{-1} = f^2[0] \, B\{z\} \, B\{z^{-1}\} \tag{12.11}$$

When we compare this with (12.9), we find that the prediction error is

$$\varepsilon_\infty^2 = \frac{1}{f^2[0]} \tag{12.12}$$

The predictor, or prediction filter, is immediately obtained either by applying (12.2) or by using the schemes of Figure 12.1, which show that its transfer function $A\{z\}$ satisfies

$$A\{z\} = 1 - B\{z\} \tag{12.13}$$

In conclusion, the whole prediction problem is solved as soon as the strong factorization of $[\Gamma\{z\}]^{-1}$ is realized.

---

(1) Note that we will systematically use the expression "the filter $B\{z\}$" for "the filter whose transfer function is $B\{z\}$."

12.2.2 Comments on These Results

(a) Strong factorization of the *z*-spectrum. It is obviously equivalent to realize the strong factorization of $[\Gamma\{z\}]^{-1}$ or that of $\Gamma\{z\}$, written as

$$\Gamma\{z\} = H\{z\}H\{z^{-1}\} \tag{12.14}$$

If we compare this with (11.118), we deduce that $H(z) = [F\{z\}]^{-1}$. Furthermore, the transfer function can be expressed as

$$B\{z\} = \frac{\alpha}{H\{z\}} \tag{12.15}$$

where α is such that $b[0] = 1$. In order to calculate the constant α, let us introduce the impulse response $h[k]$ of the filter $H\{z\}$. As

$$B\{z\} = b[0] + \sum_{k=1}^{\infty} b[k]\, z^{-k} = 1 + \sum_{k=1}^{\infty} b[k]\, z^{-k} \tag{12.16}$$

we deduce that $b[0]$ is the limit when $|z|$ tends to infinity of $B\{z\}$. Applying this result to (12.15), we find that $\alpha = h[0]$, or

$$B\{z\} = \frac{h[0]}{H\{z\}} \tag{12.17}$$

Similarly, the comparison with (12.6) yields

$$\varepsilon_\infty^2 = h^2[0] \tag{12.18}$$

This means that in prediction problems we can use the strong factorization of either $\Gamma\{z\}$ or of $[\Gamma\{z\}]^{-1}$.

(b) Boundaries of the prediction error. The prediction error must satisfy the general inequality $0 \le \varepsilon_\infty^2 \le \sigma^2$, where σ^2 is the variance of the signal $x[k]$. The lower bound means that the prediction problem is *singular*, a situation which will be analyzed in detail later. The upper bound means that the prediction vanishes. This occurs if the observed signal $x[k]$ is *white*, because then its correlation function is given by (7.1) and its *z*-spectrum is equal to σ^2. Consequently $H\{z\}$ in (12.14) is reduced to σ, which also gives $h[0] = \sigma$ and $\varepsilon_\infty^2 = \sigma^2$. This is the only case giving a prediction error equal to σ^2. To understand this point, it is enough to remember that the purpose of the filter $B\{z\}$ is to whiten the signal with the constraint $b[0] = 1$. In the case where the signal $x[k]$ is already white, $B\{z\}$ becomes useless, and the constraint $b[0]$ indicates that $B\{z\}$ is the identity filter. This means that the signal is equal to its innovation.

This allows us to understand the function of the error filter. If the power spectrum of the signal $x[k]$ is not constant, it is possible to reach a greater whiteness by using the filter $B\{z\}$, an operation that generates a reduction in the error of prediction, sometimes called residual variance.

(c) Case of non-zeromean signals. If the signal $x[k]$ has the mean value m, we can simply apply (11.16) instead of (12.1), which gives

$$\hat{x}[k] = m + \sum_{l=1}^{\infty} a[l] \{x[k-l] - m\} \tag{12.19}$$

where the impulse response $a[k]$ is still given by the inverse z-transform of (12.13) and $B\{z\}$ by (12.10) or (12.17). It is obvious that the mean value of $\hat{x}[k]$ is m and consequently the innovation of prediction, or the output of the error filter, is zeromean-valued.

(d) Application of the general theory. It is interesting to find the optimum predictor by using the general procedure introduced in the previous chapter. First, we must define the observation and estimandum signals. To avoid confusion, these signals will be called $x'[k]$ and $y'[k]$. It is clear that $y'[k] = x[k]$, which is the signal to be predicted. On the other hand, the observation is $x'[k] = x[k-1]$, and the general filter (11.68) with causality constraint takes the form

$$\hat{y}'[k] = \sum_{l=0}^{\infty} a'[l]\, x'[k-l] \tag{12.20}$$

Returning to the values of x and y, this gives

$$\hat{x}[k] = \sum_{l=0}^{\infty} a'[l]\, x[k-l-1] \tag{12.21}$$

Taking

$$a[l] = a'[l-1] \tag{12.22}$$

we can write (12.21) as

$$\hat{x}[k] = \sum_{l=0}^{\infty} a[l+1]\;\; x[k-l-1] \tag{12.23}$$

and, replacing $l+1$ by l, we obtain

$$\hat{x}[k] = \sum_{l=1}^{\infty} a[l]\, x[k-l] \tag{12.24}$$

which is (12.1). After these preliminaries we calculate the filter $A'\{z\}$, and by using (12.22), deduce the filter $A\{z\}$.

The starting point is the strong factorization of the z-spectrum of $x'[k]$. But we note that $x[k]$ and $x'[k]$ have the same autocorrelation function and thus the same spectrum $\Gamma\{z\}$ that can be factored as in (12.14), where $H\{z\}$ is a minimum phase filter. The crosscorrelation between $y'[k]$ and $x'[k]$ can be written as

$$\gamma_{y'x'}[p] \triangleq E\{y'[k]\,x'[k-p]\} = E\{x[k]\,x[k-p-1]\} = \gamma[p+1] \quad (12.25)$$

where $\gamma[p]$ is the autocorrelation function of $x[k]$. As a result the z-transform of $\gamma_{x'y'}[p]$ is

$$\Gamma_{y'x'}\{z\} = z\Gamma\{z\} = zH\{z\}\,H\{z^{-1}\} \quad (12.26)$$

Inserting this expression in (11.127), where $F\{z\}$ is replaced by $1/H\{z\}$, we obtain

$$A'\{z\} = \frac{1}{H\{z\}}\,[zH\{z\}]_+ \quad (12.27)$$

Using (12.22), we deduce that the predictor is given by

$$A\{z\} = z^{-1}[H\{z\}]^{-1}\,[zH\{z\}]_+ \quad (12.28)$$

It now remains to calculate the bracket $[\;]_+$ to arrive at the final result. Let us write $H\{z\}$ in the form

$$H\{z\} = h[0] + G\{z\} \quad (12.29)$$

As the bracket selects the causal part, we deduce that

$$[zH\{z\}]_+ = zG\{z\} \quad (12.30)$$

because $h[0]\,z$ is not causal and the first term in the expansion of $G\{z\}$ is $h[1]\,z^{-1}$. As a result, we find from (12.28) that

$$A\{z\} = [H\{z\}]^{-1}G\{z\} \quad (12.31)$$

To arrive now at (12.17), it suffices to use (12.13), which gives

$$B\{z\} = 1 - A\{z\} = \frac{H\{z\} - G\{z\}}{H\{z\}} \quad (12.32)$$

Using (12.29), we deduce that the numerator of this function is simply $h[0]$, which shows the identity between (12.17) and (12.32).

The above shows that the calculation of the predictor given in Section 12.2.1 is much simpler and more elegant than the application of the general method presented in the previous chapter.

12.2.3 Mean Square Prediction

In the previous calculations the predictor was *linear*, but in general there is no reason for the optimal predictor to be linear. Consequently, the constraint of linearity of the filter can introduce an increase in the prediction error with respect to the optimum system. As indicated in the previous chapter, this system must calculate the *regression*, as in (11.3), which for prediction can be written as

$$\hat{x}[k] = E\{x[k] \mid x[k-1],\, x[k-2],\, \ldots\} \quad (12.33)$$

and has no reason *a priori* to be linear. It is almost impossible to calculate (12.33) in general, and the calculation is only possible in some specific cases.

The simplest case obviously appears when the expected value (12.33) is *linear*, in which case we can apply the previous results. This situation occurs especially when the signal $x[k]$ to be predicted is *Gaussian*, as described in Section 7.4. This is one of the reasons for the importance of such signals. Linear regression also appears for some other signals, as for example the spherically invariant signals introduced in Section 7.6.

There is also another very important class of signals which are *a priori* neither Gaussian nor spherically invariant, for which the optimum prediction is linear: the class of *autoregressive signals*. This is a direct consequence of (9.41), which indicates that if in the regression (9.20) or (9.22) the driving noise $u[k]$ is a strict white noise, that is, a sequence of IID random variables, then the regression (12.33) is given by (9.42) which is linear. This last equation can be written as

$$\hat{x}[k] = \sum_{l=1}^{p} a[l]\ x[k-l] \tag{12.34}$$

which shows a new simplification: the optimum predictor is not only linear but also has a *finite impulse response* (FIR), meaning that the infinite sum in (12.1) is a sum of only p terms.

Let us now give an example of a non-linear optimum predictor, a situation already discussed in Example 9.4 where a signal $x[k]$ normal and AR(1) was introduced. It was shown that the signal $y[k] = x^3[k]$ was zeromean-valued and that its regression was

$$\hat{y}[k] = a^3 y[k-1] + 3a\sigma^2 y^{1/3}[k-1] \tag{12.35}$$

which is the best mean square predictor of $y[k]$. This predictor is obviously non-linear. However, it is also possible to calculate the best linear predictor, since the correlation function of $y[k]$ was also calculated in Example 9.4. By calculating its z-transform, we obtain the z-spectrum from which the strong factorization can be deduced, and this yields the optimum linear predictor by using (12.17) and (12.13).

12.2.4 Extension to Complex Signals

Using the ideas presented in Section 10.7.1, it is easy to extend the theory of linear prediction in Section 12.2.1 to complex signals. This theory is especially adapted to the case where the signal $x[k]$ is *complex circular*, which means that $E\{x[k]\ x[k']\} = 0$ for all k and k'. However, the application of the ideas of Section 10.7.2 requires that a theory of linear prediction of vector signals be developed. This is possible but outside the scope of this book. The main difficulty is in the extension of strong factorization to spectral matrices, introducing some complications not present in the scalar case.

To conclude this section let us analyze a situation using a signal introduced in Example 7.1. This signal is defined by (7.3), and (7.5) shows that it is complex circular. Furthermore, we deduce from (7.4) that this signal is weakly white, which means that it is unpredictable linearly. This is characterized by $\hat{x}[k] = 0$, or by the fact that the signal is equal to its linear innovation. However, the non-linear innovation is null, meaning that the signal can be determined without error from its past values. In other words, the prediction reaches its upper or lower bound when passing from linear to non-linear prediction.

12.3 PREDICTION ERROR

12.3.1 Calculation of the Error

The infinite past one-step prediction error is given by (12.18) where $H\{z\}$ is the filter in the strong factorization of the z-spectrum $\Gamma\{z\}$ of $x[k]$, given by (12.14). Consequently, the calculation of the prediction error is especially simple when the z-spectrum is a rational function of z. In fact by using the method indicated in Section 6.11, it is possible to write the filter $H\{z\}$ in the form

$$H\{z\} = \pm c \prod_{i=1}^{q} \left[\frac{z - z_i}{z}\right] \prod_{j=1}^{p} \left[\frac{z}{z - p_j}\right] \tag{12.36}$$

where z_i and p_j are the zeros and poles respectively. As $H\{z\}$ is causal, it can also be written as

$$H\{z\} = \sum_{k=0}^{\infty} h[k]\, z^{-k}$$

which shows that $h[0]$ is obtained from $H\{z\}$ when $|z| \to \infty$. Applying this result in (12.36), we obtain $h^2[0] = \varepsilon_\infty^2 = c^2$. For example the power spectrum (6.180) introduces the function $H\{z\}$ given by (6.184) and the prediction error is equal to 4 while the variance, obtained by integration of (6.180), is equal to 5. Similarly, the spectrum (6.180) corresponds to a signal with a variance of 36 and a prediction error of 4.

Let us now consider the general case. For this we note that the filter $H\{z\}$ has neither zero nor pole outside the unit circle and consequently the filter

$$P\{z\} = \ln[H\{z\}] \tag{12.37}$$

has all its poles inside the unit circle. It is then causal, which implies that

$$P\{z\} = p[0] + \sum_{n=1}^{\infty} p[n]\, z^{-n} \tag{12.38}$$

In the frequency domain this expression becomes

$$P(\nu) = p[0] + \sum_{n=1}^{\infty} p[n] \exp(-2\pi j n \nu) \tag{12.38'}$$

which is a Fourier series. We then have

$$p[0] = \int_{-1/2}^{+1/2} P(\nu)\, d\nu = \int_{-1/2}^{+1/2} \ln[H(\nu)]\, d\nu \tag{12.39}$$

Using the expression of the logarithm of a complex number, we deduce that

$$\ln[H(\nu)] = \ln[|H(\nu)|] + j\phi(\nu) \tag{12.40}$$

where $\phi(\nu)$ is the phase of the filter $H(\nu)$. As this filter is real, it satisfies $H(\nu) = H^*(-\nu)$, which implies that $\phi(\nu) = \phi(-\nu)$ and (12.39) becomes

$$p[0] = \int_{-1/2}^{+1/2} \ln[|H(\nu)|]\, d\nu \tag{12.41}$$

Noting that, because of (12.14), the power spectrum of $x[k]$ satisfies

$$\Gamma(\nu) = |H(\nu)|^2 \tag{12.42}$$

we deduce that

$$p[0] = \frac{1}{2} \int_{-1/2}^{+1/2} \ln[\Gamma(\nu)]\, d\nu \tag{12.43}$$

Finally, by taking $|z| \to \infty$, it results from (12.37), (12.38) and (12.18) that

$$p[0] = \ln\{h[0]\} = \frac{1}{2} \ln\{\varepsilon_\infty^2\} \tag{12.44}$$

and comparing this with (12.43), we obtain the expression of the error, which is

$$\varepsilon_\infty^2 = \exp\left\{ \int_{-1/2}^{+1/2} \ln[\Gamma(\nu)]\, d\nu \right\} \tag{12.45}$$

This expression of the error is sometimes called Kolmogoroff's formula.

We note that this error is null if the term inside the bracket tends to $-\infty$. This is in relation with the condition (6.86) ensuring strong factorization. In fact, the integral in (12.45) cannot tend to $+\infty$ if $x[k]$ is a second order signal. This results from the inequality $\ln x \leq x - 1$, and if $x[k]$ is second order, the integral of $\Gamma(\nu)$ is the variance, which is finite. However, the integral of (12.45) can tend to $-\infty$, in which case ε_∞^2 is null and (6.86) does not hold. In conclusion, the signals with a null prediction error are those for which strong factorization does not hold. This leads to a more careful analysis of the prediction error.

12.3.2 Properties of the Error

(a) Upper bound of the prediction error. Let us introduce the normalized power spectrum $\tilde{\Gamma}(\nu)$ defined by

$$\Gamma(\nu) = \sigma^2 \tilde{\Gamma}(\nu) \tag{12.46}$$

Because of (6.41), the integral of $\tilde{\Gamma}(\nu)$ is equal to one, and inserting (12.46) into (12.45), we obtain

$$\varepsilon_\infty^2 = \sigma^2 \exp\left\{ \int_{-1/2}^{+1/2} \ln[\tilde{\Gamma}(\nu)]\, d\nu \right\} \tag{12.47}$$

Using the classical inequality $\ln x \leq x - 1$, already used in (4.183), we deduce that the latter factor of (12.47) is smaller than one, or $\varepsilon_\infty^2 \leq \sigma^2$, which is the upper bound of the prediction error already indicated in 12.2.2 (**b**). This upper bound is reached if $\tilde{\Gamma}(\nu) = 1$, which means that the signal $x[k]$ is weakly white.

(b) Lower bound of the prediction error. This lower bound is in principle $\varepsilon_\infty^2 = 0$. The simplest example of a signal giving a null prediction error is the constant signal defined by (5.46). In fact, if $x[k] = X(s)$, where $X(s)$ is a zeromean-valued RV, we have $\hat{x}[k] = x[k-1]$ and the innovation is null. But the power spectrum of this signal is limited to a spectral line at the frequency 0, and does not satisfy the condition for strong factorization. In other words, (12.45) is established provided that strong factorization is possible, or that (6.86) holds, which means that $\varepsilon_\infty^2 > 0$, according to (12.45). However, if we approximate a non factorable spectrum by a sequence of factorable spectra, (12.45) is still valid at the limit, and the conclusion is that a signal with a non-factorable spectrum has a null prediction error. In this case the prediction problem is *singular* and the signal is said to be *predictable*. Expressions such as "singular" and "deterministic" are also occasionally used. The latter term is not really appropriate because of the possible confusion with deterministic signals, meaning that there is no randomness, as in (1.1). A signal which is not predictable, that is, when $\varepsilon_\infty^2 \neq 0$, is said to be *regular*.

Let us now explain the meaning of a predictable signal physically. It is a *random signal* which can be estimated linearly and without error in terms of its entire past. If the past is noted P_k, and if P_k is observed, $x[k]$ is known without error. From P_k and $x[k]$ we can reconstruct $x[k+1]$, and so on. Consequently the whole future of the signal is known from P_k. This property seems difficult to verify in practice because the verification needs an experiment of infinite duration. It is more an asymptotic property, and we will present later some of its verifications.

Predictable signals can be classified in *two categories*: those which can be determined with a finite past, and those for which the property needs an infinite past.

(c) Finite past predictable signals. Consider a stationary random signal $x[k]$ such that

$$x[k] = a_1 x[k-1] + a_2 x[k-2] + \ldots + a_p x[k-p] \tag{12.48}$$

This equation characterizes a signal that is a linear combination of past values at each time instant k. Multiplying by $x[k-n]$ and taking the expectation, we deduce that the correlation function satisfies the same equation, or

$$\gamma[n] = \sum_{q=1}^{p} a_q \gamma[n-q] \ , \qquad \forall n \tag{12.49}$$

We will deduce that this relation implies a very specific structure of the power spectrum. In fact, using (6.40'), we can write (12.49) in the form

$$\int_{-1/2}^{+1/2} \Gamma(\nu)\, e^{2\pi j \nu n} \left[1 - \sum_{q=1}^{p} a_q e^{-2\pi j \nu q} \right] d\nu = 0 \tag{12.50}$$

which must be valid whatever n. It follows that

$$\Gamma(\nu) F(\nu) = 0 \tag{12.51}$$

where $F(\nu)$ is the term inside the brackets of (12.50). This means that $\Gamma(\nu)$ can be non-null only at the frequencies ν_i such that $F(\nu_i) = 0$. Introducing the polynomial

$$P\{z\} = z^p - \sum_{q=1}^{p} a_q z^{p-q} \tag{12.52}$$

we see that

$$F(\nu) = e^{-2\pi j \nu p} P\{e^{2\pi j \nu}\} \tag{12.53}$$

and the frequencies ν_i, roots of $F(\nu)$, are then the roots of $P\{z\}$ *located on the unit circle*. The conclusion of this discussion is that a signal which is predictable with a finite past has a power spectrum limited to *spectral lines*. The amplitudes of these spectral lines are arbitrary, and their number is limited to p, since $P\{z\}$ is a polynomial of degree p. But $P\{z\}$ can only have p' ($p' < p$) roots on the unit circle, the others being located inside the circle. If this is the case, it is easy to show by using the same reasoning in reverse that $x[k]$ can be written only in terms of p' past values. As a consequence $p' = p$ and all the roots of $P\{z\}$ are located on the unit circle.

This is closely connected to the AR(p) signals studied in Chapter 9, as seen in Section 9.2.3 (**d**). In particular the signal (9.49) has exactly the same structure as (12.48) and introduces a second order recursion. It has two symmetric spectral lines, a characteristic of a sinusoid signal.

(d) Infinite past predictable signals. As a consequence of the previous discussion, these signals cannot have a spectrum containing only a finite number of spectral lines. In order to study them, we shall return to the expression of the error given by (12.45).

In this expression we see that all signals such that $\Gamma(\nu) = 0$ for frequencies belonging to an arbitrary *frequency band* $\Delta\nu$ lead to a null prediction error, and are then predictable. In fact, for frequencies which belong to $\Delta\nu$, $\ln[\Gamma(\nu)]$ is equal to $-\infty$. This is especially so for DT random signals which are *band-limited.* This property seems surprising, because most signals used in communication systems are assumed to be band-limited, and are usually not considered to be singular. It needs a more specific analysis, which will be given in a special section after the introduction of means necessary for its development. For the time being we can consider that strictly band-limited signals are only mathematical models and in practice are difficult to realize. It is well known in circuit theory that it is impossible to realize a band-limited filter with a rational transfer function. This means that a signal whose power spectrum is given by (6.80) with a rational filter $H(z)$ cannot be band-limited. This can be easily understood, as $H(z)$ has a finite number of roots and thus cannot be null in a frequency band.

But a signal can be predictable even if its power spectrum is *null at only one frequency* ν_0. As an example of this, consider the DT signal whose power spectrum is

$$\Gamma(\nu) = \exp[-1/|\nu|], \quad -\frac{1}{2} < \nu < \frac{1}{2} \tag{12.54}$$

which is represented in Figure 12.2. It is clear that $\Gamma(\nu) = 0$ only at the frequency $\nu = 0$ even if it appears practically null for $|\nu| < 0.1$. Furthermore, as $\Gamma(\nu) < 1$, the signal is second order. Inserting (12.54) into (12.45), we can see that $\varepsilon_\infty^2 = 0$. In fact, the integral of $1/|\nu|$ is infinite because of the frequency $\nu = 0$. We thus have a power spectrum which is null at only one frequency but leads to a null prediction error, or introduces a predictable signal. It is clear that there are many other similar situations.

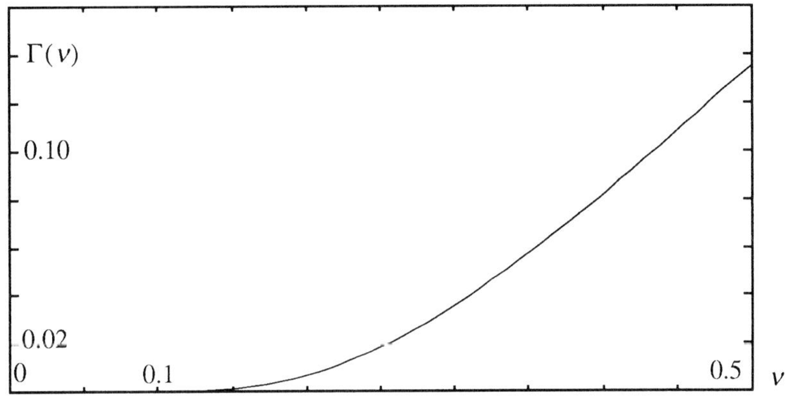

Figure 12.2 Power spectrum of (12.54).

Let us conclude this section by noting that all the power spectra leading to a null prediction error have in common the property of being null at at least one frequency. We shall show that indeed this is a *necessary condition* for a power spectrum. Let $\Gamma(\nu)$ be an arbitrary power spectrum of a second order signal. If $\Gamma(\nu)$ has no zero in the finite interval $[-1/2, +1/2]$, there is a finite lower bound of $\Gamma(\nu)$ in this interval, or $\Gamma(\nu) \geq b > 0$. This implies that $\ln[\Gamma(\nu)] \geq \ln b$ and therefore the integral is also greater than $\ln b$. The prediction error is thus greater than b and cannot be null.

12.3.3 Comparison with the Interpolation Error

If we compare (11.90) and (12.1), we note that the interpolation of a signal is similar to a prediction in which we take into account the whole past and the whole future. Note immediately that the interpolation error is certainly smaller than the prediction error, because of the contribution to the estimation of the future values of the signal. Let us at once verify this property. The interpolation error, noted here η^2, is given by (11.107). Using (12.9) we deduce that

$$[\Gamma(\nu)]^{-1} = \frac{1}{\varepsilon_\infty^2} \, |B(\nu)|^2 \tag{12.55}$$

Inserting this expression into (11.107), we find that

$$\frac{\eta^2}{\varepsilon_\infty^2} = \left[\int_{-1/2}^{+1/2} |B(\nu)|^2 d\nu \right]^{-1} \tag{12.56}$$

If we introduce the impulse response of the filter $B(\nu)$, we can write

$$B(\nu) = b[0] + \sum_{k=1}^{\infty} b[k] \exp(-2\pi jk\nu) \tag{12.57}$$

As a result we have

$$I \triangleq \int_{-1/2}^{+1/2} |B(\nu)|^2 d\nu = b^2[0] + \sum_{k=1}^{\infty} b^2[k] \tag{12.58}$$

and as $b[0] = 1$, as seen in (12.4), we deduce that $I \geq 1$, and $[\eta^2/\varepsilon_\infty^2] \leq 1$. Consequently any predictable signal is also interpolable, but there is no reason for the converse to be true.

Let us give an example of this situation. Suppose that $\Gamma(\nu) = |\nu|$. This power spectrum is represented in Figure 12.3. The corresponding variance is equal to 1/4. The prediction error can be easily calculated from (12.45) and is equal to $1/2e$. However, the interpolation error is obviously null, and we obtain a signal which is interpolable but non-predictable.

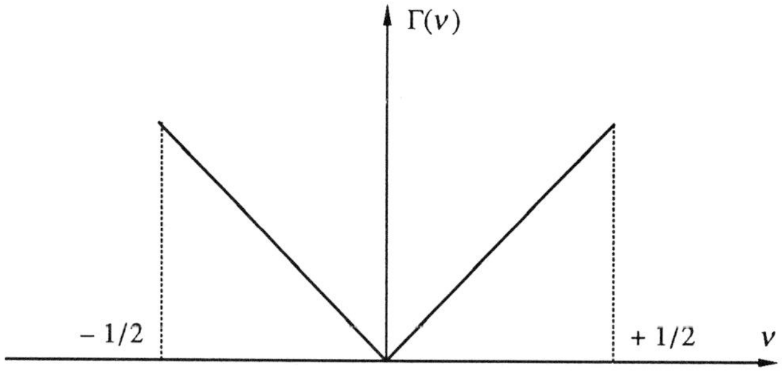

Figure 12.3 Power spectrum $\Gamma(\nu) = |\nu|$.

We may ask whether a signal can be interpolable with a *finite* past and future. Using the same reasoning as for prediction, it can be shown that this situation also needs a spectrum containing only spectral lines. But here the signal is predictable, and there is no difference between prediction and interpolation. For example, the signal (9.49) can also be written as

$$x[k] = \frac{1}{2\cos(2\pi f)} \{x[k-1] + x[k+1]\} \tag{12.59}$$

which indicates an interpolation of order one. The difference between interpolable and predictable signals can therefore only appear with an infinite past and future.

12.4 PREDICTION WITH FINITE PAST

The prediction filter in (12.1) and calculated above has in general an infinite impulse response. However, if the signal $x[k]$ is an AR(p) signal, the prediction filter takes the form (12.34), which introduces a finite impulse response, or a transversal filter. But if we use (12.34) instead of (12.1) we realize a prediction with a finite past which generally gives an error greater than ε_∞^2. The other consequence is that the innovation in this form of prediction has no reason to be white. However, for practical or technological reasons it is sometimes too complex to realize (12.1), and we will therefore now examine finite past prediction.

Using the same notation as in (9.22), we can write

$$\hat{x}_p[k] = \mathbf{a}^T \mathbf{X}[k] \tag{12.60}$$

where the components a_i, $1 \le i \le p$, of the vector $\mathbf{a}$ are the values $a[i]$ of the impulse response of the predictor. The problem now is to determine the vector $\mathbf{a}$ in such a way that the prediction error is minimum. This problem was solved in Section 10.4, and as we are working with zeromean-valued signals, the solution is given by (10.65). This equation can be written

$$\mathbb{\Gamma}\,\mathbf{a} = \mathbf{c} \tag{12.61}$$

where $\mathbb{\Gamma}$ is the covariance matrix of the observation vector, or

$$\mathbb{\Gamma} = E[\mathbf{X}\mathbf{X}^T] \tag{12.62}$$

and $\mathbf{c}$ is the vector

$$\mathbf{c} = E(x[k]\,\mathbf{X}[k]) \tag{12.63}$$

These equations are equivalent to (9.23) or (9.24) and (12.61) is the first *normal equation* (9.27). In other words the finite past one-step predictor is obtained by solving the normal equation.

The finite past innovation is defined by

$$\tilde{x}_p[k] = x[k] - \hat{x}_p[k] \tag{12.64}$$

and its variance is the prediction error given by (10.74), which can be expressed as

$$\varepsilon_p^2 = \gamma[0] - \mathbf{c}^T \Gamma^{-1} \mathbf{c} \tag{12.65}$$

Using (12.61), we obtain the most interesting expression of this error by

$$\varepsilon_p^2 = \gamma[0] - \mathbf{a}^T \mathbf{c} \tag{12.66}$$

which is also the second normal equation (9.29).

We thus arrive at the conclusion that solving a one-step prediction problem with finite past is equivalent to solving the *normal equations* encountered in the framework of AR(p) signals. This does not mean that the signal $x[k]$ to be predicted is an AR(p) signal. If this arises we have $\varepsilon_p^2 = \varepsilon_\infty^2$, but in general this is not so and we have $\varepsilon_p^2 \geq \varepsilon_\infty^2$.

Let us now summarize the procedure used to solve the prediction problem. We observe a signal $x[k]$ whose correlation function $\gamma[k]$ is known. Using only the values of this correlation function for $0 \leq k \leq p$, we construct the vector $\mathbf{c}$ defined by (9.30), or

$$\mathbf{c}^T = \{\, \gamma[1],\ \gamma[2],\ \ldots,\ \gamma[p] \,\} \tag{12.67}$$

and the matrix

$$\Gamma = T\{\, \gamma[0],\ \gamma[1],\ \ldots,\ \gamma[p-1] \,\} \tag{12.68}$$

defined by (9.33) or (9.33'). Using these elements, we solve the linear equation (12.61) which gives the filter, and obtain the corresponding error with (12.66).

Let us comment on some points in this problem.

(a) Minimum-phase property. Consider the error filter calculating $\tilde{x}_p[k]$ defined by (12.64) where $\hat{x}_p$ is given by (12.60). The transfer function of this filter is

$$B_p\{z\} = 1 - a_1 z^{-1} - a_2 z^{-2} - \cdots - a_p z^{-p} \tag{12.69}$$

where the a_is are the components of the vector $\mathbf{a}$ solution of (12.61). As with any error filter, this filter satisfies the property $b[0] = 1$, already seen in (12.4) in connection with the infinite past case. We will now show that all the zeros of $B_p\{z\}$ are necessarily located inside the unit circle, which means that $B_p\{z\}$ is a minimum-phase filter. For this, note that the error ε_p^2 is the variance of the output of $B_p\{z\}$ when the input is $x[k]$. Using (6.41) and (6.55), this means that

$$\varepsilon_p^2 = \int_{-1/2}^{+1/2} |B_p(\nu)|^2 \, \Gamma(\nu) \, d\nu \tag{12.70}$$

where $\Gamma(\nu)$ is the power spectrum of $x[k]$ and $B_p(\nu)$ the frequency response of $B_p\{z\}$ defined by $B_p(\nu) = B_p\{\exp(2\pi j \nu)\}$. Introducing the zeros of $B_p\{z\}$, and using the property $b[0] = 1$, we can write

$$B_p\{z\} = (1 - z_1 z^{-1}) \prod_{k=2}^{p} (1 - z_k z^{-1}) = (1 - z_1 z^{-1})\, B_p'\{z\} \tag{12.71}$$

where an arbitrary zero, say z_1, has been isolated. We will now show that it is impossible to have $|z_1| > 1$. For this we work by contradiction and suppose that this inequality is satisfied. Using the angular frequencies $\omega = 2\pi\nu$, we have

$$B_p(\omega) = (1 - z_1 e^{-j\omega}) B'_p(\omega) \tag{12.72}$$

Considering now the term inside the bracket, we can write the following relations

$$|1 - z_1 e^{-j\omega}|^2 = |z_1|^2 \left|\frac{1}{z_1} - e^{-j\omega}\right|^2 = |z_1|^2 \left|\frac{1}{z_1^*} - e^{j\omega}\right|^2 = |z_1|^2 \left|1 - \frac{1}{z_1^*} e^{-j\omega}\right|^2$$

Thus, by comparing the first and the last terms and as $|z_1| > 1$, we deduce that

$$|1 - z_1 e^{-j\omega}|^2 > \left|1 - \frac{1}{z_1^*} e^{-j\omega}\right|^2 \tag{12.73}$$

which means that if we replace z_1 in (12.72) by $1/z_1^*$, which is located inside the unit circle, we obtain a filter that still satisfies $b[0] = 1$ and which gives, by (12.70), a smaller error. But as ε_p^2 is the minimum error, it appears that the assumption $|z_1| > 1$ is impossible. As z_1 is an arbitrary zero, we conclude that all the zeros of $B\{z\}$ must be located inside the unit circle, and consequently $B\{z\}$ is a minimum-phase filter.

Another proof of this result will be given later, after the introduction of the reflection coefficients.

Let us now discuss the physical meaning of this result. It is in relation to the *autoregressive modeling* of an arbitrary signal. Let $x[k]$ be a signal with a unit power, or such that $E\{x^2[k]\} = 1$. The covariance matrix of the vector $\mathbf{X}[k]$ in (12.60) is given by (12.68) which has the structure (9.33') with diagonal elements equal to one. Similarly the vector $\mathbf{c}$ in (12.61) is given by (12.67). As a result the vector $\mathbf{a}$ defining the finite-past predictor is only a function of the vector $\mathbf{c}$, because if $\mathbf{c}$ is known, $\mathbb{\Gamma}$ is also known, with (12.68). As the normal equation (12.61) has a unique solution when $\mathbb{\Gamma}$ is positive-definite, we can write this solution as

$$\mathbf{a} = \mathbf{f}(\mathbf{c}) \tag{12.74}$$

The main result of this section is that if $\mathbf{c}$ is such that $\mathbb{\Gamma}$ is positive definite, $\mathbf{a}$ is such that $B\{z\}$ is a minimum phase filter. It is then possible to introduce an AR(p) signal $x'[k]$ defined by the vector $\mathbf{a}$ and driven by a white noise such that the power of $x'[k]$ is one. The first p values of the correlation function of $x'[k]$ are the same as those of $x[k]$, and nothing is said concerning the other values. We can then approximate any signal $x[k]$ by an AR(p) signal, the approximation meaning that the correlation functions $\gamma[n]$ and $\gamma'[n]$ of $x[k]$ and $x'[k]$ respectively are the same for $0 \le n \le p$. It is for this reason that the vector $\mathbf{a}$ solution of the normal equation will be called the *regression vector*, while $\mathbf{c}$ is called the *correlation vector*.

Conversely, starting from a given regression vector $\mathbf{a}$, or such that $B\{z\}$ is minimum phase, we construct an AR(p) signal with a definite correlation function, which gives a correlation vector **c** introducing a positive-definite matrix Γ. So (12.74) can be inverted and

$$\mathbf{c} = \mathbf{f}^{-1}(\mathbf{a}) \tag{12.75}$$

This means that the region of $\mathbb{R}^p$ such that **a** gives a minimum phase filter, or a stable AR(p) signal, is transformed by $\mathbf{f}^{-1}(.)$ in the region of $\mathbb{R}^p$ such that **c** introduces a positive-definite matrix. This property will be extensively studied later.

(b) Invariance by scaling effect. If we multiply the signal $x[k]$ by λ, its correlation becomes $\lambda^2\gamma[k]$. Consequently Γ and **c** are multiplied by λ^2 and the vector **a** is unchanged. This means that the prediction *filter* is invariant by a scaling operation on the signal, the reason why it is often assumed that $\gamma[0] = 1$. On the other hand, the prediction error given by (12.66) is multiplied by λ^2.

(c) Case of complex signals. In spectral analysis, complex signals are often used because it is simpler to represent a sinusoid signal by $\exp(j\omega k)$ than by $\cos(\omega k)$. If the complex signal is circular, prediction problems can be solved by using the normal equation (9.47).

(d) Singular case. This situation appears if the matrix Γ has at least one null eigenvalue. In this case the solution of (12.61) requires specific treatment, which will be analyzed later.

(e) Evolution of the prediction error. The prediction error given by (12.66) is a non-increasing function of p. This is an obvious property since if p increases, this means that the prediction is realized with a longer past, and this cannot increase the error. The following figures give the most common situations in the evaluation of ε_p^2 in terms of p. For $p = 0$ we make no prediction, and then $\varepsilon_0^2 = \sigma^2$, the variance of the signal, which is also equal to $\gamma[0]$. When p tends to infinity we obtain the error ε_∞^2 studied in Section 12.3 and given by (12.45). To simplify the presentation, the curves are presented as if p were a continuous parameter, while it is clear that p is an integer. In other words, we assume that all the points ε_p^2 are connected by one curve. In Figure 12.4 the prediction errors of *regular* signals are presented. This means that $\varepsilon_\infty^2 > 0$. Curve 1 has no specific property, but shows that for some values of p, ε_p^2 can remain constant. Curve 2 is such that the error reaches its limit for $p = r$. This means that the signal $x[k]$ is an AR(r) signal, as a result of the fact that for an AR(r) signal like (9.22) we can obtain a *white* innovation by calculating $x[k] - \mathbf{a}^T\mathbf{X}[k]$, where **a** is a vector of dimension r. As the characteristic property of the infinite past innovation is to be white, we reach this property with a finite past for AR signals.

The case of *predictable* signals is presented in Figure 12.5. Curve 1 shows a signal for which the prediction error is never null, but tends to zero when p tends to infinity. This is especially so for band-limited signals, as shown later. Curve 2 corresponds to a signal predictable with a finite past, and we have seen that this means that the power spectrum contains only r spectral lines of arbitrary amplitudes.

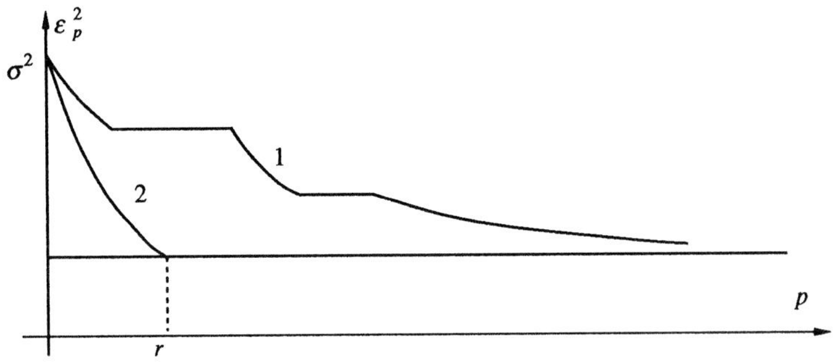

Figure 12.4 Evolution of the prediction error of regular signals.

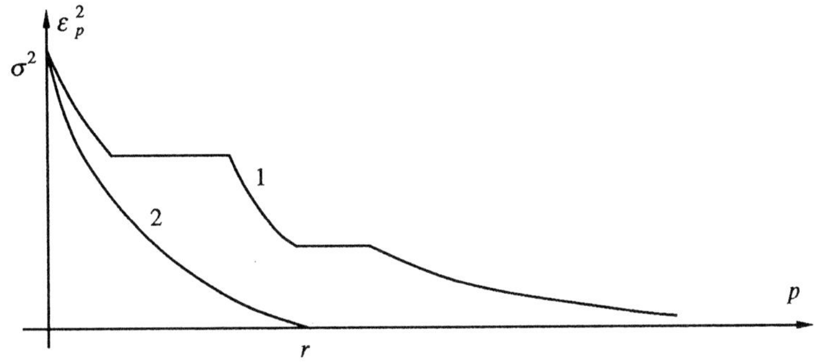

Figure 12.5 Evolution of the error for predictable signals.

12.5 SOLUTION OF THE NORMAL EQUATIONS

In this section we will calculate explicitly the solution to the normal equations (12.61) and (12.66). Starting from a correlation function $\gamma[k]$, $0 \le k \le p$, we want to determine the vector $\mathbf{a}$ and the error ε_p^2. It may seem *a priori* strange to devote a section to the solution of a linear system such as (12.61), as there are many well known standard techniques for this problem.

However, the main point of this section is that the matrix $\mathbb{\Gamma}$ is not arbitrary, because it has a Toeplitz structure. This fact has many fundamental consequences. First, it simplifies the solution of (12.61), introducing the concept of fast algorithm which was already used, for example, in connection with discrete Fourier transforms. Second, the fast, or Levinson, algorithm has a very interesting interpretation in terms of filtering, as it introduces the concept of lattice filters for prediction.

Before entering into calculations, note that although there are many ways of introducing the Levinson algorithm, the presentation below is the most appropriate to explain the origin of the meaning of a fast algorithm.

12.5.1 The Levinson Algorithm

The purpose of this section is to solve the linear system (12.61) in which **c** is a given vector and $\mathbb{\Gamma}$ a Toeplitz matrix. The starting point is a set of $p + 1$ values of a correlation function $\gamma[k]$, also noted γ_k, which defines the vector **c** by (12.67) and the matrix $\mathbb{\Gamma}$ by (9.33'). The procedure used in this solution is called an order-recursive method. Its principle consists of calculating a sequence of p vectors $\mathbf{a}_1, \mathbf{a}_2, \ldots, \mathbf{a}_p$ in such a way that the last vector of this sequence is the regression vector **a**, solution of the normal equation (12.61). At the step n of the recursion, the vector $\mathbf{a}_n$ is the solution of the normal equation

$$\mathbb{\Gamma}_n \mathbf{a}_n = \mathbf{c}_n \tag{12.76}$$

where $\mathbb{\Gamma}_n$ is the Toeplitz matrix

$$\mathbb{\Gamma}_n = T[\gamma_0, \gamma_1, \ldots, \gamma_{n-1}] \tag{12.77}$$

In this equation the notation is the same as in (9.33), and its meaning is explained in (9.33'), where p is replaced by n. Note that if $\mathbb{\Gamma}$ defined by (9.33') is positive-definite, $\mathbb{\Gamma}_n$ is also positive-definite for $n < p$. To demonstrate this, we calculate the quadratic form (3.69) with a vector **u** whose last $p - n$ components are null.

To any vector **v** of $\mathbb{R}^n$ let us associate the vector $\mathbf{v}^{(-)}$ such that if the components of **v** are v_i, $1 \le i \le n$, then $\mathbf{v}^{(-)}$ is defined by

$$\mathbf{v}^{(-)} = [v_n, v_{n-1}, \ldots, v_2, v_1]^T \tag{12.78}$$

The components of $\mathbf{v}^{(-)}$ are thus those of **v** read in reverse order. It is clear that we can write $\mathbf{v}^{(-)}$ in the form

$$\mathbf{v}^{(-)} = \mathbb{J}\,\mathbf{v} \tag{12.79}$$

where $\mathbb{J}$ is the $n \times n$ matrix defined by

$$\mathbb{J} = \begin{bmatrix} 0 & & & & 1 \\ & & & 1 & \\ & & \cdot & & \\ & \cdot & & & \\ & 1 & & & \\ 1 & & & & 0 \end{bmatrix} \tag{12.80}$$

and the matrix elements of $\mathbb{J}$ are

$$J_{kl} = \delta[k + l - n - 1] \tag{12.81}$$

where $\delta[q]$ is the Kronecker-Delta symbol. The matrix $\mathbb{J}$ satisfies

$$\mathbb{J} = \mathbb{J}^T \quad ; \quad \mathbb{J}^2 = \mathbb{I} \tag{12.82}$$

where $\mathbb{I}$ is the identity matrix.

But the main property of $\mathbb{J}$ used in what follows is that if $\mathbb{T}$ is a symmetric Toeplitz matrix, then

$$\mathbb{J}\,\mathbb{T} = \mathbb{T}\,\mathbb{J} \tag{12.83}$$

which means that $\mathbb{J}$ and $\mathbb{T}$ are commuting matrices. In order to prove (12.83), let us calculate the matrix elements of $\mathbb{J}\,\mathbb{T}$ by applying the classical rule concerning the product of matrices and given by (3.67). Using (12.81), this gives

$$[\mathbb{J}\,\mathbb{T}]_{kl} = \sum_p \delta[k + p - n - 1]\, T_{pl} = T_{n+1-k,\,l}$$

But as $\mathbb{T}$ is symmetric and Toeplitz, we have $T_{ij} = f(|i-j|)$, or

$$[\mathbb{J}\,\mathbb{T}]_{kl} = f(|n + 1 - k - l|)$$

Applying the same procedure to $\mathbb{T}\,\mathbb{J}$, we obtain

$$[\mathbb{T}\mathbb{J}\,]_{kl} = f(|k + l - n - 1|)$$

which proves the relation (12.83). The consequence of this relation is that we have

$$\Gamma_n\, \mathbf{a}_n^{(-)} = \mathbf{c}_n^{(-)} \tag{12.84}$$

In fact, starting from (12.76) and using (12.83), we obtain

$$\mathbb{J}\,\Gamma_n\, \mathbf{a}_n = \Gamma_n\, \mathbb{J}\, \mathbf{a}_n = \Gamma_n\, \mathbf{a}_n^{(-)}$$

and as $\Gamma_n\, \mathbf{a}_n = \mathbf{c}_n$, the first term of these equations is $\mathbf{c}_n^{(-)}$, which shows (12.84).

In order to solve (12.76) recursively, we introduce a partition of Γ_n in four blocks by selecting the last row and column. This gives

$$\Gamma_n = \left[\begin{array}{c|c} \Gamma_{n-1} & \mathbf{c}_{n-1}^{(-)} \\ \hline \mathbf{c}_{n-1}^{(-)T} & \gamma_0 \end{array}\right] \tag{12.85}$$

The fact that the last column contains the vector $\mathbf{c}_{n-1}^{(-)}$ is simply deduced from the structure appearing in (9.33'). To this partition we can associate the partitions of the vectors $\mathbf{a}_n$ and $\mathbf{c}_n$ given by

$$\mathbf{a}_n = \left[\begin{array}{c} \mathbf{a}_{n-1}^n \\ \hline a_n^n \end{array}\right] \quad ; \quad \mathbf{c}_n = \left[\begin{array}{c} \mathbf{c}_{n-1} \\ \hline \gamma_n \end{array}\right] \tag{12.86}$$

It is important to note that these two partitions are quite different. The partition of the vector $\mathbf{c}_n$ is simply a consequence of the definition of this vector by (12.67), where

$p = n$. On the other hand, there is no reason why the first $n-1$ components of $\mathbf{a}_n$ should be equal to the vector $\mathbf{a}_{n-1}$ calculated at the previous step of the recursion. This is why there are two indices, and the vector $\mathbf{a}_{n-1}^n$ is the vector of $\mathbb{R}^{n-1}$ obtained by selecting the first $n-1$ components of $\mathbf{a}_n$. In other words, the vectors $\mathbf{a}_n$ and $\mathbf{a}_{n-1}^n$ are defined by

$$\mathbf{a}_n = \left[a_1^n ,\ a_2^n ,\ a_3^n ,\ \ldots,\ a_{n-1}^n ,\ a_n^n \right]^T \tag{12.87}$$

$$\mathbf{a}_{n-1}^n = \left[a_1^n ,\ a_2^n ,\ a_3^n ,\ \ldots,\ a_{n-1}^n \right]^T \tag{12.88}$$

Using (12.85) and (12.86), the normal equation (12.76) can be written in two equations

$$\Gamma_{n-1}\, \mathbf{a}_{n-1}^n + a_n^n\, \mathbf{c}_{n-1}^{(-)} = \mathbf{c}_{n-1} \tag{12.89}$$

$$\mathbf{c}_{n-1}^{(-)T}\, \mathbf{a}_{n-1}^n + a_n^n\, \gamma_0 = \gamma_n \tag{12.90}$$

As Γ_{n-1} is positive-definite, we can multiply the first equation by its inverse and, by using (12.76) and (12.84), we obtain

$$\mathbf{a}_{n-1}^n = \mathbf{a}_{n-1} - a_n^n\, \mathbf{a}_{n-1}^{(-)} \tag{12.91}$$

This equation clearly shows why $\mathbf{a}_{n-1}^n$ is not equal to $\mathbf{a}_{n-1}$, as indicated just above. Using this value in (12.90), we deduce that a_n^n can be written as

$$a_n^n = \beta_n / \alpha_n \tag{12.92}$$

$$\beta_n = \gamma_n - \mathbf{a}_{n-1}^T\, \mathbf{c}_{n-1}^{(-)} \tag{12.93}$$

$$\alpha_n = \gamma_0 - \mathbf{a}_{n-1}^T\, \mathbf{c}_{n-1} \tag{12.94}$$

In these last calculations we have used the obvious relations deduced from (12.78)

$$\mathbf{u}^T\mathbf{v} = \mathbf{u}^{(-)T}\, \mathbf{v}^{(-)} \quad ; \quad \mathbf{u}^T\mathbf{v}^{(-)} = \mathbf{v}^T\mathbf{u}^{(-)} \tag{12.95}$$

We summarize below the basic ideas of the Levinson recursion and explain why this algorithm is called a *fast algorithm*.

(a) Summary of the Levinson algorithm. At the step $n-1$ of the recursion we know the vector $\mathbf{a}_{n-1}$. We can then calculate α_n, β_n, and a_n^n, given by the equations (12.92) – (12.94). From the value a_n^n we construct $\mathbf{a}_{n-1}^n$ by using (12.91), and by concatenation of $\mathbf{a}_{n-1}^n$ and a_n^n we deduce $\mathbf{a}_n$ from (12.87) and (12.88). When $n = p$ the system (12.76) is equivalent to (12.61), and we obtain the solution $\mathbf{a} = \mathbf{a}_p$.

(b) Initial conditions. Any recursive algorithm must have some initial conditions. This corresponds to the case $n = 1$, where (12.76) becomes a scalar equation written as $\gamma_0\, a_1 = \gamma_2$, or

$$a_1 = \gamma_1 / \gamma_0 \tag{12.96}$$

If the signal studied has a unit power, or if $\gamma_0 = 1$, we immediately deduce that $a_1 = \gamma_1$. Note that (12.96) is equivalent to (9.9) valid for an AR(1) signal.

(c) Fast algorithm. Let us evaluate the number of operations necessary to obtain the solution **a**. It is usual in this perspective to count only the number of multiplications (or divisions), although it is well known that this does not completely specify the complexity of an algorithm. The solution of a linear system such as (12.61) using the standard techniques requires a number of multiplications written as $O(p^3)$, which means that this number is in the order of p^3 multiplications. This especially implies that for large values of p the number increases as p^3. However, if the matrix in (12.61) is diagonal, we obtain p linear operations similar to (12.96), and consequently only p multiplications are required to solve the system.

We will now show that the Levinson algorithm requires only $O(p^2)$ multiplications, which justifies the term *fast algorithm*. To do this, let us again consider the step giving $\mathbf{a}_n$ in terms of $\mathbf{a}_{n-1}$. There are $(n-1)$ multiplications in (12.91), (12.93), and (12.94), and (12.92) introduces an additional multiplication. Consequently the corresponding step needs $3(n-1)+1$ multiplications. As it is well known that the sum of the first p integers is $O(p^2)$, we deduce that the calculation of **a** by the Levinson recursion needs $O(p^2)$ multiplications. In this calculation we ignore the coefficient 3 and the additional multiplication at each step, because for large p, $p << p^2$.

It is important to understand clearly from where the reduction of the number of multiplications arises. In reality it is a direct consequence of the *Toeplitz structure* of the matrix Γ. Because of this structure (12.84) holds, with the result that the last term of (12.91) is proportional to $\mathbf{a}_{n-1}^{(-)}$, which is already calculated. If (12.84) does not hold, it is easy to show that this last term requires the product of a matrix and a vector, which introduces n^2 multiplications, and finally an algorithm with $O(p^3)$ multiplications.

(d) Further simplification of the Levinson algorithm. Let us now show that the recursion (12.94) giving the coefficient α_n can be simplified. For this we calculate α_{n+1} given by

$$\alpha_{n+1} = \gamma_0 - \mathbf{c}_n^T \mathbf{a}_n \tag{12.97}$$

Using (12.86) this equation can be written as

$$\alpha_{n+1} = \gamma_0 - (\mathbf{c}_{n-1}^T \mathbf{a}_{n-1}^n + \gamma_n a_n^n) \tag{12.98}$$

and, applying (12.91), we obtain

$$\begin{aligned}\alpha_{n+1} &= \gamma_0 - \mathbf{c}_{n-1}^T \mathbf{a}_{n-1} + a_n^n (\gamma_n - \mathbf{c}_{n-1}^T \mathbf{a}_{n-1}^{(-)}) \\ &= \alpha_n - a_n^n \beta_n = \alpha_n - (a_n^n)^2 \alpha_n \\ &= \alpha_n \left[1 - (a_n^n)^2 \right]\end{aligned} \tag{12.99}$$

This recursion is much simpler than (12.94) and requires only two multiplications. With this simplification the complete Levinson algorithm can be written as follows:

Initial conditions: $a_1 = a_1^1 = \gamma_1 / \gamma_0$

$\alpha_1 = \gamma_0$

Step $(n-1)$ *to* n:

$\mathbf{a}_{n-1}$ and α_{n-1} are known

a_{n-1}^{n-1} is deduced from $\mathbf{a}_{n-1}$

$$\alpha_n = \alpha_{n-1}\left[1 - (a_{n-1}^{n-1})^2\right]$$

$$\beta_n = \gamma_n - \mathbf{a}_{n-1}^T \mathbf{c}_{n-1}^{(-)}$$

$$a_n^n = \beta_n / \alpha_n$$

$$\mathbf{a}_{n-1}^n = \mathbf{a}_{n-1} - a_n^n \mathbf{a}_{n-1}^{(-)}$$

$\mathbf{a}_n$ and α_n are known.

12.5.2 Interpretation and Examples

The principle of the Levinson recursion is to solve n successive normal equations, such as (12.76), of increasing order. In other words, starting from a sequence of values of a correlation function $\gamma[k]$, $0 \le k \le p$, we construct a sequence of vectors as in Table 12.1.

$\mathbf{a}_1$	$\mathbf{a}_2$	$\mathbf{a}_3$	$\cdots$	$\mathbf{a}_p$	$=$	$\mathbf{a}$
a_1^1	a_1^2	a_1^3		a_1^p	$=$	a_1
	a_2^2	a_2^3		a_2^p	$=$	a_2
		a_3^3		a_3^p	$=$	a_3
				$\vdots$		
				a_p^p	$=$	a_p

Table 12.1 Sequence of regression vectors.

This sequence of vectors constitutes a triangular table of numbers a_j^i. The vectors $\mathbf{a}_n$ are not arbitrary. In fact, as they are solutions to the normal equations such as (12.76), they enjoy all the properties of the solutions of these equations discussed in the previous section. In particular the vector $\mathbf{a}_n$ is that used in the linear prediction of $x[k]$ in terms of $x[k-1]$, $x[k-2]$, ..., $x[k-n]$, which can be written as (12.60) provided that p is replaced by n. In this prediction the mean square error is given by (12.66) where p is replaced by n. Comparing this equation with (12.97), we deduce that

$$\alpha_{n+1} = \varepsilon_n^2 \tag{12.100}$$

and the recursion (12.99) takes the form

$$\varepsilon_n^2 = \varepsilon_{n-1}^2(1 - k_n^2) \tag{12.101}$$

where the coefficient k_n, called the *reflection coefficient*, is defined by

$$k_n = a_n^n \tag{12.102}$$

As the prediction error is positive for regular signals, we deduce that

$$|k_n| < 1 \tag{12.103}$$

which is a fundamental relation very often used later.

The figures 12.4 and 12.5 can be interpreted in terms of these coefficients k_n. If $k_n = 0$ we have $\varepsilon_n^2 = \varepsilon_{n-1}^2$, which appears in curve 1 of Figure 12.4. This corresponds to the part of this curve where ε_p^2 remains constant. On the other hand, if $k_r \neq 0$ and $k_{r+1} = k_{r+2} = \ldots = 0$, we deduce that $\varepsilon_n^2 = \varepsilon_r^2$ for $n > r$, which characterizes an AR(r) signal. Let us verify this point on the Levinson recursion (12.91). It is obvious from (12.102) that if $k_n = 0$, $a_n^n = 0$ and then $\mathbf{a}_{n-1}^n = \mathbf{a}_{n-1}$, which means that the vector $\mathbf{a}_n$ is simply deduced from $\mathbf{a}_{n-1}$ by adding one zero. It is clear in (9.17) that adding null coefficients does not change the transfer function, and consequently an AR(p) signal is also an AR($p + q$) signal if the last components of the regression vector are null. This was verified in Example 9.1.

Let us now consider the *singular case* described in Figure 12.4. Suppose first that $k_n^2 = 1$. In this case $\varepsilon_n^2 = 0$ and the Levinson recursion cannot be continued. In fact, this implies that $\alpha_{n+1} = 0$, and it is then impossible to calculate a_{n+1}^{n+1} by using (12.92).

This situation corresponds to a predictable signal with a finite past, and we have seen that this implies that the polynomial defined by the vector $\mathbf{a}_n$ has all its roots on the unit circle. Finally, if $k_n \to 1$, $\varepsilon_n^2 \to 0$, and we have the case of a predictable signal, but with an infinite past.

It is clear that the recursion on the error given by (12.101) and the initial condition $\alpha_1 = \gamma_0$ give an expression of the error in terms of the reflection coefficients by

$$\varepsilon_p^2 = \gamma_0 \prod_{n=1}^{p} (1 - k_n^2) \tag{12.104}$$

The last point that must be noted is that all the successive normal problems give vectors $\mathbf{a}_n$ which introduce minimum phase filters, or are regression vectors of second order AR signals.

Example 12.1. Levinson recursion with an exponential correlation function. Let $x[k]$ be a signal with the correlation function $\gamma[k] = a^{|k|}$, as in (9.5). The initial conditions are $a_1 = a$ and $\alpha_1 = 1$.

$$\textit{Step 2} \qquad \alpha_2 = 1 - a^2$$

$$\beta_2 = a^2 - a^2 = 0$$

$$a_2^2 = 0$$

$$a_1^2 = a$$

Let us now show by induction that $a_i^i = 0$, a property which is valid for $i = 2$. If it is valid till $i - 1$, we have

$$\mathbf{a}_{i-1} = [a, 0, 0, \ldots, 0]^T$$

As the last component of $\mathbf{c}_{i-1}$ is a^{i-1}, we deduce that

$$\beta_i = a^i - a.\ a^{i-1} = 0$$

which proves the property. In conclusion, the correlation function $a^{|k|}$ introduces an AR(1) signal and a prediction error equal to $1 - a^2$. This is in accordance with the results in Section 9.1.

Example 12.2. Constant signal in a white noise. Consider a situation similar to that discussed in Example 3.2. The signal to be predicted is

$$x[k] = n[k] + A \tag{12.105}$$

where A is an RV uncorrelated with $n[k]$ which is a weak white noise with a correlation function given by (7.1). Assuming that A and $n[k]$ are zeromean-valued, the correlation function of $x[k]$ is given by

$$\gamma[k] = v_N\ \delta[k] + v_A \tag{12.106}$$

where v_A and v_N are the variance of A and $n[k]$ respectively. Introducing the signal to noise ratio defined as in (3.108) by $\rho = v_A/v_N$, we deduce that

$$\gamma[0] = v_N\ (1 + \rho) \quad ; \quad \gamma[i] = v_A\,, \quad i \neq 0 \tag{12.107}$$

The initial conditions are

$$a_1 = a_1^1 = \rho/(1 + \rho) \quad ; \quad \alpha_1 = v_N\ (1 + \rho) \tag{12.108}$$

At step 2 of the algorithm we obtain

$$\alpha_2 = v_N\ (1 + 2\rho)\ /(1 + \rho)$$

$$\beta_2 = v_A\ /(1 + \rho)$$

$$a_2^2 = \rho/(1 + 2\rho)$$

This yields

$$\mathbf{a}_2 = \frac{\rho}{1 + 2\rho}\ [1, 1]^T \tag{12.109}$$

Let us now show by induction that

$$\mathbf{a}_n = \frac{\rho}{1 + n\rho}\ [1, 1, \ldots, 1]^T \tag{12.110}$$

For this we calculate $\mathbf{a}_{n-1}$. From (12.110) we deduce that

$$a_n^n = \rho/(1 + n\rho)$$

$$\alpha_{n+1} = v_N \frac{1 + (n+1)\rho}{1 + n\rho}$$

$$\beta_{n+1} = v_A/(1 + n\rho)$$

$$a_{n+1}^{n+1} = \frac{\rho}{1 + (n+1)\rho}$$

Applying the Levinson recursion, we obtain (12.110) where n is replaced by $n + 1$, which completes the proof of this expression.

It is interesting to investigate the two extreme cases of weak or strong signal-to-noise ratio. If $\rho = 0$, there is no signal, and $\mathbf{a}_n = \mathbf{0}$, which is normal because a white noise is unpredictable. If, however, $\rho \to \infty$, we find that

$$\mathbf{a}_n = \frac{1}{n}[1, 1, \ldots, 1]^T \tag{12.111}$$

which shows that the predictor calculates

$$\hat{x}[k] = \frac{1}{n}\sum_{i=1}^{n} x[k-i] \tag{12.112}$$

which is the discrete-time time averager analog to (6.93) valid for the continuous-time case. This is quite natural, because the best mean square linear estimation of a constant RV from its samples is the time averaging.

12.5.3 Inversion of a Toeplitz Matrix

It is clear that solving the sequence of normal equations (12.76) is equivalent to calculating the inverse of the covariance matrices $\mathbb{\Gamma}_n$ which are Toeplitz. In fact, the sequence of regression vectors can be written as

$$\mathbf{a}_n = \mathbb{\Gamma}_n^{-1}\mathbf{c}_n \tag{12.113}$$

We will now explain this inversion matrix problem by analyzing the structure of $\mathbb{\Gamma}_n^{-1}$. Utilizing the same partition as in (12.85), we can write this inverse matrix in the form

$$\mathbb{\Gamma}_n^{-1} = \left[\begin{array}{c|c} \mathbb{M}_{n-1} & \mathbf{v}_{n-1} \\ \hline \mathbf{v}_{n-1}^T & m_n \end{array}\right] \tag{12.114}$$

where $\mathbb{M}_{n-1}$ is an $(n-1) \times (n-1)$ matrix, $\mathbf{v}_{n-1}$ a vector of $\mathbb{R}^{n-1}$ and m_n a scalar. Applying the relation $\mathbb{\Gamma}_n \mathbb{\Gamma}_n^{-1} = \mathbb{I}$ yields

$$\mathbb{\Gamma}_{n-1}\mathbb{M}_{n-1} + \mathbf{c}_{n-1}^{(-)}\mathbf{v}_{n-1}^T = \mathbb{I}_{n-1} \tag{12.115}$$

$$\mathbb{\Gamma}_{n-1}\mathbf{v}_{n-1} + \mathbf{c}_{n-1}^{(-)} m_n = \mathbf{0} \tag{12.116}$$

$$\mathbf{c}_{n-1}^{(-)T}\mathbf{v}_{n-1} + \gamma_0 m_n = 1 \qquad (12.117)$$

Consider first the last two equations. If we replace $\mathbf{c}_{n-1}^{(-)}$ by (12.84) in (12.116) and if we note that Γ_{n-1} is positive-definite, we obtain

$$\mathbf{v}_{n-1} = -m_n \mathbf{a}_{n-1}^{(-)} \qquad (12.118)$$

Using this expression in (12.117), and comparing it with (12.94) shows that

$$m_n = \alpha_n^{-1} \qquad (12.119)$$

If we combine all these results we finally find

$$\Gamma_n^{-1} = \left[\begin{array}{c|c} \Gamma_{n-1}^{-1} + \alpha_n^{-1}\mathbf{a}_{n-1}^{(-)}\mathbf{a}_{n-1}^{(-)T} & -\alpha_n^{-1}\mathbf{a}_{n-1}^{(-)} \\ \hline -\alpha_n^{-1}\mathbf{a}_{n-1}^{(-)T} & \alpha_n^{-1} \end{array}\right] \qquad (12.120)$$

This expression exhibits very nice properties. The most important is the structure of the last row of this matrix. According to (12.100), the last term of this row is the inverse of the prediction error ε_{n-1}^2 obtained with a predictor using a past of length $n-1$. Multiplying this row by this error, we obtain the vector $\mathbf{u}^{(-)}$ such that

$$\mathbf{u}^T = \left[1 \;\vdots\; -\mathbf{a}_{n-1}^T\right] \qquad (12.121)$$

Comparing this with (12.64) or (12.69) we see that this is the vector $\mathbf{b}_{n-1}$ characterizing the *error filter* of past $n-1$. In conclusion the last row, or the last column, of the matrix Γ_n^{-1} contains all the elements characterizing the prediction problem with past $n-1$. As a consequence it is possible to deduce recursively the sequence of matrices Γ_n^{-1} by using only the sequence of regression vectors $\mathbf{a}_n$ appearing in Table 12.1 and the sequence of errors given by (12.101). The sequence of matrices is initialized by $\Gamma_1^{-1} = \{\gamma[0]\}^{-1}$.

The structure (12.120) also yields an expression of the prediction error as a ratio of two determinants. In fact it is well known in linear algebra that the inverse of a matrix $\mathbb{M}$ is given by the expression

$$\mathbb{M}^{-1} = [\det(\mathbb{M})]^{-1}\mathrm{Adj}(\mathbb{M}) \qquad (12.122)$$

where the matrix $\mathrm{Adj}(\mathbb{M})$ is a matrix of determinants. Applying that to Γ_n we deduce that

$$[\Gamma_n^{-1}]_{nn} = [\det(\Gamma_n)]^{-1}[\mathrm{Adj}(\Gamma_n)]_{nn} \qquad (12.123)$$

Using the partition (12.85) and the value of $[\Gamma_n]^{-1}$ given by (12.120), we deduce that

$$\varepsilon_{n-1}^2 = \alpha_n = \frac{\det(\Gamma_n)}{\det(\Gamma_{n-1})} \qquad (12.124)$$

This expression is obtained by another method in a problem.

As an illustration of the previous considerations, it is possible to calculate the inverse of the matrix

$$\Gamma_p = T[1,\ a,\ a^2,\ \ldots,\ a^{p-1}] \tag{12.125}$$

which is the covariance matrix of a signal with an exponential correlation function studied in Example 12.1.

Using the previous recursion of matrices, it is easy to find that the inverse of Γ_p is

$$\Gamma_p^{-1} = \frac{1}{1-a^2}\begin{bmatrix} 1 & -a & 0 & 0 & \ldots\ldots & 0 & 0 \\ -a & 1+a^2 & -a & 0 & \ldots\ldots & 0 & 0 \\ 0 & -a & 1+a^2 & -a & \ldots\ldots & 0 & 0 \\ & & & & \cdot & & \\ & & & & \cdot & & \\ & & & & \cdot & & \\ 0 & 0 & 0 & 0 & \ldots\ldots & 1+a^2 & -a \\ 0 & 0 & 0 & 0 & \ldots\ldots & -a & 1 \end{bmatrix} \tag{12.126}$$

As (12.125) is the covariance matrix of an AR(1) signal, we find the prediction vector of such a model in the last column, and also the prediction error $1-a^2$. Note that, as indicated previously, (12.126) is no longer a Toeplitz matrix because of the structure of its diagonal elements.

12.6 THE TRIPLET a, c, k

In this section we consider only unit power signals or signals whose correlation function satisfies $\gamma[0] = 1$.

The Levinson recursion is an algorithmic method to solve (12.74), or to associate to a vector **c** of $\mathbb{R}^p$, called the *correlation vector*, another vector **a** of $\mathbb{R}^p$, called the *regression vector*. We saw in (12.75) that the inverse operation is possible, which means deducing a correlation vector from a regression vector, but the algorithm allowing this operation was not indicated. Furthermore, during the recursion the reflection coefficients k_n defined by (12.102) were introduced. These are the diagonal elements of the table of numbers in Table 12.1. It is then possible to introduce a *reflection vector* **k**, with components k_i. The purpose of this section is to study the relationships between these three vectors of $\mathbb{R}^p$:

regression vector : $\mathbf{a} = [a_1,\ a_2,\ \ldots\ldots,\ a_p]^T$

correlation vector : $\mathbf{c} = [\gamma_1,\ \gamma_2,\ \ldots\ldots,\ \gamma_p]^T$

reflection vector : $\mathbf{k} = [k_1,\ k_2,\ \ldots\ldots,\ k_p]^T$

The first point to note is that the reflection vector $\mathbf{k}$ is obviously obtained as soon as the sequence S_p of regression vectors $\mathbf{a}_n$, $1 \le n \le p$ is obtained. This is a direct consequence of (12.102). The sequence S_p is visualized in Table 12.1. The Levinson recursion is an algorithm allowing us to construct S_p from the vector $\mathbf{c}$. In the recursion the dimension of the vectors $\mathbf{a}_n$ increases by one at each step of the algorithm, and for this reason we speak of the increasing, or direct, Levinson algorithm. We will now show that the inverse procedure is also possible.

12.6.1 Inverse Levinson Algorithm

Starting from a vector $\mathbf{a}$ of $\mathbb{R}^p$, we will show that it is possible to calculate a decreasing sequence of vectors $\mathbf{a}_n$ allowing us to construct the complete sequence S_p. For this purpose it is sufficient to make an inversion of the Levinson recursion (12.91). Suppose that $\mathbf{a}_n$ is known. By using (12.87) and (12.88) we deduce $k_n = a_n^n$ and $\mathbf{a}_{n-1}^n$. Applying the matrix $\mathbb{J}$ to (12.91) gives

$$-k_n \mathbf{a}_{n-1} + \mathbf{a}_{n-1}^{(-)} = \mathbf{a}_{n-1}^{n(-)} \tag{12.127}$$

Noting that (12.91) can be written as

$$\mathbf{a}_{n-1} - k_n \mathbf{a}_{n-1}^{(-)} = \mathbf{a}_{n-1}^n \tag{12.128}$$

we obtain two equations, which yields

$$\mathbf{a}_{n-1} = \frac{1}{1-k_n^2} \left[\mathbf{a}_{n-1}^n + k_n \mathbf{a}_{n-1}^{n(-)} \right] \tag{12.129}$$

This gives the vector $\mathbf{a}_{n-1}$ in terms of the vector $\mathbf{a}_n$. The solution is always possible as soon as $k_n^2 \ne 1$, which is ensured if (12.103) is satisfied. This is assumed in this section.

12.6.2 The Relation a → c

Suppose that $\mathbf{a}$ is given. By using (12.129) we construct the table S_p of vectors $\mathbf{a}_n$. From this table we can construct the vector $\mathbf{c}$. Using (12.92) and (12.102), the relation (12.93) can be written as

$$\gamma_n = \alpha_n k_n + \mathbf{a}_{n-1}^T \mathbf{c}_{n-1}^{(-)} \tag{12.130}$$

The reflection coefficients k_n are known, as diagonal elements of the table S_p. Furthermore, α_n is deduced from (12.101), which can be written as

$$\alpha_{n+1} = \alpha_n (1 - k_n^2) \tag{12.131}$$

From these two equations we deduce an increasing recursion allowing the calculation of γ_n from $n = 1$ to $n = p$, provided that γ_1 and α_1 are known. In order to calculate these initial conditions, we use the remark made at the beginning of this section. As it is assumed that $\gamma[0] = 1$, we deduce that $\alpha_1 = 1$, which is a consequence of (12.94).

Similarly, (12.93) and (12.92) give $\gamma_1 = \alpha_1 k_1 = k_1 = a_1^1$, which is the first element of the sequence S_p. Consequently, starting from the initial conditions $\alpha_1 = 1$, $\gamma_1 = a_1^1$, we can use (12.130) and (12.131) to construct a sequence of γ_n from $n = 1$ to $n = p$, which gives the components of the vector $\mathbf{c}$.

We now have an *algorithmic procedure* to solve (12.74) and (12.75). For (12.74) the procedure is the increasing Levinson recursion, and for (12.75) there are two steps: the first is the decreasing Levinson recursion (12.129) and the second an increasing recursion giving the coefficients γ_n.

We make here a final comment concerning the application of these methods to AR(p) signals. Suppose that $x[k]$ is an AR(p) signal with a unit power, which implies that $\gamma[0] = 1$, and then defined by a regression vector $\mathbf{a}$. We have noticed that if $\mathbf{a}$ is known, the power spectrum of $x[k]$ is known, and consequently by inverse Fourier transformation the correlation function can be determined. This procedure is in general complex, and the simplest way is to use the previous calculation. As $\mathbf{a}$ is known, we can calculate the vector $\mathbf{c}$ by the recursion indicated above. This yields the values $\gamma_0 = 1$, γ_1, γ_2, ..., γ_p. In order to calculate γ_k for $k > p$, we use (9.32). It is interesting to note that (9.32) is a particular case of (12.130) when $k_n = 0$. This is quite logical if we remember that an AR(p) signal is an AR(∞) signal in which all the components a_i of the regression vector are null for $i > p$. This implies, as already indicated, that the reflection coefficients k_i are also null if $i > p$. In conclusion, the inverse Levinson algorithm is the best numerical method for calculating the entire correlation function of an AR(p) signal.

12.6.3 The Relation a $\to$ k

This relation is a direct product of the inverse Levinson recursion presented in the previous section. In fact, starting from a vector $\mathbf{a}$, this recursion allows the construction of the sequence of vectors S_p. Taking the last component of each vector $\mathbf{a}_n$ of this sequence, and using (12.102), we obtain the p components of the vector $\mathbf{k}$.

12.6.4 The Relation k $\to$ a

This relation is deduced from the direct Levinson recursion (12.91) where a_n^n is expressed as in (12.102). This gives

$$\mathbf{a}_{n-1}^{n} = \mathbf{a}_{n-1} - k_n\, \mathbf{a}_{n-1}^{(-)} \tag{12.132}$$

$$a_n^n = k_n \tag{12.133}$$

Because of its practical importance, discussed in the next section, we will analyze this relation $\mathbf{a} = \mathbf{h}(\mathbf{k})$ in more detail.

The first comment that must be made is that this relation does not imply any condition on the vector $\mathbf{k}$. This is the great difference with the Levinson recursions (direct or inverse), which are given under the condition (12.103). In reality the inverse recursion (12.129) only requires $k_n^2 \neq 1$. No such conditions appear in the recursion defined by (12.132) and (12.133).

To have some idea of the structure of the vectors $\mathbf{a}_n$ deduced from a set of coefficients k_i, we indicate in Table 12.2 the sequence S_3 of the first three vectors $\mathbf{a}_i$

$\mathbf{a}_1$	$\mathbf{a}_2$	$\mathbf{a}_3$
k_1	$k_1 - k_1 k_2$	$k_1 - k_1 k_2 - k_2 k_3$
	k_2	$k_2 - k_1 k_3 + k_1 k_2 k_3$
		k_3

Table 12.2 Sequence of vectors $\mathbf{a}_i$ in terms of reflection coefficients.

It is clear that the components of $\mathbf{a}_i$ are not simple functions of the k_is, and this complexity rapidly increases. However, the recursion allowing the change from $\mathbf{a}_n$ to $\mathbf{a}_{n+1}$ remains extremely simple. In spite of the complexity of the relation $\mathbf{a} = \mathbf{h}(\mathbf{k})$, some general features can be indicated. They are discussed in some problems and their interpretation requires the use of lattice filters presented in the next section. However, in the following example we discuss the second order case (vectors of $\mathbb{R}^2$) which shows some interesting properties.

Example 12.3 Transformation $\mathbf{a}_2 = \mathbf{h}(\mathbf{k}_2)$. Let us analyze the transformation of $\mathbb{R}^2$ into itself by the relation $\mathbf{a}_2 = \mathbf{h}(\mathbf{k}_2)$, written as

$$a_1 = k_1 - k_1 k_2 \;;\; a_2 = k_2 \tag{12.134}$$

The first point to note is that the origin and the axes $k_1 = 0$ or $k_2 = 0$ are invariant in this transformation. Let us now consider the four following straight lines

$$D_1 = [\mathbf{k} \mid k_1 = 1]$$

$$D_2 = [\mathbf{k} \mid k_1 = -1]$$

$$D_3 = [\mathbf{k} \mid k_2 = 1]$$

$$D_4 = [\mathbf{k} \mid k_2 = -1]$$

By elementary calculations we find that these lines are transformed into D'_1, D'_2, D'_3, D'_4 defined by

$$D'_1 = \{\mathbf{a} \mid a_2 = 1 - a_1\}$$

$$D'_2 = \{\mathbf{a} \mid a_2 = 1 + a_1\}$$

$$D'_3 = \{\mathbf{a} \mid a_1 = 0,\; a_2 = 1\}$$

$$D'_4 = \{\mathbf{a} \mid a_2 = -1\}$$

We note that D'_3 is limited to the point $a_1 = 0$, $a_2 = 1$. These relations are pictured in Figures 12.6 and 12.7.

The inverse transformation of (12.134) is of course

$$k_1 = a_1(1 - a_2)^{-1} \quad ; \quad k_2 = a_2 \tag{12.135}$$

and the line D'_5 defined by $a_2 = 1$ has no image in the plane $k_1\ k_2$.

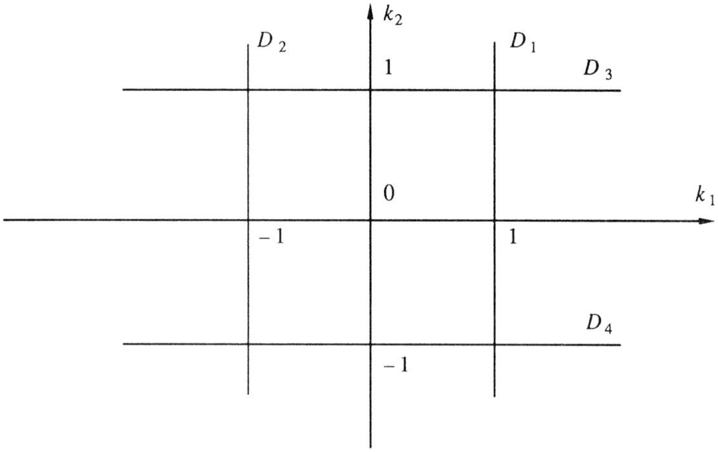

Figure 12.6 Plane $k_1\ k_2$.

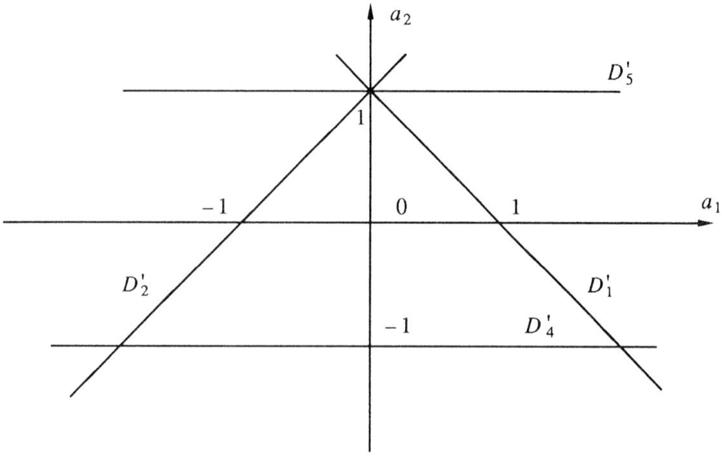

Figure 12.7 Plane $a_1\ a_2$.

12.7. LATTICE FILTERS FOR PREDICTION

Suppose that we want to realize one-step prediction of a signal $x[k]$ with an indefinite past. This means that the past can be changed according to the requirements on the prediction error. The predictor is characterized by a regression vector $\mathbf{a}$ belonging to $\mathbb{R}^p$ and its output can be written as in (12.60), or

$$\hat{x}_p[k] = \mathbf{a}^T \mathbf{X}[k] = \sum_{i=1}^{p} a_i x[k-i] \qquad (12.136)$$

The corresponding error filter gives the output

$$\tilde{x}_p[k] = x[k] - \hat{x}_p[k] = \sum_{i=1}^{p} b_i x[k-i] \tag{12.137}$$

where $b_0 = 1$ and $b_i = -a_i$. The corresponding prediction error is given by

$$\varepsilon_p^2 = \gamma[0] - \mathbf{a}^T \mathbf{c} \tag{12.138}$$

where **c** is the correlation vector studied in the previous sections and defined by (12.63) or (12.67). This error behaves as in Figures 12.4 and 12.5. Let us now describe the structures of the predictor and of the error filter. Using the same conventions as in Figures 9.2, 9.3, and 9.4, the flow graph of the prediction and error filters is given in Figure 12.8 for $p = 3$, for which the extension to an arbitrary value of p is straightforward. The realization of these filters needs only three delays, or three functions z^{-1}, and more generally, if the past is p, the number of delays z^{-1} is also p. The *transfer function* of the predictor characterized by (12.136) can be written as

$$A_p\{z\} = \sum_{i=1}^{p} a_i z^{-i} = \mathbf{a}^T \mathbf{Z}_p \tag{12.139}$$

where $\mathbf{Z}_p$ is the vector whose components are z^{-i}, $1 \le i \le p$.

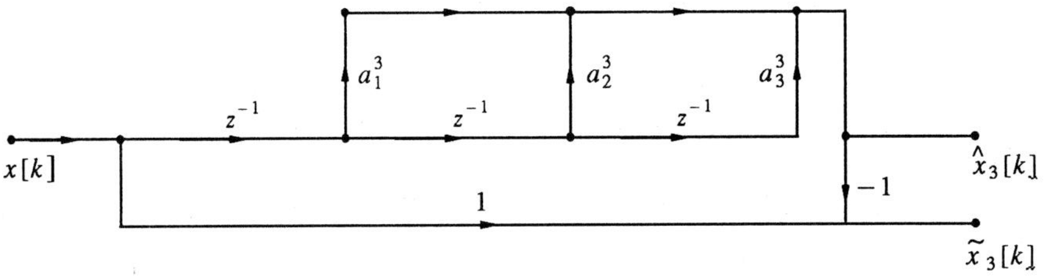

Figure 12.8 Transversal realization of prediction and error filters.

Suppose now that we want to *reduce* the prediction error by adding one more delay. This means that in the previous equations p is replaced by $p + 1$. To do this we must calculate the new regression vector $\mathbf{a}_{p+1}$, or add another column to Table 12.1. This can be realized by using the Levinson recursion studied in the previous sections. But we saw, especially from (12.91), that in this operation all the components of vector **a** are changed. Physically this means that if in Figure 12.8 we add a delay cell z^{-1}, we must change all the transversal coefficients which become a_1^4, a_2^4, a_3^4, and a_4^4. This is rather inconvenient, and we will show that, by using lattice filters, it is possible to introduce a structure where the increment of the order does not change the structure already calculated.

The predictor corresponding to the normal equation (12.76) calculates the one-step prediction with a past n, or

$$\hat{x}_n[k] = \mathbf{X}_n^T[k]\mathbf{a}_n \tag{12.140}$$

where $\mathbf{X}_n[k]$ is the vector defined by

$$\mathbf{X}_n^T[k] = \{x[k-1], x[k-2], \ldots, x[k-n]\} \tag{12.141}$$

The corresponding innovation is

$$\tilde{x}_n[k] = x[k] - \hat{x}_n[k] \tag{12.142}$$

These quantities are called *forward* prediction and innovation.

To interpret the vector $\mathbf{a}_n^{(-)}$ appearing in the Levinson recursion, let us introduce the *backward* prediction and innovation. From the observation vector (12.141) we can deduce the forward prediction, which is the estimation of $x[k]$, and the backward prediction, which is the estimation of $x[k-n-1]$. We will show that this backward prediction uses the vector $\mathbf{a}_n^{(-)}$, and can be written as

$$\hat{x}_n^{(-)}[k-n-1] = \mathbf{X}_n^T[k]\,\mathbf{a}_n^{(-)} \tag{12.143}$$

For this purpose let us introduce the random variable $y = x[k-n-1]$. The linear estimation of y in terms of $\mathbf{X}_n[k]$ can be written as

$$\hat{y} = \mathbf{h}^T\mathbf{X}_n[k] \tag{12.144}$$

where $\mathbf{h}$ is obtained by (10.65), or

$$\Gamma_n\mathbf{h} = \mathbf{v} \tag{12.145}$$

In this equation $\mathbf{v}$ is the vector $E[y\,\mathbf{X}_n]$, and using (12.141) and the definition of y, we deduce that $\mathbf{v} = \mathbf{c}^{(-)}$. Comparing (12.145) with (12.84), we deduce that $\mathbf{h} = \mathbf{a}_n^{(-)}$. Furthermore, it is important to distinguish $\hat{x}_n[k-n-1]$ and $\hat{x}_n^{(-)}[k-n-1]$. The first is the forward prediction at the time $k-n-1$, while the latter is the backward prediction at the same time. Similarly, the *backward innovation* can be expressed as

$$\tilde{x}_n^{(-)}[k-n-1] = x[k-n-1] - \hat{x}_n^{(-)}[k-n-1] \tag{12.146}$$

As the Levinson recursion introduces a relation using the vector $\mathbf{a}_n^{(-)}$, it seems interesting to translate this relation in terms of the forward and backward innovations.

Decomposing $\mathbf{X}_n[k]$ into two blocks as in (12.86), we can write

$$\hat{x}_n[k] = \mathbf{X}_{n-1}^T[k]\mathbf{a}_{n-1}^n + a_n^n\,x[k-n] \tag{12.147}$$

As a result of the Levinson recursion (12.132), it appears that

$$\hat{x}_n[k] = \mathbf{X}_{n-1}^T[k]\{\mathbf{a}_{n-1} - k_n\mathbf{a}_{n-1}^{(-)}\} + k_n\,x[k-n] \tag{12.148}$$

which can be written as

$$\hat{x}_n[k] = \hat{x}_{n-1}[k] + k_n\,\tilde{x}_{n-1}^{(-)}[k-n] \tag{12.149}$$

In fact, the backward estimation of $x[k-n]$ is deduced from (12.143) by replacing n by $n-1$. Using (12.142) and (12.149), we deduce

$$\boxed{\tilde{x}_n[k] = \tilde{x}_{n-1}[k] - k_n \tilde{x}^{(-)}_{n-1}[k - n]} \tag{12.150}$$

Apart from the time variables, this relation resembles (12.132).

Making the same calculation for the backward prediction, we can write, as in (12.147),

$$\hat{x}^{(-)}_n[k - n - 1] = \mathbf{X}^T_n[k - 1]\, \mathbf{a}^{n(-)}_{n-1} + k_n\, x[k - 1] \tag{12.151}$$

and, once again using the Levinson recursion, we obtain

$$\hat{x}^{(-)}_n[k - n - 1] = \hat{x}^{(-)}_{n-1}[k - n - 1] - k_n \tilde{x}_{n-1}[k - 1] \tag{12.152}$$

Using (12.146), we deduce that

$$\tilde{x}^{(-)}_n[k - n - 1] = \tilde{x}^{(-)}_{n-1}[k - n - 1] - k_n \tilde{x}_{n-1}[k - 1] \tag{12.153}$$

For the discussion which follows, it is more interesting to write this equation at the time $k - n$ instead of $k - n - 1$, which gives

$$\boxed{\tilde{x}^{(-)}_n[k - n] = \tilde{x}^{(-)}_{n-1}[k - n] - k_n \tilde{x}_{n-1}[k]} \tag{12.154}$$

These equations are the basis for the introduction of *lattice filters*. In fact, they allow us to pass from the innovations indexed by $n - 1$ to the innovations indexed by n, as presented in Figure 12.9. The transformation is linear and completely specified by the reflection coefficient k_n.

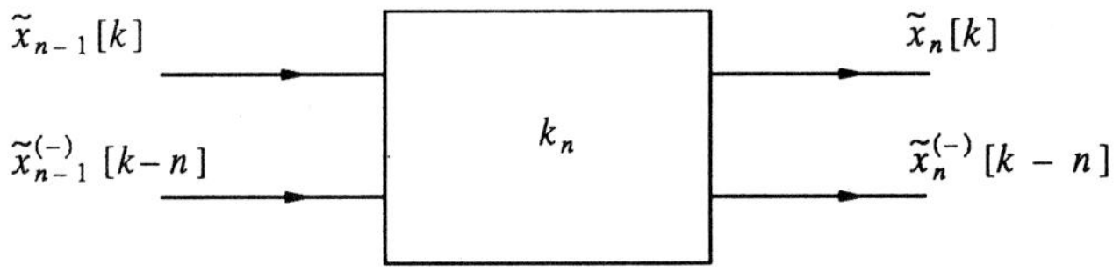

Figure 12.9 Transformation of the forward and backward innovations.

If we look inside the box represented in Figure 12.9, we arrive at the flow graph in Figure 12.10, which specifies the two equations (12.150) and (12.154). This graph shows a lattice structure characterized by the reflection coefficient k_n.

Applying the general rules of the flow graphs indicated in Figures 9.2 and 9.3, we find that the signal in node A is given by (12.150) and the signal in node B by (12.154).

To pass on to the next cell corresponding to the reflection coefficient k_{n+1}, we replace n in the equations by $n + 1$. This does not change the time dependence of the forward prediction, but as the parameter n appears simultaneously as index and in the time dependence of the backward innovation, the cell $n + 1$ must have as input $\tilde{x}^{(-)}_n[k - n - 1]$, which justifies the delay z^{-1} appearing after node B.

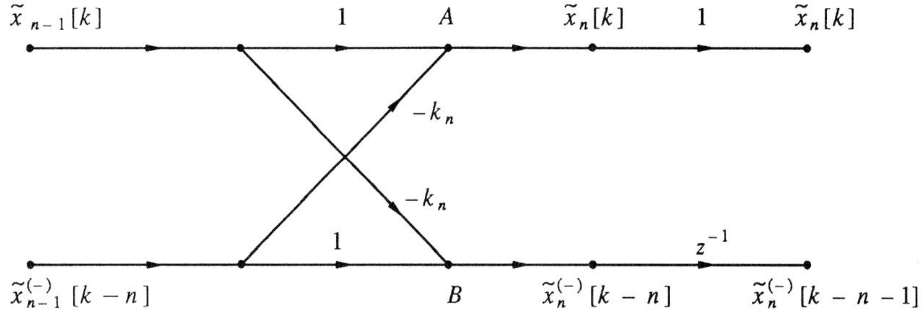

Figure 12.10 Flow graph of the lattice structure.

Combining all these elements, we give in Figure 12.11 the flow graph allowing the calculation of the forward innovation with a past $p = 3$. Extensions to higher values of p are straightforward.

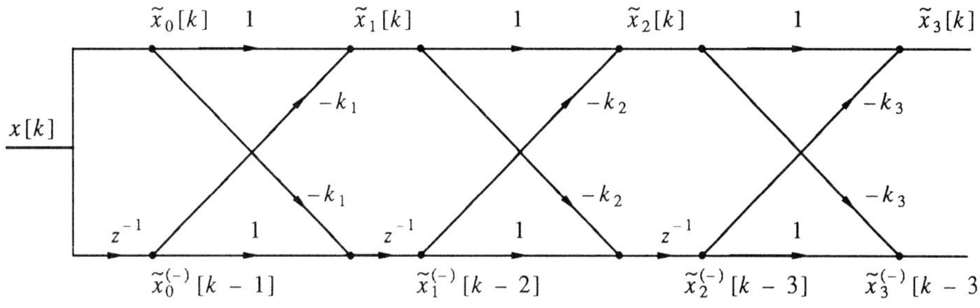

Figure 12.11 Error filter with past 3.

Let us make some comments on this lattice realization of an error filter.

(**a**) For prediction problems only the upper line is of interest and gives the forward innovations at time k for an increasing past. The lower line is a means for the calculation of these innovations, but has no physical interest in prediction problems, which of course concerns forward prediction.

(**b**) The innovation $\tilde{x}_3[k]$ in Figure 12.11 is exactly the same as that in Figure 12.8; only the structure of the filters is different.

(**c**) This can easily be verified by calculating the transfer function. Let us call $B_p\{z\}$ the transfer function of an error filter with past p. This function is given by (12.69), where the a_i s must be replaced by a_i^p. These coefficients appear in Table 12.2 for $p = 1$, 2, 3. To calculate the same transfer function with the use of lattice filters, we calculate the sum of the transfer functions corresponding to all the paths from the input to the output. Using the flow graph in Figure 12.11, we obtain the following three transfer functions:

$$B_1\{z\} = 1 - k_1 z^{-1} \tag{12.155}$$

$$B_2\{z\} = 1 - (k_1 - k_1 k_2) z^{-1} - k_2 z^{-2} \tag{12.156}$$

$$B_3\{z\} = 1 - (k_1 - k_1 k_2 - k_2 k_3) z^{-1} - (k_2 - k_1 k_3 + k_1 k_2 k_3) z^{-2} - k_3 z^{-3} \tag{12.157}$$

When we compare this with Table 12.2 we find that the transfer functions of the transverse and lattice filters are effectively the same.

(d) The number of delay cells z^{-1} is the same in the two realizations appearing in Figures 12.8 and 12.11.

(e) The principal advantage of the lattice structure is its *modular form*. This means that when passing from $\tilde{x}_p$ to $\tilde{x}_{p+1}$ we need only add another cell *without changing the previous ones*. For example, the calculation of $\tilde{x}_4[k]$ needs only an additional lattice with the reflection coefficient k_4. This was not the case when using the transverse structure, and we have noted many times that when passing from $\mathbf{a}_p$ to $\mathbf{a}_{p+1}$, all the first components of the vector $\mathbf{a}_{p+1}$ must be changed.

(f) It is clear that the lattice representation of transversal filters can be introduced without reference to prediction problems (see [Picinbono], p. 162). Furthermore, any dynamical filter can have a lattice structure, which is the origin of the term *reflection coefficient.* In fact, in the lattice representation of purely recursive filters, such as those used for the introduction of AR signals, an analogy with partial reflection, as for optical waves, can be introduced. This justifies the expression in a context unconnected with the reflection property.

(g) PARCOR coefficients. The reflection coefficients are sometimes called partial correlation (or PARCOR) coefficients. Let us explain this terminology. At first we want to show that the reflection coefficient k_{n+1} is the correlation coefficient of the two RVs $\tilde{x}_n[k]$ and $\tilde{x}_n^{(-)}[k-n-1]$. The correlation coefficient of two RVs is defined by (3.33). As we are working with zeromean RVs, this correlation coefficient can be written as a ratio, or as N/D, $D > 0$,where

$$N = E\{\tilde{x}_n[k]\,\tilde{x}_n^{(-)}[k-n-1]\} \tag{12.158}$$

$$D^2 = E\{(\tilde{x}_n[k])^2\}E\{(\tilde{x}_n^{(-)}[k-n-1])^2\} \tag{12.159}$$

As the forward and backward prediction are described by the same equations, the prediction errors are the same and we deduce from (12.100) that $D = \alpha_{n+1}$. The numerator N can also be written as

$$\begin{aligned} N &= E\,\{(x[k] - \mathbf{X}_n^T[k]\mathbf{a}_n)(x[k-n-1] - \mathbf{X}_n^T[k]\mathbf{a}_n^{(-)})\} \\ &= \gamma[n+1] - \mathbf{c}_n^{(-)T}\mathbf{a}_n - \mathbf{c}_n^T\mathbf{a}_n^{(-)} + \mathbf{c}_n^{(-)T}\mathbf{a}_n \\ &= \gamma[n+1] - \mathbf{c}_n^{(-)T}\mathbf{a}_n \end{aligned}$$

As a result of (12.93), $N = \beta_{n+1}$, and with (12.92) this shows that

$$k_{n+1} = N/D \tag{12.160}$$

This indicates that the reflection coefficient is the correlation coefficient between the forward and backward *innovations*. This is the definition of the *partial* correlation coefficient. More precisely, if we start from two RVs X and Y, their correlation coefficient is defined by (3.33). If we now add a random vector $\mathbf{V}$ correlated with X and Y, and if we calculate the innovations $\tilde{X}$ and $\tilde{Y}$ in the linear mean square estimation of X and Y in terms of $\mathbf{V}$, the correlation coefficient between $\tilde{X}$ and $\tilde{Y}$ is called the partial correlation coefficient between X and Y. This is the case of k_{n+1} which appears in (12.160).

12.8 POSITIVITY AND STABILITY

The relations between the elements of the triplet $\mathbf{a}$, $\mathbf{c}$, $\mathbf{k}$ studied in Section 12.6 were analyzed within the framework of algorithmic procedure. No specific constraint was imposed on these vectors, except to make the algorithms possible. In this perspective, the only situation making some algorithms impossible was characterized by $k_m^2 > 1$ for the direct Levinson recursion and $k_m^2 = 1$ for the inverse Levinson recursion.

However, it is clear that the vectors $\mathbf{c}$ and $\mathbf{a}$ are not arbitrary. The vector $\mathbf{c}$ must satisfy the so-called *positivity condition.* This condition ensures that the matrix $\mathbb{\Gamma}$ defined by (12.68) is positive-definite (PD), as being a covariance matrix defined by (12.62) without zero eigenvalue, in order to find a unique solution to (12.61).

Similarly, the vector $\mathbf{a}$ must satisfy a *stability condition.* This means that the purely recursive filter defined by $\mathbf{a}$ and introducing an AR(p) signal must be stable. This is equivalent to saying that the roots of the polynomial $z^p B_p\{z\}$, where $B_p\{z\}$ is defined by (12.69), are located inside the unit circle.

We shall now analyze these conditions and their consequences on the reflection vector $\mathbf{k}$.

12.8.1 Correlation Vector and Positivity

Suppose, as previously, that in (12.68) $\gamma[0] = 1$. As a result the matrix $\mathbb{\Gamma}$ is a Toeplitz matrix written

$$\mathbb{\Gamma} = T[1,\ \mathbf{c}^T] \tag{12.161}$$

and the problem is to find the condition on $\mathbf{c}$ which ensures that $\mathbb{\Gamma}$ is PD. There are many ways of doing this, but none is simple. The first uses the calculation of the eigenvalues of $\mathbb{\Gamma}$ which must be positive for a PD matrix. The second uses the result studied in Problem 3.10 indicating that $\mathbb{\Gamma}$ is PD if, and only if, all the principal minors are positive. The principal minors are the determinants of the matrices extracted from $\mathbb{\Gamma}$ by using the first n rows and columns, $n \le p$. It is obvious that the calculation of all these minors rapidly becomes difficult, for example as soon as $p > 4$. But it is not enough to calculate the minors; it is later necessary to ensure that they are positive.

We shall explain the procedure for $p = 2$. The first determinant gives the condition $\gamma_1^2 < 1$, which is the Schwarz inequality for the correlation. The third order minor is the determinant of the matrix

$$\Gamma = \begin{bmatrix} 1 & \gamma_1 & \gamma_2 \\ \gamma_1 & 1 & \gamma_1 \\ \gamma_2 & \gamma_1 & 1 \end{bmatrix} \tag{12.162}$$

Calculating this determinant, we find that the condition $\det(\Gamma) > 0$ is

$$\gamma_2^2 - 2\gamma_1^2\gamma_2 - 1 + 2\gamma_1^2 < 0 \tag{12.163}$$

Assuming that the first condition $\gamma_1^2 < 1$ is satisfied, we deduce that (12.163) gives a condition on γ_2 which can be written as

$$-1 + 2\gamma_1^2 < \gamma_2 < 1 \tag{12.164}$$

It is interesting to represent, in the plane γ_1, γ_2, the set of points such that Γ of (12.162) is PD, as appears in Figure 12.12. The domain D_c such that the matrix Γ is PD is defined by the parabola $\gamma_2 = 2\gamma_1^2 - 1$, and the straight line $\gamma_2 = 1$. This domain becomes much more complex in $\mathbb{R}^3$ and is widely unknown for $\mathbb{R}^p$, $p > 3$.

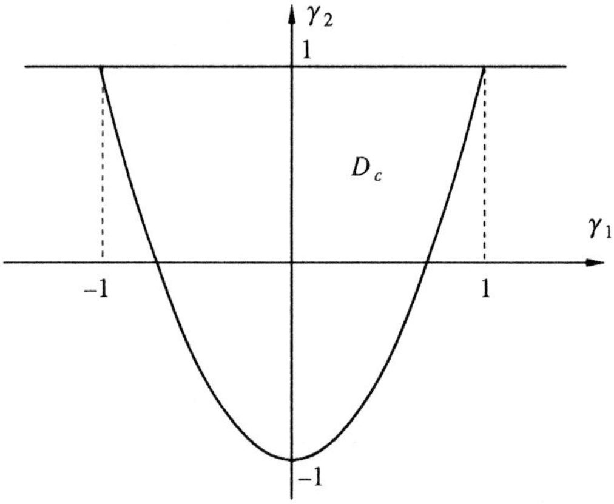

Figure 12.12 Domain D_c such that Γ is PD.

However, there are some general conditions which can be calculated. One is discussed in a problem, and shows that if $\gamma_1, \ldots, \gamma_{n-1}$ are chosen in such a way that $\Gamma_{n-1} = T\,[1, \gamma_1, \gamma_2, \ldots, \gamma_{n-1}]$ is PD, then the condition ensuring that Γ_n is PD is

$$b_n < \gamma_n < B_n \tag{12.165}$$

where the difference between the upper and lower bound is

$$B_n - b_n = 2\alpha_n = 2\varepsilon_{n-1}^2 \tag{12.166}$$

where ε_{n-1}^2 is the prediction error discussed above. This already appears in (12.164).

At this point we note the following. Taking $|\gamma_1| < 1$, there is no reason why the matrix obtained by using $\gamma_i = 0$, $2 \leq i < p$, would be PD. This is already evident for the matrix $T[1, \gamma_1, 0]$ and we note in Figure 12.12 that if $2^{-1/2} < |\gamma_1| < 1$, the point $[\gamma_1, 0]$ is not in the domain D_c which ensures the positivity of Γ in (12.162). Here the question arises: what is the condition on γ_1 ensuring the matrix such that $\gamma_2 = \gamma_3 = ... = \gamma_n = 0$ is PD, whatever n? This condition is that the function γ_i defined by

$$\gamma_0 = 1 \; ; \; \gamma_1 = \gamma_1 \; ; \; \gamma_i = 0 \, , \; i > 1 \tag{12.167}$$

is a correlation function. This is realized if its FT defined by (6.42') is non-negative, as being a power spectrum. The spectrum is

$$\Gamma(\nu) = 1 + 2\gamma_1 \cos(2\pi\nu) \tag{12.168}$$

which is positive if $|\gamma_1| < 1/2$.

12.8.2 Regression Vector and Stability

This problem is classical in the study of dynamical filters and can be stated as follows: how can we ensure that the the roots of the polynomial $z^p B_p\{z\}$ defined by the vector **a** are located inside the unit circle without calculating these roots? There are many possible criteria which are discussed in works devoted to DT linear systems. The best known criterion is the Jury criterion, which can be deduced from the famous Routh criterion valid for CT systems (see [Picinbono] p. 221). These criteria are not simple to use. As an example, for $p = 2$ it can be shown, even by explicitly calculating the roots, that these roots are inside the unit circle if the point with coordinates a_1 and a_2 is located in the domain D_a inside the triangle which is shown in Figure 12.7.

Note here that the stability or positivity criteria discussed above do not have the *modularity property* discussed in connection with lattice filters in Section 12.8 (**e**). This means that it is impossible to ensure the stability or the positivity by a condition valid for each a_i or γ_i independently of the others. This is clear in the figures showing the domains D_c and D_a. The reason for this is that the boundaries of the domains are not parallel to the axes. For example, we see in (12.164) that the condition on γ_2 depends on the value chosen for γ_1, and this property shown for $p = 2$ is quite general. We shall now see that the situation changes in the case of the reflection vector.

12.8.3 Condition on the Reflection Vector

It is shown in Appendix 12.12 that the previous conditions on stability and positivity are equivalent to

$$|k_i| < 1 \; , \; 1 \leq i \leq p \tag{12.169}$$

The proof is rather technical and does not help the reading of this section. It is more interesting to concentrate on the meaning of the result.

(a) It is shown in the proof that the roots of $z^p B_p\{z\}$ are located inside the unit circle *if and only if* (12.169) holds. This means that, starting from a vector **k** satisfying (12.169), we can construct a vector **a** by using the results of Section 12.6.4, the result of which is a stable polynomial. The converse is always true. Starting from **a**, it is possible to arrive at a vector **k** by using the inverse Levinson algorithm, which always gives a well defined solution because there is no k_i such that $k_i^2 = 1$. But this unique solution also satisfies (12.169).

(b) The condition on the k_is is also equivalent to the condition on the γ_is because there is a one-to-one correspondence between **a** and **c**. From **a** we deduce **c**, as seen in Section 12.6.2, and from **c** we deduce **a** from the normal equation. Furthermore, a stable filter defined by **a** defines a vector **c** which is a correlation vector, and thus a PD covariance matrix. We then have another proof of the minimum-phase property discussed in Section 12.4(**a**). This is entirely due to the one-to-one correspondence between the vectors **a** and **c**, ensured by (12.169).

(c) Finally, it is worth pointing out that (12.169) implies that the stability domain in the space of **k** is a hypercube, and it is clear that the condition (12.169) enjoys the modularity property, which means that the stability can be tested with each k_i independently of the others.

(d) Let us now show the practical aspect of these results. In order to know if a given vector **c** corresponds to a PD Toeplitz matrix, the simplest way is to use the Levinson recursion given in Section 12.5.1. If all the k_is defined by (12.102) satisfy (12.169), the vector **c** possesses the positivity property. If at least one k_i does not satisfy (12.169), the Toeplitz matrix is not PD.

In order to know if a given vector **a** introduces a stable purely recursive filter, it is sufficient to use the inverse Levinson algorithm described in Section 12.6.1. The conclusion is the same: **a** is stable if all the k_is satisfy (12.169) and unstable if at least one k_i is such that $k_i^2 \geq 1$. This is probably the simplest stability criterion for DT dynamical filters.

12.9 *S*-STEP PREDICTION

The concept of s-step prediction, introduced in Section 12.1, will now be studied more carefully. In fact, we shall see that our knowledge of the one-step case ($s = 1$) is very helpful in solving this problem.

12.9.1 Prediction with Infinite Past

We start from the error filter in (12.3). It transforms the stationary signal $x[k]$ into its innovation of one-step prediction with infinite past $\tilde{x}[k]$. This innovation is a white noise of variance ε_∞^2 given by (12.12) and studied in Section 12.3. The filter $B\{z\}$ corresponding to (12.3) is such that $b[0] = 1$ and is a minimum phase filter. Let us call $C\{z\}$ its inverse

$$C\{z\} = 1/B\{z\} \tag{12.170}$$

which then is causal and gives the signal $x[k]$ in terms of its innovation, or

$$x[k] = \sum_{l=0}^{\infty} c[l]\,\tilde{x}[k-l] \tag{12.171}$$

This can be written as

$$x[k] = \sum_{l=0}^{s-1} c[l]\,\tilde{x}[k-l] + \sum_{l=s}^{\infty} c[l]\,\tilde{x}[k-l] \tag{12.172}$$

The two terms of this expression are orthogonal, because of the whiteness of the innovation. Furthermore, the last term of (12.172) belongs to the Hilbert subspace spanned by the RVs $x[k-s]$, $x[k-s-1]$, ... , $x[k-s-i]$, ... A result of this is that we can write (12.172) as

$$x[k] = \tilde{x}_s[k] + \hat{x}_s[k] \tag{12.173}$$

introducing the innovation and the prediction with s steps and infinite past. In fact, $\hat{x}_s[k]$ satisfies the orthogonality principle in which $x[k] - \hat{x}_s[k]$ is orthogonal to the subspace of the signal observation from $-\infty$ to $k-s$. For $s = 1$ we again find (12.2), provided that $c[0] = 1$, which will be shown below.

We must now calculate the coefficients $c[k]$. For this we assume that the one-step prediction problem is solved, or that $B\{z\}$ is known. In fact, this is equivalent to calculating the strong factorization of the z-spectrum of the signal $x[k]$, as seen in (12.9). As $C\{z\}$ is given by (12.170) and is causal, we can write $B\{z\}\ C\{z\} = 1$, and using the property $b[0] = 1$, we find

$$[1 + b_1 z^{-1} + b_2 z^{-2} + \dots][c_0 + c_1 z^{-1} + c_2 z^{-2} + \dots] = 1 \tag{12.174}$$

where b_i and $b[i]$ are identical. This gives $c[0] = 1$. We also have $b_1 + c_1 = 0$, or

$$c[1] = -b[1] \tag{12.175}$$

Furthermore, the nullity of the coefficient of z^{-i} yields

$$c[i] = -(b[i] + b[i-1]c[1] + \dots + b[1]c[i-1]) \tag{12.176}$$

which is a recursive procedure for calculating the coefficients $c[i]$ in terms of the $b[i]$s.

In conclusion, starting from the impulse response $b[i]$ of the error filter studied in Section 12.2, we can calculate the coefficients $c[i]$ for $1 \le i \le s-1$, which allows us to obtain the innovation (or the error) of the s-step prediction with infinite past, which is

$$\tilde{x}_s[k] = \sum_{l=0}^{s-1} c[l]\,\tilde{x}[k-l] \tag{12.177}$$

The prediction filter is, of course, defined by

$$\hat{x}_s[k] = x[k] - \tilde{x}_s[k] \tag{12.178}$$

This operation is summarized in Figure 12.13 corresponding to $s = 3$. Note that the filter $B\{z\}$ in the figure can be realized either by a transversal filter or by a lattice filter, as discussed in Section 12.7. The transfer function $H_s\{z\}$ of the s-step error filter is deduced from (12.177), which gives

$$H_s\{z\} = B\{z\} \sum_{l=0}^{s-1} c[l]\, z^{-l} \tag{12.179}$$

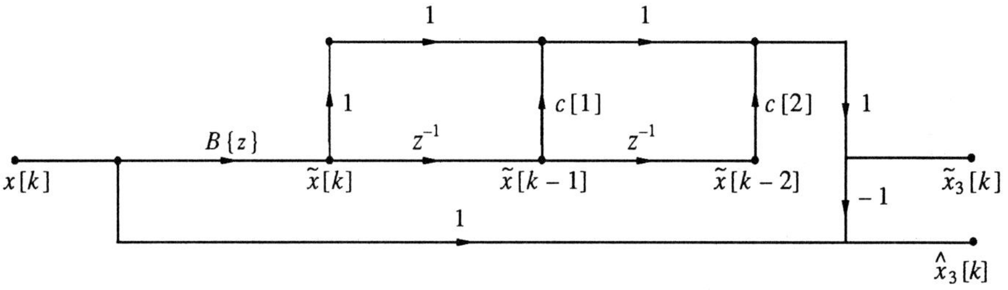

Figure 12.13. Three steps predictor and error filters.

Let us now calculate the prediction error. It is given by

$$\varepsilon_{s,\infty}^2 = E\{\tilde{x}_s^2[k]\} \tag{12.180}$$

and using (12.177), and remembering that the $\tilde{x}[i]$s are orthogonal, we obtain

$$\varepsilon_{s,\infty}^2 = \varepsilon_\infty^2 \left\{1 + \sum_{l=1}^{s-1} c^2[l]\right\} \tag{12.181}$$

where ε_∞^2 corresponds to the one-step prediction, and is calculated in Section 12.3. It is clear that the s-steps prediction error cannot be smaller than the one-step, which is quite natural.

Finally, note that $\tilde{x}_s[k]$ given by (12.177) has no reason to remain a white noise, as was the case in the one-step innovation.

Example 12.4 Some s-step predictors. Consider first the case where $x[k]$ is an AR(1) signal driven by unit white noise. This implies that $\varepsilon_\infty^2 = 1$. Using (9.6) we deduce that the strong factorization is realized with

$$B\{z\} = z^{-1}(z - a) \tag{12.182}$$

which gives $b_1 = -a$ and $b_i = 0$ for $i > 1$. This implies that $c[1] = a$ and (12.176) yields $c[i] = a\, c[i-1]$. Thus we have $c[k] = a^k$, which defines the

s-step predictor as in Figure 12.13. Its transfer function is given by (12.179) and (12.182), which yields

$$H_s\{z\} = 1 - a^s z^{-s} \tag{12.183}$$

This result can be obtained directly by noting that $\tilde{x}[k] = x[k] - ax[k-1]$. Inserting $c[l] = a^l$ in (12.177) and using (12.178) yields

$$\hat{x}_s[k] = a^s x[k-s] \tag{12.184}$$

This is also a consequence of the Markovian property of the AR(1) signal studied in Chapter 9.

The prediction error is given by (12.181) which becomes

$$\varepsilon^2_{s,\infty} = \frac{1 - a^{2s}}{1 - a^2} \tag{12.185}$$

Let us now consider the case of the MA(1) signal. Suppose that it is generated by a unit white noise through a filter with the transfer function $H\{z\} = 1 - a z^{-1}$. As a result, we have

$$B\{z\} = [1 - az^{-1}]^{-1} \quad ; \quad C\{z\} = 1 - az^{-1} \tag{12.186}$$

and the transfer function of the s-step predictor is deduced from (12.179), which gives

$$H_s\{z\} = [1 - az^{-1}]^{-1}[1 - az^{-1}] = 1 \ , \ s > 1 \tag{12.187}$$

This result is easily understood. In fact, we saw in Example 9.3 that the correlation function $\gamma[k]$ of an MA(1) signal is null for $k > 1$. This means that $x[k+1]$ is uncorrelated with the past values $x[k-i]$, $i \geq 1$, and we arrive at a null prediction. The corresponding error given by (12.181) is

$$\varepsilon^2_{s,\infty} = 1 + a^2 \tag{12.188}$$

which is the variance of the signal and indicates the null prediction. Other examples are discussed as problems.

12.9.2 Prediction with Finite Past

The procedure is similar to that described in the previous section. Starting from the observation vector $\mathbf{X}[k]$ defined by (9.21), we want to estimate $x[k]$. The basic idea is to transform the components of $\mathbf{X}[k]$ into a sequence of orthogonal RVs by using a sequence of innovations. The oldest component of $\mathbf{X}[k]$ is $x[k-p]$, and we state

$$\tilde{x}_0[k-p] \triangleq x[k-p] \tag{12.189}$$

Realizing the one-step prediction with past one, we introduce

$$\tilde{x}_1[k-p+1] \triangleq x[k-p+1] - \hat{x}_1[k-p+1] \tag{12.190}$$

and because of the orthogonality principle, $\tilde{x}_1[k-p+1]$ is orthogonal to $\tilde{x}_0[k-p]$. Furthermore, the calculation of $\tilde{x}_1[k-p+1]$ is obtained by (12.60) using the coefficient a_1^1 in Table 12.1. Following the same procedure, we introduce

$$\tilde{x}_m[k-p+m] \triangleq x[k-p+m] - \hat{x}_m[k-p+m] \tag{12.191}$$

which is orthogonal to the innovations $\tilde{x}_i[\ ]$ for $i < m$. Similarly, $\tilde{x}_m[k-p+m]$ is calculated by using (12.60) where the vector $\mathbf{a}_m$ is the mth column of Table 12.1. Finally, we arrive at

$$\tilde{x}_p[k] = x[k] - \hat{x}_p[k] \tag{12.192}$$

which is (12.64). All these innovations are deduced from the elements of Table 12.1, which can be obtained by the Levinson recursion studied in Section 12.5.1.

From the previous equations we can deduce that

$$x[k] = \sum_{l=0}^{p} c_p[l]\, \tilde{x}_{p-l}[k-l] \tag{12.193}$$

where $c_p[0] = 1$. This comes from the fact that if we replace m by $p-l$, (12.191) can be expressed as

$$x[k-l] = \tilde{x}_{p-l}[k-l] + \hat{x}_{p-l}[k-l] \tag{12.194}$$

The condition $c[0] = 1$ comes from (12.192), which is the only equation in which the term $\tilde{x}_p$ appears.

The relation (12.193) corresponds to (12.171) valid when the past is infinite. We can now decompose (12.193) as (12.172), which gives

$$x[k] = \sum_{l=0}^{s-1} c_p[l]\, \tilde{x}_{p-l}[k-l] + \sum_{l=s}^{p} c_p[l]\, \tilde{x}_{p-l}[k-l] = \tilde{x}_p^s[k] + \hat{x}_p^s[k] \tag{12.195}$$

and $\hat{x}_p^s[k]$ is the s-step prediction with past p, because the two terms are orthogonal and $\hat{x}_p^s[k]$ is only a function of $x[k-s]$, $x[k-s-1]$, ..., $x[k-p]$.

It is obvious that the error in this prediction problem is the variance of $\tilde{x}_p^s[k]$. Noting that $E\{(\tilde{x}_{p-l}[k-l])^2\}$ is the one-step prediction error with past $p-l$ given by (12.66) or (12.100), we deduce from the first term of (12.195) that

$$\varepsilon_s^2[p] = \varepsilon_p^2 + \sum_{l=1}^{s-1} c_p^2\,[l]\varepsilon_{p-l}^2 \tag{12.196}$$

The last point that must be solved is, of course, the calculation of the coefficients $c_p[l]$ in (12.193). For this we multiply this expansion by $\tilde{x}_{p-i}[k-i]$, and taking the expectation, we obtain

$$c_p[i]\, E\{(\tilde{x}_{p-i}[k-i])^2\} = E\{x[k]\tilde{x}_{p-i}[k-i]\} \tag{12.197}$$

because of the orthogonality of the innovations. The coefficient of $c_p[i]$ is the prediction error ε_{p-i}^2 deduced from the Levinson algorithm and used in (12.196). The last term of (12.197) is easy to calculate if we note that

$$\tilde{x}_{p-i}[k-i] = x[k-i] - \sum_{j=1}^{p-i} a_j^{p-i} x[k-i-j] \tag{12.198}$$

This last expression is due to the definition of the one-step prediction by (12.60), which gives

$$\hat{x}_{p-i}[k] = \sum_{j=1}^{p-i} a_j^{p-i} x[k-j] \tag{12.199}$$

where the a_n^ms are the components of the vector $\mathbf{a}_m$ which appears in the mth column of Table 12.1. Using (12.198) in (12.197), we deduce that

$$c[i] = [\varepsilon_{p-i}^2]^{-1} \left\{ \gamma[i] - \sum_{j=1}^{p-i} a_j^{p-i} \gamma[i+j] \right\} \tag{12.200}$$

where $\gamma[.]$ is the correlation function of the signal $x[k]$.

In conclusion, we note that the s-step prediction problem with finite past is completely solved with the use of the Levinson algorithm, which gives both the errors ε_m^2 and the vectors $\mathbf{a}_m$ used in the previous equations.

Let us note here that if we want to estimate $x[k+s-1]$ in terms of the observation vector $\mathbf{X}[k]$ defined by (9.21), we have to solve the same kind of problem, but the Levinson recursion must be calculated with the covariance matrix of the vector

$$\mathbf{Y}^T[k] \triangleq \{x[k-p], x[k-p+1],\ldots,$$

$$x[k-1], x[k], x[k+1], \ldots, x[k+s-2]\} \tag{12.201}$$

which is a vector of $\mathbb{R}^{p+s-1}$. We recommend the reader to rewrite the above equations in this context. In fact, in these equations the Levinson algorithm was calculated from an n x n covariance matrix, and the estimation of $x[k]$ was realized in terms of the vector

$$\mathbf{Z}^T[k] \triangleq \{x[k-p], x[k-p+1],\ldots, x[k-1]\} \tag{12.202}$$

It is obvious that for $s = 1$ the two vectors $\mathbf{Y}$ and $\mathbf{Z}$ are coincide.

12.10 THE MAXIMUM ENTROPY METHOD

Let $x[k]$ be a zeromean second order real signal and suppose that its correlation function $\gamma[k]$ is known for $0 \le k \le p$. The problem which appears is to extend this correlation function to any value of k. We already know of some cases where this is possible, the simplest being when $x[k]$ is an MA(p) signal. In fact, we saw in Section 9.3 that in this situation the correlation function is null for $k > p$, and its extension then becomes obvious: it suffices to add zeros to the sequence $\gamma[0]$, $\gamma[1]$, ..., $\gamma[p]$ to obtain the entire correlation function. However, we know that this method is not general, and we saw in (12.167) an example illustrating this point.

Another case in which the problem can be easily solved is when the signal $x[k]$ is an AR(p) signal. The knowledge of $\gamma[k]$ for $0 \le k \le p$ is equivalent to that of the variance and the vector **c**, and we saw in Section 12.6 how to pass from this vector to the regression vector **a**. Using **a** and **c** it is then possible to calculate the entire correlation function by using (9.32).

The problem of the extension of a correlation function from its first values is important in many situations. This is especially the case in *spectral analysis* when we want to use (6.42') which is a series using all the values of k.

In this section we will establish the following result: using the concept of entropy rate of a signal, it appears that the correlation function whose first p values are known and which maximizes the entropy rate is that of a *normal* AR(p) signal. This signal is completely characterized by the known vector **c**, as indicated just above.

It is not our objective here to develop the origin and properties of entropy, which are analyzed in books on information theory or statistical mechanics. We shall use only the concepts introduced in Section 4.8.1, and especially the result concerning the maximum of entropy of a vector with a given covariance matrix.

12.10.1 Entropy and Innovations

Let **x** be a random vector of components x_i. Using a procedure similar to that in the previous section concerning s-step prediction, we introduce the innovation $\tilde{x}_i$ defined by

$$\tilde{x}_i = x_i - \text{LMSE}[x_i | x_1, x_2, \ldots, x_{i-1}] \tag{12.203}$$

where the last term means the linear mean square estimation of x_i in terms of $x_1, x_2, \ldots, x_{i-1}$. As a result of the orthogonality principle, $\tilde{x}_i$ is orthogonal to (or uncorrelated with) $x_1, x_2, \ldots, x_{i-1}$, and thus to $\tilde{x}_1, \tilde{x}_2, \ldots, \tilde{x}_{i-1}$ because $\tilde{x}_{i-k}$ is a linear function of $x_1, x_2, \ldots, x_{i-k}$. It is then possible to introduce a linear transformation, called an innovation transformation, and written as

$$\tilde{\mathbf{x}} = \mathbb{A}\,\mathbf{x} \tag{12.204}$$

where $\tilde{\mathbf{x}}$ is obviously the vector whose components are the $\tilde{x}_i$s. From this procedure we find that the square matrix $\mathbb{A}$ is lower triangular and all its diagonal elements are equal to 1. Its determinant is then equal to 1. The covariance matrix of the vector $\tilde{\mathbf{x}}$ is given by

$$\mathbb{\Gamma}' = \mathbb{A}\,\mathbb{\Gamma}\,\mathbb{A}^T \tag{12.205}$$

according to Section 3.5.2(**c**).

Suppose now that the vector **x** is $N(\mathbf{0}, \mathbb{\Gamma})$. Its entropy is given by (4.178). The vector $\tilde{\mathbf{x}}$ is obviously $N(\mathbf{0}, \mathbb{A}\,\mathbb{\Gamma}\,\mathbb{A}^T)$, and its entropy is given by (4.180). However, as the determinant of $\mathbb{A}$ is equal to 1, we deduce that the last term of (4.180) is null, which means that the vectors **x** and $\tilde{\mathbf{x}}$ have the same entropy. In other words, the innovation transformation (12.104) preserves the entropy in the normal case.

Let us now calculate this entropy. This is especially simple for the vector $\tilde{\mathbf{x}}$ because it is a normal vector with a diagonal covariance matrix, as a consequence of the orthogonality of the $\tilde{x}_i$s. Let us call η_i^2 the variance of the component $\tilde{x}_i$. It is clear from the definition of the innovation that η_i^2 is also the mean square error in the estimation problem appearing in (12.203). Combining all these results and using (4.178) and (4.180), we deduce that the entropy of the vector $\mathbf{x}$ is

$$H(\mathbf{x}) = (n/2)\ln(2\pi e) + (1/2)\sum_{j=1}^{n} \ln(\eta_i^2) \tag{12.206}$$

This is the basis for introducing the *entropy rate* of a DT stationary signal $x[k]$. This quantity is defined by

$$h_x = \lim_{n\to\infty} (1/n)H(\mathbf{x}) \tag{12.207}$$

where $H(\mathbf{x})$ is the entropy of the vector of components $x[1], x[2], \ldots, x[n]$. It is clear that

$$\lim_{i\to\infty} \eta_i^2 = \varepsilon_\infty^2 \tag{12.208}$$

which is the infinite past one-step prediction error studied in the previous sections. Consequently, if the signal is not singular, or if $\varepsilon_\infty^2 \neq 0$, we deduce from (12.206) and (12.207) that

$$h_x = (1/2)\ln(2\pi e) + (1/2)\ln(\varepsilon_\infty^2) \tag{12.209}$$

This equation shows that for a normal DT signal the entropy rate is an increasing function of the prediction error ε_∞^2. For the discussion which follows it is important to remember the expression of the error given by (12.104), and to extend this expression to the infinite past case, which gives

$$\varepsilon_\infty^2 = \gamma_0 \prod_{n=1}^{\infty} (1 - k_n^2) \tag{12.210}$$

12.10.2 The Maximum of Entropy Rate

Proposition. Consider a function $\gamma[i]$, $0 \le i \le p$, such that the Toeplitz matrix $T\{\gamma[0], \gamma[1], \ldots, \gamma[p]\}$ is positive-definite. The signal $x[k]$ such that the first values of its correlation are $\gamma[i]$, $0 \le i \le p$, and giving the maximum entropy rate, is the normal AR(p) signal defined by the given $\gamma[i]$s.

Proof. The normal property is a direct consequence of the proposition in Section 4.8.1. In fact, it is impossible for the maximum of the entropy rate to come from a non-normal signal, because to each non-normal signal it is possible to associate a normal signal whose entropy rate cannot be smaller.

Let us now call $G_p(\{\gamma[i]\})$ the class of normal signals which have the same correlation function $\gamma[i]$ for $0 \le i \le p$. As a result of (12.209), maximizing the entropy rate is equivalent to maximizing the infinite past prediction error (12.210). However, all the signals of $G_p(\{\gamma[i]\})$ have the same reflection coefficients k_i for $1 \le i \le p$ and the same variance $\gamma[0]$. Consequently ε_∞^2 is maximum if all the k_is which are not determined are null. These k_is correspond to $i > p$, which characterizes the normal AR(p) signal defined by the $\gamma[i]$s. This completes the proof.

12.10.3 Physical Interpretation of the Maximum Entropy Method

It may be strange to use a criterion which maximizes the prediction error, and we shall explain this by using the very simple example where only $\gamma[0]$ is known. Thus $G_0(\{\gamma[0]\})$ corresponds to all the normal signals with the same variance $\gamma[0]$. The prediction errors of this class of signals satisfy the general rule

$$0 \le \varepsilon_\infty^2 \le \gamma_0 \tag{12.211}$$

presented in Section 12.2.2(**b**). The lower bound is reached for all normal signals of variance $\gamma[0]$ and giving a singular prediction. We have previously discussed this question, showing that there are many and various signals with this property. However, the upper bound is reached if the correlation function takes the form

$$\gamma[k] = \gamma[0]\ \delta[k] = \gamma_0\ \delta[k] \tag{12.212}$$

which completely defines the normal DT white noise of variance $\gamma[0]$. In this example it appears that maximizing the entropy or the prediction error can be understood as maximizing the whiteness of the signal. This interpretation, evident for $p = 0$, can be extended to any value of p, for example, by using the concept of spectral flatness of a power spectrum $\Gamma(\nu)$ defined by

$$f = \frac{\exp\left\{ \int_{-1/2}^{+1/2} \ln[\Gamma(\nu)]\, d\nu \right\}}{\int_{-1/2}^{+1/2} \Gamma(\nu)\, d\nu} = \frac{\varepsilon_\infty^2}{\sigma^2} \tag{12.213}$$

In fact, maximizing this flatness for a given value of σ also corresponds to maximizing the prediction error.

12.11 WOLD DECOMPOSITION AND RELATED PROBLEMS

12.11.1 Principle of Wold Decomposition

The principle of Wold decomposition has often been used without the mathematical arguments below. The first and simplest situation appears in the *finite* case, discussed in the previous section. We shall briefly recall the basic idea. Consider a signal $x[k]$

observed from $k = 1$ to $k = n$, and let $\mathbf{x}$ be the vector whose components are the $x[i]$s. These components are in general correlated, and by using successive innovations it is possible to transform these components into an orthogonal sequence. For this purpose we introduce the LMSE $\hat{x}[k]$ of $x[k]$ in terms of past values, or $x[i]$, $1 \leq i \leq k-1$. Let $\tilde{x}[k]$ be the corresponding innovation

$$\tilde{x}[k] = x[k] - \hat{x}[k] \tag{12.214}$$

It is obvious that $\tilde{x}[k]$ is a function of the $x[i]$s for $1 \leq i \leq k$ and that the successive innovations are orthogonal, as a consequence of the orthogonality principle. In fact, $\tilde{x}[k]$ is orthogonal to past values of $x[k]$ and $\tilde{x}[k-p]$ is a linear combination of these past values. Furthermore,

$$x[k] = \tilde{x}[k] + \hat{x}[k] \tag{12.215}$$

which means that $x[k]$ is a linear combination of $\tilde{x}[k]$ and $x[k-i]$, $1 \leq i \leq k-1$. Repeating the same operation for $x[k-1]$, $x[k-2]$, ..., we arrive at the following expression

$$x[k] = \sum_{i=0}^{k-1} c_k[i]\, \tilde{x}[k-i] \tag{12.216}$$

where $c_k[0] = 1$. This expression is valid for any k from $k = 1$ to $k = n$.

This equation gives an expression of the values $x[k]$ of the signal in terms of the past *innovations*. This is valid for $1 \leq k \leq n$, which explains why this equation is referred to as the finite case. But (12.216) contains the basic idea of the Wold representation, which is the expansion of a signal in terms of its innovation.

Let us now extend this idea to situations related to the infinite case. Let $x[k]$ be an AR(1) signal. This signal can be expressed as in (9.4), and conversely $u[k]$ in this equation satisfies (9.7), which gives

$$u[k] = x[k] - a\, x[k-1] \tag{12.217}$$

As $x[k] - ax[k-1]$ is orthogonal to the past values of $x[k]$, we deduce that $u[k]$ is the innovation of the one-step prediction with infinite past. So (12.217) expresses the innovation in terms of the signal and (9.4) expresses the signal in terms of its innovation, as in (12.216). The purpose of Wold decomposition is to extend this kind of expression to a broader class of stationary discrete-time signals.

12.11.2 Wold Decomposition

Let $x[k]$ be a regular discrete-time stationary signal. This means that the prediction error ε_∞^2 introduced in Section 12.2 and studied in Section 12.3 is not zero. Let $\tilde{x}[k]$ be the innovation in the infinite past one-step prediction of $x[k]$.

In the following discussion we assume that the power spectrum $\Gamma(\nu)$ of $x[k]$ is expressed in the form

$$\Gamma(\nu) = S(\nu) + \sum_i \gamma_i \delta(\nu - \nu_i) \tag{12.218}$$

where $S(\nu)$ is a non-negative function without spectral lines, and satisfying the condition (6.86) ensuring strong factorization. The last term of (12.218) exhibits the possible spectral lines and if this term is not zero we cannot apply (6.86) for the power spectrum $\Gamma(\nu)$. Finally it is worth pointing out the analogy between (12.118) and (2.14). This appears if, instead of the power spectrum, we use the spectral distribution function appearing in (6.38) or (6.39). In this case the two terms of (12.218) correspond to the continuous and discontinuous cases of the spectral distribution function and the spectral lines correspond to the discontinuities of this function. The possible singular case is not considered here. We shall discuss the two cases of spectra with or without spectral lines separately.

(a) Signal without spectral lines. In this case $\Gamma(\nu)$ satisfies (6.86) and the Wold decomposition can be written as

$$x[k] = \sum_{l=0}^{\infty} c[l]\, \tilde{x}[k-l] \quad , \ c[0] = 1 \tag{12.219}$$

This expression is a direct consequence of the results of Section 12.2. In fact the error filter $B\{z\}$ in (12.3) is given by (12.10), where $F\{z\}$ is defined by the strong factorization (11.118). Consequently, $F\{z\}$ is a minimum phase filter and $B\{z\}$ has the same property. The filtering equation (12.3) can then be inverted by using a causal filter. This is (12.219). The condition $c[0] = 1$ is a result of the definition of the innovation in (12.7). In fact, as $\hat{x}[k]$ is only a function of past values of the signal, the coefficient of $\tilde{x}[k]$ is equal to 1. The condition $c[0] = 1$ can also be deduced from (12.174) defining the inverse filter of $B\{z\}$.

(b) Signal with spectral lines. This means that there is at least one non-zero γ_i in (12.218). The Wold decomposition can be stated as follows.

Proposition. Let $x[k]$ be a regular signal with the spectrum (12.118) and $\tilde{x}[k]$ its innovation. It is then possible to write

$$x[k] = u[k] + v[k] \tag{12.220}$$

with the following properties:

(α) The signals $u[k]$ and $v[k]$ are uncorrelated.

(β) The spectrum of $u[k]$ is $S(\nu)$.

(γ) The signal $u[k]$ can be expressed as

$$u[k] = \sum_{l=0}^{\infty} c[l]\, \tilde{x}[k-l] \quad , \ c[0] = 1 \tag{12.221}$$

(δ) The spectrum of $v[k]$ contains only spectral lines, as the last term of (12.218). Consequently, $v[k]$ is singular (or deterministic, or predictable).

(ε) The decomposition (12.220) is unique.

We will not present the proof of this proposition which is rather technical and can be found in Anderson or Hannan. It is more interesting to discuss its consequences.

12.11.3 Comments on the Wold Decomposition

(a) Assumption of regularity. It is important to note that (12.220) and its properties are valid provided that the signal $x[k]$ is regular, which means that $\varepsilon_\infty^2 \neq 0$. This is not the case of some signals studied in Section 12.3 which are predictable and whose power spectrum does not contain spectral lines. However, we saw in Section 12.3 (**c**) that if a signal is predictable *with a finite past*, its spectrum contains only spectral lines. In this case (12.220) again becomes valid with $u[k] = 0$.

(b) The error formula revisited. Comparing (12.219) and (12.221), we note that the innovations of $x[k]$ and of $u[k]$ are the same. This means that the presence of the term $v[k]$ in (12.220), or of spectral lines in the spectrum of $u[k]$, does not change the error ε_∞^2. Of course the predictors of $x[k]$ and of $u[k]$ are not the same, but they give the same asymptotic prediction error.

Because of the importance of this result, we shall present some further information. Consider first the signal (12.105). Its correlation function is given by (12.106) and the corresponding power spectrum is

$$\Gamma(\nu) = v_N + v_A\,\delta(\nu) \tag{12.222}$$

It is then the sum of a constant spectrum and a spectral line at the frequency 0. Its Wold decomposition is simply given by (12.105), and as $n[k]$ is white, it is equal to its innovation, giving a prediction error equal to v_N. Let us now calculate the prediction error of $x[k]$. In the case of prediction with a finite past, the prediction error is given by (12.66). Using (12.106) and (12.110), we deduce that

$$\varepsilon_n^2 = v_N + v_A(1 + n\rho)^{-1} \tag{12.223}$$

and when $n \to \infty$, we obtain $\varepsilon_\infty^2 = v_N$, which is the variance of the white noise. This shows clearly that the spectral line changes the prediction error with finite past but not that with infinite past.

Let us now give another argument showing that a spectral line plays no role in the expression of the error (12.45). Consider the spectrum $\Gamma_b(\nu)$ represented in Figure 12.14.

It corresponds to the case of the sum of two uncorrelated signals of the same power equal to 1. The first is a unit white noise and the second a band-limited white noise in the frequency band $-b$, $+b$. It is easy to calculate the prediction error ε_∞^2 by (12.45). As a result of the properties of the logarithm function, this error tends to 1 when b tends to 0, which means that the contribution of the central band of the spectrum, which always gives a unit power, tends to 0 for the prediction error. Furthermore when b tends to 0, the spectrum of Figure 12.14 tends to the form (12.218) with a spectral line at the zero frequency. This reasoning can easily be extended to several spectral lines.

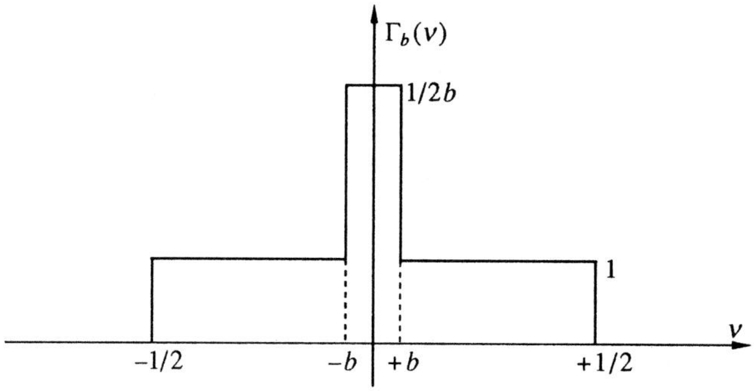

Figure 12.14. Example of power spectrum.

(c) On the sum of predictable signals. Let $a[k]$ and $b[k]$ be two uncorrelated predictable signals, and call $x[k]$ their sum. It is worth pointing out that $x[k]$ is not necessarily predictable. This is especially the case when the spectra of $a[k]$ and $b[k]$ are zero for some frequency bands in such a way that their sum is never null. On the other hand, if the spectra of the two signals contain only spectral lines, this is the same for $x[k]$ which is then also predictable.

12.11.4 Evolution of the Prediction and Interpolation Errors

The one-step and infinite past prediction error plays an important role in classifying signals between the regular and the singular. The general expression of this error is given by (12.45) and many of its properties were given throughout this chapter.

Our purpose here is to study the evolution of the error when the past is increasing, especially for singular signals. General behavior of the prediction error was described in Section 12.4(e) and we want here to present results similar to those in Figures 12.4 and 12.5 but calculated explicitly for some specific correlation functions. Furthermore, as there are some relationships between prediction and interpolation, studied at the end of Section 12.3, we will make the same type of calculations for the interpolation error.

Consider first the *prediction error*. In order to calculate ε_p^2 given by (12.66), we calculate the vector $\mathbf{a}_p$ solution of (12.76). This vector is obtained recursively from the Levinson algorithm. It is not even necessary to calculate the scalar product in (12.66), because the prediction error is obtained recursively from the reflection coefficients by using (12.101). It is this expression which is used in the computer calculations given below. In order to calculate ε_p^2 for a signal of unit power, it is enough to know the correlation function or the power spectrum, and to use the Levinson recursion.

Consider now the *interpolation error*. Let us call η_p^2 this error. This is the error in the LMSE of $x[k]$ in terms of both $x[k+i]$ and $x[k-i]$, $1 \leq i \leq p$. The calculation of this interpolation error is analyzed as a problem whose result is similar to that obtained for prediction, which means that this interpolation error can be calculated by a recursive algorithm with a complexity similar to that of the Levinson algorithm.

Let us now present some results obtained by these methods.

(a) Prediction error of (12.54). The power spectrum (12.54) is represented in Figure 12.2 and seems null for frequencies smaller than 0.1. In reality it is only null at the frequency 0. The evolution of its prediction error is represented in Figure 12.15, which shows a regular decreasing of this error, as for singular signals. In fact, the signal is only asymptotically singular because its spectrum is not reduced to spectral lines and the figure shows a slow decreasing of the error for $p > 200$. In practice, using 200 values of the past for prediction is a very long observation time.

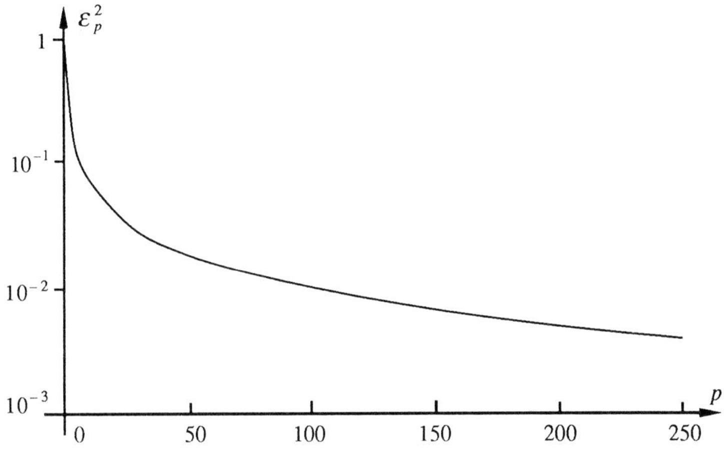

Figure 12.15 Evolution of the prediction error corresponding to the spectrum (12.54).

(b) Non-predictable and interpolable signals. In Figure 12.16 we see the prediction and interpolation errors corresponding to the power spectrum $\Gamma(\nu) = 12\nu^2$ corresponding to a signal with a unit power. It is easy to show that the prediction error ε_∞^2 given by (12.45) is $3e^{-2}$, while the interpolation error η_∞^2 given by (11.107) is zero. This property appears in Figure 12.16, showing that the asymptotic prediction error is reached for $p \approx 20$ while the interpolation error decreases regularly.

(c) Signals with band-limited spectra. We have already shown above that any signal whose power is null in a frequency band B is predictable. By using the same calculation method, it is possible to observe experimentally that the prediction error ε_p^2 decreases exponentially for p sufficiently large, or that $\varepsilon_p^2 = k\,\alpha^p$. This property is mathematically shown in [Slepian] for the case of a band-limited white signal, and can be extended to any band-limited spectrum. This exponential behavior also appears for the interpolation error.

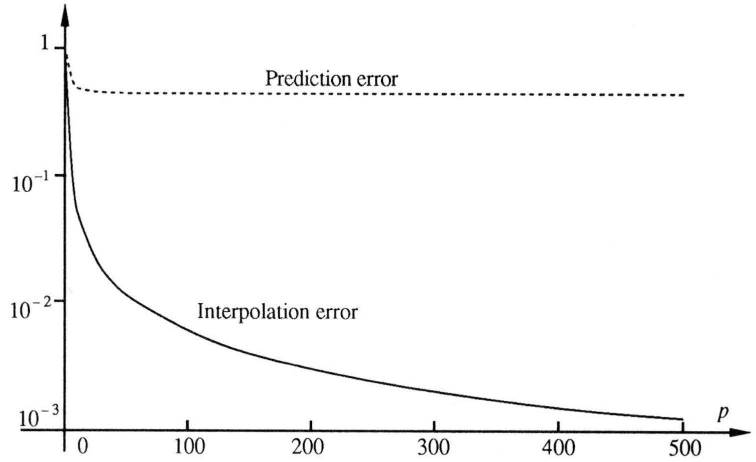

Figure 12.16 Prediction and interpolation errors corresponding to $\Gamma(\nu) = 12\nu^2$.

12.12 APPENDIX ON STABILITY

The purpose of this section is to give a proof for the condition (12.169) and extensions of this condition. The problem is to relate the locations of the roots of the polynomials $B_i\{z\}$, transfer functions of the filters calculating the one-step innovation with finite past, and the reflection coefficients which appear in the lattice representation of these filters represented in Figures 12.10 and 12.11.

As the detailed calculations are rather technical, we shall outline the main steps in the reasoning. The basic idea of the lattice representation deduced from the Levinson algorithm is the concept of *recursion*. In other words, we calculate the vector $\mathbf{a}_n$ from the vector $\mathbf{a}_{n-1}$. This idea can easily be translated to polynomials, and was used to calculate $B_i\{z\}$ for $1 \le i \le 3$ in (12.155) to (12.157). So our first objective is to establish this recursion between successive polynomials. The second idea from the concept of recursion is to present a proof by induction.

12.12.1 Recursion on Polynomials

Consider the box appearing in Figure 12.9. To simplify notation, let us introduce the signals $u[k]$ and $v[k]$ defined by

$$u_n[k] \triangleq \tilde{x}_n[k] \quad ; \quad v_n[k] \triangleq \tilde{x}_n^{(-)}[k-n] \tag{12.224}$$

Consequently, using a lattice structure similar to that in Figure 12.11, we realize a linear filter with one input, $x[k]$, and two outputs, $u_n[k]$ and $v_n[k]$. The corresponding transfer functions of these filters are $B_n\{z\}$ and $B_n^{(-)}\{z\}$ respectively. Consequently the outputs of the filters $B_{n-1}\{z\}$ and $B_{n-1}^{(-)}\{z\}$ are $u_{n-1}[k]$ and $v_{n-1}[k]$ respectively. Furthermore note that (12.224) implies that $\tilde{x}_{n-1}^{(-)}[k-n] = v_{n-1}[k-1]$. Using this result and looking at the flow graph of Figure 12.10, or at (12.150) and (12.154), we deduce a recursion on the transfer functions $B\{z\}$ which is

$$B_n\{z\} = B_{n-1}\{z\} - k_n z^{-1} B_{n-1}^{(-)}\{z\} \tag{12.225}$$

$$B_n^{(-)}\{z\} = -k_n B_{n-1}\{z\} + z^{-1} B_{n-1}^{(-)}\{z\} \tag{12.226}$$

Furthermore, we see in Figure 12.11 that

$$B_1\{z\} = 1 - k_1 z^{-1} \tag{12.227}$$

$$B_1^{(-)}\{z\} = -k_1 + z^{-1} = z^{-1} B_1\{z^{-1}\} \tag{12.228}$$

We shall show that this relation is general, or that, whatever n,

$$B_n^{(-)}\{z\} = z^{-n} B_n\{z^{-1}\} \tag{12.229}$$

This is valid for $n = 1$, from (12.228). Suppose that the property is valid for $n-1$. By using the previous recursion, we obtain

$$B_n^{(-)}\{z\} = -k_n B_{n-1}\{z\} + z^{-n} B_{n-1}\{z^{-1}\} = z^{-n}\left[B_{n-1}\{z^{-1}\} - k_n z^n B_{n-1}\{z\}\right]$$

$$= z^{-n}\left[B_{n-1}\{z^{-1}\} - k_n z B_{n-1}^{(-)}\{z^{-1}\}\right] = z^{-n} B_n\{z^{-1}\} \tag{12.230}$$

which is (12.229). Consequently, the property (12.229) is valid whatever n and its meaning is that the polynomial $B_n^{(-)}\{z\}$ can be obtained as soon as $B_n\{z\}$ is known. Physically this means that there is only one transfer function to calculate the two outputs of the lattice filter.

The polynomials $B_n\{z\}$ are polynomials of degree n in z^{-1}. This appears in (12.155) to (12.157) and can be shown by induction from (12.225) and (12.226). However, for stability problems we have to work with polynomials in z defined by

$$P_n\{z\} = z^n B_n\{z\} \tag{12.231}$$

This appears, for example, when passing from (9.17) to (9.18). In stability problems we are dealing with the location of the roots of $P_n\{z\}$ with respect to the unit circle (UC).

The first step is to transpose to the polynomials $P_n\{z\}$ the recursions valid for the $B_n\{z\}$s. As a result of (12.230), $B_n^{(-)}\{z\}$ is also a polynomial of degree n in z^{-1}, and we can again use (12.231) to introduce the polynomial

$$P_n^r\{z\} \triangleq z^n B_n^{(-)}\{z\} \tag{12.232}$$

Multiplying the recursion (11.225) by z^n, and using (12.231) and (12.232), we obtain

$$P_n\{z\} = z\,P_{n-1}\{z\} - k_n P^r_{n-1}\{z\} \tag{12.233}$$

This is the increasing order recursion on the polynomials $P_n\{z\}$. As for the Levinson recursion, there is also a decreasing order recursion. To establish this, let us first show that (12.229) can be translated to polynomials $P_n\{z\}$. In fact, using this relation, (12.231) and (12.232), we obtain

$$P^r_n\{z\} = B_n\{z^{-1}\} = z^n P_n\{z^{-1}\} \tag{12.234}$$

Using this relation in (12.233), we obtain

$$P^r_n\{z\} = z^n[z^{-1}P_{n-1}\{z^{-1}\} - k_n P^r_{n-1}\{z^{-1}\}] \tag{12.235}$$

As a result of (12.234),

$$P_{n-1}\{z^{-1}\} = z^{-(n-1)}P^r_{n-1}\{z\}\,;\, P^r_{n-1}\{z^{-1}\} = z^{-(n-1)}P_{n-1}\{z\} \tag{12.236}$$

in such a way that (12.235) becomes

$$P^r_n\{z\} = -z\,k_n P_{n-1}\{z\} + P^r_{n-1}\{z\} \tag{12.237}$$

From (12.233) and (12.237) we deduce that

$$P_{n-1}\{z\} = \frac{1}{1-k_n^2}\,z^{-1}[P_n\{z\} + k_n P^r_n\{z\}] \tag{12.238}$$

which is similar to (12.129).

Before entering into the proof of the root location theorem, let us discuss the structure of the polynomials $P_n\{z\}$ and $P^r_n\{z\}$. Let $\mathbf{a}_n$ be the vector solution of the normal equations (12.76). The corresponding filter is

$$B_n\{z\} = 1 - a_1 z^{-1} - a_2 z^{-2} - \cdots - a_n z^{-n} \tag{12.239}$$

and (12.231) gives

$$P_n\{z\} = z^n - a_1 z^{n-1} - a_2 z^{n-2} - \cdots - a_n \tag{12.240}$$

However, as a result of (12.229),

$$B^{(-)}_n\{z\} = z^{-n} - a_1 z^{-(n-1)} - a_2 z^{-(n-2)} - \cdots - a_n \tag{12.241}$$

and (12.232) yields

$$P^r_n\{z\} = 1 - a_1 z - a_2 z^2 - \cdots - a_n z^n \tag{12.242}$$

The two polynomials $P_n\{z\}$ and $P^r_n\{z\}$ are called *reciprocal,* which justifies the letter r. Furthermore, it is obvious that the transformation of a polynomial into its reciprocal is similar to the transformation of a vector $\mathbf{a}$ into $\mathbf{a}^{(-)}$ defined by (12.78).

12.12.2 Stability Criterion

Theorem. The polynomial $P_n\{z\}$ defined by a set of coefficients k_i, $1 \le i \le n$, has all its roots inside the UC if and only if $|k_i| < 1$, $1 \le i \le n$.

Proof of the 'if' part

As $P_1\{z\} = z - k_1$, the property is true for $P_1\{z\}$. Suppose that the property is true from 1 to m and suppose also that $|k_{m+1}| < 1$. We have to show that all the roots of $P_{m+1}\{z\}$ are inside the UC. As $P_m\{z\}$ begins with z^m we can write

$$P_m\{z\} = \prod_{i=1}^{m} (z - z_i) \tag{12.243}$$

which, with (12.234) gives

$$P_m^r\{z\} = \prod_{i=1}^{m} (1 - zz_i) = \prod_{i=1}^{m} (1 - zz_i^*) \tag{12.244}$$

because we are working with polynomials with real coefficients. Let ζ be an arbitrary root of $P_{m+1}\{z\}$. If $k_{m+1} = 0$, we deduce from (12.233) that the roots of P_m and of P_{m+1} are the same and there is an additional root at $z = 0$, which shows the property. Suppose that $k_{m+1} \neq 0$. In this case $\zeta \neq 0$ because it results from (12.102) and (12.240) that the constant term of $P_{m+1}\{z\}$ is $-k_{m+1}$. As $P_{m+1}\{\zeta\} = 0$, we obtain from (12.233) written for $m + 1$

$$\frac{P_m\{\zeta\}}{P_m^r\{\zeta\}} = \frac{k_{m+1}}{\zeta} \tag{12.245}$$

or with (12.243) and (12.244)

$$\prod_{i=1}^{m} \frac{\zeta - z_i}{1 - \zeta z_i^*} = \frac{k_{m+1}}{\zeta} \tag{12.246}$$

Let us now consider one term of the product, or $\alpha_i = (\zeta - z_i)(1 - \zeta z_i^*)^{-1}$. We have $|\alpha_i|^2 = N^2/D^2$ with

$$N^2 = |\zeta|^2 + |z_i|^2 - (\zeta z_i^* + z_i\zeta^*) \tag{12.247}$$

$$D^2 = 1 + |\zeta|^2 |z_i|^2 - (\zeta z_i^* + z_i\zeta^*) \tag{12.248}$$

We then deduce

$$|\alpha_i|^2 - 1 = D^{-2}\{ |\zeta|^2 + |z_i|^2 - 1 - |\zeta|^2 |z_i|^2 \}$$

$$= D^{-2}\{ (|\zeta|^2 - 1)(1 - |z_i|^2) \} \tag{12.249}$$

Let us work by contradiction. For this, suppose that $|\zeta|^2 > 1$. As $|z_i|^2 < 1$, we have $|\alpha_i|^2 > 1$, and repeating the same argument for $i = 1$ to $i = m$ we finally obtain $k_{m+1}/|\zeta| > 1$. But as $|k_{m+1}| < 1$ by construction, this is impossible if $|\zeta|^2 > 1$, so that we must have $|\zeta|^2 < 1$. In other words all the roots of $P_{m+1}\{z\}$ are inside the UC.

Proof of the 'only if' part

It is sufficient to show that if all the roots z_i of $P_m\{z\}$ satisfy $|z_i| < 1$, then $|k_m| < 1$ and all the roots of $P_{m-1}\{z\}$ are inside the UC.

For the first part we note that the last term of (12.240) is $-k_m$ as a consequence of (12.102). We then deduce from (12.243) that

$$\prod_{i=1}^{m} (-z_i) = -k_m \tag{12.250}$$

As $|z_i| < 1$, we also have $|k_m| < 1$. For the second part we start from (12.238). If ζ is a root of $P_{m-1}\{z\}$, or if $P_{m-1}\{\zeta\} = 0$, we obtain

$$\prod_{i=1}^{m} \frac{\zeta - z_i}{1 - \zeta z_i^*} = -k_m \tag{12.251}$$

and as previously we deduce that it is impossible to have $|\zeta| > 1$ and $|k_m| < 1$ simultaneously. Then, as $|k_m| < 1$, all the roots of $P_{m-1}\{z\}$ are inside the UC. Using the same result for $P_{m-1}\{z\}$, and so on, we arrive at $|k_i| < 1$, $1 \le i \le n$, which completes the proof.

12.12.3 Interpretation and Extensions

It is well known that the concept of stability can be extended by accepting some roots located in the UC. This corresponds to filters with undamped modes. It would seem natural that the stability criterion would be transformed into the relation $|k_i| \le 1$, or by accepting that some reflection coefficients satisfy $|k_i| = 1$.

This is not the case, as we will see in the following example. Consider the filter $B_2\{z\}$ given by (12.156). The corresponding polynomial $P_2\{z\}$ is

$$P_2\{z\} = z^2 - (k_1 - k_1 k_2) z^1 - k_2 \tag{12.252}$$

Suppose that $k_2 = 1$. We obtain $P_2\{z\} = z^2 - 1$, whatever the value of k_1. This polynomial has two roots located on the UC even though $k_1 > 1$. This is due to the fact that the straight line D_3 in Figure 12.6 is transformed into a point in Figure 12.7.

The complete criterion of wide-sense stability, that is, the possibility of roots on the UC, is given in [Benidir] and many other properties of polynomials are also given in [Picinbono, 1986]. Some are discussed as problems.

PROBLEMS

12.1 Let $B\{z\}$ be the transfer function of the error filter of the one-step prediction with infinite past and $b[k]$ its impulse response. The same functions $C\{z\}$ and $c[k]$ are defined for the inverse of this error filter. Give a recursive method to calculate the coefficients $c[k]$ in terms of the $b[k]$s.

12.2 Calculate the one-step prediction error with infinite past defined by (12.18) of the signal $x[k]$ whose correlation function is $\gamma[k] = \alpha a^{|k|} + \beta b^{|k|}$. For this calculation use the results of Problem 6.24.

12.3 Calculate the variance and one-step prediction error with infinite past defined by (12.18) of the signal $x[k]$ obtained by passing a unit white noise in the filter whose transfer function is $H(z) = z^2[(z - a)(z - b)]^{-1}$. Compare the results with those obtained in Problem 6.23.

12.4 Calculate the variance and one-step prediction error with infinite past of the signal $x[k]$ whose power spectrum is represented in Figure 12.3.

12.5 **Comparison between linear and non-linear prediction.** The purpose of this problem is to study the properties of the predictor (12.35). Let $x[k]$ be an AR(1) signal driven by a white noise $u[k]$ which is $N(0, 1 - a^2)$. Let $y[k]$ be the signal equal to $x^3[k]$.

(a) Calculate the best mean square prediction of $y[k]$ with infinite past and compare the result with (12.35).

(b) Calculate the corresponding prediction error.

(c) Using the correlation function of $y[k]$, calculate the prediction error resulting from the use of a linear prediction.

(d) Show that the best linear prediction cannot be realized with a finite past.

(e) Calculate the best linear prediction $\hat{x}[k]$ of the signal $x[k]$. Suppose that the prediction of $y[k]$ is taken as $\hat{x}^3[k]$. Calculate the corresponding prediction error and discuss the interest of this method compared to the two other procedures studied in the previous questions.

12.6 Let $x[k\,;s]$ be the signal $\cos[2\pi F(s)k + \Theta(s)]$, where s refers to the randomness. In this expression $F(s)$ is a random frequency satisfying $-1/2 < F(s) < 1/2$ and $\Theta(s)$ is a random phase uniformly distributed over $0, 2\pi$.

(a) Calculate the correlation function of $x[k\,;s]$ and express its power spectrum $\Gamma(\nu)$ in terms of the PDF of $F(s)$.

(b) Suppose that $F(s)$ is uniformly distributed over $-1/2, 1/2$. Calculate in this case the one-step prediction error with infinite past and show that we are in the case of null prediction.

(c) Explain the apparent contradiction between the preceding result and the point that $x[k\,;s]$ satisfies (9.49) for any s.

(d) Suppose now that $\Gamma(\nu)$ contains four spectral lines at frequencies $+\nu_1$, $-\nu_1$, $+\nu_2$ and $-\nu_2$. What is the structure of the signal $x[k\,;s]$ and why is it predictable?

12.7 Suppose that the correlation function of a random signal $x[k]$ is given by

$$\gamma[k] = [(a-b)(1-ab)]^{-1}[a(1-a^2)^{-1}a^{|k|} - b(1-b^2)^{-1}b^{|k|}]$$

Calculate the sequence of regression vectors appearing in the Levinson recursion and the corresponding prediction errors. Give an interpretation of the results and specify the structure of the signal $x[k]$.

12.8 Make the same calculation as in the previous problem when $\gamma[k]$ is given by

$$\gamma[k] = [(1-m^2)(m^4 - 2m^2\cos 2\phi + 1)\sin\phi]^{-1}[\sin[(k+1)\phi - m^2\sin[(k-1)\phi]m^k$$

12.9 **Correlation function of AR(2) signals.** Let $x[k]$ be the signal defined by the recursion $x[k] = a_1 x[k-1] + a_2 x[k-2] + u[k]$, where $u[k]$ is a second order white noise with a variance such that $E\{(x[k])^2\} = 1$.

(**a**) Calculate the reflection vector studied in Section 12.6.3. What are the values of the reflection coefficients k_i for $i > 2$?

(**b**) Deduce from the results of the previous equation the sequence of prediction errors.

(**c**) Calculate the vector **c** studied in Section 12.6.2.

(**d**) We want to calculate the correlation function $\gamma[i]$ of the signal $x[k]$. Give the values of this function for $i = 0, 1, 2$. For $i > 2$, write a second order linear recursion between $\gamma[i]$, $\gamma[i-1]$ and $\gamma[i-2]$. Calculate the solutions of this equation which are in the form cr^k, where c is an arbitrary constant. Show that there are only two possible values r_1 and r_2. Determine the general solution by using the fact that $\gamma[i]$ is known for $i = 1, 2$.

(**e**) Apply all the previous results to the case where $a_1 = 2m\cos\phi$ and $a_2 = -m^2$.

12.10 **The triplet a, c, k.** Consider a signal with unit power and the three vectors introduced in Section 12.6. We assume in this problem that $p = 3$.

(**a**) Calculate the three components of the vectors **a** and **k** in terms of those of the vector **c**.

(**b**) Calculate the three components of the vectors **a** and **c** in terms of those of the vector **k**.

(**c**) Calculate the three components of the vectors **c** and **k** in terms of those of the vector **a**.

12.11 **Other form of the Levinson recursion.** As indicated in Section 9.2.3(c), the regression vector is often written with another sign. Consequently (12.60) is written by replacing the vector **a** by the vector **A** equal to $-$**a**. Of course the definitions of Γ and **c** remain unchanged.

(**a**) Write the normal equation (12.61) in terms of **A**.

(**b**) By using the same procedure as in Section 12.5.1, write the Levinson recursion valid for the vector **A**.

(**c**) Express this recursion by using the reflection coefficients K_n defined by $K_n = A_n^n$. Write the relations between the triplet **A**, **K** and **c** by simply extending the results of Section 12.6.

(d) Find the lattice structure appearing in Figures 12.10 and 12.11 in terms of the reflection coefficients K_n.

(e) Discuss the relative interests of the two possible notations.

12.12 Prediction of a sinusoid signal. Let $x[k]$ be the signal $2^{1/2}\cos(\omega_0 k + \phi)$, where ϕ is a random phase uniformly distributed over 0, 2π.

(a) Calculate $E\{x[k]\}$ and $E\{x[k]x[k - p]\}$. Verify that this last expectation does not depend on k and can then be written as a correlation function $\gamma[p]$.

(b) Using this correlation function, calculate the vector $\mathbf{a}_2$ defining the one step predictor with past two and the corresponding error. Give an interpretation of the result obtained.

(c) Let $H_2\{z\}$ be the transfer function of the error filter corresponding to the prediction studied in the previous question. Find the zeros and the poles of $H_2\{z\}$. Is it possible to say that $x[k]$ is an AR(2) signal?

(d) We want to calculate the vector $\mathbf{a}_3$ defining the one-step predictor with past three. Make this calculation first by directly solving the normal equation and secondly by using a recursive procedure. Compare and give an interpretation of the results.

(e) Calculate the power spectrum of $x[k]$ and apply the results of the discussion presented in Section 12.3.2(c) to this spectrum.

12.13 The purpose of this problem is to find another proof of the finite past prediction error given by (12.124). The starting point is to replace the observation vector $\mathbf{x}$ by the vector $\tilde{\mathbf{x}}$ defined as in (12.204) from a sequence of successive innovations.

(a) From a simple ordering of the terms of $\mathbf{x}$, find a relation between the prediction error and the determinant of the covariance matrix Γ' of $\tilde{\mathbf{x}}$.

(b) Compare the determinants of the covariance matrices of $\tilde{\mathbf{x}}$ and $\mathbf{x}$ and deduce the relation (12.124).

12.14 Infinite past s-step prediction error of AR(2) signals. Let $x[k]$ be an AR(2) signal driven by a white noise of unit power. Let a and b be the roots of the polynomial in z appearing in the transfer function of the filter transforming the unit white noise into $x[k]$.

(a) Write in terms of a and b the difference equation verified by the signal $x[k]$.

(b) Calculate the filter $B\{z\}$ generating the innovation of the infinite past, one-step prediction.

(c) Calculate the transfer function $C\{z\}$ of the inverse filter of $B\{z\}$ and deduce the transfer function $H_s\{z\}$ of the infinite past, s-step error filter.

(d) Give the expression of the s-step prediction in terms of the signal $x[k]$ and explain the result by an extension of that obtained for AR(1) signals.

(e) Calculate the infinite past s-step prediction error.

(f) Explain the results obtained when one of the roots a or b is null.

(g) Explain all the previous results in the case of complex roots or when $a = m\exp(j\phi)$ and $b = m\exp(-j\phi)$.

12.15 Let $x[k]$ be the MA(1) signal defined by $x[k] = u[k] - au[k-1]$ where $u[k]$ is a white noise of unit power.

(a) Calculate the correlation function and the z-spectrum of $x[k]$.

(b) Calculate the infinite past and one-step prediction of $x[k]$ and the corresponding prediction error. Consider separately the cases $a < 1$ and $a > 1$.

12.16 We want to extend the previous problem to the case where $a = 1$. For this purpose we construct the sequence of one-step predictors with increasing past by using the Levinson algorithm.

(a) Calculate the correlation function of $x[k]$.

(b) Using the Levinson algorithm, deduce from this correlation function the sequence of predictors and the corresponding errors. For this calculation find by induction the general structure of the predictors.

(c) Give the general expression of $\hat{x}_p[k]$ in terms of the past values of $x[k]$. Calculate the corresponding prediction error.

(d) Explain the result which appears when p tends to infinity.

12.17 Let $x[k]$ be the MA(2) signal defined by $x[k] = u[k] - (a+b)u[k-1] + abu[k-2]$, $(a < 1,\ b < 1)$, where $u[k]$ is a white noise of unit power.

(a) Calculate the infinite past and one-step prediction filter and express this estimation in terms of the signal $x[k]$.

(b) Calculate the error in the prediction studied in the previous question.

(c) Make the same calculations for the 2-steps prediction.

(d) Explain all the previous results in the case of complex roots or when $a = m\exp(j\phi)$ and $b = m\exp(-j\phi)$.

12.18 The purpose of the following two problems is to prove the relations (12.165) and (12.166). In order to simplify the notation the determinant of a matrix $\mathbb{\Gamma}$ is simply written Γ. Assuming that the Toeplitz matrix $\mathbb{\Gamma}_n = T[\gamma_0, \gamma_1, \ldots, \gamma_{n-1}]$ is positive definite (PD), we want to determine the condition on γ_n ensuring that $\mathbb{\Gamma}_{n+1}$ is also PD.

(a) Show that the only point to prove is that $\Gamma_{n+1} > 0$.

(b) Show that Γ_{n+1} is a polynomial of degree 2 in γ_n.

(c) Suppose that we have shown that this polynomial is

$$\Gamma_{n+1} = -\Gamma_{n-1}\gamma_n^2 - 2t\gamma_n + s$$

with $t^2 + \Gamma_{n-1}s = \Gamma_n^2$. Using (12.124), show that $\Gamma_{n+1} > 0$ if and only if (12.165) and (12.166) hold. It remains now to prove the equation of question (c). This is the most difficult part and will be analyzed in the following problem.

12.19 Let $\mathbb{B}_n$ be the matrix deduced from $\mathbb{\Gamma}_{n+1}$ by suppressing the first column and the last row. Write this matrix explicitly and show that it contains only one element depending on γ_n.

(a) By using (12.120) and calculating the last element of the first column of $\mathbb{\Gamma}_{n+1}^{-1}$, show that $-\alpha_{n+1}^{-1}\, a_n^n = B_n/\Gamma_n$.

(b) By using (12.124), deduce that $(a_n^n)^2 = (B_n/\Gamma_n)^2$.

(c) Deduce from (12.99) that $\Gamma_{n+1} = \{\Gamma_n^2/\Gamma_{n-1}\} - \{B_n^2/\Gamma_{n-1}\}$.

(d) By calculating B_n from its last column, show that

$$B_n = (-1)^{n-1}\gamma_n\Gamma_{n-1} + B_n^0$$

where B_n^0 is the determinant of the matrix $\mathbb{B}_n$ in which γ_n is replaced by 0.

(e) Using the expression obtained in (**c**), write Γ_{n+1} as a polynomial of degree 2 in γ_n.

(f) Show that $(B_n^0)^2 = \Gamma_n^2 - \Gamma_{n-1}\Gamma_{n+1}^0$, where Γ_{n+1}^0 is the determinant of $T[\gamma_0, \gamma_1, \ldots, \gamma_{n-1}, 0]$.

(g) Deduce the second order equation of the preceding problem and its discriminant.

12.20 Finite past and future interpolation. The purpose of this problem is to calculate the interpolation error represented in Figure 12.16. Let $x[k]$ be a random signal and $\gamma[i]$, or γ_i, its correlation function. In order to solve the interpolation problem we will use the method analyzed in Problem 10.15. As a result, it appears that the first task is to calculate the diagonal elements of the inverse of the covariance matrix of the vector **X** appearing in the one-step prediction with finite past problem and with components $x[k-i]$, $1 \le i \le n$. Let $\mathbb{\Gamma}_n$ be the Toeplitz matrix $T[\gamma_0, \gamma_1, \ldots, \gamma_{n-1}]$ and $\mathbb{\Gamma}_n^{-1}$ its inverse. The diagonal elements of this matrix are noted d_i^n, which means that $d_i^n = [\mathbb{\Gamma}^{-1}]_{ii}$. Let $\mathbf{d}_n$ be the vector defined by $\mathbf{d}_n^T = [d_1^n, d_2^n, \ldots, d_n^n]$.

(a) Deduce from the Levinson algorithm a recursion allowing the calculation of the sequence of vectors $\mathbf{d}_n$. For this purpose it is recommended to use the vector $\mathbf{v}_n$ defined by its components $(a_1)^2, (a_2)^2, \ldots, (a_n)^2$.

(b) We want to estimate $x[k-i]$ in terms of the other components of the vector **X**. Using the results of Problem 10.15 and of the previous question, give the expressions of this estimation and of the corresponding error.

(c) We want to find the position i giving the minimum interpolation error. For this we will present several examples. Suppose that $x[k]$ is an AR(1) signal of unit power characterized by the regression coefficient a. Calculate in this case the vector $\mathbf{d}_5$ and find the positions i giving the minimum interpolation error.

(d) Make the same calculation when $x[k]$ is an AR(2) signal of unit power characterized by the regression coefficients a_1 and a_2.

(e) Make the same calculation when the correlation function of $x[k]$ satisfies: $\gamma_0 = 1$; $\gamma_3 = a$, $\gamma_i = 0$ if i is not equal to 0 or 3. Show in this case that the position $k - 3$, which corresponds to a symmetric interpolation, is not the best. Explain this result.

12.21 Some properties of lattice filters. The regression vector **a** can be expressed in terms of the reflexion vector and an example of this appears in Table 12.2. The purpose of this problem is to establish some general properties of the relation $\mathbf{a} = \mathbf{f}(\mathbf{k})$. These properties are in general proved by induction, starting from the results of Table 12.2 and can be interpreted with the lattice structure appearing in Figure 12. 11.

(a) Show that each component of the vector $\mathbf{a}_n$ is a polynomial in k_i with a degree not greater than n.

(b) Show that the powers of k_i in the polynomials introduced in (**a**) are either 0 or 1.

(c) Show that these polynomials are sums of products of k_i with the sign + if the number of k_i is odd and − if it is even.
(d) Show that if the reflexion coefficients are negative, then the components of $\mathbf{a}_n$ are also negative.
(e) Show that if $Sg(k_i) = (-1)^{i+1}$, then $Sg(a_i) = (-1)^{i+1}$.
(f) Show that the number of terms appearing in the polynomial giving a_i is $\binom{n}{i}$.
(g) Show that the total degrees of the polynomials corresponding to the components of the vectors $\mathbf{a}_n$ are

$$[2, 4, 6, \ldots, 2k, 2k-1, 2k-3, \ldots, 3, 1] \text{ for } \mathbf{a}_{2k}$$

$$[2, 4, 6, \ldots, 2k, 2k+1, 2k-1, 2k-3, \ldots, 3, 1] \text{ for } \mathbf{a}_{2k-1}$$

12.22 Some consequences of stability. Combining the results of the previous problem and the stability condition (12.169), show the following results.
(a) If the regression vector **a** corresponds to a stable filter, then $|a_i| < \binom{n}{i}$.
(b) If the regression vector **a** corresponds to a stable filter, then $\Sigma a_i < 2^n - 1$.
(c) If the regression vector **a** corresponds to a stable filter, then the magnitude of this vector is smaller than $[\binom{2n}{n} - 1]^{1/2}$.
(d) If the regression vector **a** corresponds to a stable filter, then the maximum of the magnitude is reached for two vectors corresponding to $k_i = -1$ and to $k_i = (-1)^{i-1}$ respectively and its value is $[\binom{2n}{n} - 1]^{1/2}$.
(e) Deduce from the previous results that the maximum value of the interpolation error of an AR(n) signal is given by $\varepsilon^2(n!)^2/(2n)!$, where ε^2 is the prediction error.

Chapter 13

Time Recursive Methods

13.1 INTRODUCTION

The idea of recursivity has been used several times in preceding chapters, for example, when discussing properties of autoregressive signals, where (9.20) is a recursive equation introducing a recursive system (see Figure 9.4). Similarly, the Levinson algorithm is a recursive method used to solve the normal equation such as (9.27). Recursive methods are especially interesting in signal processing problems because they are easy to implement on computers or on any digital signal processor. There is therefore a strong incentive to give a recursive form to many signal processing methods. However, it is important to delineate the difference between the two major ideas of recursivity: time and order.

It is clear that the recursive method in the Levinson algorithm has nothing to do with time. It is a way of solving normal equations of increasing order, which is why we call these procedures order recursive methods. On the other hand, the recursion in (9.20) which introduces an AR signal is a time recursive method, the only method with which we shall deal in this chapter.

Let us explain in more detail the general concept of the time recursive method. Consider once more the input-output relationship in a time-invariant linear filter using discrete-time signals. This relationship is a convolution given by (5.55) in the general case and by (3.119) if the causality constraint is introduced. It is obvious that the convolution is not a recursive operation. In fact, the knowledge of the output $y[.]$ at time instants $k-1$, $k-2$, ..., is in general without interest for calculating $y[k]$. It is then necessary to calculate at each time instant k the complete convolution, which introduces a large number of calculations. But the convolution, which is not a recursive procedure, can sometimes be transformed and written in a recursive way. The most illuminating example of this appears in the convolution (9.4) associated with an AR(1) signal. The relation (9.4) is not written in a recursive form, but can easily be transformed into (9.7) which is an especially simple recursive equation. One may wonder if this is not general, or if any causal convolution such as (3.119) can be transformed into a recursive equation.

The answer is no, at least if we impose the condition of using recursive linear equations of finite order.

Let us now define the concept of time recursivity for a system. Let S be a discrete-time (DT) system whose input and output are $x[k]$ and $y[k]$ respectively. The input-output relationship can be written as

$$y[k] = h[x\{.\}] \tag{13.1}$$

meaning that $h\{.\}$ is in general non-linear and that the output y at time k is, in principle, a function of the input x at any time instant. In this chapter we will restrict our study to *causal* systems, which means that only past inputs contribute to the output at k.

The system S is said to be *directly recursive of order* r if (13.1) can take the form

$$y[k] = f\{y[k-1], y[k-2], \ldots, y[k-r], x[k]\} \tag{13.2}$$

This means that in order to calculate y at time k we must know the input at the same instant k and use the outputs calculated at the r previous time instants $k - l$. If f is linear we return to (9.20) introducing AR signals. In the context of random signals it is clear that if the input is a strict white noise, $y[k]$ is a Markov signal of order r, as defined in Section 9.6.4. However, there is no reason for it to be linear Markovian, as defined by (9.102). The state representation studied in Section 9.5 introduces the concept of a signal which is not recursive but has a recursive representation. In fact, (9.70) shows that the state vector is directly recursive of order 1, but $y[k]$ cannot generally be written as in (13.2). This is for the same reason as that discussed in Section 9.6.5 for the Markovian property.

Let us point out here that time-recursivity has nothing to do with stationarity, or time invariance. In other words, the function $f\{.\}$ appearing in (13.2) can perfectly well be time-dependent, and written as $f_k\{.\}$. We illustrate this point with a very simple example. Suppose that we want to estimate the mean value of a signal observed from k = 1. For this we can use the sample mean defined by (4.308). At time k this mean is given by

$$y[k] = \frac{1}{k} \sum_{i=1}^{k} x[i] \tag{13.3}$$

which is not written in a recursive form. However, it is easy to show that $y[k]$ satisfies

$$y[k] = \{1 - (1/k)\}\, y[k-1] + (1/k)\, x[k] \tag{13.4}$$

which introduces a linear recursivity of order 1. The coefficients of this recursion are time-dependent. It is also possible to give (13.4) another form, which is interesting for the following discussion. By isolating the term $y[k-1]$ in (13.4) we deduce

$$y[k] = y[k-1] + (1/k)\, \{x[k] - y[k-1]\} \tag{13.5}$$

The last term of this equation is sometimes called a correcting term, and its behavior is the key point of the convergence of (13.3) already studied in Section 4.7.4(**a**).

13.2 TIME RECURSIONS FOR NORMAL EQUATIONS

Normal equations were introduced in the context of autoregressive modelling of signals in Section 9.2. However, the normal equation (9.27) also appears in the more general context of linear mean square estimation, and the solution to this problem introduces (10.65), which takes a similar form. Finally, the normal equation (12.61) was extensively examined in Section 12.5 in connection with the concepts of the recursive methods and fast algorithms. The basic purpose common to all these discussions was to solve an equation such as (12.61), or

$$\mathbb{\Gamma}\mathbf{a} = \mathbf{c} \tag{13.6}$$

where $\mathbb{\Gamma}$ and $\mathbf{c}$ are known.

In many practical problems, however, this is not the case, and both the matrix $\mathbb{\Gamma}$ and the vector $\mathbf{c}$ must be deduced from the observation. This introduces a completely new situation which must first be clearly described.

Let us again consider the one-step linear *prediction* of a signal with a finite past, extensively dealt with in Section 12.4. Suppose that $x[i]$ is a DT zeromean stationary signal observed from the initial time instant $i = 0$. The prediction at time k can still be expressed by (12.60), but we must replace $\mathbb{\Gamma}$ and $\mathbf{c}$ in (12.61) and (13.6) by their estimates deduced from the observation of the signal from 0 to $k - 1$. There are many ways of calculating these estimates which will not be discussed here. The simplest way is to adapt to our problem the method introduced for variance estimation in Section 4.8.7. As the signal is zeromean-valued, we can use the *sample variance* defined by (4.287) in order to estimate the variance, or the diagonal elements of the covariance matrix. By an extension of the terminology, we can call *sample correlation vector* the vector $\mathbf{c}_k$ defined by

$$\mathbf{c}_k = (1/k) \sum_{i=0}^{k-1} x_i \, \mathbf{X}_i \tag{13.7}$$

where $\mathbf{X}_i$ is still the past vector with components $x[i-1]$, $x[i-2]$, ..., $x[i-p]$. We shall briefly comment on this equation.

The first term in this sum is $x_0 \mathbf{X}_0$ that, strictly speaking, requires the knowledge of the signal at time instants $0, -1, -2, \ldots, -p$. If we replace these values by 0, because the observation begins at $i = 0$, we introduce a systematic bias in the correlation estimation. This bias of course tends to zero when $k \to +\infty$. To eliminate this problem, the simplest procedure is to say that the observation begins at $-p$ instead of 1. This corresponds to a translation of the beginning of the experiment. For long observations it introduces no problems, but for short observations a more complex analysis is required, which is not given here.

The second comment deals with the index of $\mathbf{c}$ in (13.7). It would be equally logical to write $\mathbf{c}_{k-1}$ instead of $\mathbf{c}_k$ in (13.7). In fact, this equation gives an estimation of the correlation vector by using only observations until time $k - 1$, because of the term x_i and of the definition of the vector $\mathbf{X}_i$. This is simply a question of notation, and we can only say that $\mathbf{c}_k$ is the correlation vector deduced from the observation until $k-1$ on order to predict the signal at time k.

The same reasoning can be introduced to estimate the covariance matrix at time $k-1$ which is then defined by

$$\Gamma_k = (1/k) \sum_{i=0}^{k-1} \mathbf{X}_i \mathbf{X}_i^T \tag{13.8}$$

Here also we can make some comments on the indices used. Strictly speaking, it would be possible to extend this sum until k instead of $k-1$ because the vector $\mathbf{X}_k$ uses the signal only until $k-1$. However, it is preferable to omit the term $\mathbf{X}_k\ \mathbf{X}_k^T$ in order to arrive at the same number of terms in (13.7) and (13.8).

Finally, observing the signal $x[i]$ from $i = -p$ to $i = k-1$ we want to estimate its value at time k by the predictor (12.60), or

$$\hat{x}[k] = \mathbf{a}_k^T\, \mathbf{X}[k] = \mathbf{a}_k^T\, \mathbf{X}_k \tag{13.9}$$

where $\mathbf{a}_k$ satisfies the normal equation

$$\Gamma_k\, \mathbf{a}_k = \mathbf{c}_k \tag{13.10}$$

We have here an example of an *adaptive procedure*. In fact, as time passes, the predictor (13.9) makes use of a time varying filter defined by the vector $\mathbf{a}_k$ which is deduced from past observations. In other words, these observations are used to calculate both the filter and its output. Adaptive filtering is a broad field of research in signal processing, and only the basic ideas are introduced here.

Before continuing, let us make a few comments. First, it is worth noting that no *optimality criterion* has been introduced. This is due to the fact that two different problems have been combined. The normal equation (13.6) gives the optimum predictor in the mean square sense when Γ and $\mathbf{c}$ are known, which is not the present situation. On the other hand, (13.7) and (13.8) are optimum estimates in the sense of Section 4.8.7, but we cannot say at this stage that the global procedure is optimum. This question will be analyzed later in the context of least square methods.

Second, it is important to note the difference in the numbers k and p. The first refers to time and is essentially variable while p is the order of the predictor which is constant. However, it is intuitively obvious that the previous calculations have a physical meaning only if $k >> p$, which means that k is sufficiently large to achieve a correct estimation of the covariance and the correlation. One can even say that if $k < p$, the problem is impossible to solve because Γ_k is not invertible. In fact, Γ_k in (13.8) is a $p \times p$ matrix and its rank is smaller or equal to k, because the dyadic matrices $\mathbf{X}_i\, \mathbf{X}_i^T$ are of rank one. Consequently, solving (13.10) needs at least $k \geq p$. However, as the dyadic matrices are non-negative definite, it is clear that if Γ_k in (13.8) is positive definite at time k, it has the same property at any posterior time. Therefore, (13.10) can be solved at all those time instants, for example by using algorithms discussed in the previous chapter.

We now arrive at the question of *recursivity*. Instead of solving (13.10) at each k, it would be interesting to calculate $\mathbf{a}_k$ when $\mathbf{a}_{k-1}$ is calculated, which introduces a time recursion, for a fixed order of p of the filter. The basic idea for this is that both (13.7) and (13.8) can be expressed in a recursive form, by using the same arguments as for (13.3). First note that the vector $\mathbf{a}_k$ in (13.10) is unchanged if we replace $\mathbb{\Gamma}$ and $\mathbf{c}$ by $\lambda\mathbb{\Gamma}$ and $\lambda\mathbf{c}$. The argument was already noted in Section 12.4(**b**). Consequently, we can suppress the factor $(1/k)$ both in (13.7) and in (13.8), which allows us to write

$$\mathbf{c}_k = \mathbf{c}_{k-1} + x_{k-1}\,\mathbf{X}_{k-1} \tag{13.11}$$

$$\mathbb{\Gamma}_k = \mathbb{\Gamma}_{k-1} + \mathbf{X}_{k-1}\mathbf{X}_{k-1}^T \tag{13.12}$$

In order to calculate $\mathbf{a}_k$ by using (13.10) we need to obtain the matrix $\mathbb{M}_k$ inverse of $\mathbb{\Gamma}_k$ or

$$\mathbb{M}_k \triangleq \mathbb{\Gamma}_k^{-1} = [\mathbb{\Gamma}_{k-1} + \mathbf{X}_{k-1}\mathbf{X}_{k-1}^T]^{-1} \tag{13.13}$$

This can be done by using an inversion procedure specified by

$$[\mathbb{A} + \mathbf{b}\mathbf{b}^T]^{-1} = \mathbb{A}^{-1} - \gamma\mathbb{A}^{-1}\mathbf{b}\mathbf{b}^T\mathbb{A}^{-1} \tag{13.14}$$

$$\gamma = (1 + \mathbf{b}^T\mathbb{A}^{-1}\mathbf{b})^{-1} \tag{13.15}$$

where the matrices $\mathbb{A}$ and $\mathbb{A} + \mathbf{b}\,\mathbf{b}^T$ are regular.

To prove this equation we multiply (13.14) by $\mathbb{A} + \mathbf{b}\,\mathbf{b}^T$, which gives a matrix $\mathbb{P}$ such that

$$\mathbb{P} = (\mathbb{A}^{-1} - \gamma\mathbb{A}^{-1}\mathbf{b}\mathbf{b}^T\mathbb{A}^{-1})(\mathbb{A} + \mathbf{b}\mathbf{b}^T)$$

$$= \mathbb{I} + \mathbb{A}^{-1}\mathbf{b}\mathbf{b}^T - \gamma\mathbb{A}^{-1}\mathbf{b}\mathbf{b}^T - \gamma\mathbb{A}^{-1}\mathbf{b}\mathbf{b}^T\mathbb{A}^{-1}\mathbf{b}\mathbf{b}^T$$

Noting that the last term of this equation can be written as $\gamma\mathbb{A}^{-1}\,\mathbf{b}(\mathbf{b}^T\,\mathbb{A}^{-1}\,\mathbf{b})\mathbf{b}^T$ and that the term in brackets is scalar, we deduce that

$$\mathbb{P} = \mathbb{I} + \mathbb{A}^{-1}\mathbf{b}\mathbf{b}^T(1 - \gamma - \gamma\mathbf{b}^T\mathbb{A}^{-1}\mathbf{b})$$

and it results from (13.15) that $\mathbb{P} = \mathbb{I}$.

Using (13.14) and (13.15) we deduce that the matrix $\mathbb{M}_k$ can be expressed as

$$\mathbb{M}_k = \mathbb{M}_{k-1} - \gamma_k\,\mathbb{M}_{k-1}\mathbf{X}_{k-1}\mathbf{X}_{k-1}^T\,\mathbb{M}_{k-1} \tag{13.16}$$

$$\gamma_k = (1 + \mathbf{X}_{k-1}^T\,\mathbb{M}_{k-1}\mathbf{X}_{k-1})^{-1} \tag{13.17}$$

These two equations show that $\mathbb{M}_k$, inverse matrix of $\mathbb{\Gamma}_k$, can be calculated recursively, without matrix inversion procedures, directly from the observation appearing in the vector $\mathbf{X}_{k-1}$.

It is now interesting to find a recursion for the filter vector $\mathbf{a}_k$ defined by (13.10), or

$$\mathbf{a}_k = \mathbb{M}_k\,\mathbf{c}_k \tag{13.18}$$

For this purpose we find another recursion for the matrix $\mathbb{M}_k$ in which the factor γ_k does not appear. We calculate $\mathbb{M}_k \mathbf{X}_{k-1}$, which, by using (13.16) and (13.17), gives

$$\mathbb{M}_k \mathbf{X}_{k-1} = \mathbb{M}_{k-1}\mathbf{X}_{k-1} - \gamma_k \mathbb{M}_{k-1}\mathbf{X}_{k-1}\mathbf{X}^T_{k-1}\mathbb{M}_{k-1}\mathbf{X}_{k-1}$$

$$= \mathbb{M}_{k-1}\mathbf{X}_{k-1}(1 - \gamma_k \mathbf{X}^T_{k-1}\mathbb{M}_{k-1}\mathbf{X}_{k-1})$$

$$= \gamma_k \mathbb{M}_{k-1}\mathbf{X}_{k-1} \tag{13.19}$$

Consequently, we deduce from (13.16) that

$$\mathbb{M}_k = \mathbb{M}_{k-1} - \mathbb{M}_k \mathbf{X}_{k-1}\mathbf{X}^T_{k-1}\mathbb{M}_{k-1} \tag{13.20}$$

By using this equation and (13.11), we can write $\mathbf{a}_k$ in the form

$$\mathbf{a}_k = (\mathbb{M}_{k-1} - \mathbb{M}_k \mathbf{X}_{k-1}\mathbf{X}^T_{k-1}\mathbb{M}_{k-1})\,\mathbf{c}_{k-1} + \mathbb{M}_k \mathbf{X}_{k-1} x_{k-1}$$

which gives

$$\mathbf{a}_k = \mathbf{a}_{k-1} + \mathbb{M}_k \mathbf{X}_{k-1}(x_{k-1} - \mathbf{a}^T_{k-1}\mathbf{X}_{k-1}) \tag{13.21}$$

Let us now comment on this equation.

(a) Recursivity. The first point to analyze is to explain in which sense the previous method is recursive. To do this we combine the basic equations of the problem. Our aim is to predict the signal x at time k from observations realized until $k-1$. The estimate is given by

$$\hat{x}[k] = \mathbf{a}^T_k \mathbf{X}[k] \tag{13.9}$$

where the filter $\mathbf{a}_k$ satisfies

$$\mathbf{a}_k = \mathbf{a}_{k-1} + \mathbb{M}_k \mathbf{X}_{k-1}(x_{k-1} - \mathbf{a}^T_{k-1}\mathbf{X}_{k-1}) \tag{13.21}$$

In this equation the matrix $\mathbb{M}_k$ satisfies

$$\mathbb{M}_k = \mathbb{M}_{k-1} - \gamma_k \mathbb{M}_{k-1}\mathbf{X}_{k-1}\mathbf{X}^T_{k-1}\mathbb{M}_{k-1} \tag{13.16}$$

with

$$\gamma_k = (1 + \mathbf{X}^T_{k-1}\mathbb{M}_{k-1}\mathbf{X}_{k-1})^{-1} \tag{13.17}$$

Suppose that the problem is solved at time $k-1$. At this time we know $\mathbf{a}_{k-1}$ and $\mathbb{M}_{k-1}$. The first step is to calculate $\mathbb{M}_k$ that is realized with (13.16) and (13.17). For this calculation we use the vector $\mathbf{X}_{k-1}$, or the values of the observations at time $k-2$, $k-3$, ..., $k-p-1$. Using $\mathbb{M}_k$ we can calculate $\mathbf{a}_k$ by the filter equation (13.21) which also uses the value of the signal at time $k-1$. Finally, the prediction is realized by (13.9) which uses the values of the signal between $k-p$ and $k-1$. At time k the signal $x[k]$ is observed and the procedure is repeated to estimate $x[k+1]$. We then have a recursive procedure.

The main feature of this procedure is that the vector $\mathbf{a}_k$ thus calculated satisfies the normal equation (13.10) which is solved without an inversion matrix procedure, in general a complex task. Similarly, as the calculations above are given without approximations, the vector $\mathbf{a}_k$ in (13.21) is the exact solution of (13.10). We have thus solved the same problem by two completely different methods.

(b) Initialization. Any recursive procedure must start from an initial point. If we do not want to make approximations, we can start the algorithm at the first instant k_i such that the matrix $\mathbb{\Gamma}_k$ defined by (13.8) is positive definite. At this time k_i we can calculate the inverse matrix $\mathbb{\Gamma}_i^{-1}$ or $\mathbb{M}_i$ and the vector $\mathbf{a}_i$ solution of (13.10). So the procedure needs at least one matrix inversion at the initial time. This matrix inversion can be avoided, if some approximations are accepted, as indicated below.

(c) Complexity. The algorithm (13.21) giving the filter is rather simple, as essentially it implies the product of a vector by a matrix. On the other hand, the algorithm giving $\mathbb{M}_k$ is much more complex, because of the presence of the product of several matrices. It is obviously this step of the procedure which needs most calculation.

(d) Innovation. The innovation in prediction problems is given by (12.64). Extending this expression in the present context, we can introduce an innovation defined by

$$\tilde{x}_k = \tilde{x}[k] = x[k] - \mathbf{a}_k^T \mathbf{X}[k] \tag{13.22}$$

This innovation appears in the algorithm giving the filter, which can then be written as

$$\mathbf{a}_k = \mathbf{a}_{k-1} + \mathbb{M}_k \, \mathbf{X}_{k-1} \tilde{x}_{k-1} \tag{13.23}$$

The last term can then be considered as an error term.

(e) Convergence. For a strictly stationary signal observed during a very long period of time, this recursive procedure is perhaps of less interest. In fact, it is clear that under very general ergodic conditions the vector $\mathbf{c}_k$ and the matrix $\mathbb{\Gamma}_k$ converge to a vector $\mathbf{c}$ and a matrix $\mathbb{\Gamma}$ respectively. In this case we return to the situation analyzed in the previous chapter where these quantities were known. However, these calculations are the basis of adaptive procedures, especially interesting in the real situations where stationarity cannot be taken for granted.

(f) Extensions to other estimation problems. It is easy to extend these calculations valid for prediction to other estimation problems. We shall simply outline the main ideas here. The basic equation for solving linear estimation problems with zeromean signals is (10.69), which can be written as

$$\hat{y} = \mathbf{a}^T \mathbf{x} \tag{13.24}$$

where $\mathbf{a}$ is the solution of

$$\mathbb{\Gamma} \, \mathbf{a} = \mathbf{c} = E(y \, \mathbf{x}) \tag{13.25}$$

It is then possible to extend all previous calculations by replacing x_i by y_i in (13.7). It is sometimes said that y_i is a reference signal for realizing the adaptive procedure.

13.3 RECURSIVE LEAST SQUARE METHODS

Let us reconsider the problem of the previous section consisting of predicting a signal at time k from observations taken at times from $-p$ to $k-1$, by using a linear system such as (13.9), where the vector $\mathbf{a}_k$ of $\mathbb{R}^p$ is to be determined. To make this determination we will use the least square method, which has a long practical history.

The basic idea of this method is to associate to the unknown vector $\mathbf{a}_k$ the quantity T_k defined by

$$T_k \triangleq \sum_{i=0}^{k-1} (x_i - \mathbf{a}_k^T \mathbf{X}_i)^2 \tag{13.26}$$

The physical meaning of this quantity is as follows. The term $x_i - \mathbf{a}_k^T \mathbf{X}_i$ is the prediction error of the signal at time i when the vector $\mathbf{a}_k$ is used, and T_k is then the sum of the square of these errors. This term is obviously a function of the vector $\mathbf{a}_k$ and the least square method consists of finding the vector $\mathbf{a}_k$ such that T_k is minimum. The solution is later used to predict x_k which is not yet observed.

To find the solution, we expand the squares in (13.26) in the form

$$(x_i - \mathbf{a}_k^T \mathbf{X}_i)^2 = \mathbf{a}_k^T \mathbf{X}_i \mathbf{X}_i^T \mathbf{a}_k - 2\,\mathbf{a}_k^T \mathbf{X}_i x_i + (x_i)^2 \tag{13.27}$$

which allows us to write T_k as

$$T_k = \mathbf{a}_k^T \Gamma_k \mathbf{a}_k - 2\,\mathbf{a}_k^T \mathbf{c}_k + t_k \tag{13.28}$$

where Γ_k is given by (13.8) without the factor $(1/k)$, and $\mathbf{c}_k$ by (13.7). Finally t_k is

$$t_k = \sum_{i=0}^{k-1} (x_i)^2 \tag{13.29}$$

As Γ_k is positive definite, T_k is a function with one minimum obtained by making null the derivative, which immediately yields (13.10). We can avoid the calculation of the derivative by simply noting the analogy between the present problem and that encountered in Section 10.4 on mean square estimation. If we suppose that the signals are zeromean-valued, we can state $c = 0$ and $\alpha = 0$ in (10.60) and (10.61). The optimum estimate $\hat{y}_L(\mathbf{x})$ then takes the form $\mathbf{a}^T \mathbf{x}$ and $\mathbf{a}$ is chosen to minimize the mean square error ε^2 defined by

$$\varepsilon^2 \triangleq E\{[\hat{y}(\mathbf{x}) - y]^2\} = \mathbf{a}^T \Gamma_x \mathbf{a} - 2\mathbf{a}^T \mathbf{c} + \sigma_y^2 \tag{13.30}$$

where $\mathbf{c}$ is the vector $\Gamma_{xy} = E(\mathbf{x}y)$. The minimum is obtained by the projection theorem, or the orthogonality principle, and yields (10.66) which is equivalent to (13.10). Consequently this normal equation gives the minimum of T_k and defines the optimum filter with the least square criterion.

Starting from this point, we can repeat the previous discussion without any changes, and the recursive solution (13.21) associated with (13.16) and (13.17) introduces the recursive least square estimation of $\mathbf{a}_k$. This gives an optimality criterion which was not introduced earlier.

13.4 INTRODUCTION TO ADAPTIVE FILTERING

It was indicated above that the algorithm (13.21) giving the vector $\mathbf{a}_k$ introduced the idea of adaptive filtering. Roughly speaking this means that the prediction filter characterized by $\mathbf{a}_k$ is not constant but can be adapted to the information extracted from the observation. We shall extend this concept in several ways, giving some ideas on a broad field of signal processing methods explained in more detail in Honig, Widrow, and Haykin.

The first point to note concerns the stationarity of the signals. All previous calculations referred to stationary signals. For example, it is clear that (13.7) is an estimation of the correlation vector $\mathbf{c} = E(x_k \mathbf{X}_k)$ for large values of k if stationarity is taken for granted. This appears because all the terms $x_i \mathbf{X}_i$ in (13.7) are considered as having the same importance. However, in many situations of practical importance it is not realistic to assume that the signals observed are stationary. Thus not all the terms $x_i \mathbf{X}_i$ can be considered as having the same importance. It seems logical to assume that for large values of k the term $x_{k-1} \mathbf{X}_{k-1}$ is more significant than the term $x_0 \mathbf{X}_0$. A method to be considered at this juncture is to limit (13.7) and use instead the relation

$$\mathbf{c}_k = \frac{1}{N} \sum_{i=1}^{N} x_{k-i} \mathbf{X}_{k-i} \tag{13.31}$$

This is the discrete-time version of the time averager in (6.93). It corresponds to a linear filtering similar to that shown in Figure 6.6.

This elementary expression immediately introduces the basic dilemma of adaptive systems. If N is very large, we return to (13.7), which means that we have a fairly precise measurement of the vector $\mathbf{c}$, as discussed in Section 6.6. However, for non-stationary signals we have a poor adaptivity to the variations of the properties of the signal. On the other hand, if N is small, it is possible to follow these variations but the precision in the measurement of $\mathbf{c}$ is also small. In practice it is always necessary to make a compromise between these two opposite requirements of precision and adaptivity, and no general rule can *a priori* be imposed. It depends completely on the experimental situation or on the purpose of the system.

The time averaging in (13.31) is simple to realize, but cannot be easily written in a recursive form, an objective of many adaptive systems. In order to arrive at this recursivity, we replace (13.11) by

$$\mathbf{c}_k = \alpha\, \mathbf{c}_{k-1} + x_{k-1} \mathbf{X}_{k-1}\ , \quad |\alpha| < 1 \tag{13.32}$$

where α is a damping factor. In comparison with (9.7), this replaces (13.31) with an expression such as (9.4). The past values $x_{k-i} \mathbf{X}_{k-i}$ are weighted by the factor α^i, which decreases with the amplitude of the past characterized by i. The same procedure can be used for (13.12), which becomes

$$\mathbb{\Gamma}_k = \alpha\, \mathbb{\Gamma}_{k-1} + \mathbf{X}_{k-1}\, \mathbf{X}_{k-1}^T \tag{13.33}$$

Using (13.32) and (13.33) instead of (13.11) and (13.12), we can use exactly the same procedure to calculate (13.13) and the vector $\mathbf{a}_k$. The calculations are similar, as discussed in a problem. However, the general conclusion remains the same, and we thus obtain an algorithm taking into account the fact that there is a time evolution of the statistics of the signal. Furthermore, if $\alpha = 1$, we return to the previous situation valid for the stationary case.

For some signal processing problems the algorithm (13.21) remains too complex to realize and various approximations have been suggested. The simplest consists of replacing the matrix $\mathbb{M}_k$ by a simple scalar, which gives

$$\mathbf{a}_k = \mathbf{a}_{k-1} + \mu_k (x_{k-1} - \mathbf{a}_{k-1}^T \mathbf{X}_{k-1})\, \mathbf{X}_{k-1} \tag{13.34}$$

It is even possible to assume that μ does not depend on k. It is clear that this algorithm is especially simple because it only needs the calculation of one scalar product at each step. By using methods outside the scope of this book, it is possible to show that, with appropriate conditions on the coefficients μ_k and the signal $x[k]$, the algorithms (13.21) and (13.34) have the same limit. However, it is clear that (13.21) gives, at each k, the exact solution to (13.10), which is obviously not the case with (13.34). Furthermore, for nonstationary signals, it is possible to use (13.34) with a constant value of μ, which introduces a situation similar to that encountered in the least square method with damping factor α. Readers interested in this topic may wish to consult more specialised works.

13.5 KALMAN FILTERING

In principle Kalman filtering is a time-recursive method for solving the same problem as Wiener filtering in Section 11.5. In that section we showed that in order to solve the Wiener-Hopf equation or to calculate the Wiener filter, it was necessary to realize strong factorization of the spectrum of the observation, which is not always an easy task. In the Kalman approach, the causality constraint is preserved but, to avoid the factorization problem, the signals are described by a state representation as studied in Section 9.5. Let us point out here that the assumption introduced in Wiener and Kalman filtering are not contradictory. In fact, we saw that strong factorization can be easily realized for rational spectra, which means that the signals belong to the category of ARMA signals studied in Section 9.4. But it is well known that dynamical filters, which are the basis of ARMA representation, can be described by a state variable method. This means that the class of signals for which the Wiener filter can be easily calculated is the same as that used to introduce Kalman filtering.

However, the great difference between the two procedures is due to the time recursivity. The concept of recursivity plays no role in Wiener filtering, while it is the basic idea used to introduce Kalman filtering, and it is for this reason that we discuss the method in this chapter.

13.5.1 Statement of the Problem

Our starting point is the internal representation described by (9.70) and (9.71). In the following the state equation is written as

$$\mathbf{x}_{i+1} = \mathbb{A}\,\mathbf{x}_i + \mathbf{b}u_i \tag{13.35}$$

The state vector $\mathbf{x}_i$ at time i is a vector of $\mathbb{R}^n$. The random signal u_i is scalar, and thus the vector $\mathbf{b}$ is also a vector of $\mathbb{R}^n$. The dynamical matrix $\mathbb{A}$ is an $n \times n$ matrix, and we suppose that $\mathbb{A}$ and $\mathbf{b}$ are time-independent. This assumption of stationarity simplifies the equations, but can be removed when considering non-stationary systems.

The observation is the signal y_i defined by

$$y_i = \mathbf{c}^T \mathbf{x}_i + v_i \tag{13.36}$$

where $\mathbf{c}$ is a constant vector of $\mathbb{R}^n$ and v_i a random signal.

The basic statistical assumptions facilitating the solution of the problem are that the signals u_i and v_i are zeromean valued and jointly weakly white. To explain this expression, let us consider the vector random signal $\mathbf{w}_i$ defined by

$$\mathbf{w}_i = [\, u_i\,, v_i\,]^T \tag{13.37}$$

Its matrix correlation function is defined by (5.14) and we assume that it satisfies

$$\gamma\,[i,j\,] = \mathbb{\Gamma}\,\delta\,[i,j\,] \tag{13.38}$$

where $\mathbb{\Gamma}$ is the covariance matrix defined as in (3.51) by

$$\mathbb{\Gamma} = \begin{bmatrix} \gamma_u & \gamma_{uv} \\ \gamma_{vu} & \gamma_v \end{bmatrix} \tag{13.39}$$

with $\gamma_{uv} = \gamma_{vu}$. As a result u_i and v_i can be correlated, which is described by the term γ_{uv}.

For the following calculations we assume that $\mathbb{A}$, $\mathbf{b}$, $\mathbf{c}$ and $\mathbb{\Gamma}$ *are known*, which completely defines the second order statistical properties of $\mathbf{x}$ and y.

Let us now introduce the problem we want to solve. We observe the signal from time 0 to i, and this observation is described by the vector $\mathbf{y}_i$ with components y_0, y_1, ..., y_i. Note that at each new observation the number of components of this vector increases by one. Starting from the observation vector, we want to solve two problems. The first is the one-step linear prediction of y_i, written as $\hat{y}_i$. It is the best linear mean square estimation (LMSE) of y_{i+1} in terms of $\mathbf{y}_i$. The second is the state estimation, written $\hat{x}_{i|i}$, which is the best LMSE of $\mathbf{x}_i$ in terms of $\mathbf{y}_i$. We will see that these two problems are closely interconnected.

The basic method to solve these problems is a recursive one, as explained next.

13.5.2 Prediction Filter

The prediction of the signal y_i can be written as in 10.5, or

$$\hat{y}_{i+1} = \mathrm{LMSE}[\, y_{i+1} \mid \mathbf{y}_i\,] \tag{13.40}$$

We recall at this point that the expression LMSE means a projection, as studied in Section 10.2.2 and expressed by (10.31). As we are using only *linear* estimation, this projection takes the form of (10.45), where H_C is the space defined by (10.60) with $c = 0$. As LMSE in (13.40) means a projection, we will systematically use the properties of this operation, especially its linearity and orthogonal decomposition specified by (3.149). Using (13.36), we can write

$$\hat{y}_{i+1} = \mathrm{LMSE}[\mathbf{c}^T \mathbf{x}_{i+1} + v_{i+1} \mid \mathbf{y}_i\,] \tag{13.41}$$

Noting that v_{i+1} and $\mathbf{y}_i$ are uncorrelated because of (13.38), we deduce from the linearity of the projection that

$$\hat{y}_{i+1} = \mathbf{c}^T \hat{\mathbf{x}}_{i+1} \tag{13.42}$$

where

$$\hat{\mathbf{x}}_{i+1} = \mathrm{LMSE}[\mathbf{x}_{i+1} \mid \mathbf{y}_i\,] \tag{13.43}$$

This quantity is the prediction of the state at time $i + 1$ in terms of the vector observation $\mathbf{y}_i$. The method which follows consists of solving (13.43) recursively and deducing the prediction of y with (13.42). For this we use (10.12) written as

$$y_i = \hat{y}_i + \tilde{y}_i \tag{13.44}$$

where the innovation $\tilde{y}_i$ is orthogonal, or uncorrelated, to the vector $\mathbf{y}_{i-1}$. Consequently the vector $\mathbf{y}_i$ equal to $[\mathbf{y}_{i-1},\ y_i]$ spans the same linear space as the vector $[\mathbf{y}_{i-1},\ \tilde{y}_i]$, and the latter introduces two orthogonal components. As a result we can write (13.43) in the form

$$\hat{\mathbf{x}}_{i+1} = \mathrm{LMSE}[\mathbf{x}_{i+1} \mid \mathbf{y}_{i-1}, \tilde{y}_i] \tag{13.45}$$

and using the state equation (13.35), we obtain

$$\hat{\mathbf{x}}_{i+1} = \mathrm{LMSE}[\mathbb{A}\,\mathbf{x}_i + \mathbf{b}u_i \mid \mathbf{y}_{i-1}, \tilde{y}_i] \tag{13.46}$$

Using the two properties of the projection recalled just above, we can write this expression as

$$\hat{\mathbf{x}}_{i+1} = \mathrm{LMSE}[\mathbb{A}\,\mathbf{x}_i \mid \mathbf{y}_{i-1}] + \mathrm{LMSE}[\mathbf{b}u_i \mid \mathbf{y}_{i-1}] + \mathrm{LMSE}[\mathbf{x}_{i+1} \mid \tilde{y}_i] \tag{13.47}$$

The second term of the right hand side of this equation is null. In fact $\mathbf{y}_{i-1}$ is a linear function of the u_is up to u_{i-2}, as it results from (13.35) and (13.36), and the u_is are uncorrelated. Similarly $\mathbf{y}_{i-1}$ is a linear function of the v_is up to v_{i-1}, as it results from

(13.36), and u_i and v_j are uncorrelated for $i \neq j$. Furthermore, the first term causes the term $\hat{\mathbf{x}}_i$ defined by (13.43) to appear. Finally, as we are using linear estimation, the last term depends linearly on $\tilde{y}_i$, which allows us to write

$$\hat{\mathbf{x}}_{i+1} = \mathbb{A}\hat{\mathbf{x}}_i + \mathbf{K}_i \tilde{y}_i \tag{13.48}$$

where $\mathbf{K}_i$ is a vector of $\mathbb{R}^n$ called the *gain* of the filter. Using (13.44) and (13.42), we can now express $\tilde{y}_i$ in terms of y_i as

$$\tilde{y}_i = y_i - \mathbf{c}^T\hat{\mathbf{x}}_i \tag{13.49}$$

and, inserting this expression in (13.48), we obtain

$$\boxed{\hat{\mathbf{x}}_{i+1} = (\mathbb{A} - \mathbf{K}_i\mathbf{c}^T)\,\hat{\mathbf{x}}_i + \mathbf{K}_i\, y_i} \tag{13.50}$$

which is the basic recursion of Kalman filtering. This recursion defines $\hat{\mathbf{x}}_{i+1}$ recursively provided $\mathbf{K}_i$ is known. Before beginning this calculation, let us make some comments.

First, remember that our starting problem is the prediction of y_i, which is given by (13.42). Second, there is an analogy between the recursive solution of this prediction problem and the solution presented in the context of recursive least squares. This appears when comparing (13.50) and (13.21), or (13.42) and (13.9). As in this problem, we will now see that the calculation of the gain of the filter is the most complex part of Kalman filtering.

We start from the definition of the gain which appears in (13.48), and is

$$\text{LMSE}[\,\mathbf{x}_{i+1} \mid \tilde{y}_i] = \mathbf{K}_i \tilde{y}_i \tag{13.51}$$

To calculate $\mathbf{K}_i$ we use the basic expression of LMSE, which is (10.69), and expressing the correlations appearing in this expression, we deduce that

$$\mathbf{K}_i = r_i^{-1}\,\mathbf{k}_i \tag{13.52}$$

where

$$r_i = E(\tilde{y}_i^2) \tag{13.53}$$

$$\mathbf{k}_i = E(\tilde{y}_i\,\mathbf{x}_{i+1}) \tag{13.54}$$

It is obvious that r_i is the one-step linear prediction error discussed at length in the previous chapter.

The basic approach to calculating these quantities is to express them in terms of the innovation $\tilde{\mathbf{x}}_i$ of the state vector. We begin from (13.36) and (13.42), which give

$$\tilde{y}_i = \mathbf{c}^T\mathbf{x}_i + v_i - \mathbf{c}^T\hat{\mathbf{x}}_i = \mathbf{c}^T\tilde{\mathbf{x}}_i + v_i \tag{13.55}$$

This shows that there is the same relation between the innovations $\tilde{\mathbf{x}}_i$ and $\tilde{y}_i$ as between the state vector $\mathbf{x}_i$ and the signal y_i. The statistical properties of the innovation vector $\tilde{\mathbf{x}}_i$ are characterized by its covariance matrix $\mathbb{P}_i$ defined by

$$\mathbb{P}_i \triangleq E\,[\tilde{\mathbf{x}}_i \tilde{\mathbf{x}}_i^T] \tag{13.56}$$

This matrix is also called the *error matrix* of the state prediction. Using (13.55), we then deduce that

$$r_i = E\{[\mathbf{c}^T \tilde{\mathbf{x}}_i + v_i][\tilde{\mathbf{x}}_i^T \mathbf{c} + v_i]\} \tag{13.57}$$

and noting that $\tilde{\mathbf{x}}_i$, linear function of the v_is up to v_{i-1}, is orthogonal to v_i, we deduce

$$r_i = \mathbf{c}^T \mathbb{P}_i \mathbf{c} + \gamma_v \tag{13.58}$$

where γ_v appears in (13.39).

The calculation of $\mathbf{k}_i$ is realized in a similar way. Writing $\mathbf{x}_i$ as $\hat{\mathbf{x}}_i + \tilde{\mathbf{x}}_i$, we deduce from (13.35) that

$$\mathbf{k}_i = E\{[\mathbf{c}^T \tilde{\mathbf{x}}_i + v_i][\mathbb{A}\,\mathbf{x}_i + \mathbf{b}u_i]\} = E\{[\mathbf{c}^T \tilde{\mathbf{x}}_i + v_i][\mathbb{A}\,\hat{\mathbf{x}}_i + \mathbb{A}\,\tilde{\mathbf{x}}_i + \mathbf{b}u_i]\} \tag{13.59}$$

This equation can be greatly simplified because of various orthogonality relations. By definition $\tilde{\mathbf{x}}_i$ and $\hat{\mathbf{x}}_i$ are orthogonal. Similarly, $\tilde{\mathbf{x}}_i$ and u_i are orthogonal because $\tilde{\mathbf{x}}_i$ is a function of $\mathbf{y}_{i-1}$ which depends linearly on the u_is up to $i-1$. We then obtain

$$\mathbf{k}_i = \mathbb{A}\,\mathbb{P}_i \mathbf{c} + \gamma_{uv}\,\mathbf{b} \tag{13.60}$$

where γ_{uv} appears in (13.39).

As a result of these calculations, the gain $\mathbf{K}_i$ can be calculated when the covariance matrix $\mathbb{P}_i$ is known. Later we will establish a recursion concerning this matrix.

13.5.3 Estimation of the State

This estimation can be written as

$$\hat{\mathbf{x}}_{i|i} = \text{LMSE}[\mathbf{x}_i |\, \mathbf{y}_i] \tag{13.61}$$

Using the same decomposition as in (13.45), this can be written as

$$\hat{\mathbf{x}}_{i|i} = \text{LMSE}[\mathbf{x}_i |\, \mathbf{y}_{i-1}, \tilde{y}_i] \tag{13.62}$$

From the orthogonality between $\mathbf{y}_{i-1}$ and $\tilde{y}_i$, and from (13.43), we find that

$$\hat{\mathbf{x}}_{i|i} = \hat{\mathbf{x}}_i + \mathbf{H}_i\,\tilde{y}_i \tag{13.63}$$

where $\mathbf{H}_i$ is a gain similar to $\mathbf{K}_i$, and defined by

$$\mathbf{H}_i = r_i^{-1}\,\mathbf{h}_i \tag{13.64}$$

$$\mathbf{h}_i = E(\tilde{y}_i\,\mathbf{x}_i) \tag{13.65}$$

It appears from (13.63) that the problems of state estimation and state prediction are interconnected, and the estimation $\hat{\mathbf{x}}_{i|i}$ can immediately be deduced from the results of the previous section which calculates $\hat{\mathbf{x}}_i$ recursively, provided that $\mathbf{h}_i$ in (13.65) is known.

This quantity is calculated by using the same procedure as in (13.53), which gives, as for (13.60),

$$\mathbf{h}_i = E\{[\mathbf{c}^T\tilde{\mathbf{x}}_i + v_i][\hat{\mathbf{x}}_i + \tilde{\mathbf{x}}_i]\} = \mathbb{P}_i\mathbf{c} \tag{13.66}$$

13.5.4 Recursion on the Error Matrix

This matrix is given by (13.56), and we calculate its recursion from that of the innovation $\tilde{\mathbf{x}}_i$. From (13.35) and (13.48) we obtain

$$\tilde{\mathbf{x}}_{i+1} \triangleq \mathbf{x}_{i+1} - \hat{\mathbf{x}}_{i+1} = \mathbb{A}\tilde{\mathbf{x}}_i + \mathbf{b}u_i - \mathbf{K}_i\,\tilde{y}_i \tag{13.67}$$

Using (13.55), this gives

$$\tilde{\mathbf{x}}_{i+1} = (\mathbb{A} - \mathbf{K}_i\mathbf{c}^T)\tilde{\mathbf{x}}_i + \mathbf{b}u_i - \mathbf{K}_i\,v_i \tag{13.68}$$

We must now calculate the square of this expression to obtain a recursion for $\mathbb{P}_i$ defined by (13.56). Using the orthogonality between $\tilde{\mathbf{x}}_i$ and u_i or v_i, this gives

$$\mathbb{P}_{i+1} = (\mathbb{A} - \mathbf{K}_i\mathbf{c}^T)\,\mathbb{P}_i\,(\mathbb{A} - \mathbf{K}_i\mathbf{c}^T)^T + \gamma_u\mathbf{b}\mathbf{b}^T + \gamma_v\mathbf{K}_i\mathbf{K}_i^T - \gamma_{uv}(\mathbf{b}\mathbf{K}_i^T + \mathbf{K}_i\mathbf{b}^T) \tag{13.69}$$

This expression can be reorganized in the form

$$\mathbb{P}_{i+1} = \mathbb{A}\mathbb{P}_i\mathbb{A}^T - \mathbf{K}_i(\mathbf{c}^T\mathbb{P}_i\mathbb{A}^T + \gamma_{uv}\mathbf{b}^T) - (\mathbb{A}\mathbb{P}_i\mathbf{c} + \gamma_{uv}\mathbf{b})\mathbf{K}_i^T + \mathbf{K}_i(\mathbf{c}^T\mathbb{P}_i\mathbf{c} + \gamma_v)\mathbf{K}_i^T + \gamma_u\mathbf{b}\mathbf{b}^T \tag{13.70}$$

Using (13.60), (13.58) and (13.52), we obtain

$$\mathbb{P}_{i+1} = \mathbb{A}\,\mathbb{P}_i\,\mathbb{A}^T - r_i^{-1}\,\mathbf{k}_i\mathbf{k}_i^T + \gamma_u\mathbf{b}\mathbf{b}^T \tag{13.71}$$

where we note that r_i and $\mathbf{k}_i$ are still functions of $\mathbb{P}_i$ as seen in (13.58) and (13.60). It is clearly a non-linear recursion on the $n \times n$ matrix $\mathbb{P}_i$ and is the more complex computational part of the Kalman filtering procedure.

13.5.5 Summary of the Complete Recursion

In order to facilitate the understanding of Kalman filtering, we shall give a summary of the above calculations indicating the recursive procedure used.

Signal model

$$\mathbf{x}_{i+1} = \mathbb{A}\,\mathbf{x}_i + \mathbf{b}u_i \tag{13.35}$$

$$y_i = \mathbf{c}^T\mathbf{x}_i + v_i \tag{13.36}$$

A priori knowledge

The quantities $\mathbb{A}$, $\mathbf{b}$ and $\mathbf{c}$ characterizing the model are known as well as the second order statistical properties of u_i and v_i characterized by γ_u, γ_v and γ_{uv}.

Recursion on the error matrix

Suppose that $\mathbb{P}_i$ is calculated. To obtain $\mathbb{P}_{i+1}$ we use the recursion

$$\mathbb{P}_{i+1} = \mathbb{A}\,\mathbb{P}_i\,\mathbb{A}^T - r_i^{-1}\,\mathbf{k}_i\mathbf{k}_i^T + \gamma_u\mathbf{b}\mathbf{b}^T \tag{13.71}$$

$$r_i = \mathbf{c}^T\,\mathbb{P}_i\mathbf{c} + \gamma_v \tag{13.58}$$

$$\mathbf{k}_i = \mathbb{A}\,\mathbb{P}_i\,\mathbf{c} + \gamma_{uv}\,\mathbf{b} \tag{13.60}$$

Note that these recursions are independent of the observations, and thus can be calculated before the experiment.

Prediction of the signal

Suppose that $\hat{\mathbf{x}}_i$ is calculated. We then have

$$\hat{\mathbf{x}}_{i+1} = (\mathbb{A} - \mathbf{K}_i\mathbf{c}^T)\,\hat{\mathbf{x}}_i + \mathbf{K}_i\,y_i \tag{13.50}$$

$$\mathbf{K}_i = r_i^{-1}\,\mathbf{k}_i \tag{13.52}$$

$$\hat{y}_{i+1} = \mathbf{c}^T\,\hat{\mathbf{x}}_{i+1} \tag{13.42}$$

It is obvious that this prediction depends on the observation up to time i through y_i and $\hat{\mathbf{x}}_i$.

Estimation of the state

Suppose, as above, that $\hat{\mathbf{x}}_i$ is known. We then have

$$\hat{\mathbf{x}}_{i|i} = \hat{\mathbf{x}}_i + \mathbf{H}_i\,\tilde{y}_i \tag{13.63}$$

$$\tilde{y}_i = y_i - \mathbf{c}^T\hat{\mathbf{x}}_i \tag{13.49}$$

$$\mathbf{H}_i = r_i^{-1}\,\mathbf{h}_i \tag{13.64}$$

$$\mathbf{h}_i = \mathbb{P}_i\mathbf{c} \tag{13.66}$$

Initializing the procedure

This is the last point, not previously discussed. As seen in the summary, there are two kinds of recursions: those independent of the observations, which can be calculated *a priori*, and those depending on the observation y_i. In the first category we essentially have to determine the covariance matrix $\mathbb{P}_0$. There are many ways of doing this, as, for example, to choose the identity or the null matrices. But the most logical is to say that at time 0 there is no estimation, which leads to the assumption that

$$\mathbb{P}_0 = E\,[\mathbf{x}_0\,\mathbf{x}_0^T] \tag{13.72}$$

This matrix appears in (9.74) and (9.75), and we can then choose the matrix $\mathbb{P}_0$, solution of the equation

$$\mathbb{P}_0 = \mathbb{A}\,\mathbb{P}_0\,\mathbb{A}^T + \gamma_u\mathbf{b}\mathbf{b}^T \tag{13.73}$$

For scalar state equations we return to (9.8) valid for AR(1) signals.

The last point is the state $\mathbf{x}_0$ at time 0. We know that it is a random vector with the covariance matrix defined by (13.73), but the exact value of $\mathbf{x}_0$ is unknown. If we have no *a priori* knowledge, we can take the null vector or make a random choice of $\mathbf{x}_0$ taking into account our knowledge of its statistical properties.

13.5.6 Additional Comments on Kalman Filtering

The above discussion is essentially oriented towards the time recursivity of Kalman filtering. We shall now make some additional comments to explain Kalman filtering a little further.

(a) Extension to vector cases. Although the observation y_i above was a scalar signal, there is no difficulty in adapting the calculations to the case where it becomes a vector signal. The only change appears in (13.36) that becomes

$$\mathbf{y}_i = \mathbb{C}\mathbf{x}_i + \mathbb{D}\mathbf{v}_i \tag{13.74}$$

where $\mathbb{C}$ is an $m \times n$ matrix and $\mathbf{v}_i$ a random vector. The previous equations remain the same provided that $\mathbf{c}^T$ is replaced by $\mathbb{C}$ and (13.39) is adapted to the fact that $\mathbf{v}$ is a vector. It would be useful for the reader to calculate this in full.

(b) Extension to the non-stationary case. It is not difficult to convert the above expressions to the case where both the parameters of the model ($\mathbb{A}$, $\mathbf{b}$, $\mathbf{c}$) and those of the signals ($\mathbb{\Gamma}$) depend on time. Because of this it is sometimes said that Kalman filtering is well adapted to non-stationary signals which have a state representation.

(c) Complexity. Looking at the equations in the previous section, it appears that the most complex part of the recursive procedures is the calculation of the error matrix $\mathbb{P}_i$. At each step of the recursion the most complex calculation is, of course, that of the matrix $\mathbb{A}\,\mathbb{P}_i\,\mathbb{A}^T$. We shall present later a method to reduce this complexity.

(d) Comparison with Wiener filtering. The Wiener filtering procedure studied in Section 11.5 is established for stationary signals, and can only be compared with the Kalman procedure if the stationarity is taken for granted, as in the previous calculation, and when the effect of the time origin disappears. We can then expect that Kalman filters tend to Wiener filters when $i \to \infty$. However, the identity of the two approaches is not easy to prove. To understand this point, let us study the asymptotic aspect of the Kalman procedure. For sufficiently large values of i all the parameters of the filter become time-independent, and we can suppress the letter i in all the equations in the summary 13.5.5, except for the model (13.35) and (13.36). So the asymptotic covariance matrix satisfies

$$\mathbb{P} = \mathbb{A}\,\mathbb{P}\,\mathbb{A}^T + r^{-1}\mathbf{k}\,\mathbf{k}^T + \gamma_u \mathbf{b}\mathbf{b}^T \tag{13.75}$$

$$r = \mathbf{c}^T\mathbb{P}\,\mathbf{c} + \gamma_v \tag{13.76}$$

$$\mathbf{k} = \mathbb{A}\,\mathbb{P}\,\mathbf{c} + \gamma_{uv}\,\mathbf{b} \tag{13.77}$$

It is not easy to calculate a positive definite matrix $\mathbb{P}$ satisfying these non-linear equations and allowing us to calculate the one-step predictor of y_i with infinite past discussed in the previous chapter and defined recursively by (13.42) where $\hat{\mathbf{x}}_i$ satisfies

$$\hat{\mathbf{x}}_{i+1} = (\mathbb{A} - r^{-1}\mathbf{k}\mathbf{c}^T)\,\hat{\mathbf{x}}_i + r^{-1}\mathbf{k}y_i \tag{13.78}$$

and where $\mathbf{k}$ and $\mathbf{r}$ result from the above equations.

In other words, the difficulties encountered in Wiener filtering stemming from the calculation of strong factorization appear in asymptotic Kalman filtering in the calculation of the covariance matrix of the error. Some examples illustrating this point are discussed as problems at the end of the chapter.

(e) Innovation representation. Using (13.48) and (13.49), we can write

$$\hat{\mathbf{x}}_{i+1} = \mathbb{A}\hat{\mathbf{x}}_i + \mathbf{K}_i\,\tilde{y}_i \tag{13.79}$$

$$y_i = \mathbf{c}^T\,\hat{\mathbf{x}}_i + \tilde{y}_i \tag{13.80}$$

These equations show that the signal y_i can be expressed in terms of its innovation as in the Wold decomposition in the previous chapter. When the asymptotic situation is reached, $\mathbf{K}_i$ becomes a constant vector $\mathbf{K}$ and $\tilde{y}_i$ is a stationary white noise. In other words, (13.79) and (13.80) describe the filter inverse of the error filter generating the innovation from the signal.

It is also possible to interpret these equations as an internal representation of the signal y_i with a state vector $\hat{\mathbf{x}}_i$ and a white driving noise which is the innovation of y_i. This establishes the relation between the internal and the external representation, which needs strong factorization.

13.5.7 Complexity Reduction

We shall briefly return to the complexity problem outlined in (**c**) above. There is abundant literature on this subject, and we intend here only to mention some fundamental ideas rather than to cover such a broad field of research.

For ease of presentation, we assume that the initial error matrix $\mathbb{P}_0$ is taken as the null matrix instead of the solution of (13.73). In fact, this equation is not easy to solve and we know that the initial conditions are not important when i is sufficiently large. The same property was discussed in the case of AR(1) signals in Section 9.2.3 (**e**). As a result of this assumption, and from (13.58), (13.60) and (13.71),

$$r_0 = \gamma_v \ ; \ \mathbf{k}_0 = \gamma_{uv}\mathbf{b} \tag{13.81}$$

and

$$\mathbb{P}_1 = \gamma_v^{-1}(\gamma_u\gamma_v - \gamma_{uv}^2)\,\mathbf{b}\mathbf{b}^T = \gamma_u(1-c^2)\,\mathbf{b}\mathbf{b}^T \tag{13.82}$$

where c is the correlation coefficient between u and v defined by (3.33).

This shows that if $\mathbb{P}_0 = \mathbb{O}$, $\mathbb{P}_1$ is a rank one matrix. The same situation was encountered in the least square method, as a result of (13.12). It is thus appropriate to work with the *increments* of functions when passing from $i-1$ to i. Let us consider an arbitrary quantity M_i (scalar, vector or matrix). By definition, ΔM_i is

$$\Delta M_i \triangleq M_i - M_{i-1} \tag{13.83}$$

It is obvious from our initial assumption that $\Delta \mathbb{P}_1 = \mathbb{P}_1$. Similarly, we can write

$$\Delta r_i = \mathbf{c}^T \Delta \mathbb{P}_i \mathbf{c} \tag{13.84}$$

$$\Delta \mathbf{k}_i = \mathbb{A} \Delta \mathbb{P}_i \mathbf{c} \tag{13.85}$$

because the last terms of (13.58) and (13.60) are constant. Consider now the more complex equation (13.71). This gives

$$\Delta \mathbb{P}_{i+1} = \mathbb{A} \Delta \mathbb{P}_i \mathbb{A}^T + \mathbb{T}_i \tag{13.86}$$

where the matrix $\mathbb{T}_i$ is

$$\mathbb{T}_i \triangleq r_{i-1}^{-1} \mathbf{k}_{i-1} \mathbf{k}_{i-1}^T - r_i^{-1} \mathbf{k}_i \mathbf{k}_i^T \tag{13.87}$$

In order to eliminate the negative powers, we return to (13.52), which gives

$$\mathbb{T}_i = r_{i-1} \mathbf{K}_{i-1} \mathbf{K}_{i-1}^T - r_i \mathbf{K}_i \mathbf{K}_i^T \tag{13.88}$$

Using the relations

$$\mathbf{K}_{i-1} = \mathbf{K}_i - \Delta \mathbf{K}_i \; ; \quad r_i = r_{i-1} + \Delta r_i \tag{13.89}$$

we can write (13.88) in the form

$$\begin{aligned}
\mathbb{T}_i &= r_{i-1} (\mathbf{K}_i - \Delta \mathbf{K}_i)(\mathbf{K}_i - \Delta \mathbf{K}_i)^T - (r_{i-1} + \Delta r_i) \mathbf{K}_i \mathbf{K}_i^T \\
&= r_{i-1} (\Delta \mathbf{K}_i \Delta \mathbf{K}_i^T - \Delta \mathbf{K}_i \mathbf{K}_i^T - \mathbf{K}_i \Delta \mathbf{K}_i^T) - \Delta r_i \mathbf{K}_i \mathbf{K}_i^T \\
&= r_{i-1} (\Delta \mathbf{K}_i \Delta \mathbf{K}_i^T - \Delta \mathbf{K}_i \mathbf{K}_i^T - \mathbf{K}_i \Delta \mathbf{K}_i^T) - \mathbf{K}_i \mathbf{c}^T \Delta \mathbb{P}_i \mathbf{c} \mathbf{K}_i
\end{aligned} \tag{13.90}$$

Let us now calculate $\Delta \mathbf{K}_i$. We can write

$$\begin{aligned}
\Delta \mathbf{K}_i &\triangleq \mathbf{K}_i - \mathbf{K}_{i-1} = (r_{i-1} \mathbf{K}_i - r_{i-1} \mathbf{K}_{i-1}) r_{i-1}^{-1} = \\
&[(r_i - \Delta r_i) \mathbf{K}_i - r_{i-1} \mathbf{K}_{i-1}] r_{i-1}^{-1}
\end{aligned} \tag{13.91}$$

As a result of (13.52) we have

$$\Delta \mathbf{k}_i = \Delta [r_i \mathbf{K}_i] = r_i \mathbf{K}_i - r_{i-1} \mathbf{K}_{i-1} \tag{13.92}$$

in such a way that (13.91) becomes

$$\begin{aligned}
\Delta \mathbf{K}_i &= (\Delta \mathbf{k}_i - \Delta r_i \mathbf{K}_i) r_{i-1}^{-1} = (\mathbb{A} \Delta \mathbb{P}_i \mathbf{c} - \mathbf{c}^T \Delta \mathbb{P}_i \mathbf{c} \mathbf{K}_i) r_{i-1}^{-1} \\
&= (\mathbb{A} - \mathbf{K}_i \mathbf{c}^T) \Delta \mathbb{P}_i \mathbf{c} r_{i-1}^{-1}
\end{aligned} \tag{13.93}$$

Using this expression in (13.90), we obtain

$$\mathbb{T}_i = r_{i-1}^{-1}(\mathbb{A} - \mathbf{K}_i\mathbf{c}^T)\Delta\mathbb{P}_i\mathbf{c}\mathbf{c}^T\Delta\mathbb{P}_i^T(\mathbb{A} - \mathbf{K}_i\mathbf{c}^T)^T - (\mathbb{A} - \mathbf{K}_i\mathbf{c}^T)\Delta\mathbb{P}_i\mathbf{c}\mathbf{K}_i^T$$

$$- \mathbf{K}_i\mathbf{c}^T\Delta\mathbb{P}_i^T(\mathbb{A} - \mathbf{K}_i\mathbf{c}^T)^T - \mathbf{K}_i\mathbf{c}^T\Delta\mathbb{P}_i\mathbf{c}\mathbf{K}_i \tag{13.94}$$

Noting that $\Delta\mathbb{P}_i = \Delta\mathbb{P}_i^T$ because $\mathbb{P}_i$ is a symmetric matrix, and after grouping the terms, we deduce from (13.86) and (13.94) that

$$\Delta\mathbb{P}_{i+1} = (\mathbb{A} - \mathbf{K}_i\mathbf{c}^T)(\Delta\mathbb{P}_i + r_{i-1}^{-1}\,\Delta\mathbb{P}_i\mathbf{c}\mathbf{c}^T\Delta\mathbb{P}_i)(\mathbb{A} - \mathbf{K}_i\mathbf{c}^T)^T \tag{13.95}$$

From this expression we find that if we have

$$\Delta\mathbb{P}_{i+1} = s_i\mathbf{a}_i\mathbf{a}_i^T \tag{13.96}$$

then we have

$$\Delta\mathbb{P}_{i+2} = s_{i+1}\,\mathbf{a}_{i+1}\,\mathbf{a}_{i+1}^T \tag{13.97}$$

with

$$\mathbf{a}_{i+1} = \mathbb{A}\mathbf{a}_i - r_{i+1}^{-1}(\mathbf{c}^T\mathbf{a}_i)\,\mathbf{k}_{i+1} \tag{13.98}$$

$$s_{i+1} = s_i + s_i^2(\mathbf{c}^T\mathbf{a}_i)r_i^{-1} \tag{13.99}$$

Similarly, as a result of (13.96), (13.84) and (13.85),

$$r_{i+1} \triangleq r_i + \Delta r_{i+1} = r_i + (\mathbf{c}^T\mathbf{a}_i)^2 s_i \tag{13.100}$$

$$\mathbf{k}_{i+1} \triangleq \mathbf{k}_i + \Delta\mathbf{k}_{i+1} = \mathbf{k}_i + s_i\mathbb{A}\mathbf{a}_i\mathbf{a}_i^T\mathbf{c} = \mathbf{k}_i + s_i(\mathbf{c}^T\mathbf{a}_i)\mathbb{A}\mathbf{a}_i \tag{13.101}$$

Let us now group all these equations in a recursive procedure. Suppose that r, s, $\mathbf{k}$, and $\mathbf{a}$ are known up to time i. We then deduce that

$$\begin{aligned}
r_{i+1} &= r_i + (\mathbf{c}^T\mathbf{a}_i)^2 s_i \\
s_{i+1} &= s_i + s_i^2 r_i^{-1}(\mathbf{c}^T\mathbf{a}_i) \\
\mathbf{k}_{i+1} &= \mathbf{k}_i + s_i(\mathbf{c}^T\mathbf{a}_i)\mathbb{A}\mathbf{a}_i \\
\mathbf{a}_{i+1} &= \mathbb{A}\mathbf{a}_i - r_{i+1}^{-1}(\mathbf{c}^T\mathbf{a}_i)\,\mathbf{k}_{i+1} \\
\mathbf{K}_{i+1} &= r_{i+1}^{-1}\mathbf{k}_{i+1}
\end{aligned}$$

which shows that the Kalman gain $\mathbf{K}_i$ can be calculated recursively by a set of recursions in which the more complex numerical operation appears for the calculation of $\mathbb{A}\mathbf{a}_i$, which is much simpler than (13.71).

The last point concerns the initial values of the algorithm. These values are deduced from (13.81) and (13.82), which gives

$$r_0 = \gamma_v \ ; \ s_0 = \gamma_v^{-1}(\gamma_u \gamma_v - \gamma_{uv}^2) \ ; \ \mathbf{k}_0 = \gamma_{uv} \mathbf{b}$$

$$a_0 = \mathbf{b}$$

In conclusion, after a rather tedious calculation, we arrive at a recursive algorithm which is simpler than (13.71). However, it should be noted that part of this simplification is a consequence of the fact that the initial value of $\mathbb{P}_i$ is the null matrix.

PROBLEMS

13.1 Starting from (13.32) and (13.33), show that the recursion of the vector $\mathbf{a}_k$ is still given by (13.21). Calculate the recursion corresponding to (13.16) and (13.17) and indicate what are the changes due to the presence of the damping factor α.

13.2 Starting from (13.35) and (13.74), write down the complete recursion defining the Kalman filtering corresponding to a vector-valued signal $\mathbf{y}$.

13.3 Consider the signal y_i whose state representation is given by (13.35) and (13.36) with the following parameters. The state vector is scalar and the noises u_i and v_i are equal with a variance γ. The system is characterized by $A = a$, $b = 1$, and $c = a$, where a is a real parameter. We want to make a prediction about the signal y_i by using the Kalman filtering.

(a) Write the equations defining the filtering calculating the one-step prediction of y_i with infinite past.

(b) Calculate the gain K_i of the Kalman filter and deduce the prediction-estimation of the state x_{i+1}.

(c) Deduce the expression of the prediction of y_{i+1} in terms of its past values.

(d) In order to explain the result obtained, calculate the transfer function of the filter defined by (9.73). What is the particularity of the signal y_i?

(e) Calculate the steady form of the error defined by (13.71) where $P_{i+1} = P_i$ and deduce that the state vector can be determined without error, which means that its estimation is singular. Deduce the prediction error of the signal y_i.

(f) Show that this final result could be obtained without any calculation from the state and observation equations.

13.4 **Comparison between the Kalman and Wiener filters.** We want to solve by the Kalman procedure the statistical filtering problem studied in Section 11.5.3(**b**). The first step is to express the problem of causal filtering of an AR(1) signal in a white noise in the form of state and observation equations. The signal satisfies the difference equation (9.7)

$$s[i] = as[i-1] + u[i]$$

where $u[i]$ is the driving noise with variance γ_u. Taking $v[i] = u[i-1]$, this equation can be written as

$$s[i + 1] = as[i] + v[i]$$

which is similar to (13.35). Consequently the state vector is scalar and can be taken as equal to the signal. The observation equation is (11.32), where x is replaced by y. The model is then defined by the parameter a and the variances γ_u and γ_n of the driving and observation noises assumed to be uncorrelated. The problem is then reduced to the state estimation presented in Section 13.5.3.

(a) Write in full the equations giving the solution to the problem.

(b) In order to compare the solution with that obtained with the Wiener filtering, we assume that the steady state is reached, which implies in (13.71) that $P_{i+1} = P_i = P$. Calculate P and show that there always exists only one positive solution to this problem. Find where the similar step in the Wiener procedure is.

(c) Express the solution in terms of P and of the other parameters of the problem and show that the Kalman filter leads to the same structure as in (11.156).

13.5 Consider the signal described by the state representation (13.35) and (13.36), where

$$\mathbb{A} = \begin{bmatrix} a_1 & a_2 \\ 1 & 0 \end{bmatrix} \; ; \; \mathbf{b} = \begin{bmatrix} 1 \\ 0 \end{bmatrix} \; ; \; \mathbf{c}^T = [a_1, a_2] \; ; \; u_i = v_i$$

(a) Show that the components of the state vector $\mathbf{x}[i + 1]$ are $y[i]$, $y[i - 1]$.

(b) Deduce that the error in the estimation prediction of the state is zero, which means that this estimation is singular.

(c) Deduce the structure of the Kalman filtering and calculate the prediction (13.42).

(e) Explain the result obtained by calculating the transfer function defined by (9.73).

(f) Show that there is no other non-negative solution than zero to (13.71).

Chapter 14

Matched Filters

14.1 INTRODUCTION

The concept of matched filters was originally introduced within the framework of signal to noise ratio (SNR) maximization. The basic idea is as follows. Let $s(t)$ be a deterministic signal of finite energy and $n(t)$ a second order zeromean noise. The SNR at t_0 is defined by

$$[\mathrm{SNR}]_t \triangleq \frac{s^2(t)}{\sigma^2} \tag{14.1}$$

where σ^2 is the variance of the noise. By filtering both $s(t)$ and $n(t)$ in a linear filter of frequency response $G(\nu)$ defined by (6.51), we modify the SNR, and the problem is to find the filter such that at a given time instant t_0 the SNR is maximum. We will see later that this problem can be easily solved and that the optimum filter depends on the Fourier transform $S(\nu)$ of $s(t)$, on the power spectrum $\Gamma(\nu)$ of the noise and on the time instant t_0 arbitrarily chosen. This filter is called the *matched filter* because it is matched to the properties of the signal and of the noise.

However, the idea of the matched filter can be introduced into a much broader framework as follows. There seems to be no reason to limit the concept to *linear* filters, so that matched filtering can be extended to non-linear systems with, of course, the trade-off of more analytical complexity.

In the same way, the idea of optimal filtering appears in many other situations with results very similar to those when the SNR is maximized. This is, for example, the case of signal amplitude estimation, and (4.278) is an example of a matched filter. Here the problem was to to minimize the variance with a constraint specified by (4.271). Exactly the same equations appear in the method of minimum variance spectral estimation (see [Kay], p. 370).

The purpose of this chapter is to show that several problems usually presented independently have in reality the same nature, which is why we use the term matched filtering, although this expression is usually restricted to SNR maximization.

14.2 THE CLASSICAL MATCHED FILTER

14.2.1 General Theory

Let $\mathbf{x}$ be a zeromean real random vector belonging to $\mathbb{R}^n$ with the covariance matrix Γ. By linear filtering of $\mathbf{x}$ we mean an operation characterized by

$$y = \mathbf{h}^T\mathbf{x} \tag{14.2}$$

where the vector $\mathbf{h}$ defines the filter. As a result of (14.2), y is a zeromean RV whose variance is

$$\sigma_y^2 = \mathbf{h}^T\Gamma\mathbf{h} \tag{14.3}$$

Let $\mathbf{s}$ be a given vector, called the signal vector. The signal-to-noise ratio at the output of this filter is defined by

$$[\text{SNR}] = \frac{(\mathbf{h}^T\mathbf{s})^2}{\mathbf{h}^T\Gamma\mathbf{h}} \tag{14.4}$$

and is the ratio between the square of the output of the filter when the input is $\mathbf{s}$ and the variance of the output noise. The problem is to find the filter, called the matched filter, which maximizes the SNR.

The solution is obvious if we use various results given in the previous chapters. Suppose that the matrix Γ is positive definite (the case where some eigenvalues are null will be discussed later). With this assumption we can write (14.4) in the form

$$[\text{SNR}] = \frac{(\mathbf{h}^T\Gamma\Gamma^{-1}\mathbf{s})^2}{\mathbf{h}^T\Gamma\mathbf{h}} \tag{14.5}$$

Using (3.131), we can introduce the scalar product between two vectors of $\mathbb{R}^n$ defined by

$$(\mathbf{u},\ \mathbf{v})_\Gamma \triangleq \mathbf{u}^T\Gamma\mathbf{v} \tag{14.6}$$

With this scalar product we can write (14.5) in the form

$$[\text{SNR}] = \frac{(\mathbf{h}^T\Gamma^{-1}\mathbf{s})_\Gamma^2}{(\mathbf{h},\ \mathbf{h})_\Gamma} \tag{14.7}$$

As a result of the Schwarz inequality, the maximum value of the SNR is

$$d^2 = (\Gamma^{-1}\mathbf{s},\ \Gamma^{-1}\mathbf{s})_\Gamma = \mathbf{s}^T\Gamma^{-1}\mathbf{s} \tag{14.8}$$

which is obtained with the filters

$$\mathbf{h}_o = \alpha\,\Gamma^{-1}\mathbf{s} \tag{14.9}$$

An arbitrary constant α appears because the SNR does not change if $\mathbf{h}$ is replaced by $\alpha\mathbf{h}$. Note the analogy between (14.9) and the vector defined by (4.278) and obtained in another context.

It is possible to extend the above reasoning to the *complex case*, replacing the transposition operation indicated by the letter T with the Hermitian operation indicated by H. In this case the scalar product (14.6) becomes

$$(\mathbf{u},\ \mathbf{v})_{\mathbb{\Gamma}} = \mathbf{u}^H \mathbb{\Gamma} \mathbf{v} \tag{14.10}$$

Defining the SNR by

$$[\text{SNR}] = \frac{|\mathbf{s}^H \mathbf{h}|^2}{\mathbf{h}^H \mathbb{\Gamma} \mathbf{h}} \tag{14.11}$$

and repeating the same reasoning, we again arrive at (14.9). However, the same problem appears as for estimation in the complex case (see Section 10.7). It is possible to show that simple transposition to the complex case of the method valid for the real case does not necessarily give the maximum SNR, and is valid only for some particular structures of the noise as for example, the circularity. As the discussion closely resembles that in Section 10.7.2, we shall not repeat it.

14.2.2 Principal Properties of the Matched Filter

In this section we shall investigate the most important properties of the matched filter, leaving others of less interest to be discussed as problems.

(a) Matched filter in white noise. As we are dealing only with second order properties, the white noise assumption means that the covariance matrix of $\mathbf{x}$ takes the form $\sigma^2 \mathbb{I}$. Inserting this expression in (14.9) we deduce that the matched filter can be written as

$$\mathbf{h}_o = \beta \mathbf{s} \tag{14.12}$$

where β is an arbitrary constant. Assuming that $\beta = 1$, we deduce that the output of the matched filter is simply

$$y = \mathbf{s}^T \mathbf{x} = \sum_{i=1}^{n} s_i x_i \tag{14.13}$$

If the index i refers to time, this output is sometimes said to be obtained by a *correlator receiver*. In fact, the last term of (14.13) realizes a kind of correlation between the signal x_i and the signal s_i.

The maximum value of the SNR is given by (14.8), and with $\mathbb{\Gamma} = \sigma^2 \mathbb{I}$ we obtain

$$d^2 = \mathbf{s}^T \mathbf{s} / \sigma^2 \tag{14.14}$$

As a consequence, all the signals with the same energy $E = \mathbf{s}^T \mathbf{s}$ exhibit the same maximum SNR, and the only way to increase the SNR is to increase this energy. From a geometric point of view, the signals with the same energy are located on a hypersphere of $\mathbb{R}^n$. It is, of course, possible to select some specific signals on this sphere, which is a signal design problem, but not on the basis of increasing SNR.

(b) Matched filter and likelihood ratio. The matched filter is used in signal detection theory, which is not dealt with in this book, but in which the likelihood ratio (LR) plays a very important role. We shall define this function and show its relationship to the matched filter. Consider two distinct probability density functions $p_0(\mathbf{x})$ and $p_1(\mathbf{x})$. The LR $L(\mathbf{x})$ is defined by (4.211), or

$$L(\mathbf{x}) \triangleq \frac{p_1(\mathbf{x})}{p_0(\mathbf{x})} \tag{14.15}$$

We have already met this in Problem 2.5 and in Section 4.8, when discussing the concept of sufficient statistic. We assume that $L(\mathbf{x})$ is well defined, which means that there is no point $\mathbf{x}$ where $p_0(\mathbf{x}) = 0$ and $p_1(\mathbf{x}) \neq 0$.

In signal detection problems the PDF $p_0(\mathbf{x})$ describes the random vector $\mathbf{x}(\omega)$ when there is only noise, while $p_1(\mathbf{x})$ corresponds to the situation of signal plus noise. We will assume that $p_0(\mathbf{x})$ introduces a zero mean value, while the mean value associated with $p_1(\mathbf{x})$ is $\mathbf{s}$. This is especially the case in the detection of a *deterministic* signal $\mathbf{s}$ in an additive noise for which we have $p_1(\mathbf{x}) = p_0(\mathbf{x} - \mathbf{s})$. In the following discussion we shall assume that the vector $\mathbf{s}$ in the matched filter (14.9) is precisely the mean value corresponding to the PDF $p_1(\mathbf{x})$.

Let us now consider the Hilbert subspace H_L defined by (10.60) with $c = 0$, in which the scalar product of two functions $S(\mathbf{x})$ and $S'(\mathbf{x})$ of $\mathbf{x}$ is defined with the PDF $p_0(\mathbf{x})$. This means that (10.26) must be written as

$$(S, S') = E_0[SS'] \triangleq \int S(\mathbf{x})\, S'(\mathbf{x})\, p_0(\mathbf{x})\, d\mathbf{x} \tag{14.16}$$

Finally, let us call $T(\mathbf{x})$ the output of the matched filter with $\alpha = 1$ given by (14.2) and (14.9). It can be expressed as

$$T(\mathbf{x}) = \mathbf{s}^T \Gamma^{-1} \mathbf{x} \tag{14.17}$$

We will now prove that

$$T(\mathbf{x}) = \mathrm{Proj}[L(\mathbf{x}) | H_L] \tag{14.18}$$

which means that the output of the matched filter $T(\mathbf{x})$ is the best linear mean square estimation of the LR.

To prove (14.18) we have to show that

$$L(\mathbf{x}) - T(\mathbf{x}) \perp\!\!\perp H_L \tag{14.19}$$

where the orthogonality has the same meaning as in (10.46) where H_C is replaced by H_L. The orthogonality relation (14.19) can be stated by saying that whatever the vector $\mathbf{h}$ of $\mathbb{R}^n$ we must have

$$L(\mathbf{x}) - T(\mathbf{x}) \perp\!\!\perp \mathbf{h}^T\mathbf{x} \tag{14.20}$$

Using the scalar product (14.16), this can be expressed as

$$E_0\left\{ \mathbf{h}^T\mathbf{x}[T(\mathbf{x}) - L(\mathbf{x})] \right\} = 0 \tag{14.21}$$

It suffices now to calculate the two terms of this equation. The first is

$$E_0\left[\mathbf{h}^T\mathbf{x}\,\mathbf{x}^T\mathbb{\Gamma}^{-1}\mathbf{s}\right] = \mathbf{h}^T\mathbf{s} \tag{14.22}$$

The second is calculated by using (14.15) and (14.16), which gives

$$E_0[\mathbf{h}^T\mathbf{x}L(\mathbf{x})] = E_0\left[\mathbf{h}^T\mathbf{x}\,\frac{p_1(\mathbf{x})}{p_0(\mathbf{x})}\right] = \int \mathbf{h}^T\mathbf{x}\,p_1(\mathbf{x})\,d\mathbf{x} \tag{14.23}$$

Using the assumption that the mean value associated with $p_1(\mathbf{x})$ is $\mathbf{s}$, the last term of (14.23) is also equal to $\mathbf{h}^T\mathbf{s}$, which gives (14.21) or (14.20).

(c) Matched filter and linear mean square estimation. In this paragraph we will write $T(\mathbf{x})$ given by (14.17) in a different form using the concepts of LMSE discussed in Section 10.4.

Suppose that the vector $\mathbf{s}$ in (14.17) is a unit vector, or satisfies $\mathbf{s}^T\mathbf{s} = 1$. This does not reduce the generality, because we have seen in (14.9) that the matched filter is defined with an arbitrary amplitude α. By taking $\alpha = (\mathbf{s}^T\mathbf{s})^{-1/2}$, we arrive at a unit vector $\mathbf{s}$ in (14.9), and then in (14.17). The unit vector $\mathbf{s}$ belongs to the space $\mathbb{R}^n$. Let R_2 be the subspace of $\mathbb{R}^n$ orthogonal complement of $\mathbf{s}$, the orthogonality being defined by the standard scalar product in $\mathbb{R}^n$, or the scalar product (14.6) with $\mathbb{\Gamma} = \mathbb{I}$. Any vector $\mathbf{v}$ of $\mathbb{R}^n$ can be decomposed in a unique way as

$$\mathbf{v} = v_1\mathbf{s} + \mathbf{v}_2 \tag{14.24}$$

where $\mathbf{v}_2$ belongs to R_2 or is orthogonal to $\mathbf{s}$. As a result of this property, and as $\mathbf{s}$ is a unit vector, we have

$$v_1 = \mathbf{s}^T\mathbf{v} \tag{14.25}$$

which is the component of $\mathbf{v}$ in the direction $\mathbf{s}$.

Introducing the vector

$$\mathbf{r} \triangleq \mathbb{\Gamma}^{-1}\mathbf{s} \tag{14.26}$$

the output $T(\mathbf{x})$ of the matched filter can be written as

$$T(\mathbf{x}) = \mathbf{r}^T\mathbf{x} \tag{14.27}$$

Let us now apply the decomposition (14.24) to the vector $\mathbf{r}$ and to the random vector $\mathbf{x}$, which gives

$$\mathbf{r} = r_1\mathbf{s} + \mathbf{r}_2\ , \quad r_1 = \mathbf{s}^T\mathbf{r} \tag{14.28}$$

$$\mathbf{x} = x_1\mathbf{s} + \mathbf{x}_2\ , \quad x_1 = \mathbf{s}^T\mathbf{x} \tag{14.29}$$

Using the orthogonality between $\mathbf{s}$ and $\mathbf{r}_2$ or $\mathbf{x}_2$, the function $T(\mathbf{x})$ takes the form

$$T(\mathbf{x}) = T_1(\mathbf{x}) + T_2(\mathbf{x}) \tag{14.30}$$

with

$$T_1(\mathbf{x}) = r_1 x_1 \tag{14.31}$$

$$T_2(\mathbf{x}) = \mathbf{r}_2^T \mathbf{x}_2 \tag{14.32}$$

Note that r_1 defined by (14.28) can be calculated with (14.26), which gives

$$r_1 \triangleq \mathbf{s}^T\mathbf{r} = \mathbf{s}^T\Gamma^{-1}\mathbf{s} = d^2 \tag{14.33}$$

where d^2 is defined by (14.8).

Furthermore, as $\mathbf{r}_2$ is orthogonal to $\mathbf{s}$, we deduce from (14.29) that

$$T_2(x) = \mathbf{r}_2^T \mathbf{x} \tag{14.34}$$

It results from these preliminary calculations that the output $T(\mathbf{x})$ can be expressed as

$$T(\mathbf{x}) = d^2[x_1 - u(\mathbf{x})] \tag{14.35}$$

where

$$u(\mathbf{x}) = -(1/d^2)\,\mathbf{r}_2^T \mathbf{x} \tag{14.36}$$

We will now show that $u(\mathbf{x})$ can be interpreted as LMSE. Let H_2 be the subspace defined by (10.77) where S is the subspace R_2 introduced above as the subspace of $\mathbb{R}^n$ orthogonal complement to $\mathbf{s}$. It is obvious that $u(\mathbf{x})$ belongs to H_2. Let us now show that

$$u(\mathbf{x}) = \mathrm{Proj}[x_1 | H_2] \tag{14.37}$$

where the projection is taken in the same sense as in (10.78) or (10.79). Note also that x_1 is the scalar random variable defined by the last term of (14.29). To prove (14.37) we can express the orthogonality principle in (10.79), or write

$$E\{[x_1 - u(\mathbf{x})]\,\mathbf{x}^T\mathbf{h}_2\} = 0 \tag{14.38}$$

whatever $\mathbf{h}_2 \in R_2$. Using (14.35), this gives

$$E[T(\mathbf{x})\,\mathbf{x}^T\mathbf{h}_2] = 0 \tag{14.39}$$

and with (14.27), this becomes

$$E\,[\mathbf{r}^T\mathbf{x}\,\mathbf{x}^T\mathbf{h}_2] = \mathbf{r}^T\Gamma\mathbf{h}_2 = 0 \tag{14.40}$$

Using (14.26), we see that this gives $\mathbf{s}^T\mathbf{h}_2 = 0$, which is valid for any vector $\mathbf{h}_2$ of R_2. This completes the proof of (14.37).

Let us now interpret this result. Using (10.78), we see that $u(\mathbf{x})$ can be written as

$$u(\mathbf{x}) = \hat{x}_1 \tag{14.41}$$

where $\hat{x}_1$ is the LMSE of x_1 in terms of $\mathbf{x}_2$, because any element of H_2 can be written as $\mathbf{h}_2^T\mathbf{x} = \mathbf{h}_2^T\mathbf{x}_2$, as in (14.34). Consequently, the output of the matched filter given by (14.18) or (14.35) can also be written as

$$T(\mathbf{x}) = d^2(x_1 - \hat{x}_1) = d^2\tilde{x}_1 \tag{14.42}$$

introducing the innovation in the LMSE problem above. This is a very simple expression.

Analyzing (14.42) in more detail, we see that the term x_1 is given by the last term of (14.29) and is the output of the matched filter used when the noise is white, and expressed by (14.13). As a result, the term $\hat{x}_1$ is the correction which must be applied to the white noise matched filter when the true covariance matrix is taken into consideration.

Finally, as the variance of $T(\mathbf{x})$ defined by (14.17) is equal to d^2 defined by (14.8), we deduce that the variance of $\tilde{x}_1$, or the error in the LMSE problem, is equal to $1/d^2$.

Let us conclude this section by noting that the term $\hat{x}_1$ in (14.42) enjoys a property of Noise Alone Reference (NAR property). To explain this point, let us suppose that the observation $\mathbf{x}$ takes the form

$$\mathbf{x} = \mathbf{n} + k\mathbf{s} \tag{14.43}$$

where $\mathbf{n}$ is a zeromean noise while $\mathbf{s}$ is a deterministic unit vector and k an amplitude factor. If we apply (14.29), we immediately obtain

$$\mathbf{x} = (n_1 + k)\,\mathbf{s} + \mathbf{n}_2 \tag{14.44}$$

which shows that the vector $\mathbf{n}_2$ is independent of k, which can even be equal to zero. In other words, the term $\mathbf{n}_2$, which is the observation used in our LMSE problem, is only due to the noise, even though a signal is present, and whatever its amplitude. This constitutes the NAR property. It is interesting to compare this result with the method of unbiased variance estimation discussed in Section 4.8. The geometric interpretation of (4.311) also introduces the subspace of $\mathbb{R}^n$ orthogonal to the signal $\mathbf{s}$.

14.2.3 Matched Filter in Time

Let us return to the initial problem stated in (14.1), and use the previous results to solve it. Consider a deterministic signal $s(t)$ of finite energy, and let us call $S(\nu)$ its Fourier transform. Suppose also that there is a noise $n(t)$ which is stationary, zeromean valued, and let $\Gamma(\nu)$ be its power spectrum. The output at time t_0 of a filter with frequency response $G(\nu)$ is

$$y(t_0) = \int G(\nu)\, S(\nu)\, \exp(2\pi j \nu t_0)\, d\nu \tag{14.45}$$

Assuming that the signals and filter are real, this quantity is real, which is ensured by the Hermitian property similar to (6.24). On the other hand, the variance of the output noise is given by (6.96), or

$$\sigma_y^2 = \int |G(\nu)|^2\, \Gamma(\nu)\, d\nu \tag{14.46}$$

Consequently, the output signal to noise ratio defined by (14.4) can be expressed as

$$[\mathrm{SNR}]_{t_0} = \frac{\left| \int ab^* d\nu \right|^2}{\int aa^* d\nu} \tag{14.47}$$

where

$$ab^* = G(\nu)\ S(\nu)\ \exp(2\pi j\nu t_0) \tag{14.48}$$

$$aa^* = G(\nu)G^*(\nu)\Gamma(\nu) \tag{14.49}$$

Note that the square of the modulus in the numerator of (14.47) is due to the fact that $y(t_0)$ is real although the terms under integration are complex. The structure (14.47) leads to the use of the Schwarz inequality for integrals, which shows that

$$[\mathrm{SNR}]_{t_0} \le \int S^*(\nu)[\Gamma(\nu)]^{-1}S(\nu)\,d\nu \tag{14.50}$$

The maximum is obtained when $a(\nu) = k\ b(\nu)$. Using the values of a and b extracted from (14.48) and (14.49), we find that the maximum value D^2 of the SNR is

$$D^2 = \int \frac{|S(\nu)|^2}{\Gamma(\nu)}\,d\nu = \int S^*(\nu)\,[\Gamma(\nu)]^{-1}S(\nu)\,d\nu \tag{14.51}$$

This last expression is written to show a formal analogy with d^2 given in (14.8). This maximum value is obtained with the filter characterized by

$$G_o(\nu) = k\,[\Gamma(\nu)]^{-1}S^*(\nu)\exp(-2\pi j\nu t_0) \tag{14.52}$$

where k is an arbitrary constant.

Let us now briefly indicate the most significant properties of this *matched filter*.

(a) The previous calculation is valid both in continuous and discrete time. The only difference concerns the boundaries of the integral in the frequency domain, as appears in (6.5) or (6.6).

(b) Let us examine the matched filter for white noise, or WNMF. In this case $\Gamma(\nu) = \alpha$ and, mixing this constant with k, we can write, for simplification, that

$$G(\nu) = S^*(\nu)\exp(-2\pi j\nu t_0) \tag{14.53}$$

while the corresponding SNR is

$$D^2 = (1/\alpha)E \tag{14.54}$$

where E is the energy of the signal. This expression is similar to (14.14), and we can thus deduce the same consequences.

The impulse response of the matched filter (14.53) is the inverse Fourier transform of $G(\nu)$, and, using the classical properties of the Fourier transformation, we deduce that

$$h(t) = s(t_0 - t) \tag{14.55}$$

This shows that the impulse response of the WNMF is simply the signal $s(t)$ transformed by a time reversal and a time translation of t_0.

(c) Let us now discuss the output of the WNMF. Using the classical convolution relation, we can write the input-output relationship as

$$y(t) = \int x(\theta)\, s(t_0 - t + \theta)\, d\theta \tag{14.56}$$

In particular, for $t = t_0$, we obtain

$$y(t_0) = \int x(\theta)\, s(\theta)\, d\theta \tag{14.57}$$

which is (14.13). In other words, at the time instant t_0 where the SNR is maximum, the output of the matched filter is identical to that of a correlator receiver.

(d) Let us now calculate the output of the WNMF when only the signal is present. In this case we replace $x(t)$ in (14.56) by $s(t)$, which yields

$$y_S(t) = \int s(\theta)\, s(\theta - t + t_0)\, d\theta \triangleq \gamma_s(t - t_0) \tag{14.58}$$

where $\gamma_s(\tau)$ is called the correlation function of the deterministic signal $s(t)$. The Fourier transform of this correlation function is

$$\Gamma_s(\nu) = |S(\nu)|^2 \tag{14.59}$$

It is worth pointing out that this correlation function is introduced in a non-random context. However, as $\Gamma_s(\nu)$ is non-negative, its Fourier transform is a non-negative definite function, as introduced in Section 5.2.3. Using the properties discussed here, we deduce that $\gamma_s(\tau) \le \gamma_s(0)$, which shows that in the presence of signal only, the output of the WNMF is maximum at the time instant t_0 arbitrarily chosen.

(e) It is possible to use this last property for other purposes than signal to noise maximization. If we want to obtain the maximum value of the SNR at t_0 in the presence of white noise, all signals with the same energy E are equivalent, as is seen from (14.54). In the class of signals of energy E, however, there is a large variety of possible correlation functions. As the matched filter is also used in another context, especially to measure the position of a signal, it is sometimes useful to design the signal in such a way that its correlation function $\gamma_s(\tau)$ is concentrated in a very short time interval. This is a specific signal design problem, not analyzed here.

14.2.4 Singular Case

The problem of SNR maximization solved by the matched filter becomes singular if it is possible to obtain an infinite value of the maximum SNR. This means that either d^2 of (14.8) or D^2 of (14.51) become infinite.

Considering the first case, we write d^2 in a more convenient form using the covariance basis introduced in Section 3.5.2 **(e)**. It is the orthonormal basis using the eigenvectors of the covariance matrix $\mathbb{\Gamma}$. With this basis the real matrix $\mathbb{\Gamma}^{-1}$ can be expressed as

$$\mathbb{\Gamma}^{-1} = \sum_{i=1}^{n} \frac{1}{\lambda_i} \mathbf{u}_i \mathbf{u}_i^T \tag{14.60}$$

The signal $\mathbf{s}$ can be written as

$$\mathbf{s} = \sum_{i=1}^{n} s_i \mathbf{u}_i \tag{14.61}$$

where s_i is the composant of $\mathbf{s}$ corresponding to the unit vector $\mathbf{u}_i$ and defined by $\mathbf{s}^T\mathbf{u}_i$. Using these expressions in (14.8) gives

$$d^2 = \sum_{i=1}^{n} \frac{s_i^2}{\lambda_i} \tag{14.62}$$

As $\mathbf{s}$ is a vector of finite energy, which means that

$$\sum_{i=1}^{n} s_i^2 = E < +\infty \tag{14.63}$$

we deduce that the only way to obtain an infinite value of d^2 is that there is at least one null eigenvalue, for example λ_k, and that $s_k \neq 0$. In fact, we must eliminate the vectors $\mathbf{u}_i$ for which both λ_i and s_i are null, and which do not contribute to d^2. The corresponding matched filter can be written as

$$T(\mathbf{x}) = \mathbf{u}_k^T \mathbf{x} \tag{14.64}$$

because the variance of $T(\mathbf{x})$ defined by (14.3) becomes $\mathbf{u}_k^T \Gamma \mathbf{u}_k$, and then is null while $\mathbf{u}_k^T \mathbf{s} = s_k$, the numerator of (14.4), is not null.

The same conclusion appears for D^2 in the discrete-time case, since we have

$$D^2 = \int_{-1/2}^{+1/2} \frac{|S(\nu)|^2}{\Gamma(\nu)} d\nu \tag{14.65}$$

The way to give an infinite value to D^2 corresponds to the situation in which there is some frequency band $\Delta\nu$ where $S(\nu) \neq 0$ and $\Gamma(\nu) = 0$. This is similar to that of singular estimation of a signal in noise, discussed in Example 11.1.

The continuous-time case is more complex because

$$D^2 = \int_{-\infty}^{+\infty} \frac{|S(\nu)|^2}{\Gamma(\nu)} d\nu \tag{14.66}$$

This expression can, in fact, become infinite although $|S(\nu)|^2 / \Gamma(\nu)$ is always finite, an effect due to the infinite frequencies. It can be said that this case of singular detection is only a mathematical singularity, because no physical device can use arbitrary high frequencies, and there is always a band limitation which must be introduced.

14.3 AMPLITUDE ESTIMATION

We shall now see that the matched filter (14.17) can be introduced in a completely different context, corresponding to the signal amplitude estimation. This is a frequent problem in communication systems. Amplitude modulation of a signal $s(t)$ consists of multiplying this signal by a, giving $a\,s(t)$, where the interesting information is contained in a. In vector notation we can use the vector $\mathbf{s}$ instead of the function $s(t)$. In the presence of additive noise $\mathbf{n}$, the observation becomes

$$\mathbf{x} = \mathbf{n} + a\mathbf{s} \tag{14.67}$$

where a must be estimated from $\mathbf{x}$. Note that in this problem the estimandum a is non-random, and thus this problem cannot be solved with MSE methods as in Chapter 10.

Suppose that $\mathbf{n}$ is a zeromean random vector with a known covariance matrix $\mathbb{\Gamma}$. The problem becomes similar to that considered in Section 4.8.7 for mean estimation. Because of the importance of the result, we will briefly recapitulate here. Our purpose is to find a linear estimate of a written as

$$\hat{a}(\mathbf{x}) = \mathbf{h}^T\mathbf{x} \tag{14.68}$$

in such a way that the estimator is unbiased, or satisfies

$$E[\hat{a}(\mathbf{x})] = a \tag{14.69}$$

and that its variance

$$V = \mathbf{h}^T\mathbb{\Gamma}\mathbf{h} \tag{14.70}$$

is minimum. In (14.67) we assume that the mean value of $\mathbf{n}$ is null, which transforms (14.69) into

$$\mathbf{h}^T\mathbf{s} = 1 \tag{14.71}$$

and $\mathbb{\Gamma}$ is the covariance matrix of noise $\mathbf{n}$.

The problem now is to find the filter minimizing the quadratic form (14.70) with the constraint (14.71). This is obtained by using again the Schwarz inequality, as in (4.275) and (4.276), which gives the optimum estimate by

$$\hat{a}(\mathbf{x}) = d^{-2}\mathbf{s}^T\mathbb{\Gamma}^{-1}\mathbf{x} = d^{-2}T(\mathbf{x}) \tag{14.72}$$

where d^2 is given by (14.8) and $T(\mathbf{x})$ by (14.17). The corresponding minimum variance is

$$V_{\min} = d^{-2} \tag{14.73}$$

In other words, this variance is the inverse of the maximum SNR calculated in the previous section. In particular, this minimum variance becomes null in the singular case studied above.

Let us now examine the same problem in a different context. Suppose that noise $\mathbf{n}$ is $N(0, \mathbb{\Gamma})$. As a result, $\mathbf{x}$ becomes $N(a\,\mathbf{s}, \mathbb{\Gamma})$, and its PDF is given by (4.82), or

$$p(x;\, a) = k\exp\{-(1/2)Q(\mathbf{x})\} \tag{14.74}$$

where the quadratic for $Q(\mathbf{x})$ is

$$Q(\mathbf{x}) = (\mathbf{x} - a\mathbf{s})^T\,\mathbb{\Gamma}^{-1}(\mathbf{x} - a\mathbf{s}) \tag{14.75}$$

In order to estimate a from the observation, we can use the *maximum likelihood* estimator. This is the number $a(\mathbf{x})$ such that $p(x;\ a)$, considered as a function of a, is maximum. This is equivalent to minimizing $Q(\mathbf{x})$, or

$$Q(\mathbf{x}) = a^2\mathbf{s}^T\Gamma^{-1}\mathbf{s} - 2a\mathbf{s}^T\Gamma^{-1}\mathbf{x} + \mathbf{x}^T\Gamma^{-1}\mathbf{x} \tag{14.76}$$

The minimum of this polynomial in a is obviously obtained for $\hat{a}(\mathbf{x})$ given by (14.72). This means that when noise $\mathbf{n}$ is normal, it is equivalent to using the maximum likelihood method and minimizing the variance with the constraint (14.71) ensuring a null bias.

Before leaving this section, it is worth making a comparison between the results of the previous section and these results. It is clear in (14.4) that the SNR does not vary when changing $\mathbf{h}$ into $k\mathbf{h}$. This is why the constant α appears in (14.9). In (14.72) this no longer holds because of the constraint (14.71). The maximization of the SNR (14.4) can be interpreted as a minimization of the variance in the denominator, provided that the numerator is constant, or

$$\mathbf{h}^T\mathbf{s}\mathbf{s}^T\mathbf{h} = c \tag{14.77}$$

The constant c is arbitrary, while in (14.17) it is equal to 1, and the constraint (14.77) is quadratic in $\mathbf{h}$, while it is linear in (14.71). This leads to a more general analysis of matched filtering.

14.4 GENERALIZED MATCHED FILTERS

We are still working with linear systems, for which the output is given by (14.2). The output variance V is given by (14.3) or (14.70), and the purpose is to find a filter $\mathbf{h}$ giving the minimum variance under various constraints. Let us first examine the most interesting constraints for application, which can be imposed separately or simultaneously.

(a) Directional constraint. This is the constraint (14.71), often written in the complex case as

$$\mathbf{h}^H\mathbf{s} = 1 \tag{14.78}$$

In this case the variance which must be minimized is

$$V = \mathbf{h}^H\Gamma\mathbf{h} \tag{14.79}$$

The expression "directional" comes from its use in *spatial signal processing*. This was touched upon in Section 5.2.5 (**b**), and we shall now look at it more closely. Instead of using two sensors as in Figure 5.5, we can use n sensors, and their outputs are the components of the vector $\mathbf{x}$. By linearly combining these components, we realize an output like (14.2), and $\mathbf{h}$ characterizes a particular antenna structure. A point source at infinity generates plane waves, as in Figure 5.5. It is then possible to associate to this source a vector $\mathbf{s}$ whose components are the outputs of the sensors when only this source, and no noise, is present. In other words, the condition (14.78) means that the gain of the antenna in the direction of this source is equal to 1. If there are various other sources of noise, the problem is to find a vector **h** minimizing the global noise characterized by a covariance matrix $\mathbb{\Gamma}$, provided that the gain corresponding to the source **s** is maintained equal to 1.

The same problem arises in *spectral analysis* with the method of minimum variance spectral estimation. In this case, we are looking for a filter with a frequency response equal to 1 for a given frequency and minimizing the noise power due to the other frequencies. The problem can easily be stated in the same form.

(b) Subspace constraint. This constraint means that the vector **h** minimizing the variance must belong to a subspace S of $\mathbb{R}^n$ if **x** is real, or of $\mathbb{C}^n$ if **x** is complex. This constraint was widely used above, and the simplest example is given in (10.91), meaning that **h** must be orthogonal to a given vector **s'**. For, example, in the antenna problem the constraint

$$\mathbf{h}^T\mathbf{s}' = 0 \tag{14.80}$$

means that the source associated to the vector **s'** must be eliminated.

Note that a double directional constraint can be transformed into a directional and a subspace constraint. In fact, the condition

$$\mathbf{h}^T\mathbf{s} = 1 \quad ; \quad \mathbf{h}^T\mathbf{s}' = 1 \tag{14.81}$$

is equivalent to

$$\mathbf{h}^T\mathbf{s} = 1 \quad ; \quad \mathbf{h}^T(\mathbf{s} - \mathbf{s}') = 0 \tag{14.82}$$

(c) Power constraint. This constraint is expressed as

$$\mathbf{h}^T\mathbb{K}\,\mathbf{h} = 1 \tag{14.83}$$

and an example was given by (14.77). This also appears in the problem of maximizing the SNR in the presence of random signals. If the covariance matrices of the signal and noise are $\mathbb{K}$ and $\mathbb{\Gamma}$ respectively, the SNR becomes

$$[\mathrm{SNR}] = \frac{\mathbf{h}^T\mathbb{K}\,\mathbf{h}}{\mathbf{h}^T\mathbb{\Gamma}\mathbf{h}} \tag{14.84}$$

which is a ratio of two quadratic forms, sometimes called a Rayleigh ratio. The problem is then to find the filter giving the maximum value of this ratio, and (14.4) is a particular case of (14.84) corresponding to $\mathbb{K} = \mathbf{s}\,\mathbf{s}^T$.

After this general introduction we shall now give the solutions to various problems of matched filtering.

14.4.1 Matched Filter with Directional and Subspace Constraints

The problem is exactly the same as in the previous section, with the additional subspace constraint. It can be stated as follows: find the vector **h** minimizing the quadratic form (14.79) such that

$$\mathbf{h}^T\mathbf{s} = 1 \text{ and } \mathbf{h} \in S \qquad (14.85)$$

To solve this we use the scalar product (14.6) extensively. Note first that if S is $\mathbb{R}^n$, or if the last constraint (14.85) disappears, we return to solution (14.72), written here in the form

$$\mathbf{h}_o = d^{-2}\mathbf{r} \qquad (14.86)$$

where d^2 is given by (14.8) and $\mathbf{r}$ by (14.26). Let us now apply the projection theorem to this vector $\mathbf{r}$. Using the orthogonality deduced from the scalar product (14.6), we can introduce the subspace $S^\perp$ orthogonal complement to S. Consequently, $\mathbf{r}$ can be written, as in (3.150), in the form

$$\mathbf{r} = \mathbf{r}_S + \mathbf{r}_S^\perp \qquad (14.87)$$

where $\mathbf{r}_S$ and $\mathbf{r}_S^\perp$ belong to S and $S^\perp$ respectively. The same decomposition can be applied to any vector $\mathbf{h}$ and the second constraint (14.85) implies that $\mathbf{r}_S^\perp = \mathbf{0}$. Let us now write the first in the form

$$\mathbf{h}^T\mathbf{s} = \mathbf{h}^T\Gamma\Gamma^{-1}\mathbf{s} = (\mathbf{h},\ \mathbf{r})_\Gamma = 1 \qquad (14.88)$$

As $\mathbf{h}$ belongs to S, or as $\mathbf{r}_S^\perp = \mathbf{0}$, we deduce from (14.87) that the directional constraint can take the form

$$(\mathbf{h},\ \mathbf{r}_S)_\Gamma = 1 \qquad (14.89)$$

We then return to the same kind of problem as without constraint: find a vector belonging to S which minimizes the variance and satisfies (14.89). Applying again the Schwarz inequality to (14.89) yields

$$1 = (\mathbf{h},\ \mathbf{r}_S)_\Gamma^2 \le (\mathbf{h},\ \mathbf{h})_\Gamma\ (\mathbf{r}_S, \mathbf{r}_S)_\Gamma \qquad (14.90)$$

The solution is similar to (14.86) and can be written as

$$\mathbf{h}_{o,1} = (\mathbf{r}_S, \mathbf{r}_S)_\Gamma^{-1}\ \mathbf{r}_S = (\mathbf{r}_S^T\Gamma\mathbf{r}_S)^{-1}\mathbf{r}_S \qquad (14.91)$$

The minimum variance obtained with this filter is deduced from the last term of (14.90), and can be written as

$$V_1 = (\mathbf{r}_S\ \Gamma\mathbf{r}_S)^{-1} = (\mathbf{r}_S, \mathbf{r}_S)_\Gamma^{-1} \qquad (14.92)$$

Let us calculate the loss in performance introduced by the subspace constraint. Using (14.87) and the orthogonality of the vectors $\mathbf{r}_S$ and $\mathbf{r}_S^\perp$, we deduce that

$$(\mathbf{r},\ \mathbf{r})_\Gamma = (\mathbf{r}_S, \mathbf{r}_S)_\Gamma + (\ \mathbf{r}_S^\perp,\ \mathbf{r}_S^\perp\ )_\Gamma \qquad (14.93)$$

Recalling that the minimum variance without constraint V_m is given by (14.73), and noting that

$$d^2 \triangleq \mathbf{s}^T\Gamma^{-1}\mathbf{s} = \mathbf{r}^T\Gamma\mathbf{r} = (\mathbf{r},\ \mathbf{r})_\Gamma \qquad (14.94)$$

we deduce from (14.93) that

$$(1/V_m) = (1/V_1) + (\mathbf{r}_S^{\perp}, \mathbf{r}_S^{\perp})_{\Gamma} \tag{14.95}$$

This last term is of course non-negative and the expression gives the increase of the variance due to the constraint introduced. Note the analogy between this expression and (10.54) or (10.89) giving the increase of the estimation error when a linear constraint is introduced in MSE problems. If there is no constraint at all, $\mathbf{r}_S^{\perp} = \mathbf{0}$, and $V_1 = V_m$.

Let us now consider the calculation of the constrained matched filter defined by (14.91). The main problem is, of course, to calculate $\mathbf{r}_S$ given by (14.87). This expression means that

$$\mathbf{r}_S = \text{Proj}[\mathbf{r} | S] \tag{14.96}$$

and we arrive at a situation similar to that encountered in LMSE problems with linear constraint and characterized by (10.82). Here also we must note that the projection in (14.96) corresponds to a projection with the scalar product (14.6) defined by $\mathbb{\Gamma}$. In conclusion, to calculate $\mathbf{r}_S$ defining the constrained matched filter, we must first calculate the unconstrained matched filter $\mathbf{r}$, and secondly find its projection ($\mathbb{\Gamma}$) onto the subspace S defining the constraint. This can be achieved by applying the orthogonality principle meaning that for any $\mathbf{u}$ belonging to S, we must have

$$(\mathbf{u}, \mathbf{r} - \mathbf{r}_S)_{\Gamma} = 0 \tag{14.97}$$

This method can be applied to the constraint (14.80) equivalent to (10.91), and a result similar to (10.96) is obtained.

14.4.2 Matched Filter with Power and Subspace Constraints

The problem now is to minimize the variance (14.70) with the constraint (14.83), provided that $\mathbf{h}$ belongs to the linear subspace S.

We will first assume that $\mathbb{K}$ is symmetric and non-negative definite. This means that $\mathbb{K}$ is a covariance matrix, as shown in Section 3.5.2. Taking again the scalar product (14.6), the constraint (14.83) can be written as

$$(\mathbf{h}, \mathbb{\Gamma}^{-1}\mathbb{K}\mathbf{h})_{\Gamma} = 1 \tag{14.98}$$

The matrix $\mathbb{\Gamma}^{-1}\mathbb{K}$, product of two non-negative definite matrices, no longer has a reason to have this property. Nevertheless, the operator $\mathbb{\Gamma}^{-1}\mathbb{K}$ is non-negative definite when used with the scalar product (14.6). In order to verify this point, it is sufficient to note that for every vector $\mathbf{u}$ we have

$$(\mathbf{u}, \mathbb{\Gamma}^{-1}\mathbb{K}\mathbf{u})_{\Gamma} \stackrel{\Delta}{=} \mathbf{u}^T\mathbb{K}\mathbf{u} \geq 0 \tag{14.99}$$

Let us now call λ_i and $\mathbf{v}_i$ the eigenelements of the operator $\mathbb{\Gamma}^{-1}\mathbb{K}$. As this is non-negative with the scalar product (14.6) we deduce that $\lambda_i \geq 0$ and the eigenvectors are orthogonal ($\mathbb{\Gamma}$). If they are normalized, we have

$$(\mathbf{v}_i, \mathbf{v}_j)_{\Gamma} = \mathbf{v}_i^T \Gamma \mathbf{v}_j = \delta_{ij} \tag{14.100}$$

It is then possible to expand the vector **h** in terms of the vectors $\mathbf{v}_i$, which gives

$$\mathbf{h} = \sum_{i=1}^{n} h_i \mathbf{v}_i \tag{14.101}$$

and the variance V defined by (14.70) becomes

$$V = (\mathbf{h}, \mathbf{h})_{\Gamma} = \sum_{i=1}^{n} |h_i|^2 \tag{14.102}$$

Similarly, the constraint (14.98) can be written as

$$1 = (\mathbf{h}, \Gamma^{-1}\mathbb{K}\mathbf{h})_{\Gamma} = \sum_{i=1}^{n} \lambda_i |h_i|^2 \tag{14.103}$$

and, calling λ_M the greatest eigenvalue of $\Gamma^{-1}\mathbb{K}$, we deduce

$$1 \le \lambda_M \sum_{i=1}^{n} |h_i|^2 = \lambda_M V \tag{14.104}$$

This shows that

$$V \ge V_2 \triangleq (1/\lambda_M) \tag{14.105}$$

and the minimum is reached when the vector **h** can be written as

$$\mathbf{h}_2 = \frac{1}{\sqrt{\lambda_M}} \mathbf{v}_M \tag{14.106}$$

In this expression $\mathbf{v}_M$ is the normalized (Γ) eigenvector of $\Gamma^{-1}\mathbb{K}$ corresponding to the greatest eigenvalue. It is easy to verify that $\mathbf{h}_2$ satisfies (14.98).

Let us now examine what appears when we add the subspace constraint meaning that **h** belongs to a subspace S. Let us call Proj(Γ) the projection operator onto S appearing in the decomposition (14.87) and introducing the matrix $\mathbb{P}$. As any projection matrix satisfies $\mathbb{P}^2 = \mathbb{P}$, the subspace constraint can be written as

$$\mathbf{h} = \mathrm{Proj}(\Gamma)[\mathbf{h}|S] \triangleq \mathbb{P}\mathbf{h} \triangleq \mathbf{h}_S = \mathbb{P}\mathbf{h}_S \tag{14.107}$$

It follows that the constraint (14.98) can be written as

$$(\mathbb{P}\mathbf{h}_S, \Gamma^{-1}\mathbb{K}\mathbb{P}\mathbf{h}_S)_{\Gamma} = 1 \tag{14.108}$$

As any projection operator is symmetric, we obtain

$$(\mathbf{h}_S, \mathbb{P}\Gamma^{-1}\mathbb{K}\mathbb{P}\mathbf{h}_S)_{\Gamma} = 1 \tag{14.109}$$

With this formulation we arrive at the same problem as previously, that is, minimizing the variance $V = (\mathbf{h}_S, \mathbf{h}_S)$ with the constraint (14.109). For this, let us call μ_j and $\mathbf{w}_j$ the eigenelements of $\mathbb{P}\Gamma^{-1}\mathbb{K}\mathbb{P}$. Using the same procedure as above, we deduce that

$$\mathbf{h}_3 = \frac{1}{\sqrt{\mu_M}}\mathbf{w}_M \tag{14.110}$$

where μ_M is the greatest eigenvalue and $\mathbf{w}_M$ the corresponding eigenvector. Furthermore, the minimum value of the variance is

$$V_3 = (1/\mu_M) \tag{14.111}$$

Let us now verify that $\mathbf{w}_M$ and then $\mathbf{h}_3$ belong to S. We start from the eigenequation

$$\mathbb{P}\Gamma^{-1}\mathbb{K}\mathbb{P}\mathbf{w} = \mu\,\mathbf{w} \tag{14.112}$$

Note that the matrix $\mathbb{I} - \mathbb{P}$ makes a projection onto the subspace orthogonal (Γ) complement to S, and then, if $\mu \neq 0$, we have

$$\mathbf{w}_S^{\perp} \overset{\Delta}{=} (\mathbb{I} - \mathbb{P})\,\mathbf{w} = \mu^{-1}(\mathbb{I} - \mathbb{P})\,\mathbb{P}\Gamma^{-1}\mathbb{K}\mathbb{P}\,\mathbf{w} = \mathbf{0} \tag{14.113}$$

because the product $(\mathbb{I} - \mathbb{P})\,\mathbb{P}$ is null. As μ_M is necessarily positive, we deduce that $\mathbf{w}_M$ belongs to S.

As in the case of a directional constraint, it is of interest to calculate the loss due to the additional subspace constraint. We deduce from (14.104) and (14.111) that

$$\frac{1}{V_2} - \frac{1}{V_3} = \lambda_M - \mu_M \tag{14.114}$$

It is clear that this quantity is non-negative. In fact λ_M and μ_M are the maximum values of $(\mathbf{u}, \Gamma^{-1}\mathbb{K}\mathbf{u})_\Gamma$ and $(\mathbf{u}, \Gamma^{-1}\mathbb{K}\mathbb{P}\mathbf{u})_\Gamma$ respectively. As this last quantity can also be written as $(\mathbb{P}\mathbf{u}, \mathbb{P}\Gamma^{-1}\mathbb{K}\mathbb{P}\mathbf{u})_\Gamma$, its maximum is smaller than or equal to that of $(\mathbf{u}, \Gamma^{-1}\mathbb{K}\mathbf{u})_\Gamma$.

14.4.3 Additional Comments on Matched Filtering

(a) Justification of the expression of matched filters. As the two constraints (14.71) and (14.83) seem quite different, we may ask why the same term "matched filter" is used for both. In fact, the solutions also seem quite different when we compare $\mathbf{h}_0$ and $\mathbf{h}_2$ obtained without subspace constraints or $\mathbf{h}_1$ and $\mathbf{h}_3$ obtained with this constraint. This difference is more apparent than real, and we will see that constraint (14.71) can be considered as a particular case of constraint (14.83), except for a sign ambiguity. Suppose that $\mathbb{K}$ is a rank one matrix. As it is symmetric it can be written as $\mathbb{K} = \mathbf{s}\mathbf{s}^T$, and constraint (14.83) becomes

$$|\mathbf{h}^T\mathbf{s}|^2 = 1 \tag{14.115}$$

Comparing this with (14.71), we observe that the only difference is the result of a sign ambiguity, which also appears in the values of $\mathbf{h}_1$ and $\mathbf{h}_3$. To complete the comparison, let us verify that the two filters $\mathbf{h}_1$ and $\mathbf{h}_3$ are the same, apart from this phase factor. For this it is sufficient to verify that $\mathbf{w}_M$ appearing in (14.110) is proportional to $\mathbf{r}_S$ in (14.91). In fact, if this point is verified, we can say that

$$\mathbf{h}_3 = \pm\, \mathbf{h}_1 \tag{14.116}$$

because the other terms in (14.91) or (14.110) are only normalization factors ensuring the condition $|\,\mathbf{h}^T\mathbf{s}\,| = 1$.

Starting from the value of $\mathbb{K}$, we deduce that

$$\mathbb{P}\Gamma^{-1}\mathbb{K}\mathbb{P} = \mathbb{P}\Gamma^{-1}\mathbf{s}\,\mathbf{s}^T\mathbb{P} \tag{14.117}$$

As $\mathbb{P}\Gamma^{-1}\mathbf{s} = \mathbf{r}_S$, we obtain

$$(\mathbf{r}_S\,\mathbf{s}^T\mathbb{P})\,\mathbf{r}_S = \mathbf{r}_S\,(\mathbf{s}^T\mathbb{P}\,\mathbf{r}_S) = (\mathbf{s}^T\mathbf{r}_S)\,\mathbf{r}_S \tag{14.118}$$

which shows that $\mathbf{r}_S$ is an eigenvector of $\mathbb{P}\Gamma^{-1}\mathbf{s}\,\mathbf{s}^T\,\mathbb{P}$ with the eigenvalue $\mathbf{s}^T\mathbf{r}_S$. But as this matrix is still of rank one, this is the only non-null eigenvalue, and it is therefore the greatest. This completes the proof.

In conclusion, even if the constraints of direction or of power appear different, they give the same filter, which justifies the term of matched filter.

(b) Unicity of the matched filters. It is clear that the filters $\mathbf{h}_0$ and $\mathbf{h}_1$ are defined without any ambiguity and are the unique solution of the problems. This is because there is only one solution to the equation (14.26) when Γ is invertible, and because the projection $\mathbf{r}_S$ is also unique.

This is no longer necessarily the case for the filters $\mathbf{h}_2$ and $\mathbf{h}_3$. There is first a sign indetermination. But also, even if the greatest eigenvalues λ_M or μ_M are unique, they can correspond to several eigenvectors $\mathbf{v}_i$ or $\mathbf{w}_j$. This appears when these maximum eigenvalues are degenerated. In this case there is no unique vector associated with the greatest eigenvalue, but a subspace S_M, and any vector of this subspace can be taken to construct the filters $\mathbf{h}_2$ or $\mathbf{h}_3$. A choice between these vectors can be made by using considerations other than those discussed here.

(c) Some examples of constraints or of similar problems. In spatial signal processing the constraint $|\,\mathbf{h}^T\mathbf{s}\,| = 1$ means that the gain of the processor is fixed for one source direction characterized by the vector s, while the constraint $\mathbf{h}^T\mathbf{u} = 0$ means, on the contrary, that the system cancels the direction corresponding to the vector $\mathbf{u}$. That is why we say that it is a null constraint which appears when we want to completely cancel a particular jammer. Suppose now that we impose P directional constraints specified by P vectors $\mathbf{s}_i$ and Q null constraints corresponding to Q vectors $\mathbf{u}_i$. After having chosen one of the directional constraints, for example $\mathbf{h}^T\mathbf{s}_1 = 1$, the $P - 1$ other constraints can be written as $\mathbf{h}^T\mathbf{v}_i = 0$ where $\mathbf{v}_i = \mathbf{s}_i - \mathbf{s}_1$. The P directional constraints are therefore equivalent to one directional and $P - 1$ null constraints, and the global problem is characterized by one directional and $P + Q - 1$ null constraints. This corresponds to a subspace constraint because the vector $\mathbf{h}$ must be orthogonal to $P + Q - 1$ given vectors.

The subspace constraint also appears in some projection problems. Suppose that the filter vector $\mathbf{h}$ must satisfy the condition

$$\mathbb{P}[\mathbf{h} \mid T] = \mathbf{v} \tag{14.119}$$

where $\mathbb{P}$ is the standard projection on a given subspace T and $\mathbf{v}$ a given vector of T. Let us call $\{\mathbf{u}_i\}$ a set of m orthonormal (in the ordinary sense) vectors, generating the subspace T. As a result, we can write the matrix associated with $\mathbb{P}$ in the form

$$\mathbb{P} = \sum_{i=1}^{m} \mathbf{u}_i \mathbf{u}_i^T \, , m < n \tag{14.120}$$

Condition (14.119) can then be written as

$$\sum_{i=1}^{m} \mathbf{u}_i \mathbf{u}_i^T \; \mathbf{h} = \mathbf{v} \tag{14.121}$$

and, as the vectors $\mathbf{u}_i$ are orthonormal, we deduce that

$$\mathbf{u}_i^T \mathbf{h} = \mathbf{u}_i^T \mathbf{v} = c_i \, , 1 \le i \le m \tag{14.122}$$

This gives a set of m conditions similar to m directional constraints. As previously, this is equivalent to one directional constraint $\mathbf{h}^T \mathbf{u}_1 = c_1$ and $m - 1$ null constraints $\mathbf{h}^T(c_i \mathbf{u}_1 - c_1 \mathbf{u}_i) = 0$.

Similar comments can be made on quadratic constraints. For example, the constraint appearing in (14.77) can be written as $\mathbf{h}^T \mathbb{P} \mathbf{h} = 1$, where $\mathbb{P}$ is the projector $\mathbf{s}\mathbf{s}^T$ onto the subspace generated by the vector $\mathbf{s}$. Similarly the constraint

$$\sum_{i=1}^{m} | \mathbf{h}^T \mathbf{s}_i |^2 = 1 \tag{14.123}$$

can easily take the form of (14.83), and the ordinary matched filter maximizing the signal-to-noise ratio (14.4) is a particular case of (14.123).

(d) Extensions to the complex case. The previous results can easily be translated to the complex case with the restriction indicated after (14.11). However, a phase ambiguity appears such that the vector in (14.106) can be multiplied by $\exp(j\phi)$ without changing the fact that it gives the minimum variance. This also appears in (14.110).

(e) Extensions to non-linear cases. As in MSE problems, the theory of matched filtering can be extended to the class of Volterra filters, using a method similar to that described in Section 10.6. This is interesting in detection problems when the likelihood ratio cannot be used because of its complexity. It can be approximated when some higher-order moments are known, and one can, for example, show that the linear-quadratic matched filter is the best approximation of $L(\mathbf{x})$, extending to linear-quadratic filters the property (14.18) (see [Picinbono, 1990]).

In conclusion, minimizing the variance of a system, linear or not, is a problem which in principle is quite different to MSE problems. However, the geometric approach, frequently used in this book, shows a close analogy in their solution. The interest of geometric methods appears in many other problems concerning random signals and systems.

PROBLEMS

14.1 The quantity d^2 defined by (14.8) gives the maximum value of the signal to noise ratio which can be obtained with a matched filter. This quantity depends on the covariance matrix of the noise and of the signal $\mathbf{s}$. Suppose that $\mathbf{s}$ can be chosen in the class of signals with a given energy or satisfying $\mathbf{s}^T\mathbf{s} = E$. For a given value of the noise covariance matrix and of E, find the signals giving the maximum value to d^2.

14.2 We want to extend the principle of matched filtering to quadratic filters. In this case (14.2) is replaced by (4.251) where $\mathbb{M}$ is symmetric. The signal and the noise are assumed to be zeromean, normal and defined by the covariance matrices $\mathbb{\Gamma}_s$ and $\mathbb{\Gamma}_n$ respectively. The criterion (14.2) is replaced by the deflexion criterion. The deflexion C is defined as the ratio N/D with $N = \{E(y)\}^2$ and $D = V(y)$, where E means the expectation calculated with $\mathbb{\Gamma}_s$ and V means the variance calculated with $\mathbb{\Gamma}_n$.

(a) Calculate the deflexion C.

(b) By using the results of Sections 4.8.6 and 4.8.7, show that C can be expressed by a ratio similar to (14.7).

(c) Calculate the matrix $\mathbb{M}_O$ which maximizes the deflexion and its maximum value.

Abbreviations and Symbols

AR	Autoregressive
AR(p)	Autoregressive signal of order p
ARMA	Autoregressive and moving average
ARMA (p, q)	Autoregressive (order p) and moving average (order q) signal
AS	Analytic signal
CT	Continuous-time
DF	Distribution function
DFT	Discrete Fourier transform
DT	Discrete time
FIR	Finite impulse response
FT	Fourier transform
IID	Independent and identically distributed
LMSE	Linear mean square estimation
LT	Laplace transform
M(1)	Markovian of order 1
M(p)	Markovian of order p
MA	Moving average
MA(q)	Moving average of order q
MSE	Mean square estimation
$N(m, \sigma^2)$	Normal distribution, mean m, variance σ^2
$N(\mathbf{m}, \mathbb{\Gamma})$	Normal distribution, mean $\mathbf{m}$, covariance matrix $\mathbb{\Gamma}$
NND	Non-negative definite

PD	Positive definite
PDF	Probability density function
RV	Random variable
SI	Spherically invariant
ZT	Z-transform
*	Complex conjugate
$\mathbf{v}$	Column vector
$\mathbb{M}$	Matrix
T	Transpose of a matrix
H	Hermitian conjugate of a matrix
$m \wedge n$	Minimum of m or n

Bibliography

BOOKS

ANDERSON, T.W., *Time Series Analysis*, John Wiley and Sons, New York, 1970.

BLANC-LAPIERRE, A. et PICINBONO, B., *Fonctions aléatoires*, Masson, Paris, 1981.

BRÉMAUD, P., *An Introduction to Probabilistic Modeling*, Springer-Verlag, New York, 1988.

BRILLINGER, D., *Time Series Data Analysis and Theory*, Holt, Rinehart and Winston, New York, 1975.

CRAMÉR, H. and LEADBETTER, M., *Stationary and Related Stochastic Processes*, John Wiley and Sons, New York, 1967.

DOOB, J.L., *Stochastic Processes*, John Wiley and Sons, New York, 1953.

FELLER, W., *An Introduction to Probability Theory and its Applications*, Vol. 1, John Wiley and Sons, New York, 1968, Vol. 2, John Wiley and Sons, New York, 1971.

GALLAGHER, R., *Information Theory and Reliable Communications*, John Wiley and Sons, New York, 1968.

GARDNER, W.A., *Introduction to Random Processes with Applications to Signals and Systems*, McGraw-Hill, New York, 1990.

GRAY, R. and DAVISSON, L., *Random Processes: A Mathematical Approach for Engineers*, Prentice-Hall, Englewood Cliffs, 1986.

HANNAN, E.J., *Multiple Time Series*, John Wiley and Sons, New York, 1970.

HAYKIN, S., Adaptive Filter Theory, Prentice-Hall, Englewood Cliffs, 1991.

HELSTROM, C., *Probability and Stochastic Processes for Engineers*, Second Edition, Macmillan Publishing Company, New York, 1991.

HONIG, M.L. and MESSERSCHMITT, D.G., *Adaptive Filters Structures, Algorithms and Applications*, Kluwer Academic Publishers, Boston, 1984.

KAY, S.M., *Modern Spectral Estimation, Theory and Applications*, Prentice-Hall, Englewood Cliffs, 1988.

LOÈVE, M., *Probability Theory*, 4th edition, Springer Verlag, New York, 1978.

MILLER, K., *Complex Stochastic Processes, An Introduction to Theory and Applications*, Addison-Wesley, Reading, 1974.

MILLER, K., *Multidimensional Gaussian Distributions*, John Wiley and Sons, New York, 1964.

MONZIGO, R.A. and MILLER, T.W., *Introduction to Adaptive Arrays*, John Wiley and Sons, New York, 1980.

MORTENSENS, R.E., *Random Signals and Systems*, John Wiley and Sons, New York, 1987.

OCHI, M., *Applied Probability and Stochastic Processes*, John Wiley and Sons, New York, 1990.

OPPENHEIM, A. and WILLSKY, A., with YOUG, I., *Signals and Systems*, Prentice-Hall, Englewood Cliffs, 1983.

PAPOULIS, A., *Probability, Random Variables, and Stochastic Processes*, Second Edition, McGraw-Hill, New York, 1984.

PFEIFFER, P., *Probability for Applications*, Springer-Verlag, New York, 1990.

PICINBONO, B., *Principles of Signals and Systems: Deterministic Signals*, Artech House, London, 1988.

SNYDER, D.L., *Random Point Processes*, John Wiley and Sons, New York, 1975.

STARK, H. and WOODS, J.M., *Probability, Random Processes and Estimation Theory for Engineers*, Prentice-Hall, Englewood Cliffs, 1986.

STROBACH, P., *Linear Prediction Theory*, Springer-Verlag, Berlin, 1990.

WIDROW, B. and STEARNS, S.D., *Adaptive Signal Processing*, Prentice Hall, Englewood Cliffs, 1985.

WONG, E., *Stochastic Processes in Information and Dynamical Systems*, McGraw-Hill, New York, 1971.

YAGLOM, A.M., *An Introduction to the Theory of Stationary Random Functions*, Dover Publications, New York, 1962.

PAPERS

BENIDIR, M. and PICINBONO, B., "Extensions of Stability Criterion for ARMA Filters," *IEEE Trans. on Acoustics, Speech and Signal Processing*, Vol. ASSP 35, pp. 425-432, 1987.

PICINBONO, B. and BENDJABALLAH, C., "Photoelectron Shot Noise," *Journal of Mathematical Physics*, Vol. 12, pp. 2166-2176, 1970.

PICINBONO, B. and BENIDIR, M., "Somes Properties of Lattice Autoregressive Filters," *IEEE Trans. on Acoustics, Speech and Signal Processing*, Vol. ASSP 34, pp. 342-349, 1986.

PICINBONO, B. and DUVAUT, P., "Geometrical Properties of Optimal Volterra Filters for Signal Detection," *IEEE Trans. Inform. Theory*, Vol. IT 36, p. 1061-1068, 1990.

SLEPIAN, D., "Prolate Spheroidal Wawe Functions, Fourier Analysis and Uncertainty, V. The Discrete Case," *Bell System Technical Journal*, Vol. 57, pp. 1371-1430, 1978.

Index

D

E

F

G

N

Q

R

T

U

V

W

Z